Natural and Engineered Pest Management Agents

ACS SYMPOSIUM SERIES **551**

Natural and Engineered Pest Management Agents

Paul A. Hedin, EDITOR
U.S. Department of Agriculture

Julius J. Menn, EDITOR
U.S. Department of Agriculture

Robert M. Hollingworth, EDITOR
Michigan State University

Developed from the Conference on Natural and Derived
Pest Management Agents sponsored
by the Division of Agrochemicals,
Snowbird, Utah,
August 9–14, 1992

American Chemical Society, Washington, DC 1994

Library of Congress Cataloging-in-Publication Data

Conferences on Natural and Derived Pest Management Agents (1992: Snowbird, Utah)

Natural and engineered pest management agents / Paul A. Hedin, Julius J. Menn, Robert M. Hollingworth, [editor].

p. cm.—(ACS symposium series, ISSN 0097–6156; 551)

"Developed from the Conference on Natural and Derived Pest Management Agents sponsored by the Division of Agrochemicals, Snowbird, Utah, August 9–14, 1992."

Includes bibliographical references and index.

ISBN 0–8412–2773–X

1. Natural pesticides—Congresses.

I. Hedin, Paul A. (Paul Arthur), 1926– . II. Menn, Julius J. III. Hollingworth, Robert M., 1939– . IV. American Chemical Society. Division of Agrochemicals. V. Title. VI. Series: ACS symposium series; 551.

SB951.145.N37C66 1992
632′.96—dc20 93–41197

CIP

The paper used in this publication meets the minimum requirements of American National Standard for Information Sciences—Permanence of Paper for Printed Library Materials, ANSI Z39.48–1984. ♾

PRINTED IN THE UNITED STATES OF AMERICA

Foreword

THE ACS SYMPOSIUM SERIES was first published in 1974 to provide a mechanism for publishing symposia quickly in book form. The purpose of this series is to publish comprehensive books developed from symposia, which are usually "snapshots in time" of the current research being done on a topic, plus some review material on the topic. For this reason, it is necessary that the papers be published as quickly as possible.

Before a symposium-based book is put under contract, the proposed table of contents is reviewed for appropriateness to the topic and for comprehensiveness of the collection. Some papers are excluded at this point, and others are added to round out the scope of the volume. In addition, a draft of each paper is peer-reviewed prior to final acceptance or rejection. This anonymous review process is supervised by the organizer(s) of the symposium, who become the editor(s) of the book. The authors then revise their papers according to the recommendations of both the reviewers and the editors, prepare camera-ready copy, and submit the final papers to the editors, who check that all necessary revisions have been made.

As a rule, only original research papers and original review papers are included in the volumes. Verbatim reproductions of previously published papers are not accepted.

M. Joan Comstock
Series Editor

Contents

INDEXES

Preface

SINCE ANCIENT TIMES, PEOPLE HAVE RECOGNIZED that plants contained substances that could promote health and cure illnesses, and they observed that some plants resisted pests better than others. Through trial and error, people discovered the use of these substances and developed procedures for their extraction and use. With the advent of modern chemistry, the structures of many of these biologically active agents became known, and the systematic studies of natural products that protected plants from pests became a recognized activity within the field of chemistry.

For a period of time, the use of natural products for the protection of plants waned with the increased use of large-scale synthesis and production of synthetic chemical compounds. During recent years, there has been a growing recognition that many of these compounds display a variety of adverse effects, and their use is often circumscribed. The need to control pests is even greater today, and attention is being refocused on natural products because they often have a reduced impact on the host and the environment, they may have desired specificity, and they may serve as prototype models.

We have had a continuing interest in biologically active agents that protect plants from pests. Following this interest, we organized three special conferences on these topics in 1984, 1987, and 1992. Books based on the presentations at all three meetings have been compiled and published.

This book is based on presentations given at the 1992 conference. The organization of the book follows that of the conference and includes sections on natural product pesticides, peptides and neuropeptides, natural and engineered viral agents, evolving approaches to pesticide discovery emphasizing biochemistry and computer-aided design, and registration of biopesticides.

We are moving into a new era in which crops will be protected from a broad spectrum of pests, including plants, insects, and diseases, by more selective technologies. This book will contribute to the understanding and the subsequent adoption of additional criteria and research strategies for the control of pests based on natural products and will be of interest to industry, the academic and federal research sectors, and the agencies responsible for regulation of natural products.

Acknowledgments

We thank the participants of the conference and the contributors to this book, without whom the project would not have been possible. We acknowledge financial aid received from the U.S. Department of Agriculture National Competitive Research Initiative Grants Program and the U.S. Department of Agriculture Office of Technology Transfer that was used for organization of the conference. Finally, we thank the Agricultural Research Service of the U.S. Department of Agriculture for support associated with organization of the conference and compilation of the book.

PAUL A. HEDIN
U.S. Department of Agriculture
Mississippi State, MS 39762

JULIUS J. MENN
U.S. Department of Agriculture
Beltsville, MD 20705

ROBERT M. HOLLINGWORTH
Michigan State University
East Lansing, MI 48824

September 28, 1993

Innovation in Discovery

The Career of Toshio Fujita

A special award recognizing Toshio Fujita's contribution to biological, computational, and synthetic chemistry was presented to him at the 1992 conference by the Division of Agrochemicals of the American Chemical Society. The award was presented at a session held in Fujita's honor entitled "Evolving Approaches to Pesticide Discovery".

Toshio Fujita is held in such high regard by his colleagues for numerous reasons. His career is typified by innovative research and creative teaching in his laboratory. Fujita provided novel ways of looking at biological activity to generations of biological chemists, many of whom view the world in terms of π, σ, and other physical–chemical parameters developed by Fujita, Hansch, and others.

Fujita is Professor Emeritus of Bioregulation and Pesticide Chemistry, Kyoto University, Kyoto, Japan. He earned his B.S. degree in the Department of Agricultural Chemistry of Kyoto University in 1951 and was appointed as an instructor of that department the same year, serving until 1964. On the basis on his research from 1951 to 1961, he received a D.Sc. from the university in 1962. He was promoted to lecturer in 1964, associate professor in 1966, and professor in 1981.

Fujita is a member of the Japan Society for Bioscience, Biotechnology, and Agrochemistry (formerly the Agricultural Chemical Society of Japan), the Chemical Society of Japan, the Pharmaceutical Society of Japan, and the Pesticide Science Society of Japan, as well as the American Chemical Society. He served as President of the Pesticide Science Society of Japan in 1985 and 1986. He received Agricultural Chemical Society Awards in 1967 for promoting research activity and in 1989 for his research performance. In addition, he received the Pesticide Science Society Award in 1979, and, as recently announced, the American Chemical Society International Award in Pesticide Chemistry Research for 1993.

Since his undergraduate years, his research career has focused on the structure–activity relationships of bioactive compounds, starting with the auxin-type plant-growth regulators and herbicides. In 1961, he joined Corwin Hansch at Pomona College in Claremont, California, as a Postdoctoral Research Fellow. From 1961 to 1963, he and Hansch developed a novel procedure to quantitatively analyze structure–activity relationships of agrochemicals and medicines (QSAR). QSAR is regarded as a forerunner to the development of a number of recent computer-aided

Toshio Fujita

drug design methodologies. Two early seminal papers (Fujita, Hansch, and Muir[1]), introducing QSAR, were selected as Citation Classics in 1982 and 1986 by the Institute of Scientific Information in Philadelphia, Pennsylvania.

Fujita is the author of more than 250 papers and has been the editor or co-editor of five books on QSAR methodologies and applications to structure–activity optimization. His major interests are QSAR and SAR and their applications to a wide variety of agrochemicals and medicines. Upon retirement from Kyoto University in 1992, he began working at the Fujitsu Kansai Systems Laboratory to construct a knowledge-based system named EMIL. This system uses the rapidly expanding power of computer systems to analyze databases of biologically active agrochemicals and medicines as a creative tool in compound design. His chapter in this book provides ample illustration that Toshio Fujita is still making innovative contributions.

BRUCE D. HAMMOCK
University of California
Davis, CA 95616

July 6, 1993

[1]Hansch, C.; Fujita, T. *J. Am. Chem. Soc.* **1964,** *86,* 1616–1626. Fujita, T.; Iwasa, J.; Hansch, C. *J. Am. Chem. Soc.* **1964,** *86,* 5175–5180.

Natural Product Pesticides

Chapter 1

Development of Natural Products and Their Derivatives for Pest Control in the 21st Century

Paul A. Hedin[1], Julius J. Menn[2], and Robert M. Hollingworth[3]

[1]Crop Science Research Laboratory, Agricultural Research Service, U.S. Department of Agriculture, Mississippi State, MS 39762
[2]Plant Sciences Institute, Agricultural Research Service, U.S. Department of Agriculture, Beltsville, MD 20705
[3]Pesticide Research Center, Michigan State University, East Lansing, MI 48824

This overview chapter and those that follow explore the emerging technologies for crop protection that are accelerated by the development of new natural pest management agents and materials derived from natural models. We are moving into a new era in which crops will be protected from a broad spectrum of pests including weeds, insects, and diseases by these new biotechnologies. The development of these agents, along with genetically altered pest resistant crops, will provide new means by which crops will be protected from pests in future years. Also addressed here is the evolving approach to pesticide discovery, such as computer-aided design and strategies based on understanding of the underlying biochemical target systems. Finally, the monitoring of these products in the environment, and as a related process, the devising of strategies to regulate them with an appropriate balance between safety and the encouragement of innovation are discussed.

Over the past 25 years, there has been much activity directed to chemical work on the isolation and identification of a wide array of biologically active natural products that in some way affect the behavior, development and/or reproduction of pests such as insects, diseases and weeds. However, with regard to crop plants, agronomists, entomologists, and other agricultural scientists have generally depended on traditional methods for selecting resistant varieties with adequate yield properties. Even though various biologically active compounds have been identified, only in a limited percentage of situations was chemical guidance the leading factor in screening for these properties. Agents such as pheromones, antifeedants, and insect and plant growth regulators, to name a few, have found limited commercial application. In

0097–6156/94/0551–0002$06.00/0

contrast, synthetic analogues based on the activity of some natural products such as the avermectins have found a wider market.

However, the day of biologically active natural products entering the mainstream of crop protection may have come. Environmental concerns are constricting synthetic chemical pesticide usage. Growing resistance of pests to present pesticides has given added urgency to the search for better, safer compounds, and improved delivery systems. The need to treat crops precisely has also provided additional opportunities for the use of natural products. It should be stated, however, that resistance can be expected to develop rapidly to toxic natural products that act on a single target site.

Natural Product Pesticides

Some biologically active natural products can be considered as bioregulators, the name modeled in part on the plant growth hormones that were shown to regulate various plant growth and development processes. The levels required for activity of these compounds are often very low. For example, pheromones are active at picogram levels and below. Insect neurohoromones have been found to be just as active physiologically, and these are just two examples of endogenous bioactive compounds that may be further exploited in control strategies.

The elucidation of the often complex structure and stereochemistry of natural products becomes more and more feasible as access to powerful instrumental techniques rapidly increases. It is thrilling to realize that so much power and capability has been developed, particularly in the past decade.

With regard to the pest species that need to be controlled, they can most broadly be classified as those attacking animals, plants, and their products. Those pests that attack animals are often arthropods, usually insects. Other pests of animals are microbes and nematodes. The protection of plants from pests can be divided into at least five categories; control of (1) mammals and other vertebrates, (2) insects and other arthropods, (3) nematodes, (4) diseases, and (5) weeds. While pest control has traditionally been accomplished over the past 50 years with pesticides of synthetic origin, this was not the case prior to the development of synthetic pesticides such as DDT. In recent years, a number of natural products from fermentation such as the avermectins have been finding a niche in the market place, because of lower costs of production and increasing environmental concerns about synthetic chemical pesticides. An effort has also been made to identify those natural products such as azadirachtin, which have been found useful for the control of a broad diversity of pests.

Natural products provide leads for the synthesis of pesticides with desirable properties. Often, from an economic standpoint, only synthesis will provide adequate quantities to justify the development of a product. There are a few examples of identification of a natural product that has led to a marketable product. They include the pyrethroids, azadirachtin, the β-methoxyacrylate fungicides, insect pheromones and hormones, and the avermectins.

Peptides and Neuropeptides

Among the newer categories of potential pest control agents are the biologically active peptides and neuropeptides that appear to have great promise for control of insects, weeds, and diseases because of their very potent hormone-like activities. Their structures and the genes encoding them are already known in several instances (1).

Early work in this century suggested a functional relationship between the neural and endocrine systems in insects. This led to the concept of neurosecretion. Subsequent discoveries made using vertebrate systems illustrated the central role of peptides in regulating behavior and maintaining homeostasis (2). Breakthroughs in technology, and the continuing discovery of proteinaceous factors in insect neural tissue, and most recently, in plants which have important physiological activities, stimulated numerous efforts to isolate and identify insect (and plant) peptides (1). Coupled with these elements is the need to develop alternative insect control and crop protection methods which are environmentally acceptable and economically viable. All of these forces have generated great interest in peptides as leads to insect control agents. Efforts are underway to isolate peptides and their genes, analyze sequences and structure-function relationships, explore the use of natural vectors and the insertion of constructs and foreign genes into these vectors (particularly viruses). Studies are also in progress to increase virulence, understand the role of plant protective peptides, to identify and understand the molecular biology of peptide (especially neuropeptide) biogenesis, action and clearance, and to improve delivery.

The physiological processes regulated by neuropeptides are extensive in insects. These neuropeptide-regulated processes have been discovered at an ever-increasing pace and are now being followed by the purification and sequencing of the regulatory agents and their genes (3). Since many neuropeptides control essential life processes, mechanisms for blocking their synthesis or action, or enhancement of their degradation may be necessary for developing pest control agents based on this technology. However, the danger of non-target effects may be high. Investigations into sites of synthesis and release continue to reveal complexity in the insect neuroendocrine system. The rapidly increasing knowledge in this area suggests that useful prototypes for the design of selective pest control agents will emerge in the near future.

Insects respond to the presence of bacteria in their body cavity with a combination of hemocyte-mediated sequestration processes and the induced synthesis of several families of bacteria-elicited, hemolymph proteins and peptides (1). Among the antibacterial peptides synthesized by various insect species are the cecropins, diptericins, and defensin-like peptides. The structures of these peptides and genes encoding them, and the specificity and mode of action of the peptides are known (Dunn, P., Purdue University, Personal Communication, 1992). Soluble fragments of bacterial cell wall peptidoglycan have been shown to elicit the synthesis of cecropin-like peptides by body fat, other tissues and a cell line derived from the tobacco hornworm, *Manduca sexta*. Antibacterial peptides are also synthesized and released into the lumen by the metamorphosing midgut of non-infected M. sexta. This has been interpreted as a prophylactic mechanism protecting the insect from infection during gut restructuring.

Venoms of insect predators serve as a vast reservoir of toxins which attack key proteins in excitable membranes. Targets for venom toxins include voltage-gated ion channels and components of the neurotransmitter excitation-secretion process. Using venom toxins as pharmacological probes, it has become apparent that pronounced phylogenetic differences exist in these target proteins. Such differences indicate that a high potential exists for development of engineered insect-specific toxins (Adams, M., University of California, Riverside, Personal Communication, 1992).

Tentoxin is a cyclic tetrapeptide derived from the fungus Alternaria alternata that causes disruption of chloroplast development in most of the major weed species in soybean and of johnsongrass in corn while having no effect on either of these crop species. The major impediment to the development of tentoxin as a herbicide is its limited availability because of low biosynthetic yields. Chemical synthesis is possible and the production of analogues has led to meaningful structure/activity relationships. To date, no analog has provided greater activity or specificity than the parent molecule. Molecular genetic manipulation is considered the best strategy for the increase in cost effective production. The success of cloning of the biosynthetic genes may determine the eventual deployment of these alternative biorational pesticides (Lax, A., USDA, New Orleans, LA, Personal Communication, 1992).

Some antifungal peptides have sufficiently broad spectrum activity against opportunistic plant pathogens to provide resistance against the soft rot diseases that damage most major crops. Research has focused on peptides that were sufficiently small to be amenable to chemical synthesis of both the peptide and the corresponding coding sequence for the gene. This attribute permitted the research to start with naturally occurring peptides and then, using principles of rational design, move on to unique peptides (4).

Natural and Engineered Viral Agents

Although insects are infected by a number of viruses, members of the family Baculoviridae, the nuclear polyhedrosis viruses, and the granulosis viruses are the most promising candidates for use in insect population management because of their many desirable characteristics. They are found only in Arthropods, frequently cause devastating epizootics in natural populations, and infect only the larval feeding stage in Lepidoptera normally with lethal effect (5). In addition, the infectious virion is embedded in a protein body that protects it to some extent from solar inactivation. The alteration of the viruses through recombinant DNA techniques to increase their effectiveness is a key goal of this line of work.

Nuclear polyhedrosis viruses have been extensively researched and are used more often for microbial control of insect pests than any of the other viruses. Their generally high degree of host specificity and ability to produce epizootics are among their most favorable attributes. The broader host ranges of a few NPV's offer hope for using a single virus to control several key pests and expands the potential market for these viruses and is a challenge for mass production (6). The rather rapid inactivation of NPV's by sunlight and the longer time required to kill pests, compared with chemical pesticides, are undesirable attributes. In the future, the addition of feeding stimulants, UV protectants, and natural or chemical enhancers to increase infectivity of viruses may make viruses more competitive with synthetic chemicals for insect control.

Many insect species are known to be susceptible to one or more isolates of granulosis viruses (7), however they have been sparingly used for insect control. Nevertheless, many of the susceptible species are economically important Lepidoptera, they are family, genus, or species specific, and they are generally non-hazardous to non-target species. In the case of the codling moth, they have been produced in large quantities using in vivo methods. Their use is compatible with traditional control methods, and they have been genetically engineered for increased potency.

Several approaches have been taken to improve insect baculoviruses as biospecific pesticides. Among these is the deletion of a viral gene, egt, which encodes an enzyme allowing the virus to block the molting of its host by inactivating ecdysteroids, insect molting hormones. Insects infected with viruses lacking functional egt cause less crop damage because they attempt to molt, gain less weight, and die sooner than insects infected with the wild-type virus. A second improvement involves expression of a neurotoxin gene, Tox34. Viruses expressing Tox34 paralyze insects during infection thus limiting crop damage (8).

Viral insecticides must be active in formulations that protect the virus from light degradation. The optimum use of viral insecticides in an insect pest management system requires application systems that deliver the product to the target efficiently and economically. Viruses have typically been applied as sprays by air or ground equipment using conventional application equipment. However, this equipment was designed for fast acting contact poisons. Since viral "insecticides" are slow-acting stomach poisons (actually infectious agents), current research is being centered around spray equipment that will selectively target spray droplets and unique delivery systems (9).

Design of Pesticides

The development of effective compounds for pest control most often evolves from the identification of a biologically active natural product whose structure is not conducive for synthesis or microbial production. While the approaches to a synthetic compound vary, they normally utilize some elements of modeling. Examples consist of various structural modifications of "lower-ordered" lead structures including simple exchanges of bioisosteric groupings, and more drastic skeletal changes such as ring openings and closures. These structural modifications which could apply to starting structures are not necessarily limited within compound series showing similar biological modes of action, but are also sometimes extended to those for bioactive compound series of similar structures regardless of the differences in biological modes of action.

Successful design of novel herbicides based on specific inhibition of selected enzyme targets requires careful consideration of the choice of the target, mechanism of action, design of potent inhibitors, delivery of the inhibitor to the target, and metabolic fate of the inhibitor. Validated targets can be identified by obtaining chemical leads or by genetic methods. Genetic approaches include studies of conditionally lethal bacterial and plant mutants and use of antisense technology. In the absence of chemical leads with known sites of action, targets for chemical validation may be selected by the following criteria: the target is essential to plants and, preferably, inhibition leads to multiple deleterious effects; the target is present in plants, but not in mammals; the target has low intracellular concentration, i.e., has potential for low use rates; and the proposed inhibitors of the target are available.

Potent inhibition of the selected target may still not produce an effective herbicide. Studies of the uptake, translocation, and metabolism of the inhibitor are needed to determine if the cause of poor in vivo performance is due to these factors or to an intrinsically poor target. Without full appreciation of each of these aspects of herbicide design, the chances for success with the target-site directed approach are reduced (Rendina, A., DuPont de Nemours and Co., Newark, DE, Personal Communication, 1992).

Another approach involves some genetic alteration of the plant. For example, the herbicide glyphosate is currently used for the control of a wide spectrum of weeds. Because glyphosate is a non-selective, post-emergent herbicide, one limitation of its use in a weed management system is that it cannot come into contact with normal crop plants during the growing season. Two general methods are available for the development of herbicide-tolerant crops using genetic modification: target-site modification and metabolic inactivation. Using a combination of mutagenesis, X-ray crystallography, and physical methodologies, the glyphosate/phosphonenolpyruvate and shikimate-3-phosphate binding sites of EPSPS have been identified. Glyphosate-tolerant EPSPS's which maintain a high catalytic efficiency have also been identified. In addition, a glyphosate-degrading enzyme has been used to provide in planta tolerance, thus allowing the illustration of both approaches for imparting glyphosate-tolerance to crops (10).

Another approach involves elucidating the role of substrate conformation in key biological processes. In a model study, the primary step in the biosynthesis of all N-linked glycoproteins was found to be catalyzed by the membrane associated enzyme oligosaccharyl transferase and involved the co-translational transfer of a complex carbohydrate from a pyrophosphate donor to the side-chain nitrogen of an asparagine. The primary peptide sequence requirements for the process are simple and include a minimum -Asn-Xaa-Ser/Thr- tripeptide recognition motif. This enzyme-catalyzed transformation is intriguing in that it represents a rare demonstration of nucleophilic behavior at an amide nitrogen, and furthermore, the selectivity is even more surprising when one considers the potential for competing functionally in the complex peptidyl substrates in which the "reactive" asparagine is localized (11).

To provide a convenient means for selective bacterial marking of engineered microorganisms and for semiquantitative bacterial population determination, reversible, metal-based switching sites were introduced into the enzyme, rat pancreatic trypsin. Carefully chosen metal complexes down-regulate enzymatic activity and act as noncompetitive, allosteric enzyme inhibitors. The variant trypsins retain good catalytic activity and possess K_i-values for certain copper complexes in the range 0.1-20 μM. Computer-based methods for designing metal-switching properties into enzymes have been developed (Haymore, B., Monsanto Co., St. Louis, MO, Personal Communication, 1992).

Registration of Biopesticides

It is realized that unless these natural and derived pest management agents, however developed, can be sold in the marketplace, the underlying research may be viewed by many as having been of no practical consequence. Therefore, along with work directed to their development, parallel efforts to secure approval for their use must also be pursued. The U.S. Environmental Protection Agency (USEPA) has encouraged the development of pest management techniques that offer alternatives to chemical control. Many natural products, engineered microbials and viruses, and biologically-derived pest management agents differ significantly from "conventional" chemical pesticides in their chemistry, complexity, and mode of action. However, such differences are difficult to categorize in terms of relative risk and there are few regulatory precedents. This has created difficulties for the regulatory community and the decision-making process has sometimes appeared slow. Progress is being made as more biopesticides are registered. Because certain uses of semiochemicals present comparatively low risk compared with conventional chemicals owing to their target specificity, unique modes of action, and low volume used, the USEPA has exempted these uses from FIFRA (Plimmer, J., ABC Laboratories, Columbia, MO, Personal Communication, 1992).

The public has become increasingly focused on the issue of pollution prevention. The USEPA is now shifting its perspective from being a neutral evaluator to being an active seeker of ways to promote safer approaches to pest control. EPA has begun to realign its regulatory programs to promote the development, registration, marketing, and use of safer pesticides, such as the biologicals. These "environmental-friendly" pesticides can be easily divided into two basic groups: naturally-occurring microbials and engineered viruses, and biochemicals and engineered transgenic plants. The registration process for these two pesticide groupings is governed by different regulatory schemes and policy issues. The new focus on "lower risk" pesticides as well as the loss of "minor uses" through the re-registration process will result in greater reliance on integrated pest management, which is USEPA's challenging goal to the pesticide-producing community (Lindsay, A., Office of Pesticide Programs, USEPA, Washington, DC, Personal Communication, 1992).

Products developed using the tools of recombinant DNA technology are subject to special regulations that affect not only the final product approval but all phases of field research. Many products may be subject to regulation by more than one federal agency, and most will be subject to both federal and state controls. Close consulation with regulatory officials early in the research and development process is the key to insuring that research activities are focused not just on product development but also in developing the data base necessary for initial field testing and for final product approval. Managing the regulatory process involves creating a framework for responding to public concerns raised by individual citizens, public interest groups and the press (Davis, J., Crops Genetics International, Hanover, MD, Personal Communication, 1992).

Technologies based on new knowledge of biological ecosystems such as those involving insect pheromones, genetically modified organisms, or suppression of biological populations have been developed by support from the public sector in response to the demand for alternative pest control methods. They are usually most suited for limited niche markets where commercial sales are limited, or they represent entirely different biological approaches where significant risk must be taken by business before profitable markets are created. These newly developed techniques are regulated by the USEPA under FIFRA and USDA's Animal and Plant Health Inspection Service under the Plant Pest Act. However, the regulatory procedures can become an important obstacle which can impede commercial development.

As these new agents become available, they should be used as part of a management system that will embrace increasingly species-specific agents used very sparingly in an integrated program, and that involves environmental management to minimize the incidence and impact of pests along with the use of a multiplicity of pest control agents. The trend toward species specificity in pest control agents will select for those that have substantial commercial potential against major economic pests, so that the costs of registration can be recovered through sales. Unless the regulatory process is relaxed or is subsidized, the control of moderately severe, sporadic or geographically limited pests will remain the province of broad-spectrum agents.

The development of natural and derived pest management agents seems to be a highly desirable strategy for obtaining government approval and public support so that crops, animals, and the public can be protected from pests. In most cases, these new agents have met with much scrutiny, and they hopefully will gain public support because they appear to be environmentally friendly.

Literature Cited

1. Kelly, T. J.; Masler, E. P.; Menn, J. J. In Pesticides and Alternatives: Innovative Chemical and Biological Approaches to

Pest Control; Casida, J., Ed.; Elsevier, Amsterdam, 1990, pp 283-297.
2. Thorpe, A.; Duve, H. In Stimulus-Secretion Coupling in Neuroendocrine Systems, Current Topics in Neuroendocrinology; Ganten, D.; Pfaff, D., Eds.; Springer-Verlag: New York, NY, 1988, Vol. 9; pp 185-230.
3. Dave, H.; Thorpe, A. In Neurohormones in Invertebrates; Thorndyke, M. C.; Goldsworthy, G. J., Eds.; Cambridge University, 1988; pp 115-139.
4. Soravia, E. S.; Martini, G.; Zasloff, M. F.E.B.S. Lett. 1988, 288, 337.
5. Vaughn, J. L.; Dougherty, E. M. In Viral Insecticides for Biological Control; Maramorosch, K.; Sherman, K. E., Eds.; Academic Press, Inc.: Orlando, FL, 1985, pp 569-633.
6. Harper, J. D. In Epizootiology of Insect Diseases; Fuxa, J. R.; Tanada, Y., Eds.; John Wiley and Sons, Inc.: New York, NY, 1987, pp 473-496.
7. Crook, N. E. In Viruses of Invertebrates; Kurstak, E., Ed.; Marcel Dekker, Inc.: New York, NY, 1991, pp 73-110.
8. Tomalski, M. D.; Miller, L. K. Nature, 1991, 352, 82.
9. Groner, A. In Safety of Microbial Insecticides; Laid, M.; Lacey, L. A.; Davidson, E. W., Eds.; CRC Press: Boca Raton, FL, 1989, pp 135-148.
10. Barry, G.; Kishore, G.; Padgette, S.; Taylor, M.; Kolacz, K.; Weldon, M.; Re, D.; Elchholtz, D.; Fincher, K.; Hallas, L. In Biosynthesis and Molecular Regulation of Amino Acids in Plants; Singh, B. K.; Flores, H. E.; Shannon, J. C., Eds.; American Society of Plant Physiologists: 1992, pp 139-145.
11. Imperiali, B.; Shannon, K. L.; Rickert, K. W. J. Am. Chem. Soc., 1992, 114, pp 7942-7944.

RECEIVED August 19, 1993

Chapter 2

Natural Products as Sources of Pest Management Agents

Koji Nakanishi

Department of Chemistry, Columbia University, New York, NY 10027

Terrestrial and marine animals, plants, insects, and microorganisms produce numerous secondary metabolites. Some are for self-defense, symbiosis, sexual attraction, or to fulfill a variety of other purposes while some are simply metabolic degradation products. The structures of most such Natural Products are extremely cleverly designed so that not only are the skeletons constructed but even each functional groups are arranged in a manner to exert maximal effect. The design of natural products are very ingenious and "imaginative" and it is impossible for scientists to cope with Nature in terms of designing unprecedented bioactive structures. In the following are summarized some examples taken from our past and present studies; some were left at the stage of isolation and structure determination without further investigations, while more recent studies are being investigated beyond the stage of mere structural elucidation. No doubt a great number of natural products published in the literature and described with no mention of activity or simply listed to be "antibiotic" have to be reinvestigated. However, the most difficult is how to assay them, and this depends a great deal on a better understanding of the mode of bioactivity.

With the explosive advancement in biology and biochemistry in recent years, it is not surprising that Natural Products lost its glamorous status when structure determination was at its peak up to the 1970's. The situation has, however, changed. Previously, when a new structure was determined, it was simply regarded as providing a lead to more active compounds by synthesizing a myriad of derivatives. After 20 years of rapid advancement, application of modern molecular biology and/or biochemical techniques have now become routine, and it has become possible to start probing into the very difficult multidisciplinary question of what constitutes the basis of bioactivity, i.e., why is a compound active? The solution of this lies in understanding the interaction between a bioactive substrate and its receptor, often membrane proteins, on a truly molecular structural basis. Thus, it is not surprising that natural products chemistry is facing a revival. Natural product structures can now be re-investigated from a totally different viewpoint which includes a rationalization of their mode of action. However, such an approach has just started, and requires collaborative efforts between all disciplines, chemists, biochemists, biologists, pharmacologists, biophysicists, clinicians, theorists, ethologists, etc. The artificially

0097–6156/94/0551–0011$06.00/0

divided disciplines in the past now have to be recombined to understand life and utilize natural products. It opens a totally new field that is extremely challenging.

A Plant Hormone that Promotes Cell Arrest in the G_2 Stage of the Cell Cycle.

Gap-2 (G2) refers to the stage in the mitotic cycle after cells have doubled their DNA content and which precedes mitosis (Fig. 1). Cells deprived of energy source will be predominantly arrested in the G1, the stage before DNA synthesis. An understanding and control of the mitotic cell cycle obviously has a direct bearing with cancer research. Natural substances that regulates this cellular proliferation are still poorly characterized glycoproteins called chalones. In contrast to the biopolymeric chalones, Evans and Van't Hoff characterized the physiological parameters of a small molecule in the cotyledons of the garden pea that promotes cell arrest in G2 after germination; many of the responses of this factor resembled those of the chalones (*1*).

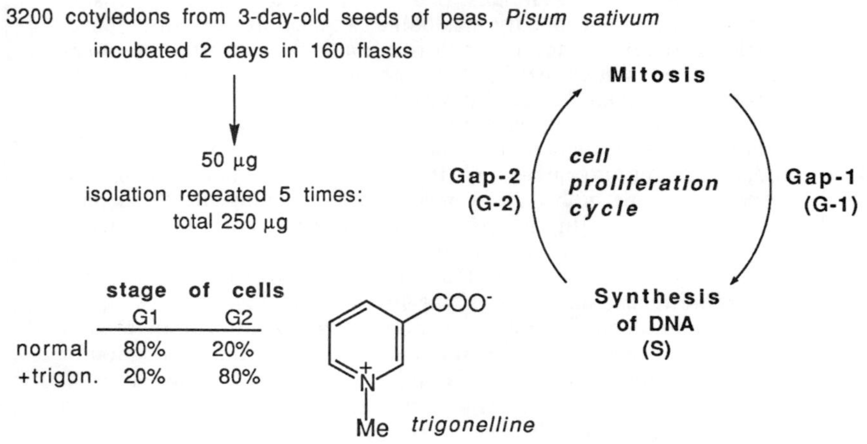

Figure 1. Mitotic cycle, isolation and assay of N-methylnicotinic acid (*2,3*).

Aseptic excision of 3200 seedlings of garden peas, incubation in 160 culture flasks for 2 days, and assay-monitored isolation yielded 50 μg of pure G2 factor, λmax 265 nm; 5 repetitions gave a total of 250 μg of a very polar material. The factor turned out to be none other than N-methylnicotonic acid (trigonelline), a coupound isolated in 1895 and synthesized in 1896. After many failures to measure the MS, satisfactory HRMS was obtained with a sample presumably containing a mixture of the zwitterionic and the more volatile protonated form. FTIR, 80 MHz ^{1}H-NMR and a 9-day measurement of its ^{13}C-NMR clarified the structure (*2, 3*).

Seed Germination Inhibitor.

Extracts of plants collected in East Africa were tested for seed antigermination activity using a simple lettuce seed assay performed in a Petri dish or test tube. Extracts of the East African herb described in Fig. 2 resulted in the isolation and characterization of 4-hydroxyisoxazole ("triumferol"), a simple but new compound (*4*).

***Triumfetta rhomboidea* (Tillaceae), East African medicinal herb**

500 g air-dried leaves

↓

210 mg triumferol
(4-hydroxyisoxazole)

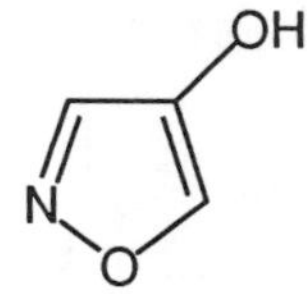

antigermination against lettuce seeds

Figure 2. Seed antigermination factor from East African plant.

Although many 3- and 3,5-disubstituted-4-hydroxyisozoles had been synthesized and tested for antigermination activity (*5*), the parent compound itself had never been synthesized. This is probably because all conventional syntheses failed and hence it may not have been clear whether it could exist at all. It was isolated, hence we knew it existed; a specific synthesis was therefore devised (*4*). What is its activity due to?

Phytoalexins from Sweet Popato infected by Pathogenic Fungi.

Infection of plants by pathogens lead to the rapid production of antifungal compounds for self-defense. The molecular mechanism of this rapid and efficient defense is an important problem but is still obscure. When the sweet potato *Ipomea batatas* is is invaded by the pathogenic black rot fungus *Ceratocystis fimbriata* and other fungi, ipomeamarone, a bitter sesquiterpenoid, rapidly accumulates in the infected tissue. The structure of this first phytoalexin was elucidated in 1956 by Kubota and Matsuura (*6*). As shown in Fig. 3 and 4, inoculation of the roots with fungi or treatment with $HgCl_2$, incubation for 3 days and extraction yielded 7-hydroxycostal, a new class of phytoalexins (*7*), and ten new farnesol-derived ipomeamarones (*8*).

Inoculation of roots of sweet potato (*Ipomea batatas*) with spore suspension of black rot fungus *Ceratocystis fimbriata* and other pathogenic fungi or $HgCl_2$.

Incubation for 3 days, 30°, extraction with $CHCl_3$/MeOH → ***many phytoalexins !***

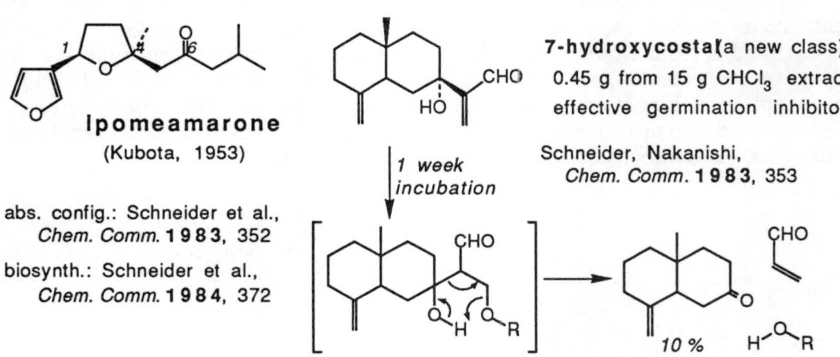

Figure 3. 7-Hydroxycostal, a new class of phytoalexin.

The absolute configuration of ipomeamarone was in conflict, but this was settled unambiguously by chemical correlations. (*9*). The 7-deoxyaldehyde was also produced but this was devoid of activity. It was therefore proposed that the antifungal activity of 7-hydroxycostal is due to the production of acrolein; indeed, a one week incubation led to a 10% production of the bycyclic ketone. In view of the multitude of phytoalexins produced upon inoculation or injuring, this may provide an attractive and simple means of producing new antifungal agents in general.

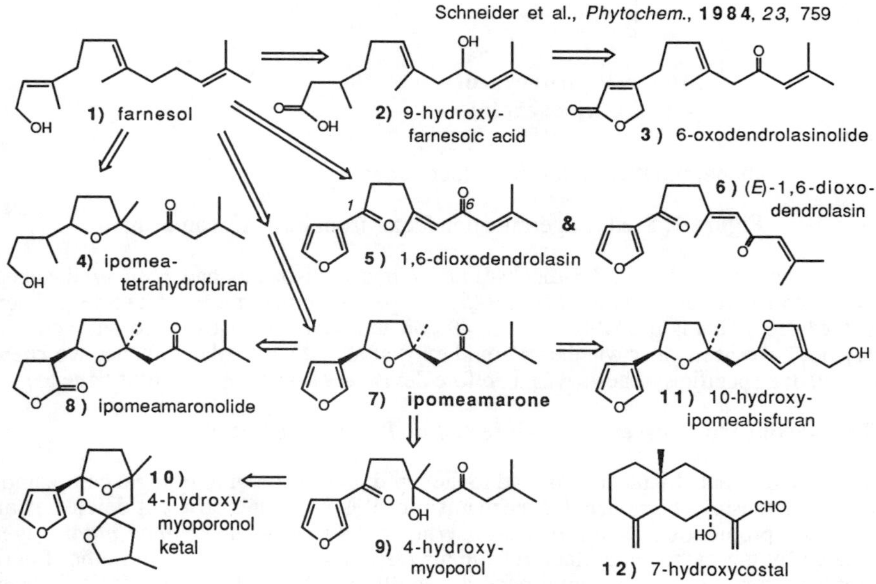

Figure 4. New phytoalexins produced by fungal inoculation or $HgCl_2$ injuring.

Cabenegrins, Anti-snake Venom Factors from *Cabeca ne negra.*

This problem evolved in a most unusual style, the general picture being still enigmatic and requiring further investigation. Dr. Laszlo Darko whom I had never met walked into my office and produced a bottle containing a colorless solution claiming it was an extract of an Amazonian plant "*Cabeca de negra*", and was used by plantation workers in the jungle as an antidote against snake and spider venoms. He had obtained it from a hunter who did not want to disclose the identity of the plant. Two-years of extraction, monitored by injections into mice the 2.5-fold lethal dose of the venom from a common South American snake Fer de lance (*Bothrops atrox*) indeed yielded two phytoalexins, cabenegrins A-I and II (Fig. 5). The minimal dosage required for survival was 2.0-2.8 mg/kg of A-I or A-II (*10*).

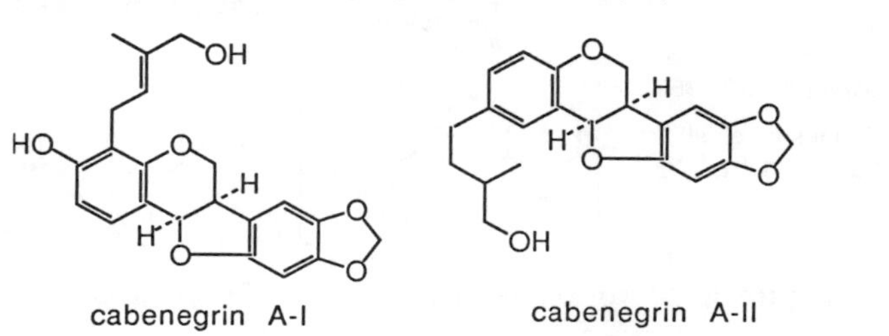

Figure 5. Cabenegrins A-I and A-II, anti-snake venom factors.

Injection of 2.5 mg/kg of the snake venom into a 9 kg male beagle dog led to hypotension, followed by respiratory and cardiac arrest. However, injection of 1 mg/kg of A-I 15 min prior to i.v. injection of the venom restored all three symptoms after 1 hour; converseley, injection of venom followed by injection of cabenegrin also reversed the toxic cardiovascular effects. However, when A-I and -II were tested against some other snake venoms (at a snake venom laboratory in Okinawa) they were ineffective. The cabenegrins must be specific for the Fer de lance venom (and spider venom?). The colorless solution contained a more potent antidote, but the studies were discontinued because Dr. Vick who was performing the assays was retiring. Furthermore, we found that at least ten plants exist in South America called by the name of "Cabeca de negra". Not only would studies of the mode of action be interesting, but from a more pragmatic view point, further investigations to collect more material and characterize the more potent antidote(s) are necessary.

The Brassinosteroids, Plant Growth Promoters.

Following the isolation of the first steroidal plant growth promoter, brassinolide from rape pollen (*11*), 24-methylenebrassinolide and castasterone were characterized (Fig. 6). Fractionation of cat-tail pollen (25 kg), monitored by the standard rice lamina joint bending assay, led to the fourth brassinosteroid typhassterol; 1.7 mg from 25 kg of pollen (*12*).

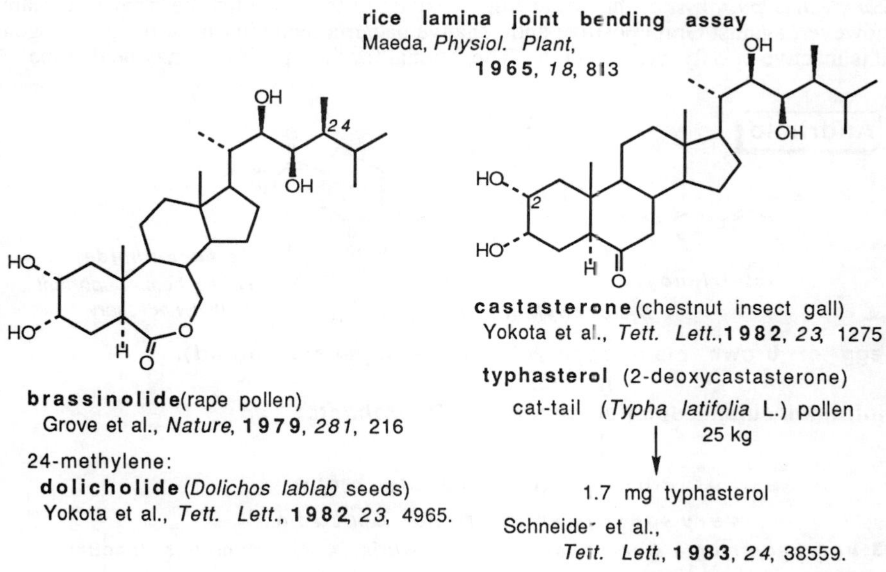

Figure 6. Brassinolisteroids, promising plant growth promoters.

The contents of these plant growth promoters are minuscule, but all are relatively simple to synthesize in quantities. The low LD_{50}, the potent plant growth activity, the ease in synthesizing analogs, growth promoting activity in some fungi as well as in plants, etc. make them attractive candidates for field use in agriculture. Note that typhasterol retains potent plant-growth activity despite the lack of the 2-hydroxy function. Plant physiology is far behind mammalian physiology, and no receptor for a plant hormone has yet been cloned (to my knowledge); moreover, the mode of action is still very obscure in most cases. Clearly, further research is required.

Intracellular Symbionts as a Source of New Antibiotics.

Some insects, e.g., planthoppers, leafhoppers, aphids, and beetles, carry intracellular symbiotic microorganisms in the fat body which are transmitted to the next generation through the eggs. Incubation of the eggs at 35°C gives rise to insects with many defects, *i.e.*, covered with mold, fail to reproduce, greatly reduced in size, etc. The chemical basis of these phenomena remained unaccounted for, due to difficulties associated in culturing these anaerobic organisms, until the successful incubation studies carried out by Nasu, Noda and coworkers (*13,-15*). In the case of the brown planthopper *Nilaparvata lugens,* it has been found that yeast-like symbionts biosynthesize a cholesterol intermediate from mevalonic acid and provide it to the insects which lack steroid biosynthetic capabilities; the planthopper then converts the intermediate into cholesterol, the precursor to the molting hormone, 20-hydroxyecdysone (*17*).

Insects lack immune systems and hence depend on survival by other means such as secretion of defense substances, cuticle formation, and rearing of symbionts. Thus selective killing of symbionts lead to infection by microorganisms and hence nymphs covered with mold. The systematic screening of symbiont cultures from eggs of several insects against crop pest microorganisms led to the characterization of a new antibiotic andrimid from *Nilaparvata lugens* (brown planthopper, collected in Thailand)(*18-20*). As described in Fig. 7, it is highly specific against *Xanthomonas campestris pv. oryzae,* the white blight pathogen that invades the Thai rice plant; however, against Gram positive and negative bacteria including most plant pathogens it is inactive or only weakly active. The rational for this specificity may be that the

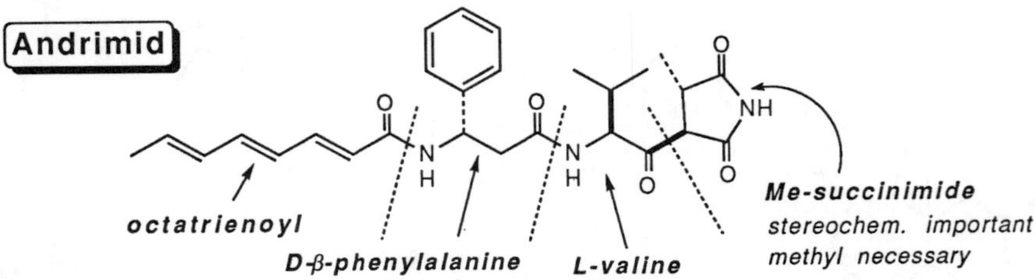

egg of brown planthopper *Nilaparvata lugens* (Thailand).

⇩

Intramolecular bacterial symbiont *Enterobactor sp*

⇩

***andrimid* (from culture broth)**

***very specific* antimicrobial spectrum**

3 *Agrobacterium* 4 *Corynebacterium* 9 *Erwinia*, 34 *Pseudomonas* inactive

Out of 21 *Xanthomonas*
only active against *X. campestris* pv. *oryzae* (blight pathogen) at MIC 0.1μg/ml
and a few others; also *Proteus vulgaris* MIC 0.2 μg/ml

Figure 7. Andrimid, a potent and specific antibiotic from an intracellular symbiont.

symbiont kills the pathogen invading the rice-plant in order to preserve its food source. Several other symbionts were also examined but so far they have only given rise to known antibiotics: *Sogatella furcifera* (planthopper) tested against *Corynebacterium* (canker of tomato) gave pyoluteorin and 2,4-diacetylphloroglucinol; *P. stali* (stinkbug) tested aginst *Salmonella* gave polymixin E_1; and *S. micado* (ambrosia beetle) tested against *Corynbacterium* yielded monoacetyl phloroglucinol and 2,4-ddiacetyl phloroglucinol. Symbiont cultures should provide an unexplored source for the discovery of new bioactive products.

Insect Antifeedants.

An obvious defense mechanism for plants against insects, birds, and animals is to produce feeding deterrents or antifeedants. Furthermore, it is not surprising that plants growing in tropical rain forests are exposed to many more natural hazards and hence are better developed in producing antifeedants. During my connection with the International Centre of Insect Physiology and Ecology, Nairobi, Kenya (1969-1977), we carried out searches for insect antifeedants, which was developed into a systematic and intensive survey by Isao Kubo (*21, 22*). Of the close to 30 antifeedants that were isolated, warburganal and azadirachtin are the two most potent (Fig. 8).

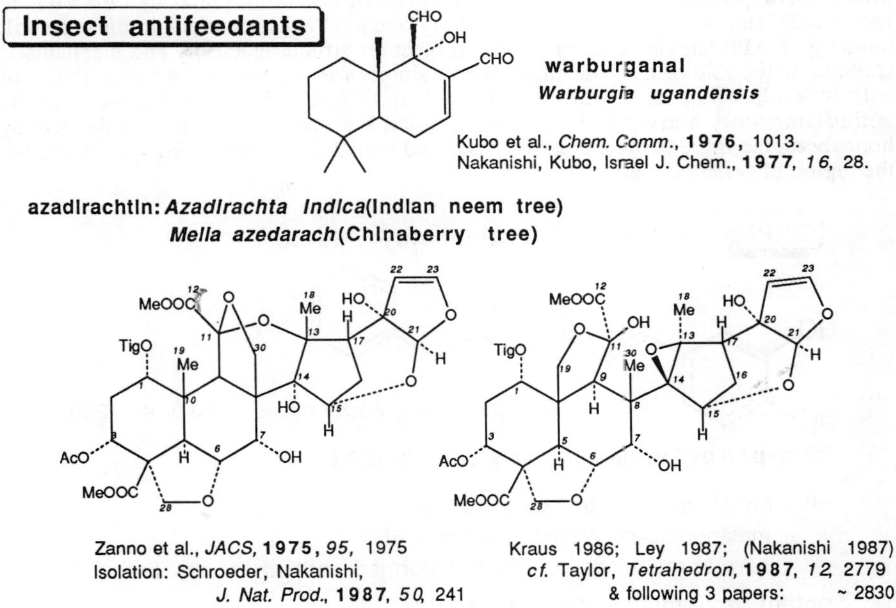

Figure 8. Warburganal and azadirachtin, two potent insect antifeedants.

Warburganal was isolated from the leaves and bark of *Warburgia stuhlmannii* and *W. ugandensis,* plants that are widely used as food spice and folk medicine in East Africa. Electrophysiological studies show that it irreversibly blocks the taste buds of *Spodoptera exempta* (African army-worm)(*23*, also *Isr. J. Chem..* quoted in Figure 8) so that when the army-worm is first placed onto a corn leaf treated with warburganal for 1 hour and then transferred to an untreated leaf, it starves to death. Warburganal exhibits potent molluscidal and antibiotic activity (e.g., *Candida utilis* 3.1 μg/ml (*21*). It is contained in "TADE", a spice traditionally accompanying sashimi, but upon injection exhibits acute toxicity (LD_{50} 20.4 mg/kg against mice) and cytotoxicity (KB 0.01 μg/ml). Many syntheses of warburganal have been published. Electrophysiology tests for insect antifeedants are easy to set up and readily give impulse signals that reflect the responses visually and semiquantitatively.

Azadirachtin was first isolated as a potent antifeedant against the desert locust by Morgan (*24*) who also elucidated partial structures (*25*). A structure was proposed in 1975 by us (Zanno et al., Figure 8) which was subsequently revised by Kraus (*26*), Ley (*27*) and later ourselves (*28*). Azadirachtin, isolated from the Indian neem tree

Azadirachta indica A. Juss is more than an antifeedant and exerts profound physiological effects on insects, the details of which are under study in various laboratories. It is by far the best known "antifeedant", and because ot its ready availability from the plant, it is attracting great interest; several companies dealing with azadirachtin have been formed in India and the US.

CAPE, an Antiinflammatory and Preferential Cytotoxic Agent from Propolis.

Propolis, the brown gum made by honeybees on beehives has long been a popular folk medicine in Eastern Europe, the Near East and other countries, and is alleged to possess antibiotic, antiinflammatory and tumor-growth-arresting properties. Fractionation monitored by cytotoxicity assays yielded the simple but unknown caffeic acid phenethyl ester, which can be synthesized in a one-step reaction without protection/deptrotection (Fig. 9)(*29*). CAPE exhibits preferential cytotoxicity on tumor cells and is a potent antiinflammatory agent. Pharmaceutical studies are still ongoing. CAPE should be an intriguing subject for structure/activity and mechanistic studies. Is the role of the phenethyl moiety simply for increasing the permeability of caffeic acid, which is a well-documented antibiotic? What is the basis of its antiinflammatory activity? It should be noted that since propolis is collected by honeybees, the ingredients are not constant and would differ depending on the season, the region of collection, etc.

CAPE Caffeic Acid Phenethyl Ester

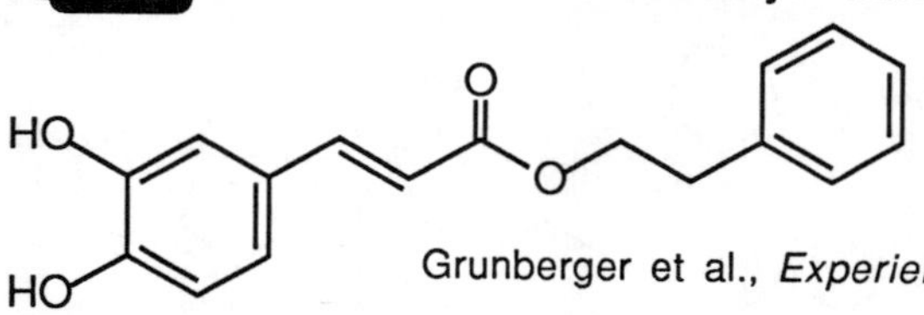

Grunberger et al., *Experientia,* **1988**, *44*, 230

from **propolis** dark-brown wax of beehives

differential cytotoxicity observed in
melanoma and breast carcinoma of:
normal rat/human *versus* transformed rat/human cell lines

potent anti-inflammatory agent

Synthesis caffeic acid + phenethyl alcohol

TsOH ↓ reflux in benzene, 3 days
water removed by Dean-Stark trap

CAPE 40% YIELD

Figure 9. CAPE from propolis

Phytoecdysteroids.

At the time of the structure elucidation of ecdysone **1** (*30*)(Fig. 10), the precursor of the insect molting hormone which Butenandt and Karlson had isolated (25 mg from 500 kg of silkworm pupae, a brilliant and exciting achievement), and Horn and coworkers had isolated 2 mg of 20-hydroxyecdysone **2** from 1 ton of crayfish waste (*31*), our group and that of Takemoto's had isolated large quantities of similar compounds which exhibited potent molting activity, the ponasterones **3** and inokosterone **4** (*32, 33*)(Fig. 10). 20-Hydroxyecdysone **2** is the representative and universal molting hormone of insecta and crustaceans, although other indogenous ecdysteroid molting hormones have recently been found. Approximately 70 ecdysteroids have since been isolated from plants. The discovery of phytoecdysteroids obviously had an impact on insect physiology because it made these hormones readily accessible.

Compared to 20-hydroxyecdysone **2**, muristerone (**5** in Fig. 10 with 14-β-OH) and ponasterone A **3** have 10-fold increased affinities to the ecdysteroid receptor of cultured *Drosophila* Kc-H cells (*34*). Treatment of muristerone with NaI-TMSCl in acetonitrile under argon resulted in removal of the 14-OH to yield **5**, which exhibited an unexpected 5-8 fold further increase in the affinity for the ecdysteroid receptor, the strongest of all known ecdysteroids so far; this 14-deoxymuristerone **5** is also active in chromosome puffing and imaginal disk binding (*34*). 14-Dehydroxylation of various other ecdysteroids have resulted in a clearcut structure/activity relationship in the affinity for the ecdysteroid receptor; in Fig. 10, **6**, the number *n* by parenthesized OH indicates an *n*-fold increase in affinity upon *removal* of that OH, whereas *n* by non-parenthisized OH indicates an *n*-fold increase with *introduction* of that OH (Stonard, R.; Trainor, D. A.; Nakanishi, K.; Cherbas, P. unpublished).

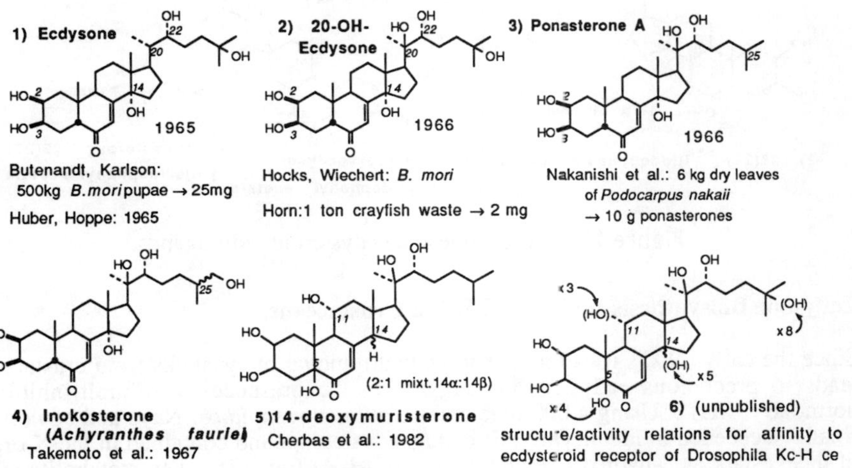

Figure 10. Ecdysteroids.

Preecdysone and Ecdysteroid Radioligands.

The prothoracid glands (PTG) of insects produce ecdysone which is then oxidized to 20hydroxyecdysone in peripheral tissues. Gilbert and coworkers found that incubation of tobacco hornworm (*Manduca sexta*) PTG in the presence of a hemolymph protein fraction (HPF) increased the ecdysone cohntent 8-fold, thus an unknown substance X which is converted into ecdysone must be present (Fig. 11). Usage of RIA antisera H-2 and H-22, which specifically recognize side-chain and ring A modified ecdysteroids, respectively, showed substance X to be ring A modified. Microscale chemical and spectroscopic methods, including CD measurements that showed the presence of an extra sdaturated carbonyl group (Cotton effect at 287 nm), showed substance X t be a 2:1 mixture of 3-dehydroecdysone **1** and 2-dehydroecdysone. This led to the intriguing question as to whether these were preecdysones or their storage forms (*35*). It has recently been shown that indeed in crustaceans, 3-dehydroecdysone is a precursor to ecdysone (Y. Naya et al., in press; see below).

A 4 step derivatization of inokosterone (Fig. 10, **4**) yielded 26-iodoponasterone which turned out to be an ecdysteroid with the strongest affinity to the ecdysteroid receptor (K_d 3.8x10^{-10} M, 20-fold of 20-hydroecdysone) after 14-deoxymuristerone

(50-80 fold, Figure 9, **5**) (*36, 37*); the radioactive analog **2** (*36*) has been used widely in locating the ecdysteroid receptor and used in its receent cloning (*38*). For the purpose of investigating the tertiary structure of the ecdysteroid receptor the potent photoaffinity labeled ecdysteroids **3** and **4** linked to a new phenyl acetate moiety, which readily reacts with hindered secondary OH groups (i.e., 26-OH in inokosterone) have also been prepared (*39*)(Fig. 11).

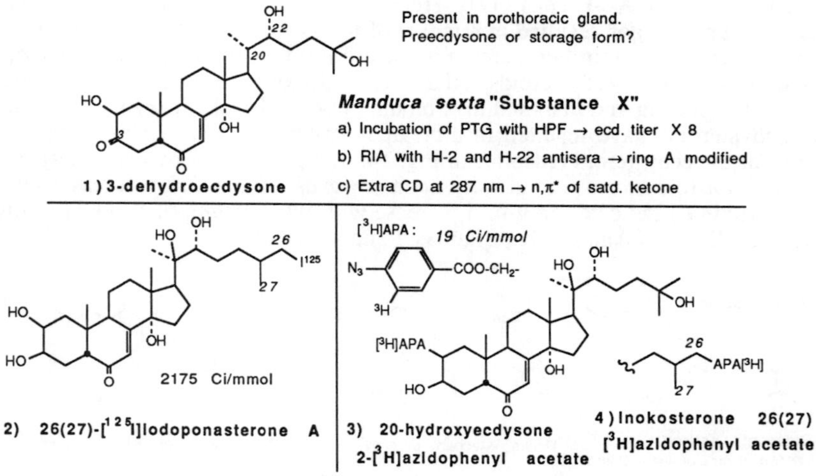

Figure 11. Preecdysone and ecdysteroid radioligands.

Ecdysone Biosynthesis Inhibitor (EBI) of Crustaceans.

Since the early 1900's it had been known that removal of eyestalks from crustaceans leads to precocious ecdysis, thus suggesting the presence of a "molt-inhibiting hormone (MIH)". Using 4,000 blue crabs *Callinectes sapidus* , Naya and coworkers finally succeeded in identifying this endogenous compound contained in the X organ of theeyestalk as 3-hydroxy-l-kynurenine (3-OH-K)(Fig. 12), a key metabolite of L-tryptophan (*40*). It was further found that during the flow of 3-OH-K to the Y-organ complex (YOC) located at the stem of the eyestalk, the site of ecdysone biosynthesis, it was converted into xanthurenic acid, the real ecdysone biosynthesis inhibitor. Xanthurenic acid interferes with the hydroxylation of cholestrol to ecdysosne by interacting with the P-450 system, but does not block the hydroxylation step leading from ecdysone to 20-hydroxyecdysone (*41, 42*). More specifically, the 8-OH of xanthurenic acid coordinates with the P-450 central iron (*42*).

Extensive physiological studies performed recently on the American crayfish (*Procambarus clarkii*) have clarified the following (Naya, Y.; Ikeda, M. *Pure & Appl. Chem.*, in press): (i) 3-dehydroecdysone (Fig. 11, **1**) is the precursor of ecdysone in crustaceans; (ii) 3-dehydroecdysone is not produced upon incubation of the Y organ homogenate, presumably due to destruction of membranes involved in its biosynthesis, but is formed when the total Y organ complex is incubated; (iii) xanthurenic acid also inhibits the ecdysone biosynthesis in the silkworm upon addition to the PTG culture with a detergent. Oligopeptides with molt inhibitory activities have recently been reported for crustaceans (*43, 44*); there clearly must be a close relation between these putative MIHs and EBI but currently this is not clear.

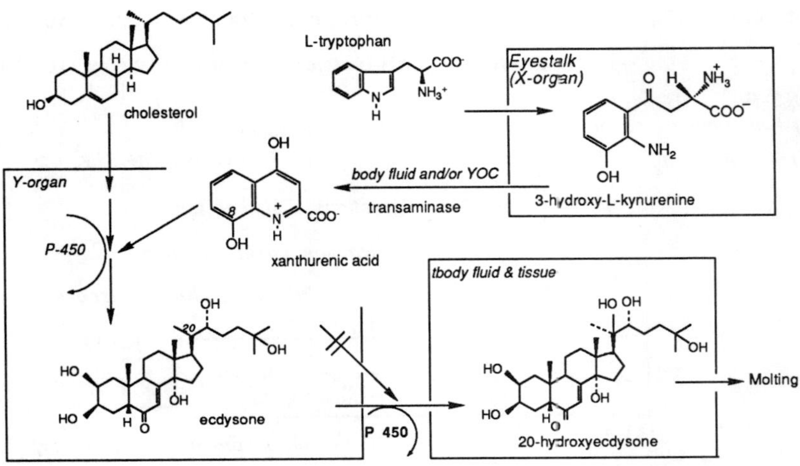

Figure 12. Ecdysone Biosynthesis Inhibitor of Crustaceans (Adapted from ref. 42).

Philanthotoxin, Noncompetitive Antagonist of Glutamate and nACh Receptors.

The glutamate receptors (Glu-R) are implicated in neurological disorders such as Alzheimer's, Huntington's, Parkinson's diseases, epilepsy, ishemic damage due to strokes, as well as in learning and memory. In recent years many subtypes of the Glu-Rs are being rapidly cloned and are attracting great interest in various fields. The venom of the Egyptian solitary digger wasp *Philanthus triangulum*, first noted by T. Piek (*45*), is an antagonist of transmission at the quisqualate-sensitive glutamate synapses (QUIS-R) of locust leg muscles. The major component is a polyamine with the butyryl / tyrosyl / thermospermine sequence, philanthotoxin-433 (PhTX-433)(*46, 47*). In attempts to obtain radioactive (^{3}H or ^{125}I) photoaffinity-labeled analogs with higher biological activity, close to 100 analogs have been synthesized systematically by dividing the molecule into 4 zones and by making single to multiple changes in respective zones (*48, 49*). Although it was initially hoped that the philanthotoxins be selective for the QUIS-R, it was found that they inhibit both N-methyl-D-aspartate (NMDA)(*50, 51*) and kainate receptors (*52*) as well; in addition, they also inhibited nicotinic acetylcholine (nACh) receptors noncompetitively(*53*). Certain types of analogs were more selective for QUIS-R while others were more selective for nACh-R (*54*). The results of electrophysiological assays using locust muscle for the QUIS-R (*55*) and [^{3}H]perhydrohistrionicotoxin displacement assays using*Torpedo* electric organ for the nACh-R (*54*) are summarized below; the NMDA-R assays were performed by binding studies with [^{3}H]MK-801 using rat brain cortex (*54*) and injection of rat or chick brain RNA into *Xenopus* oocytes (*50, 51*) but results are not given in Fig. 13 because they were similar to those of nACh-R (*54*).

The nonspecificity of the PhTX analogs make them poor ligands for receptor identification. However, the availability of many radioligands with photoaffinity labels at various sites make them suitable probes for isolating receptors from relatively pure receptor sources. More importantly, the distribution of photoaffinity groups along a long molecule make the analogs unique "rulers" for probing the tertiary structures and the mode of action of receptor channels (Fig. 14).

Since no Glu-R protein has as yet been secured even in a minuscule amount for structural studies, our current effort is focused on the much more readily available nACh-R (Fig. 14), composed of 5 subunits forminga channel, *i.e.*, α, α, β, γ, and δ.

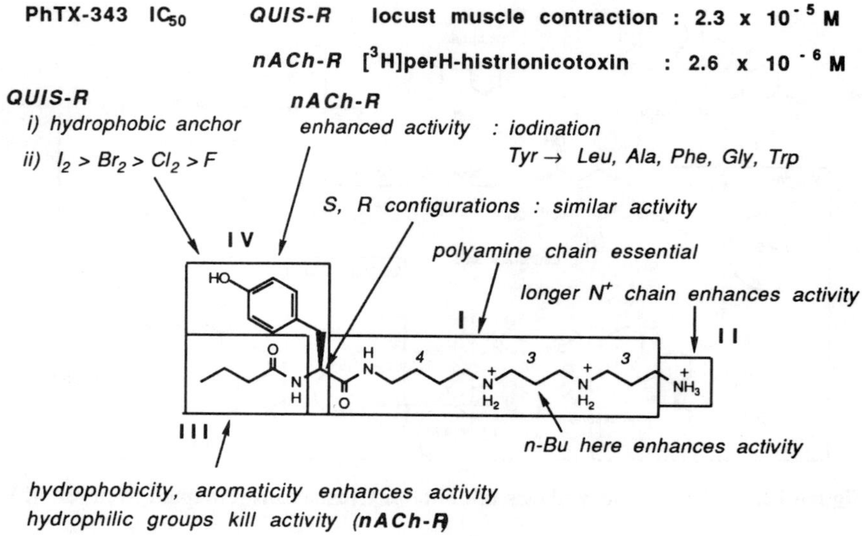

Figure 13. PhTX-433 Structure–activity relationships.

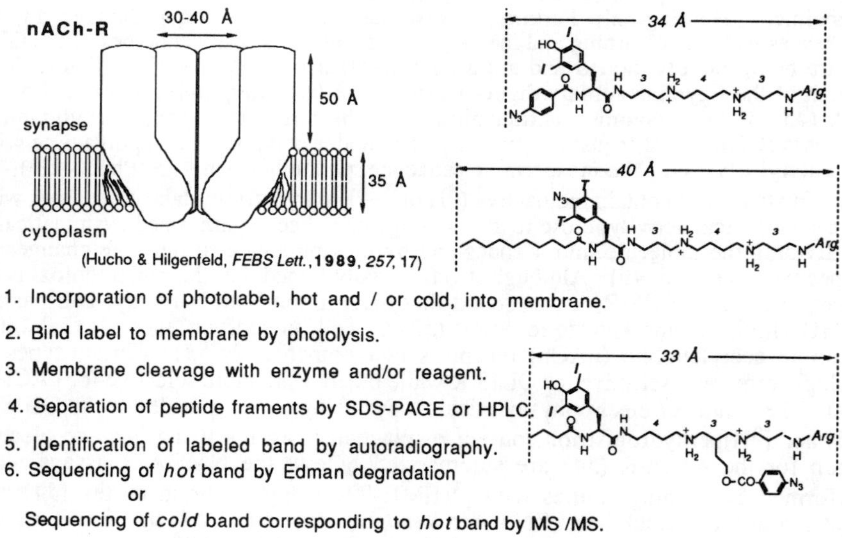

Figure 14. Photoaffinity labeling of nACh-R with PhTX analogs.

The first α-subunit of nACh-R was cloned in 1982 by Numa (*56*) and elucidation of the remaining three followed. Extensive structural studies are being continued by Changeux (*57*), Hucho (*58*), Karlin (*59*), Numa (*60*), Unwin (*61*) and others. These results are summarized in Fig. 13 (*58*). The gross structure of this complex receptor is known, but details are unclear on a molecular structural basis. Photoaffinity labeled PhTX analogs such as those depicted in Fig. 14 have been prepared, and have been shown to irreversibly block the QUIS-R upon irradiation, thus showing that they are binding properly (*62*). These analogs have been bound to pure nACh-R and the protocol outlined in Fig. 14 is being pursued. In order to streamline the tedious separation of cleaved membrane protein fragments, solid phase photoaffinity labeling, cleavage, and tandem MS is under study.

Acknowledgements. The studies were supported by NIH AI 10187. I am most grateful to all coworkers, both at Columbia University and at the Suntory Institute for Bioorganic Research, Osaka.

Literature Cited.

1. Evans, L. S.; Van't Hof, J. *Exp. Cell Res.* **1974**, *87*, 259.
2. Lynn, D. G.; Nakanishi, K.; Patt, S.L.; Occolowitz, J. L.; Almeida, S.; Evans, L. S. *J. Am. Chem. Soc.* **1978**, *100*, 7759.
3. Evans, L. S.; Almeida, M. S.; Lynn, D. G.; Nakanishi, K. Science **1979**, *203*, 1122.
4. Kusumi, T.; Chang, C. W.; Wheeler, M.; Kubo, I.; Nakanishi, K.; Naoki, H. *Tetrahedron Lett.* **22**, *36*, 3451.
5. Kashima,C.;Yamamoto,Y.;Tsuda, Y.;Omote,Y.*Bull.Chem.Soc.Jap.* **1976**, *49*, 1047.
6. Kubota, T,; Matsuura, T.*Chem. Ind. (London)* **1956**, 521.
7. Schneider, J. A.; Nakanishi, K. *Chem. Commun.*. **1983**, 353.
8. Schneider, J. A.; Lee, J.; Naya, Y.; Nakanishi, K.; Oba, K.; Uritani, I. *Phytochem.* **1984**, *23*, 759.
9. Schneider, J. A.; Yoshihara, Y.; Nakanishi, K. *Chem. Commun.* **1983**, 352.
10. Nakagawa, M.;Nakanishi, K.; Darko, L. L.; Vick, J. A. *Tetrahedron Lett.* **1982**, *23*, 3855.
11. Grove, M. D.; Spencer, G. F.; Rohwedder, W. K.; Mandava, N.; Worley, J. F.; Warthen, J. D. Jr.; Steffens, G. L.; Flippen-Anderson, J. L.; Cook, J. C. *Nature (London)* **1979**, *281*, 216.
12. Schneider,J.A.;Yoshihara,K.;Nakanishi,K.;Kato,N.*Tetrahedron Lett.*.**1983**, 3859.
13. Noda, H; Saito, T. *Appl. Ent. Zool.* **1979**, *14*, 64.
14. Kusumi, K.; Suwa, T.;Kita, H.; Nasu, S.; *ibid.*, **1980**, *15*, 129; **1981**, *16*, 88.
15. Noda, H.; Wada, K.; Saito, T. J. Insect Physiol. **1979**, *25*, 443.
16. Fredenhagen, A.; Kenny, P.; Kita, H.; Komura, H.; Naya, Y.; Nakanishi, K.; Nishiyama, K.; Sugiura, K.; Sugiura, M.; Tamura in *Pesticide Science and Biotechnology*; Greenhalgh, R.; Roberts, T. R. Ed.; Proc. IUPAC 6th Internat. Congress Pesticide Chem.; Blackwell Scientific Publications, Oxford, **1987**, 101.
17. Eya, B. K.; Kenny, P. T. M.; Tamura, S. Y.; Ohnichi, M.; Naya, Y.; Nakanishi, K.; Sugiura, M. *J. Chem. Ecology*, **1989**, *15*, 373.
18. Fredenhagen, A.;Tamura, S. Y.;Kenny, P.T. M.; Komura, H.; Naya, Y.; Nakanishi, K.; Nishiyama, K.; Sugiura, M.; Kita, H. *J. Am. Chem. Soc.* **1987**, *109*, 4409.
19. Synthesis: McWhorter, *W.;* Fredenhagen, A.; Nakanishi, K.; Komura, H. *Chem. Commun*, **1989**, 299.
20. Kenny, P. T. M.; Tamura, S. Y.; Fredenhagen, A.; Naya, Y.; Nakanishi, K.; Nishiyama, K.; Sugiura, M.; Kita, H.; Komura, H. *Pestic. Sci.* **1989**, *27*, 117.
21. Kubo, I.; Nakanishi, K. In *Host Pland Tesistance to Pests:* Hedin, P. A. Ed.; ACS Symposium Series No. 62, Am Chem. Soc.: Washington, D.C., **1977**, 165.
22. Kubo, I.; Nakanishi, K. In *Advances in Pesticide Science* , Part 2: Geissbuhler, H. Ed.; Pergamon Press, New York, 1979, 284.
23. Ma, W. -C. Physiol. Entomol. **1977**, *2*, 199.
24. Butterworth, J. H.; Morgan, E. D. *Chem. Commun.* **1968**, 23.
25. Butterworth, J. H.; Morgan, E. D.; Percy, G. R. *JCS Perkin Trans. I,* **1973**, 2445.
26. Kraus, W.; Bokel, M.; Klenk, A.; Pohnl, H. *Tetrahedron Lett.* **1985**, *26*, 6435.
27. Broughton, H. B.; Ley, S. V.; Slawin, A. M. Z.; Williams, D. J.; Morgan, E. D. *Chem. Commun.* **1986**, 46.
28. Turner, C. J.; Tempesta, M. S.; Taylor, R. B.; Zagorski, M. G.; Termini, J. S., Schroeder, D.; Nakanishi, K. *Tetrahedron* **1987**, *43*, 2789.
29. Grunberger, D.; Banerjee, R.; Eisinger, K.; Oltz, E. M.; Efros, L.; Caldwell, M.; Estivez, V.; Nakanishi, K. *Experientia* **1988**, *44*, 230.
30. Huber, R.; Hoppe, W. *Chem. Ber.* **1965**, *98*, 2403.
31. Hampshire, F.; Horn, D. H. S. *Chem. Commun.* **1966**, 37.

32. Nakanishi, K.; Koreeda, S.; Sasaki, S.; Chang, M. L. Hsu, H. Y. *Chem. Commun.* **1966**, 915.
33. Takemoto, T.; Ogawa, S, Nishimoto, N. *Yakugaku Zasshi* **1967**, *87*, 325.
34. Cherbas, P.;Trainor,D. A.; Stonard, R.;Nakanishi, K. *Chem. Commun.* **1982**, 1307.
35. Warren, J. T.; Sakurai, S.; Roountree, D. B.; Gilbert, L. I.; Lee, S. S.; Nakanishi, K. *Proc. Natl. Acad. Sci. USA*, **1988**, *85*, 958.
36. Cherbas, P.; Cherbas, L.; Lee, S. S.; Nakanishi, K. *Proc. Natl. Acad. Sci. USA* **1988**, *85*, 2096.
37. Lee, S. S.; Nakanishi, K.; Cherbas, P. *Chem. Commun.* **1991**, 51.
38. Koelle, M. R.; Talbot, W. S.; Segraves, W. A.; Bender, M. T.; Cherbas, P.; Hogness *Cell* **1991**, *67*, 59.
39. Boehm, M. F.; Nakanishi, K.; Cherbas, P. *Chem. Commun.* **1991**, 52.
40. Naya, Y.; Kishida, K.; Sugiyama, M.; Murata, M.; Miki, W.; Ohnishi, M., Nakanishi, K. *Experientia* **1988**, *44*, 50.
41. Naya, Y.; Ohnishi, M.; Ikeda, M.; Miki, W.; Nakanishi, K. *Proc. Natl. Acad. Sci. ISA*. **1989**, *86*, 6826.
42. Naya, Y.; Miki, W.; Ohnishi, M.; Ikeda, M.; Nakanishi, K. *Pure & Appl. Chem.* **1989**, *61*, 465.
43. Webster, S. G.; Keller, R. *J. Comp. Physiol.* **1986**, *B 156*, 617.
44. Chang, E. S.; Bruce, M. J.; Newcomb, R. W. *Gen. Comp. Endocr.* **1987**, *65*, 56.
45. Piek, T.; Mantel, P.; Engels, E. *Comp. Gen. Pharmacol.* **1971**, *2*, 317.
46. Eldefrawi, A.T.; Eldefrawi, M.E.; Konno, K.; Monsour, K.; Nakanishi, K.; Oltz, E.; Usherwood,P. *Proc. Natl. Acad. Sci.* **1988** *85*, 4910-4913.
47. Piek, T.; Fokkens, R. H.; Karst, H.;`Kruk, C.; Lind, A.; Van Marle, J.; Nakajima, T.; Nibbering, N. M. M.; Shinozaki, H.; Spanjer, W.; Tong, Y. C. In *Neurotox '88:* Lunt, G. G., Ed.; Elsevier: Amsterdam, **1988**, 61.
48. Goodnow, R.; Konno, K.; Niwa, M.; Kallimopoulos, T.; Bukownik, R.; Lenares, D.; Nakanishi, K. *Tetrahedron* **1990**, *46*, 3267-3286.
49. Nakanishi, K.; Goodnow, R.; Konno, K.; Niwa, M.; Bukownik, R.; Kalimopoulos, T.A.; Usherwood, P.; Eldefrawi, A.T.; Eldefrawi, M.E. *Pure & Appl. Chem.* **1990**, *62*, 1223-1230.
50. Brackley, P.; Goodnow, R.; Nakanishi,K.; Sudan, H.L.; Usherwood, P.N.R. *Neuroscience Lett.* ***1990***. *114*, 51-56.
51. Ragsdale, D.; Grant,D.B.; Anis, N.A.; Eldefrawi, A.T.; Eldefrawi,M.E.; Konno, K.; Miledi, R. *J. Pharmacol. Exp. Therap.*, **1989**, *251*, 156-163.
52. Goodnow, R.A. ; Bukownik, R.; Nakanishi, K.; Usherwood, P.N.R.; Eldefrawi, E.T.; Anis, N.A. ; Eldefrawi, M.E. *J. Med. Chem.* **1991**, *34*, 2389.
53. Rozenthal, R.; Scoble, G.T.; Albuquerque, E.X.; Idriss, M.; Sherby, S.; Saattelle, D.B.; Nakanishi, K.; Konno, K.; Eldefrawi, A.T.; Eldefrawi, M.E. *J.Pharmacol. Exp. Therap.* **1989**, *249*, 123-130.
54. Anis, N.; Sherby, S.; Goodnow, R.; Niwa, M.; Konno, K.; Kallimopoulos, T.;Bukownik, R.; Nakanishi, K.; Usherwood, P.N.R.; Eldefrawi, A.T; Eldefrawi, M.E. *J. Pharmacol. Exp. Ther.* **1990**, *254*, 764.
55. Bruce, M.; Bukownik, R.; Eldefrawi, A.T.; Eldefrawi, M.E.; Goodnow, R.; Kalimopoulos, T.; Konno, K.; Nakanishi, K.;Niwa, M. ; Usherwood, P.N.R. *Toxicon*, **1990**, *28*, 1333-1346.
56. Noda, M.; Takahashi, H.; Tanabe, T.; Toyosato, M.; Furutani, Y.; Hirose, T.; Asai, M.; Inayama, S.; Miyata, T.; Numa, S. *Nature* **1982**, *299*, 793.
57. Changeux, J.-P. *Fidia Research Foundation Neuroscience Award Lectures*, Vol. 4, Raven Press, New York, **1990**, 21-168.
58. Hucho, F.; Hilgenfeld, R. *FEBS LETT.* **1989**, *257*, 17.
59. Karlin, A. *The Harvey Lectures,* Series 85, Wiley-Liss, New York, **1991**, 71-107.
60. Imoto, K.; Busch, C.; Sakman, B.; Mishima, M.; Konno, T.; Nakai, J.; Bujo, H.; Mori, Y.; Fukuda, K.; Numa, S. *Nature* **1988**, *335*, 645.
61. Toyoshima, Ch.; Unwin, N. *Nature* **1988**, *335*, 247.
62. Goodnow, R. A.; Nakanishi, K.; Sudan, H. L.; Usherwood, P. N. R. *Neurosci. Lett.* **1991**, *125*, 62.

RECEIVED October 11, 1993

Chapter 3

Microbial Secondary Metabolites as a Source of Agrochemicals

Richard J. Stonard[1], Stephen W. Ayer[2], John J. Kotyk[1], Leo J. Letendre[1], Carolyn I. McGary[3], Thomas E. Nickson[1], Ngo LeVan[4], and Paul B. Lavrik[1]

[1]Agricultural Group, Monsanto Company, 700 Chesterfield Parkway, Chesterfield, MO 63198
[2]Institute for Marine Biosciences, National Research Council of Canada, Halifax, Nova Scotia B3H 3Z1, Canada
[3]Bristol Myers-Squibb, Evansville, IN 47721
[4]Sandoz Agro Ltd., Crop Protection Research, CH–4002 Basel, Switzerland

The isolation and structure elucidation of four novel microbial secondary metabolites are described. Diabroticin A ($C_{16}H_{28}N_6O_6$) and B ($C_{16}H_{28}N_6O_5$) are *Bacillus* derived pyrazines possessing potent activity against the southern corn rootworm. Arginidiene ($C_{14}H_{21}N_5O_5$) and phytophthoradiene ($C_{13}H_{17}N_3O_6$) are weakly fungicidal dipeptides incorporating an unusual cyclohexadiene moiety.

Several of the top 20 pharmaceuticals in gross sales in 1992 were natural products or close derivatives thereof (eg. Mevacor, Ceclor). Of the top 20 crop chemicals in this same period, only one class of compounds, the pyrethroids, are based on a natural substance, the plant metabolite pyrethrin. Three active areas of chemistry have the potential to impact this situation within the next ten years: the methoxyacrylate family of fungicides based upon the fungal metabolite, strobilurin; Basta, a post-emergent broad-spectrum herbicide based upon the *Streptomyces viridochromogenes* metabolite phosphinothricylalanylalanine; and avermectin, and its semi-synthetic derivatives, insecticides originating from *Streptomyces avermitilis*.

Several reasons are cited for the success of natural products in health care as compared to agriculture. The most prevalent of these are; their frequent lack of amenability to cost effective total synthesis, or that they are too structurally complex to serve as models for analogue synthesis programs; the absence of environmental stability; insufficient efficacy (typically too specific/ too narrow a spectrum of activity); and the high cost-of-goods of the natural substance *per se*. The latter is a particularly significant distinction between crop chemicals and pharmaceuticals. For example, the average costs per treatment in 1992 for some of the more common agrochemicals were: cotton caterpillar control, $10/acre; wheat foliar disease control, $17/acre; and broad and narrow leaf weed control in

0097–6156/94/0551–0025$06.00/0

corn, \$17/acre. For comparative purposes, perhaps the least expensive fermentation product, penicillin at \$50/lb, as a hypothetical agrochemical would have to be used in the range of 40g/acre or less in order to be economically viable. This use rate is within the range of only a few modern crop chemicals.

While significant hurdles to commercialization exist, natural products offer several potential advantages including: target specificity; novel mechanisms of action; and unique structures upon which to base synthesis efforts. Consequently, several major agrochemical companies have been or are presently engaged in natural product screening programs. In this chapter we report on the discovery in our laboratories of a new class of natural insecticides derived from micro-organisms, the diabroticins, and two structurally related microbial secondary metabolites, arginidiene and phytophthoradiene, possessing antifungal activity.

Diabroticins

Microbial secondary metabolites have proven to be a modestly significant source of insecticidally active substances. Avermectin, a metabolite of *Streptomyces avermitilis,* and its derivatives have found utility as commercial pest control agents (*1*). Others such as pyrrolnitrin have served as the basis for the development of novel synthetic insecticides (*2*). As part of an investigation of antiinsectan metabolites produced by micro-organisms, we identified a new family of water soluble pyrazine-amino acid conjugates produced by *Bacillus subtilis* having potent activity against the southern corn rootworm, *Diabrotica undecimpunctata*. Herein, we describe the isolation, characterization and biological activity of the most abundant members of this family, diabroticin A (**1**) and diabroticin B (**2**).

The diabroticins were isolated from fermentation broths of *B. subtilis* (isolate ALB102) obtained from a soil sample collected in an oat field in South Dakota. Bioassay guided purification versus *D. undecimpunctata* afforded **1**, 13 mg/L, and **2**, 3 mg/L, as colorless oils. The 0.45μm filtered broth was purified by open column, gradient cation

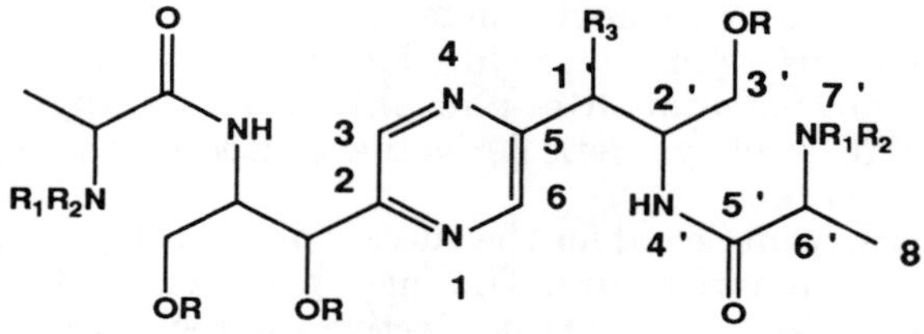

1 R, R_1, R_2 = H, R_3 = OH
2 R, R_1, R_2, R_3 = H
3 R, R_1 = Ac, R_2 = H, R_3 = OAc
4 R, R_1 = Ac-d_3, R_2 = H, R_3 = OAc-d_3
5 R = H, R_1, R_2 = =CHN$(CH_3)_2$, R_3 = OH
6 R, R_1 = H, R_2 = 2,4-dinitrophenyl, R_3 = OH
11 R, R_1 = Ac, R_2, R_3 = H
12 R, R_3 = H, R_1, R_2 = =CHN$(CH_3)_2$

exchange chromatography (IRC 50 NH_4^+ form; 0-5N NH_4OH) followed by repeated cation exchange HPLC (Whatman SCX). Crude broths showed significant activity against Colorado potato beetle, boll weevil, mosquito larvae, *Staphylococcus aureus* and *Micrococcus lutea,* but were inactive against European corn borer, *Escherichia coli, B. subtilis* and *Pseudomonas aeruginosa*. Diabroticin A was also isolated (7 mg/L from 1.8L of filtrate) from fermentation broths of *B. cereus* (isolate B739) isolated from a soil sample taken from a cantaloupe field in Georgia.

Mass spectral analysis (EI, CI, +/- FAB) of diabroticin A (**1**, $[\alpha]_D^{23}$ (*c* 2.1 in MeOH) -6.7) failed to detect a molecular ion or characteristic mass fragments. The molecular formula of **1**, $C_{16}H_{28}N_6O_6$ (MW 400 Da, 6 sites of unsaturation), was deduced on the basis of mass spectral analysis of acetylated, dimethylamidine, and hydrogenated derivatives (Table I). A UV maximum at 276 nm in conjunction with a downfield singlet resonance at δ8.67 in the ^1H NMR spectrum (Table II) and a ^{13}C NMR spectrum (Table III) displaying resonances at δ145.0 and δ157.2 suggested the presence of a heteroaromatic moiety. Absorption bands at 1660 and 1560 cm^{-1} in the FTIR spectrum in combination with a ^{13}C NMR APT resonance at δ173.52 (qC) indicated the presence of a secondary amide functionality. The ^1H NMR and ^{13}C NMR spectra of **1** in D_2O contained resonances for 9 protons and 8 carbons, respectively, denoting a high degree of molecular symmetry.

Acetylation of **1** afforded hexa-acetyldiabroticin A (**3**, Table I). Derivatization with Ac_2O-d_6 (**4**) and analysis by FABMS and ^1H NMR confirmed the synthetic origin of all six acetyl groups. The ^1H NMR spectrum of **3** in $CDCl_3$ (Table II) exhibited two acetyl methyl singlet resonances at δ2.19 (12H) and δ2.10 (6H) and two D_2O-exchangeable two proton doublet resonances at δ7.06 (J=10.0 Hz) and δ6.97 (J=9.0 Hz). In conjunction with the molecular formula and IR data, this indicated the presence of four hydroxyl and two primary amine groups in **1**. The presence of the two primary amine groups was confirmed by the reaction of **1** with *N,N*-dimethylformamide dimethyl acetal (DMF/DMA) to give the bisdimethylamidine derivative **5** (see Table I). In addition, treatment of **1** with Sanger's reagent led to the formation of **6** with incorporation of two 2,4-dinitrophenyl units.

The proton decoupled ^{15}N NMR (40.5 MHz, δ in ppm upfield of internal nitromethane at 0 ppm) spectrum of **1** contained three signals at δ -61.7, δ –263.9 and δ –341.9 indicating, when taken in conjunction with the molecular formula, the presence of two aromatic, two amide and two amine nitrogen atoms, respectively. Hydrogenation of **1** resulted in the uptake of three moles of hydrogen (Table I). Subsequent acetylation of the hydrogenation product, **7**, gave an octaacetyl derivative, **8**, having a protonated molecular ion at m/z 743 Da (+FABMS; confirmed by a parallel Ac_2O-d_6 acetylation experiment), thereby confirming that the heteroaromatic ring was six membered and contained two nitrogen atoms.

Table I. HRFAB Mass Spectral Data for Derivatives of Diabroticin A

Derivative	Formula	ΔM mmu
Hexa-acetyl (**3**)[a]	$C_{28}H_{40}N_6O_{12}$	2.4
Bisdimethylamidine (**5**)	$C_{22}H_{38}N_8O_6$	3.6
Hexahydro (**7**)[b]	$C_{16}H_{34}N_6O_6$	0.5

[a] Ac_2O/4-pyrrolidinopyridine, rt, 24 hr [b] H_2/PtO_2/TFA rt, 1 atm, 4 hrs, 3 moles absorbed.

Table II. ^{1}H NMR Data for Diabroticin A (1), B (2) and Hexa-acetyldiabroticin A (3)

position	diabroticin A (**1**)[a]	hexacetyl-diabroticin A (**3**)[b]	diabroticin B (**2**)[c]
3	8.67 (2H, s)[e]	8.46 (2H, s)[e]	8.58 (1H, d,1.3)
6			8.45 (1H, d,1.3)
1´	4.97 (2H, d, 7)	5.64 (2H, d, 10.5)	3.13 (1H, dd, 14, 4) 2.86 (1H, dd,14,10)
2´	4.28 (2H, brq, 7)	4.76 (2H, dddd, 10.5, 10, 3, 3)	4.24 (1H, m)
3´	3.76 (4H, d, 7)	4.61 (2H, dd, 12, 3) 4.17 (2H, dd, 12, 3)	3.73 (1H, dd, 11,1.5) 3.88 (1H, dd, 11, 6)
4´		7.06 (2H, d, 10,)	
6´	3.94 (2H q, 7)	4.50 (2H, dq, 9, 7)	3.54 (1H, q, 7)[d]
7´		6.97 (2H, d, 9)	
8´	1.44 (6H, d, 7)	1.24 (6H, d, 7)	1.31 (3H, d, 7)
1´´			4.75 (1H, d, 8.5)
2´´			4.16 (1H, m)
3´´			3.63 (2H, m)
6´´			3.58 (1H, q, 7)[d]
8´´			1.34 (3H, d, 7)
OAc		2.19 (12H, s)	
NAc		2.10 (6H, s)	

[a] ^{1}H NMR spectra were recorded at 300 MHz (D_2O) or [b]at 400 MHz in $CDCl_3$ or [c]at 500 MHz in CD_3OD. [d]Assignments of 6´ and 6´´ may be interchanged. Assignments were aided by DQCOSY (*3*) experiments. *J* values are in hertz, and chemical shifts, δ in ppm relative to HOD at δ 4.63 ppm. [e] Signals correspond to position 3 and 6.

Compound **7** showed a UV maximum at 222 nm and disappearance of the aromatic protons observed at δ 8.67 in **1.** Thus, the twelve heteroatoms present in **1** could be accounted for by 4 -OH, 2 -HNCO-, 2 $-NH_2$ and 2-ArN and the six sites of unsaturation could be ascribed to two amide groups and one six membered aromatic ring containing two nitrogen atoms.

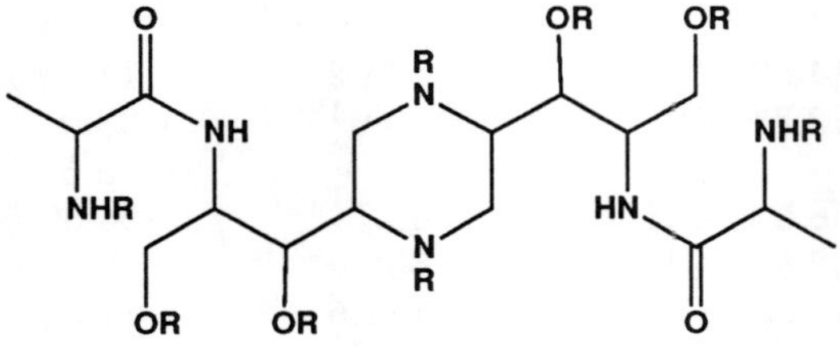

7 R = H
8 R = Ac

The structure of diabroticin A (**1**) was established by a series of NMR experiments. The COSY spectra of **1** and of hexa-acetyldiabroticin A (**3**) showed essentially the same proton-proton correlations, fully consistent with two isolated spin systems as illustrated in partial structures **A** and **B** for hexacetyldiabroticin A (**3**). The two amide proton doublet resonances at δ7.06 (H_b) and δ6.97 (H_c) in **3** showed strong correlations with a methine resonance at δ4.76 (H_e, dddd) and a methine resonance at δ4.50 (H_g, dq), respectively. In addition, observation of a correlation between the aromatic proton resonance at δ8.67 in **1** and the methine doublet resonance at δ4.97 (H_d , not resolved in 1D - ^1H NMR) established the connection of the heteroaryl group as shown in **A**.

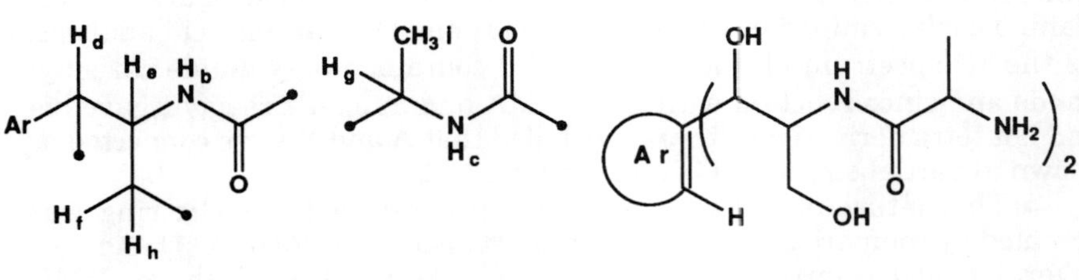

A **B** **C**

Determination of the side chain structure was facilitated by two acid hydrolyses and subsequent derivatizations of the reaction products. Treatment of **1** with 2N-HCl in the presence of trace amounts of methanol at 85°C for 3 hr followed by acetylation with

Table III. ^{13}C NMR Data for Diabroticin A (1), B (2) and Hexa-acetyldiabroticin A (3)

carbon no.	diabroticin A (1)[a]		hexa-acetyl-diabroticin A (3)[b]	diabroticin B (2)[a]	
2	157.2	C	151.4	164.5	C
5				155.4	C
3,6	145.0	CH	144.2	144.0[c], 146.0[c]	CH
1´	73.8[d]	CH	70.9[d]	38.0	CH_2
1´´				73.6	CH
2´, 2´´	58.5	CH	49.7	54.3[c], 58.4[c]	CH
3´, 3´´	62.5	CH_2	63.2	65.5[c], 62.0[c]	CH_2
5´, 5´´	173.5	C	171.6	173.5, 173.5	C
7´, 7´´	51.9	C	47.8	52.0, 52.0	C
8´, 8´´	19.4	CH_3	21.0	26.0[c], 19.8[c]	CH_3
$CH_3\underline{C}$=O			171.5,170.6,169.5,169.4		
$\underline{C}H_3C$=O			15.0, 23.3		

[a] ^{13}C NMR were recorded at 75 MHz in D_2O or, [b] in $CDCl_3$. [c] Assignments of carbons 3/6, 2´/2´´, 3´/3´´, 8´/8´´ may be interchanged. Assignments for **1** and **2** were aided by APT (*4*) experiments. Chemical shifts, δ in ppm downfield from TMS. [d]Signals represent positions 1´ and 1´´.

Ac_2O/4-PP resulted in the formation of compounds **9** and **10**. Compound **9** exhibited a protonated molecular ion at *m/z* 511 Da (+FABMS) . The ^1H NMR spectrum of **9** contained an aromatic proton singlet resonance at δ8.62 and three acetyl methyl resonances (δ2.17, δ2.07 and δ1.92), consistent with the formation of a symmetrical degradation product of **1** lacking two alanine residues. More vigorous acid hydrolysis (6N HCl/110°C/24 hr) of **1** resulted in the formation of two mole equivalents of alanine as determined by quantitative PTH and OPA amino acid analysis. As the IR spectrum of diabroticin A (**1**) contained only amide carbonyl bands and since reaction with DMF/DMA and Sanger's reagent gave bis and not tetra derivatives, it was concluded that **A** and **B** were connected as shown in partial structure **C** for diabroticin A (**1**).

The nature of the six-membered aromatic heterocyclic ring was revealed by comparison of spectral data between diabroticin A (**1**) and several model compounds (see Table IV). As mentioned above, NMR analysis of compound **1** showed one aromatic nitrogen resonance at δ -61.8, two aromatic carbon resonances both adjacent to nitrogen (δ144.96, δ157.15) and one aromatic proton resonance at δ8.67, consistent with a 2,5-disubstituted pyrazine. The 2,6-disubstituted pyrazine possibility was ruled out on the basis of the equivalence of the 2 new acetyl methyl signals

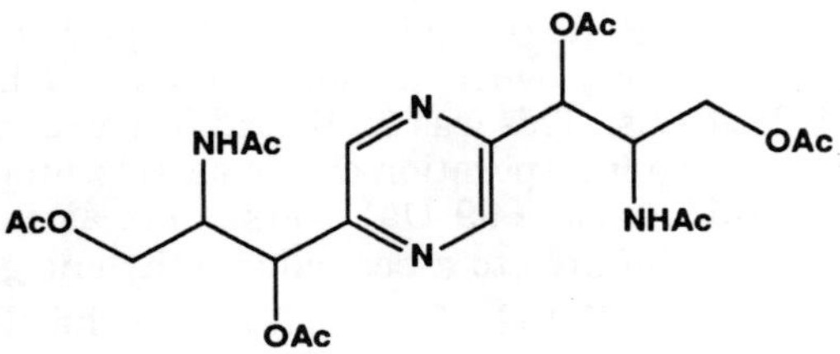

9

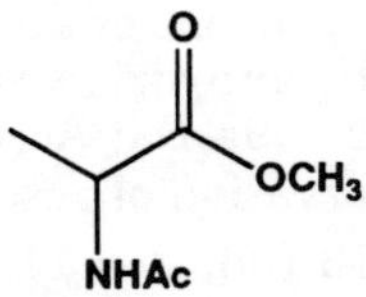

10

Table IV. NMR[f] and UV Comparison of 1 and Related Model Compounds

	UV λmax	^{15}N NMR[a]	^{13}C NMR[b]	^{1}H NMR[b]
Pyrimidine	240 nm	-89[g]	122, 160,157[e]	7.36,8.78, 9.26[d]
Pyridazine[g]	247 nm	+17[g]	128, 153	7.54, 9.24[c]
2,6-Dimethylpyrazine[g]	275 nm	+53 (t), +56 (s_{br})[h]	145 d, 153 s	8.3 s
2,5-Dimethylpyrazine[g]	274 nm	-52[h]	145 d, 153 s	8.3 s
Fructosazine (*5*)	276 nm	ND[i]	145 d, 158 s	8.7 s
Diabroticin A (**1**)	275 nm	-61	145 d, 157 s	8.67 s

[a] 40.5 MHz, nitromethane as external reference at 0ppm in D_2O. [b] See Table 2 and 3 for detail. [c] H-4, H-3, resp., $J_{3,4}$=4.7 Hz, $J_{4,5}$ =9.0 Hz. [d] H-2, H-4, H-5, resp., $J_{4,5}$=5.0 Hz. [e] C-5, C-2, C-4,6 resp. [f] In D_20 unless indicated. Aromatic resonances only. [g] $CDCl_3$. [h] CD_3OD. [i] Not determined.

observed in ^{1}H NMR spectra of **8**, resulting from the acetylation of the hydrogenation product (**7**) of **1**. Furthermore, 2,6-disubstituted pyrazines show two discrete aromatic nitrogen signals in the proton decoupled ^{15}N NMR spectrum (Table IV). Consequently, the structure of diabroticin A is concluded to be as shown in **1**. Confirmation of this regiochemical assignment was provided by the structure determination of diabroticin B as discussed below.

As was the case with diabroticin A (**1**), diabroticin B (**2**) failed to yield a molecular ion or useful mass fragments by either EI, CI or +/- FABMS. The molecular formula of **2**, $C_{16}H_{28}N_6O_5$ (MW 384 Da, 6 sites of unsaturation), was deduced by mass spectral analysis (+FABHRMS) of bisdimethyl-amidine (**12**) and pentacetyl derivatives (**11**). Comparison of the molecular formulae, UV, ^{1}H NMR and ^{13}C NMR spectra clearly indicated that **1** and **2** differed only by the presence of a single hydroxyl group on one of the two side chains. This was confirmed by treatment with acetic anhydride resulting in the incorporation of five acetyl units (CIMS, *m/z* 594 Da; Ac_2O-d_6, CIMS, *m/z* 609 Da) versus six as in **1**. The observation of resonances attributable to a benzylic methylene group at δ3.13 (dd, J = 14, 4 Hz) and δ2.86 (dd, J = 14,10 Hz) in the ^{1}H NMR spectrum and at δ38.0 in the ^{13}C NMR APT spectrum of **2**, assigned the missing hydroxyl in **2** to one of the two benzylic positions (i.e. C-1′). Consequently, the structure of diabroticin B was concluded to be as shown in **2**, above.

Further evidence for the presence of two intact alanine residues in diabroticin B (**2**) was provided by acid hydrolysis (6N HCl/110°C/24 hr) followed by quantitative PTH and OPA amino acid analysis. Treatment of **2** with DMF-DMA resulted in formation of a bisdimethylamidine adduct (FABMS, *m/z* 494 Da) as predicted. Confirmation of the 2,5-pyrazine substitution pattern in diabroticin A (**1**) and B (**2**) was facilitated by the unsymmetric nature of **2**. Two chemically nonequivalent aromatic proton resonances were observed in the ^{1}H NMR spectrum at δ8.58 and δ8.45 as 1.3 Hz doublets consistent with $J_{3\text{-}6}$ in mono-substituted pyrazines (*6*). 2-Methylpyrazine and other mono-substituted pyrazines show long range coupling constants of $J_{3\text{-}6}$ = 1.33-1.54 Hz, $J_{3\text{-}5}$ = 0.01-0.35 Hz and $J_{5\text{-}6}$=2.4-2.9 Hz.

Isolated diabroticin A (**1**) was active against first instar larvae of *D. undecimpunctata* at 2-4 ppm in an *in vitro* diet incorporation assay(*7*). Similarly, diabroticin B (**2**) was active at 25-50 ppm. No effect was observed on egg hatch with either compound, and **1** did not deter feeding of *D. undecimpunctata* first instar larvae in a choice assay (*8*). The toxicity of the diabroticins to mammals has not been determined.

Several metabolites related to **1** and **2** have been isolated from fermentations of *B. subtilis* (isolate ALB 102) but have not been fully characterized. NMR analysis suggests that they differ from **1** in their amino acid substitutions.

Arginidiene and Phytophthoradiene

Several microbial metabolites have found utility in crop agriculture as control measures for plant pathogenic fungi, mainly in Japan (eg. blasticidin, polyoxin and validamycin). Significant patent activity has occurred during the past few years surrounding a new class of methoxyacrylate fungicides based upon the fungal metabolite, strobilurin (*9*). These compounds have not been commercialized to date.

During the course of screening for new fungicides from micro-organisms we encountered two closely related dipeptides with modest activity principally against *Phytophthora* spp. Herein we describe the isolation and structure elucidation of a novel fungicidal dipeptide, arginidiene (**13**), and briefly the related compound, phytophthoradiene (**14**). Both molecules contain a previously unreported amino diacid moiety and yet interestingly, arginidiene was isolated from *Bacillus megaterium* while phytophthora-diene was isolated from a *Streptomyces* sp.

Arginidiene was isolated from the cell-free filtrate of a 10 liter fermentation of *B. megaterium* (isolate B18252), obtained from a sand and soil sample collected in San Diego, California. The cell free fermentation broth was loaded onto a 600 X 100 mm column of AG50W-X8 cation exchange resin and eluted with 2L of 10% (v/v) NH_4OH. The active fraction was concentrated *in vacuo* and chromatographed on a 150 X 30 mm RP flash column eluting with 250 ml each of water, 1:1 methanol:water and methanol. The active water fractions were combined, concentrated *in vacuo* and further purified on an LH-20 column (900 X 25 mm) using 25% methanol in water. The active fractions (*ca.* 5 grams of hydroscopic material) were again combined and purified on an LH-20 column (500 X 25 mm) using a water mobile phase. The bioactive fractions were then combined, lyophilized and dissolved in 8 mL of 0.25M ammonium formate, pH 9.1, and further purified on a QAE-Sephadex column (750 X 20 mm) which had been previously equilibrated in the starting buffer. The column was washed with 150 mL of 0.25M ammonium formate, pH 9.1, followed by a gradient made from 200 mL each of pH 9.1 and pH 6.3 ammonium formate. The active fractions were combined and crystallized from water to yield 18 mg of **1**. The sterile fermentation broth of *B. megaterium* B18252 was initially observed to be active against *Phytophthora megasperma*, the causative agent of soybean root rot, in an agar-based *in vitro* assay. Subsequent testing of purified arginidiene demonstrated *in vitro* activity against *Botrytis cinerea*, the causative agent of eggplant gray mold.

High resolution FABMS analysis of arginidiene (**13**) provided a protonated molecular ion of *m/z* 340.1656 Da corresponding to a molecular formula of $C_{14}H_{21}N_5O_5$ (ΔM 3.5 mmu, 6 sites of unsaturation). Treatment of **13** with 6N HCl for 12 hours produced arginine and isophthalic acid, accounting for all of the carbon atoms present in the molecule. ^{1}H NMR and ^{13}C NMR spectra of **13** (Table V) contained several resonances which could be readily assigned, except for the amide carbon, by comparison with spectra for arginine. An APT spectrum clearly indicated that the remaining carbons could be accounted for by the presence of a cyclohexadiene ring containing one aliphatic and two olefinic methines, one aliphatic methylene and two quaternary olefinic carbons, rather than the fully aromatic isophthalic acid obtained as a product of hydrolysis. The geminal coupling constant of 23 Hz observed for the two aliphatic methylene protons (δ3.19, δ3.28 ppm) is consistent with the characteristically large couplings observed in 1,4-cyclohexadiene rings (see partial structure **D**). Additional support for this assignment was provided by the observation of large (6.8 and 8.3 Hz) homo-allylic coupling constants between the methylene protons and the aliphatic methine proton at δ5.4 ppm (H1, see **D**). The allylic juxtaposition of the olefinic methine protons (δ6.52, δ6.54 ppm) and the aliphatic methylene protons was further suggested by the observed coupling constant (1.9 Hz, see **D**).

Table V. ^1H and ^{13}C NMR Data for Arginidiene (13) and Phytophthoradiene (14)

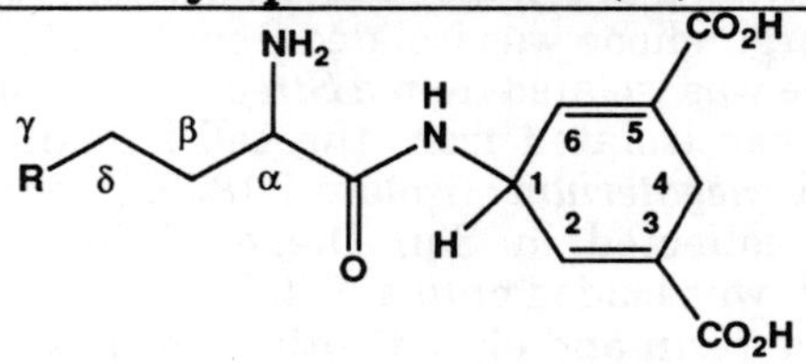

13 R = $CH_2NHC(=NH)NH_2$

14 R = $CONH_2$

Carbon/ Proton	^{13}C NMR δ ppm (mult)[a]		^1H NMR δ ppm[a] (mult)	^{13}C NMR δ ppm[a]	^1H NMR δ ppm[a]
1	44.1	CH	5.40 ddd, J=3,4,8.3,6.8	48.5	5.35m
2, 6	125.3, 125.4	CH	6.52, 6.54 m	130.2, 130.3	6.68m
3, 5	135.0, 135.2	C		138.9, 139.0	
4	25.3	CH_2	3.19 ddt, J=23,8.3,1.9 3.28 ddt, J=23,6.8,1.9	29.5	3.12m 3.18m
Acid	173.4	C		170.9[b]	
Amide	166.8	C		161.9	
α	51.4	CH	4.12 t, J=6.5	55.5	4.03t, J=7
ß	26.4	CH_2	2.10 m	28.	
9	2.23	q,J=7			
γ	21.9	CH_2	1.84 m	32.8	2.51t, J=7
δ	38.7	CH_2	3.41 td, J=6.4,1.7	170.9	
Guanidine	155.2	C			

[a]Proton and ^{13}C NMR spectra were recorded in D_2O at 300 and 75 MHz respectively. Carbon multiplicities were determined using the APT (*4*) experiment.[b]Assigments were made based upon a series of HMBC and HMQC experiments.

Assignment of the protonated carbons was facilitated by a proton-detected heteronuclear one bond correlation (HMQC) experiment.

Assignment of key carbon resonances and the determination of the point of attachment of the arginine and cyclohexadiene groups was accomplished by the analysis of three proton-carbon HMBC experiments which differed in their $^nJ_{CH}$ values (see partial structure **E**). The arginine carbonyl carbon (δ 166.8 ppm) was assigned based upon its correlation with the arginine α-methine (δ 4.12 ppm) and ß-methylene (δ 2.10 ppm) protons. The cyclohexadiene carbonyl carbons were assigned based upon their correlations to the olefinic protons.

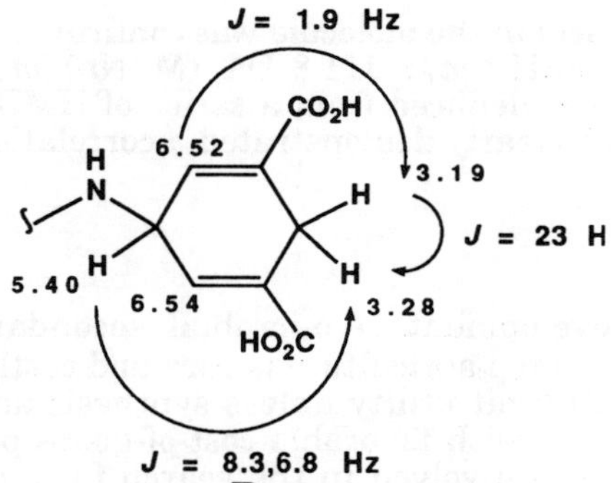

D

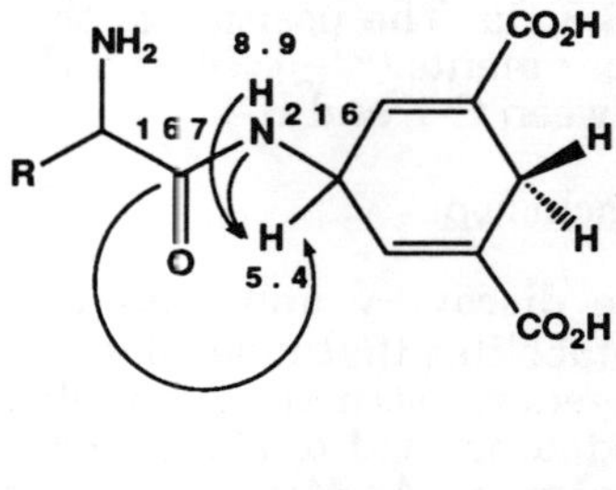

E

Observation of a correlation between the methine proton (H1, δ 5.4ppm) and the amide carbonyl carbon (δ 166.8 ppm) in an HMBC spectrum obtained with $^{n}J_{CH}$ = 4 Hz demonstrated that the cyclohexadiene moiety was connected to the arginine residue as shown in **13** (see **E**). This was confirmed by a COSY experiment performed in pH 4.5 10% D_2O in which H1 was coupled to a previously unobserved proton resonance at δ 8.9 ppm which was assigned to the amide proton (see **E**). Further evidence for this attachment was provided by a ^{1}H-^{15}N HMBC experiment in which H1 correlated to an amide nitrogen (δ 216 ppm, see **E**).

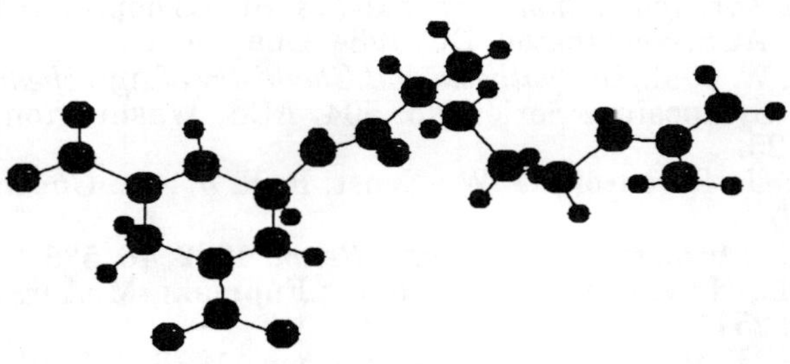

F

The structure proposed for arginidiene (**13**), based upon chemical and spectroscopic evidence, was subsequently confirmed by single crystal X-ray analysis (see **F**).

Shortly after the elucidation of arginidiene we isolated (5-10 mg/L) a closely related molecule from a fungicidally active (*in vitro* activity was observed versus *P. infestans* and *P. megasperma*) fermentation broth of *Streptomyces* sp. (isolate A16240) isolated from a soil sample collected in a wooded savannah in Boual, Central West Africa. The structure of phytophthoradiene (**14**) was readily deduced by comparison of spectroscopic information with data gathered on arginidiene (Table V). Hydrolysis of **14**, as described for **13** above, afforded isophthalic acid ($(M-H)^-$ m/z 165 Da) and glutamic acid ($(M+H)^+$ m/z 148 Da). The presence of glutamine as opposed to glutamic acid in the molecule was confirmed by ion spray mass spectral analysis; $(M+H)^+$ m/z 311.8 Da; $(M+Na)^+$ m/z 333.8 Da. The position of the ring was deduced from a series of HMBC experiments ($^nJ_{CH}$ = 4 or 8 Hz) which clearly demonstrated a correlation between C-1' and H-1.

Conclusion

The discovery and subsequent development of microbial secondary metabolites that have direct utility in crop agriculture is rare and costly. In general structural complexity will limit utility unless synthesis and modelling lead to simplified analogues with favorable cost-of-goods per treatment. At Monsanto we have been involved in the search for crop chemicals of natural origin for several years. While numerous active compounds have been characterized, the diabroticins and the recently reported herbicide, herboxidiene (*10*), are by far the most highly active that we have discovered to date.

Acknowledgments

We would like to thank Nancy Biest, Minhtien Tran, Barb Reich, Mike Prinsen, Jay Pershing, Steve Sims, John Greenplate and Margann Miller-Wideman for biological support, and Hideji Fujiwara and Tom Solsten for performing the HRMS analyses.

Literature Cited

1. Campbell, W. C.; Burg, R. W.; Fisher, M. H.; Dybas, R. A. In *Pesticide Synthesis Through Rational Approaches*, ACS Symposium Series No. 255; ACS,Washington, DC, 1984; Chapter 1.
2. Addor, R. W. et al., In *Synthesis and Chemistry of Agrochemicals III,* ACS Symposium Series No. 504, ACS, Washington, DC, 1992; Chapter 25.
3. Piantini, U.; Sorenson, O. W.; Ernst, R. R. *J. Am. Chem. Soc.* **1982**, 104, 6800.
4. Patt, S. L.; Shoolery, J. N. *J. Magn. Reson.* **1982**, 46, 535.
5. Tchushida, H.; Komoto, M.; Kato, H.; Fujimaki, M. *Agr. Biol. Chem.* **1973**, 11, 2571.
6. Cox, R. H.; Bothner-By, A. A. *J.Phys.Chem.* **1968**, *5*, 1646-1649.
7. Marrone, P.G.; Ferri, F. D.; Mosley, T. R.; Meinke, L. J. *J. Econ. Entomol.* **1985**, 78, 290.
8. Higgans, R. A.; Pedigo, L. P. *J. Econ. Entomol.* **1968**, *72*, 238-244.
9. Beautement, K; Clough, J. M.; de Fraine, P. J.; Godfrey, C. R. A. *Pestic. Sci.* **1991**, *31*, 499-519.
10. Isaac, B. G.; Ayer, S. W.; Elliott, R. C.; Stonard, R. J. *J. Org. Chem.* **1992,** 57, 7220.

RECEIVED September 21, 1993

Chapter 4

Role of Natural Products in Pesticide Discovery

The β-Methoxyacrylate Fungicides

John M. Clough, David A. Evans, Paul J. de Fraine, Torquil E. M. Fraser, Christopher R. A. Godfrey, and David Youle

ZENECA Agrochemicals, Jealott's Hill Research Station, Bracknell, Berkshire RG12 6EY, United Kingdom

The development of a new class of synthetic agricultural fungicides based on the naturally occurring β-methoxyacrylates is described. A consideration of structure-activity relationships and key physical properties has led to a series of tricyclic compounds which still incorporate the β-methoxyacrylate toxophore. The best examples of these compounds combine high levels and a broad spectrum of systemic fungicidal activity with good crop safety.

As the technical and commercial requirements for new agrochemicals become more and more demanding, it is increasingly difficult to discover new classes of compounds with sufficient potential to justify the escalating costs of development. There is therefore a continuing need for scientists in industry and academia to develop new strategies for pesticide discovery.

Five standard approaches to the generation of new areas for agrochemical (and pharmaceutical) research are commonly cited: random screening; speculative synthesis; imitative ("me-too") chemistry; biorational design; and finally the exploitation of natural products (*1*). Although this last approach has found much commercial success in the development of novel insecticides (for example the pyrethroids), there are still relatively few examples of herbicides and fungicides which have been discovered in this way. This account describes the development of a new class of fungicides derived from the strobilurin family of natural products and related compounds, all of which are derivatives of β-methoxyacrylic acid.

Natural Products in Fungicide Discovery

Natural products can be exploited in a number of different ways in the development of new fungicides: they can be used as products in their own right or as leads for the design of novel synthetic products. Alternatively, they can be used to highlight novel modes of action (*2*).

0097–6156/94/0551–0037$06.00/0

The most important natural products in use as commercial fungicides are illustrated in Figure 1. Individually, these are minor products, each with a narrow spectrum of activity and application predominantly on rice, vegetable and fruit outlets. Taken together, natural products now command less than 1% of the total fungicide market, which was estimated in 1991 to account for 21% of the $26.8 billion total for agrochemicals (*1*). The relative unimportance of natural products, when compared to their synthetic counterparts, can be attributed to a number of factors, including limitations in their spectrum of activity and, in certain instances, the onset of resistance.

Although the activity of natural products can in principle be improved by chemical modification, this approach relies heavily on the ready availability of sufficient quantities of the natural starting materials and the development of appropriate synthetic methodology. A far more promising and effective strategy is to use a knowledge of the structure of a fungicidal natural product as the starting point for the synthesis of simpler compounds with optimised physical, biological and environmental properties. Recent examples of natural products which have been looked at in this way are hadacidin (*3*), thiolutin (*4*), pyrrolnitrin (*5*) (which has recently led to the development of fenpiclonil and CGA173506 as seed treatments) and, the subject of this paper, the β-methoxyacrylates strobilurin A and oudemansin A (Figure 2).

The Fungicidal β-Methoxyacrylates

We first became interested in strobilurin A and oudemansin A as potential leads for synthesis in the early 1980's. Since then, the number of structurally related compounds reported in the literature has risen steadily. There are now eleven known strobilurins and their structures are illustrated in Figure 3. They are all reported to be fungicidal. Each member of this family of natural products incorporates a methyl β-methoxyacrylate group, linked at its α-position to a phenylpentadienyl unit, and all but the parent compound strobilurin A carry either one or two additional substituents on the benzene ring. The strobilurins therefore vary in complexity from strobilurin A to strobilurin E, the latter containing relatively elaborate spiro-ketal functionality. Two new oudemansins, B and X, have also been isolated, and their structures, together with the structurally related β-methoxyacrylamide myxothiazol, which is also fungicidal, are depicted in Figure 4.

From a structural point of view, the β-methoxyacrylates looked relatively attractive and amenable to synthesis. Significantly, they have a mode of action [inhibition of mitochondrial respiration by binding at a specific site on cytochrome b (*6*, *7*)] which is not shared by any other known class of fungicide. This novel mode of action led us to believe that there would be no cross-resistance between the β-methoxyacrylates and other fungicides. Furthermore, a detailed knowledge of the mode of action meant that an *in vitro* assay for activity could be established, which could potentially provide useful structure-activity information. For all of these reasons, this family of compounds appeared to be a good starting point to investigate the fungicidal activity in this area. Accounts of the early phases of our work have already been published (*8*, *9*).

Simple Analogues of the Natural Products

Although active against a range of fungi growing on agar, strobilurin A, when applied as a foliar spray, showed no useful activity against fungi growing on plants in the glasshouse. This was found to be due to a combination of its photochemical instability and relatively high vapour pressure, through which it is rapidly lost from a leaf surface

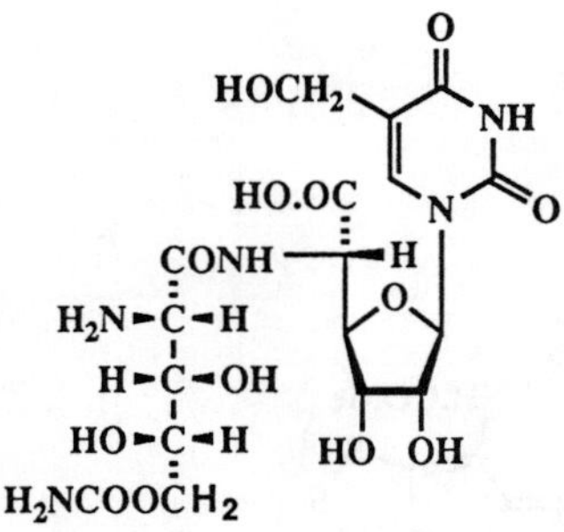

Polyoxin B

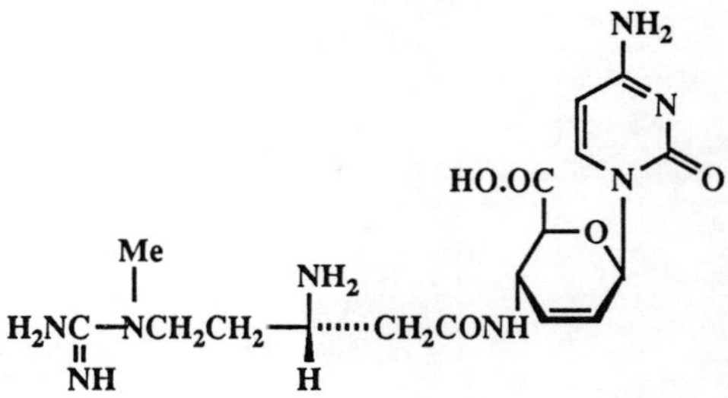

Blasticidin S

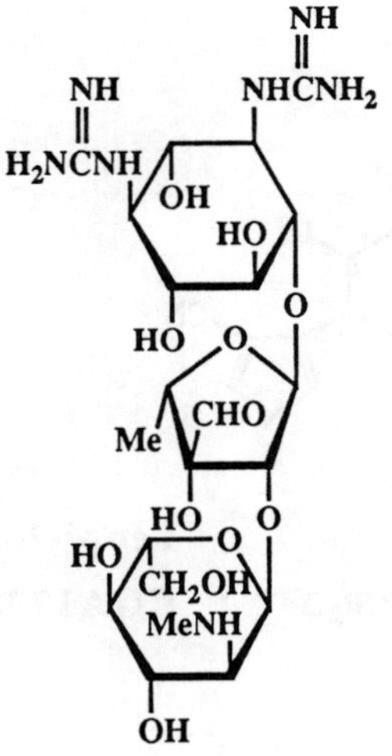

Streptomycin

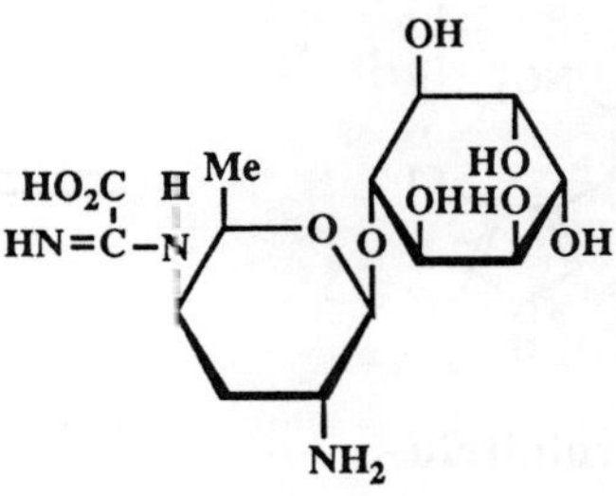

Kasugamycin

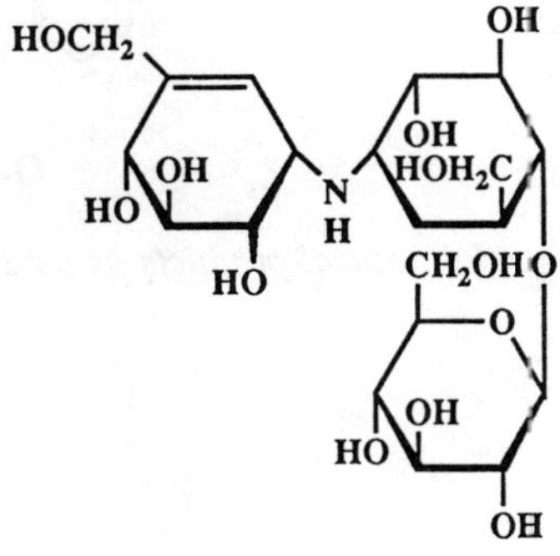

Validamycin A

Figure 1. Natural products in use as fungicides.

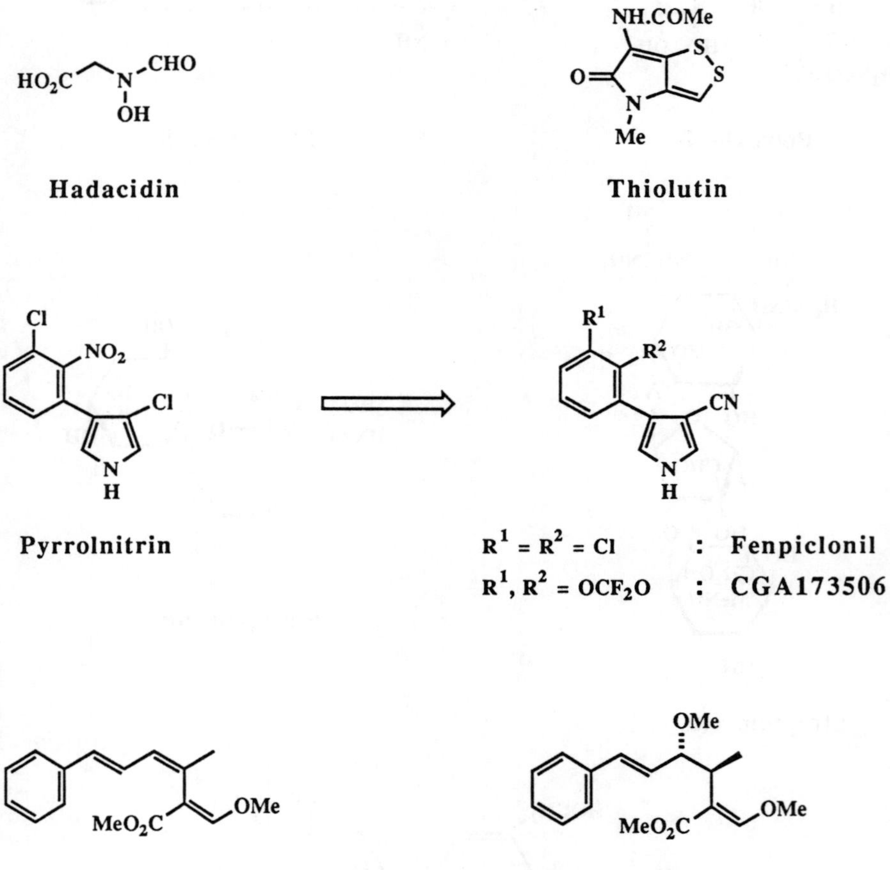

Figure 2. Fungicidal natural products as leads for synthesis.

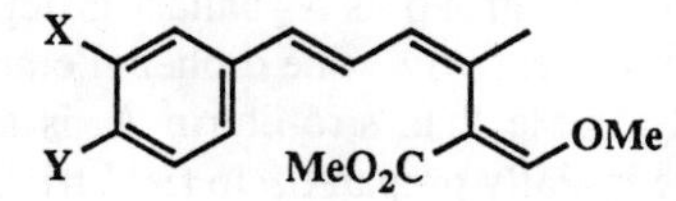

X	Y	STROBILURIN	
H	H	A	(1969)
MeO	Cl	B	(1977)
H	MeO	X	(1983)
HO	H	F	(1990)
MeO	H	H	(1990)

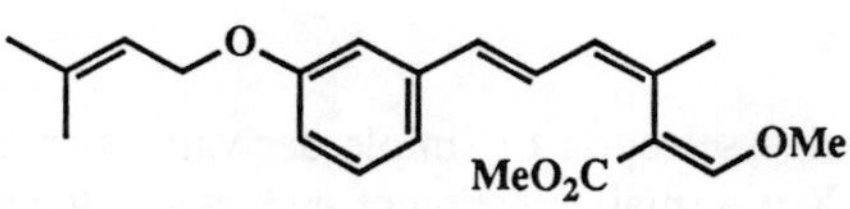

Strobilurin C (1983)

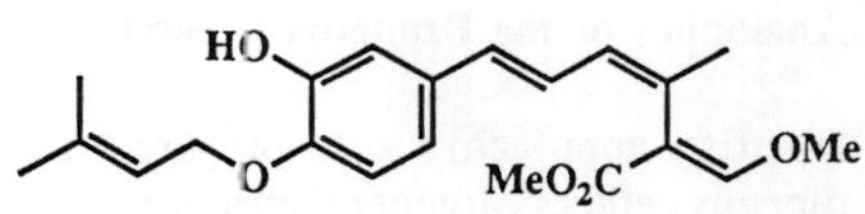

Strobilurin F (1990)

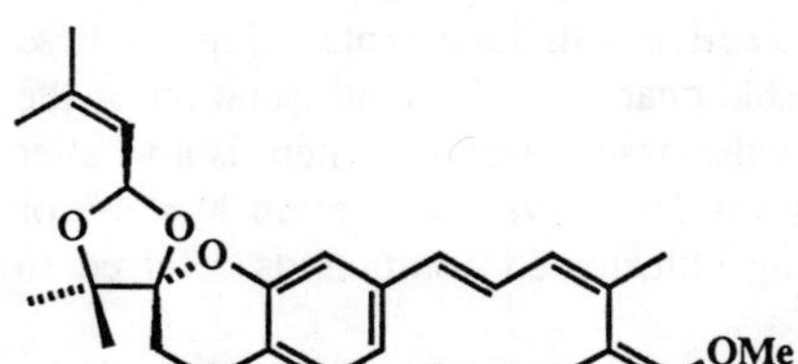

Strobilurin E (1989)

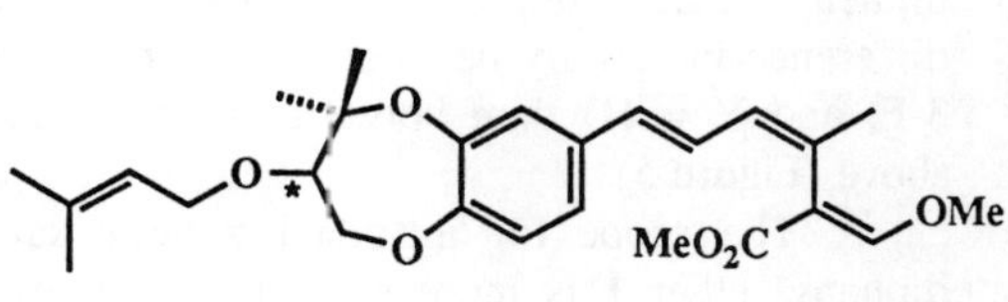

Strobilurin G (1990)

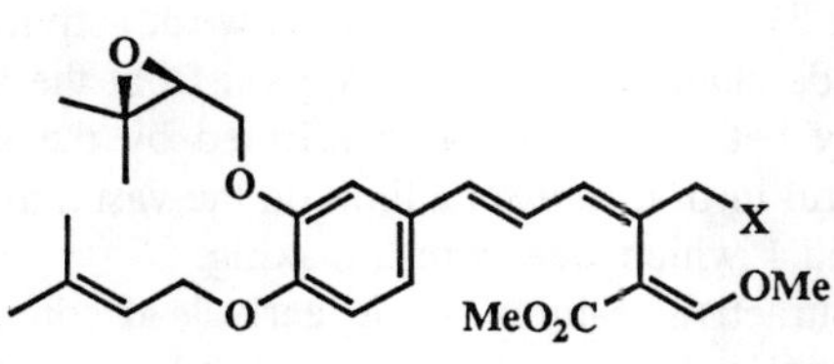

X = H Strobilurin D (1990)
X = OH Hydroxystrobilurin D (1988)

Figure 3. The strobilurins, showing the year when first described. Note that two different compounds have been designated "strobilurin F". The asterisk marks a centre of unknown configuration in strobilurin G.

(*8*, *9*). We were able to synthesize analogues which have both appropriate physical properties for use in agriculture and increased fungicidal activity *in vivo*. The key to the success of this work was the identification of the β-methoxyacrylate group as the toxophore, important for activity, and the phenylpentadienyl group as a "ballast group" which could be modified in a variety of ways without loss of activity. The diphenyl ether **1**, less volatile and considerably more photochemically stable than strobilurin A, is an example of one of the most active analogues which were initially prepared. In field trials, it gave moderate levels and a good spectrum of fungicidal activity, but also produced unacceptable crop damage in some cases (on wheat, barley and vines for example) (*9*). We therefore followed up this work with a programme of synthesis based on the lead compound **1** with the aim of discovering analogues with higher fungicidal activity and improved crop safety.

Analogues of the Diphenyl Ether 1

Our first approach was to prepare a representative selection of simple derivatives of **1**, diphenyl ethers of general structure **2** in which X is a small substituent such as a halogen or a methyl or methoxyl group and Y is hydrogen (*9*). The activity of the chlorinated compounds **2** wherein X = Cl and Y = H is representative: although the 3- and 4-chloro-derivatives, when applied as foliar sprays, are highly active against a variety of fungi growing on plants in the glasshouse, the 2-chloro-derivative is only weakly active and the 2, 6-dichloro-derivative (**2**, X,Y = 2,6-di-Cl) is less active still. One explanation for these results is that a 2-substituent causes an unfavourable change in the conformation of the diphenyl ether. Consistent with this explanation is the observation that there is a smaller difference in activity between the 2- and the 3-fluoro-derivatives (**2** wherein X = 2-F or 3-F, and Y = H) than between the corresponding chlorinated compounds referred to above (Figure 5).

The scope for introducing large substituents onto the phenoxy-group of the diphenyl ether **1** is more restricted. For example, while the tricyclic compound **3** is highly fungicidal, more active than **1**, its regioisomer **4** is a very poor fungicide(See Fig. 13). The fact that **3** can be accommodated at the active site is perhaps not surprising in view of the fungicidal activity of the more elaborate strobilurins such as strobilurin E and strobilurin G (Figure 3). Nevertheless, from the weak activity of **4**, it is evident that the shape of the large side-chain is critical for a good fit at the active site. This very clear difference in activity between **3** and **4**, confirmed by the synthesis of further related compounds, was useful insofar as it set a limit on the vast number of potential derivatives of the lead compound **1** which were worth making.

One of the attractive features of our early lead compound, the simple diphenyl ether **1** [log P (octanol/water) = 3.3 ; water-solubility = 30ppm], was its systemicity. Systemicity is a key attribute of modern fungicides: it improves field performance through redistribution of the compound within plant tissues after spraying, and potentially allows the farmer to apply the compound either as a foliar eradicant or as a seed treatment. The first evidence for the systemic activity of **1** came from indicator tests in the glasshouse in which the compound, when applied as a root drench, protected both cereal and broad-leaved crops from infection by fungal spores which were applied as a foliar spray 24 hours later. (The diphenyl ether **1** also gave good control in parallel tests in which it was applied as a foliar spray). In further tests, solutions of **1**, applied as small spots at a site near the base of a leaf of a growing barley plant, gave protection from infection by *Erysiphe graminis* f. sp. *hordei* over the surface of that leaf between the site of application

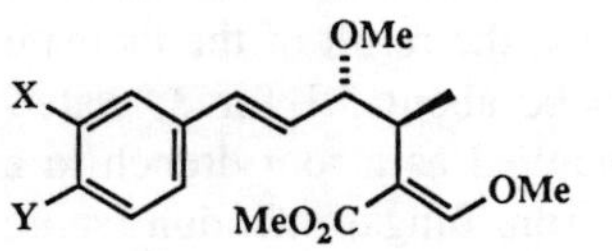

X	Y	OUDEMANSIN	
H	H	A	(1979)
MeO	Cl	B	(1983)
H	MeO	X	(1990)

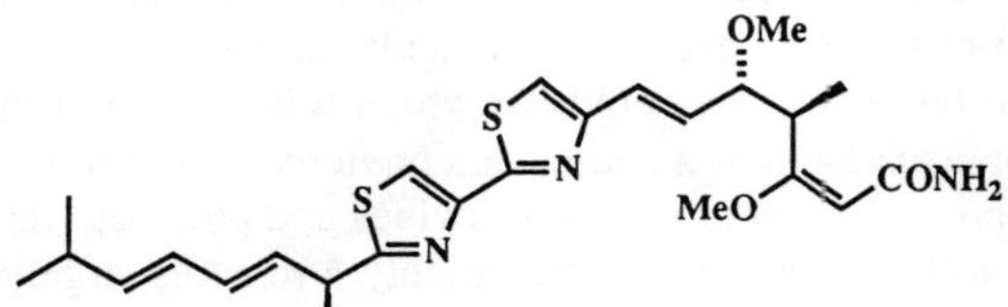

Myxothiazol (1980)

Figure 4. The oudemansins and myxothiazol.

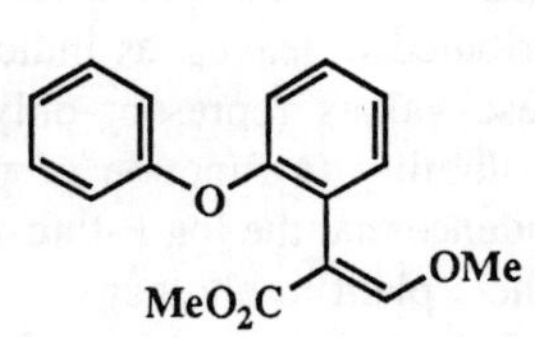

1

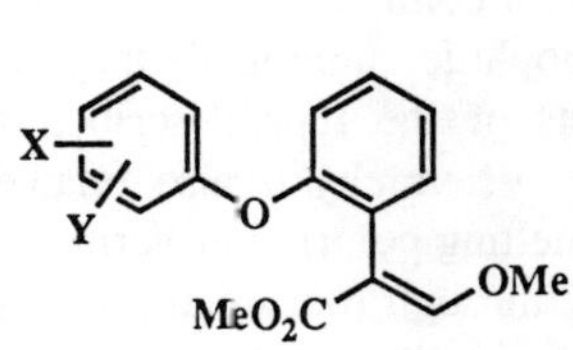

2

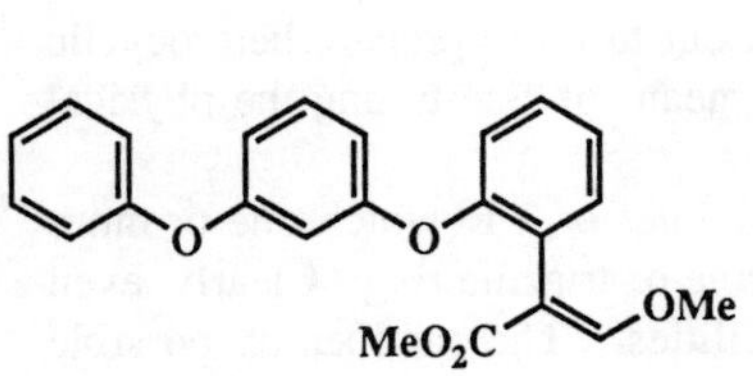

3

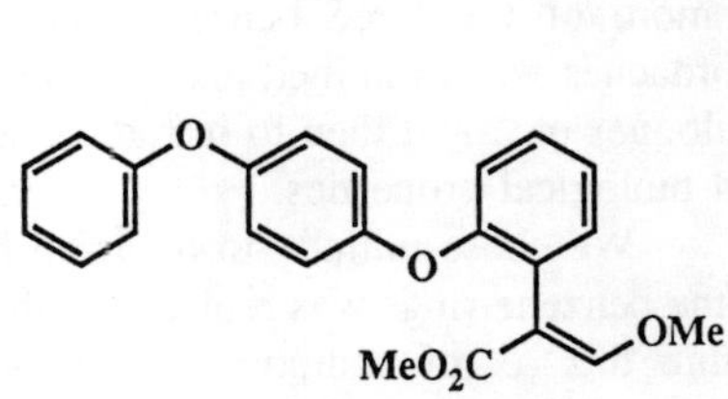

4

Figure 5. Aryl ether analogues of the strobilurins.

and the leaf tip. The fact that there was no disease control between the site of application and the base of the leaf indicated that this effect is due to xylem translocation (apoplastic movement) and not phloem translocation (symplastic movement) or vapour activity. Further studies with a ^{14}C-labelled sample of **1** confirmed that the material moves from the site of application towards the tip of the leaf.

An important and predicted penalty which followed from the addition of an extra phenoxy-group to **1** to give **3** was the loss of systemic activity, the result of the increase in partition coefficient [log P (octanol/water) estimated to be about 5.1 for **3**; water-solubility (measured) = 0.03ppm]. Thus, solutions of **3**, applied as a root drench to a range of cereal and broad-leaved crops, give no protection from fungal infection (spore suspension applied as a foliar spray, 24 hours later), while application of the same solutions *via* foliar spray provides excellent fungicidal activity.

In contrast, one favourable consequence which followed from the weak systemic activity of **3** was the observation that it causes less phytotoxicity than the simple diphenyl ether **1**, especially on broad-leaved crops such as vines and potatoes (its phytotoxicity on wheat and barley was still unacceptable). Presumably **3** remains largely near the surface of the leaf, at or close to the site of application, and so is not able to exhibit its intrinsic phytotoxicity. Nevertheless, since systemicity is a highly desirable property for an agricultural fungicide, we set about finding ways in which the log P might be reduced in analogues of **3**. Our experience with β-methoxyacrylates which were prepared and tested earlier in this programme of synthesis showed that systemic activity through root uptake generally occurs for compounds which have octanol/water log P's below about 3.5, while slightly more lipophilic compounds may still be redistributed in leaves, as indicated by leaf-spotting tests of the type described above. These values represent only rough guidelines, since systemicity is also related to water-solubility (a function of partition coefficient and melting point). Furthermore, there is evidence that the log P-threshold for systemic activity through root uptake varies from one host plant to another.

We reasoned that the partition coefficient of **3** could be lowered by the introduction of one or more suitable substituents, by replacement of one or both of the oxygen links between rings with suitable alternative groups, and/or by replacement of one or more of the three benzene rings with appropriate heterocycles. Each of these approaches was examined and the most fruitful turned out to be to prepare heterocyclic analogues of **3** and then to introduce substituents as a means of fine-tuning the physical and biological properties.

We chose initially to focus on heterocyclic analogues of **3** in which one or more of the benzene rings was replaced with a pyridine, diazine or triazine ring. Clearly, even within this set of compounds there are many possibilities. The number of possible analogues of **3** in which one benzene ring has been replaced by pyridine, for example, is 11, while 18 diazines (3 pyrazines, 8 pyrimidines and 7 pyridazines) and 14 regioisomeric triazines are possible. If target compounds in which either two or all three benzene rings are replaced by these heterocycles are considered, the scope for synthesis multiplies dramatically. Therefore, compounds which had a representative range of physical properties (log P, pK_a) were initially chosen for synthesis. While these properties could be predicted with some accuracy, it was, of course, not known at the outset which modifications to the tricyclic compound **3** would be accommodated at the active site, or would be beneficial for binding. This only emerged as examples were prepared and tested and structure-activity relationships were established.

Synthesis of Tricyclic Acrylates

Compound **3**, the first highly active tricyclic acrylate to be made, was prepared on a 0.4 mol scale for field trials by the pathway shown in Figure 6. An Ullmann ether coupling was the key step in the construction of the intermediate phenoxyphenoxyphenylacetate **5** which was then converted stereospecifically into **4** *via* Claisen condensation with methyl formate followed by *O*-methylation of the resulting β-hydroxyacrylate.

Alternatively, compounds of this type could be prepared in a more convergent way from the intermediate **6**. The syntheses shown in Figure 7 were chosen to illustrate this chemistry because they demonstrate that regardless of whether **6** or the corresponding (*Z*)-isomer **7** is used, only the required (*E*)-isomeric Ullmann ether product is isolated (*10*). For example, when the reaction between **7**, 3-methylphenol and potassium carbonate was monitored by GC, it could be seen that isomerisation of the β-methoxyacrylate double bond occurred smoothly at 50°C, and clearly preceded the Ullmann coupling which took place only at much higher temperatures. Stereomutation to the thermodynamically-favoured (*E*)-isomer presumably occurs through reversible Michael addition of the phenolate to the acrylate group of **7** (*11*). Although the yields in Ullmann ether couplings of the type shown in Figure 7 are not high (the best examples gave yields of about 60%), this disadvantage was clearly outweighed by the fact that the readily-available intermediate **6** could be treated with a variety of phenols to produce a range of compounds for biological testing.

Representative tricyclic compounds which incorporate either one or two heterocyclic rings are shown in Figure 8 and were prepared by related chemistry (Figures 9 to 12). As expected, coupling could be achieved under milder conditions whenever the electrophilic partner was activated by one or more nitrogen atoms in the ring. Again, whenever possible, convergent intermediates were used to prepare the first small samples for testing for fungicidal activity. Examples are the phenol **8** [Figures 9 (*12*) and 10 (*13*)], the chloro-1,3,5-triazine **9** [Figure 9 (*12*)], the chloropyrimidine **10** [Figure 10 (*13*)], the phenol **11** [Figure 11 (*14*)] and the phenol **12** [Figure 12 (*15*)]. Many of these intermediates are crystalline solids which are readily prepared on a multigram scale.

The phenol **8**, a crystalline solid (melting point 125-6°C), which was easily prepared on a mole scale, was a particularly useful intermediate (*12*, *13*). It reacts under basic conditions with a variety of electrophilic heterocycles, such as chlorinated 1,3,5-triazines [Figure 9 (*12*)] and 4,6-dichloropyrimidine [Figure 10 (*13*)], but its reaction with less electrophilic partners is limited by its tendency to lactonise under basic conditions. Target compounds which were formally derived from the phenol **8** and less electrophilic heterocycles could often be obtained by using heterocycles containing one or more additional activating chlorine atoms which could be removed by reduction at a later step [see Figure 9 (*12*)].

Biological Activity of Tricyclic Acrylates

The fungicidal activity of the representative compounds shown in Figure 8 is compared with that of the original tricyclic lead compound **3** in Figure 13. All the results shown are from 24-hour protectant tests in the glasshouse, that is, tests in which the plant is treated first with the acrylate, either as a foliar spray or as a root drench, and then, 24 hours later, with a spore suspension of the fungal pathogen as a foliar spray.

Several points are worth noting. Firstly, as predicted, only the most hydrophilic examples have systemic activity through root uptake. The pyridinyloxypyrimidine **18**

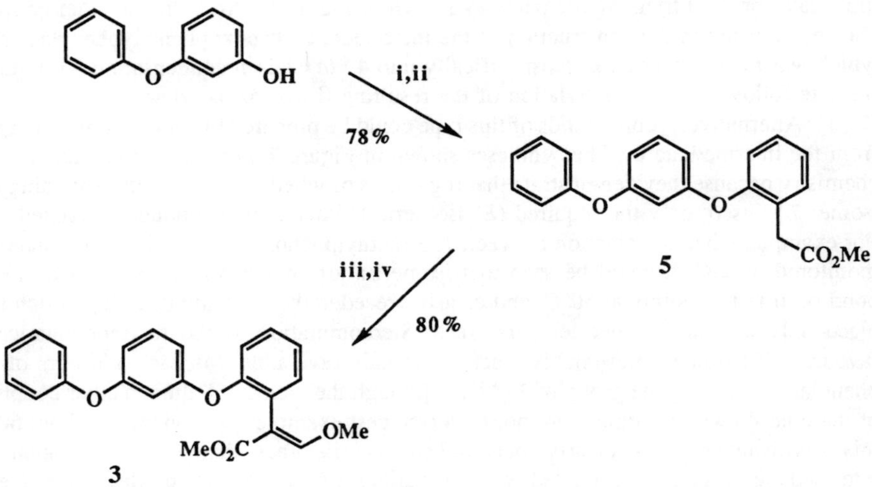

Reagents : i, $2\text{-}Cl\text{-}C_6H_4\text{-}CH_2CO_2H/K_2CO_3$/cat.CuCl/130-140°C/24h.
ii, $MeOH/H^+$
iii, HCO_2Me/NaOMe/diglyme/20°C
iv, Me_2SO_4/K_2CO_3/DMF/20°C

Figure 6. Synthesis of compound **3**.

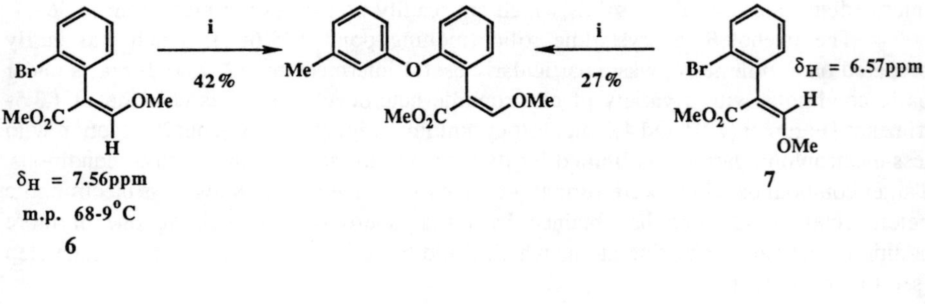

Reagents : i, $3\text{-}Me\text{-}C_6H_4\text{-}OH/K_2CO_3$/cat.CuCl/170°C/4h.

Figure 7. Synthesis of (E)*-isomeric Ullmann ether product.*

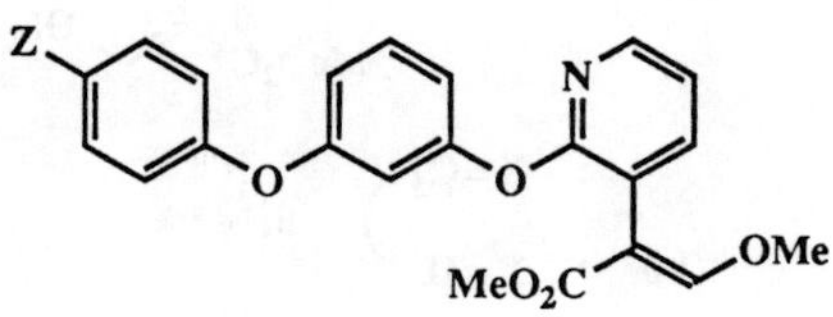

Z=H : 13
$Z=NO_2$: 14

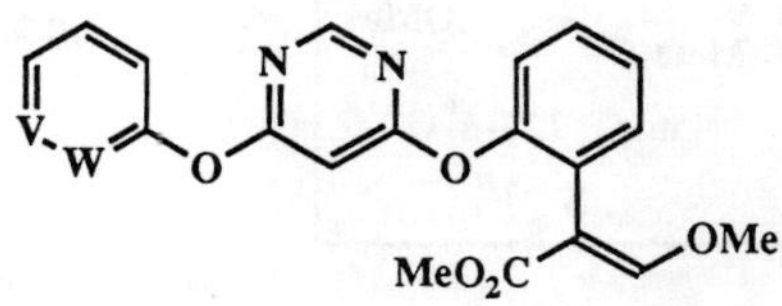

V=N ; W=C-Cl : 18
V=C-Cl ; W=N : 19

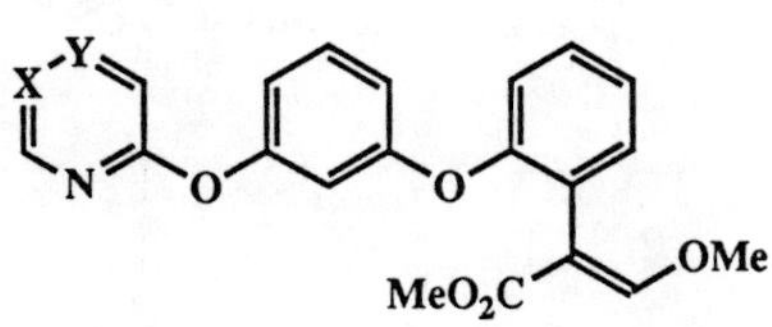

X=Y=CH : 15
X=CH ; Y=N : 16
X=N ; Y=CH : 17

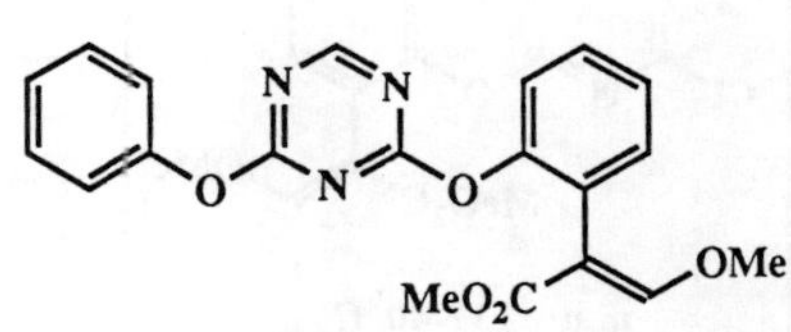

20

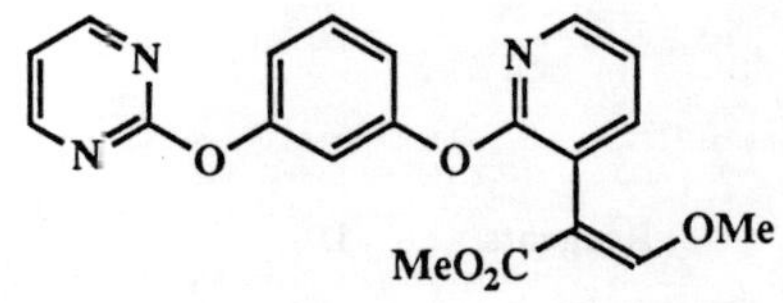

21

Figure 8. Representative tricyclic acrylates.

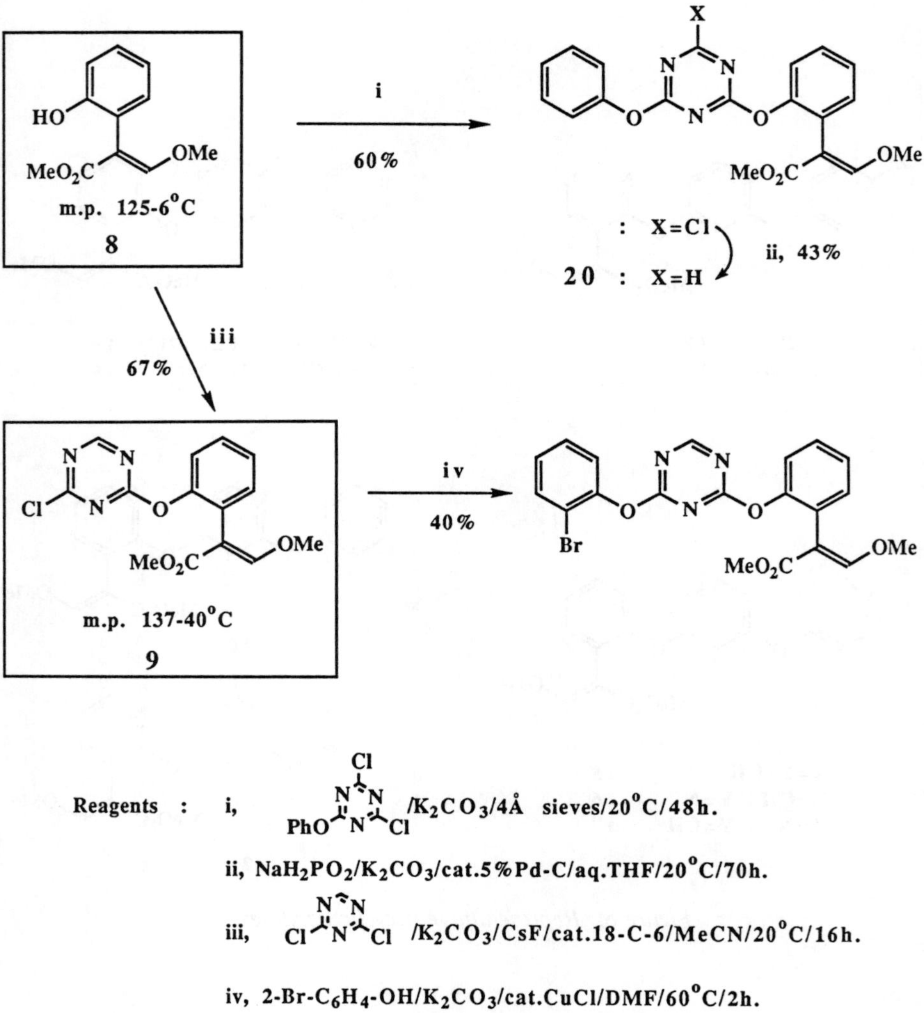

Figure 9. Synthesis of a triacyclic acrylate.

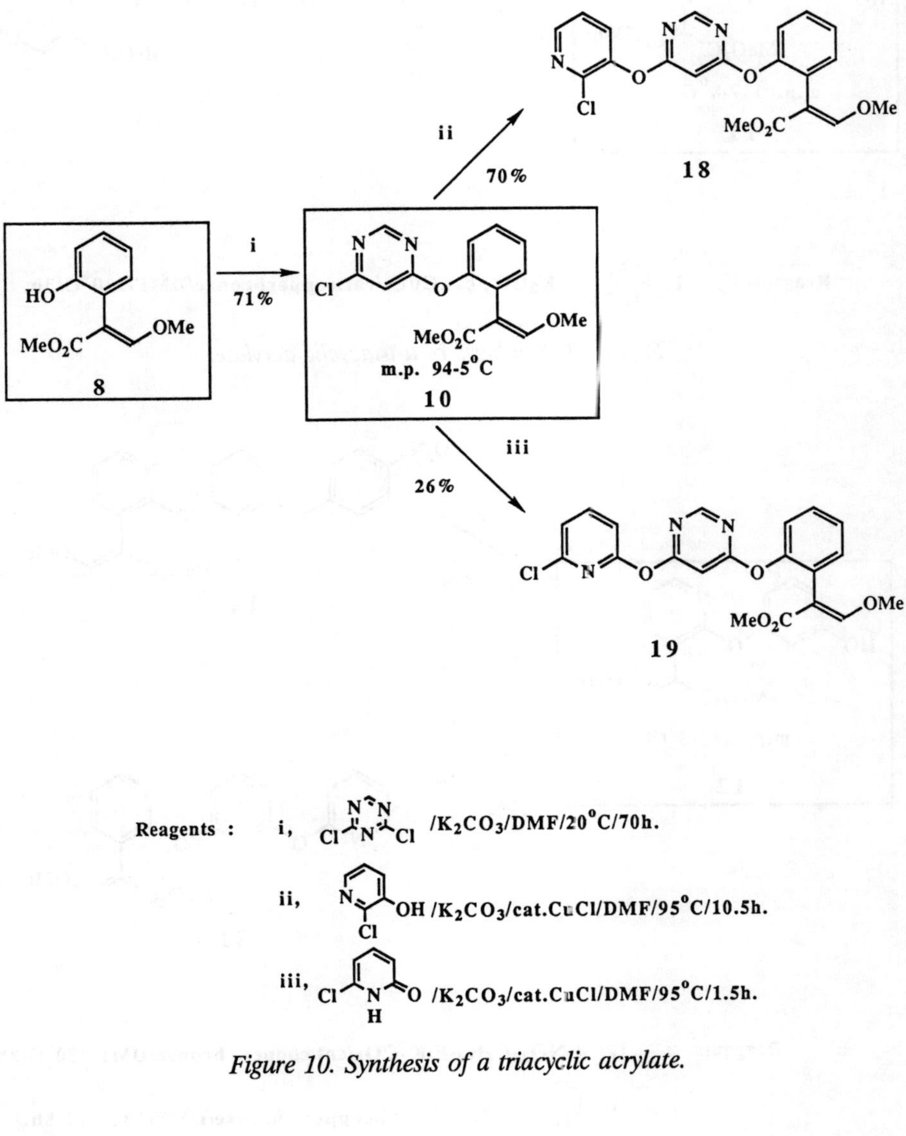

Figure 10. Synthesis of a triacyclic acrylate.

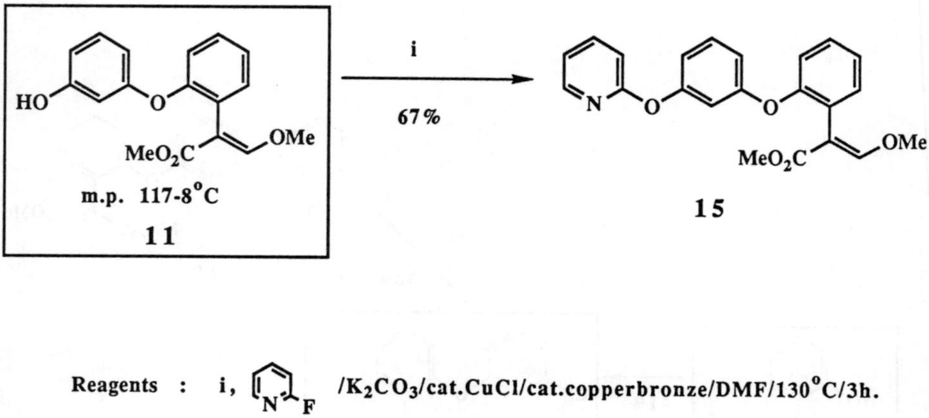

Figure 11. Synthesis of a triacyclic acrylate.

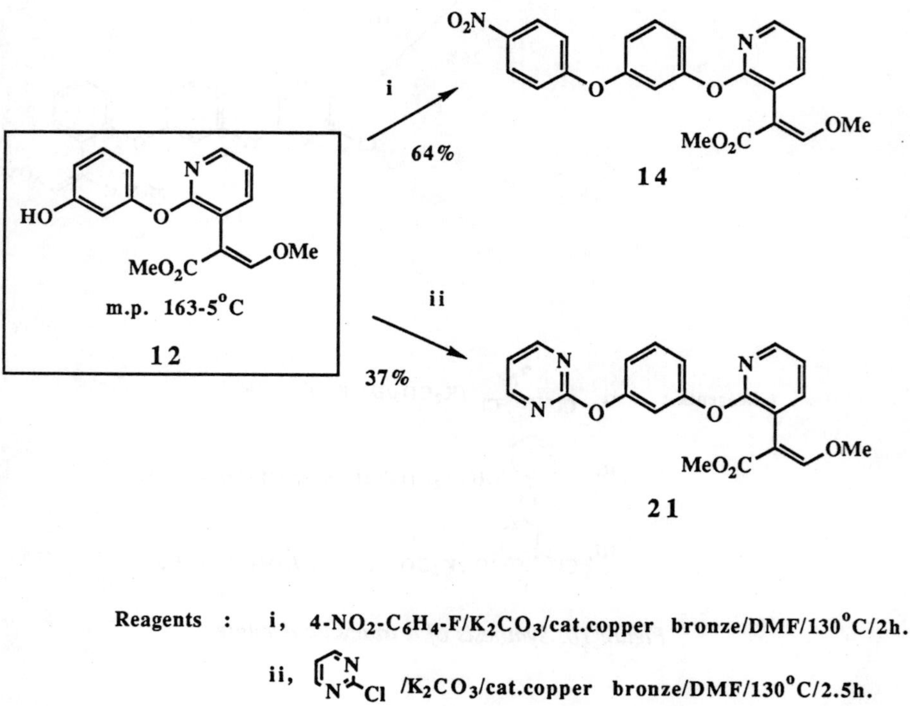

Figure 12. Synthesis of a triacyclic acrylate.

Compound	logP	PUCCRE		PLASVI		VENTIN		PYRIOR	
		Foliar	Root	Foliar	Root	Foliar	Root	Foliar	Root
1	3.3	*	*	*	-	*	*	**	*
3	+5.1	**	-	**	-	**	-	**	-
4	+5.1	-	-	-	-	-	-	-	-
13	4.1	**	-	**	-	**	-	**	-
14	+4.1	*	-	**	-	**	-	-	-
15	+3.5	*	-	**	-	**	-	*	-
16	3.2	*	-	**	-	**	-	≥*	-
17	3.0	-	-	**	-	*	-	-	-
18	2.4	**	*	**	*	**	-	**	**
19	+2.2	**	-	**	-	**	-	*	*
20	2.1	**	*	**	-	**	*	-	-
21	2.0	-	-	-	-	*	*	-	-

KEY : - Weak activity, * Moderate activity, ** Activity ≥ standard,
log P measured in octanol/water (+ estimated value).

PUCCRE = *Puccinia recondita*, brown rust on wheat
PLASVI = *Plasmopara viticola*, downy mildew on vines
VENTIN = *Venturia inaequalis*, apple scab
PYRIOR = *Pyricularia oryzae*, rice blast

The following standards were used:

PUCCRE, Foliar/Root	-	Cyproconazole
PLASVI, Foliar	-	Captafol
PLASVI, Root	-	Metalaxyl
VENTIN, Foliar	-	Hexaconazole
VENTIN, Root	-	(5*RS*, 6*RS*)-6-Hydroxy-2,2,7,7-tetramethyl-5-(1,2,4-triazol-1-yl)octan-3-one (see reference 16)
PYRIOR, Foliar	-	Tricyclazole
PYRIOR, Root	-	Pyroquilon

Untreated plants had at least 60% disease coverage in all tests.

Figure 13. Fungicidal activity of representative β-methoxyacrylates.

(*13*), for example, is a highly mobile compound, and this was confirmed in leaf-spotting tests versus *Erysiphe graminis* f. sp. *hordei* on barley. Furthermore, although the best compounds are highly active, with useful levels of movement in plant tissue, others, quite unpredictably, have a narrower spectrum of activity or are uniformly weak fungicides. The 1,3,5-triazine **20** (*12*), for example, gives only poor control of *Pyricularia oryzae* despite its excellent activity against other fungi, and the pyrimidinyloxyphenoxypyridine **21** (*15*), although superficially satisfying the requirements for activity in terms of shape, size and physical properties, is a relatively weak fungicide across the whole spectrum of pathogens shown.

Finally, the phytotoxicity which we had observed earlier with the diphenyl ether **1** is also a problem with some of the tricyclic compounds. Nevertheless, certain compounds or families of compounds which were prepared were found to cause no significant damage to a range of cereal and broad-leaved crop plants. Further details of these results will be published shortly.

Conclusions

We have described the discovery of a new class of synthetic agricultural fungicides inspired by a family of naturally-occurring β-methoxyacrylates. The natural products themselves have physical properties which make them unsuitable for use in agriculture. However, having identified the structural features required for activity, we were able to prepare analogues with improved stability and physical properties. Early synthetic compounds were still insufficiently active to be cost-effective fungicides, and some also caused damage to crops. Further work led to a series of tricyclic compounds, the best of which combine a broad spectrum of systemic fungicidal activity with good crop safety.

Acknowledgements

We wish to thank our many colleagues at ZENECA Agrochemicals who have participated in this project, particularly Vivienne Anthony, Brian Baldwin, David Bartlett, Ian Streeting and Tom Wiggins.

Literature Cited

1. Evans, D. A., and Lawson, K. R., *Pestic. Outlook*, 1992, *3*, 10.
2. Koller, W., in *Target Sites for Fungicide Action*, CRC Press, 1991, p.255.
3. Dunn, C. L., and Klein, S. P., in *Fungicides for Crop Protection, BCPC Monograph No. 31*, 1985, 407; Wade, M., Highwood, D. P., Dunn, C. L., Moncorge, and J-M., Perugia, G., *ibid.*, 1985, 455.
4. Dell, I., Godfrey, C. R. A., and Wadsworth, D.J., in *Synthesis and Chemistry of Agrochemicals III*, Baker, D. R., Fenyes, J. G., and Steffens, J. J., Eds., ACS Symposium Series No. 504; American Chemical Society: Washington, DC, 1992, p.384.
5. Nyfeler, R., and Ackermann, P., *ibid.*, p.395.
6. Becker, W. F., von Jagow, G., Anke, T., and Steglich, W., *FEBS Letts.*, 1981, *132*, 329.
7. Mansfield, R. W., and Wiggins, T. E., *Biochim. Biophys. Acta*, 1990, *1015*, 109.
8. Beautement, K., Clough, J.M., de Fraine, P. J., and Godfrey, C. R. A., *Pestic. Sci.*, 1991, *31*, 499.

9. Clough, J. M., de Fraine, P. J., Fraser, T. E. M., and Godfrey, C. R. A., in *Synthesis and Chemistry of Agrochemicals III*, Baker, D. R., Fenyes, J. G., and Steffens, J. J., Eds., ACS Symposium Series No. 504; American Chemical Society: Washington, DC, 1992, p.372.
10. EP 307,101 (1987, ICI).
11. GB 2,248,615 (1990, ICI).
12. EP 405,782 (1989, ICI).
13. EP 393,861 (1989, ICI).
14. EP 307,103 (1987, ICI).
15. EP 312,243 (1987, ICI).
16. de Fraine, P. J., and Worthington, P. A., *Pestic. Sci*, 1986, *17*, 343.

RECEIVED April 30, 1993

Chapter 5

Advances in Research and Development of Avermectins

H. Mrozik

Merck Research Laboratories, P.O. Box 2000, Rahway, NJ 07065

Avermectins have potent biocidal activities against a wide spectrum of nematodes, insects, and arachnids. The semisynthetic derivative ivermectin (22,23-dihydro-avermectin B1) has been in wide use as a broad spectrum endectocide in animals for the last decade. The major agricultural pesticidal uses of Abamectin (avermectin B1) currently are as a miticide in crops such as citrus, pear, deciduous tree nuts, ornamental plants, cotton, vegetables and strawberries. Chemical modifications have been carried out with the aim of increasing its insecticidal spectrum, residual activities and chemical stabilities. Emamectin (MK-244, 4"-deoxy-4"-epi-N-methylaminoavermectin B1), one of many 4"-substituted analogs, has greatly increased potency against lepidoptera larvae. Mode of action studies recently progressed towards the identification of an avermectin binding protein with a molecular weight of approximately 50 kD.

The avermectins are a group of closely related macrocyclic lactones with exceedingly potent activities against helminths and arthropods.[1,2] They are produced as a mixture of eight components by fermentation of the microbe *Streptomyces avermitilis*.[3] Their chemical structure[4] is related to the milbemycins,[5,6] which were described first in 1974 by workers at Sankyo as very potent miticides and insecticides for crop protection. In 1975 Merck scientists discovered the structurally closely related avermectins as highly potent endo and ectoparasiticides with a wide spectrum of activities mainly for animal, but also certain human applications.[7] Similar interesting anthelmintic activities were subsequently also described for 13-deoxyavermectin aglycones and for the milbemycins by the Merck group.[8] Ivermectin, the 22,23-dihydro derivative of avermectin B1,[9] is a chemically modified derivative of this group of natural products that has found wide use as a systemic antiparasitic agent against endo and ectoparasites of animals.[10] Ivermectin is also used as a treatment for human filarial worm infections (*Onchocerca volvulus*, River Blindness).[11] The scientific and commercial success of ivermectin

0097–6156/94/0551–0054$06.00/0

as an endectocide for animal health attracted the interest of organic chemists to the total synthesis of the interesting structures,[12,13] and of biochemists to the investigation of the novel mode of action.[14,15] It also encouraged many drug companies to search for their own avermectin or milbemycin.

Antiparasitic Efficacies

Although the concern of this publication is agricultural pest managment, this article will review some of the animal and human health data, since the major impact of the avermectins lies in that area, and crop protection is of secondary importance. Currently two avermectin derivatives, ivermectin and avermectin B1, are commercially available for animal health applications. A third one, doramectin, is under development by Pfizer (FIGURE 1). Avermectin B1, with the generic name abamectin, is the major and most important product of the fermentation of *Streptomyces avermitilis*. It also serves as the starting material for the chemical conversion to ivermectin via selective hydrogenation of the 22,23 double bond. Milbemycin derivatives have only recently become commercially available. Milbemycin A_3/A_4 has just been introduced in Japan for crop protection, milbemycin D is used in Japan as an anthelmintic agent, milbemycin A_3/A_4 5-oxime is marketed by Ciba-Geigy as an anthelmintic exclusively for dogs, and moxidectin is being developed by American Cyanamid as a broad spectrum animal health drug (FIGURE 2).

Ivermectin has been on the market since 1980 as an endectocide for animals. Its special advantage over conventional anthelmintic agents is its wide spectrum against gastrointestinal and systemic parasites as well as against many ectoparasitic insects with a single application. Ivermectin is used at the unprecedented low dose of 200 ug/kg in cattle, sheep and horses. The small dose allows for easy formulation in oral, parenteral, and topical applications. It is used in swine at 300 ug/kg, the higher dose being required due to higher metabolism in this species. In dogs it is used under the name Heartgard 30 for the prevention of heartworm infections at the low dose of 6 ug/kg, and is effective with one single monthly application. It has been used by several hundred thousand humans with a single dose every six months at 50 to 200 ug/kg to alleviate the most damaging symptoms of onchocerciasis, or river blindness. It is highly efficacious against a wide spectrum of parasitic nematodes and many ectoparasitic species from the grub, lice, mite, tick and bot families. Representatives of each of these, for instance, are responsible for economic losses in cattle or sheep. No activity, however, has been observed against flatworms, protozoa, bacteria or fungi.

Ivermectin has revolutionized the treatment of man for onchocerciasis in African river regions, where a good part of the adult population becomes blind due to irritation of the eys caused by the microfilaria, and where children are seen to guide their blind fathers to the fields for their work. Most parasites have complicated life cycles with obligate intermediate hosts. Onchocerciasis, as an example, is spread by the "Black Fly" which lives in fast flowing rivers and picks up *Onchocerca volvulus* microfilariae from human blood while feeding. After molting in the insect the infective larvae are then reintroduced into man, where they migrate through the lymph system and finally settle as adults in nodes under the skin. The females deliver microfilaria by life birth, which then spread via the bloodstream over the body. They cause particular damage to the eyes, leading in severe cases of repeat infections to blindness. Eventually the microfilaria are picked up by the flies for a new cycle. It was found that mass treatment of whole villages in Africa so drastically reduces the number of microfilaria in human blood that the subsequent infection rate was greatly reduced.[16] Although ivermectin possesses

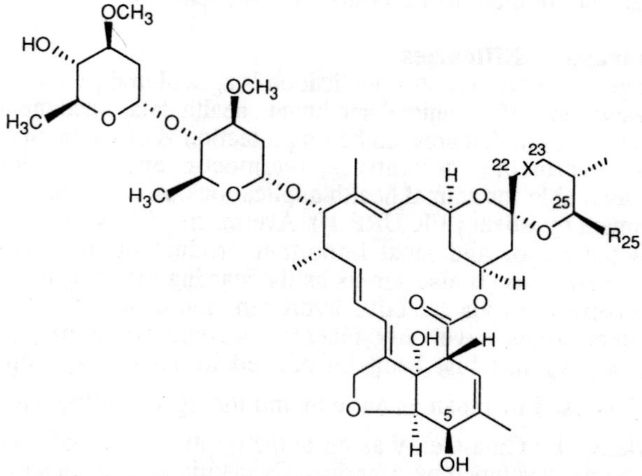

Ivermectin: X = $-CH_2CH_2-$ R_{25} = $CH(CH_3)CH_2CH_3$ and $CH(CH_3)_2$
Abamectin: X = $-CH{=}CH-$ R_{25} = $CH(CH_3)CH_2CH_3$ and $CH(CH_3)_2$
Doramectin: X = $-CH{=}CH-$ R_{25} = Cyclohexyl

Figure 1. Avermectins for animal and human uses.

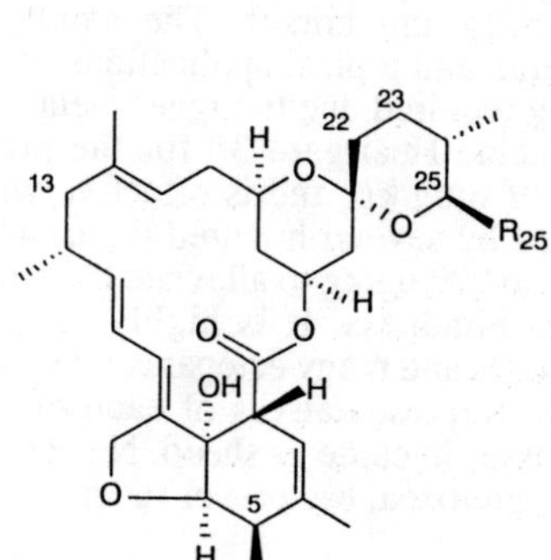

Milbemycin D: R_5 = ━OH R_{25} = $-CH(CH_3)_2$
Milbemycin A_3/A_4 5–Oxime: R_5 = =NOH R_{25} = CH_3 and C_2H_5
Moxidectin: R_5 = ━OH R_{23} = $=NOCH_3$ R_{25} = 

Figure 2. Milbemycins for animal health.

considerable activities against the major gastrointestinal worm infections of man, it is judged to be insufficiently efficacious for this application.

Agricultural Applications

With a good part of development costs and safety testing absorbed by the animal applications, it was possible to introduce abamectin, the natural precurser of ivermectin, for agricultural uses.[17,18] One derivative of abamectin is currently under development for crop protection, and very recently milbemycin A_3/A_4 has become available in Japan for agricultural uses. Avermectin B1, or abamectin, possesses in laboratory assays particularly high toxicity towards mites with an LC_{90} of .02 to .03 ppm, while it is generally less toxic against insects. For example lepidoptera larvae such as the corn earworm (*Heliothis zea*) and the southern armyworm (*Spodoptera eridania*) require 1.5 and 6.0 ppm respectively to achieve an LC_{90}. The toxic effects are mainly caused by feeding of the insects on treated foliage. Abamectin is a stomach poison, and is less effective on contact. There are no ovicidal activities, and the onset of activity coincides with the beginning of the feeding of the larvae. As suggested by the assay results, the primary commercial use of abamectin is as an acaricide. In combination with paraffinic oil it is used against citrus rust mites at 13.5 to 54 g ai/ha on citrus fruit, spider mites on ornamentals and on cotton at 9 to 22 g ai/ha, the twospotted spider mite, European red mite, pear rust mite on various crops such as tomatoes, canteloupe, strawberries, pears and similar high value crops due to its considerable cost. Leafminers are controlled on ornamentals and vegetables, and psylla are significantly reduced on pears. Varying potency of .02 to 6.0 ppm against lepidoptera allows select uses for tomato pinworm control at 11 to 22 g ai/ha, and for control of the diamondback moth, which is highly susceptible. Abamectin is considered safe for the environment due to a rapid breakdown in UV light through photooxidation. Its halflife in the environment in bright sunlight is 4 to 24 hours. It is tightly bound to the soil, which prevents it from being washed into aquatic bodies. It's halflife in soil is 20 to 40 days. All residue studies in crops point to rapid degradation. It is, however, acutely toxic to fish at 3 to 40 ppb, but accumulation in aquatic organisms is low, so that its concentration is not likely to be multiplied through a food chain. Honeybees are very sensitive, but in practice little effect is seen due to the rapid depletion of abamectin residue.[19]

Structural Modifications of Avermectins

Avermectins have been chemically converted by selective hydrolysis of either one or two of the oleandrose glycoside bonds to monosaccharides and aglycones. Deoxygenation of the 13-hydroxy group gave the 13-deoxyaglycones, which are structurally closely related to the milbemycins (FIGURE 3). Conversely, somewhat costly chemical and microbiological procedures exist to introduce a 13-hydroxy group into milbemycins and to attach the sugars to yield avermectins.

The modification of the multifunctional molecule was largely guided by chemistry. It was discovered early on that derivatization of the 4"-hydroxy group retains the high anthelmintic activity against the sheep nematode *Trichstrongylus colubriformis* in a gerbil *in vivo* assay, but that the free 5-hydroxy group was required for activity. A number of 4"-esters and carbamates were highly active anthelmintics. Selective protection of the 5 and 7 hydroxy groups is possible, and subsequent reaction of the 4"-hydroxy group with the fluorinating reagent diethylaminosulfur trifluoride (DAST), for instance, gives via an internal oxonium ion the expected 4"-deoxy-4"-fluoro analog plus a ring contracted isomer (FIGURE

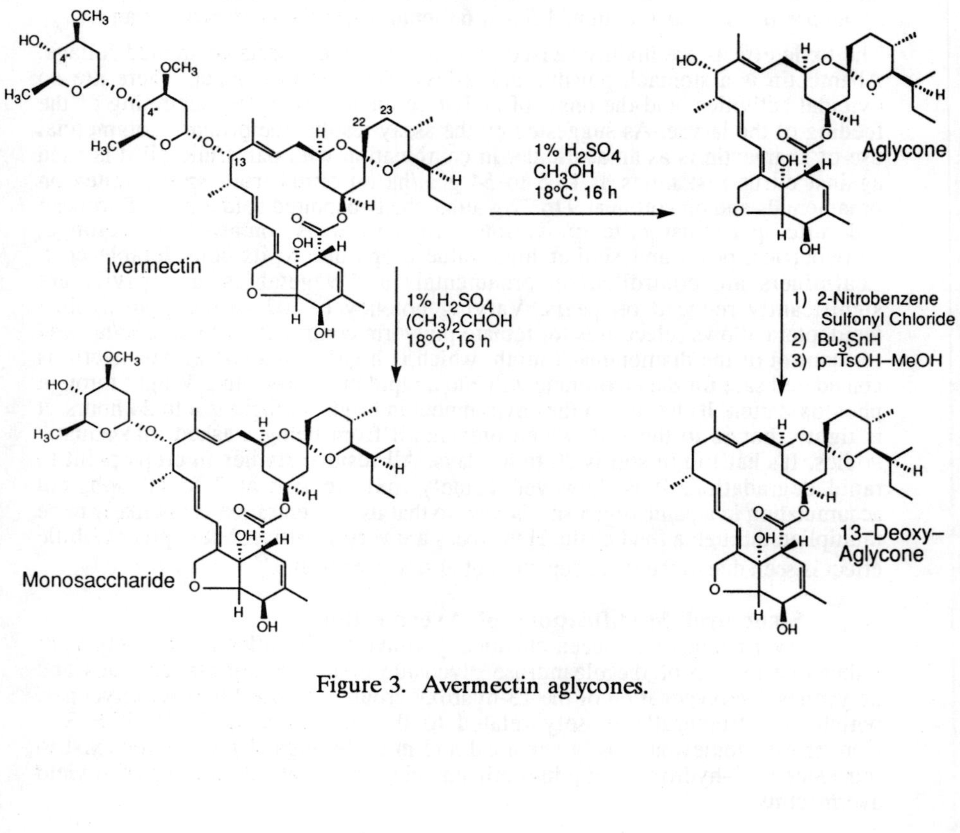

Figure 3. Avermectin aglycones.

4). The fluoro analogs, however, had somewhat reduced biological activities. On the other hand, 4"-epi-hydroxy, oxo, cyano, even deoxy, and particularly methyl analogs have good activities in the twospotted spider mite and southern armyworm assays. One also often observes with a minor modification a shift in spectrum, as seen with the southern army worm activities of the two epimeric 4"-methyl derivatives (EC_{90} = 0.5 and 8.0 ppm, respectively).

4"-Amino Substituted Second Generation Avermectins

For a number of reasons we were interested in the introduction of a basic amino group into the avermectin molecule. This should change its physical properties, make it more polar, and result in a different tissue distribution. In addition, most of the antibacterially active macrolide antibiotics contain an aminosugar. After suitable protection and deprotection, reductive amination of a 4"-oxo intermediate gave the 4"-amino-4"-deoxy-analog as major reaction product (FIGURE 5). When testing these new amino analogs we discovered that they had reduced potency in our brine shrimp assay and also in the twospotted spider mite assay in comparison with abamectin. A substantial increase in efficacy, however, was discovered in the southern armyworm assay, where the EC_{90} of the amino and epi-amino analogs were 80 and 400 fold lower, respectively, than those of abamectin. Since we had observed earlier better activities in the southern armyworm assay also for certain monosaccharides, we prepared their 4'-amino derivatives as well as the 13-amino aglycone, but did not observe further improvement of biological activities. We then looked at alkylamino analogs and found the monomethyl analog the most promising of the group. Further modifications with acylamino, aroylamino, sulfonamido, hydrazino, amidino, cyanamido, azido, carbamino, ureido, or hydrazono substituents represent mostly potent derivatives, and some of these are of considerable interest as endectocides for animal health, but none surpassed the epi-monomethyl derivatives in potency against lepidoptera species (TABLE 1).

4"-Deoxy-4"-epi-N-methylaminoavermectin B1,[20] or MK 244, with the generic name emamectin, is currently under development as an agricultural insecticide, particularly against lepidoptera species (FIGURE 6). It is reserved initially for high value crops, since it will be a rather expensive compound. The foliar ingestion toxicities of MK 244 for a number of important insects such as tobacco hornworm, cabbage looper, beet armyworm, fall armyworm, Colorado potato beetle are from .003 to .03 ppm (LC_{90}), for Mexican been beetle and two-spotted spider mite .2 to .3 ppm, but for bean aphid as high as 20 ppm.[21] The methylamino compound MK-244 showed topical potency superior to the amino analog against the three lepidoptera species of southern armyworm, tobacco budworm, and corn earworm. MK-244 is considerably more potent in a foliar residue test against southern armyworm and tobacco budworm than thiodicarb or fenvalerate.

Structural Modifications of the Spiroketal of Avermectins

More recently we, as well as others, directed our attention to the modification of the spiroketal ring, in particular the 24 and 25 substitutions. Avermectin B_2 is a fermentation product distinguished by an axial hydroxy group at the 23 position from abamectin, which contains a 22,23-double bond. This was a readily available starting material for the desired reaction sequence (FIGURE 7). To this end suitably 4",5-diprotected avermectin B2 was oxidized to the 23-oxo analog, reacted via the 22,23-en-23-ol trimethylsilyl ether, and the 22,23-oxide to

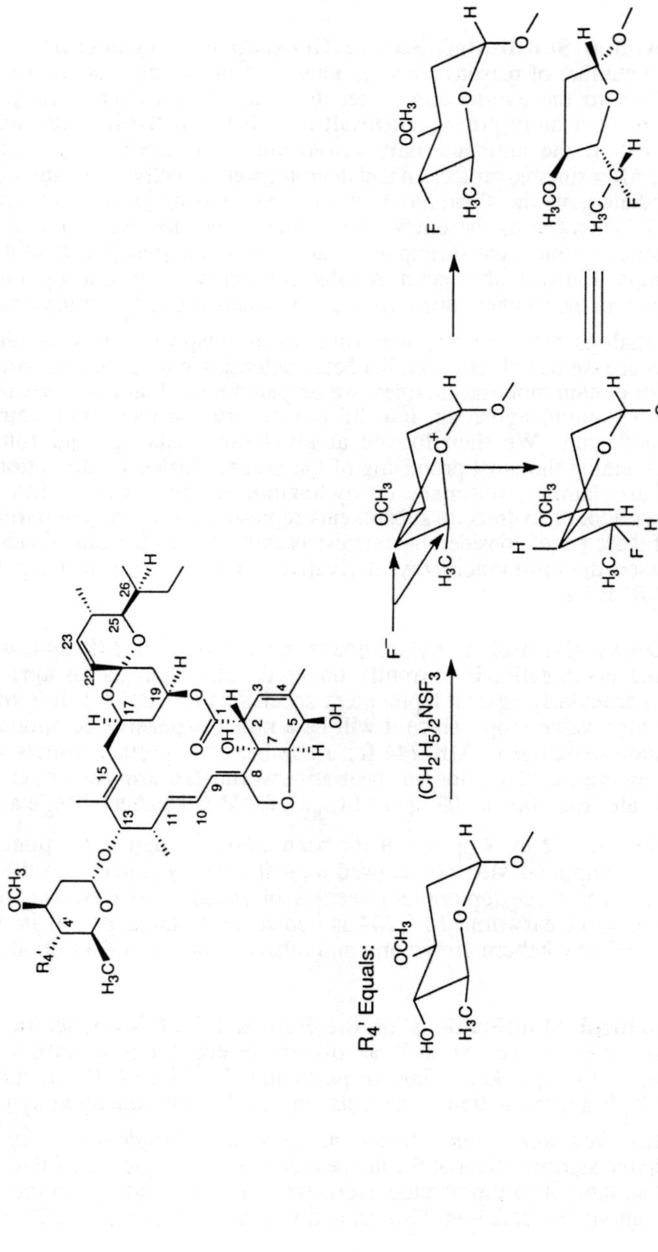

Figure 4. 4"–deoxy–4"–fluoroavermectin.

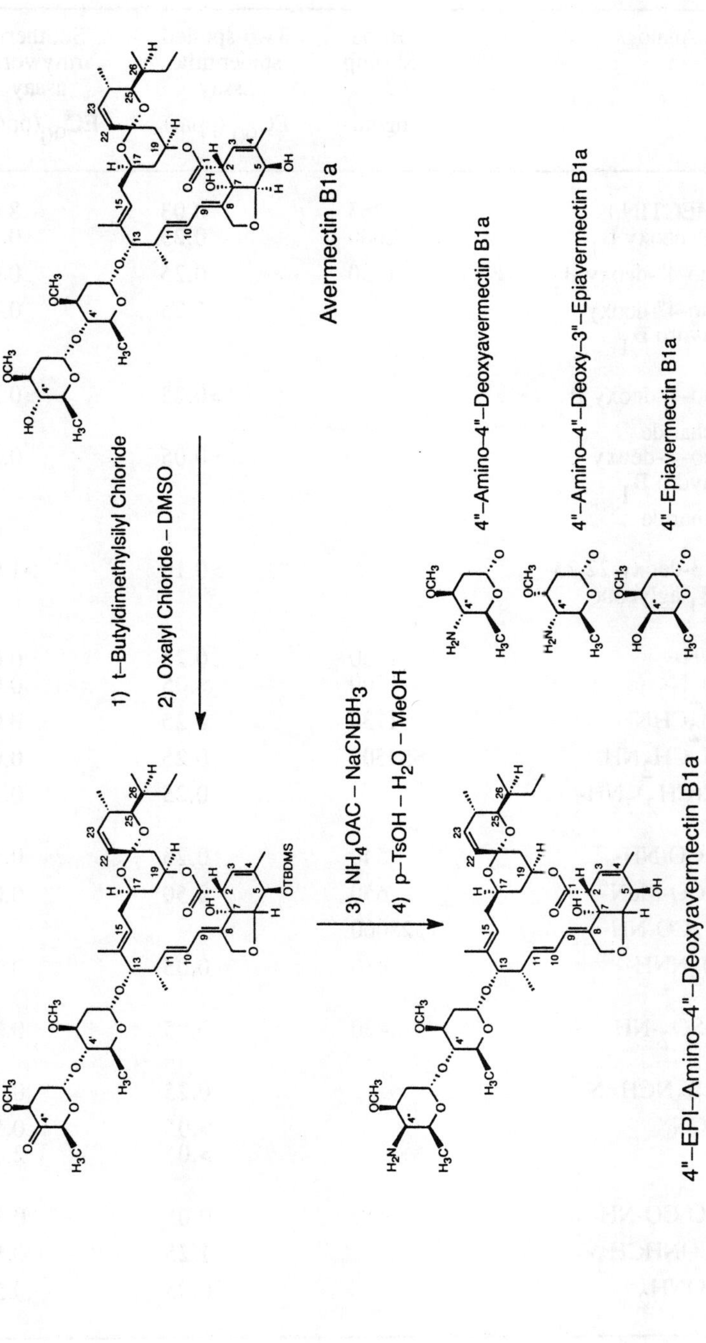

Figure 5. Synthesis of 4"–amino–4"–deoxyavermectin derivatives.

Table 1

Biological Activities of Aminosubstituted Avermectin Derivatives

Avermectin Analogs	Brine Shrimp IC_{100} ng/ml	Two-spotted spider mite assay EC_{90} (ppm)	Southern armyworm assay EC_{90}(ppm)
B1 (ABAMECTIN)	263	0.03	8.000
4"-amino-4"-deoxy B_1	2600	0.25	0.100
4"-epi-amino-4"-deoxy B_1	1730	0.25	0.020
4"-epi-amino-4"-deoxy-22,23-dihydro B_1		1.25	0.500
4'-epi-amino-4'-deoxy B_1 monosaccharide		>0.25	>0.500
4'-epi-amino-4'-deoxy-22,23-dihydro B_1 monosaccharide		>0.05	0.100
13-amino-13-deoxy-22,23-dihydro-B_1 aglycone		>0.10	>1.000
4"-epi-MeNH-	1730	0.25	0.004
4"-epi-$(Me)_2N$-	1300	>.05	0.020
4"-epi-$(Me)_2CHNH$-	1730	0.25	0.020
4"-epi-$C_6H_5CH_2NH$-	>55500	0.25	0.020
4"-epi-$H_3C(CH_2)_7NH$-		0.25	0.100
4"-epi-CH_3CO-NH-	540	0.25	0.500
4"-epi-CH_3CO-MeN-	650	0.50	0.050
4"-epi-C_6H_5CO-NH-	28000		
4"-epi-$(Me)_2NNH$-	430	0.05	0.100
4"-epi-CH_3SO_2-NH-	430	0.05	0.100
4"-epi-$(CH_3)_2NCH$=N-		0.25	0.100
4"-epi-HN-CN		>.05	0.500
4"-epi-N_3		>.05	2.000
4"-epi-CH_3O-CO-NH-	430	0.05	0.100
4"-epi-$NHCONHCH_3$-		1.25	0.500
4"=N-$NHCONH_2$		0.05	0.500

4" –Deoxy–4"-epi-Methylaminoavermectin B1 Benzoate Salt

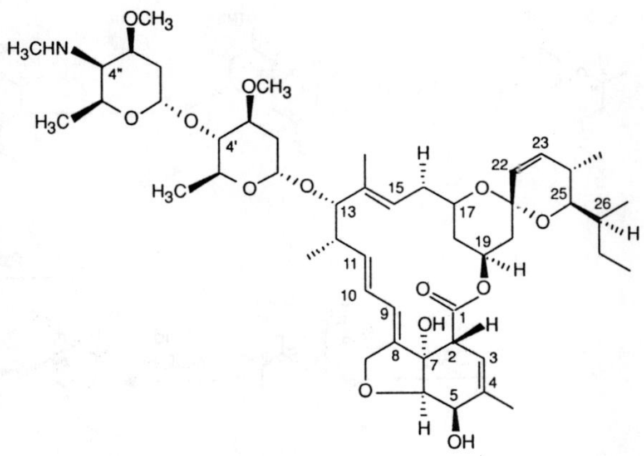

Figure 6. MK–244 / Emamectin.

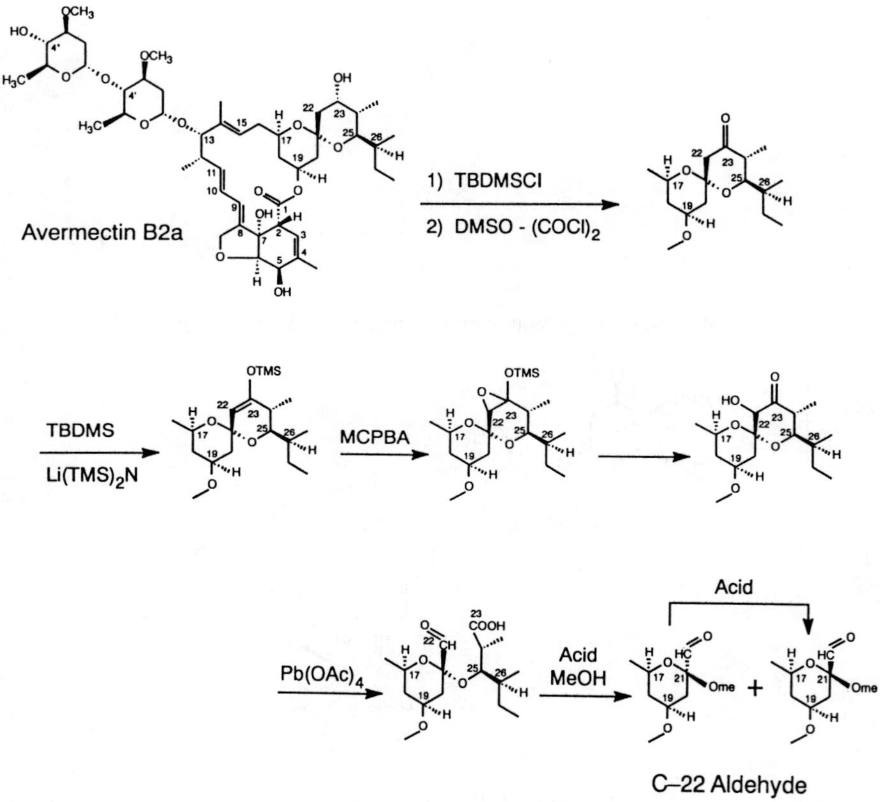

TBDMS = tert-butyldimethylsilyl
TMS = trimethylsilyl
MCPBA = metachloroperbenzoic acid
DMSO = dimethylsulfoxide
$(COCl)_2$ = oxalyl chloride
$Pb(OAc)_4$ = lead tetraacetate

T.L. SHIH *et al.*, Tet. Letters 1990, 31, 3525-3528

Figure 7A. Degradation of avermectin B2 to the starting materials for C24/25 modified avermectins. (Reprinted with permission from reference 22. Copyright 1990 Pergamon Press Ltd.)

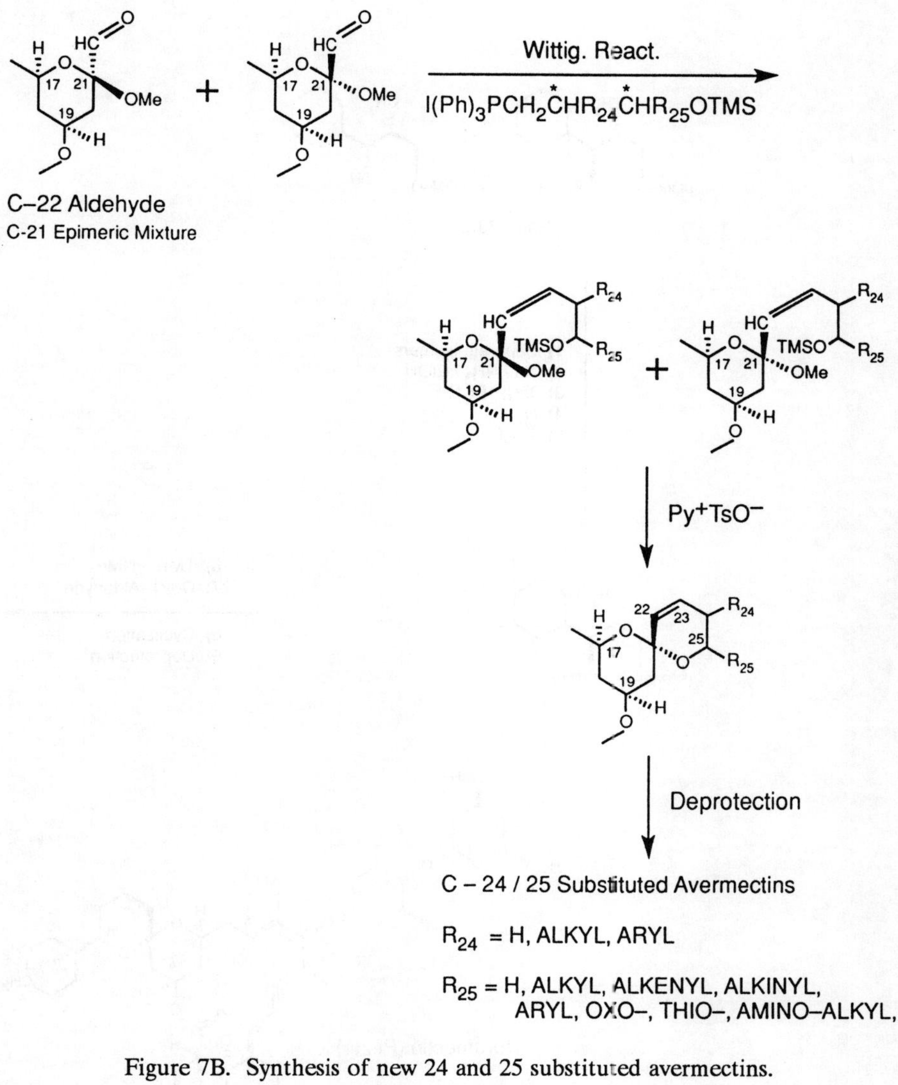

Figure 7B. Synthesis of new 24 and 25 substituted avermectins.

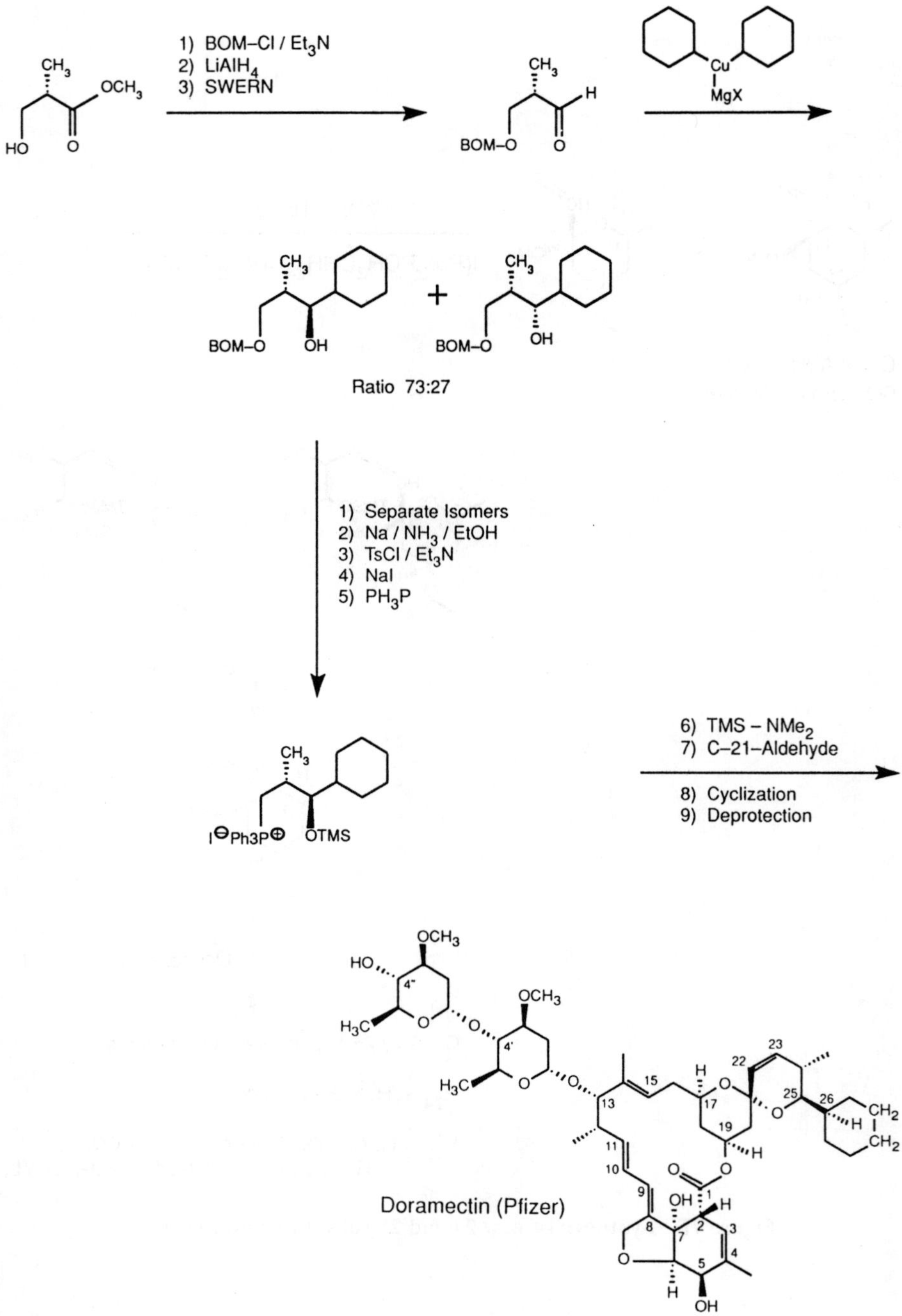

Figure 7C. Chiral synthesis of 25–DES–(2–butyl)–25–cyclohexyl avermectin B1a.

the rearranged 22-hydroxy-23-ketone. This is then cleaved to the 22,23-seco compound, which gave upon methanolysis the desired C-22 aldehyde intermediate for further synthetic modifications.[22] The epimeric mixture of this aldehyde is now used for a new carbon-carbon bond formation with a Wittig reagent containing the desired C-24 and C-25 substituents. The C-21 epimeric mixture of avermectin B1 precursors is then equilibrated with pyridinium tosylate to the thermodynamically more stable natural avermectin B1 analog. After deprotection a wide variety of C-24 and C-25 substituted analogs with natural and epimeric stereochemistry are obtained.[23] This approach lends itself also to the chiral syntheses of analogs with predetermined stereochemistry at C-24 and C-25. Starting with methyl (S)-(+)-3-hydroxy-2-methylpropionate and reaction with a cyclohexylcuprate synthon gives the chiral Wittig reagent needed for the synthesis of 25-des-(2-butyl)-25 cyclohexylavermectin B1. This compound was also obtained via directed biosynthesis and is currently under development by Pfizer. The activities of the new compounds in the brine shrimp assay suggest that none of them are more active than avermectin B1. Additional testing in an *in vivo* anthelmintic model assay confirmed anthelmintic activities for these compounds. In a definitive sheep test against a spectrum of gastrointestinal nematodes again little difference from avermectin B1 was shown in potency and spectrum. In a twospotted spider mite laboratory assay the 25-phenyl analog, for instance, had activity comparable to that of abamectin. Although many of these compounds are potent avermectin analogs, they did not show any significant improvement over the parent compound. We used a related synthetic scheme to prepare derivatives where the 6,6-spiroketal of natural avermectins was modified to a 6,5-system (FIGURE 8). A series of reactions starting with a C-22-aldehyde and a stabilized Wittig or Horner-Emmons olefination reagent, subsequent reduction of the double bond, reduction of the carbonyl to an epimeric mixture of alcohols, cyclization and deprotection gave a series of analogs as epimeric mixtures at C-24, which were separated chromatographically. Either one of these stereoisomers could also be prepared by stereospecific reduction of the carbonyl group with a chiral oxazaborolidine-borohydride complex.[24] A wide variety of these derivatives were prepared in yields from 30 to 76 % from the avermectin B_2 intermediate (not optimized). These 6,5-spiroketal compounds are potent anthelmintic and insecticidal compounds, but no significant advantages over abamectin could be found in subsequent tests.

Additional C-25-substituted avermectins were also obtained by directed biosynthesis. Incorporation studies revealed that natural avermectins are produced in the fermentation medium from 7 acetate and 5 propionate building blocks, which account for all the carbon atoms except for C-25 and the attached C-25-substituent.[25] These atoms are not labelled by acetate or propionate. Instead it was found that isoleucine or 2-methylbutyrate are incorporated into the 2-butyl group of avermectin B1a.[26] Subsequently it was shown that addition of 2-methylpentanoate and 2-methylhexanoate to the avermectin fermentation gave additional products corresponding to mono and bis homologs of avermectin B1a.[27] A systematic approach by scientists at Pfizer using a mutant devoid of branched chain 2-oxo acid dehydrogenase, which blocks the formation of 2-methylbutyrate, and addition of a wide variety of carboxylic acids lead to avermectin derivatives with modified C-25-side chains.[28,29]

Mode of Action and Avermectin Binding Site

Many pharmacologic effects of avermectin B1 in a number of different animals and tissue preparations have been described.[14] Nematodes are paralyzed rapidly without causing hypercontraction or flaccid paralysis.[30] Signal transmission from ventral interneurons to excitatory motorneurons of *Ascaris* are blocked.[30] A

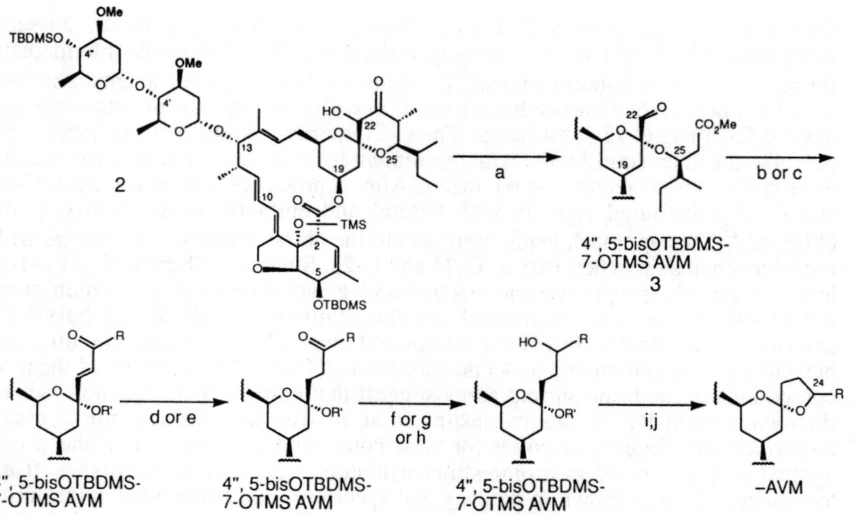

a) $Pb(OAc)_4$, MeOH, pyr; **b)** $(MeO)_2P(O)CH_2C(O)R$, LiCl, DIEA; **c)** $Ph_3PCHC(O)R$; **d)** 9 eq $Na_2S_2O_4$, 18 eq $NaHCO_3$; 1:1 $PhH:H_2O$, reflux; **e)** $Pd(PPh_3)_4$, nBu_3SnH; **f)** $NaBH_4$; **g)** RMgBr; **h)** BH_3SMe_2, oxazaborolidine, 0° C; **i)** 4:1 PPTS:TsOH; **j)** HF.pyr

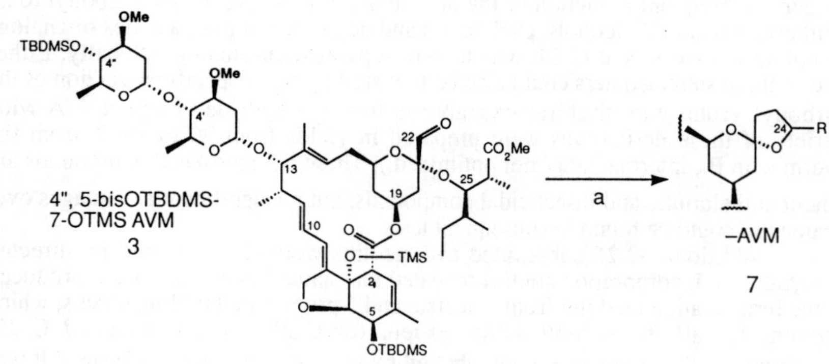

R	Yield (%, 3 → 7)	R	Yield (%, 3 → 7)
H	62	CH_2OMe	57
Me	42	CH_2OPh	43
i-Pr	68	Ph	38
t-Bu	76	(*p*-F)Ph	30
n-C_8H_{17}	41	(*p*-MeO)Ph	42
c-C_6H_{11}	59	2-Furyl	25
CH_2OH	48	OMe	56

Figure 8. Top: 25–NOR–6,5–spiroketal avermectins. Bottom: 25–NOR–24-substituted avermectin analogs. (Table reprinted with permission from reference 24. Copyright 1992 Pergamon Press Ltd.)

reversible increase of chloride ion permeability of GABA sensitive fibers of the extensor tibiae muscle of the locust *Schistocerca gregaria* is observed at nanomolar concentrations. In contrast an irreversible inhibition of GABA sensitive and insensitive muscle fibers of *Schistocerca gregaria* is observed at micromolar concentrations.[31] A reversible opening of crayfish stomach chloride channels is observed at subpicomolar concentrations, and irreversible opening of crayfish stomach chloride channels at 10 pmol or higher.[32] Avermectin binds specifically to a number of chloride channel proteins but its binding site is distinct from that of all other effector molecules.[33] A high affinity binding site was described for a membrane preparation of *C. elegans*, a free living nematode, which is known to be very sensitive to avermectins.[34] Recently the protein containing the binding site was solubilized without much destruction as suggested by the almost identical dissociation constants for ivermectin of 0.14 and 0.18 nM, respectively.[35] The binding affinity of several avermectin derivatives closely correlates to their biological activities. The highly potent ivermectin, the water soluble avermectin B1 phosphate, and avermectin B_2 have the lowest K_I values. In contrast the biologically inactive octahydroavermectin B_1 does not inhibit the binding of tritiated ivermectin. Aglycones and monosaccharides with intermediate activities have intermediate inhibition constants.[34]

For the further identification of the binding proteins a photoaffinity probe was constructed from suitably protected 4"-aminoavermectin B1. 4-Azidosalicylic acid was attached to the aminogroup through a beta-alanyl-omega-aminocaproyl spacer. After iodination to the 4-Azido-3-^{125}iodosalicylic acid derivative it was found suitable for the radioactive labelling of avermectin binding proteins.[36] The specific binding of ^{125}I-azido-AVM in the dark to *C. elegans* membranes occurs with a K_D = 0.136 nM, which is comparable to that of ivermectin. It also is inhibited by ivermectin.[37] Frozen *C. elegans* membrane bound receptor proteins are obtained from the free living nematode grown in liquid culture with *E. coli* as food source. After flotation on 35 % sucrose, washing with 0.1 M NaCl the worms were homogenized in HEPES buffer in the presence of protease inhibitors. Centrifugation at 1000 x G and subsequent centrifugation of the supernatent at 28000 x G gave a pellet containing the membrane bound receptor proteins, which was resuspended in HEPES plus protease inhibitors in a concentration of 5 mg/ml protein. The solution was dialysed and frozen. The frozen membranes were thawed, diluted, and stirred with TRITON X for 1 hour at 0° C for solubilization of the proteins. Centrifugation and filtration gave the TRITON X solubilized protein solution containing the avermectin binding site.

After further dilution the protein mixture was incubated with excess ^{125}I-azido-AVM in the dark, and the unbound ^{125}I-azido-AVM was removed with charcoal. Photolysis with UV light and precipitation of the proteins with methanol gave a mixture containing the ^{125}I labelled avermectin binding proteins.[37] The Coomassie stained gel after electrophoresis shows the bands of *C. elegans* proteins after incubation and photolabelling in the presence of 0.0, 0.2, 0.8, 2.0, and 20 nM unlabelled ivermectin. The ^{125}I labelled avermectin binding proteins are, however, not visible in the Coomassie stained gel due to their very low concentrations, and all lanes are identical in this gel. The autoradiogram of the same gel (FIGURE 9) shows three protein bands labelled by ^{125}I-azido-AVM with molecular weights of 53, 47 and 8 kDa. With increasing concentrations of unlabelled ivermectin from left

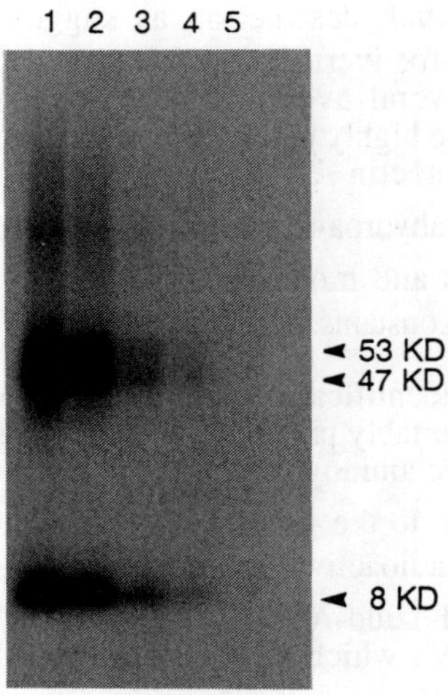

Ivermectin Concentration

Lane 1	0
Lane 2	2×10^{-10}M
Lane 3	8×10^{-10}M
Lane 4	2×10^{-9}M
Lane 5	2×10^{-8}M

Figure 9. Photoaffinity labeling of the *C. Elegans* avermectin receptor. (Reprinted with permission from reference 37. Copyright 1992 National Academy of Sciences.)

to right the binding of ^{125}I-azido-AVM decreases.[37] The autoradiogram of a similar gel shows the binding experiment in lane 1 with ^{125}I-azido-AVM only, in lane 2 and 3 after addition of high level cold ivermectin, which inhibits the binding of ^{125}I-azido-AVM and thus the labelling of the binding proteins. Lane 4 and 5 show addition of high concentrations of 3,4,8,9,10,11,22,23-octahydro avermectin B_1, a close analog of ivermectin, which is known to be completely inactive and is not binding to the avermectin receptor. Consequently it does not interfere with the labelling of the avermectin binding proteins. A single major avermectin binding protein with a molecular weight of approximately 47 kDA was subsequently detected in Drosophila head membranes.[37]

The cloning and structure determination of these binding proteins, which are presumably part of an avermectin sensitive chloride channel, should enhance our understanding of the mode of action of avermectins and milbemycins.

REFERENCES

(1) Burg, R. W; Miller, B. M; Baker, E. E.; Birnbaum, J; Currie, S. A.; Hartman, R; Kong, Y-L; Monaghan, R. L.; Olsen, G; Putter, I; Tunac, J. B.; Wallick, H.; Stapley, E. O.; Oiwa, R.; Omura, S. Avermectins, New Family of Potent Anthelmintic Agents: Producing Organism and Fermentation. Antimicrob. Agents Chemother. **1979**, 15, 361-67.

(2) Egerton, J. R.; Ostlind, D. A.; Blair, L. S.; Eary, C. H.; Suhayda, D.; Cifelli, S.; Fiek, R. F.; Campbell, W. C. Avermectins, New Family of Potent Anthelmintic Agents: Efficacy of the B_{1a} Component. Antimicrob. Agents Chemother. **1979**, 15, 372-00.

(3) Miller, T. W.; Chaiet, L.; Cole, D. J.; Cole, L. J.; Flor, J. E.; Goegelman, R. T.; Gullo, V. P.; Joshua, H.; Kempf, A. J.; Krellwitz, W. R.; Monaghan, R. L.; Ormond, R. E.; Wilson, K. E.; Albers-Schonberg, G.; Putter, I. Avermectins, New Family of Potent Anthelmintic Agents: Isolation and Chromatographic Properties. Antimicrob. Agents Chemother. **1979**, 15, 368-71.

(4) Albers-Schonberg, G.; Arison, B. H.; Chabala, J. C.; Douglas, A. W.; Eskola, P.; Fisher, M. H.; Lusi, A.; Mrozik, H.; Smith, J. L.; Tolman, R. L. Avermectins. Structure Determination J. Am. Chem. Soc. **1981**, 103, 4216-4221.

(5) Mishima, H.;.Ide, J; Muramatsu, S.; Ono, M. Milbemycins, a New Family of Macrolide Antibiotics: Structure Determinations of Milbemycins D, E, F, G, H, J and K. J. Antibiot. **1983**, 36, 980-990.

(6) Takiguchi, Y.; Mishima, H.; Okuda, M.; Terao, M.; Aoki, A.; Fukuda, R. Milbemycins, a New Family of Macrolide Antibiotics: Fermentation, Isolation and Physico-Chemical Properties. J. Antibiot. **1980**, 33,1120-1127.

(7) Ivermectin and Avermectin, W. C. Campbell ed., Springer-Verlag, **1989**, 1-363.

(8) Chabala, J. C.; Fisher, M. H.; Mrozik, H H. 13-Halo and 13 -Deoxy C076 Compounds. **1979**, US. Patent No. 4,171,314. Putter, I. Milbemycin Compounds as Anthelmintic Agents. **1979**, US. Patent No. 4,144,352.

(9) Chabala, J. C.; Mrozik, H.; Tolman, R. L.; Eskola, P.; Lusi, A.; Peterson, L. H.; Woods, M. F.; Fisher, M. H. Ivermectin, a New Broad-Spectrum Antiparasitic Agent. J. Med. Chem. **1980**, 23, 1134-1136.

(10) Campbell,W. C.; Fisher, M. H.; Stapley, E. O.; Albers-Schonberg, G. Jacob, T. A. Ivermectin: A Potent New Antiparasitic Agent. Science **1983**, 221, 823-828.

(11) Taylor, H. R.; Greene, B. M. The Status of Ivermectin in the Treatment of Onchocerciasis. Am. J. Trop. Hyg. **1989**, 41, 460-466.

(12) Blizzard, T.; Fisher, M. H.; Mrozik, H.; Shih, T. L. Avermectins and Milbemycins. In Lukacs G (ed) Recent Progress in the Chemical Synthesis of Antibiotics, Springer-Verlag Berlin Heidelberg, **1990**, 65-102.

(13) Davies, H. G; Green, R. H. Avermectins and Milbemycins Part II. Chem. Soc. Rev. **1991**, 20, 271-339.

(14) Turner, M. J.; Schaeffer, J. M. Mode of Action of Ivermectin, in ref. 7, 73-88.

(15) Fisher, M. H.; Mrozik, H. The Chemistry and Pharmacology of Avermectins. Annu. Rev. Pharmacol. Toxicol. **1992**, 32, 537-53.

(16) Taylor, H. R.; Pacque, M.; Munoz, B.; Greene, B. M. Impact of Mass Treatment of Onchocerciasis with Ivermectin on the Transmission of Infection. Science, **1990**, 250, 116-118.

(17) Putter, I.; MacConnell, J. G.; Preiser, F. A.; Haidri, A. A.; Ristich, S. S.; Dybas, R. A. Avermectins: Novel Insecticides, Acaricides and Nematicides from a Soil Microorganism. Experientia **1981**, 37, 963.

(18) Lasota, J. A.; Dybas, R., A. Avermectins, a Novel Class of Compounds: Implications for Use in Arthropod Pest Control. Annu. Rev. Entomol. **1991**, 36, 91-117.

(19) Wislocki, P. G.; Grosso, L. S.; Dybas, R. A. Environmental Aspects of Abamectin Use in Crop Protection. See Ref 7, 182-200.

(20) Mrozik, H.; Eskola, P.; Linn, B. O.; Lusi, A.; Shih, T. L.; Tischler, M.; Waksmunski, F. S.; Wyvratt, M. J.; Hilton, N. J.; Anderson, T. E.; Babu, J. R.; Dybas, R. A.; Preiser, F. A.; Fisher, M. H. Discovery of Novel Avermectins with Unprecedented Insecticidal Activity. Experientia **1989**, 45, 315-6.

(21) Dybas, R. A.; Hilton, N. J.; Babu J. R.; Preiser, F. A.; Dolce, G. J. Novel Second-Generation Avermectin Insecticides and Miticides for Crop Protection. In Novel Microbial Products for Medicine and Agriculture, ed. A. L. Demain, G. A. Somkuti, J. C. Hunter-Cevera, H. W. Rossmoore, **1989**, 203-12, New York: Elsevier.

(22) Shih, T. L.; Mrozik, H; Holmes, M. A.; Fisher, M. H. Cleavage of the Spiroketal Portion of Avermectin B_{2a}. Tetrahedron Lett. **1990**, 31, 3525-3528.

(23) Shih, T. L.; Mrozik, H; Holmes, M. A.; Fisher, M. H. A Wittig Approach to Novel C24 and C25-substituted Avermectins. Tetrahedron Lett. **1990**, 31, 3529-3532.

(24) Meinke, P. T.; O'Connor, S. P.; Mrozik, H.; Fisher, M. H. Synthesis of Ring-Contracted 25-nor-6,5-Spiroketal-Modified Avermectin Derivatives. Tetrahedron Lett. **1992**, 33, 1203-1206.

(25) Cane, D. E.; Liang, T. C.; Kaplan, L.; Nallin, M. K.; Schulman, M. D.; Hensens, O. D.; Douglas, A. W.; Albers-Schonberg, G. J. Biosynthetic Origin of the Carbon Skeleton and Oxygen Atoms of the Avermectins. Am. Chem. Soc. **1983**, 105, 4110-12.

(26) Chen, T. S.; Arison, B. H.; Gullo, V. P.; Inamine, E. Further Studies on the Biosynthesis of the Avermectins. Am. Chem. Soc. 186 Meet., **1983**, Abstr. MBTD 28.

(27) Chen, T. S.; Inamine, E. S.; Hensens, O. D.; Zink, D.; Ostlind, D. A. Directed Biosynthesis of Avermectins. Arch. Biochem. Biophys. **1989**, 269, 544-547.

(28) Hafner, E. W.; Holley, B. W.; Holdom, K. S.; Lee, S. E.; Wax, R. G.; Beck, D.; McArthur, H. A. I.; Wernau, W. C. Branched Chain Fatty Acid Requirement for Avermectin Production by a Mutant of Streptomyces avermitilis Lacking Branched-Chain 2-Oxo Acid Dehydrogenase Activity. J. Antibiot. **1991**, 44, 349-356.

(29) Dutton, C. J.; Gibson, S. P.; Goudie, A. C.; Holdom, K. S.; Pacey, M. S.; Ruddock, J. C.; Bu'Lock, J. D.; Richards, M. K. J. Novel Avermectins Produced by Mutational Biosynthesis. J. Antibiot. **1991**, 44, 357-365.

(30) Kass, I. S.; Wang, C. C.; Walrond, J. P.; Stretton, A. O. W. Avermectin B_{1a}, a Paralyzing Anthelmintic that Affects Interneurons and Inhibitory Motoneurons in Ascaris. Proc. Natl. Acad. Sci. USA **1980**, 77, 6211-6215.

(31) Duce, I. R.; Scott, R. H. Actions of Dihydroavermectin B_{1a} on Insect Muscle. Br. J. Pharmacol. **1985**, 85, 395-401.

(32) Zufall, F.; Franke, C.; Hatt, H. J. The Insecticide Avermectin B_{1a} Activates a Chloride Channel in Crayfish Muscle Membrane. J. Exp. Biol. **1989**, 142, 191-205.

(33) Pong, S. S.; Wang, C. C. The Specificity of High Affinity Binding of Avermectin B_{1a} to Mammalian Brain. Neuropharmacololgy **1980**, 19, 311-317.

(34) Schaeffer, J. M.; Haines, H. W. Avermectin Binding in *Caenorhabditis elegans*. A Two-State model for the Avermectin Binding Site. Biochem. Pharmacol. **1989**, 38, 2329-2338.

(35) Cully, D. F.; Paress, P. S. Solubilization and Characterization of a High Affinity Ivermectin Binding Site from *Caenorhabditis elegans*. Mol. Pharmacol. **1991**, 40, 326-332.
(36) Meinke, P. T.; Rohrer, S. P.; Hayes, E. C.; Schaeffer, J. M.; Fisher, M. H.; Mrozik, H. Affinity Probes for the Avermectin Binding Proteins. J. Med. Chem. **1992**, 35, 3879-3884.
(37) Rohrer, S. P.; Meinke, P. T.; Hayes, E. C.; Mrozik, H.; Schaeffer, J. M. Photosffinity Labelling of Avermectin Binding Sites from *Caenorhaditis elegans* and *Drosophila melanogaster*. Proc. Natl. Acad. Sci. USA **1992**, 89, 4168-4172.

RECEIVED April 30, 1993

Chapter 6

Hydantocidin and Cornexistin

New Herbicidal Antibiotics

Shuji Takahashi[1], Mutsuo Nakajima[1], Takeshi Kinoshita[2], Hideyuki Haruyama[2], Soji Sugai[3], Toyokuni Honma[3], Sadao Sato[2], and Tatsuo Haneishi[4]

[1]Fermentation Research Laboratories, [2]Analytical and Metabolic Research Laboratories, [3]Agricultural Chemical Research Laboratories, and [4]International Drug Development Department, Sankyo Company, Ltd., 1–2–58, Hiromachi, Shinagawa-ku, Tokyo 140, Japan

Hydantocidin and cornexistin are new compounds with herbicidal activities isolated from the culture broth of *Streptomyces hygroscopicus* SANK 63584 and *Paecilomyces variotii* SANK 21086, respectively. Hydantocidin(**1**), $C_7H_{10}N_2O_6$ (MW 218), formed colorless needles which were water soluble, and **1** was found to be composed of hydantoin and ribose with a spiro structure. Cornexistin(**2**), $C_{16}H_{20}O_6$ (MW 308), formed colorless needles which were water insoluble, and **2** was found to have a nonadride structure.

Herbicidal activities of **1** are more pronounced against perennial plants than annual plants, and this activity is greater than bialaphos and almost equal to glyphosate. **2** is useful for postemergence weed control with selective protection of corn. Neither **1** nor **2** had activities against gram-positive and negative bacteria and fungi. Both compounds were essentially non-toxic in mice. LD_{50}s of **1** and **2** were greater than 1 g/kg (p.o.) and 100 mg/kg (i.p.).

During the course of screening for herbicidal antibiotics from metabolites of microorganisms, we found several compounds which exhibited herbicidal activity. Among those found were herbicidins which had been previously reported to be nucleoside antibiotics[1)]. Also, we isolated two previously unreported compounds, hydantocidin(**1**) and cornexistin(**2**), which are produced by *Streptomyces hygroscopicus* SANK 63584 and *Paecilomyces variotii* SANK 21086, respectively (Fig. 1). In this paper, we first describe the isolation and determination of structure for hydantocidin and cornexistin and then report the herbicidal activities of these compounds

Fermentation, Isolation and Structure Determination of Hydantocidin[2)]

The producing organism of hydantocidin was identified as *Streptomyces hygroscopicus* SANK 63584 from microscopic characteristics and chemical analysis of the cell wall[2)].

0097–6156/94/0551–0074$06.00/0

We found that production of hydantocidin was very limited when shaking liquid culture equipment was employed but production was increased using a malted rice culture as seed culture. 2 L of a malted rice culture was transferred into a 600 L fermenter containing 300 L of production medium as shown in Fig. 2. Hydantocidin was detected by a germination test of chinese cabbage. The minimun inhibition concentration in the germination test was 0.4 μg/ml. The maximum production of hydantocidin in fermentation broth was determined to be 4 μg/ml after 139 hrs.

After fermentation was completed, the cultured filtrate was passed through a Diaion HP-20 column as described in Fig. 2. In this time, co-producing antibiotics, including hygromycin and azalomycin, as well as a considerable quantity of unrelated impurities, were retained in this column. The collected eluate was adsorbed on a carbon column. Hydantocidin was subsequently eluted with 30% MeOH, and the active fractions were purified by several columns including CHP-20P, Avicel and Dowex 50 (Ca-type). For further purification, the active fractions were rechromatographed on Diaion CHP-20P and then crystallized from acetone.

Hydantocidin was obtained as colorless needles. Hydantocidin was a water soluble, neutral compound and had no characteristic UV absorption. Its molecular formula was established by high resolution FAB-MS spectra at 219 $(M+H)^+$ as $C_7H_{10}N_2O_6$ and the structure of hydantocidin was determined as **1** by detailed IR and NMR spectra[2].

In order to confirm the stereochemistry and reconfirm the structure of hydantocidin, we performed a total synthesis of the compound and all its stereoisomers[3-5]. We envisioned that a condensation of tartaric aldehyde and hydantoin could be employed in the synthesis of hydantocidin. The condensation of the D-tartaric aldehyde derivative(**3**) and acetylhydantoin(**4**) gave compound **5**, and after treatment with p-TsOH, two dehydroxy spiro compounds (**6** and **7**) were obtained. Stereoisomers at the C1 position were also observed. Compound(**8**), which is the 1'-*N*-CBZ derivative of compound(**6**) with *S*-configuration at the C1-position, was oxidized with OsO_4 to form the diol derivative(**10**). This reaction stereoselectively oxidized the double bond to yield hydroxy groups which were opposite to the *N*-CBZ group. Compound(**10**) gave hydantocidin by a deprotection reaction with ceric ammonium nitrate(CAN). The optical rotation of synthesized hydantocidin was identical with that of the natural compound. Therefore, the absolute configuration of natural hydantocidin was assigned 1*S*, 2*S*, 3*R* and 4*R* by total synthesis (Fig. 3). By using L-tartaric aldehyde and/or other reagents in place of OsO_4, we synthesized the other 15 stereoisomers of hydantocidin in a similar way.

The herbicidal activities of the synthetic stereoisomers of hydantocidin against annual plants are shown in Table 1. The synthetic hydantocidin had the same activity as the natural compound, and hydantocidin which had been epimerized at the C1-isomer was less active than the natural compound. The other 14 stereoisomers had no activity. These results suggested that the configuration of the hydroxy groups on the ribose ring play an important role in herbicidal activity.

Isolation and Structure Determination of Cornexistin[6]

Cornexistin is a nonadride compound having a single anhydride structure in an organic solvent and having dicarboxylic acid structure in an aqueous solution. Thus, the nature of cornexistin is that of a neutral compound in an organic solvent and that of an acidic compound in aqueous solution. We can easily isolate cornexistin by taking advantage of these characteristics[6].

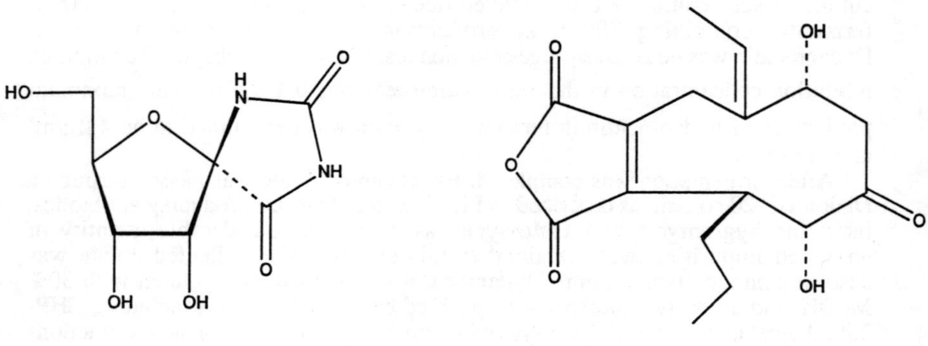

Hydantocidin Cornexistin

Fig. 1. Structures of Hydantocidin and Cornexistin

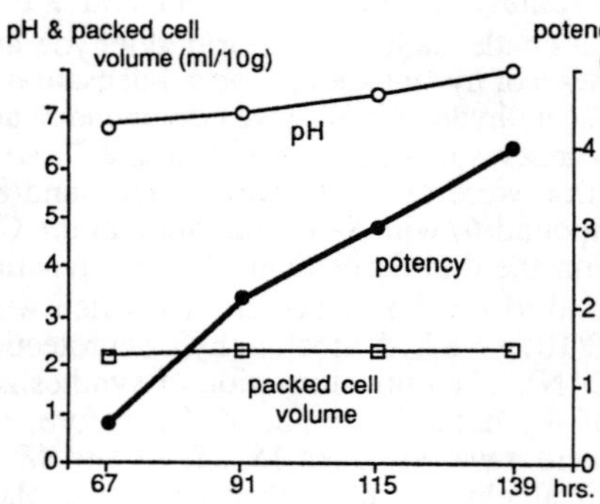

Medium	(%)
Glucose	3.0
Pressed yeast	1.0
Soybean meal	3.0
$CaCO_3$	0.4
$MgSO_4$	0.2
CB-422	0.02
pH 7.2	

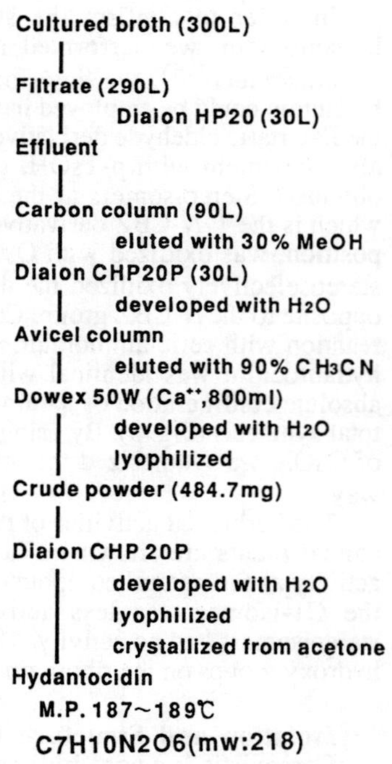

Fig. 2. Fermentation and Purification of Hydantocidin Produced by *Streptomyces hygroscopicus* SANK 63584

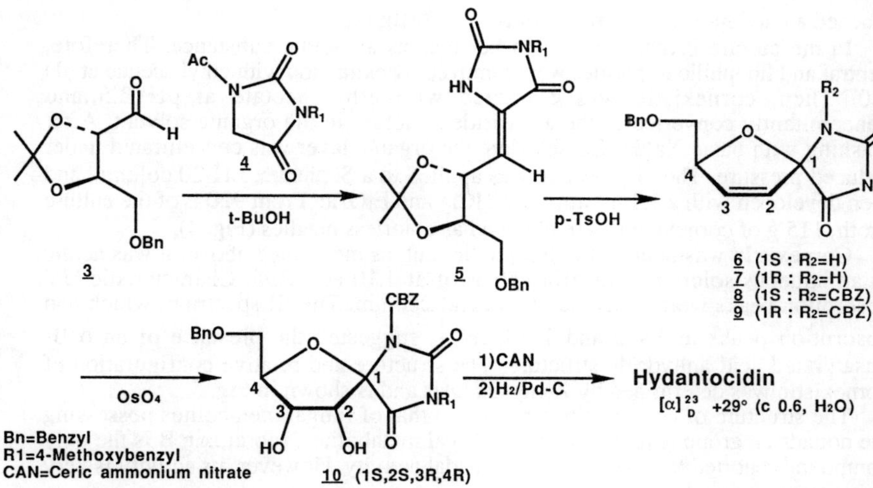

Fig. 3. Total Synthesis of Hydantocidin

Table 1. Herbicidal activities of stereoisomers of hydantocidin against annual mono- and dicotyledonous weeds (1000 ppm)

	Hydantocidin (synthetic)	Epi-hydantocidin (C1-epimer)	Other 14 Isomers
Barnyardgrass	5	1	
Black grass	5	2	
Crabgrass, large	5	4	0~1
Foxtail, green	5	4	
Johnsongrass	5	0	
Wild oat	4	0	
Cocklebur, common	5	4	
Lambsquarters, common	4	2	
Morningglory, tall	5	2	0~1
Night shade, black	5	3	
Velvetleaf	5	4	
Wild mustard	5	4	

Cornexistin was isolated from the culture broth of *Paecilomyces variotii* SANK 21086. This strain was newly isolated from a deer dung sample collected at Nojak, Alberta, Canada.

The fermentation process was directly monitored by HPLC analysis of cornexistin itself. After 161 hr, in a 600 L fermenter, the production of cornexistin reached a maximum concentration of about 260 μg/ml.

In the culture broth, cornexistin behaves as an acidic substance. Therefore, neutral and lipophilic impurities were removed by extraction with ethyl acetate at pH 8.0. Then, cornexistin was extracted with ethyl acetate at pH 2.5,.and concomitantly converted to the anhydride structure in the organic solvent. After washing with basic Na_2HPO_4 solution, the organic layer was concentrated under reduced pressure. The oily residue was applied to a Sephadex LH-20 column, and then developed with a 1:1 mixture of $CHCl_3$ and EtOAc. From 920 L of the culture broth, 115 g of cornexistin were obtained as colorless needles (Fig. 4).

Cornexistin was neutral and lipophylic, but, as mentioned above, it was acidic in an aqueous solution with pKa values of at 4.10 and 5.95. Characteristic UV absorption peaks were observed at 238 and 280 nm. The IR spectrum, which had absorption peaks at 1820 and 1761 cm^{-1}, suggested the presence of an α,β-unsaturated acid anhydride structure. The structure and relative configuration of cornexistin was determined by X-ray analysis and is shown in Fig. 5.

The structure of cornexistin is related to that of fungal metabolites possessing the nonadride group (Fig. 6). Of these fungal metabolites, rubratoxin B is the only compound reported to demonstrate herbicidal activity. However, its activity is very weak compared to that of cornexistin.

The Herbicidal Activities of Hydantocidin and Cornexistin

The herbicidal activities were obtained from an experiment in which the plant seeds were covered with soil of 10 mm depth in flowerpots, and cultivated in a greenhouse for 2 weeks. Herbicidal effects were examined by foliar spraying with 500 ppm solution, and each sample was evaluated on a 0 to 5 scale 2 weeks after treatment.

The herbicidal activities of hydantocidin and cornexistin against annual monocotyledonous plants are shown in Table 2. The results are also compared with those of glyphosate and bialaphos. Hydantocidin and cornexistin had potent herbicidal activity against most of the monocotyledonous plants, almost or great as glyphosate, and stronger than bialaphos.

The herbicidal activities of hydantocidin and cornexistin against dicotyledonous plants are shown in Table 3. Hydantocidin and cornexistin also had potent herbicidal activities against dicotyledonous plants, but the killing effects on these plants were non-selective. One notable exception was that of corn seedings which were able to tolerate cornexistin.

The herbicidal activity of hydantocidin against perennial weeds was compared with that of glyphosate and bialaphos. The results are shown in Table 4. In general, most samples were not strongly effective against perennial weeds. However, hydantocidin exhibited greater activity than glyphosate and bialaphos. Since hydantocidin similarly damaged roots and treated leaves, and also inhibited the growth of roots and buds, hydantocidin was considered to be translocated throughout the plant symplastically.

A field test of hydantocidin is shown in Fig. 7. The right side shows untreated weeds. The left side shows weeds 3 weeks after treatment with hydantocidin by foliar spraying at 4 kg/ha. Hydantocidin was effective on various weeds.

The pot tests of cornexistin are shown in Fig. 8. The control, which contained 9 types of weeds as well as corn, is shown on the left. The picture on the right shows corn growing in association with weeds 25 days after planting. These

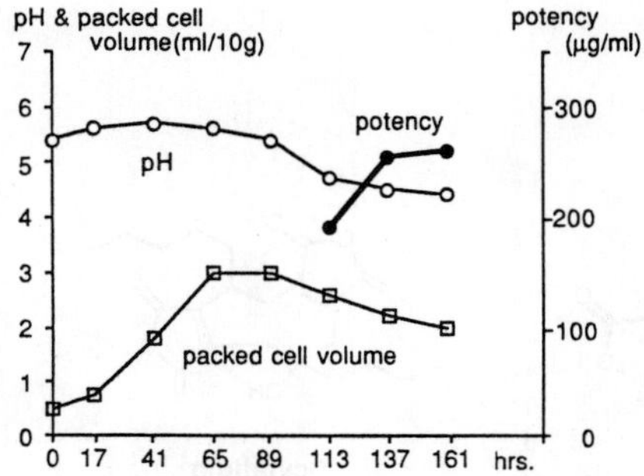

Cultured broth
| filtered
Filtrate (920L)
| pH8.0 ethyl acetate
Aqueous layer
| pH2.5 ethyl acetate
Ethyl acetate layer
| washed with 0.1M Na_2HPO_4 and NaCl-water
| dried on Na_2SO_4
| concentrated
Oil
| Sephadex LH-20 column (2.5L)
| $CHCl_3$: EtOAc (1:1)
Active fractions
| concentrated
| crystallized from CH_2Cl_2
Cornexistin (115g)

Medium	(%)
Glycerol	5.0
Fresh poteto	5.0
Yeast extract	0.5
Malt extract	0.5
CB-422	0.01
pH 6.0	

Fig. 4. Fermentation and Purification of Cornexistin Produced by *Paecilomyces variotii* SANK 21086

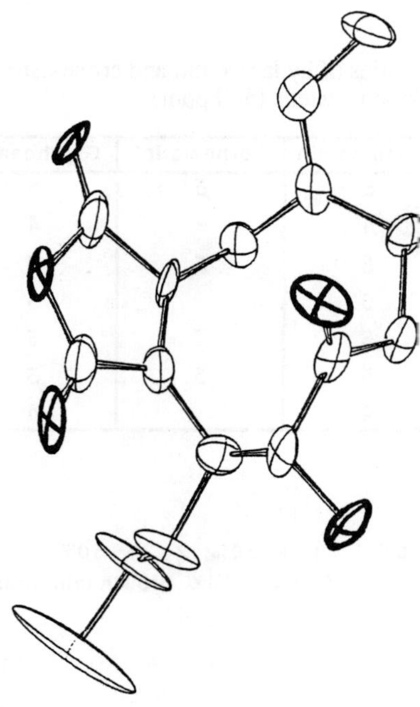

Fig. 5. X-Ray Structure of Cornexistin

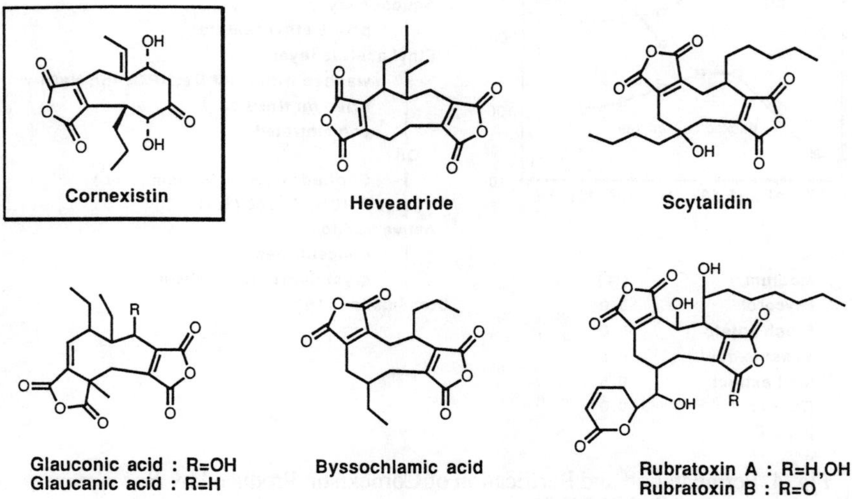

Fig. 6. Structure of Nonadride Group Compounds

Table 2. Herbicidal activities of hydantocidin and cornexistin against annual monocotyledonous weeds (500 ppm)

	Hydantocidin	Cornexistin*	Glyphosate	Bialaphos
Barnyardgrass	5	5	5	3
Black grass	4	-	4	2
Crabgrass, large	5	5	5	5
Foxtail, giant	5	-	5	4
Foxtail, green	5	5	5	5
Johnsongrass	5	5	5	3
Wild oat	4	-	4	3

* ; separated assay

- ; not tested

Herbicidal rating ; 0 : 0~10% , 1 : 11~30%, 2 : 31~50% , 3 : 51~70%, 4 : 71~90% , 5 : 91~100% growth inhibition

Table 3. Herbicidal activities of hydantocidin and cornexistin against annual dicotyledonous weeds (500 ppm)

	Hydantocidin	Cornexistin*	Glyphosate	Bialaphos
Cocklebur, common	5	5	5	5
Jimsonweed	5	-	5	5
Lambsquarters, common	4	-	5	3
Morningglory, tall	4	5	4	4
Night shade, black	5	5	5	5
Pigweed, redroot	4	-	5	5
Prickly sida	5	-	5	5
Ragweed, common	5	5	4	5
Velvetleaf	5	5	4	3
Wild mustard	5	-	5	5

Table 4. Herbicidal activity of hydantocidin against perennial weeds

	Hydantocidin		Glyphosate		Bialaphos	
	500ppm	250ppm	500ppm	250ppm	500ppm	250ppm
Bermudagrass	1	0	3	2	3	2
Quack grass	3	1	3	1	1	0
Field bindweed	5	5	3	2	2	1
Horsenettle	5	3	3	3	5	3
Nutsedge, purple	4	3	3	1	3	2
Nutsedge, yellow	5	5	5	5	4	2

Control rating ; 0 : no effect, 5 : 100% kill

Fig. 7. Pictures of control (right) and treated with hydantocidin (left) in field test.

Fig. 8. Pictures of control (left) and treated with cornexistin (right) in pot test.

Fig. 9. Pictures of control (top) and treated with cornexistin (bottom) field test.

plants were sprayed with cornexistin (500 g/ha) on day 9, and excellent control of annual weeds is demonstrated.

The field test of cornexistin in Hokkaido, Japan, is shown in Fig 9. On the left is shown the untreated control 7 weeks after planting. A variety of weed species grew naturally in the control crop, but not in the crop treated with cornexistin. The photo on the right shows the test field in which plants had been treated 1 month previously by foliar spraying of cornexistin at a concentration of 500 g/ha. These results demonstrated that cornexistin can selectively exhibits herbicidal activity against annual dicotyledonous plants. However, phytotoxicity against the corn was observed only at high doses. Herbicidal activity of cornexistin against monocotyledonous plants such as corn was less, compared with that of chemical herbicides such as sulfonyl urea compounds.

Hydantocidin and cornexistin had no activity against gram-positive and negative bacteria and fungi at 1 mg/ml using the paper-disc agar diffusion assay, and showed very low toxicity with LD_{50} values of greater than 1 g/kg (p.o.) and 100 mg/kg (i.p.) in mice. In comparative acute toxicity studies, cornexistin was less toxic than rubratoxin B.

In conclusion, hydantocidin and cornexistin show non-selective, broad spectrum herbicidal activity against annual plants including mono- and dicotyledonous weeds. In addition, hydantocidin had potent herbicidal activity against perennial plants, and cornexistin may be useful for postemergence weed control with selective protection of corn.

References

1) Terahara, A.; T. haneishi, M. Arai, T. Hata, H. Kuwano & C. Tamura: The revised structure of herbicidins. *J. Antibiotics* **35**, 1711-1714, (1982)
2) Nakajima, M.; K. Itoi, Y. Takamatsu, T. Kinoshita, T. Okazaki, K. Kawakubo, M. Shindo, T. Honma, M. Tohjigamori & T. Haneishi: Hydantocidin, a new compound with herbicidal activity from *Streptomyces hygroscopicus*. *J. Antibiotics* **44**, 293-300, (1991)
3) Mio, S.; R. Ichinose, K. Goto & S. Sugai: Synthetic studies on (+)-hydantocidin (1); a total synthesis of (+)-hydantocidin, a new herbicidal metabolite from microorganism. *Tetrahedron* **47**, 2111-2120, (1991)
4) Mio, S.; M. Shiraishi, S. Sugai, H. Haruyama & S. Sato: Synthetic studies on (+)-hydantocidin (2): aldol addition approaches toward the stereoisomers of (+)-hydantocidin. *Tetrahedron* **47**, 2121-2132, (1991)
5) Mio, S.; M. Ueda, M. Hamura, J. Kitagawa & S. Sugai: Synthetic studies on (+)-hydantocidin (4): synthesis of stereoisomers of (+)-hydantocidin. *Tetrahedron* **47**, 2145-2154, (1991)
6) Nakajima, M.; K. Itoi, Y. Takamatsu, S. Sato, Y. Furukawa, K. Furuya, T. Honma, J. Kadotani, M. Kozasa & T. Haneishi; Cornexistin, a new fungal metabolite with herbicidal activity. *J. Antibiotics* **44**, 1065-1072, (1991)

RECEIVED April 30, 1993

Chapter 7

Advances in the Use of Brassinosteroids

Horace G. Cutler

Richard B. Russell Research Center, Agricultural Research Service, U.S. Department of Agriculture, P.O. Box 5677, Athens, GA 30613

The brassinosteroids are a unique class of plant growth regulators that have the potential to increase yields in economic and horticultural crops. Their original discovery occurred almost simultaneously in Japan and America, respectively in *Distylium racemosum*, "Isunoki," an evergreen tree and in canola (*Brassica napus*) pollen. Extracts from 227 kg canola pollen, gathered by bees, gave America the lead in the field and the structure of brassinolide was determined. There followed an intense search for sources of brassinosteroids in plants and by chemical synthesis, especially in Japan and Europe. The discovery that 24-epibrassinolide was active in field tests and was more readily available from synthesis led to the compound becoming the focus of attention in laboratory and field experiments. Consequently, 24-epibrassinolide and brassinolide were examined as fungal hormones, to produce fruiting bodies, as antiecdysis compounds in insects and especially as yield enhancers in field and horticultural crops. By 1994 it is projected that Japan will have treated more than 23,000 hectares of wheat in China where yields, so far, have been increased 8-15%.

History has a unique way of fashioning events, especially in science, and sometimes significant discoveries are made independently at almost precisely the same time. But the odd feature is that these events occur, even in the age of rapid communication, without any connections having been made. Perhaps one of the best examples of this is the almost simultaneous discovery of the differential calculus by Sir Isaac Newton and Gottfried Leibniz in the latter part of the 17th century. While they worked independently, one in England and the other in Germany, the presentation of their work was so close that a dispute went on for years as to whose discovery came first. It is now conceded that Newton was a few months ahead of Leibniz though, of course, their synthesis was based on the work of all the mathematicians who had gone before them.

An almost identical situation occurred with the discovery of the brassinosteroids and the logic which brought about their discovery had only one common link, the curiosity to find compounds that regulated plant growth with a view to increasing agricultural yields. Following World War II, there had been a burst of energy in the agricultural chemical industry and this had been brought about in part by the fact that the world was in the process of rebuilding and that a growing population would need a food supply that was both plentiful and inexpensive. The catalyst that gave rise to this industry was the small, but important, handful of natural products that had been discovered in the 1930's and 1940's. These included indole-3-acetic acid (Figure 1) upon which the synthetic homolog indole-3-butyric acid (*1*) was based (it is still used today for rooting plant cuttings) and in the early 1950's the production of gibberellic acid (GA_3) (Figure 2) which was discovered by the Japanese in the late 1920's but hidden in the Japanese literature because few translations, if any, were made until after World War II. Again, GA_3 increased yields in certain crops and is used today in at least twenty-seven different horticultural and agronomic situations (*1*). One example is Thompson seedless grape where vines sprayed at 7-10 days after flowering with 20-40 mg/liter increase yields by ~ 150%. Prior to the use of GA_3 vines had been girdled by cutting a ring around the base of the trunks at approximately a week following flowering and even the most careful operator occasionally killed plants by slicing through the vascular cambium. By 1955 the arsenal of natural products consisted of indole-3-acetic acid, GA_3, cytokinin and ethylene. The latter was produced by ripening fruits and found practical in use in shipping green bananas in the sealed holds of banana boats from Central and South America, and the West Indies. By the time the bananas reached their destination, they were well on the way to being a nice yellow color for immediate market use. But all these materials had the following significant properties. They were active at exceedingly low concentrations, and in many cases treatments at 10^{-5} M induced the desired response; they were non-toxic at the concentrations used; and they left no apparent residue in the treated product. It was believed that nature contained many such compounds and that it was only necessary to look in the right places to discover them. Consequently, in the early 1950's a number of bioassays were developed to evaluate plant extracts for plant growth regulatory activity in the hope of finding new biologically active natural products.

Brassinosteroid Research in the US and Japan.

It seemed obvious to certain researchers, including J. W. Mitchell, who worked for the Agricultural Research Service, that plant organs that grew rapidly should be sources for growth stimulating substances, and among these were seeds and pollen. So Mitchell actively pursued bioassay systems and made pollen extracts in his search for new plant growth regulators. Unfortunately, there was a lag time between his initial verbal report to an evening meeting of the Potomac Division of the Plant Physiology Society at the National Arboretum in the winter of early 1963, when he exhibited time-lapse movies showing the effects of rape pollen extract (*Brassica napus* L.) on bean second node extension, and his formal publication in 1970 in *Nature* (*2*). And the lag time is essential in understanding the historical development and division of opinion as to whether the primacy of discovery of the brassinosteroids first occurred in the United States, or Japan.

Events in Japan had followed a slightly different course. For years scientists worldwide had observed that insect induced galls in plants grew rapidly. Therefore, it was surmised that the spurt in growth was, most probably, under the influence of a plant growth regulatory substance produced either by the insect, or by the plant, that was extremely potent. If the substance could be isolated it could have important agronomic and commercial implications. And while many projects were mounted, the literature surrounding specific natural products that controlled galling was sparse until the research carried out in Japan on an evergreen tree, *Distylium racemosum* Sieb. et Zucc, commonly called "Isunoki," came into play. It transpired that the very young leaves of the tree, when attacked by the aphid *Neothoracaphis yanonis* in the spring, rapidly produced 1-2 mm galls which, in two to three months, grew to ~ 1 cm. Doubtless this observation had been made for years but, at some point in the 1960's, work began in earnest to isolate the "*Distylium* factor." Another vital point in this work was the selection and development of a bioassay system and in 1965 the circumstances surrounding the rice-lamina inclination assay were published (*3*). Surprisingly, this assay and the second internode bean assay used by Mitchell were specific for detecting the brassinosteroids and without either the discovery of this class of compounds would not have been possible at the time because other assays were insufficient to detect their activity. Early work showed that the active material occurred not only in galls but also in healthy *Distylium* leaves. Incredibly, 430 kg of fresh leaves were harvested in 1966 (*4*), extracted with methanol to give a neutral fraction which was soluble in ether, and active in the rice-lamina assay. On chromatography this fraction gave 751 μg of *Distylium* factor A_1, 50 μg of factor A_2 and 236 μg of B: all had biological activity. But, obviously, there was not enough material for chemical work, so the biological and isolation results were published in 1968 (*5*).

Bees as pollen harvesters. American research approached the problem of collecting enough rape (canola) pollen in an ingenious way to obtain enough material for extraction. At the time, in the 1960's, the health food industry was starting to collect pollen from bee hives and some of the pollen was also used in baked products. It was also known that a healthy bee colony can collect about 34 kg of pollen during a season (*6*). In addition, the maximum distance that a bee can fly for harvesting is 2 miles but, like all other creatures, they prefer to forage close to home. If you put all these ingredients together and place the hives in the middle of a flowering canola field in Canada, pure canola pollen can be gathered. Special traps are placed at the hive entrance to collect the pollen so that the pure pollen pellets fall onto a wire screen and these can be suitably harvested. Approximately 227 kg were harvested in this manner. Finally, after solvent extraction, several chromatographic separations (both open column and high pressure liquid chromatography) ~ 4 mg of brassinolide (Figure 3) crystals were obtained from a

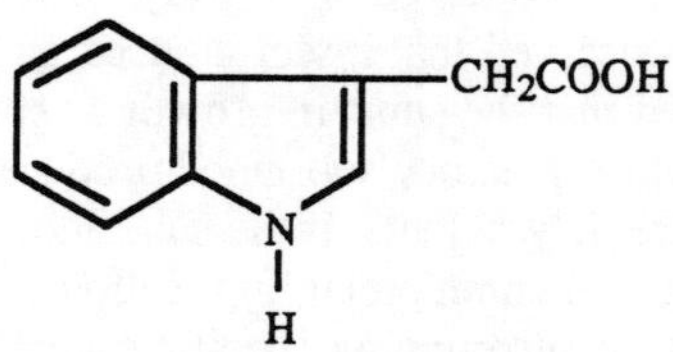

Figure 1. Indole-3-acetic acid (*Rhizopus suinis*).

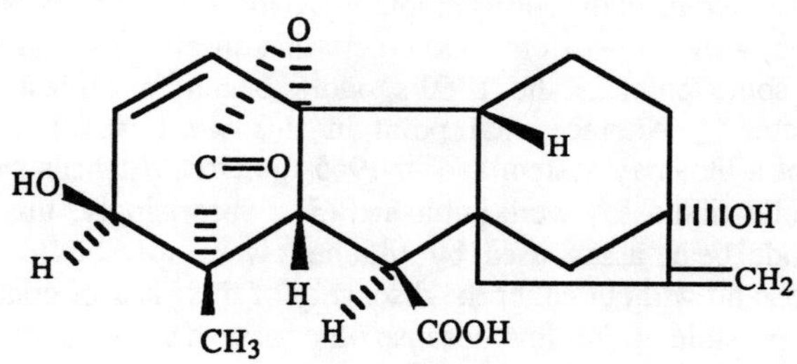

Figure 2. Gibberellic acid (*Gibberella fujikuori*).

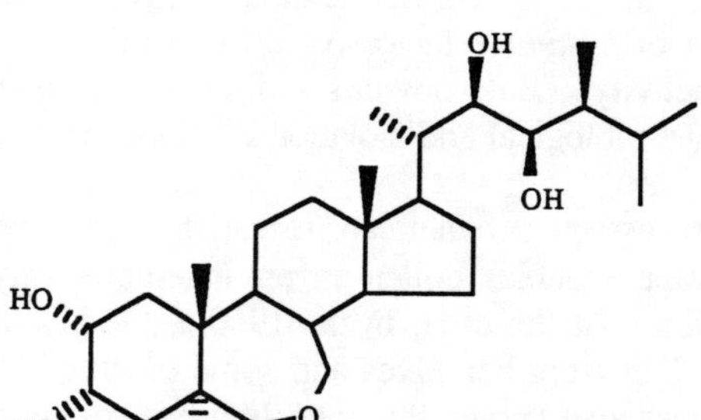

Figure 3. Brassinolide.

methanol fraction though this represented only approximately one-sixth of the pollen sample. That is, 23 mg of crystals would have been found in 227 kg of pollen (*7*).

Again, two pieces of serendipity occurred. First, the material was suitably crystalline for X-ray analysis and, second, the necessity for synthesizing heavy atom derivatives for X-ray determination had only been superseded in ~1973.

Brassinosteroid Production. As soon as the structure for brassinolide had been proved, an immediate search followed to find systems that would give greater quantities for practical applications. This followed two routes: examination of organisms that produced not only brassinolide but, hopefully, other brassinosteroids possessing higher specific activity and, concomitantly, chemical synthesis.

Fermentation and Plant Sources. It was hoped that fermentation of a suitable microorganism would yield sufficient quantities of brassinolide for industrial use but although a project was apparently mounted the yields were extremely poor (personal communications), and none of the information was published. Also, while there have been rumors of gene splicing to produce brassinosteroids in bacteria no official publications support this endeavor. The reasons for taking the fermentation approach are logical because all the major plant growth regulators, indole-3-acetic acid, abscisic acid, cytokinins (Figure 4), gibberellic acid and ethylene have been found in microorganisms and higher plants (*8*). Indole-3-acetic acid and gibberellic acid were originally discovered in microorganisms and the yields were relatively large. Their yields from plant sources were minuscule. Conversely, abscisic acid and cytokinins were discovered in higher plants (kinetin was initially found in stale fish sperm) in low quantities and later in microorganisms in much higher amounts. For example, 93 mg/L of abscisic acid have been isolated from cultures of *Botrytis cinerea* irradiated at 366 nm (*9*). Thus, the genetic mechanism for producing these plant growth regulators seems to have been preserved during the course of evolution from microorganisms to the higher plants. Furthermore, the isoprene precursors for the production of the brassinosteroids are, presumably, operative in microorganisms.

At the present time, over sixty types of brassinosteroids have been detected in plants but of these only thirty-one have been characterized (*10*) (Tables I, II, and III).

Table I. Brassinosteroid Sources of Green Algae and Gymnosperms

Green Algae	
Hydrodyction retriculatum	24-epicastasterone 24-ethylbrassinone
Gymnosperms	
Pinus thunbergii pollen	2-deoxycastasterone
Picea sitchensis shoots	Castasterone Typhasterol
Pinus sylvestris cambium	Castasterone Brassinolide

Table II. Brassinosteroid Sources of Monocotyledonous Plants and Dicotyledonous Plants

Monocotyledonous Plants	
Oryza sativa shoots	Castasterone Dolichosterone
Zea mays pollen	Castasterone Typhasterol Teasterone
Typha latifolia pollen	2-Deoxycastasterone
Dicotyledonous Plants	
Alnus glutinosa pollen	Brassinolide Castasterone
Castanea crenata insect galls	Brassinolide 6-Deoxodihydrocastasterone Brassinone
Castanea crenata stems, leaves flowers	Castasterone 6-Deoxodihydrocastasterone
Pharbitis purpurea fruit	Castasterone Brassinone
Thea sinensis leaves	Brassinolide Castasterone Typhasterol Teasterone
Vicia faba pollen	Brassinolide Castasterone 24-Epibrassinolide

Generally, they are found in such small amounts that processing plant material to isolate sufficient quantities is not yet practical. Among the highest yielding sources are seeds. With few exceptions, all these isolations have been carried out in Japanese laboratories and the obvious academic and industrial support for brassinosteroid research and development in Japan will play a very decisive role in crop production, as we shall discuss later. One system that may present a commercial source of brassinosteroids is tissue culture. Crown gall cells from *Catharanthus roseus D. Don* (*Vinca rosea* L.) have yielded brassinolide and castasterone at ~ 30-40 μg/kg fresh weight of tissue and while this does not compare to the yield for brassinolide from rape pollen (~ 100 μg/kg) the material is easier to obtain and handle. The crown gall cells were generated from *C. roseus* through transformation with a strain of *Agrobacterium tumefaciens* (A208) carrying the Ti-plasmid, pTi-T37, to produce *C. roseus* (V208) and this cell line is characterized by the biosynthesis of the amino acid nopaline as opposed to octopine in other cell lines. The titre of brassinosteroids was increased by the addition of plant growth

Table III. Brassinosteroid Sources of Immature Seed Plants

Immature Seed		
Dolichos lablab	Dolicholide - 200 μg/kg: equivalent to brassinolide in rape pollen)	
	Dolichosterone	
	Homodolicholide	
	Homodolichosterone	
	6-deoxodihydrocastasterone	
	6-deoxodihydrodolichosterone	
	Brassinolide	
	Castasterone	
Vicia faba	Brassinolide	
	Castasterone	
	24-Epibrassinolide	
Phaseolus vulgaris	Brassinolide	3,24-Diepicastasterone
	Castasterone	1β-Hydroxycastasterone
	Dolicholide	3-Epi-1α-hydroxycastasterone
	Dolichosterone	25 Methyldolichosterone
	6-Deoxodihydrocastasterone	2-Epi-25-methyldolichosterone
	6-Deoxodihydrodolichosterone	2,3-Diepi-25-methyldolichosterone
	6-Deoxodihydrohomodolichosterone	2-Deoxy-25-methyldolichosterone
	Typhasterol	3-Epi-2-deoxy-25-methyldolichosterone
	Teasterone	6-Deoxodihydro-25-methyldolichosterone
	2-Epicastasterone	23-O-β-D-glucopyranosyl-25-methyldolichosterone
	3-Epicastasterone	23-O-β-D-glucopyranosyl-2-epi-25-methyldolichosterone
	2,3-Diepicastasterone	

regulators, including indole-3-acetic acid, or naphthalene acetic acid, or 2,4-dichlorophenoxyacetic acid to the cell cultures (*11*).

While increasing yields from plant or microbial sources have not yet been fully successful, the discovery of new brassinosteroids is significant because of the questions raised. Are any of these congeners more or less active than either brassinolide or 24-epibrassinolide? Are they specific for select active sites in the plant? Do they have a synergistic effect with each other, or other endogenous plant growth regulators? At least from the chemical perspective they offer new templates for synthesis with, of course, synthesis of analogs not presently found in nature, for other exploratory work.

Synthesis of brassinosteroids. One of the limiting factors to the use of brassinosteroids in the field has been the limited availability of sufficient quantities of chemical. Apart from proving the structure and therefrom the biological activity, the first synthesis of brassinolide isomers was reported in 1979 (*12*). The starting material was ergosterol tosylate and the final products, which contained the steroid nuclei of brassinolide, had the opposite configuration at C22, C23, and C24 in one case, but in the other only the orientation of the methyl group at C24 was different. Later, brassinolide was synthesized from sterols and their degraded products including brassicasterol, bisnor dinorcholenic acid, pregnenolone, and stigmasterol (*13*). Brassinolide, castasterone, teasterone, and typhasterol have also been synthesized relatively easily from the intermediate (22R,23R,24S)-3α,5-cyclo-22,23-diacetoxy-5α-ergostan-6-one (*14*). It transpired, during the course of evaluating the biological properties of various brassinosteroids, that 24-epibrassinolide (Figure 5) elicited responses much like brassinolide in bioassays (*15*). Coincidentally, brassicasterol was determined to be present at levels of 10-20% in the sterol fraction of rapeseed oil and using solvent extraction, followed by recrystallization, the compound could be obtained in high yield (*16*). A straightforward series of reactions consisted of treating brassicasterol 3-O-mesylate with sodium carbonate to give the isoform which was oxidized with chromic acid to yield the 3,5-cyclo-6-ketone. Acid isomerization of this compound gave the 2,22-dien-6-one which, when treated with catalytic amounts of the oxidant, osmium tetroxide and N-methylmorpholine-N-oxide, yielded a 3:5 ratio of 2α,3α,22R,23R-tetrol and 2α,2α,22S,23S-tetrol, respectively. The mixture was separated and each compound was subjected to Baeyer Villiger oxidation giving 24-epibrassinolide and the 22S,23S-epimer (*17*), respectively.

A recent chemical scheme uses a chinchona alkaloid derivative as the starting material to produce 2α,3α,22R,23R epibrassinolide, as a major product (*17*). Although 24-epibrassinolide is only about 10% as active in certain bioassays, including radish and tomato, its activity in field trials is approximately equal to brassinolide and it is an attractive candidate for industrial development because of the relative ease with which it can be synthesized.

Another synthetic route leads from stigmasterol in the first step to produce 3β-acetoxy-bisnor-cholenic acid followed by steps to the next critical intermediate, 2α,3α-isopropylidenedioxy-6-ethylenedioxy-bisnor-5α-cholanal (*18*). Presumably, this will be a relatively efficient reaction.

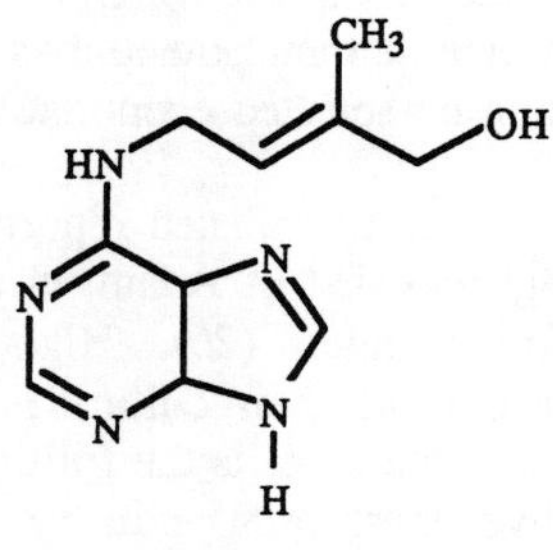

Figure 4. *trans*-Zeatin (*Pseudomonas syringae* pv. *savastanoi*).

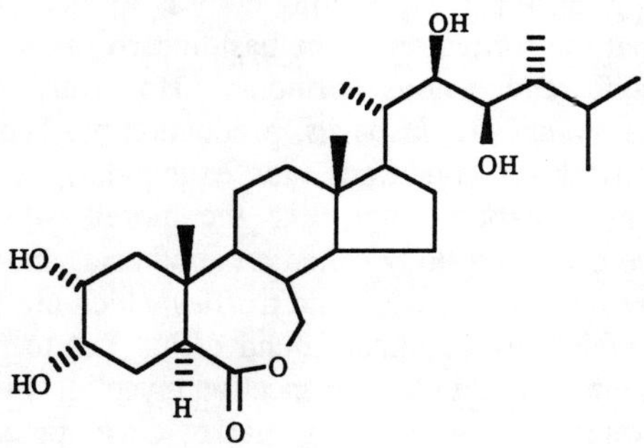

Figure 5. 24-Epibrassinolide.

Brassinolide is now available in a highly pure form in milligram quantities and 24-epibrassinolide and homobrassinolide in gram quantities from Beak Consultants, Brampton, Ontario, Canada and it is probably only a questions of time before other brassinosteroids are on the market, at which point other research avenues will be opened. But the commercial availability of large quantities also means that synthetic derivatives can now be made for other disciplines, including medicine.

Applications for the Brassinosteroids.

We now turn our attention to those areas for which some of the brassinosteroids are finding application. These include their potential as fungal growth regulators, insecticides, and field application on selected economic and horticultural crops.

Fungal Growth Regulators. The last published report concerning the effects of brassinosteroids on higher fungi was that of Adam, et al. showing results which held great practical promise for the future (*20*). Higher fungi have two distinct periods in their life cycle, insofar as their culture is concerned. One is the mycelial, or filamentous stage and the other is the fruiting stage. It is the fruiting stage upon which the lucrative mushroom industry depends to supply soup manufacturers, and processed food markets with *Agaricus campestris bisporus*. The mycelial stage, relative to the basidiocarp, is not readily visible but remains just below the soil surface. At some point during the life cycle of the fungus, the mycelium begins to aggregate and the fruiting body is formed (*19*). It is thought, but not yet proved, that the formation of the basidiocarp in the *Basidiomycetes* is in response to a specific endogenous hormone. However, in the case of the common mushroom, *A. campestris bisporus*, production has been streamlined and is carried out under rigorous conditions, for example, in limestone caves in Pennsylvania. The more exotic fungi, like the morels which belong to the *Discomycetes*, produce a fruiting body called an apothecia, and this is commonly referred to as a sponge mushroom, while the truffles which are also *Discomycetes* produce an ascocarp which grows underground (*19*). But to reiterate, all these fruiting bodies have their genesis in their respective mycelia. In addition there are numerous other gastronomic fungi awaiting market development of which some seventeen *Boletus* sp. are included. In each case, the mycelial stage is relatively easy to grow in the laboratory but the difficulty has been to induce the mycelium to produce fruiting bodies. Obviously, the first group to accomplish and patent the process will have a thriving business.

Three brassinosteroids have been used to speed up the life cycle of fungi *in vitro*. These are brassinolide, 24-epibrassinolide, and 22S,23S-homobrassinolide all of which increased mycelial growth by a factor of 2-3 and induced earlier sporocarp formation in *Psilocybe cubensis* and *Gymnopilus purpuratus* (*20*). *P. cubensis* forms part of the group known as the sacred mushrooms employed satisfactorily in religious ceremonies by Mexican Indians. One of the biologically active compounds contained in *Psilocybe* is psilocybin which has been isolated, synthesized and used in the study of schizophrenia (*19*). The effects with 22S,23S-homobrassinolide included a one-third increase in biomass, a one-quarter decrease in the number of weeks to produce fruiting bodies and a quadrupling in the number of fruiting bodies, compared to controls, with 10^{-2} ppm treatment. The exact role of the brassinosteroids in this process has not been determined. Whether the effect is to stimulate the mycelial growth and compress the life cycle or to induce mycelial aggregation has yet to be elucidated. Unfortunately, this work has been terminated because of a lack of funds and was the victim of financial restructuring of Germany following the collapse of the German Democratic Republic (correspondence with Dr. G. Adam, 1992). But the rewards for practically producing luxury fungi are too much of a temptation for the work to lie fallow for long (Table IV).

Table IV. Fungi Awaiting Commercial Development

Discomycetes	
Morchella conica	"Morels"
M. hybrida	
M. vulgaris	
Tuber aestivium	"Truffles"
T. rufum	
Basidiomycetes	
Boletus appendiculatus	
B. auranticus	*B. erythropus*
B. badius	*B. granulatus*
B. bovus	*B. luridus*
B. castaneus	*B. luteus*
B. chrysenteron	*B. queletu*
B. cynescens	*B. scaber*
B. edulis	*B. subtomentosus*
B. elegans	*B. viscidus*

Insect Growth Regulators. The structural connections between the brassinosteroids and the ecdysteroids are apparent and have led to some detailed examinations of the antiecdysteroid effects of brassinosteroids. The fact that both groups of compounds are exceedingly active in their respective domains makes them interesting models for testing in reciprocal systems and for structural modifications to pursue structure activity relationships. Again, only those brassinosteroids that are available in relatively abundant amounts have been evaluated and these include brassinolide, 22S,23S-homobrassinolide and 22S,23S-homocastasterone (Figure 6). These are similar in structure to ecdysone, 20-hydroxyecdysone and ponasterone A (Figure 7).

Examination of the brassinosteroids and ecdysteroids used in insect experiments shows that the major differences are as follows: 1) The juncture between the A and B rings is *cis* with the ecdysteroids and *trans* with the brassinosteroids. 2) The B ring in many, but not all of the brassinosteroids, is

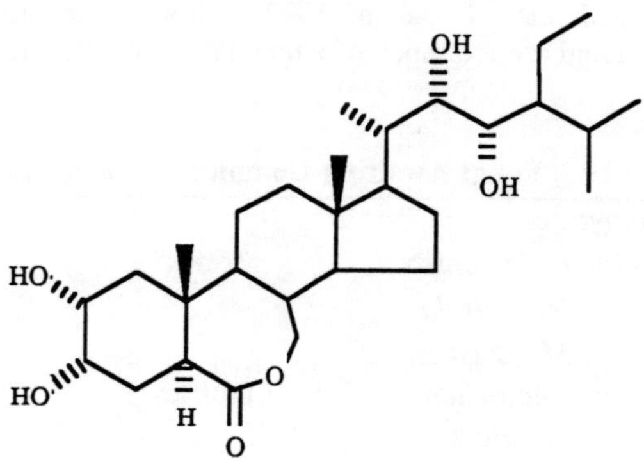

22S,23S-Homobrassinolide

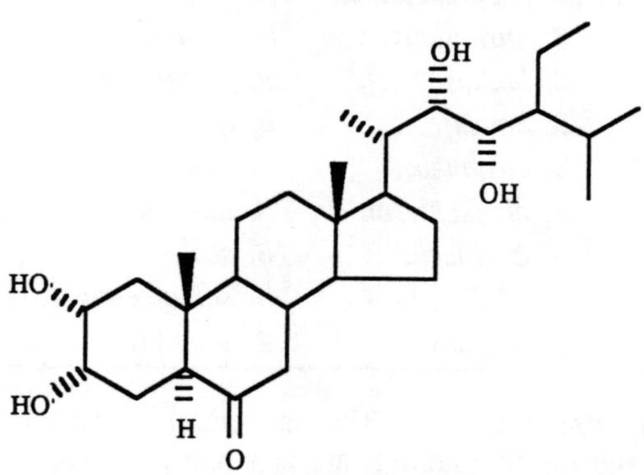

22S,23S-Homocastasterone

Figure 6. Brassinosteroids with antiecdysteroid activity.

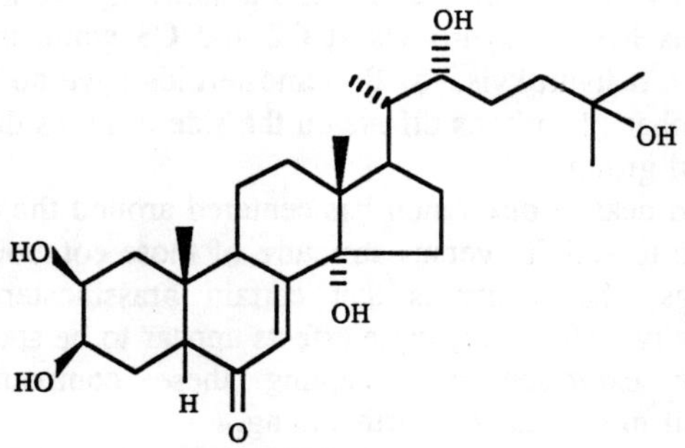

Ecdysone

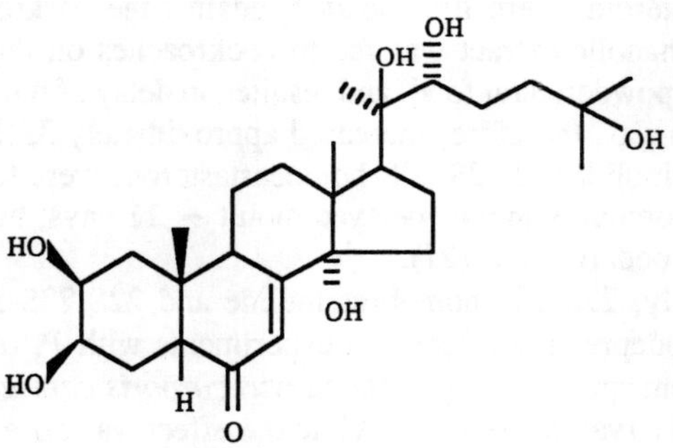

20-Hydroxyecdysone

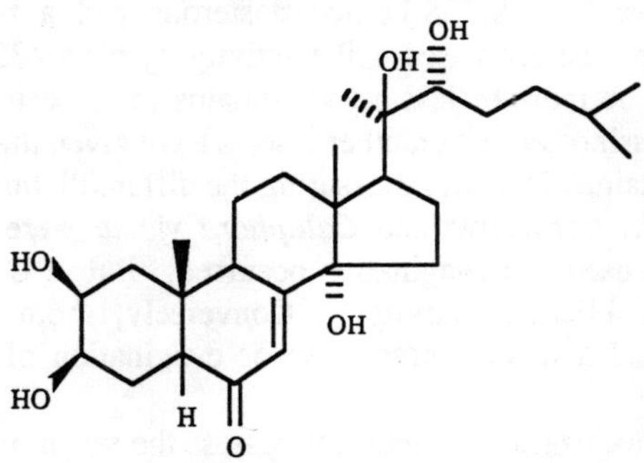

Ponasterone

Figure 7. Ecdysteroids.

seven membered and contains an extra oxygen: this is a function of the lactonic B ring. Both brassinolide and 22S,23S-homobrassinolide have a lactonic B ring; 22S,23S-homocastasterone is an example of six membered B ring and contains a ketonic function. 3) Ecdysone, 20-hydroxyecdysone, and ponasterone have an enone B ring and they more resemble 22S,23S-homocastasterone. 4) The A ring of these ecdysteroids have β hydroxyls at C2 and C3 while the A ring of the brassinosteroids have α hydroxyls. 5) Brassinosteroids have no hydroxyls on the D ring. 6) The number of carbons differs on the side chain as do the positions of hydroxyl and methyl groups.

While a good deal of discussion has centered around the definition of true ecdysteroids relative to activity versus structure of those compounds that may be regarded as analogs, the reality is that certain brassinosteroids do possess antiecdysteroid activity. However, these effects appear to be species specific and caution should be exercised in accepting these compounds as general antiecdysteroids until more data are forthcoming.

Early experiments were conducted using crude extracts of rape pollen, from which the brassinosteroids were first isolated, against the cockroach, *Periplaneta americana*. The ethanolic extract was fed to cockroaches on the 11th day of the last larval instar, on powdered rat food, and resulted in delay of moult by ~ 10 days. The length of instar was, therefore, increased approximately 33%. Subsequently, 22S,23S-homobrassinolide and 22S,23S-homocastasterone were fed to cockroaches and only 22S,23S-homobrassinolide delayed moult ~ 11 days, but only at the rate of 50 mg/4 grams food/10 larva (*21*).

More recently, 22S,23S-homobrassinolide and 22S,23S-homocastasterone have shown a neurodepressing effect. In experiments with *P. americana*, it was noted that the efferent spike activity of the *nervus corporis cardiaci* II was affected in isolated brains of larval last instars. While the effect was dose depended for the two brassinosteroids, and was equivalent to that of 20-hydroxyecdysone, a threefold concentration increase in 22S,23S-homocastasterone and a tenfold increase in 22S,23S-homobrassinolide decreased spike activity by 50% (*22*). Whether other brassinosteroids will be more or less active remains to be seen.

Effects of brassinosteroids on other insects have given different results. For example, the data obtained in assays measuring the differentiation of imaginal disks in two flies, *Phormia terraenova* and *Calliphora vicina* were similar when 20-hydroxecdysone was used and evagination occurred. But in *C. vicina*, neither of the brassinosteroids elicited activity. Conversely, both castasterone and homodolidolide caused a modest promotion or evagination of imaginal disks in *Phormia* (*22*).

22S,23S-Homocastasterone does not possess the seven numbered B ring, as does 22S,23S-homobrassinolide, and is generally considered to be a more active antiecdysis agent because of this feature (*vide infra*, where a threefold increase in 22S,23S-homocastasterone relative to 20-hydroxyecdysone induced an effect as opposed to a tenfold increase in 22S,23-homobrassinolide).

Economic and Horticultural Crop Growth Regulators.

The topics discussed up to this point have considered the brassinosteroids from their potential as fungal, or insect growth regulating agents. But there exists a large body of work which can no longer be placed in the category of purely experimental evaluation and that is the effects of the brassinosteroids on economic and horticultural crop production. Some of the information is sketchy and some is substantial.

The American experience with the brassinosteroids, specifically the brassins which were a mixture, has been well documented and one is struck by the singular lack of success in field trials, with few exceptions. In 1974-75, soybean and barley seeds were treated with brassins (the brassinosteroids were not synthesized until 1979) both in North America and Brazil to see if yields from small seed were equal to those from large seed, but no differences were found. In 1975, soybean seed were treated prior to planting at 7 locations in the United states with brassins to determine whether an increase in yield would result. Again in 1975, barley seed was treated, sorted into 3 sizes and planted. There were no results of economic importance. When the synthetic materials became available in 1979, field trials were conducted on beans, corn, lettuce, peppers, tomatoes, and radishes. These were treated at time of soil emergence with 0.01 ppm brassinosteroid. Lettuce heads were significantly increased by 25-32%, radishes ~ 20%. Peppers were increased by 9% and beans by 6%, but these increases in fruit weight in these two crops were not statistically different. There were no increases in corn or tomato yields. But by this time the American project was coming to a close and no further field work was planned. Sadly, another two years of experiments would have shown that the time of application was critical to increasing yields (*7*). For example, when seeds were soaked with solutions of brassins there was little effect on plant growth while in later experiments, increases in yields were noted in radishes, leafy vegetables and potatoes when young seedlings were treated, in greenhouse experiments (*7*).

It was approximately at this point that the Japanese, spurred on by the American disclosure of the synthesis of the brassinosteroids, went through a logarithmic effort in both the evaluation and synthesis of these plant growth regulators. The first inkling that something big was afoot, insofar as field applications and results were concerned, came to light through personal communications in 1988 (Cutler, personal communications) when it appeared that wheat yields had been increased ~ 15% in both China and Russia. The latter results were only alluded to but were finally revealed, albeit in a brief form, in the summer of 1990. In fact, the brassinosteroids, specifically 24-epibrassinolide and homobrassinolide, had been used to treat not wheat, but barley, lucerne, and some horticultural crops in Russia. Barley yields after treatment with 24-epibrassinolide were dramatically increased up to 25% with applications of 50 or 100 μg per hectare in 500 L of water. Increases in lucerne seed were also noted: 16% with homobrassinolide and up to 26% with epibrassinolide (*23*). Russia had obtained its brassinolides, if the sources are correct, from Japan.

The extent of the Japanese work in China also became evident in the summer of 1990. By that time, Nobuo Ikekawa had synthesized sufficient quantities of 24-epibrassinolide to test large areas of wheat in China with his colleague Yu-Ju Zhao. Over a six year period, they had applied 24-epibrassinolide on 3,333 hectares of wheat using concentrations of 0.1 to 0.001 ppm (*24*). The results consistently showed an increase of up to 15% in wheat, a significant figure that could well alter future wheat markets throughout the world. Since 1990, further trials have been conducted on wheat in China and the total area now treated, as a result of the 1990-1991 season, has been increased to a total of 4,000 hectares. Again, yields have been increased 8-15% during this period. The upshot is that a grant has been made by the Chinese Academy of Sciences to Japan to apply 24-epibrassinolide to wheat and the total land use will be increased to over 23,000 hectares during the next 3 years (1992-1994). The amount of 24-epibrassinolide needed for these experiments is 100 g. Horticultural crops have also been treated in China with 24-epibrassinolide and these include corn, tobacco, watermelon and cucumber. All exhibited increased yields in 1990 and 1991 though hard figures are not yet available. Grape, strawberry, orange, rape, eggplant, and sesame have also been treated at time of flowering, which appears to have been a critical time for application, and increased yields were due to increased fruit set, though in grape there was increased fruit weight, sugar content and maturity. Increased mushroom growth was also observed in plants treated with 24-epibrassinolide and this serves to confirm the work mentioned earlier with *P. cubensis* and *G. purpuratus* (*20*). Sesame yields were increased 17% (*25*). Undoubtedly, the figures from these trials will be published later.

It has been stated that the success of 24-epibrassinolide may be attributed to the lack of good cultural practice in China and that yields could easily be increased ~ 10% by proper use of fertilizer. While the point is well taken, it must be argued that if a few grams of a plant growth regulator can replace several tons of fertilizer then a more efficient agriculture exists. By analogy, it is far more economical to spray GA_3 on grapevines to increase yields than to pay a girdler who may take days to do the job.

Conclusion.

In reviewing the history of the brassinosteroids it is now obvious that these natural products, especially 24-epibrassinolide, have a future in agriculture for practical application. The fact that 24-epibrassinolide has very low toxicity, 1 g/kg (oral) for the mouse 2 g/kg (oral and dermal) for the rat, and was negative in the Ames test for mutagenicity, in addition to the very low amounts used per hectare, indicates that it has a bright future. It is also interesting to note that towards the end of the American experience with the brassinosteroids there were increases of ~ 20% in certain horticultural crops in field tests, but it was precisely at this point that research ceased. The one lesson to be learned from the brassinosteroid experience is that a long term commitment has to be made in a research endeavor of this sort. In 1979, the Japanese had no hectares under test with the brassinosteroids. By 1994 they will have over 23,000 hectares treated with 24-

epibrassinolide in China alone with, one feels certain, operations going on elsewhere. And the environmental considerations augur well for the brassinosteroids for they are natural products that are target specific, apparently biodegradable, with very high specific chemical activity. If yields are consistently higher following brassinosteroid treatment then it means that reduced land mass can be used to produce the same amounts of food and this will entail the use of less fuel and pesticides. In any event, it is obvious that the world food supply will become, in this case, dependent upon the importation of Japanese agrochemicals for success.

Acknowledgement. I am grateful to Dr. N. Ikekawa for the preliminary results of his trials with 24-epibrassinolide in China on various economic and horticultural crops.

Literature Cited.

1. Cutler, H. G.; Schneider, B. A. In *Plant Growth Regulator Handbook of the Plant Growth Regulator Society of America* 3rd Edition; Plant Growth Regulator Society of America. **1990**.
2. Mitchell, J. W.; Mandava, N.; Worley, J. F.; Plimmer, J. R.; Smith, M. V. *Nature* **1970**, *255*, 1065.
3. Maeda, E. *Physiol. Plant.* **1965**, *18*, 813.
4. Abe, H.; Marumo, S. In *Brassinosteroids*; Cutler, H. G.; Yokota, T.; Adam, G.; eds; *ACS Symposium Series No. 474*; American Chemical Society; Washington, DC 1991; p. 19.
5. Marumo, S.; Hattori, H.; Abe, H.; Nonoyama, Y.; Munakata, K. *Agric. Biol Chem.* **1968**, *32*, 528.
6. Dadant, M.G. In *The Hive and the Honeybee*; Grout, R.A.; ed; Dadant and Sons, Hamilton, Illinois; 1949; pp 203-212.
7. Steffens, G. L. In *Brassinosteroids*; Cutler, H. G.; Yokota, T.; Adam, G.; eds; *ACS Symposium Series No. 474*, American Chemical Society; Washington, DC 1991; p. 9.
8. Cutler, H. G. *CRC Critical Reviews in Plant Sciences*. **1988**, *6*, 323.
9. Marumo, S.; Kohno, E.; Natsume, M.; Kanoh, K. *Proc. 14th Ann. Meeting Plant Growth Regulator Soc. America*; 1987, 146.
10. Kim, S.-K. In *Brassinosteroids*; Cutler, H. G.; Yokota, T.; Adam, G.; eds; *ACS Symposium Series No. 474*, American Chemical Society; Washington, DC 1991; p. 26.
11. Sakurai, A.; Fujioka, S.; Saimoto, H. In "Brassinosteroids", Cutler, H. G.; Yokota, T.; Adam, G.; eds; *ACS Symposium Series No. 474*, American Chemical Society; Washington, DC 1991; p. 97.
12. Thompson, M. J.; Mandava, N. B.; Flippen-Anderson, J. L.; Worley, J. F.; Dutky, S. R.; Robbins, W. E.; Lusby, W. *J. Org. Chem.* **1979**, *44*, 5002.
13. Marumo, S. *Proc. 14th Ann. Meeting Plant Growth Reg. Soc. America.* **1987**, *14*, 174.

14. Aburatani, M.; Takeuchi, T.; Mori, K. *Agric. Biol. Chem.* **1987**, *51*, 1909.
15. Takatsuto, S.; Ikekawa, N. *Chem. Pharm. Bull.* **1984**, *32*, 2001.
16. Tsuda, K.; Sakai, K.; Ikekawa, N. *Chem. Pharm. Bull.* **1961**, *9*, 835.
17. Ikekawa, N.; Zhao, Y.-J. In *Brassinosteroids*; Cutler, H. G.; Yokota, T.; Adam, G.; eds; *ACS Symposium Series No. 474*, American Chemical Society; Washington, DC 1991; p. 281.
18. McMorris, T. C.; Donaubauer, J. R.; Silveira, M. H.; Molinski, T. F. In *Brassinosteroids*; Cutler, H. G.; Yokota, T.; Adam, G.; eds; *ACS Symposium Series No. 474*; American Chemical Society; Washington, DC 1991; p. 41.
19. Alexopoulos, C. J. In *Introductory Mycology*; J. Wiley and Sons, Inc., New York, 1962, pp. 613.
20. Adam, G.; Marquardt, V.; Vorbrodt, H. M.; Hörhold, C.; Andreas, W.; Gartz, J. In *Brassinosteroids*; Cutler, H. G.; Yokota, T.; Adam, G.; eds; *ACS Symposium Series No. 474*; American Chemical Society, Washington, DC 1991; p. 83-84.
21. Richter, K.; Koolman, J. In *Brassinosteroids*; Cutler, H. G.; Yokota, T.; Adam, G.; eds; *ACS Symposium Series No. 474*; American Chemical Society, Washington, DC 1991; p. 265-272.
22. Richter, K. *Naturwissenschaften.* **1991**, *78*, 138.
23. Khripach, V. A.; Zhabinskii, V. N.; Litvinovskaya, R. P. In *Brassinosteroids*; Cutler, H. G.; Yokota, T.; Adam, G.; eds; *ACS Symposium Series No. 474*; American Chemical Society, Washington, DC 1991; p. 53.
24. Ikekawa, N.; Zhao, Y.-J. In *Brassinosteroids*; Cutler, H. G.; Yokota, T.; Adam, G.; eds; *ACS Symposium Series No. 474*; American Chemical Society, Washington, DC 1991; p. 284-290.
25. Ikekawa, N. Unpublished results 1992.

RECEIVED April 30, 1993

Chapter 8

Advances in Neem and Azadirachtin Chemistry and Bioactivity

Dale J. Hansen, John Cuomo, Mamunur Khan, Rex T. Gallagher, and William P. Ellenberger

AgriDyne Technologies, Inc., 417 Wakara Way, Salt Lake City, UT 84108

Utilization of azadirachtin is rapidly becoming an accepted method for the disruption of insect molting, and thus is an effective insect growth regulator for most insects during the larval and pupal stages of development. Azadirachtin and related compounds are derived from the mevalonic acid pathway in the neem tree (*Azadirachta indica)* and other members of the Meliaceae family. Azadirachtin is a complex tetranortriterpenoid with 16 chiral carbon centers, too complex to synthesize commercially. Aflatoxins B_1 and B_2 are usually present in neem extracts so caution must be used to hygienically control the level to insure safe use. Growth regulator mechanisms may involve interactions between azadirachtin and ecdysone, with either synergistic, replacement, or antagonistic mechanisms possible. We have compared structural and biochemical properties of the two.

Advances in the chemistry of the neem tree (*Azadirachta indica* A. Juss.) and other species of the Meliaceae have been rapid as a result of interest in the unique characteristics of their limonoids and other secondary metabolites. This is especially true for azadirachtin, the final compound in the biosynthetic scheme. The search for new, powerful, non-toxic insecticides has become a great priority for the pesticide industry, and azadirachtin is currently a leading candidate. Although recent reviews have been published (*1-4*), more precise information is rapidly being generated. In this chapter we will review present information on the chemical properties of neem compounds and specifically on the chemistry of azadirachtin and its use to control insect molt.

Neem Compounds

Chemical investigations of the neem tree, *A. indica,* have yielded many interesting and structurally varied secondary metabolites. Tetranortriterpenoids (limonoids) are the

0097–6156/94/0551–0103$07.75/0

major and most characteristic class of compounds in neem. We will limit our discussion to the unique diterpenoids and triterpenoids of neem with special emphasis on the limonoids.

Diterpenoids. Among the class of terpenes, tricyclic diterpenes have been found in the greatest abundance, especially those having an aromatic C-ring with oxygen functionalities at C-3 and C-7 (Figure 1).

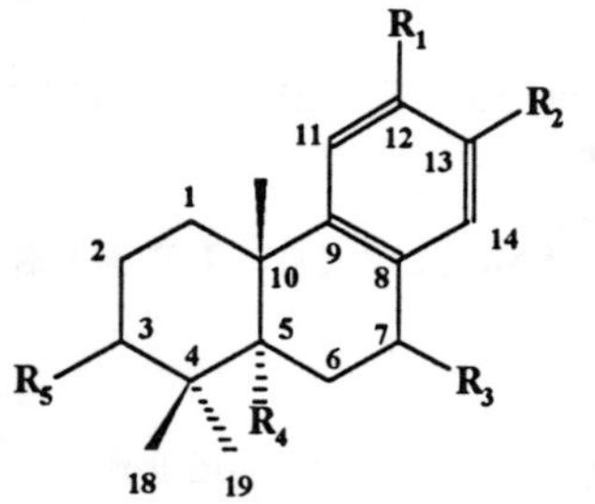

	R_1	R_2	R_3	R_4	R_5
1 Sugiol	OH	Me	=O	H	H_2
2 Methyl nimbiol	OMe	Me	=O	H	H_2
3 Nimbidiol	OH	OH	=O	H	H_2
4 Nimbosone	OMe	OMe	H_2	H	H_2
5 Nimbonone	Et	OMe	=O	H	H_2
6 Nimbonolone	Et	OMe	H_2	H	=O
7 Nimolinin	OH	i-Pr	=O	H	=O
8 Nimosone	OH	i-Pr	=O	OH	H_2
9 Methyl nimbionone	OH	i-Pr	=O	H	H_2

Figure 1. Neem tricyclic diterpenes.

Two new isomeric diterpenoids, nimbonone, **5**, and nimbonolone, **6**, have been isolated from the neutral fraction of the stem bark of *A. indica* (*5*). The unique feature of these terpenes is the presence of an ethyl group at C-12. Nimosone, **8**, nimbosone, **4**, methyl nimbiol, **2** and methyl nimbionone, **9**, have recently been isolated from the stem bark of *A. indica* along with sugiol, **1**. Structures of the new compounds have been elucidated as 12-hydroxy-8,11,13-abietatriene-3,7-dione, 13-acetyl-12-methoxy-8,11,13-podocarpatriene, 12-methoxy-13-methyl-8,11,13-podocarpatriene-7-one and 12,13-dimethoxy-8,11,13-podocarpatriene-3,7-dione, respectively, through chemical and spectral studies (*6*).

Two new tricyclic diterpenes, nimolinin, **7** (*7*), and nimbandiol, **3** (*8*), have been isolated from root bark. Various tricyclic diterpenes reported in neem possess a wide range of biological activities which include antitumor, antibiotic and insecticidal properties (*6*).

Triterpenoids and Their Derivatives. Triterpenoids and their derivatives represent the most characteristic types of compounds present in *A. indica* (Figure 2.) These compounds can be divided into two major classes, triterpenoids (protomeliacins) and meliacins (limonoids or tetranortriterpenoids).

Triterpenes. Seven tetracyclic triterpenes (protomeliacins) have so far been reported from various parts of the neem tree (Figure 3.) These include meliantriol, **16** (*9*), nimbocinone, **17** (*10*), and nimolinone, **18** (*11*), which are euphol/tirucallol derivatives, and azadirachtol, **19** (*12*), and azadirachnol, **20** (*13*), which are apo-euphol/apotirucallol derivatives. A new tetracyclic terpenoid, azadirol, **21**, has been isolated from neem fruit along with the known triterpenoid kulactone, **22**, which was previously isolated from *Melia azedarach L.* (*14*).

Meliacins. Meliacins (limonoids) are derivatives of triterpenes in which four side chain carbon atoms have been lost and the remaining carbon atoms have been

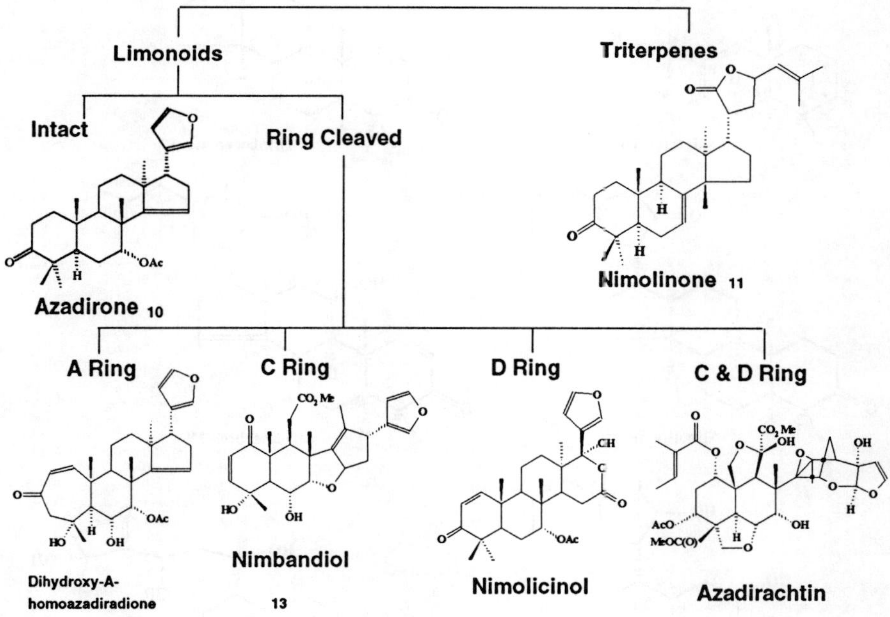

Figure 2. Neem triterpenoids

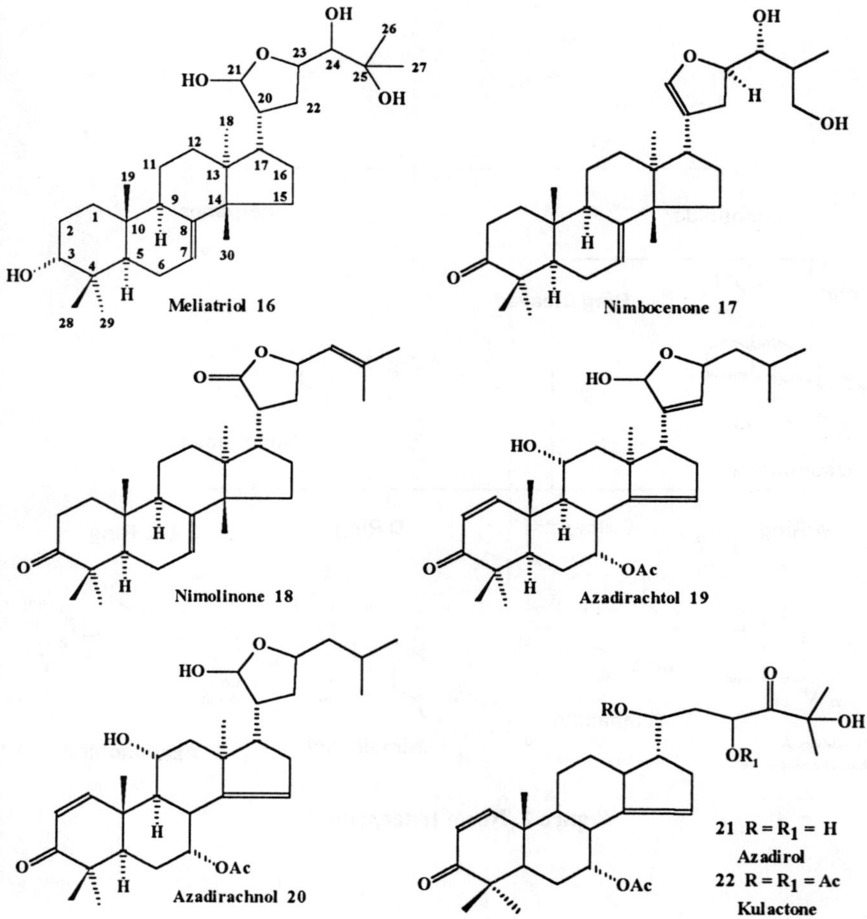

Figure 3. Neem triterpenes

cyclized into a furan ring. One or more of their ring systems may have also undergone oxidative opening. All the known meliacins possess an oxygen at C-7, a double bond in ring D, and a methyl group at C-8. Biosynthetically the intact triterpenes are considered to be precursors of limonoids because of their systematic oxidation patterns (Figure 4) (*15, 16*).

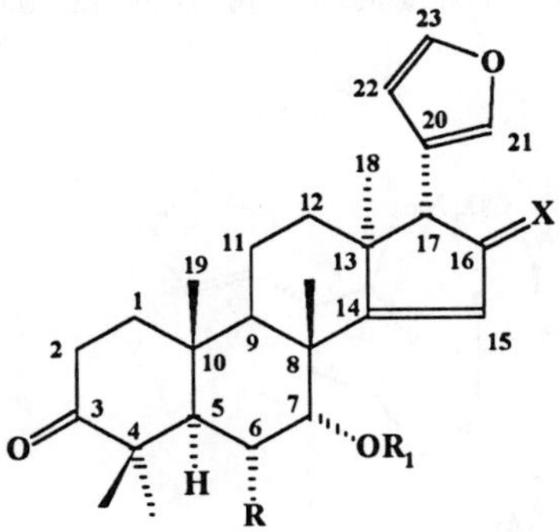

23 X=O, R_1=Ac, R=H Azadiradione
24 X=O, R_1=COPh, R=H Azadiradione benzoate
10 X=H,H, R_1=Ac, R=H Azadirone
25 X=H,H, R_1=H, R=OAc Meldenin

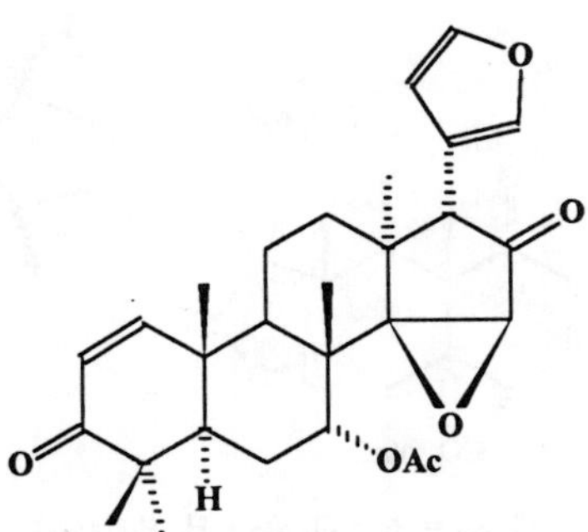

26 Epoxyazadiradione

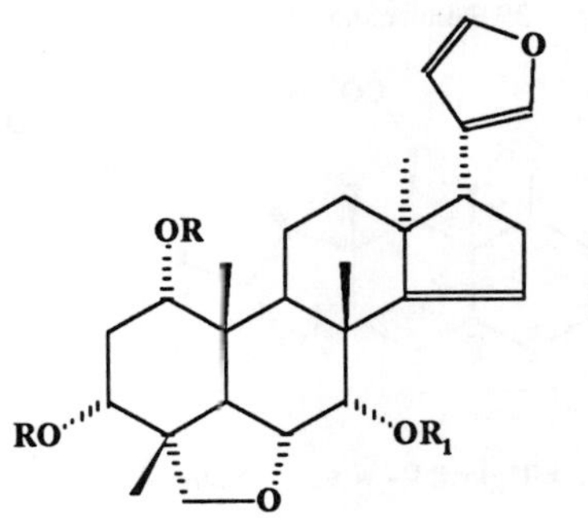

27 R=R_1=H Vilasinin
28 R=R_1=Ac Vilasinintriacetate

Figure 4. Limonoids with intact ring systems

Limonoids with Intact Ring Systems. These are also closely related to the tetracyclic triterpenoids. Representatives of these compounds are azadirone, **10**, azadiradione, **23**, epoxyazadiradione, **26** (*17*), azadiradione-7-benzoate, **24**, meldenin, **25** (*18*), vilasinin, **27**, and vilasinintriacetate, **28** (*19*). These limonoids have an intact D ring, but have been oxidized to include an epoxide ring at C-14 and C-15.

Ring Opened Limonoids This characteristic group of terpene-derived compounds consist of those compounds in which one or more of the rings has been opened or altered by oxidation. However, no B-ring-opened compounds have yet been identified in neem. *A. indica* characteristically contains C-ring-opened limonoids of the nimbin class (Figure 5.) Representative compounds of this class include salannin, **31** (*20, 21*), desacetylsalannin, **32** (*22*), nimbandiol, **29** (*22*), nimbenene, **30** (*23*), desacetyl-nimbinolide, **33**, and desacetylisonimbinolide, **34** (*22*).

The only example of an A-ring-opened compound is 4a,6a-dihydroxy-A-homoazadiradione, **12,** reported by Bruhn et al. (*23*). Nimolicinol, **14,** is among the few examples of the D-ring-opened limonoids (*24*).

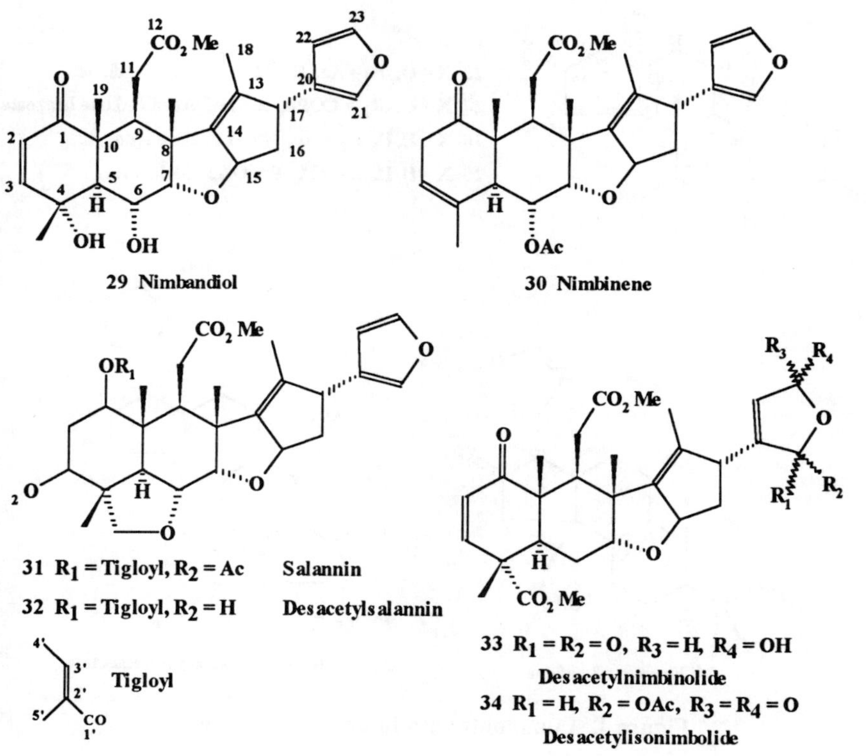

29 Nimbandiol

30 Nimbinene

31 R_1 = Tigloyl, R_2 = Ac Salannin

32 R_1 = Tigloyl, R_2 = H Desacetylsalannin

Tigloyl

33 $R_1 = R_2 = O$, R_3 = H, R_4 = OH
Desacetylnimbinolide

34 R_1 = H, R_2 = OAc, $R_3 = R_4 = O$
Desacetylisonimbolide

Figure 5. Ring opened limonoids

The most highly oxidized limonoids in neem are modified in both the C and D rings. Azadirachtin, **15,** (*25, 26*), the most studied member of this structural class, contains 16 oxygen atoms and has 16 chiral centers. In addition to their own unique structural modifications, members of the class have side-chain substitutions similar to those observed in compounds already discussed. Other representative examples of this class are shown in Figure 6 (*25-27).* These compounds, although found in various plant tissues, are predominantly in the neem kernel. Azadirachtin is the most abundant member of this group, and under ideal natural conditions can reach levels of 0.6% to 0.7% on a dry kernel-weight basis, but typically is present in levels of 0.2% to 0.3%.

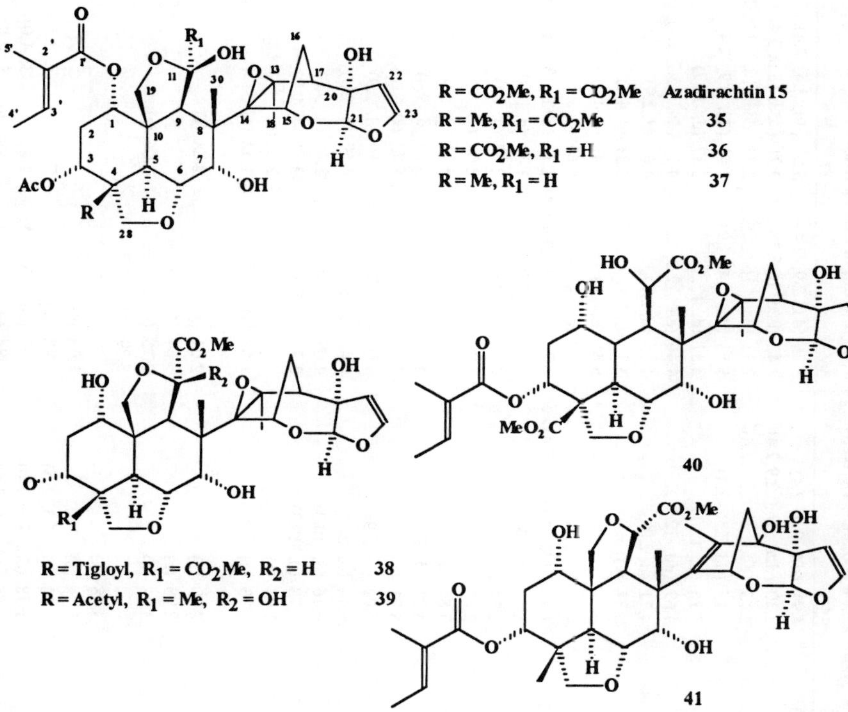

Figure 6. Neem limonoids modified in both the C and D rings.

Spectral Properties of Tetranortriterpenes. ^{1}H- and ^{13}C-NMR data for some selected neem compounds are included in Tables 1-4.

Chemical Synthesis of Azadirachtin

Total Synthesis. The complexity of azadirachtin can best be illustrated by the number of academic groups, and publications, aimed at the total synthesis of this molecule (*28-39*). Despite this massive effort, over nearly a decade, no synthesis is near completion. At this point, the only progress has been in model studies approaching the furopyran, **I** (*28-35*), and decalin, **II** (*36-38*), portions of the molecule. Completion of this work is still years away, and when finished will furnish azadirachtin costing orders of magnitude more than the natural product.

Table 1. ^{1}H-NMR Data on Azadirachtin and Some Derivatives [δ H, m, J (Hz)] in $CDCl_3$/TMS

Proton	Azadirachtin, 15	38	35	36	37
1	4.75 (dd, J=2.9, 3.1)	3.52 (ddd, 2.3, 2.7, 5.8)	4.91 (dd, 2.9, 2.4)	5.36 (dd, 2.9, 2.4)	5.01 (dd, 3.4, 2.4)
2a	2.34 (ddd, J=16.7,2.9,2.7)	3.32 (ddd, 16.3,2.3,2.8)	2.30 (ddd, 16.6,2.9,2.4)	2.42 (ddd, 16.6,2.9,2.4)	2.21 (ddd, 18.0, 3.4, 2.9)
2b	2.13 (ddd, J=16.7,3.1,2.9)	2.06 (dddd, 16.3,2.7,2.8,1.0)	2.13 (ddd, 16.6,2.9,2.4)	2.33 (ddd, 16.6,2.9,2.4)	2.26 (ddd, 18.0, 2.9, 2.4)
3	5.50 (dd, J=2.7, 2.9)	5.53 (dd, 2.8.2.8)	5.15 (dd, 2.9, 2.4)	5.53 (dd, 2.4, 2.9)	5.41 (dd, 2.9, 2.4)
5	3.35 (d, J=12.5)	3.33 (d, 12.7)	3.16 (d, 12.7)	3.37 (d, 12.2)	3.17 (d, 12.7)
6	4.60 (dd, J=12.5, 2.7)	4.55 (dd, 12.7.2.6)	4.69 (dd, 12.7, 2.0)	4.46 (dd, 12.2, 2.9)	4.11 (dd, 12.7, 2.9)
7	4.75 (d, J = 2.7)	4.72 (d, 2.6)	4.71 (d, 2.0)	4.65 (d, 2.9)	4.66 (d, 2.9)
9	3.34 (s)	3.19 (d, 1.3)	3.34 (s)	3.1 (d, 4.4)	3.21 (d, 4.4)
11		4.47 (d, 1.3)		5.41 (d, 4.4)	5.41 (d, 4.4)
15	4.67 (d, J=3.4)	4.58 (d, 3.9)	4.67 (d, 3.4)	4.58 (d, 3.4)	4.63 (d, 3.4)
16a	1.73 (ddd, J=13.0,3.4,5.1)	1.65 (ddd, 12.9, 3.9, 5.3)	1.70 (ddd, 13.0,3.4,5.4)	1.70 (ddd, 13.0,3.4,5.4)	1.73 (ddd, 13.0, 3.4, 5.4)
16b	1.31 (d, J=13.0)	1.33 (d, 12.9)	1.33 (d, 13.0)	1.32 (d, 13.0)	1.31 (d, 13.0)
17	2.38 (d, J=5.1)	2.36 (d, 5.3)	2.34 (d, 5.4)	2.37 (d, 5.4)	2.40 (d, 5.4)
18	2.01 (s)	2.04 (s)	2.05 (s)	1.99 (s)	2.03 (s)
19a	3.63 (d, J=9.6)	3.49 (d, 9.4)	3.85 (d, 9.3)	3.75 (d, 8.8)	3.71 (d, 9.0)
19b	4.15 (d, J=9.6)	3.95 (d, 9.4)	4.21 (d, 9.3)	4.09 (d, 8.8)	3.75 (d, 9.0)
21	5.65 (s)	5.66 (s)	5.71 (s)	5.66 (s)	5.72 (s)
22	5.05 (d, J=2.9)	5.03 (d, 2.9)	5.03 (d, 2.9)	5.05 (d, 2.9)	5.10 (d, 2.9)
23	6.46 (d, J=2.9)	6.43 (d, 2.9)	6.44 (d, 2.9)	6.45 (d, 2.9)	6.50 (d, 2.9)
28	4.08 (d, J=9.0)	3.83 (d, 9.0)	4.16 (d, 10.3)	4.09 (d, 8.8)	3.75 (d, 9.0)
28	3.76 (d, J=9.0)	4.04 (d, 9.0)	3.78 (d, 10.3)	3.75 (d, 8.8)	3.71 (d, 9.0)
29			1.94 (s)		2.03 (s)
30	1.74 (s)	1.45 (s)	1.73 (s)	1.64 (s)	1.32 (s)
1-OH		3.41 (dd, 5.8, 1.0)			
7-OH	3.02 (br s)	3.29 (br s)	2.80 (s)	2.64 (s)	2.68 (s)
11-OH	5.05 (s)	5.00 (s)	2.98 (s)	3.84 (s)	
20-OH	3.11 (br s)	2.78 (br s)	3.24 (s)	2.75 (s)	3.06 (s)
29 OMe	3.76 (s)	3.76 (s)		3.79 (s)	
12 OMe	3.68 (s)	3.76 (s)	3.68 (s)		
OAc	1.95 (s)	1.94 (s)	1.86 (s)	1.93 (s)	
3'	6.93 (qq, J=7.0, 1.5)	6.95 (qq, 7.0, 1.3)	6.92 (qq, 7.3, 1.5)	6.98 (qq, 7.3, 1.5)	7.01 (qq, 7.3, 1.5)
4'	1.78 (dq, J=7.0, 1.1)	1.79 (dq, 7.0, 1.3)	1.78 (dq, 7.3, 1.0)	1.77 (dq, 7.3, 1.1)	1.82 (dq, 7.3, 1.0)
5'	1.85 (dq, J=1.5, 1.1)	1.84 (dq, 1.3, 1.3)	1.85 (dq, 1.5, 1.0)	1.84 (dq, 1.5, 1.1)	1.89 (dq, 1.5, 1.0)

Table 2. ^{13}C -NMR Data on Azadirachtin and Some Derivatives [δ C, m] in $CDCl_3$/TMS

Carbon	Azadirachtin , 15		38		35		36		37	
1	70.51 d		69.37 d		70.90 d		71.58 d		72.29 d	
2	29.37 t		32.09 t		27.99 t		28.95 t		28.94 t	
3	66.99 d		67.69 d		70.32 t		66.22 d		71.31 d	
4	52.52 s		53.34 s		42.44 s		51.49 s		43.18 s	
5	37.06 d		35.18 d		35.47 d		35.81 d		35.29 d	
6	73.79 d		74.32 d		72.81 d		73.29 d		73.33 d	
7	74.37 d		73.68 d		74.87 d		75.04 d		73.66 d	
8	45.59 s		44.04 s		45.08 s		42.75 s		42.30 s	
9	44.69 d		43.80 d		44.71 d		46.97 d		47.78 d	
10	50.19 s		51.24 s		49.96 s		47.32 s		48.78 s	
11	104.10 s		79.48 d		103.99 s		99.57 d		100.51 d	
12	171.70 s		173.52 s		171.76 s					
13	68.53 s		66.59 s		68.50 s		66.69 s		66.63 s	
14	69.69 s		69.43 s		69.90 s		69.39 s		69.65 s	
15	76.43 d		76.16 d		76.38 d		76.02 d		76.80 d	
16	25.06 t		25.08 t		24.97 t		24.37 t		25.18 t	
17	48.67 d		48.99 d		48.98 d		47.11 d		48.36 d	
18	18.49 q		18.56 q		18.42 q		17.77 q		18.76 q	
19	69.07 t		71.43 t		69.97 t		68.18 t		71.20 t	
20	83.55 s		83.71 s		83.62 s		81.47 s		83.51 s	
21	108.70 d		109.16 d		108.72 d		107.54 d		108.80 d	
22	107.30 d		107.53 d		107.33 d		107.08 d		107.54 d	
23	147.00 d		146.79 d		146.82 d		144.88 d		146.90 d	
28	72.99 t		73.30 t		76.88 t		71.50 t		76.38 t	
29	173.20 s		174.12 s		18.95 q		172.88 s		18.91 q	
30	21.33 q		21.35 q		21.14 q		19.85 q		20.99 q	
CO_2Me	53.52 q	52.72 q	53.01 q	52.63 q	53.20 q		51.13 q			
AcO	169.50 s	20.88 q			170.09 s	20.88 q	168.84 s	19.57 q	170.21 s	20.93 q
1'	166.10 s		167.08 s		166.34 s		165.22 s		166.75 s	
2'	128.60 s		128.49 s		128.56 s		127.37 s		128.74 s	
3'	137.50 d		138.86 d		137.59 d		136.98 d		137.96 d	
4'	14.29 q		14.69 q		14.34 q		13.38 q		14.37 q	
5'	11.94 q		12.09 q		11.96 q		10.86 q		11.94 q	

Table 3. ^{1}H -NMR Data on Selected Ring -opened Limonoids [δ H, m, J (Hz)] in $CDCl_3$/TMS

Proton	Nimbandiol, 29	Nimbinene, 30	Desacetyl-nimbinolide, 33	Salannin, 31	Desacetyl-salannin, 32	Desacetyliso-nimbinolide, 33
1				4.97 (br s)	5.02 (t, 3)	
2	5.75 (d, J=10.2)	2.90 (dddq, 19.5, 5.5, 1.5, 1)	5.83 (d, 10.1)	6.33 (d, 10.1)		5.81 (d, 10.0)
2		3.01 (dddq 3, 2, 1)			2.08 (ddd, 3, 3, 1.6)	
3	6.53 (d, J=10.2)	5.60 (dddq, 1, 5.5, 3, 1)	6.41 (d, 10.1)	4.79 (br s)	3.88 (ddd, 3, 3, 10)	6.39 (d, 10.0)
5	2.66 (d, J=11.5)	3.30 (d, 12.5)	3.31 (d,11.6)	2.81 (d, 5)	2.73 (d, 13)	3.30 (d, 11.5)
6	4.24 (ddd, J=7.5,3.5,11.5)	5.36 (dd, 12.5, 3.5)	3.92 (dd, 11.6, 3.1)	3.98 (dd, 12, 3)	4.01 (dd, 4, 13)	3.93 (dd, 11.5, 3.0)
7	3.99 (d, J=3.5)	4.08 (d, 3.5)	4.08 (d, 3.1)	4, 19 (br s)	4.18 (d, 4)	4.10 (d, 3.0)
9	2.70 (dd, J=6, 4)	2.74 (t, 6)	2.56 (dd, 4.0, 4.5)	2.75 (dd, 6, 3,)	2.64 (dd, 10, 3)	2.55 (dd, 5.0, 5.5)
11	2.91 (dd, J=16.5, ?)	2.82 (dd, 16.5, 6)	2.89 (dd, 16.8,4.0)			2.89 (dd, 16.0, 5.0)
11	2.18 (dd, J=16.5, 4)	2.26 (dd, 16.5, 6)	2.21 (dd, 16.8, 4.5)			2.21 (dd, 16.0, 5.0)
15	5.48 (dddq, J=7, 7, 1, 1.8)	5.50 (dddq, 6.8, 8, 1, 2)	5.51 (m)	5.45 (t, 9)	5.39 (br t, 7)	5.51 (m)
16	2.16 (dd, J=12, 7)	2.17 (dd, 12, 6.8)	2.31 (dd, 12.0, 5.6)			2.29 (dd, 12.0, 5.5)
16	2.02 (ddd, J=8, 12, 7)	2.04 (ddd, 8.5, 12, 8)	2.09			2.12 (ddd, 12.0, 7.5, 7.0)
17	3.65 (d, J=8)	3.64 (d, ?1)	3.53 (d, 7.7)	3.60 (br d, 9)	3.62 (br d, 8)	3.55 (d, 7.5)
18	1.67 (d, J=1.8)	1.68 (dd, 2, 1,)	1.84 (s)	1.69 (s)	1.64 (s)	1.80 (s)
19	1.16 (s)	1.12 (s)	1.18 (s)	0.98 (s)	0.97 (s)	1.19 (s)
21	7.33 (m)	7.32 (m)		7.32 (m)	7.31 (m)	5.78 (m)
22	6.26 (m)	6.28 (m)	6.97 (m)	6.30 (m)	6.27 (m)	5.91 (m)
23	7.21 (m)	7.22 (m)	6.01 (m)	7.27 (m)	7.25 (m)	
28				3.58 3.70	4.13 (d, 7)	
29	1.58 (s)	1.78 (m)	1.22 (s)	1.22 (s)	1.18 (s)	1.22 (s)
30	1.30 (s)	1.38 (s)	1.28 (s)	1.30 (s)	1.30 (s)	1.28 (s)
OH	4.29 (s)		3.73 (m)			
OH	3.14 (d)		2.02 (m)			
OMe	3.63 (s)	3.57 (s)	3.70 (s)	3.24	3.20	3.71 (s)
OMe			3.69 (s)	3.69 (s)		3.69 (s)
OAc		2.14	2.03	1.95 (s)	2.02 (s)	
Tigloyl (Senecioxy) Group				2.03 (s)		
3' (2')			6.96 (q, 9)	6.91 (qd, 7, 1)	5.7 (d, 1.0)	
4' (4')			1.81 (d, 9)	1.85 (d, 7)	1.9	
5' (5`)			1.95 (s)	1.92 (s)	2.2	

Table 4. ^{13}C -NMR Data on Selected Ring -Opened Limonoids

Carbon	Nimbandiol, 21	Nimbinenene, 30	Desacetyl-Nimbinolide, 33	Desacetyl-Salannin, 31	Desacetyliso-salannin, 32	nimbinolide 33
1	202.49s	212.30s	202.4	77.06	76.98	202.0
2	124.74d	39.95t	126.2	29.86	30.41	126.3
3	151.88d	126.16d	148.1	71.75	72.41	148.4
4	70.95s	136.87s	39.1	39.81	39.46	39.1
5	50.18d	43.16d	48.9	39.16	38.81	50.0
6	67.31d	69.23d	65.9	70.56	70.82	66.1
7	87.40d	85.25d	86.4	84.94	85.76	87.6
8	47.94s	47.86s	47.9	48.26	48.93	47.9
9	38.68d	37.02d	39.8	39.16	39.46	39.8
10	47.65s	49.47s	47.5	40.64	40.91	47.5
11	34.03t	33.54t	34.6	29.86	30.51	34.1
12	173.53s	173.56s	175.4	171.96	172.52	175.4
13	134.98s	135.24s	131.1	136.20	137.97	131.1
14	146.49s	146.49s	145.0	145.81	146.40	145.0
15	86.78d	86.94d	87.8	87.08	87.79	88.0
16	41.34t	41.41t	39.8	41.91	41.15	39.8
17	49.59d	49.57d	52.1	48.58	49.33	52.2
18	12.74q	12.84q	13.1	12.21	12.21	13.0
19	17.39q	21.58q	17.1	18.76	19.87	17.1
20	126.66s	126.89s	137.0	126.31	127.05	162.6
21	138.88d	139.01d	169.9	137.99	138.64	96.9
22	110.38d	110.48d	142.0	109.82	110.59	120.8
23	142.91d	143.14d	96.9	142.13	142.77	171.0
28			172.0	77.76	76.58	172.0
29	23.50q	14.27q	20.5	26.72	30.51	20.6
30	15.83q	17.13q	16.1	16.07	16.88	16.1
CO_2Me	51.52q	51.65q	53.0, 53.1	50.58	50.39	52.9, 53.1
OAc		170.73s, 21.13q		169.36, 20.00		
1'				165.78	166.18	
2'				128.32	128.50	
3'				137.99	138.64	
4'				14.23	14.58	
5'				11.11	13.08	

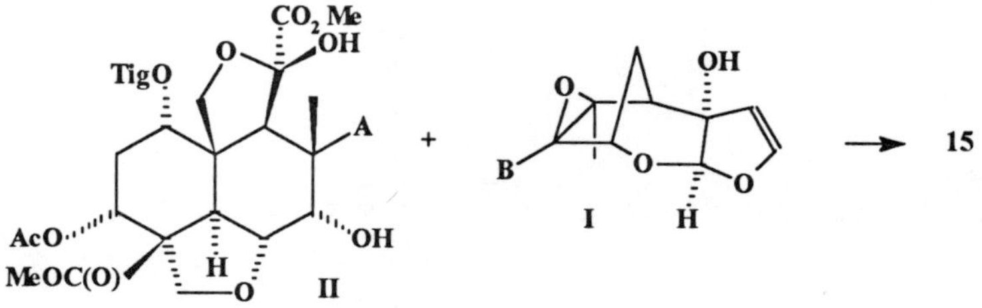

Partial Synthetic Studies. In contrast to the total synthesis efforts, there has been considerable synthetic activity starting from the natural product, **15** (azadirachtin), that is commercially interesting. Nearly all of the synthetic transformations involve initial derivatization at the highly reactive C-22,23 enol-ether double bond, thus stabilizing the molecule. For example, it has been known for many years that the hydrogenation product 22,23-dihydroazadirachtin, **42** (*40, 41*), is more stable than azadirachtin. The derivative 2',3',22,23-tetrahydroazadirachtin, **43,** is also more stable than **15** and was recently patented (*42*). It is interesting to note that the 22,23 double bond is not important for insecticidal activity. Azadirachtin, **42** and **43** all have similar biological activity (*21, 43-45*).

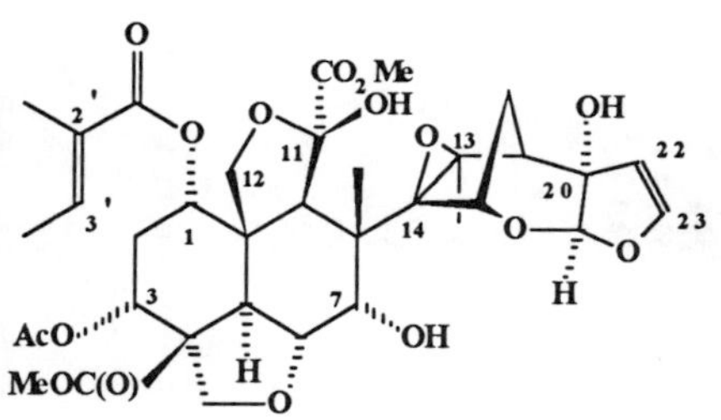

15

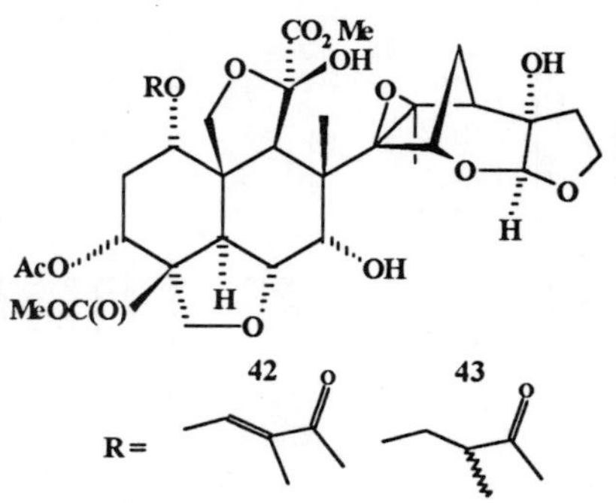

Most of the elegant synthetic studies of Ley and co-workers involve either additions to the enol-ether double bond, or reduction of this bond, giving dihydroazadirachtin (**42**) prior to any other synthetic steps. Thus, acetic acid (but no other carboxylic acids) can be added across the double bond to give **44** as a 2:1 (a:b) mixture (*46, 47*). Addition of bromine in an alcohol gives the bromo ethers, **45, 46**, or **47** as C-23 epimers (*48, 49*). Tri-n-butyl tin hydride reduction of the bromo ethers affords the ethers, **48, 49,** and **50**. The natural product 22,23-dihydro-β-methoxyazadirachtin (**51**) was synthesized by separating the β-isomer from the mixture **45** prior to reduction.

Another example of the reactivity and instability of the 22,23-enol-ether double bond is illustrated below. When the enol-ether was first hydrogenated (giving **42**) and then treated with strong acid, a rearrangement involving the C-7 hydroxyl and the C-

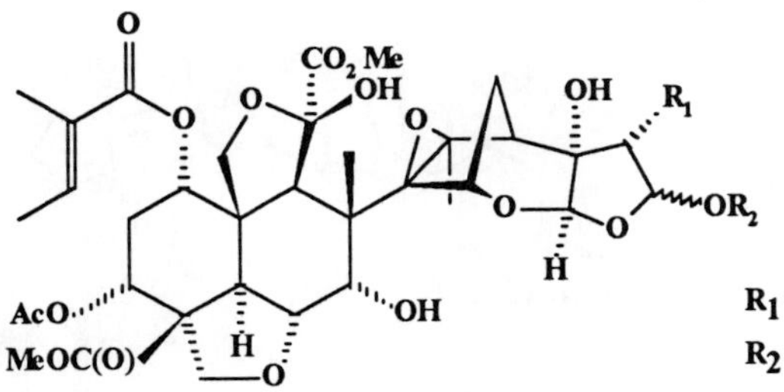

	44	45	46	47	48	49	50	51
R_1 =	H,	Br,	Br,	Br,	H,	H,	H,	H
R_2 =	Ac,	Me,	Et,	i-Pr,	Me,	Et,	i-Pr,	β–Me

13,14 epoxide occurred, yielding **52** (*48*). A similar rearrangement was also observed (yielding **53**) when the methoxy analog, **45,** was treated with acid. But azadirachtin itself could not be rearranged with acid, only decomposition products were obtained.

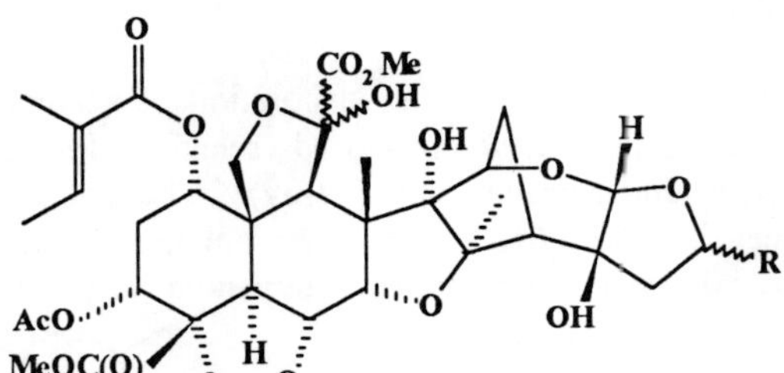

	52	53
R=	H	OMe

There are only a few known reactions where the 22,23 double bond remains intact, and these involve simple protection or deprotection to the alcohol functions of azadirachtin. For example, the acetate at C-3 can be removed with sodium methoxide in methanol at room temperature in 2.5 hours (*21*) (yielding **54**). The hydroxyl at C-11 can be selectively monoacetylated (*21, 41, 49*) (acetic anhydride reflux, 15 min., giving **55**), or monomethylated (*21*) (iodomethane, silver oxide, giving **56**). The remaining two hydroxyls at C-20 and especially C-7 are much more difficult to access synthetically. Nakanishi reported that in order to methylate all three alcohols (yielding **57**) it was necessary to react azadirachtin with 15 equivalents of KOH and 30 equivalents of iodomethane in DMSO *(40)*.

In order to remove the tiglate on C-1, it is again necessary to first reduce the 22,23 double bond. Yamasaki (*21*) found that the tiglate of dihydroazadirachtin could be removed by treatment with osmium tetroxide and sodium periodate to make the dihydroxy derivative, **58**. This derivative was decomposed to 1-detigoyl-22,23-dihydroazadirachtin, **59**, with aqueous sodium bicarbonate. Ley (*50*) has also prepared **59** by treating **42** with sodium carbonate, sodium periodate, and potassium permanganate. It is also possible to carbomethoxylate both the C-11 and C-20 alcohols of **42** by heating it in a large excess of neat dimethyl pyrocarbonate (yielding **60**). The C-7 hydroxyl is so hindered that this treatment does not affect it.

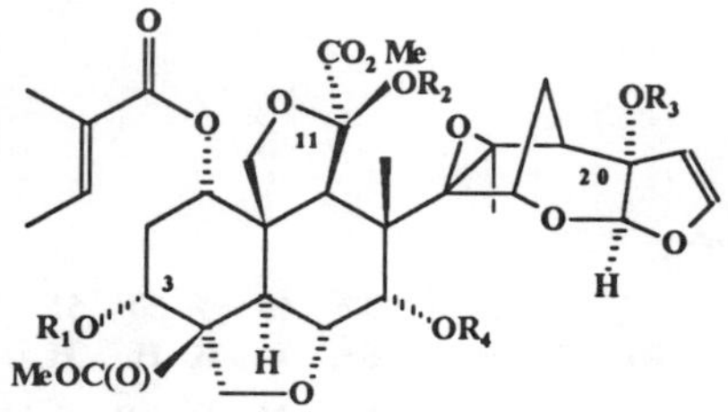

	54	55	56	57
R_1 =	H,	Ac,	Ac,	Ac
R_2 =	H,	Ac,	Me,	Me
R_3 =	H,	H,	H,	Me
R_4 =	H,	H,	H,	Me

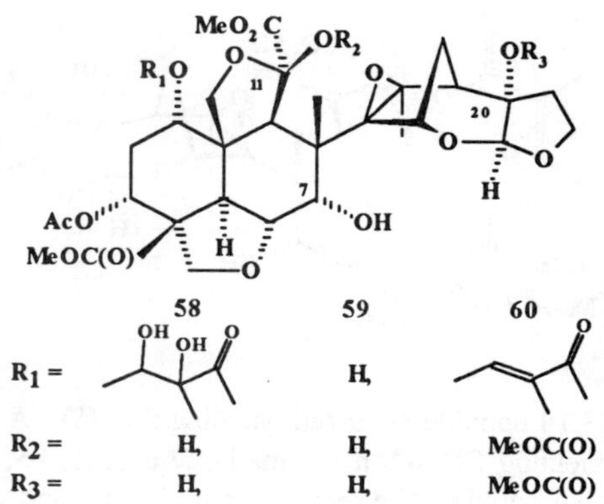

The Biogenesis of Azadirachtin

The biosynthetic pathway in neem for the elaboration of azadirachtin is unknown. However, several researchers have proposed schemes for the biogenesis of tetranortriterpenoids including azadirachtin (*51-54*). The syntheses begin with the tetracyclic triterpene tirucallol, **61**, or its C-20 epimer, euphol. The proposed rearrangement of tirucallol to afford a tetranortriterpenoid is shown in Figure 7 and

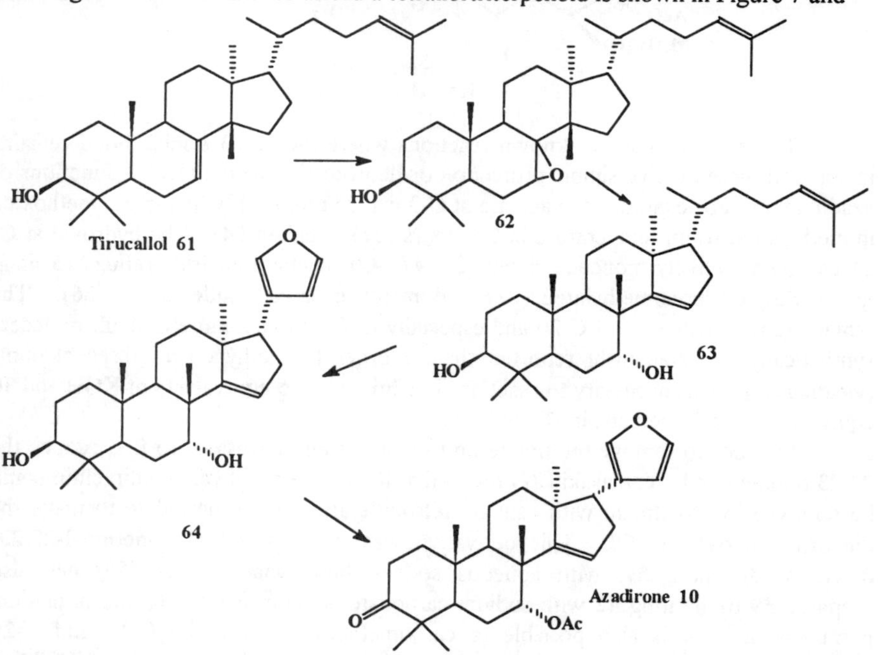

Figure 7. A proposed scheme for conversion of triterpenoids into limonoids

involves the migration of the methyl group (C-30) on C-14 to C-8 and transfer of the double bond from the C-ring to the D-ring through a C-ring epoxide intermediate.

Subsequent oxidative degradation of the eight-carbon C-17 side chain to a furan ring affords a tetranortriterpenoid, **64**. Since tirucallol (or euphol) has yet to be isolated from any part of the neem tree, support of this theory was obtained by the *in vivo* conversion of radiolabeled tirucallol and euphol into the limonoid nimbolide in the leaves of the neem tree (*55*).

The isolation of several intact tetracyclic triterpenes in which oxidation of the C-17 side chain has occurred (meliatriol, **16**, nimbocenone, **17**, nimolinone, **18**, azadirachtol, **19**, and others) from various parts of the neem tree would suggest that partial formation of the limonoid furan ring precedes the rearrangement of the carbon skeleton. The skeletal rearrangement and complete furan ring formation is first seen in the limonoid compounds azadirone, **10**, azadiradione, **23** and epoxy-azadiradione, **26**.

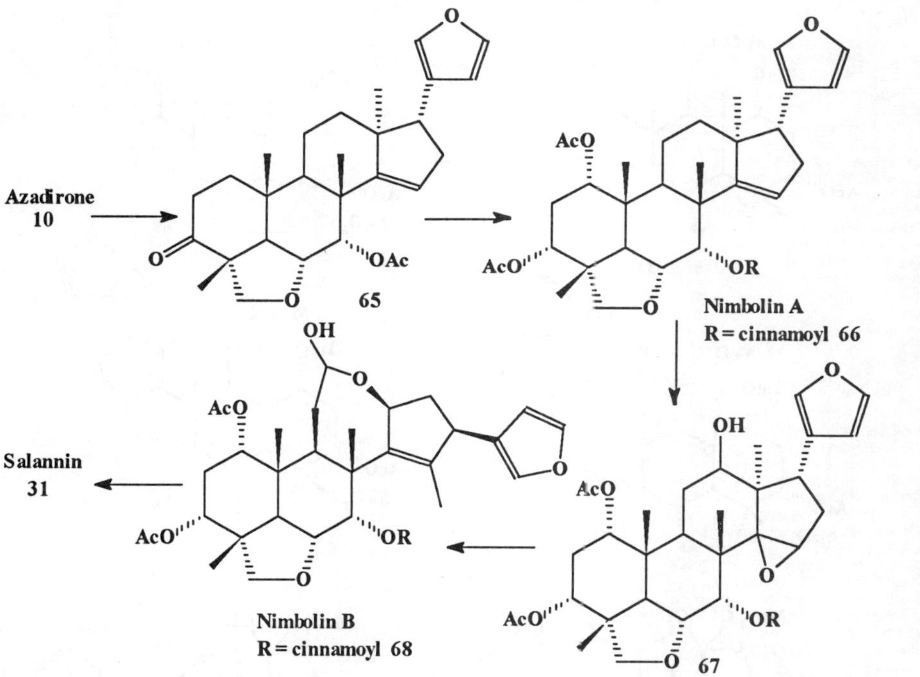

Figure 8. A proposed scheme for modifications to the C-ring and side chain additions to limonoids.

To obtain salannin, **31,** from an azadirone-like intermediate (Figure 8), the α-methyl substituent at C-4 and the C-6 carbon must be oxidized and then cyclized to form a five-membered tetrahydrofuran ring, **65**. Various stages of this type of oxidation pattern can be observed in compounds isolated from neem (meldenin, **25**, nimbinene, **30**, desacetylnimbinolide, **33**, nimbandiol, **29**, and others). Conversion of the A-ring to a 1,3-diol with esterification would give rise to nimbolin A, **66**. The transformation of nimbolin A to nimbolin B, **68** could arise via the oxidation of C-12 and epoxidation of the 14,15 double bond (*56*) followed by cleavage of the 12,13 bond with simultaneous opening of the protonated epoxide to generate 12-CHO and 15-OH functionalities (*57*). Ring opening of the hemiacetal of nimbolin B, followed by rotation about the 8,14 bond and nucleophilic displacement by the 7-α-OH on the protonated 15-OH, coupled with transesterification of the C-1 acetate to tiglate and oxidation of C-12 to methyl ester would afford salannin, **31**.

The conversion of salannin to azadirachtin, **15**, would require a series of oxidations and an intramolecular rearrangement. The oxidations of the methyl group at C-4, the C-19 methyl and C-11 methylene groups could occur in many combinations before or after the rearrangement, as evidenced by the various azadirachtin analogs isolated from neem. The intramolecular rearrangement could be

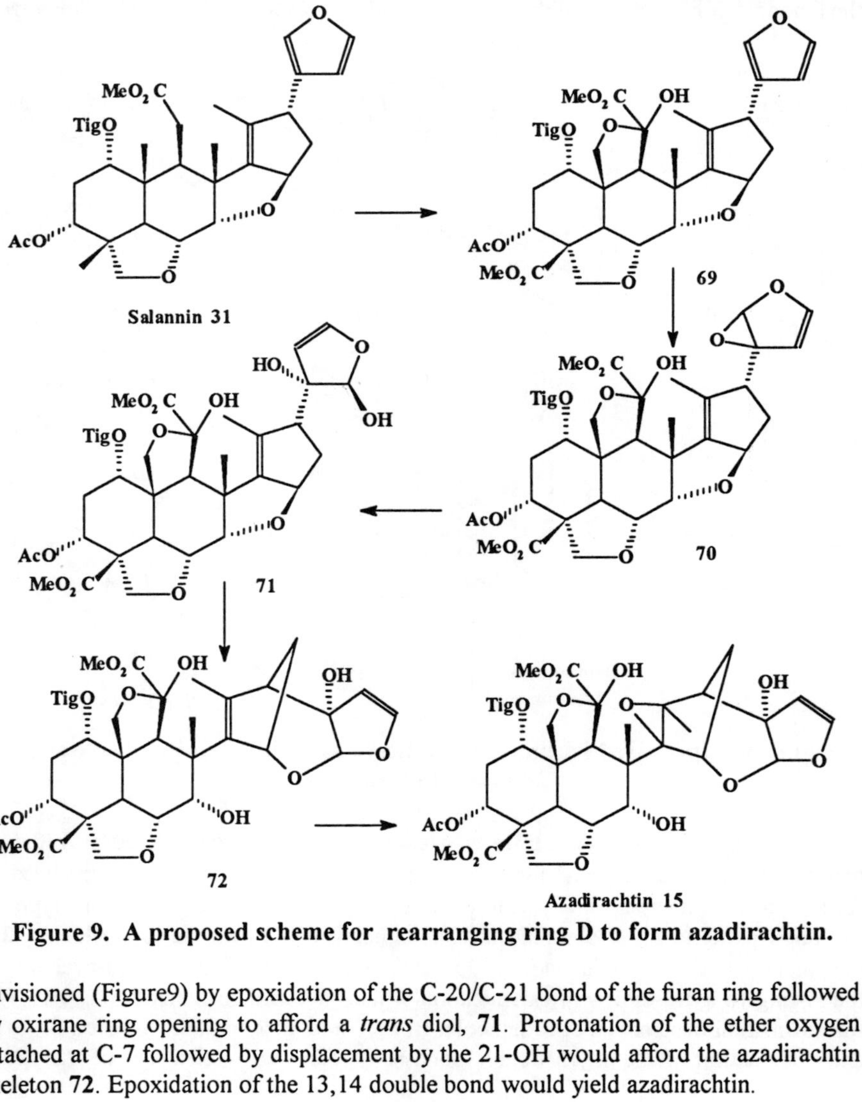

Figure 9. A proposed scheme for rearranging ring D to form azadirachtin.

envisioned (Figure9) by epoxidation of the C-20/C-21 bond of the furan ring followed by oxirane ring opening to afford a *trans* diol, **71**. Protonation of the ether oxygen attached at C-7 followed by displacement by the 21-OH would afford the azadirachtin skeleton **72**. Epoxidation of the 13,14 double bond would yield azadirachtin.

Aflatoxins in Neem

Conditions at harvest during the rainy season may result in fungal infection of the moist seed. If significant *Aspergillus flavus* invasion occurs, then production of aflatoxins may follow. Poor handling of moist seed can substantially increase aflatoxin production and may create a health hazard. Since the aflatoxins are extracted and concentrated along with the active insecticidal principles obtained from processing the seed, this can be a cause for major concern, as the aflatoxins are potent toxins and carcinogens.

Attention to hygienic practices during harvesting, handling and storage can minimize the aflatoxin problem. Thus, speedy depulping of harvested fruits, drying to moisture contents of 14% or below, and storage of the dried seed with protection from the sun and rain, are becoming recognized as key factors to help eliminate the presence of aflatoxins.

A most promising approach to controlling or minimizing aflatoxin contamination is due to recent advances in the processing of the seed extracts. It is now possible to efficiently remove aflatoxins to levels below that of public health concern. Figure 10 shows a typical result for a comparison of the levels of aflatoxins B_1 and B_2 in hygienically treated seed extracts versus non-hygienically treated extracts.

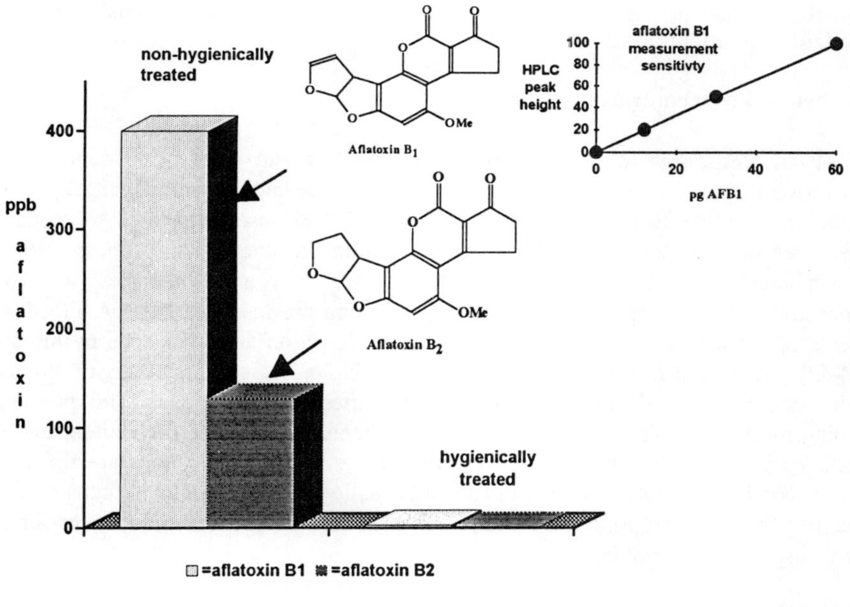

Figure 10. Comparison of aflatoxin levels in hygienically and non-hygienically treated neem seed.

It is clear that with a combination of good seed harvesting and storage practices and hygiene, coupled with the processing advances discussed above, aflatoxin contamination of pesticidal extracts and ingredients need no longer be seen as a barrier to the full exploitation of neem.

Chemical Properties of Azadirachtin

The molecular structure of azadirachtin has been firmly established as **15** by numerous studies. An initially proposed structure (*49*) of this highly oxidized tetranortriterpenoid was subsequently revised to **15** on the basis of much high-resolution NMR spectroscopic data (see, for example, *40, 50, 58*).

Major features of this architecturally interesting, 16-oxygen-atom-containing molecule include: the A-B *trans*-fused rings; the epoxide ring at positions 13,14; the terminal dihydrofuran ring; a tigoyl side chain at position 1; three free hydroxyl groups at positions 7, 11 and 20; and the possession of 16 chiral carbon centers.

Although much spectroscopic data has been published on azadirachtin, it is scattered throughout the literature, and proof of the absolute purity of the substance has been deficient. Thus, the absorption maximum of azadirachtin has been quoted as occurring at 217 nm in methanol, with an extinction coefficient value of 9,700 (*59*). Huang and Morgan (*60*) claim a maximum in the UV at 212 nm, but do not give an extinction coefficient. Since the absorption value and extinction coefficient need to be known accurately for the pure substance for quantitative analytical procedures based on, e.g. HPLC and UV detection, we addressed this problem. By repeated column chromatography, we obtained a chromatographically homogeneous substance (single peak on C-18 HPLC) which had a UV absorption maximum at 212 nm in methanol, agreeing with the value published by Huang and Morgan (*60*). Further, the extinction coefficient of this pure material was found to be 12,160, a value substantially higher than that quoted in the *Merck Index* (*59*). We believe that this establishes a new standard of purity for azadirachtin.

Biochemical Mechanisms in Insects

Observed Responses to Azadirachtin. Azadirachtin and other neem compounds cause two major observable responses in insects: 1) feeding deterrence, and 2) molt inhibition. Both effects have been noted in numerous insect species. Antifeedant effects are more species-specific than molt inhibition. In other words, concentrations of azadirachtin required to prevent feeding vary widely among species, whereas concentrations that stop growth and disrupt molting are less variable. Antifeedant effects of azadirachtin are partly due to sensory detection and avoidance by insects (*61-63*). Although many organic syntheses and biological structure-activity studies have been done to identify biologically active sites in the molecule and possible binding mechanisms, little is known about the chemistry of either the antifeedant or molting effects. Many studies have utilized assays which fail to separate the two effects, confounding interpretation of such information. This section will focus on the chemistry of molt disruption and on potential interaction with endogenous compounds regulating the insect molting process.

Insect Endocrine Scheme. This discussion of molt disruption begins with a summary of endogenous regulatory processes, followed by information on disruption caused by azadirachtin analogs. The insect brain produces a protein, prothoracotropic hormone (PTTH), which via the corpus cardiacum stimulates the prothoracic glands to initiate synthesis of a major class of insect endocrine hormones, the ecdysteroids. Juvenile hormones (JHs), are primarily produced in the corpora allata. The relative titers of ecdysteroids and juvenile hormones during the larval/nymph or pupal phase, orchestrate the molting process. In fact, these two hormones regulate cascades of protein production that modify many aspects of insect development. High titers of ecdysteriod in combination with high titers of JH result in larval-to-larval molts, while high titers of ecdysteroid in the presence of lower titers of JH result in larval-to-pupal molts. Molt-inducing spikes of ecdysteroid biosynthesis are rapidly reversed through a feedback system that shuts down PTTH production. Target tissues include the salivary glands, gut, fat body, larval cuticle, as well as imaginal discs that develop into adult structures. Although ecdysone can be found throughout the insect, biochemical studies have shown an abundance in gut tissues and in developing cuticular or imaginal discs, where it is converted to the more active isomer, 20-β-hydroxyecdysone. Insects cannot biosynthesize sterols (*64*), but must ingest them in their diets, either from plant, animal, or microbial sources. Insects control their internal ecdysteroid titers by diet changes, or physiologically through ecdysteroid conjugation and oxidation to render them hormonally inactive. In adults, juvenoid hormones are absent or are present in low concentrations while ecdysteroids are the major hormone class.

During the past few years, considerable effort has been directed at identifying the cascade of protein synthesis initiated following ecdysone treatment of insects. Most of the work has been done in *Drosophila* because of the p-element transformation system and the extensive genetics database available. It has long been known that application of ecdysone to *Drosophila* resulted in the development of large chromosome puffs (Balbiani rings), indicating binding to the chromosomes. An ecdysone receptor protein has been isolated and the gene (EcR) encoding it has been cloned and characterized (*65*). EcR protein, in the presence of active ecdysteroids, binds to *Drosophila* DNA and triggers expression at each developmental stage marked by a pulse of ecdysone. Gene transcription begins after the protein binds and is then followed by further gene activation and gene repression. Several of the activated genes in the cascade have also been isolated (*66-68*). The gene coding for PTTH has been cloned and characterized from *Bombyx* (*69*) and an antibody to PTTH has been made (*70*), both polyclonal and monoclonal antibodies (*71, 72*) have been made to 20-β–hydroxyecdysone, and an affinity label for 20-β-hydroxyecdysterone has been prepared (*73*). Such advances now make it possible to determine the similarity of ecdysone and azadirachtin-induced biological changes in insects.

Azadirachtin Mediated Modifications of Insect Morphology and Biochemistry. Most azadirachtin and ecdysone research has been done using *Bombyx, Drosophila, Locusta, Heliothis, Ostrinia, Rhodnius, Manduca* species, and others. These organisms have been studied for genetic, economic, or anatomical reasons and we will not attempt to differentiate all azadirachtin responses here on a species-by-species basis, except to elucidate salient points relative to growth and molt disruption relative to azadirachtin. Morphological responses have been summarized (*3, 4*) and can usually be categorized by molting pattern, body type, and/or insect taxonomic order. Insects that progress through nymph stages respond somewhat differently than those which develop via larval instars through a pupal stage. Insects requiring a blood meal or having aquatic larval stages respond still differently.

Azadirachtin causes a decrease of insect weight gain, disrupts and delays molting regulated by ecdysone, and often causes death to both larval and pupal stages. It also disrupts the proper shedding of old body capsules during the molting process.

Azadirachtin, applied topically to final-instar larvae of lepidopteran insects, affects oogenesis and reproductive maturation in subsequent female moths (*74*). Moths obtained from such treated larvae developed fewer mature oocytes, possibly as a result of interference of azadirachtin with vitellogenin synthesis and/or its uptake by developing oocytes. Larval treatments also cause decreased viability of emerging larvae from affected eggs.

Azadirachtin is one of the few inhibitory compounds known to alter pupal development. Pupae often die before or during eclosion; however, if they do survive to become adults, they usually have aberrant morphology such as crinkled wings, bent thoraxes, or other deformations (*75*).

Tritiated 22,23-dihydroazadirachtin accumulates in locust insect tissues and becomes tightly bound following an initial flush from the body of the insect during the first 24 hours following administration (*76*). Once bound, it cannot be removed by administering additional non-labeled compound and appears to accumulate in two major regions of the insects. Quantitative determination of the radioactive label shows attachment to neurosecretory cells and the corpus cardiacum (*77*). It also accumulates in body tissues, primarily in the gut, malpighian tubules, and ovaries (*2*). When expressed on a whole-insect basis, 0.5 μg of 22,23-dihydroazadirachtin per gram of insect tissue remain chemically unaltered and bound for long time periods, suggesting very tight binding to receptor sites or that insects lack necessary enzymes for metabolism.

Topical applications, ingestion by feeding, or hemolymph injections of azadirachtin reduces hemolymph ecdysteroid titers and/or delays the appearance of ecdysteroid peak(s) in many insect species (*78-81*). This is thought to occur because PTTH production by the brain is inhibited. Azadirachtin does not directly inhibit the *in vitro* synthesis of ecdysone when added to incubation media with previously untreated prothoracic glands (*78*), and similar results have been observed *in vivo* with *Tenebrio molitor* pupae (*82*). In addition, acetylated pregnadiene derivatives selectively and irreversibly inhibit production of ecdysteroids by blocking ecdysone synthesis (*83*). Such compounds behave as JH analogs and stimulate juvenile behavior even though they do not alter JH biochemical mechanisms. Titers of JH are elevated in relation to ecdysone since biosynthesis is blocked and typical ecdysteriod peaks fail to be produced in a timely manner throughout the life cycle. It is apparent that although azadirachtin prevents ecdysteroid production by blocking the production of PTTH, it does not cause juvenile behavior.

The ecdysone 20-monooxygenase enzyme complex (mixed-function oxidase), which converts ecdysone into 20-β-hydroxyecdysone, is inhibited by azadirachtin (*84, 85*). This enzyme complex, isolated from *Gryllus bimaculatus,* has an obligate requirement for NADPH and is also competitively inhibited by its product, 20-hydroxyecdysone. When tissues of *Calliphora vicina* and *Neobellieria bullata* are tested using fenarimol, a pyrimidine-containing cytochrome P-450 inhibitor, 20-hydroxyecdysone synthesis is inhibited while ecdysone biosynthesis is normal (*86*). Freshly ecdysed *Tenebrio molitor* pupae injected with 1 μg of azadirachtin per gram of tissue show a selective depletion of 20-hydroxyecdysone, whereas ecdysone levels remain practically unchanged in comparison with control pupae (*82*). The non-steroidal ecdysone agonist RH5849 (1,2-dibenzoyl-1-*tert*-butylhydrazine) was also found to inhibit the cytochrome P-450-dependent ecdysone 20-monooxygenase activity in the midgut of the last-instar larvae of the tobacco hornworm, *Manduca sexta* L. (Figure 11) (*87*).

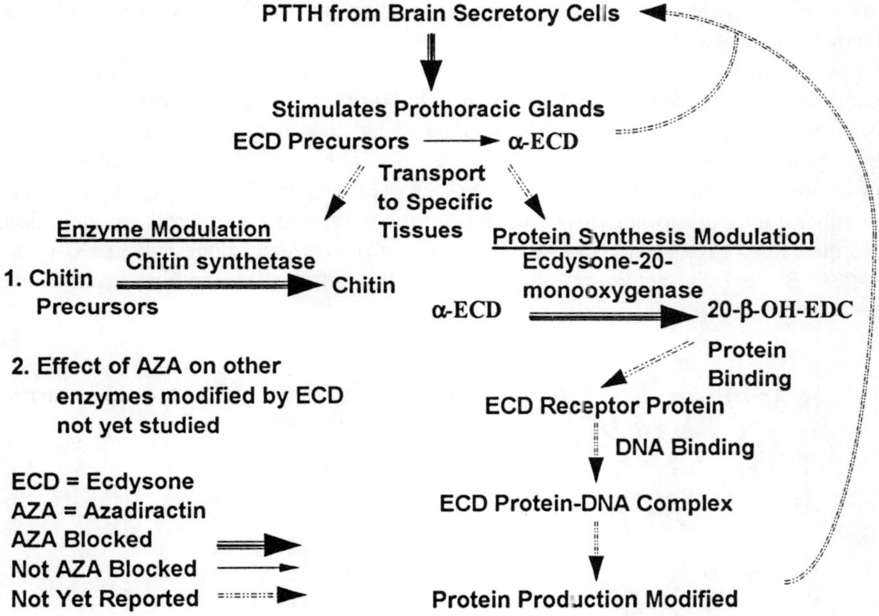

Figure 11. Proposed effects of azadirachtin on ecdysone-mediated pathways.

Chitin synthesis in wing pads of 1-day-old fifth-instar larvae of *Locusta* is stimulated *in vitro* by 20-hydroxyecdysone and makisterone A (*88*). This chitin biosynthesis system requires integrity of the microtubular system and the chitin-synthetase complex. Chitin synthesis can be inhibited in this test system by azadirachtin as well as by colcemid, a microtubular poison, or by diflubenzuron, a chitin synthetase inhibitor. It is also of interest to note that RH5849 stimulates chitin synthesis in wing discs of *Plodia interpunctella* (*89*).

Structure-Activity Relationships. Azadirachtin structure-activity studies for molt disruption show that both major portions of azadirachtin are required for activity (*90*). Components of the A and B rings of the original triterpenoid are required as well as those of the five-membered rings of the dihydrofuran-containing moiety. Structure-activity summaries have been recently published (*1, 2*) and we will confirm and extend previous findings. Several substituents play a major role in molt inhibition, most notably those at carbons 1, 3, 6, 20, and possibly the epoxide bridge at carbons 12 and 13. Figure 12 gives a list of published analogs with molt inhibitory activity. Limonoid molecules such as salannin, (**31**), which contain a furan ring and are bridged with an oxygen between carbon positions 7 and 13, do not cause molt inhibition. None of the bridged salannin analogs have a hydroxyl at C-20. Both azadirachtin and ecdysone analogs lose activity when they are extensively modified or are hindered at the 20-hydroxy position. Hydroxyl substitution at C-17 greatly reduces insect molt-inhibiting activity. Ester substitutions at positions C-1 and C-3 seem unimportant unless they are sufficiently large to preclude proper binding within the receptors. In fact, laboratory assays show the highest levels of growth inhibition and molt disruption when positions C-1 and C-3 contain only a hydroxyl group (*91*). Under these conditions, activity is increased approximately 20 times over azadirachtin. The double bond at positions 22,23 of the dihydrofuran ring does not appear to contribute to biological activity. However, stability is greatly improved when the double bond at positions 22,23 has been saturated (*43*). Substitutions on C-7, C-9, C-10, and C-11 possibly contribute to activity by conferring the appropriate structural conformation. No molecules have been made with variations in the oxygen bridge between C-6 and C-28. Biologically active molecules possess this tetrahydrofuran ring and inactive ones do not.

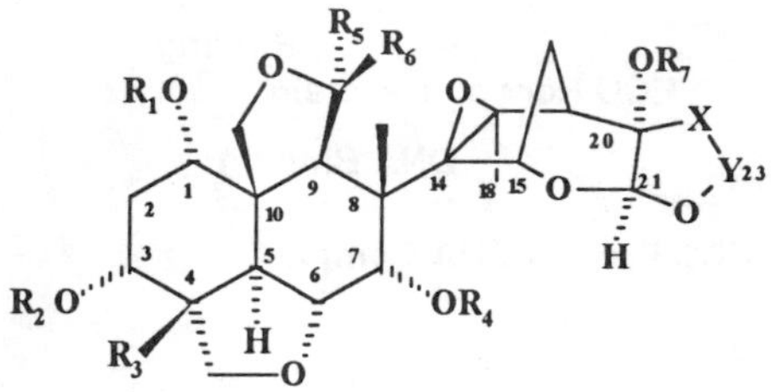

R_1 = H, Tiglate, 2-methylbutyrate, cinnamate
R_2 = H, Acetate, Tiglate, 2-methylbutyrate
R_3 = Me, CO_2Me
R_4 = H, Me
R_5 = H, CO_2Me
R_6 = H, OH, OAc, OMe
R_7 = H, Me
X-Y = CH=CH, CH_2-CH_2, CH_2-CHR_8, CH-α-Br-CHR_8
R_8 = α or β-OMe, OEt, O-*i*-Pr, OAc

Figure 12. Azadirachtin analogs with molt inhibitory activity.

Since physiological evidence has accumulated recently to suggest that azadirachtin inhibits at least some reactions involving 20-β-hydroxyecdysone, **73**, we compared 3-dimensional models of the two molecules. Suspecting that azadirachtin may adhere to insect proteins through hydrogen bonding similar to 20-β-hydroxyecdysterone, and knowing that the 20-hydroxy and 3-hydroxy positions of ecdysone are critical for biological activity, we superimposed the two molecules at these positions to look for similarities. Using Sybyl molecular modeling software (version 5.4 by Tripos), relaxed bond distances between oxygen atoms at C-3 and C-20 were found to be similar in both molecules (see Figure 13). Although the A and B rings of ecdysone and azadirachtin cannot be superimposed since these rings are *cis*-fused in ecdysone and *trans*-fused in azadirachtin, critical oxygen atoms of each molecule assume nearly the same spatial orientation. The C-3 to C-20 distance in both molecules is about 11 Å. When both molecules were superimposed, oxygen atoms at three positions (C-3, C-6, and C-20) were all less than 0.15 Å within being coincident. The oxygen atom of azadirachtin at C-1 and of ecdysone at C-2 were also within 0.47 Å. Such similar positioning suggests that azadirachtin and ecdysone may bind to one or more common receptor sites in insects. This is supported by information already

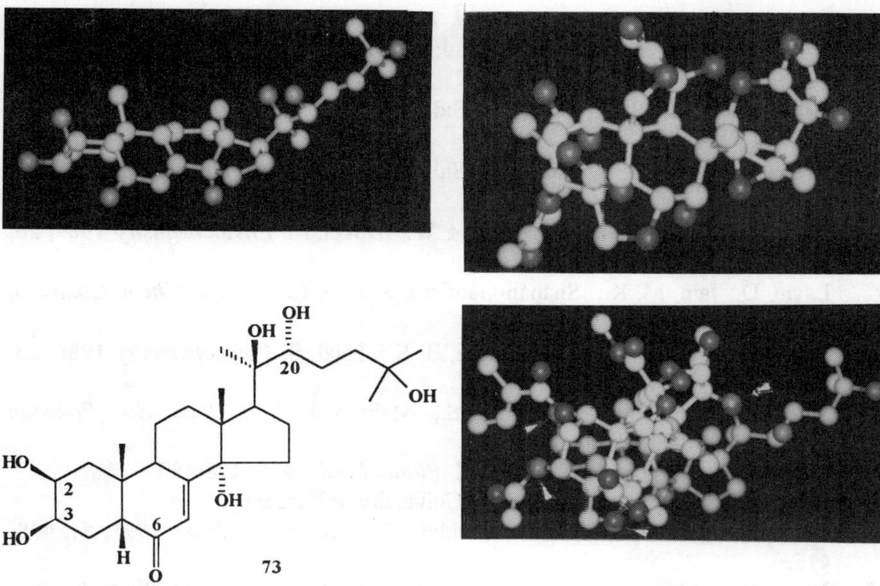

73

Molecular models were generated using Tripos Sybyl software, version 5.4. Hydrogens are not shown. Top left, 20-β-hydroxyecdysone (73), top right, azadirachtin, bottom right, superimposed. White arrows indicate oxygen atoms of both molecules that occupy similar positions. Starting clockwise from the top left corner distances are:

Azadirachtin	20-β-Hydroxyecdysone	Distance apart in Å
C-1	C-2	0.47
C-3	C-3	0.13
C-6	C-6	0.12
C-20	C-20	0.11

Figure 13. Three-dimensional molecular models of azadirachtin and ecdysone.

discussed. Further work is needed to more accurately determine additional mechanisms and reactions involved and to investigate reaction kinetics.

Literature Cited

1. Lee, S. M.; Klocke, J. A.; Barnby, M. A.; Yamasaki, R. B.; Balandrin, M. F. In *Naturally Occurring Pest Bioregulators*; ACS Symposium Series 449; ACS Washington, DC, 1991; pp. 293-304.
2. Rembold, H. In *Insecticides of Plant Origin;* Comstock, M. J., Ed.; ACS Symposium Series 387; ACS: Washington, DC, 1989; pp 150-163.
3. Saxena, R. C. In *Insecticides of Plant Origin;* Comstock, M. J., Ed.; ACS Symposium Series 387; ACS: Washington, DC, 1989; pp 110-135.

4. Schmutterer, H. *Annu. Rev. Entomol.* **1990,** *35*, 271-297.
5. Ara, I.; Siddiqui, B. S.; Faizi, S. Siddiqui, S. *Phytochemistry* **1989,** *28*, 1177-1180.
6. Ara, I.; Siddiqui, B. S.; Faizi, S.; Siddiqui, S. *J. Nat. Prod.* **1988,** *51*, 1054-1061.
7. Ara, I.; Siddiqui, B. S.; Faizi, S.; Siddiqui, S. *J. Nat. Prod.* **1989,** *52*, 1209-1213.
8. Majumder, P. L.; Maiti, D. C.; Kraus, W.; Bokel, M. *Phytochemistry* **1987,** *26*, 3021-3023.
9. Lavie, D.; Jain, M. K.; Shapan-Gabrieth, S. R. *J. Chem. Soc. Chem. Commun.* **1967,** 910-914.
10. Siddiqui, S.; Mahmood, T.; Siddiqui, B. S.; Faizi, S. *Phytochemistry* **1986,** *25*, 2183-2185.
11. Siddiqui, S.; Siddiqui, B. S.; Faizi, S.; Mahmood, T. *J. Chem. Soc. Pakistan* **1986,** *8*, 34-36.
12. Siddiqui, S.; Siddiqui, B. S.; Faizi, S. *Planta Med.* **1985,** 478-480.
13. Faizi, S., Ph.D., Dissertation, **1985**, University of Karachi.
14. Siddiqui, S.; Siddiqui, B. S.; Ghiasuddin; Faizi, S. *J. Nat. Prod.* **1991,** *54*, 408-415.
15. Jones, P. S.; Ley, S. V.; Morgan, E. D.; Santafianos, D. In *Phytochemical Pesticides;* Jacobson, M., Ed.; CRC Press: Boca Raton, FL., 1988; pp 19-45.
16. Taylor, D. A H. In *Chemistry and Chemical Taxonomy of the Rutales;* Waterman, P. G. and Michael F. Grundon, M. F., Eds..; Academic Press: London, 1983; pp 353-375.
17. Lavie, D.; Levy, E. C.; Jain, M. K. *Tetrahedron* **1971,** *27*, 3927-3939.
18. Connolly, J. D.; Handa, K. L.; McCrindle, R. *Tetrahedron Lett.* **1968,** 437-440.
19. Pachapurkar, R. V.; Kornule, P. M.; Narayanan, C. R. *Chem. Lett.* **1971,** 357-359.
20. Henderson, C. R.; McCrindle, R.; Malera, A.; Overton *Tetrahedron* **1968**, *24*, 1525-1528.
21. Yamasaki, R. B.; Klocke, J. A. *J. Agric. Food Chem.* **1987,** *35*, 467-471.
22. Kraus, W.; Cramer, R. *Chem. Ber.* **1981,** *114*, 2375-2381.
23. Bruhn, A.; Bokel, M.; Kraus, W. *Tetrahedron* **1984,** *25*, 3691-3692.
24. Siddiqui, S.; Faizi, S.; Siddiqui, B. S.; *Heterocycles* **1984,** *22*, 295-2988.
25. Govindachari, T. R.; Sandhya, G.; Raj, S. P. G. *J. Nat. Prod.* **1992,** *55*, 596-601.
26. Klenk, A.; Bokel, M.; Kraus, W. *J. Chem. Soc., Chem. Commun.* **1986,** 523-524.
27. Rembold, H. In *Economic and Medicinal Plant Research;* Academic Press: London, 1989; Vol. 3, pp 57-72.
28. Anderson, J. C.; Ley, S. V.; Santafianos, D.; Sheppard, R. N. *Tetrahedron* **1991,** *47*, 6813-6850.
29. Anderson, J. C.; Ley, S. V. *Tetrahedron Lett.* **1990,** *31*, 431-432, 3437-3440.
30. Ley, S. V; Santafianos, D.; Blaney, W. M.; Simmonds, M. S. J. *Tetrahedron Lett.* **1987,** *28*, 221-224.

31. Pflieger, D.; Muckenstrum, B. *Tetrahedron Lett.* **1987,** *28,* 1519-1522.
32. Nishikimi, Y.; Iimori, T.; Sodeoka, M.; Shibasaki, M. *J. Org. Chem.* **1989,** *54,* 3354-3359.
33. Brunetière, A. P.; Leclaire, M.; Bhatnagar, S.; Lallemand, J. Y.; Cossy, J. *Tetrahedron Lett.* **1989,** *30,* 341-344.
34. Jalali-Naini, M.; Lallemand, J. Y. *Tetrahedron Lett.* **1986,** *27,* 497-500.
35. Jauch, J.; Schurig, V. *Tetrahedron Lett.* **1991,** *32,* 4587-4690.
36. Kolb, H. C.; Ley, S. V. *Tetrahedron Lett.* **1991,** *32,* 6187-6190.
37. Ley, S. V.; Somovilla, A. A.; Broughton, H. B.; Craig, D.; Slawin, A. M. Z.; Toogood, P. L.; Williams, D. J. *Tetrahedron* **1989,** *45,* 2143-2164.
38. Brasca, M. G.; Broughton, H. B.; Craig, D.; Ley, S. V.; Somovilla, A. A.; Toogood, P. L .*Tetrahedron Lett.* **1988,** *29,* 1853-1856.
39. Chan, T. H.; Schwerdtfeger, A. E. *J. Org. Chem.* **1991,** *56,* 3294-3298.
40. Turner, C. J.; Tempesta, M. S.; Taylor, R. B.; Zagorski, M. G.; Termini, J. S.; Schroeder, D. R.; Nakanishi, K. *Tetrahedron* **1987,** *43,* 2789-2803.
41. Morgan, E. D.; Butterworth, J. H.; Percy, G. R. *J. Chem. Soc., Perkin 1* **1972,** 2445-2450.
42. Klocke, J. A.; Yamasaki, R. B. US Patent 5,001,149, March 19, 1991.
43. Barnby, M. A.; Yamasaki, R. B.; Klocke, J. A. *J. Econ. Entomol.* **1989,** *82,* 58-63.
44. Rembold, H. In *Advances in Invertebrate Reproduction 3*; Engles, W.; Clark, W. H., Jr.; Fischer, A.; Olive, P. J. W.; Went, D. F., Eds.; Elsevier: Amsterdam, New York, 1984; pp 481-491.
45. Rembold, H.; Forster, H.; Czoppelt, Ch.; Rao. P. J.; Sieber, K. -P. In *Natural Pesticides from the Neem Tree (Azadirachta indica A. Juss) and Other Tropical Plants;* Proc. 2nd Internatl. Neem Conf., (Rauischholzhausen, 1983); Schmutterer, H.; Ascher, K. R. S., Eds.; German Agency for Technical Cooperation: Eschborn, Germany, 1984; pp 153-162.
46. Ley, S. V.; Anderson, J. C.; Blaney, W. M.; Jones, P. S.; Lidert, Z.; Morgan, E. D.; Robinson, N. G.; Santafianos, D.; Simmonds, M. S. J.; Toogood, P. L. *Tetrahedron* **1989,** *45,* 5175-5192.
47. Ley, S. V.; Anderson, J. C.; Blaney, W. M.; Lidert, Z.; Morgan, E. D.; Robinson, N. G.; Simmonds, M. S. J *Tetrahedron Lett.* **1988,** *29,* 5433-5436.
48. Bilton, J. N.; Jones, P. S.; Ley, S. V.; Robinson, N. G.; Sheppard, R. N. *Tetrahedron Lett.* **1988,** *29,* 1849-1852.
49. Zanno, P. R.; Miura, I.; Nakanishi, K.; Elder, D. L. *J. Am. Chem. Soc.* **1975,** 97, 1975-1977.
50. Bilton, J. N.; Broughton, H. B.; Jones, P. S.; Ley, S. V.; Lidert, Z.; Morgan, E. D.; Rzepa, H. S.; Sheppard, R. N.; Slawin, A. M. Z.; Williams, D. J. *Tetrahedron* **1987,** *43,* 2805-2815.
51. Arigoni, D.; Barton, D. H. R.; Corey, E. J.; Jeger, O.; Caglioti, L.; Sukh Dev.; Ferrini, P. G.; Glazier, E. R.; Melera, A.; Pradhan, S. K.; Schaffner, K.; Sternhell, S.; Templeton, J. F.; Tobinaga, S. *Experientia* **1960,** *16,* 41-84.
52. Siddiqui, S.; Siddiqui, B.S.; Faizi, S.; Mahmood, T. *J. Nat. Prod.* **1988,** *51,* 30-43.
53. Jones, P. S.; Ley, S. V.; Morgan, E. D.; Santafianos D. In *Phytochemical*

Pesticides, Volume 1, The Neem Tree; Jacobson, M., Ed.; CRC Press, Boca Raton, FL; 1988; pp 19-46 (and references cited therein).
54. Connolly, J. D. In *Chemistry and Chemical Taxonomy of the Rutales, Annual Proceedings of the Phytochemical Society of Europe, Number 22*; Waterman, P. G.; Grundon, M. F. Eds.; Academic Press; London, 1983; pp 175-213.
55. Ekong, D. E. U.; Ibiyemi, S. A.; Olagbemi, E. O. *J. Chem. Soc., Chem. Commun* **1971**, 1117-1118.
56. Taylor, D. A. H. In *Fortschritte der Chemie Organischer Naturstoffe*, Herz, W.; Griseback, H.; Kirby, G. W. Eds.; Springer-Verlag, Vienna, 1984; pp 1-102.
57. Rajab, M. S.; Bentley, M. D.; Fort, R. C., Jr. *J. Nat. Prod.* **1988**, *51*, 1291-1293.
58. Kraus, W.; Bokel, M.; Bruhn, A.; Cramer, R.; Klaiber, I.; Klenk, A.; Nagl, G.; Pohnl, H.; Sadlo, H.; Vogler, B. *Tetrahedron* **1987**, *43*, 2817-2830.
59. *The Merck Index*; Budavari, S.; O'Neil, M. J.; Smith, A.; Heckelman, P. E., Eds.; Eleventh Edition; Merck & Co., Inc.: Rahway, N J, 1989; pp 142-143.
60. Huang, H. -P.; Morgan, E. D. *J. Chromatog.* **1990**, *519*, 137-143.
61. Blaney, W. M.; Simmonds, M. S. J. *J. Insect Physiol.* **1990**, *36*, 743-756.
62. Schoonhoven, L. M. In *Natural Pesticides from the Neem Tree (Azadirachta indica A. Juss). Proc. 1st Internatl. Neem Conf. (Rottach-Egern, 1980);* Schmutterer, H.; Ascher, K. R. S; Rembold. H. Eds.; German Agency for Technical Cooperation; Eschborn, Germany, 1982; pp 105-108.
63. Simmonds, M. S. J.; Blaney, W. M. In *Natural Pesticides from the Neem Tree (Azadirachta indica A. Juss) and Other Tropical Plants. Proc. 2nd Internatl. Neem Conf. (Rauischholzhausen, 1983)*; Schmutter, H.; Ascher, K. R. S. Eds.; German Agency for Technical Cooperation; Eschborn, Germany, 1984; pp 163-179.
64. Clayton, R. B. *J. Lipid Res.* **1964**, *5*, 3-19.
65. Koelle, M. R.; Talbot, W S.; Segraves, W A.; Bender, M. T.; Cherbas P.; Hogness, D. S. *Cell* **1991**, *67*, 59-77.
66. Jones, C. W.; Dalton, M. W.; Townley, L. H. *Genetics* **1991**, *127*, 535-543.
67. Thompson, C. B.; Wang, C. -Y.; Ho, I. -C.; Bohjanen, P. R.; Petryniak, B.; June, C. H.; Miesfeldt, S.; Zhang, L.; Nabel, G. J.; Karpinski, B.; leiden, J. M. *Mol. Cell. Biol.* **1992**, *12*, 1043-1053.
68. Urness, L. D.; Thummel, C. S. *Cell* **1990**, *63*, 47-61.
69. Kawakami, A.; Kataoka, H.; Oka, T.; Mizoguchi, A.; Kimura-Kawakami, M.; Adachi, T.; Iwami, M.; Nagasawa, H.; Suzuki, A.; Ishizaki, H. *Science* **1990**, *247*, 1333-1335.
70. Mizoguchi, J.; Oka, T.; Kataoka, H.; Nagasawa, H.; Suzuki, A.; Ishizaki, H. *Develop. Growth Different.* **1990**, *32*, 591- 598.
71. Shirai, Y.; Aizono, Y,; Iwasaki. T, *Appl. Entom. Zool.* **1991**, *26*, 335-342.
72. Spindler-Barth, M.; Turberg, A.; Spindler, K. D. *Arch. Insect Biochem. Physiol.* **1991**, *16*, 11-18.
73. Strangmann-Diekmann, M.; Klone, A.; Ozyhar, A.; Kreklau, F.; Kiltz, H. -H.; Hedtmann, U.; Welzel, P.; Pongs, O. *Europ. J. Biochem.* **1990**, *189*, 137-143.
74. Tanzubil, P. B.; McCaffery, A. R. *Entomol. Exp. Appl.* **1990**, *57*, 115-121.
75. Tanzubil, P. B.; McCaffery, A. R. *Crop Prot.* **1990**, *9*, 383-386.
76. Rembold, H.; Muller, T.; Subrahmanyam, B. *Z. Naturforsch.* **1988**, *43c*, 903-907.

77. Subrahmanyam, B.; Rembold, H. *Cell Tissue Res.* **1989**, *256*, 513-517.
78. Barnby, M. A.; Klocke, J. A. *J. Insect Physiol.* **1990**, *36*, 125-131.
79. Barnby, M. A.; Klocke, J. A. *J. Insect Physiol.* **1987**, *33*, 69-75.
80. Dorn, A.; Rademacher, J. M.; Sehn, E. *J. Insect Physiol.* **1986**, *32*, 231-238.
81. Redfern, R. E.; Kelly, T. J.; Borkovec, A. B.; Hayes, D. K. *Pest. Biochem. Physiol.* **1982**, *18*, 351-356.
82. Marco, M. -P.; Pascual, N.; Belles, X.; Camps, F.; Messeguer, A. *Pest. Biochem. Physiol.* **1990**, *38*, 60-65.
83. Burger, A.; Colobert, F.; Hetru, C.; Luu, B. *Tetrahedron* **1988**, *44*, 1141-1152.
84. Liebrich. W.; Dürnberger. B. B.; Hoffmann, K. H. *Comparat. Biochem. Physiol. A: Comparat. Physiol.* **1991**, *99*, 597-602.
85. Liebrich, W.; Hoffmann, K. H. *J. Comparat. Physiol. B: Biochem. Syst. Environ. Physiol.* **1991**, *161*, 93-99.
86. Kulcsár, P.; Darvas, B.; Brandtner, H.; Koolman, J.; Rees, H. H. *Experientia* **1991**, *47*, 261-263.
87. Keogh, D. P.; Smith, S. L. *Biochem. Biophys. Research Comm.* **1991**, *176*, 522-527.
88. Cassier, P.; Papillon, M. *Canadian J. Zool.* **1991**, *69*, 29-38.
89. Oberlander, H.; Silhacek, D. L.; Leach, E.; Ishaaya, I.; Shaaya, E. *Archives Insect Biochem. Physiol.* **1991**, *18*, 219-228.
90. Rembold, H. in *Focus on phytochemical pesticides. Vol I. The Neem Tree;* Jacobson, M. Ed.; CRC Press, Boca Raton, FL, 1988; pp 47-67.
91. Rembold, H.; Forster, H.; Czoppelt, C. In *Natural Pesticides from the Neem Tree (Azadirachta indica A. Juss) and Other Tropical Plants. Proc. 3rd Internatl. Neem Conf., (Nairobi, Kenya, 1986)*; Schmutterer, H.; Ascher, K. R. S., Eds.; German Agency for Technical Cooporation: Eschborn, Germany, 1987; pp 149-160.

RECEIVED April 30, 1993

Chapter 9

Ryanoid Chemistry and Action

Phillip R. Jefferies and John E. Casida

Pesticide Chemistry and Toxicology Laboratory, Department of Entomological Sciences, University of California, Berkeley, CA 94720

Ryanodine and 9,21-dehydroryanodine are the principal insecticidal components and toxicants among 11 identified ryanoids in the botanical insecticide ryania. Their biological activity is attributed to block of the calcium release channel, which is conveniently measured as inhibition of $[^3H]$ryanodine binding in muscle and nerve preparations. Structure-activity relationships based on 10 natural ryanoids and 45 of their derivatives help define the conformation of the ryanodine binding site and the structural requirements for insecticidal activity and selective toxicity.

Ryania the Botanical Insecticide

Ryania was recognized about 50 years ago as an insecticide (1) but was known much earlier as a toxicant for mammals (2,3). The current commercial insecticide is the ground stem-wood of Ryania speciosa Vahl (Flacourtiaceae) from Trinidad supplied by Agrisystems International (Windgap, PA). The roots are even more insecticidal but for reasons of convenience, conservation and regrowth the stem-wood is used. This genus is also found in the Amazon basin and adjacent parts of South America (3).

Ryania is effective against corn borer, codling moth, other lepidopterous insects, and pests of stored food (4). The extent and scope of use is limited by cost and supply. Ryania acts best in the field by ingestion causing flaccid paralysis and can be effective at a few grams of contained ryanoids per acre (5). Although the toxicity of ryania is enhanced by methylenedioxyphenyl compounds (6) it is used without addition of a synergist. Ryania is of moderate to low toxicity to mammals on ingestion (oral LD_{50} 150-2500 mg/kg) and it has low persistence (3,4). However, its chronic toxicity is only partially defined (4) and its detailed metabolism and environmental fate are unknown.

0097–6156/94/0551–0130$06.00/0

Natural Ryanoids and Their Degradation Products (Figure 1)

The natural ryanoids are conveniently isolated from ryania by wet chloroform extraction and rotary chromatography on silica gel with chloroform/methanol/aqueous methylamine followed by reverse phase HPLC with aqueous methanol (2,7,8). Ryanodine (**1**) and dehydroryanodine (**2**) are the major ryanoids, each making up 480-700 ppm (w/w) relative to the stem-wood, with seven other ryanoids contributing 10-64 ppm (w/w) each (8).

Ryanodine Series. Ryanodine (**1**) is an ester of ryanodol (**6**), a complex bridged diterpene heptol. More specifically, **1** is ryanodyl 3-(pyrrole-2-carboxylate). It was isolated in 1948 (2) and identified in 1967 (9,10). Structure elucidation involved a brilliant series of degradations based on ryanodol (**6**) formed from **1** with base, anhydroryanodine (**13**) with acid, and 4,12-seco,4,12-dioxoryanodine (**15**) with periodate (9,11). The structure of **6** was confirmed (except for the stereochemistry at C-2) by X-ray crystallography of the mono-p-bromobenzyl ether (which was found surprisingly at the 4-position) (10). Synthesis of **6** has been achieved by a multistage sequence proceeding through reductive cyclization of the 1,2-epoxide of the alcohol component of **13** with interesting use of carbonates, orthoacetates and orthocarbonates as protecting groups (12). The C-3 hydroxyl of **6** is very hindered so that esterification to **1** could not be effected. 3-Deoxyryanodol (**7**) is also a natural product discussed later. Dehydroryanodine (**2**), reported in 1984, usually exceeds **1** in amount but was overlooked in early studies because it is more soluble in the recrystallization solvents and was apparently lost in the purification steps (13-15). Minor compounds in the ryanodine series are 18-hydroxy-**1** (**3**) (14), 8-oxo-10-deoxy-9,10-dehydro-**1** (4) (7), and ryanodyl 3-(pyridine-3-carboxylate) (**5**) (16).

8_{ax}- or 9_{ax}-Hydroxy-10-epiryanodine Series. Compounds in the 8_{ax}-hydroxy series have the following substituents: 10-(O-methyl) (**8**) (7,17), 18-hydroxy-10-(O-methyl) (**9**) (17), 10-(O-methyl)-9,21-dehydro (**10**) (7,17) and 9,21-dehydro (**11**) (17,18). Another minor 10-epiryanoid is 9_{ax}-hydroxy compound **12** (7,17).

Anhydroryanodine and 4,12-Diketo Series. The ryania constituent 9-hydroxyanhydroryanodine (**14**) is closely related to **13**, the major acid-degradation product of **1** referred to above (7). Diketo compounds **15** and **16** are oxidation products of **1** and **2**, respectively, used in characterization and derivatization studies.

Cinncassiol (3-Deoxyryanodol) Series. Nonester ryanoids from genera of Lauraceae include **6** from Persea indica where it occurs together with cinnzeylanol (**7**) (19,20). Alcohol **7** and its 10-acetate occur naturally in Cinnamomum zeylanicum Nees (21) and C. cassia (Chinese cinnamon) (22) in the latter case along with sets of ryanoids with skeletons corresponding to either the anhydro series **13** (22) or the diketo type

Ryanodine series

1 ryanodine (480 ppm)
2 9,21-dehydroryanodine (700 ppm)
3 18-HO-1 (trace)
4 8-oxo-10-deoxy-9,10-dehydro-1 (trace) (B)

5 R = pyridine-3-carboxylate (58 ppm) (G)
6 ryanodol R=OH
7 cinnzeylanol R=H

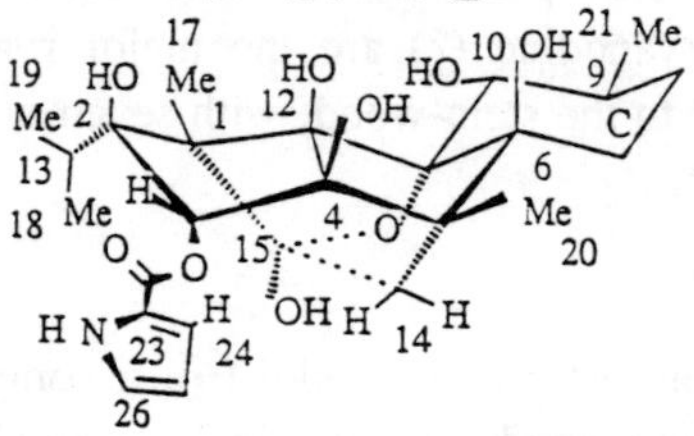

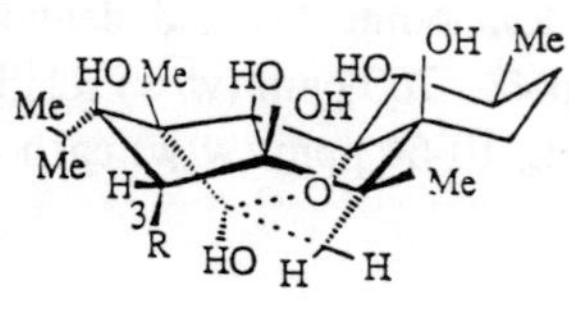

8ax- or 9ax-Hydroxy-10-epiryanodine series (C-ring modification of 1)

8 (33 ppm) (A)
9 18-HO-8 (10 ppm) (E)

10 R = Me (32 ppm) (D)
11 R = H (64 ppm) (F)

12 (17 ppm) (C1)

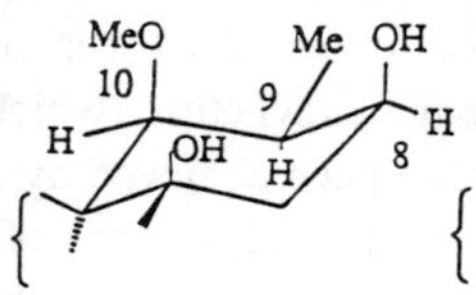

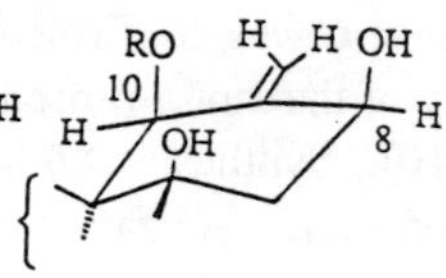

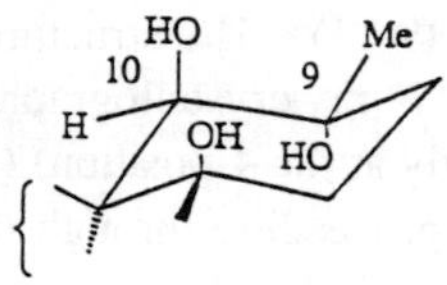

Anhydroryanodine and 4,12-diketo series (pyr=pyrrole-2-carboxylate)

13 R = H
14 R = OH (17 ppm) (C2)

15
16 9,21-ene

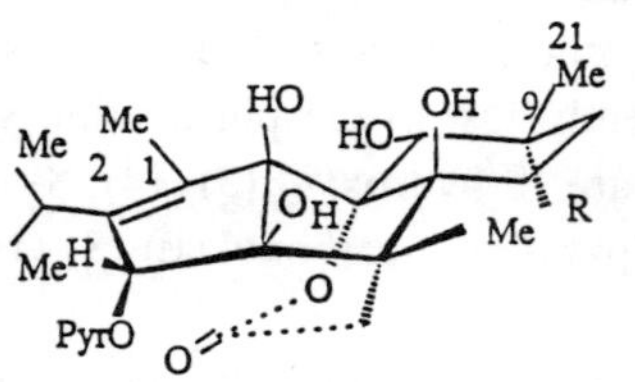

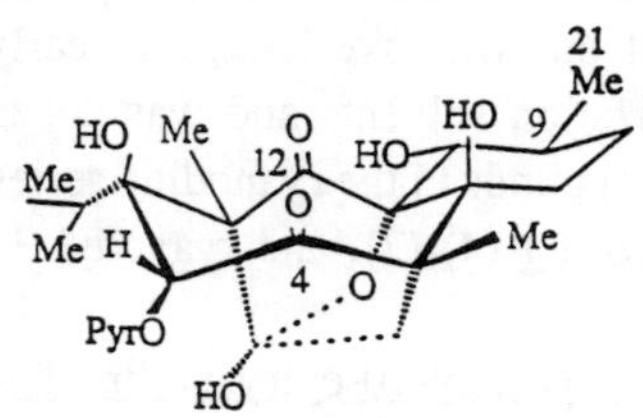

Cinncassiol series (other than 7)

17 Cinncassiol D1

18 Cinncassiol E

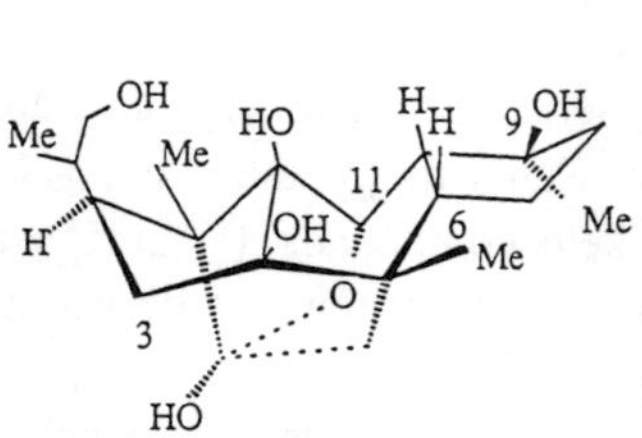

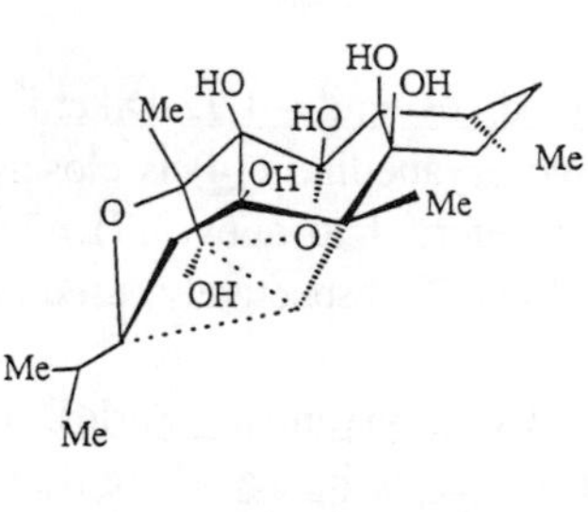

Figure 1. Natural ryanoids (**1–12**, **14**, **17**, and **18**) and their degradation products (**13**, **15**, and **16**). Ppm values or trace indicates content in a representative sample of ryania. **A–G** are literature designations for the natural products.

15 (23). In addition the cinncassiol-D series (e.g., 17) (24) and cinncassiol-E (18) (25) were obtained with modified ryanoid skeletons in which the C-ring is five membered. Details of the biosynthesis of ryanoids and cinncassiols are not known since, although they are clearly isoprenoid, cyclization of the hypothetical geranylgeranyl precursor involves linking some unusual positions (9).

NMR Spectroscopy. Ryanoids have relatively simple ^{1}H NMR spectra since most of the protons form isolated spin systems. For 2 the cyclohexane ring protons are well separated and easily identified from coupling patterns whereas in the 1 series these protons can usually be identified from COSY spectra which also serve to distinguish the secondary methyls. The carbon spectra are complicated by the large number of quaternary carbons with weak signals. By the use of double resonance techniques all the carbon shifts of 1 and 2 have been assigned (15,26) with important applications in structure elucidation (14,17). Lacking a crystal structure for a pyrrole ester, the conformation of this residue relies on spectral data. A conformation with H-bonding between the carbonyl and 4-hydroxyl accounts for the spectral data in chloroform but not in more polar solvents where nOe's are detected between H-24 of the pyrrole and the 20-methyl protons, an isopropyl methyl and the neighboring H-14 (Figure 1, compound 1). These contacts indicate that the carbonyl and H-3 are almost coplanar with H-24 near the cleft between the isopropyl and ketal groups (17).

Structural assignments for the other natural ryanoids are based partially on degradations but primarily on NMR assignments (7,14,17).

Derivatizations

Derivatives Based on Ryanodine (Figure 2). Bromination gives the tribromopyrrole derivative 19 (27,28). Alkylation occurs at the pyrrole for the butyl (20) and benzyl (21) derivatives (14). Diazomethane gives specifically **4-O-Me-1** (22) on treatment of its borate (29). Methylation with methyl iodide occurs at the pyrrole, the ketal hydroxyl and then hydoxyls at either C-4 or C-6 and perhaps C-10 (23-26) with structural assignments using the carbon resonances in the environment of the methoxyl groups and applying the well known differential shifts of the α and β carbons in relation to the parent alcohol (14). Acetylation using dicyclohexylcarbodiimide occurs at the C-10 hydroxyl to give 27 (14). Selective oxidation under Swern conditions gives the 10-ketone (28) (14) with greatly improved yields with excess reagent (29). Reduction of 28 with borohydride gives epimeric alcohol 29 as its borate from which it can be recovered with methanol (29).

Derivatives Based on Dehydroryanodine (Figure 3). Catalytic reduction of the vinyl group of 2 regenerates 1 and also gives **9-epi-1** (30) (7,13,15) with some rearrangement to isodehydroryanodine (31) (7) which is also formed in good yield using a Pd catalyst alone (29). Oxidation of 2 with the Swern reagent under the

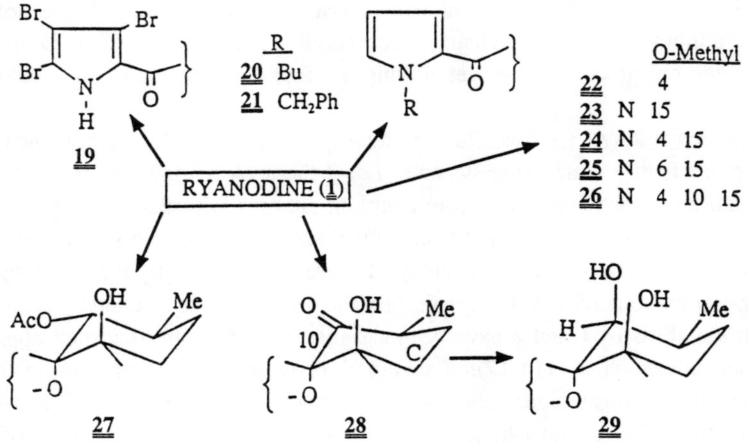

Figure 2. Derivatives of ryanodine pyrrole and cyclohexane rings.

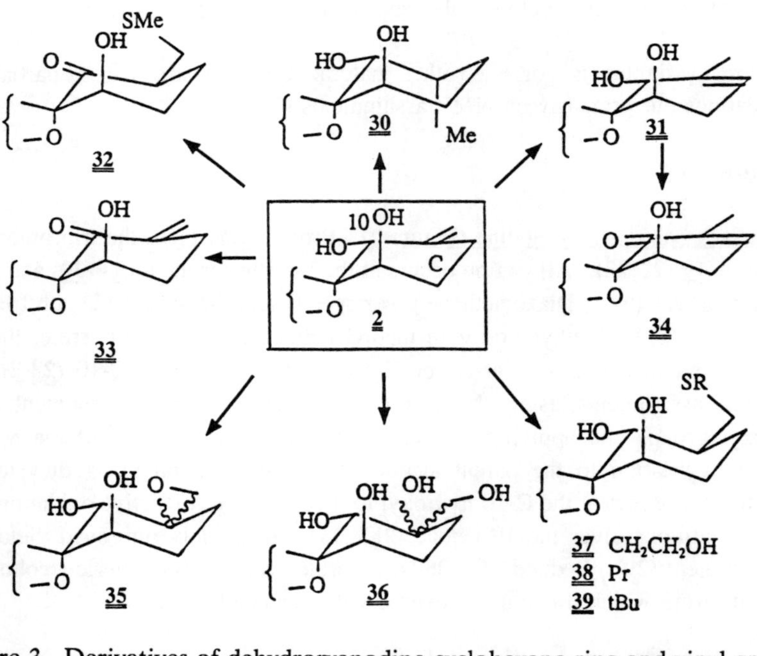

Figure 3. Derivatives of dehydroryanodine cyclohexane ring and vinyl group.

conditions originally reported gives much thiomethylketone **32** (29) although the unstable vinyl ketone **33** can be isolated when a molar portion of the reagent is used (29). Cyclohexenone **34** does not appear to be formed in this reaction but can be prepared in high yield by Swern oxidation of **31** (29). Functionalization of the vinyl group provides access to other C-ring derivatives such as epoxides **35** (14) and diols **36** (14). Thiols also add to the vinyl group such that thioethanol readily yields **37** and thioethers **38** and **39** are obtained with AIBN catalysis (30).

Derivatives Based on 4,12-seco-4,12-Dioxoryanodine (Figure 4). Our recent studies (30) establish that 4,12-diketo compound **15** with borohydride gives diol **40**, 4H-ketal **41** and 12H-ketal **42**. Nucleophilic addition products of **15** or **16** are **43** and **44** with hydroxylamine, **45** and **46** with methoxylamine, **47** with benzyloxylamine, **48** with hydrazine and its benzoyl derivative **49**, **50**-**52** with phenylhydrazine, and **53** and **54** with semicarbazide (30). More vigorous conditions with hydroxylamines give a series of oximes (**55**-**58**) two of which are reduced with sodium borohydride to hydroxylaminoketals **59** and **60** (30). Structural assignments are made on the basis of changes in chemical shifts in the region of C-4 and C-12 (30).

Radioligands and Photoaffinity Probes (Figure 5). [^{3}H]**1** of high specific activity has been prepared from **2** (13,28) or tribromopyrrole derivative **19** (28) by catalytic reduction with tritium. O_{10}-3- or 4-Substituted-benzyloxycarbonyl-β-alanyl derivatives with ^{125}I (31) or N_3 (32,33) are proposed alternative radioligands and photoaffinity probes, respectively. The latter conversions used the carbodiimide-mediated coupling first established with **27** (14).

Biological Activity

Calcium Release Channel. This channel mediates calcium release in excitation-contraction coupling in skeletal (voluntary) and cardiac (involuntary) muscle. The channel is localized in the triad junction or foot in the gap between the the terminal cisternae region of the sarcoplasmic reticulum and the transverse tubule (34). The native structure of the calcium release channel has been analyzed in two dimensions by electron microscopy (35). It consists of 4 identical protomers, each of 5037 amino acids with a known primary sequence (36-38) and emerging knowledge on the surface topography (39). The properties of the channel are characterized for mammalian skeletal muscle as above, cardiac muscle (40), brain (41,42) and liver (43). It is generally regulated by various effector molecules including calcium, magnesium and ATP and is also modulated by caffeine. The calcium release channel is implicated in malignant hyperthermia and possible death induced by inhalation anaesthesia in humans. A similar stress-induced abnormality in pigs, although simpler than the human syndrome, can be related to a single mutation of the calcium release channel (44).

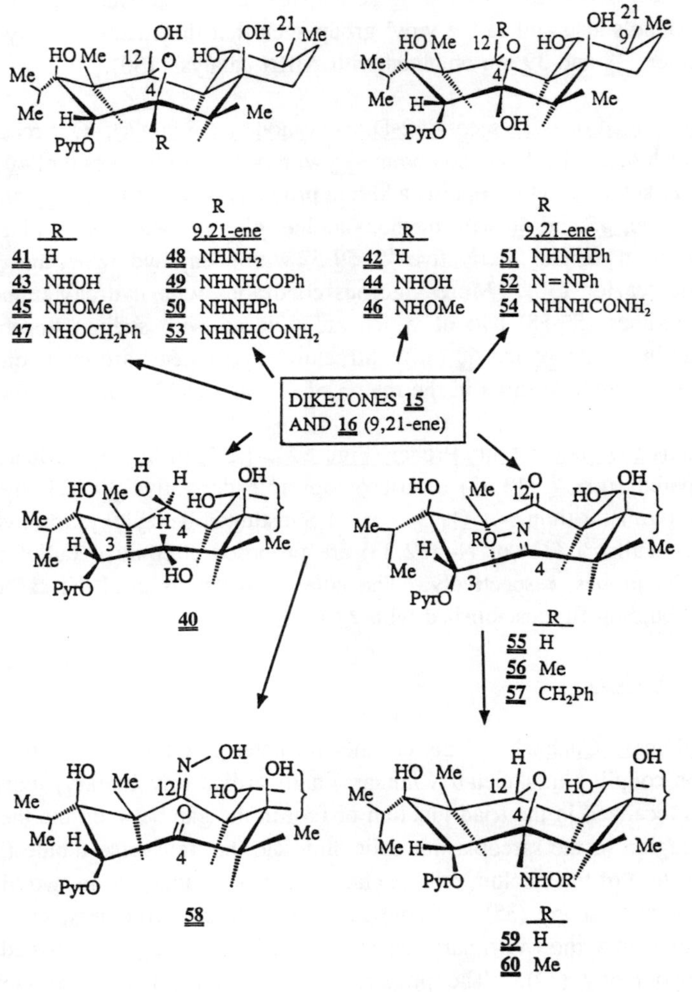

Figure 4. Partial structures showing derivatives based on diketones **15** and **16** leading to modifications in the skeleton.

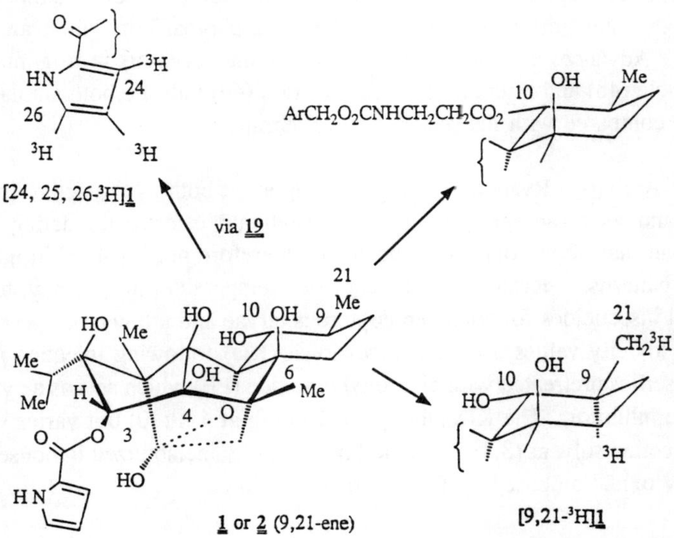

Figure 5. Derivatizations to form candidate radioligands and photoaffinity probes.

Ryanodine Receptor. The [^{3}H]ryanodine receptor, first reported in 1985 (45,46), is a region within the calcium release channel. Combination of 1 with this site leads to contracture or decline in contracture response depending on the muscle type and activity. [^{3}H]Ryanodine has been utilized in the isolation of the calcium release channel and in studying its varied pharmacological and toxicological interactions. The potency of ryanoids in blocking the [^{3}H]ryanodine receptor of rabbit skeletal muscle is a good predictor of their toxicity to mice, establishing the toxicological relevance of this assay (14,45). The IC_{50} of 1 in this assay is 10-12 nM (8,14,30). Ryanodine binds to four sites in the channel from cardiac muscle, with affinities ranging from nano- to micromolar concentrations and exhibiting negative cooperativity (40). The position of bound 1 in the receptor is not known and current information about the site comes largely from consideration of ryanoid structure-activity relationships discussed later. Studies with photoaffinity labels are essential in this area. Advances in understanding [^{3}H]ryanodine receptors in frog muscle (47), lobster muscle (48) and insect muscle and/or brain (49) indicate both similarities and differences compared with the receptor in vertebrates.

Insecticidal Activity. Ryanodine and 2 are important both as the active ingredients of ryania and as prototypes for studying candidate insecticides acting at a site different than that of any other insecticide and therefore not involved in many cross resistance patterns. Ryanoids 1 and 2 are comparable in potency to several commerical insecticides for some lepidopterous larvae and a few other pests. In this review the toxicity values used are knockdown (KD) following injection into adult female houseflies pretreated with piperonyl butoxide (PB) which acts as a cytochrome P_{450}-oxidase inhibitor. The KD_{50} for 1 is 0.19 μg/g at 4 h (30) but varies somewhat with the injection solvent (8,14). Ryanodine is rapidly metabolized in houseflies (14) probably by oxidation based on the synergism findings.

Selective Toxicity. A useful insecticide must be effective and safe, i.e. selectively toxic. Ryania as used provides a favorable selectivity (4). However, 1 is a very toxic compound with a mouse intraperitoneal (IP) LD_{50} of 0.1 mg/kg (14) and rabbit and dog intravenous LD_{50}s of 0.025 to 0.075 mg/kg, respectively (3). Selective toxicity has been examined comparing mouse IP LD_{50}s and housefly or cockroach injected KD_{50}s (with PB for the insects) with the finding that 1 is relatively nonselective, compound 6 is highly selective for insects versus mammals, and the analogs or derivatives may be more toxic to either houseflies or mice. It is not clear whether the selectivity is due to species-dependent differences in distribution, detoxification, or properties of the calcium release channel target.

Structure-Activity Relationships

Two systems are used here for evaluating structure-activity relationships (Table I) (8,14,29,30). The receptor data is for rabbit muscle sarcoplasmic reticulum and the insecticidal activity is for housefly KD with PB synergism. General trends based on both receptor and toxicity results are discussed below. Differences in the two systems can be discerned by inspection of the data. Although not tabulated, anhydroryanodine derivatives **13** and **14** are essentially inactive in both assays.

Effect of 3-Substituent. The pyrrolecarboxylate confers much higher potency than any other 3-substituent examined. The pyridinecarboxylate (**5**) is of very low activity. Hydrolysis to alcohol **6** destroys almost all receptor potency but retains much of the KD activity. 3-Deoxy-**6** (**7**) and its acetate are reported to be insecticidal (20,21). Bromination or alkylation of the pyrrole to give **19** or **20** greatly reduces activity. It is not possible as yet to esterify **6** at C-3 to examine other ester substituents.

Effect of 18-Hydroxyl Group. Comparing **3** with **1** and **9** with **8**, introduction of hydroxyl at C-18 greatly reduces potency.

Effect of Substitution of the C-Ring. Considerable structural variation is allowed in the C-ring with retention of biological activity. Introduction of the 9,21 double bond has little effect on activity (**2** vs **1** and **10** vs **8**). Oxygenation of the 9-position as in epoxide **35** or diol **12** reduces receptor potency much more than housefly toxicity and triol **36** is essentially inactive. Diol **11** has low receptor potency but high fly toxicity and on methylation to **10** much of the receptor potency is restored with similar values, as expected, for saturated analog **8**. Acetyl at the 10-position (**27**) greatly reduces potency. 10-Keto-**1** (**28**) and 10-epi-**1** (**29**) have high fly activity and good receptor potency. 9-Epi-**1** (**30**) retains 25% of the receptor potency of **1**. Isodehydroryanodine (**31**) and the corresponding 10-ketone (**34**) have relatively high insecticidal activity not shared by the thiomethylketone (**32**) analog of 10-ketoryanodine (**28**). Thiol adducts **37**-**39** are essentially inactive.

Effect of Alkylation. Alkylation at nitrogen or C-4 or conversion to di-, tri- or tetramethyl derivatives successively reduces activity (**21**-**26** vs **1**), in contrast to the enhanced receptor potency on methylation at C-10 converting **11** to **10**.

C-4 and C-12 Ketals (**41**-**54**, **59** and **60**). These compounds with a modified skeleton retain high potency when there is a polar group at C-4 and a small group at C-12, i.e. compounds **42**, **43**, **48** and **59**. Two of the adducts (**43** and **48**) dissociate during receptor assay.

C-4 and C-12 Diketones, Diol and Monooximes. Diketones **15** and **16** have only about 1-4% of the activity of **1** and **2**. Diol **40** has surprisingly low potency in view of its close relationship to ketal **42**. The 4-methoxime (**56**) and 12-oxime (**58**) are more potent than oximes **55** and **57**, respectively.

Table I. Ryanoid Structure–Activity Relationships for Inhibition of the Calcium Release Channel (Ryanodine Receptor) and for Knockdown of Houseflies Expressed as Potency Relative to Ryanodine

No.	Substituents				Potency relative to ryanodine = 100%: Receptor	Housefly
Effect of 3-substituent						
1	pyrrole-2-carboxylate				100	100
5	pyridine-3-carboxylate				0.7	12
6	hydroxyl				0.03	25
19	tribromopyrrole-2-carboxylate				<1	
20	N-butylpyrrole-2-carboxylate				<1	
Effect of 18-hydroxylation						
3	18-HO-1				11	
9	18-HO-8				<0.5	8
Effect of substitution of the C-ring						
	C-10	C-9	C-21	C-8		
2	OH	9,21-ene			110	90
8	OMe_{ax}			OH_{ax}	27	53
10	OMe_{ax}	9,21-ene		OH_{ax}	11	57
11	OH_{ax}	9,21-ene		OH_{ax}	1.8	53
12	OH_{ax}	OH_{ax}			2	26
27	OAc				7	18
28	=O				30	58
29	OH_{ax}				65	76
30	OH	epi			25	
31	OH	8,9-ene				65
32	=O		SMe			5
34	=O	8,9-ene				20
35	OH	epoxide			7	30
36	OH	OH	OH		2	<1
37	OH		SCH_2CH_2OH		<1	5
38	OH		SPr		<1	5
39	OH		StBu		<1	2

Table I.—*Continued*

Effect of alkylation							
	N	C-15	C-4	C-6	C-10		
21	CH_2Ph					3.2	
22			OMe			12	19
23	Me	OMe				3.6	<1
24	Me	OMe	OMe			<1	<1
25	Me	OMe		OMe		<1	<1
26	Me	OMe	OMe		OMe	<1	<1
C-4 and C-12 Ketals							
	C-4		C-12				
41	H		OH			28	27
42	OH		H			62	270
43	NHOH		OH			14*	270
44	OH		NHOH			9	20
45	NHOMe		OH			23	13
46	OH		NHOMe				2
47	$NHOCH_2Ph$		OH			13	2
48	$NHNH_2$		OH			11*	50
49	NHNHCOPh		OH			6	13
50	NHNHPh		OH			5	<2
51	OH		NHNHPh			2.6	<1.5
52	OH		N=NPh			4.7	6
53	$NHNHCONH_2$		OH			10	10
54	OH		$NHNHCONH_2$			6	<4
59	NHOH		H			50	200
60	NHOMe		H			20	7
C-4 and C-12 Diketones, Diol and Monooximes							
	C-4		C-12				
15	=O		=O			1.6	0.85
16	=O		=O	9,21-ene		0.8	4
40	H(OH)		H(OH)			5	
55	=NOH		=O			0.76	2.7
56	=NOMe		=O			0.75	10
57	$=NOCH_2Ph$		=O			1.3	1.0
58	=O		=NOH			1.7	8

*Dissociates in receptor assay.

Acknowledgements

This study was supported in part by National Institutes of Health grant PO1 ES00049. Some of this research was presented at the Symposium Natural and Derived Pest Management Agents, Division of Agrochemicals, American Chemical Society, Snowbird, Utah, August 1992.

Literature Cited

(1) Pepper, B.P.; Carruth, L.A. *J. Econ. Entomol.* **1945**, *38*, 59-66.
(2) Rogers, E.F.; Koniuszy, F.R.; Shavel, J., Jr.; Folkers, K. *J. Am. Chem. Soc.* **1948**, *70*, 3086-3088.
(3) Crosby, D.G. In *Naturally Occurring Insecticides*; Jacobson, M., Crosby, D.G., Eds.; Dekker: New York, 1971; pp 198-205.
(4) Penick, S.B. Ryania a Botanical Insecticide, *Tech. Bull.* **1953**, 18 pp.
(5) Casida, J.E.; Pessah, I.N.; Seifert, J.; Waterhouse, A.L. In *Pesticide Science and Biotechnology*; Greenlagh, R., Roberts, T.R., Eds.; Blackwell: Oxford, England, 1987; pp 177-182.
(6) Reed, J.P.; Filmer, R.S. *J. Econ. Entomol.* **1950**, *43*, 161-164.
(7) Ruest, L.; Taylor, D.R.; Deslongchamps, P. *Can. J. Chem.* **1985**, *63*, 2840-2843.
(8) Jefferies, P.R.; Toia, R.F.; Brannigan, B.; Pessah, I.; Casida, J.E. *J. Agric. Food Chem.* **1992**, *40*, 142-146.
(9) Wiesner, K. *Adv. Org. Chem.* **1972**, 8, 295-316.
(10) Srivastava, S.N.; Przybylska, M. *Acta Crystallogr.* **1970**, *B26*, 707-715.
(11) Kelly, R.B.; Whittingham, D.J.; Wiesner, K. *Can. J. Chem.* **1951**, *29*, 905-910.
(12) Deslongchamps, P.; Belanger, A.; Berney, D.J.F.; Borschberg, H.-J.; Brousseau, R.; Doutheau, A.; Durand, R.; Katayama, H.; Lapalme, R.; Leturc, D.M.; Liao, C.-C.; MacLachlan, F.N.; Maffrand, J.-P.; Marazza, F.; Martino, R.; Moreau, C.; Ruest, L.; Saint-Laurent, L.; Saintonge, R.; Soucy, P. *Can. J. Chem.* **1990**, *68*, 115-192.
(13) Waterhouse, A.L.; Holden, I.; Casida, J.E. *J. Chem. Soc., Chem. Commun.* **1984**, 1265-1266.
(14) Waterhouse, A.L.; Pessah, I.N.; Francini, A.O.; Casida, J.E. *J. Med. Chem.* **1987**, *30*, 710-716.
(15) Waterhouse, A.L.; Holden, I.; Casida, J.E. *J. Chem. Soc., Perkin Trans. 2* **1985**, 1011-1016.
(16) Jefferies, P.R.; Toia, R.F.; Casida, J.E. *J. Nat. Prod.* **1991**, *54*, 1147-1149.
(17) Jefferies, P.R.; Lam, W.-W.; Toia, R.F.; Casida, J.E. *J. Agric. Food Chem.* **1992**, *40*, 509-512.
(18) Humerickhouse, R.A.; Paschal, J.; Elvey, T.; Berry, D.; Besch, H.R., Jr. *Pharmacologist* **1989**, *31*, 185 (Program and Abstracts ASPET, Abstract 363).

(19) González-Coloma, A.; Hernandez, M.G.; Perales, A.; Fraga, B.M. *J. Chem. Ecol.* **1990**, *16*, 2723-2733.
(20) González-Coloma, A.; Cabrera, R.; Castañera, P.; Gutiérrez, C.; Fraga, B.M. *Phytochem.* **1992**, *31*, 1549-1552.
(21) Isogai, A.; Murakoshi, S.; Suzuki, A.; Tamura, S. *Agric. Biol. Chem.* **1977**, *41*, 1779-1784.
(22) Yagi, A.; Tokubuchi, N.; Nohara, T.; Nonaka, G.; Nishioka, I.; Koda, A. *Chem. Pharm. Bull* **1980**, *28*, 1432-1435.
(23) Kashiwada, Y.; Nohara, T.; Tomimatsu, T.; Nishioka, I. *Chem. Pharm. Bull.* **1981**, *29*, 2686-2688.
(24) Nohara, T.; Kashiwada, Y.; Murakami, K.; Tomimatsu, T.; Kido, M.; Yagi, A.; Nishioka, I. *Chem. Pharm. Bull.* **1981**, *29*, 2451-2459.
(25) Nohara, T.; Kashiwada, Y.; Nishioka, I. *Phytochem.* **1985**, *24*, 1849-1850.
(26) Krishnamurthy, V.V.; Casida, J.E. *Magn. Reson. Chem.* **1987**, *25*, 837-842.
(27) Fairhurst, A.S. *J. Labelled Compd.* **1971**, *7*, 133-136.
(28) Sutko, J.L.; Thompson, L.J.; Schlatterer, R.G.; Lattanzio, F.A.; Fairhurst, A.S.; Campbell, C.; Martin, S.F.; Deslongchamps, P.; Ruest, L.; Taylor, D.R. *J. Labelled Compd. Radiopharm.* **1986**, *23*, 215-222.
(29) Jefferies, P.R.; Casida, J.E. Unpublished results.
(30) Jefferies, P.R.; Lehmberg, E.; Lam, W.-W.; Casida, J.E. *J. Med. Chem.*, submitted.
(31) Mais, D.E.; Bowling, N.; Watanabe, A.M. *Biochem. Biophys. Res. Commun.* **1992**, *183*, 462-467.
(32) Gerzon, K.; Bidasee, K.R.; Besch, H.R., Jr.; Humerickhouse, R.; Ruest, L.; Sutko, J.L. Presented at the National Amer. Chem. Soc. Meeting, San Francisco, CA, April, 1992; Abstract 138.
(33) Gerzon, K.; Besch, H.R.; Humerickhouse, R.A. Presented at Regional Amer. Chem. Soc. Meeting, Indianapolis, IN, May 1991; Abstract 342.
(34) Fleischer, S.; Inui, M. *Annu. Rev. Biophys. Biophys. Chem.* **1989**, *18*, 333-364.
(35) Radermacher, M.; Wagenknecht, T.; Grassucci, R.; Frank, J.; Inui, M.; Chadwick, C.; Fleischer, S. *Biophys. J.* **1992**, *61*, 936-940.
(36) Takeshima, H.; Nishimura, S.; Matsumoto,T.; Ishida, H.; Kangawa, K.; Minamino, N.; Matsuo, H.; Ueda, M.; Hanaoka, M.; Hirose, T.; Numa, S. *Nature* **1989**, *339*, 439-445.
(37) Marks, A.R.; Tempst, P.; Hwang, K.S.; Taubman, M.B.; Inui, M.; Chadwick, C.; Fleischer, S.; Nadal-Ginard, B. *Proc. Natl. Acad. Sci. USA* **1989**, *86*, 8683-8687.
(38) Zorzato, F.; Fujii, J.; Otsu, K.; Phillips, M.; Green, N.M.; Lai, F.A.; Meissner, G.; MacLennan, D.H. *J. Biol. Chem.* **1990**, *265*, 2244-2256.
(39) Marks, A.R.; Fleischer, S.; Tempst, P. *J. Biol. Chem.* **1990**, 265, 13143-13149.

(40) Pessah, I.N.; Zimanyi, I. *Mol. Pharmacol.* **1992**, *39*, 679-689.

(41) McPherson, P.S.; Campbell, K.P. *J. Biol. Chem.* **1990**, *265*, 18454-18460.

(42) McPherson, P.S.; Kim, Y.-K.; Valdiva, H.; Knudson, C.M.; Takekura, H.; Franzini-Armstrong, C.; Coronado, R.; Campbell, K.P. *Neuron* **1991**, *7*, 17-25.

(43) Shoshan-Barmatz, V. *FEBS Lett.* **1990**, *263*, 317-320.

(44) MacLennan, D.H.; Phillips, M.S. *Science* **1992**, *256*, 789-794.

(45) Pessah, I.N.; Waterhouse, A.L.; Casida, J.E. *Biochem. Biophys. Res. Commun.* **1985**, *128*, 449-456.

(46) Fleischer, S.; Ogunbunmi, E.M.; Dixon, M.C.; Fleer, E.A.M. *Proc. Natl. Acad. Sci. USA* **1985**, *82*, 7256-7259.

(47) Lai, F.A.; Liu, Q.-Y.; Xu, L.; El-Hashem, A.; Kramarcy, N.R.; Sealock, R.; Meissner, G. *Am. J. Physiol.* **1992**, 263 (Cell Physiol. 32), C365-C372.

(48) Seok, J.-H.; Xu, L.; Kramarcy, N.R.; Sealock, R.; Meissner, G. *J. Biol. Chem.* **1992**, 267, 15893-15901.

(49) Lehmberg, E.; Casida, J.E. Unpublished results.

RECEIVED April 30, 1993

Chapter 10

Natural Products as Sources of Potential Agrochemicals

Muraleedharan G. Nair

Bioactive Natural Products Laboratory, Department of Horticulture and Pesticide Research Center, Michigan State University, East Lansing, MI 48824

Bioassay-directed isolation, purification and structure elucidation resulted in several natural products from plants and microbes with potential agricultural application. Compounds described include the fungicides faerifungin and the spartanamicin.

The agricultural and medicinal effects of many natural products from plants and microbes have long been known and used in our day-to-day life. Even today, more than 75% of the world population relies on plants and plant extracts (1). There are a number of useful agrochemicals and prescription drugs worldwide that are derived from plants and microbes. In spite of this natural treasure available to man, many agrochemicals and pharmaceuticals available for consumer use are synthetic compounds and some pose a serious threat to our environment and community. It is also important to note that many of these agricultural and pharmaceutical chemicals are showing their ineffectiveness in pest management and medicine due to resistance. The natural products decompose rapidly and they pose very little threat to our environment.

Several approaches exist to discover new agrochemicals. These include (a) natural product models, (b) biochemical modification of known agrochemical, (c) biochemical synthesis and screening, (d) empirical synthesis and screening and (e) allelochemical approach. The Bioactive Natural Products Laboratory in the Department of Horticulture at Michigan State University is actively involved in the discovery and identification of biologically active natural products from plants and microbes for agricultural and pharmaceutical applications. These compounds may provide useful templates to produce more active agrochemical and pharmaceutical products with less environmental and human risks.

0097–6156/94/0551–0145$05.00/0

BIOASSAYS

Since bioassays are the important preliminary step in the discovery of active natural products some of the routine bioassays carried out for the isolation and fractionation of bioactive compounds are:

Antifungal bioassay. Known amounts of the pure test compounds were dissolved in DMSO and serial dilutions prepared in the same solvent. A 20 μl aliquot of each solution was mixed with 2 ml of Emmons liquid medium seeded with ca. 2 X 10^3 CFU/ml of the test organism. The inoculated tubes were vortexed and incubated at 26°C. Similarly, inoculated tubes without test compounds served as controls. Depending on the growth characteristics of the test species, results were recorded after 2-4 days. The lowest concentration of the test compound that totally inhibited growth of test organism was recorded as the minimum inhibitory concentration (MIC) for that species.

Antibacterial assay. The antibacterial activity of all test compounds was evaluated by the same procedure as in the antifungal assay except that Mueller-Hinton broth was used as the medium and the test organism inoculum was 10^4 CFU each. The inoculated tubes containing test compounds were incubated at 37°C for 24 h and scored for growth of each test organism. The MIC for each species represents the lowest concentration of the test compound at which complete inhibition of growth occurred.

Insecticidal assay. The bioassay for insecticidal properties was conducted on 4th instar mosquito larvae, *Aedes aegypti*, reared from the mosquito eggs (University of Davis California Straw, courtesy of Dr. David Grant) and on brine shrimp, *Artemia salina* Leach (obtained locally). For the mosquitocidal assay, 10 larvae were placed in 975 μl distilled water and 25 μl of test compounds in DMSO were added and mixed at room temperature. The number of dead larvae was recorded at 2, 4 and 24 h intervals. The control tube containing 10 larvae received 25 μl of DMSO alone and mortality was recorded as in the case of test compounds.

Brine shrimp assay. The eggs of brine shrimp were placed in artificial sea water prepared by dissolving 38 g of sea salt/L of water and left at 24°C for 48 h. The larvae were then transferred into test tubes containing sea water and the experiment was conducted as with the mosquito larvae.

Herbicidal assays. Phytotoxic activity was evaluated in terms of seedling growth inhibition. Stock solutions (1 mg/ml methanol) of purified compounds were applied to Whatman #1 filter paper in plastic petri dishes (60-mm x 15-mm) to obtain doses ranging from 6.25 to 200 g/dish. Pure methanol was also added to a control plate. The plates were left open to allow the methanol to evaporate completely (20 min). Ten indicator seeds (curly cress and barnyardgrass) were then placed on the filter paper and distilled water (1.5 ml) was added. The plates were wrapped with parafilm to prevent drying and incubated in the dark (26°C, for 72 hr). After incubation, seedling root length was measured and compared to the control. The experiments were designed as randomized complete blocks, with three replicates. Post- and pre-emergence assays with weed and crop species were conducted in the greenhouse using 2 weeks old velvet leaf, cress, lettuce, redroot pigweed, sudan grass, corn and soybean grown in styrofoam trays. The plants were treated with the test compounds at two rates of 5 and 10 lbs ha^{-1} as spray applications. The pre-emergence assays were conducted as sprays on the surface of the soil with the same species of plants at the same rates two days after planting the seeds.

Nematicidal assay. Nematicidal activity was carried out on *Panagrellus redivivus*, *Caenorhabditis elegans* (cultures purchased from Dr. Bert Zuckerman, University of Massachusetts, Amherst and maintained at MSU) and *Heterodera glycines* (MSU culture). Both *P. redivivus* and *C. elegans* were reared in vials containing 5 ml of NG-medium [NaCl 3g, bacto peptone 2.5g and phosphate buffer 25 ml (11.968 g/88ml of KH_2PO_4, 2.088g/12 ml of K_2HPO_4, cholesterol, 1 ml containing 5mg/ml in EtOH, 1 ml of 1 M $CaCl_2$, 1 ml of 1M $MgSO_4$) /L]. *H. glycines* cultures were prepared by grinding soybean cysts in saline and used immediately. The aliquots of the nematode suspension (45 μl) containing 30-50 nematodes at various developmental stages were transformed into each well (0.7 cm diameter x 1.0 cm deep) of a 96-well Corning flat-bottomed tissue culture plate. Test compounds (5 μl) was added to each well and mixed gently. The inoculated plates were kept at ca. 100% relative humidity. The nematodes were examined with an inverted microscope at 40X for mortality and mobility at 4, 24 and 48h.

MICROBIAL PRODUCTS

Many microorganisms, especially actinomycetes, have been investigated for crop and animal protection. Several antibiotics discovered earlier as antimicrobial agents are now known to have toxic effects on many forms of life. For example, valinomycin produced by *Streptomyces fulvissimus* and *S. roseochromogenes* was first discovered as an antibacterial substance (2), but subsequent investigations have revealed its broad spectrum insecticidal, nematicidal and miticidal activities (3). Among the leading herbicides from actinomycetes are cycloheximide, geldanamycin, nigericin, bialaphos, herbimycins, and herbicidins (4-9).

Many polyene macrolide antibiotics are produced by a variety of soil borne Streptomyces spp. (10,11). One of these, mycoticin, a 1:1 mixture of mycoticins A and B, is produced by *Streptomyces ruber* (ATCC 3348) and was first described by Burke et al. (12) and later characterized by Wasserman and co-workers (13). Flavofungin, obtained from *Streptomyces flavofungini*, was found to be a 9:1 mixture of mycoticins A and B and showed different biological properties from mycoticin (14-16). Mycoticin, for example, showed only antifungal activities whereas flavofungin exhibited antifungal and antiviral activity (15). Neither compound showed antibacterial activity.

Antifungal and Nematicidal Compounds from *Streptomyces griseus*

Streptomyces griseus var. autotrophicus, MSU-32058/ATCC 53668, was isolated from the soil sample of a fairy ring in an old lawn in Lansing, Michigan. Cultures of *S. griseus* var. autotrophicus were grown in 2 L baffle-bottomed Erlenmeyer flasks, each containing 400 ml of A-9 medium (peptone 4 g, glucose

10 g, 'Brer Rabbit' green label molasses 20 g, distilled H_2O 1 L). The inoculated flasks were placed on a rotary shaker at 150 rpm at 26°C for 7 days. Larger batches were grown in a 130 L fermenter containing 100 L of A-9 medium, aerated at 100 L/min and stirred at 100 rpm at 26°C for 5 days. The culture broth was centrifuged and the mycelia extracted with MeOH. The MeOH extract was concentrated and the faeriefungin (**1**) (yellow needle-like crystals) was filtered off. The mother liquor was dried under reduced pressure and the resulting crude extract was purified by chromatography to afford compounds **2**, **3** and **4**.

Faeriefungin (**1**), a polyol polyene macrolide lactone antibiotic (17) is an inseparable 1:1 mixture of faeriefungin A, $C_{36}H_{58}O_{10}$ and faeriefungin B, $C_{37}H_{60}O_{10}$. It is an isomer to mycoticin and flavofungin, but possesses different physiochemical and biological properties (17-19). Preliminary cytotoxicity studies with human erythrocytes and rat liver epithelial cells indicated that faeriefungin and amphotericin B have comparable toxicity. ^{1}H-NMR studies indicated that faeriefungin is a mixture of two compounds, **A** and **B**, and they differ in the attachment of a H or a CH_3 at C_{33} (17). *Aspergillus*, *Fusarium*, *Microsporum*, *Trichophyton*, and *Alternaria* spp. were completely inhibited by faeriefungin at 3.2 ug/ml, *Candida* spp. at 5.5 ug/ml and *Pythium*, *Phialophora*, and *Leptosphaeria* spp. and some selected gram negative penicillin-resistant strains of *Neiseria gonorrhaeae* were inhibited at 16.0 ug/ml. At a concentration of 100 ppm, faeriefungin caused 100% mortality of mosquito larvae (*Aedes aegypti*, Rockfeller strain) and free-living nematodes (*Panagrellus redivivus*) (Table I). Faeriefungin showed good activity against asparagus (*Asparagus officinalis* L.) pathogens *Fusarium oxysporum* and *Fusarium moniliforme* under greenhouse conditions (19). Unlike the related polyene macrolides, faeriefungin is crystalline and stable with broad spectrum antimicrobial and insecticidal activity.

Compounds **2-4** (patent application for **2**, **3**, and **4** has been filed, and therefore structures are not available) showed potent nematocidal and mosquitocidal activities. These compounds are chemically characterized aromatic nitro compounds and the structural details will be published elsewhere soon. Compounds **2**, **3** and **4** showed 100 % mortality between 0.1 and 1 μg/ml concentrations when tested on nematodes, *Panagrellus redivivus*, *Caenorhabditis elegans* and *Heterodera glycines in vitro* at 24 h (Table II). Mosquitocidal assay on *Aedes aegyptii* gave 100% mortality at 6.25 μg/ml for compounds **2** and **4** while 62.5 μg/ml gave 100% mortality for compound **3**. This is the first report of the nematicidal and mosquitocidal activities for compounds of this nature.

Antifungal Compounds from *Micromonospora spartanea*

Spartanamicins A (**5**) and B (**6**), two antifungal antibiotics, were produced by a culture of *Micromonospora spartanea* strain No. MSU-43907 (ATCC 53803), isolated from a potted soil containing asparagus (*Asparagus officinalis* L.) plants (20). *M. spartanea* was fermented in YMG (yeast extract 4 g/l, maltose 10 g/l, glucose 4 g/l) medium for one week. The fermentation broth was centrifuged and the mycelial cake was extracted with $CHCl_3$-MeOH (4:1, v/v) mixture (20). No

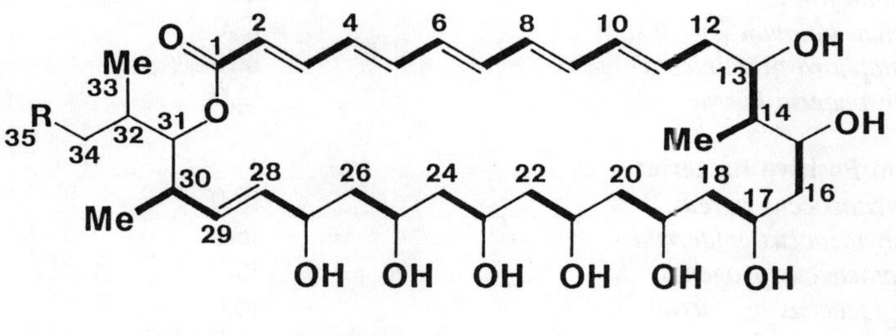

1 **Faeriefungin** A R = H
B R = CH_3

Table I. Biological activity of faeriefungin

Organism	MIC (μg/ml)
Fungi:	
Aspergillus fumigatus	3.2
A. flavus	3.2
A. niger	3.2
Candida albicans	5.5
C. tropicalis	5.5
Cryptococcus neoformans	5.5
Microsporum canis	3.2
Trichophyton rubrum	3.2
Alternaria solani	3.2
Fusarium oxysporum	3.2
F. moniliforme	3.2
Pythium ultimum	12.0
Phialophora graminicola	12.0
Leptosphaeria korrae	12.0
Gram Positive Bacteria:	
Staphylococcus aureus	16.0
Staphylococcus epidermidis	16.0
Streptococcus pyogenes	16.0
Streptococcus agalactiae	16.0
Streptococcus faecalis	16.0-32.0
Streptococcus pneumoniae	16.0
Listeria monocytogenes	32.0
Gram Negative Bacteria:	
Escherichia coli	n.a.
Enterobacter aerogenes	n.a.
Klebsiella pneumoniae	n.a.
Proteus mirabilis	n.a.
Pseudomonas aeruginosa	n.a.
Neisseria gonorrhoeae	16.0-64.0
Neisseria meningitidis	32.0-64.0
Branhamella catarrhalis	64.0
Haemophilus influenzae	64.0-128.0
Insects:	
Aedes aegypti (L.C.$_{50}$)	100
Panagrellus redivivus (L.C.$_{50}$)	100

SOURCE: Reprinted with permission from Reference 17. Copyright 1989.

Table II. Nematicidal and mosquitocidal activities of compounds **2-4**

Organism	LD_{100} in $\mu g/ml$					
	2		**3**		**4**	
	2h	24h	2h	24h	2h	24h
Panagrellus redivivus	1	0.1	1	0.1	5	0.1
Caenorhabditis elegans	1	0.1	1	0.1	5	0.1
Heterodera glycines	1	0.1	1	0.1	5	0.1
Aedes egyptii (mosquito)	*	6.25	*	6.25	*	6.25

* mortality was not recorded

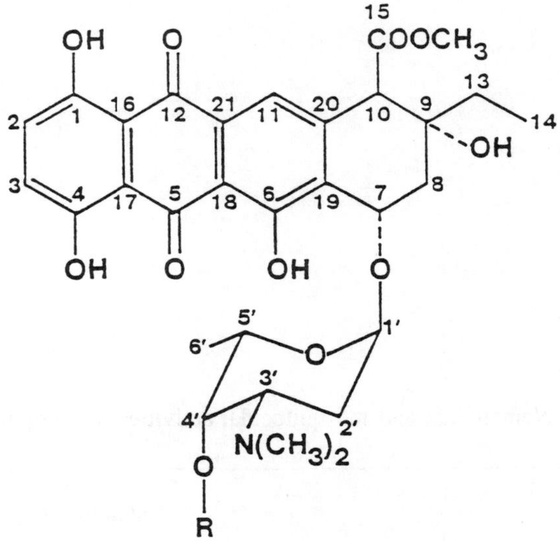

5 Spartanamicin B R =

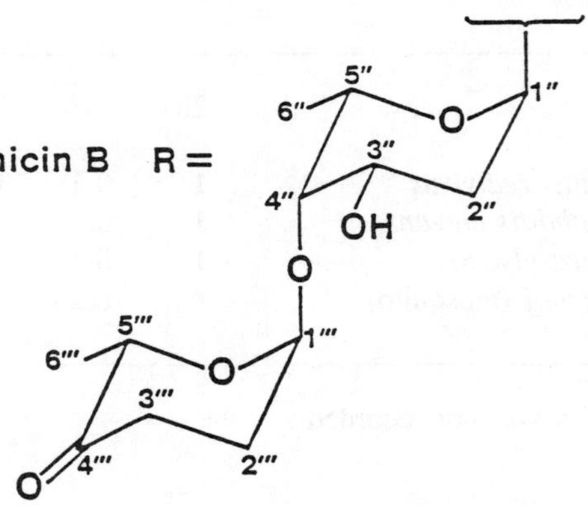

6 Spartanamicin A R =

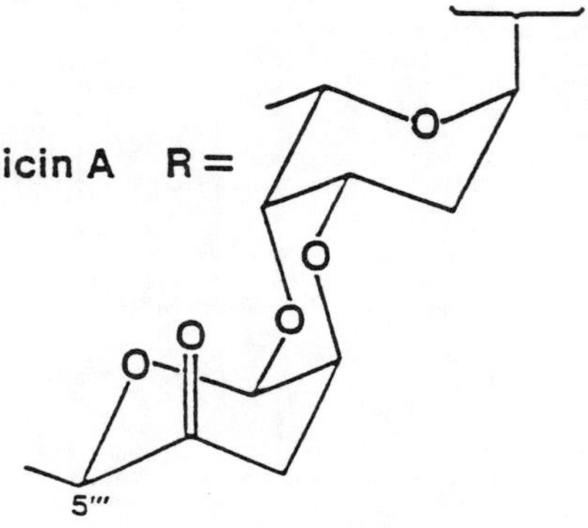

antibiotic was detected in the cell free fermentation broth. The crude extract was analyzed by TLC plates and after marking the respective spots, the plates were incubated with *C. albicans*, *A. fumigatus* and *Cladosporium* spp. Only two zones of inhibition were seen on the plate. The high Rf compound was less active than the lower Rf one. Further analysis of these active bands indicated that these compounds as spartanamicin A and B, for the high and low Rf, respectively. Since spartanamicin B was found to be much more active than it's analogue A, further study of biological activity was carried out only on spartanamicin B.

The minimum inhibitory concentration for spartanamicin B on *Candida albicans* and *Aspergillus, Cladosporium, Cryptococcus, Rhodotorula* and *Staphylococcus* spp. ranged from 0.2 to 1 μg/ml (Table III). It was not active against the *Staphylococcus aureus*, *Escherichia coli* and *Citrobacter* spp. but some strains of *S. aureus* were sensitive.

Spartanamicin A, **5**, an orange amorphous powder $C_{42}H_{51}O_{16}N$ with mp. 174-176°C has the same anthracycline aglycone moiety as spartanamicin B, **6**, a red orange amorphous solid, $C_{42}H_{54}O_{16}N$ with mp. 159-161°C. Structures of both compounds were elucidated by spectral and chemical means (20).

PLANT PRODUCTS

Magnolia virginiana

The plant family Magnoliaceae is well known for its host plant patterns of certain insect groups (21). To identify ecologically significant plant compounds from members of this family, we have examined the phytochemistry of *Magnolia virginiana* L. This tree is native to the eastern and southeastern United States and has been introduced to other parts of the world as an ornamental (22). *Magnolia* species have been investigated for compounds possessing pharmacological, antimicrobial, and pesticidal activities [23-27]. Antimicrobial activity of phenolic constituents of *M. grandiflora* L. has been reported earlier (26). Other species of *Magnolia* investigated for antimicrobial and pharmacologically active compounds are *M. officinalis*, *M. liliflora* and *M. obovata* (23-25, 27). Similar antimicrobial neolignans have been reported from the roots of *Sassafras randaiense* (28). 4',5-diallyl-2-hydroxy-3-methoxybiphenyl ether has been reported from the bark of *M. henryi* (29) and a monohydroxy ether, 4',5-diallyl-2-hydroxybiphenyl ether from *S. randaiense* (28). Magnolol, was previously isolated from the seeds of *M. grandiflora* (29) as well as from the bark of *M. henryi* (30).

Anti-microbial and insecticidal bioassay-directed work on *M. virginiana* leaves resulted in the characterization of three active compounds, a novel biphenyl ether, 4,4'-diallyl-2,3'-dihydroxybiphenyl ether (**7**), 3,5'-diallyl-2'-hydroxy-4-methoxybiphenyl (**8**) and 5,5'-diallyl-2,2'-dihydroxybiphenyl (**9**) (31). All three compounds and their methoxy analogues, **10-12**, respectively, were very toxic to brine shrimp and mosquito larvae and showed strong anti-fungal and anti-bacterial

Table III. Minimum inhibitory concentration (MIC) of Spartanamicin B, (6), against certain bacteria and fungi

Organisms	MIC (μg/ml)
Aspergillus fumigatus	0.4
A. niger	0.6
A. flavus	0.6
Penicillium spp.	0.2
Cladosporium spp.	0.4
Candida albicans	0.2
C. quilliermondii	0.8
Cryptococcus neoformans - N-2	0.8
C. neoformans- N-3	0.6
C. neoformans- G-3	1.0
C. neoformans serotype-C	0.8
C. neoformans serotype-D	0.8
Rhodotorula rubra	0.8
R. glutinis	0.4
*Citrobacter species**	not active
*Pseudomonas aeruginosa**	100
*Klebsiella pneumoniae**	100
*Escherichia coli**	not active
Staphylococcus aureus ATCC 29213*	0.8
S. aureus ATCC 25923*	0.8
S. aureus (multiple drug resistant)*	not active

N = encapsulated, mucoid strains with neurotropic tendencies.
G = non-encapsulated, dry-pasty, highly virulent strain.
* = Bacterial strains; all others are fungi.

SOURCE: Reprinted with permission from Reference 20. Copyright 1992.

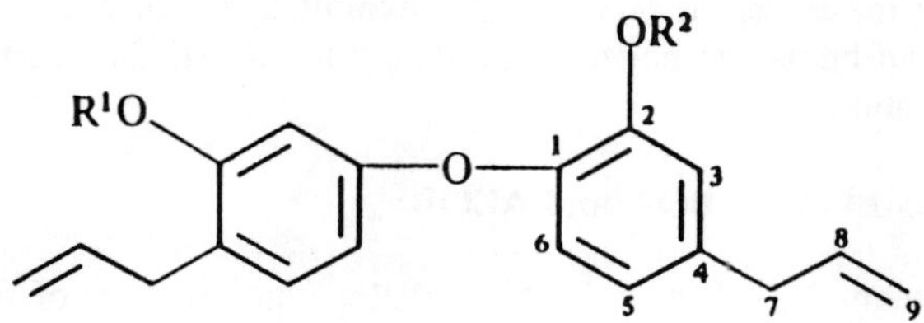

7 R^1 = H, R^2 = H
10 R^1 = Me, R^2 = H
11 R^1 = Me, R^2 = Me

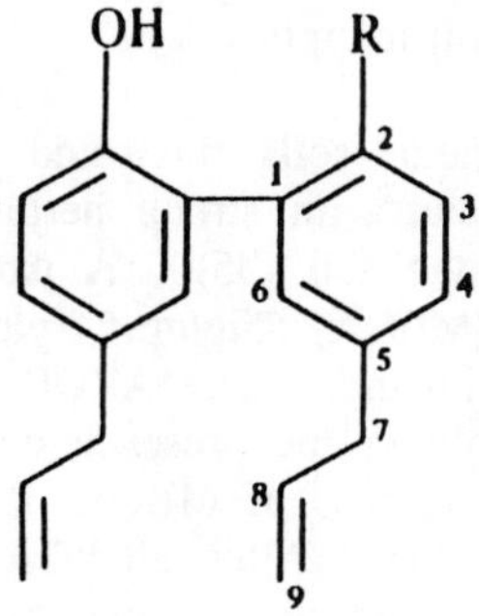

9 R = OH
12 R = OMe

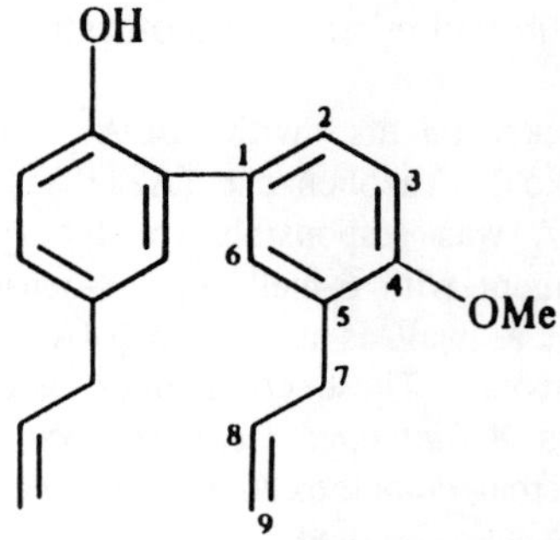

8

activities (Table IV). The biphenyl ether, compound **7**, and its methylated product showed slightly better broad spectrum activity (Table IV) than compounds **11** and **12**. All test compounds showed similar toxicity to mosquito larvae and brine shrimp. At 100 ppm concentration, compounds **7-12** gave 100% mortality for both mosquito and brine shrimp larvae within 30 min. which was similar to the control compound, valinomicin. At 10 ppm concentration, compounds **7** and **12** gave 100 % mortality to both test species within 2 h and **9** - **11** produced the same result in 12 h. None of the test compounds showed any activity towards nematodes. This is the first report of such neolignans showing insecticidal activity. Since these natural products also exhibited reasonable activity against various plant and human pathogens they should be investigated further for their potential application.

Allelochemicals BOA, DIBOA and AZOB

Barnes and Putnam (32,33) examined the allelopathic potential of residues and aqueous extracts of rye, and reported that benzoxazinones, 2,4-dihydroxy-1,4(2H)-benzoxazin-3-one (DIBOA) and 2,3-benzoxazolinone (BOA) (Figure 1) are responsible for the toxicity shown by rye residue. Patrick and Koch (34) conducted a study of the toxic substances produced as a result of microbial decomposition of plant residues, including rye and concluded that unless decomposition of the residues occurred, no toxic substances were present in the soil extract. Once in the soil, the benzoxazinones produced by rye would be susceptible to microbial transformation by various soil microbes.

Our experiments with BOA and different field soils indicated that 2,2'-oxo-1,1'-azobenzene (AZOB) (**13**), a compound with strong herbicidal activity, was responsible for the rye toxicity in the soil (35). A parallel experiment with 6-methoxy-2,3-benzoxazolinone (MBOA) (Figure 1) yielded AZOB, as well as it's mono-(MAZOB) (**14**) and dimethoxy-(DIMAZOB) (**15**) derivatives. These compounds were produced only in the presence of soil bacteria *Acinetobacter calcoaceticus* (36,37). In the case of MBOA, it was shown that demethoxylation precedes the oxidation step. Although BOA and DIBOA were leached out of rye residues, there were no detectable amounts of the bio-transformation products in the soil. When BOA was mixed with soil and rye residue, either under field conditions or in vitro, AZOB was detected. Levels of free BOA in the soil were greatly reduced by incubation with rye residue. AZOB was more toxic to curly cress (*Lepidium sativum* L.) and barnyardgrass (*Echinochloa crusgalli* L.) than either DIBOA or BOA (35).

Chemical Signal Molecule for Vesicular Arbuscular Mycorrhiza (VAM)

There are several published reports about plant factors that stimulate VAM hyphal growth *in vitro* and also the pre-colonization phase of the VAM forming fungi (38-42). It has been suggested (43,44) that the quantity rather than the presence of specific compounds in the root exudates is responsible for stimulation of fungal

Table IV. Minimum inhibitory concentrations (MIC) of compounds **7-12**

Organism	Compounds (ug/ml)					
	7	**8**	**9**	**10**	**11**	**12**
Fungi						
Candida albicans	10	10	10	50	25	25
Aspergillius flavus	10	10	10	75	50	50
Gleosporum sp.	25	25	25	25	25	25
Rhizoctonia sp.	25	25	25	25	25	25
Bacteria						
Streptococcus aureus	10	10	10	25	15	15
Staphylococcus epidermidis	10	10	10	20	10	10
Escherichia coli	20	20	20	50	25	25
Insect*						
Aedes aegypti (Mosquito larvae)	10	10	10	10	10	10
Crustacea*						
Artemia salina) (Brine shrimp)	1	1	1	1	1	1
Nematode*						
Caenorhabditis elegaus	NA	NA	NA	NA	NA	NA
Panagrellus redivivus *(Axenic)*	NA	NA	NA	NA	NA	NA

NA = not active; *activity measured at 12 h.

SOURCE: Reprinted with permission from Reference 31. Copyright 1991.

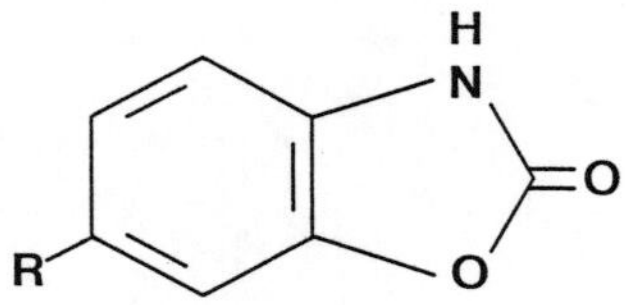

BOA: R = H
MBOA: = OMe

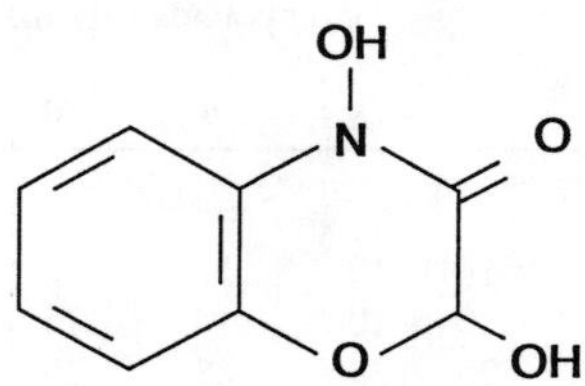

DIBOA

Figure 1. Structure of BOA, MBOA and DIBOA.

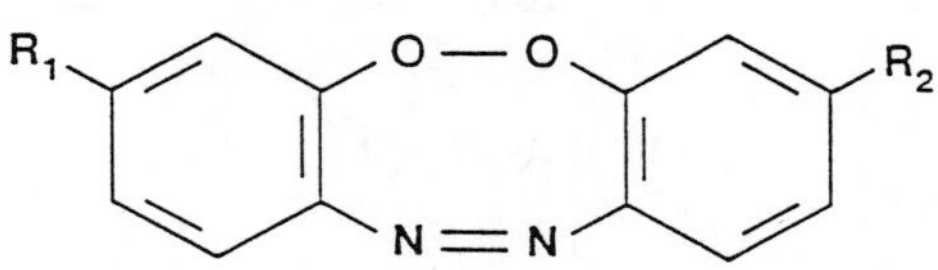

13 AZOB: $R_1 = R_2 = H$
14 MAZOB: $R_1 = H, R_2 = OMe$
15 DIMAZOB: $R_1 = R_2 = OMe$

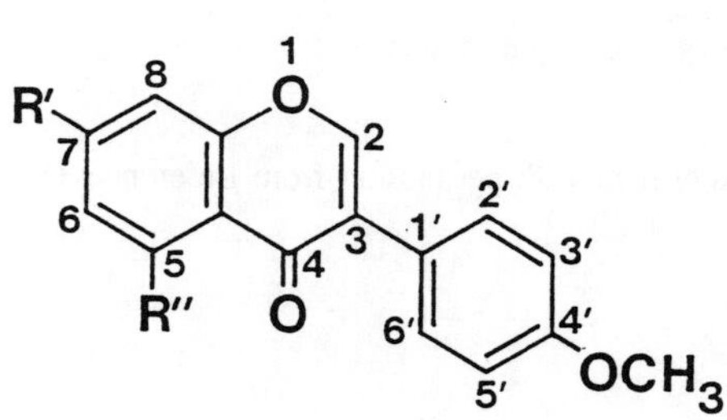

16 Formononetin R′ = OH, R″ = H
17 Biochanin A R′ = R″ = OH

growth and VAM root colonization. Viable spores of most VAM fungal species readily germinate on distilled water and there is no evidence that they require any specific host factors. However, certain components of root exudates or plant cells may act as signal molecules capable of inducing hyphal growth, branching, differentiation, and host penetration (38, 41, 45).

Elias et al. (39) indicated the presence of a transient VAM fungus stimulating factor in exudates from phosphorus-deprived young white clover seedlings. This study also indicated that the quality of the exudate is important in stimulating VAM hyphal growth. Formononetin has been found as a stress metabolite in soybean (46) and in greater quantities in clover root extracts than any other nod gene inducing flavone (47). Its concentration in clover plants is reduced by seedling age, light intensity, fertilization (48) and plant pathogens.

White clover (*Trifolium repens* L. c.v. Ladino) plants (2 weeks old) were used for all exudate and extract collections for this study. Plants were grown in sterile square glass staining dishes containing Hoagland solution with and without phosphorus as previously described (39). The VAM stimulatory compounds isolated from clover roots grown under phosphate stress were characterized as formononetin (7-hydroxy-4'-methoxyisoflavone) (**16**) and biochanin A (5,7-dihydroxy-4'-methoxyisoflavone) (**17**) (49). At 5 ppm, these compounds stimulated hyphal growth *in vitro* and root colonization by the VAM fungus *Glomus fasciculatum*. The permethylated products of the two compounds were inactive. These findings suggest that the isoflavonoids studied may act as signal molecules in the VAM symbiosis. Formononetin and biochanin A at 5 ppm concentrations gave increased VAM colonization and growth of white clover plants under growth chamber conditions (50). The effect was more prominent for compound **16** than **17**. Another interesting activity of formononetin and biochanin A was their ability to reduce herbicide injury (51). Application of VAM stimulatory compounds **16** and **17** at 5 ppm solutions to field soils containing toxic levels of imazaquin and pendimethalin reduced herbicide injury in corn and sorghum under growth chamber conditions (51). These results indicate that the isoflavonoids, formononetin and biochanin A, are involved in the stimulatory effects of clover roots towards the VAM fungi. However, the mode of action of these compounds as signal molecules in VAM symbiosis requires further studies.

LITERATURE CITED

1. Abelson, P. H. *Science*. 1990; 247, 513.
2. Brockmann, H.; Schmidt-Kastner, G. *Chem Ber*. 1955; 88, 57.
3. Patterson, E. L.; Wright, D. P., Jr. *Process for controlling insects, nematodes, and mites using valinomycins*. US Patent 3520973, 1970.
4. Wallen, V. R.; Sutton, M. D.; Skolko, A. J. *Phytopathology*. 1950; 40,156.
5. Norman, A. G. *Soil Sci. Soc. Proc*. 1959; 23, 368.

6. Arai, M.; Haneishi, T.; Kitahara, N.; Enokita, R.; Kawakubo, K.; Kondo, Y. *J. Antibiotic*. 1976;29,863.
7. Omura, S.; Gwai, Y.; Takahashi, Y.; Sadakane, N.; Nagakawa, A.; Oiwa, H.; Hasegawa, Y.; Ikai, T. *J. Antibiotic*. 1979; 32, 255.
8. Heisey, R.; Putnam, A. R. *J. Nat. Prod*. 1986; 49, 859.
9. Mishra, S. K.; Keller, J. E; Miller, J. R.; Heisey, R. M.; Nair, M. G.; Putnam, A. R. *J. Ind. Microbiol*., 1987; 2, 267.
10. Gopalakrishnan, K. S; Narasimhachari, N; Joshi, V. B.; Thirumalachar, M. J. *Nature*. 1968; 218, 597.
11. Wagman, G. H; Testa, R. T.; Patel, M.; Marquez, J. A.; Oden, E. M.; Waitz, J. A.; Weinstein, M. J. *Antimicrob. Agents. Chemother*. 1975; 7, 457.
12. Burke, R. C.; Swartz, J.; Chapman, S. S.; Huang, W. *J.Invest. Dermatol*., 1954; 23, 163.
13. Wasserman, H. H.; VanVerth, J. E.; McCaustland, D. J.; Borowitz, I. J.; Kamber, B. *J. Amer. Chem. Soc*. 1967; 89, 1535.
14. Uri, J.; Bekesi, I. *Nature*. 1958; 181, 908.
15. Schneider, M. A.; Stilbans, E. B.; Rachkovskaya, L. A.; Poltorak, V. A. *Antibiotiki*. 1983; 28, 352.
16. Bognar, R.; Brown, B. O.; Lockley, W. J. S.; Makleit, S.; Toube, T. P.; Weedon, B. C. L.; Zsupan, K. *Tetrahedron Lett*. 1970; 471.
17. Nair, M. G.; Putnam, A. R.; Mishra, S. K.; Mulks, M. H.; Keller, J. F.; Taft, W. H.; Miller, J. R.; Zhu, P-P.; Meinhart, D.; Lynn, D. G. *J. Nat. Prod*. 1989; 52, 797.
18. Mulks, M. H.; Nair, M. G.; Putnam, A. R. *Antimicrobial Agents and Chemotherapy*. 1990; 34, 1762.
19. Smith, J.; Putnam, A. R; Nair, M. G.; *J. Agric. Food Chem*. 1990; 38, 1729.
20. Nair, M. G.; Mishra, S. K.; Putnam, A. R.; Pandey, R. C. *J. Antibiotic*. 1992; 45, 613.
21. Scriber, J. M. (1986) *Oecologia*. **71**, 94.
22. Little, E. L. (1980) *The Audubon Society Field Guide to North American Trees (Eastern Region),* p 714. Alfred A. Knopf, New York.
23. Rao, K. V. and Davis, T. L., (1982) *J. Nat. Prod*., **45**, 283.
24. Kimura, I., Kimura, M., Yoshizaki, M., Yanada, K., Kadota, S., and Kikuchi, T. *Planta Medica*. 1983; **48**, 43.
25. Kamikado, T., Chang, C.-F., Murakoshi, S., Sakurai, A., and Tamura, S. (1975) *Agr. Biol. Chem*, **39**, 833.
26. Clark, A. M., El-Feraly, F. S., and Li, W.-S. (1980) *J. Pharm. Sci*., **70**, 951.
27. Watanabe, K., *Yaoxue Tongbao*, **20**, 522 (1985), (in Chinese) Chem. Abs. **104**:179748.
28. El-Feraly, F. S., Cheatham, S. F., and Breedlove, R. L. (1983) *J. Nat. Prod*., **46**, 493.
29. El-Feraly, F. S., and Li, W. S., (1978) *Lloydia*, **41**, 442.

30. Fujita, M., Hideji, I., and Sashida, Y. (1973) *Yakugaku Zasshi*, **93**, 429 (in Japanese) *Chem Abs*. **79**:52926g.
31. Nitao, J. K.; Nair, M. G.; Johnson, K. S.; Thorogood, D. L.; Scriber, J. M. *Phytochemistry*. 1991; 30, 2193.
32. Barnes, J. P.; Putnam, A. R. *J. Chem. Ecol*. 1983; 9, 1045.
33. Barnes, J. P.; Putnam, A. R. *Weed Sci*. 1986; 34, 384.
34. Patrick, Z. A.; Koch, L. W. *Can. J. Bot*. 1958; 36, 621.
35. Nair, M. G.; Whitenack, C. J.; Putnam, A. R. *J. Chem. Ecol*. 1990; 16, 353.
36. Chase, W. R.; Nair, M. G.; Putnam, A. R. *J. Chem. Ecol*. 1991; 17, 9.
37. Chase, W. R.; Nair, M. G.; Putnam, A. R.; Mishra, S. K. *J. Chem. Ecol*. 1991; 17, 1575.
38. Becard, G.; Piche, Y. *New Phytol*. 1989; 112, 77.
39. Elias, K. S.; Safir, G. R. *Applied Environmental Microbiology*. 1987; 53,1928.
40. Gianinazzi-Pearson, V.; Branzanti, B.; Gianinazzi, S. *Symbiosis*. 1989; 7,243.
41. Mosse, B. *Can. J. Bot*. 1988; 66, 2533.
42. Paula, M. A.; Siqueira, J. O. 1990. *New Phytol*. 115:69-75.
43. Graham, J. H. Mycologia. 1982; 74, 831.
44. Graham, J. H.; Leonard, R. T.; Menge, J. A. *Pl. Physiol*. (Lancaster). 1980; 68, 548.
45. Becard, G.; Piche, Y. *Can. J. Bot*. 1990; 68, 1260.
46. Osman, S. F.; Fett, W. F. *Phytochemistry*. 1983; 22, 1921.
47. Djordjevic, M. A.; Redmond, J. W.; Batley, M.; Rolfe, B. G. *EMBO J*. 1987; 6, 1173.
48. Rossitier, R. C.; Beck, A. B. *Aust. J. Agric. Res*. 1966; 17, 447.
49. Nair, M. G.; Siqueira, J. O.; Safir, G. R. *Applied and Environmental Microbiology*. 1991; 57, 434.
50. Siqueira, J. O; Safir, G. R.; Nair, M. G. *The New Phytologist*. 1991; 118, 87.
51. Siqueira, J. O.; Safir, G. R.; Nair, M. G. *Plant and Soil*. 1991; 134, 233.

RECEIVED September 27, 1993

Chapter 11

Insecticidal and Antifeedant Activities of Plant Compounds

Potential Leads for Novel Pesticides

Pierre Escoubas, Labunmi Lajide, and Junya Mizutani

Research and Development Corporation of Japan (JRDC), Eniwa RBP, Eniwa-Shi, Megumino Kita 3–1–1, Hokkaido 061–13, Japan

We have examined a variety of plants, chosen for their traditional medicinal or insect control uses, for insecticidal and insect antifeedant properties. Our test insects are crop pests such as the tobacco cutworm *Spodoptera litura*, medically important insects (e.g the yellow fever mosquito *Aedes aegyptii*) or wood-destroying organisms such as the termite *Reticulitermes speratus*.

Using innovative methods, as well as classical bioassays, we have studied feeding deterrency, larval growth inhibition and acute toxicity of a number of Nigerian plant extracts.

The results of our survey of bioactivies, as well as phytochemical studies are presented. Plants such as *Xylopia aethiopica* (Annonaceae), *Aframomum melegueta* (Zingiberaceae), *Aristolochia albida* (Aristolochiaceae), *Zanthoxylum xanthoxyloides* (Rutaceae), *Dichapetalum barteri* (Dichapetalaceae) and *Detarium microcarpum* (Leguminosae), have proven to be interesting sources of bioactive compounds.

In their often quoted *Science* paper, Balandrin et al. (1) emphasize the importance of natural products as sources of useful therapeutic and commercial products. They also state the fact that plant resources in that respect, are still largely underexploited.

In the past decade there has been renewed interest in the potential of natural products as sources of drugs or pesticides, due to some shortcomings of the "synthesis-only" approach. Natural products are therefore considered to provide a vast potential for the discovery of novel bioactive structures, new modes of actions or leads for the synthesis of interesting compounds (2).

Certainly, in the case of the insecticides, past experience has shown this approach to be successful, with synthetic pyrethroids as the best example. Other commercially useful botanical pesticides include nicotine, pyrethrum, rotenone and several other alkaloids. Similarly, other natural compounds modifying feeding behavior or inhibiting the growth of insect larvae, are considered viable alternatives to acute toxins, for insect control (3). One of the most successful examples so far is Neem, extracted from the seeds of the tree *Azadirachta indica*.

0097–6156/94/0551–0162$06.00/0

Based on these assumptions, we have started investigating plants of various origins, for their insecticidal, antifeedant, or growth-inhibitory properties against several insect pests. Among random and non-random type search strategies, we have opted for a selective approach, relying either on ecological observation or on traditional use of the plants (4). Medicinal plants, usually containing a variety of bioactive substances, were thus considered a good choice as sources of interesting material. They are still widely used in Africa and are therefore readily available (5). Plants in temperate areas have also evolved chemical defenses against herbivores, and many temperate plant species have yet to be investigated in that respect.

In this context, we present here the results of our activity screening of a series of plants from Nigeria, as well as plants collected in Hokkaido, Japan, against several insect pests.

MATERIAL AND METHODS

Sample collection: Nigerian plants were collected in Billiri, Northern Nigeria, and Ogbomoso Local Government area (Southwestern Nigeria). Japanese plants were collected in Nopporo and Tomakomai forest parks, Hokkaido, Japan.

Preparation of the samples: Samples were dried and various extractions were performed, using non-polar (hexane) and polar solvents (methanol), as well as different temperatures (room T^{o}., and boiling solvent). When the amount of material was limiting, 5% aqueous methanol was used. When possible, different parts of the plants (roots, bark, stems, leaves) were extracted separately. Solvents were removed under vacuum, the samples weighted, and an initial dilution to 10mg extract/ml was prepared in either methanol or acetone, for the bioassays.

Isolation of the active constituents was done using a bioassay-guided procedure. Depending on the extract, various combinations of chromatographic methods and solvent systems were used. Basic procedures included thin layer chromatography (analytical and preparative), as well as standard column chromatography. Structures were elucidated by a combination of NMR and MS spectrometric techniques. Details of the analyses will be reported elsewhere.

BIOASSAYS

Four different bioassays were performed with each plant extract:

1- **Mosquito Larval toxicity**: Ten second-instar larvae of *Aedes aegyptii* were placed in wells containing an aqueous solution of plant extract (100 µg/ml) in two replicates, and their mortality recorded after 24H (27°C, 16:8 L/D). Controls received solvent only (100 µl, acetone or methanol).

2- **Leaf-disk Antifeedant Bioassay**: Ten 1.0 cm^2 sweet potato leaf disks were placed in marked wells in an agar-coated petri dish. Five disks were alternatively treated with 10 µl of plant extract (100 µg/cm^2) or solvent (acetone or methanol). Five third instar *Spodoptera litura* (Lepidoptera Noctuidae) larvae per dish and three dishes were used per treatment. The treated dishes were placed in an incubator at 27°C and 75-80% RH for 16-18 hours in darkness. The leaf surface consumed was measured with a video camera interfaced to a personal computer as described earlier (6). The feeding index was calculated as I= %T / (%T + %C) (%T = % of treated disks consumed, %C = % of control disks consumed). An arbitrary level of I<20 was used as the criteria to determine effective feeding deterrents (7).

3- **Larval Growth-inhibition Bioassay**: To test for chronic growth inhibition we incorporated the plant extracts in artificial diet (0.1% w/w) using cellulose powder as an inert carrier. Control batches fed on diet containing cellulose and solvent (acetone or methanol) only. Twenty neonate *Spodoptera litura* larvae fed on the treated and control diets for seven days (27°C, 75% RH, 16:8 L/D), and their weights were recorded. Weight differences were analysed using a non-parametric statistical test (Kolmogorov-Smirnov, $p<0.05$).

4- **Termite Antifeedant Choice Bioassay**: Two filter paper disks (2cm diam.) were treated with 25 µl of extract or solvent (acetone or methanol) in three replicates (250 µg/disk), and placed on moistened sand in 55 mm diam. petri dishes, with 30 *Reticulitermes speratus* workers. After 2 weeks, the amount of feeding on each disk was recorded using video image analysis (as described above). The same antifeedant index as in the leaf-disk bioassay was calculated.

5- **TLC Bioautography bioassay**: A feeding deterrent bioassay using *S. litura* 4th instar larvae was used in the identification of *Skimmia japonica* compounds as reported earlier (8). Briefly, after migration of the sample on a 10 X 5 cm silica gel TLC plate, the dried plate was covered with a layer of agar-based artificial diet and presented to the larvae. After one night, plates were retrieved and the location of uneaten areas recorded, affording the Rf values of the active components. These results were then used in the isolation of the compounds, monitored by TLC.

RESULTS

The results of the screening are presented in Table 1 and were classified as follows:

	+++	++	+
Mosquito bioassay:	$M>80\%$	$50\%<M<80\%$	$30\%<M<50\%$
Leafdisk bioassay	$I<10$	$10<I<20$	$20<I<30$
Termite bioassay	$I<10$	$10<I<20$	$20<I<30$
Growth bioassay	$GI>50\%$ (significant/control $p<0.05$)	$50\%<GI<30\%$	$30\%<GI<20\%$

(M = % mortality after 24H., I = antifeedant index, GI = % of growth reduction/control w/w)

Plants which possessed activity against one or more of the test insect species, have therefore been chemically investigated for the presence of active constituents.

Bioassay-directed fractionations of the crude extracts have led to the isolation and identification of the compounds shown in Figures 1 and 2. Nigerian medicinal plants investigated so far are:

- *Aristolochia albida* (Aristolochiaceae)
- *Xylopia aethiopica* (Annonaceae)
- *Aframomum melegueta* (Zingiberaceae)
- *Zanthoxylum xanthoxyloides* (Rutaceae)
- *Detarium microcarpum* (Leguminosae)

Table 1. Screening results. MOSQ = *Aedes aegyptii* toxicity bioassay, TERM = *Reticulitermes speratus* antifeedant activity, LEAF / GROW = *Spodoptera litura* antifeedant / growth-inhibition bioassays

FAMILY	NAME	PART	CODE	MOSQ	TERM	LEAF	GROW
ANNONACEAE	*Xylopia aethiopica*	pod	T16	-	+++	-	++
			T17	+	+++	-	-
			T18	-	+++	-	-
			T19	-	+++	-	-
			T20A	-	++	-	-
			T20B	-	+++	-	-
			T21	++	+	-	-
		seed	T22A	-	+++	-	-
			T22B	-	-	-	++
			T23	+++	-	-	-
ANACARDIACEAE	*Lannea barteri*	root	LB-A	-	-	-	-
		stem	LB-B	-	-	+	-
		leaves	LB-C	-	+	-	-
ARISTOLOCHIACEAE	*Aristolochia albida*	leaves	AALV1		-	-	-
			AALV2		-	-	-
			AALV3		-	-	-
		root	AALR1		+++	+++	+++
			AALR2		-	-	-
			AALR3		-	-	-
			AALR4		-	-	-
COCHLOSPERMACEAE	*Cochlospermum planchonii*	leaves	CoP-A	+	-	-	
COMBRETACEAE	*Anogeissus leoicarpa*	leaves	AL-A	-	+	-	-
	Guiera senegalensis	leaves	GS-A	-	++	-	-
	Terminalia glaucescens	root	TG-A	-	-	-	-
		stem	TG-B	-	++	-	-
		leaves	TG-C	++	-	-	-
			TG-D	-	-	-	-
DICHAPETALACEAE	*Dichapetalum barteri*	root	DB-A	+++	+++	-	++
		stem	DB-B	-	-	-	-
		leaves	DB-C	-	-	-	-
EUPHORBIACEAE	*Croton zambesicus*	seed	CZ-A(Hex)	+++	-	-	-
			CZ-B(MeOH)	-	-	-	-
		root	CZ-C	-	+++	-	-
		leaves	CZ-D	-	-	-	-
		stem	CZ-E	-	-	-	-
	Euphorbia sp.	root	EU-A	-	++	-	-
	Hymenocardia acida	root	HA-A	-	+++	-	-
		stem	HA-B	-	-	-	-
		leaves	HA-C	-	-	-	-

(Continued on next page.)

Table 1.—Continued

FAMILY	NAME	PART	CODE	MOSQ	TERM	LEAF	GROW
LEGUMINOSAE	*Baphia nitida*	wood	BN-A	-	+	-	-
	Cassia nodosa	leaves	CN-A(Hex)	+	-	-	-
		leaves	CN-B(MeOH)	-	-	-	-
	Detarium microcarpum	root	DM-A	-	+ +	-	-
		leaves	DM-B	+ + +	-	-	+
	Parkia biglobosa	bark	PB-A	-		-	-
	Parkia clappertoniana	fruit	PL8	-		-	-
	Piliostigma thonningii	seed	PT-A(Hex)	+ +	+ +	-	-
			PT-B(MeOH)	-	-	-	-
		stem	PT-C	-	-	+	-
		leaves	PT-D	+ + +	-	-	-
	Tephrosia elegans	leaves	TE-A(Hex)	+ + +	-	-	-
			TE-B(MeOH)	-	-	-	-
		leaves	T1	+	-	-	-
			T2A	-	-	-	-
			T2B	-	-	-	-
		trichomes	T2C	+ +	-	-	-
		root	DB1		-	-	-
		leaf	DB2		-	-	-
	Uraria picta	leaves	UP-A	-	-	-	+ +
	Waltheria indica	leaves	WI-A	+ + +	-	-	-
			WI-B	-	-	-	-
MALVACEAE	*Hibiscus cannabinus*	root	HB-A	+	+	-	-
		leaves	HB-B	-	-	-	-
	Sida cordifolia	seed	SC	+		-	-
MELIACEAE	*Pseudocedrela kotchyii*	leaves	PsK-A	-	-	-	-
			PsK-B	-	-	-	-
MORINGACEAE	*Moringa oleifera*	leaves	T24A	-	-	-	-
			T24B	-	-	-	-
			T24C	-	-	-	-
		trichomes	T24D			-	-
		root	MO1			-	-
		leaf	MO2			-	-
		bark	MO3			-	-
		seed(def)	MO4			-	-
OLEACEAE	*Ximenia americana*	leaves	XA-A	-	-	-	-
POLYGONACEAE	*Polygonum limnatum*	aerial p.	PL7	-		-	-
		seed	PL-A	-		-	-

RUBIACEAE	*Crossopteryx febrifuga*	root	CF-A	-	-	-	-
		stem	CF-B	-	-	-	-
		leaves	CF-C	-	-	+	-
	Fadogia argestistis	leaves	T6	-	+	-	-
			T7	-	-	-	-
		root	T8	-	+	-	+
			T9	-	-	-	+ +
			T10	-	+ + +	-	-
	Gardenia erubescens	stem	GE-A	-	-	-	-
		leaves	GE-B	+	+	-	-
	Pavetta crassipes	leaves	PV1	-			-
			PV2	+ +			-
RUTACEAE	*Clausena anisata*	root	CA-A	-	+ + +	+ + +	+ + +
		leaves	CA-B	-	-	+	-
	Zanthoxylum xanthoxyloides	root	T3(CHCl3)	+ + +	+ + +	-	-
			T4(MeOH)	+ + +	+ + +	-	-
			T5(aqMeOH)	+ + +	+ + +	-	-
SAPOTACEAE	*Butyrospermum paradoxum*	root	BP-A	-	-	-	-
		stem	BP-B	-	-	-	-
		leaves	BP-C	-	-	-	+ +
SCROPHULARIACEAE	*Buchnera hispida*	stem	BH-A	-	-	-	-
		leaves	BH-B	-	-	-	-
SIMAROUBACEAE	*Balanites aegyptica*	root	BA3	-		-	-
			BA4	-		-	-
TILIACEAE	*Grewia mollis*	leaves	GM-A	-	-	-	-
VERBENACEAE	*Vitex doniana*	root	VD-A	-	+ + +	+ + +	+ +
		stem	VD-B	-	+ +	+ +	-
		leaves	VD-C	+ + +	-	-	-
			VD-D	-	-	-	-
VITACEAE	*Cissus populnea*	root bark	CP-A	-	-	-	-
ZINGIBERACEAE	*Aframomum melegueta*	seed	T11	+ + +	+ + +	-	-
			T12	-	-	-	-
		pod	T13	-	+ + +	-	-
			T14	-	+ + +	-	-
			T15A	-	+	-	-
			T15B	-	+	-	-

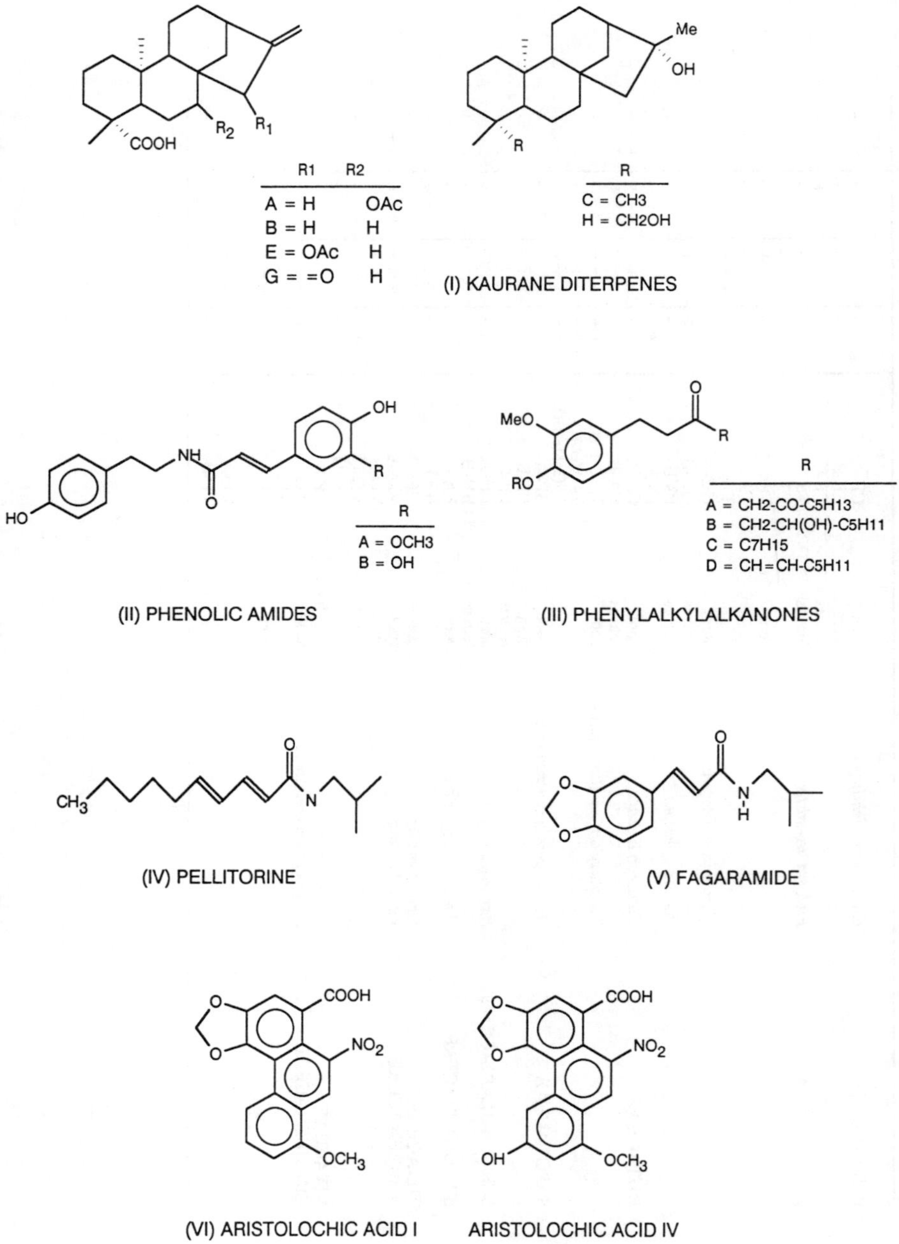

Figure 1. Bioactive compounds isolated from Nigerian medicinal plants.

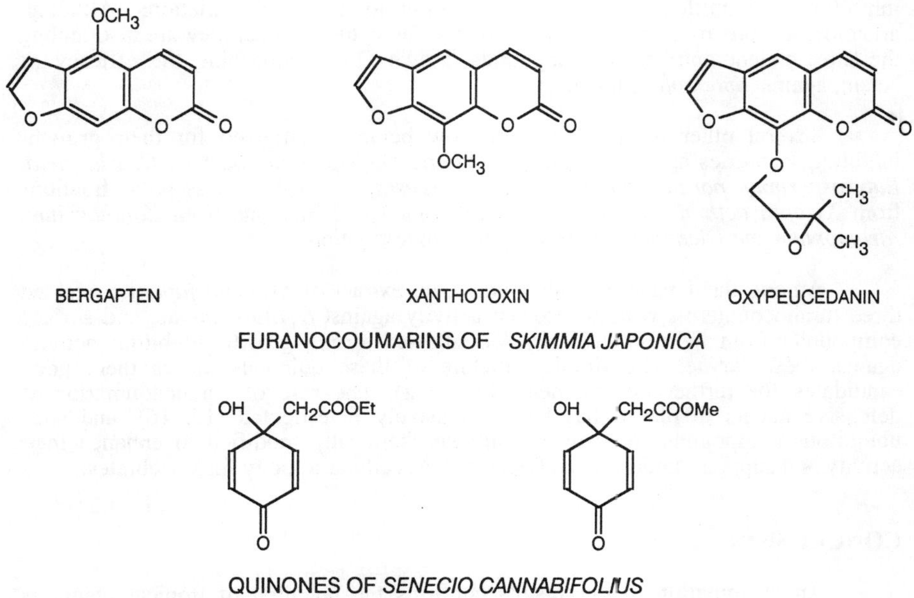

Figure 2. Antifeedant and growth-inhibiting compounds for *Spodoptera litura*, isolated from Japanese plants.

In addition, two Japanese species have also been investigated:

- *Skimmia japonica* (Rutaceae)
- *Senecio cannabifolius* (Compositae)

Phytochemical investigation of the most active plants extracts has led to the isolation of their active constituents. In particular, compounds possessing high termite-feeding deterrency have been discovered (Figure 1).
Kaurane diterpenes (**I**) and phenolic amides (**II**) from *Xylopia aethiopica*, as well as phenethylalkylketones (**III**) from *Aframomum melegueta* are now being patented as leads to new wood-protecting agents. Kaurane diterpenes have previously been found to deter feeding of Lepidopteran larvae (9) and aphids (10) and may thus have a rather broad spectrum of activities, and represent an interesting new class of active compounds.

Multiple activities have been observed for extracts of *Zanthoxylum xanthoxyloides*, a widely used medicinal plant. The principal active compounds are the isobutylamides pellitorine (**IV**) and fagaramide (**V**), which display termite and lepidopteran feeding deterrent activities. The insecticidal activity of isobutylamides isolated from various plant families, has been previously investigated (11) and their level of activity has warranted patent applications. We have demonstrated also that these compounds may act as feeding deterrents in addition to their toxicity properties.

Aristolochia albida, another medicinal plant, has yielded several aristolochic acids (**VI**), with acids I and IV showing extremely strong feeding-deterrency against *S. litura*, as well as larval growth-inhibition activity. At present, our investigation of the potential of these compounds continues, including a study of structure-activity relationships among a series of synthetic homologs. Preliminary results point to the importance of the carboxylic group for feeding deterrency, and a separation of growth inhibition and antifeedant activity in relation to structural variations. Although aristolochic acids by themselves are known to be cytotoxic (12) they are also among the most potent antifeedant compounds known (13), appoaching the efficacy of Neem, against *Spodoptera* larvae (14).

Several other plant species are now being investigated for their growth-inhibiting activities against *S. litura*, including *Dichapetalum barteri*, *Uraria picta*, *Butyrospermum paradoxum* and *Fadogia argestitis* as well as less polar fractions from *Xylopia aethiopica*. Mosquito larval toxicity of fractions from *Zanthoxylum*, *Aframomum* and *Dichapetalum* is also under investigation.

Among the Japanese plants studied, an extract of *Skimmia japonica* yielded three furanocoumarins with antifeedant activity against *S. litura* larvae, and surface compounds from *Senecio cannabifolius* showed strong growth-inhibiting activity against these larvae. The simple structure of these quinones makes them good candidates for further development (Figure 2). The role of furanocoumarins as defensive agents in plants, has been extensively investigated (15, 16), and such ubiquitous compounds could also easily be chemically modified to enhance their activity and suppress undesirable effects such as cellular toxicity for vertebrates.

CONCLUSION

An examination of our results confirms the potential of tropical plants as sources of novel biologically active compounds. Although phytochemical investigations may have been conducted previously on the plants we selected, few studies have concentrated on the bioactivity-related isolation of compounds. Moreover, most of the anti-insect properties of these plants are described here for the first time.

Structure-activity studies as well as preliminary investigation of the mode of action of these compounds will hopefully enable us to select the most promising leads, possessing either original structures or novel target sites.

It should be noted that any work dealing with natural compounds as sources of new pesticidal agents, should consider positive results as an indication of potential in a chemical series, rather than trying to discover compounds usable *per se*. Few plant extracts can measure up to the modern synthetic pesticides, and they often have undesirable properties such as quick breakdown or toxic effects. Bioactive natural compounds should therefore be considered models which can serve for the development of series of more efficient, biodegradable, safer pesticides.

ACKNOWLEDGMENTS

We thank Mr. S. L. Kela (Balewa University, Nigeria), Dr. M. O. Akanbi (Forestry Research Institute, Nigeria), as well as Dr. H. Nakai (Hokkaido University, Japan) for their help in plant collection and identification, and Miss Y. Matsuzawa for technical help. This work was funded by the Research and Development Corporation of Japan (JRDC) under the ERATO program.

LITERATURE CITED

1. Balandrin, M. F.; Klocke, J. A.; Wurtele, E. S.; Bollinger, W. H. *Science* **1985**, *228*, 1154.
2. Kulkarni, R. A. *Pure and Appl. Chem.* **1986**, *58*, 917.

3. Klocke J. A In "*Economic and Medicinal Plant Research Vol. 3*"; Wagner, H.; Hikino, H.; Farnsworth, N. R., Eds. Academic Press: CA, 1989; p 103.
4. Cox P. In *Bioactive Compounds from Plants*; Ciba Foundation Symposium Series 154; J. Wiley & Sons: UK, 1990; p 47.
5. Sofowora, A. *Medicinal Plants and Traditional Medicine in Africa*; John Wiley & Sons: UK, **1982**.
6. Escoubas, P.; Lajide, L.; Mizutani, J. *Ent. Exp. et Appl.* **1993**, *66*, 99.
7. Alkofahi, A.; Rupprecht, J. K.; Anderson, J. E.; McLaughlin, J. L.; Mikolajczak, K. L.; Scott, B. A. In *Insecticides of Plant Origin*; Arnason, J. T.; Philogene, B. J. R.; Morand, P., Eds.; ACS Symposium Series 387; American Chemical Society: Washington D.C., 1989; p 25.
8. Escoubas, P.; Fukushi, Y.; Lajide, L.; Mizutani, J. *J. Chem. Ecol.* **1992**, *18*, 1819.
9. Waiss, A. C., Jr.; Chan, B. G.; Elliger, C. A.; Garret, V. H. *Naturwissenschaften* **1977**, *64*, 341.
10. Warthen, J. D., Jr.; Morgan, E. D. In *CRC Handbook of Natural Pesticides, Vol VI Insect Attractants and Repellents*; Morgan, E. D.; Mandava, N. B., Eds. CRC: FL, 1990; p 23.
11. Miyakado, M.; Nakayama, I.; Ohno, N. In *Insecticides of Plant Origin*; Arnason, J. T.; Philogene, B. J. R.; Morand, P., Eds.; ACS Symposium Series 387; American Chemical Society: Washington D.C., 1989; p 173.
12. Saxena, B. P.; Koul, O.; Tikku, K.; Atal, C. K.; Suri, O. P; Suri, K. A. *Indian J. of Exp. Biol.* **1979**, *17*, 354.
13. Bernays, E. A.; Chapman, R. F. *Ecol. Entomol.* **1977**, *2*, 1.
14. Lajide, L.; Escoubas, P.; Mizutani, J. *J. Agric. Food Chem.* **1993** (In press - April 1993).
15. Berenbaum, M. *Rec. Adv. Phytochem.* **1985**, *19*, 139.
16. Klocke, J. A.; Balandrin, M. F.; Barnby, M. A.; Yamasaki, B. In *Insecticides of Plant Origin*; Arnason, J. T.; Philogene, B. J. R.; Morand, P., Eds.; ACS Symposium Series 387; American Chemical Society: Washington D.C., 1989; p 136.

RECEIVED October 3, 1993

Chapter 12

Aphicidal Activity of Cuticular Components from *Nicotiana tabacum*

R. F. Severson[1], R. V. W. Eckel[2], D. M. Jackson[3], V. A. Sisson[3], and M. G. Stephenson[4]

[1]Phytochemical Research Unit, Richard B. Russell Agricultural Research Center, Agricultural Research Service, U.S. Department of Agriculture, P.O. Box 5677, Athens, GA 30613
[2]Department of Entomology, North Carolina State University, Raleigh, NC 27695
[3]Crops Research Laboratory, Agricultural Research Service, U.S. Department of Agriculture, P.O. Box 1555, Oxford, NC 27565
[4]Nematodes, Weeds and Crops Research, Coastal Plain Experiment Station, Agricultural Research Service, U.S. Department of Agriculture, P.O. Box 748, Tifton, GA 31793

Most *Nicotiana* species have multicellular, glanded leaf trichomes which may produce chemical secretions containing diterpenes and/or sugar esters with C_2 to C_{10} acyl moieties. These components affect tobacco aphids, *Myzus nicotianae* Blackman, in several ways, including influencing the acceptance or rejection of plants for colonization by alate migrant aphids, and the survival and fecundity of alate and apterous aphids. Cuticular diterpenes and sucrose esters were isolated from the cuticular extracts of aphid resistant and susceptible *N. tabacum* genotypes. These compounds were applied topically to the backs of apterous aphids. LC_{50}'s (dose per aphid which kills 50% of the test population after 48 hrs.) of the isolates were; α- and ß-4,8,13-duvatriene-1,3-diols, 15.7 μg; α- and ß-4,8,13-duvatrien-1-ols, 6.4 μg; *cis*-abienol, 7.5 μg; sucrose esters (6-O-acetyl-3,3,4-tri-O-acylsucrose with C3 to C7 acyl groups), 0.25 μg; and *cis*-abienol plus α- and ß-4,8,13-duvatrien-1-ols (1:3), 6.0 μg.

Aphids are insect pests of fruit, vegetable, grain and row crops. They cause economic losses by reducing yield and quality, and by transmitting certain plant viral diseases (*1*). Some aphid species, such as the blackmargined aphid, *Monellia caryella* (Fitch), are host specific, while others, like the green peach aphid, *Myzus persicae* (Sulzer), have a wide range of host plants (*2*). The selection of the host plant is usually made by winged (alate) adults which may use physical or cuticular chemical leaf traits to accept or reject the plant as a suitable host. If found acceptable, the aphid will probe into the leaf and feed on

0097–6156/94/0551–0172$06.00/0

the phloem sap (*3, 4*). An aphid on an acceptable host plant may produce several nymphs per day. Thus, plant colonization begins an exponential population growth which is difficult to control with insecticides or biological agents. Persistent use of synthetic aphicides to control aphid population explosions has resulted in the development of resistance to these compounds. Aphid species have developed resistance to several classes of insecticides (*1*).

Recently, a change from a green to a red morph has occurred in the tobacco aphid, *Myzus nicotianae* Blackman. Although, the red morph of the tobacco aphid is often associated with an increase in insecticide resistance, this color change has not been directly linked to use of insecticides. However, since the red aphid develops faster, has a higher reproduction rate and tolerates higher temperatures, it is more difficult to control than the green form (*5*).

Thus the dependence on synthetic aphicides for the control of *M. nicotianae* is undesirable, and other insect management methods must be developed. One part of aphid management should be the use of natural plant chemicals which deter host plant acceptance (nonpreference host plant resistance) by the adult aphid, or are toxic to adults and nymphs (*6*).

For the past 30 years, the U.S. tobacco, *Nicotiana tabacum* L., and *Nicotiana* species germplasm collections have been evaluated for tobacco aphid infestations in small field plots in North Carolina, South Carolina, Kentucky, Tennessee and Georgia (*7, 13*). Aphid resistance observed with some tobacco genotypes may be due to high levels of pyridine alkaloids (*8*). In 1982 Johnson and Severson (*11*) reported that chemicals produced by leaf trichomes played an important role in determining resistance of different tobacco introductions (TI) to the green morph of the tobacco aphid. In this paper we report on the effect of different leaf surface chemistries of *N. tabacum* germplasm on the colonization of tobacco by the red morph of the tobacco aphid, and discuss studies conducted to determine mechanisms of resistance. We also discuss the effects of specific cuticular isolates on the survival and fecundity of adult tobacco aphids.

Materials and Methods

In 1990, 1991 and 1992, eighteen *N. tabacum* genotypes were grown in replicated field plots (3 replications of 12 plants) under typical flue-cured tobacco production practices at the Crops Research Laboratory, Oxford, NC and at the University of Georgia Coastal Plain Experiment Station, Tifton, GA. Six weeks after transplantation, five leaves (12-18 cm in length) were collected from each tobacco type from two of the replications. One 2-cm diameter disc was cut from the center of each leaf and the discs were dipped 8 times into a scintillation vial containing 10 ml of methylene chloride (Burdick and Jackson distilled in glass grade). The vial was sealed with a teflon lined cap, cooled in dry-ice, and stored in the freezer (-18°C) until analysis. Subjective ratings (0, no aphids to 7, maximal infestation) were made of natural aphid infestations on all three replications (*10*).

The cuticular component extracts were analyzed by capillary gas chromatography using a slightly modified method of Severson et al. (*13*). An internal standard (92.5 μg of tricosanol) was added to the vial containing the cuticular extract and the sample was mixed. About a 0.75 cm² equivalent of the sample from tobaccos with secreting trichomes or 1.5 cm² from tobaccos with nonsecreting trichomes was transferred to a 1-ml tapered reacti-vial. The solvent was removed under a stream of nitrogen, 50 μl of 1:1 N-O-bis(trimethylsilyl)-trifluroacetamide: dimethylforamide (Pierce Chemical Company) was added, the vial was capped, and was heated at 76°C for 45 min to convert hydroxylated components to trimethylsilylethers. After cooling, the sample was transferred to a 100 μl micro autosampler vial and the vial was capped. A 1 μl portion was analyzed on a Hewlett Packard 5890 gas chromatograph equipped with a splitless injector (purged activation time 1.0 min; injector temp. 250° C), flame ionization detector (detector temp 350° C), and a 7673A auto sampler using a 0.5 mm id X 20 m bonded SE54 fused silica capillary column (hydrogen flow rate 18 ml/min). A temperature program of 100° C to 160° C at 15° C/min and 160° C to 300° C at 5° C/min was used. Internal standard methodology was used to calculate component levels.

Cuticular extracts used for the isolation of components were obtained from different *N. tabacum* genotypes that were grown in blocks of 100 plants. Cuticular components were extracted from plant tops using methylene chloride as previously described (*15*). The extracts from the following tobaccos were used to isolate specific components using the methodology reported by Severson et al.(*16*): α- and ß-4,8,13-duvatriene-1,3-diols (DVT-diols) (97+% by GC, 3:1 α-diol to ß-diol) from NC 2326; α- and ß-4,8,13-duvatrien-1-ols (DVT-ols) (98+% by GC, 9:1 α-ol to ß-ol) from PD 1097; a sucrose ester fraction (SE, 6-O-acetyl-2,3,4-tri-0-acyl-sucrose) (97+%by GC) from PD 1097; labda-12,14-dien-8α-ol (*cis*-abienol) (98+% as monohydrate, mp 67°C) from NFT; and labda-13-ene-8α,15-diol(labdenediol) (99+%, mp 126-128°C) from PD 964. The *cis*-abienol plus DVT-ol mixture was isolated from the cuticular extract from TI 1223 as follows. About 3.3 g of the extract was partitioned between 200 ml each of hexane and 80% methanol-water. The hexane solubles (2.4 g) in about 10 ml of hexane were added with stirring to 100 ml of acetone and placed in the cold (3°C) overnight. The solid which precipitated was removed by filtration and the solvent was removed from the acetone solubles to yield 1.3 g which was placed on a 50 g basic alumina (activity 1) column and eluted with 250 ml of hexane. The *cis*-abienol plus DVT-ol fraction (95+% by GC) was eluted with 500 ml of 1:3 hexane-methylene chloride.

A range of concentrations of the chemicals isolated above in 75% acetone-water were applied to the dorsum of apterous (wingless) adult aphids. One ul of the each of the concentrations was placed on the backs of 60 adults on TI 1112 leaves in individual petri dishes. After 24 hours, survival and nymphal production were monitored.

Results and Discussion

Table I compares the relative field plot aphid ratings with the composition and levels of the trichome-produced cuticular components. The aphid susceptible cultivars NC 2326 (flue-cured) and KY 14 (burley) produced similar levels of DVT-diols. The DVT-diols are the major cuticular components of flue-cured and burley tobacco types (*12*). The aphid-resistant lines either have nonsecreting trichomes, and very low levels of cuticular components, or glanded trichomes which produce heavy exudates consisting of DVT-ols, and\or *cis*-abienol and labdenediol and\or SE (see *12, 14, 18* for structures). The TI 1112 and breeding

Table I. Average Levels of Cuticular Components (1990-1991) and Average Infestation Ratings for the Tobacco Aphid (1990-1992) in Field Plots at Oxford, NC and Tifton, GA

		Cuticular Components ($\mu g/cm^2$)[b]				
Tobacco entry	Relative[a] aphid rating	DVT-ols	DVT-diols	*Cis* abienol	Labdene diol	SE
NC 2326	100.00	0.6	34.5	----[c]	----	----
KY 14	92.30	0.6	22.4	----	----	----
TI 1112	10.45*	----	0.4	----	----	----
I-35	20.32*	----	0.3	----	----	----
TI 1024	21.30*	----	0.6	----	----	----
TI 1406	20.40*	----	0.3	----	----	----
TI 1068	60.89*	2.0	39.5	13.2	1.0	21.3
NFT	41.90*	----	0.9	12.5	0.9	6.3
Red Rus	49.87*	0.1	2.3	22.0	2.2	13.5
TI 1223	45.37*	21.1	1.5	4.3	0.3	4.4
TI 1623	43.24*	23.7	4.2	10.8	0.3	8.7
PD 1097	66.57	30.3	24.4	----	----	11.3
KY Blk	66.13*	24.4	19.8	----	----	----
TI 1656	42.49*	24.7	1.6	----	----	6.2
TI 1687	30.98*	26.7	1.9	----	----	7.9
TI 550	65.95*	1.8	36.2	----	----	11.3
TI 698	51.23*	3.7	41.2	----	----	37.6
TI 998	42.00*	3.5	54.8	----	----	28.3

[a] 3 replications of 12 plant plots, Oxford, NC and Tifton, GA. Rated 0-7, avg. of 1990, 1991, and 1992 ratings, Relative Rating = (avg. rating entry X/avg. rating of NC 2326) x 100.

[b] DVT-ols = α & β-4, 8, 13-Duvatrien-1-ols
DVT-diols = α & β-4,8,13-Duvatriene-1,3-diols
Cis-Abienol = Labda-12, 14-dien-8α-ol
Labdenediol = Labda-13-ene-8α,15-diol
SE = 6-0-acetyl-2,3,4-tri-o-acylsucrose with methyl butyric and methyl valeric acyl moieties.

[c] Absent or below detection levels.

* Indicates significant difference ($P \leq 0.05$) from NC2326 (SDA, PROC GCM, Dunnetts curtailed *t* test) (20)

line I-35 have simple, nonglanded trichomes and TI 1024 and TI 1406 have nonsecreting glanded trichomes (*12*). TI 1112, I-35, TI 1024, TI 1406, NFT, Red Russian, TI 1223, TI 1623, TI 1656 and TI 1687 produce low levels of the DVT-diols. The cuticular extracts of TI 1068, TI 1223 and TI 1623 had detectable levels of all the duvane and labdane diterpenes and SE. The major surface chemicals of NFT and Red Russian were *cis*-abienol and SE. The DVT-ols were the major cuticular components of TI 1223, TI 1623, PD 1097, KY Black, TI 1656 and TI 1687. High SE producers were TI 1068, TI 698 and TI 998.

Shown in Table II are data from experiments where alate aphids were given a choice in a petri dish between two leaf discs of NC 2326 and two leaf discs of one of the test entries in Table I. Compared to NC 2326, TI 1112, I-35, TI 1024, NFT, Red Russian, TI 1223, and TI 1068 had significantly fewer alates after 24 hours and fewer nymphs deposited after 72 hours. However, in no-choice feeding tests, apterous aphids on TI 1112, I-35, and TI 1024 had similar survival and reproductive rates relative to NC 2326. Both survival and reproduction were reduced on NFT, Red Russian, TI 1223 and TI 1068 in no-choice feeding experiments. Jackson et al (*13*) proposed that the major mechanism of resistance of the nonsecreting entries is nonpreference by alate aphids, and that the entries with cuticular chemistries similar to TI 1223 and TI 1068 have both nonpreference and antibiotic components which contribute to their resistance.

To investigate the antibiotic role of these components, the different cuticular components were isolated and bioassayed topically on apterous adult aphids. The antibiotic effects of the cuticular components are illustrated in Figure 1 as dose response curves determined by Probit analysis. Commercial nicotine (97+% purity) was included as a control aphicide. At doses up to 50 μg, the labdenediol did not depress survival, and these data are not included in Figure 1. LC_{50}'s (expressed here as the lethal dose per aphid required to kill 50% of the test population) for the DVT-ols (LC_{50}=6.4 μg), *cis*-abienol (LC_{50} = 7.5 μg), *cis*-abienol plus DVT-ol isolate (LC_{50} = 6.0 μg) were comparable to that of nicotine (LC_{50} = 2.5 ug). The SE fraction (LC_{50} =0.25 μg) was about 10 times more active than nicotine and 70 times more active than the DVT-diols (LC_{50} = 15.7 μg).

These studies clearly demonstrate a relationship between aphid resistance in *N. tabacum* and differences in the levels and composition of trichome exudates. Tobaccos which are resistant to the green morph of the tobacco aphid are also resistant to the red form (*10, 11, 13*). This indicates that both color forms of the tobacco aphid have similar response to plant surface chemistry. It also appears that, as seen in the tobacco budworm and tobacco hornworm (*12,17-19*), the DVT-diols are preference compounds used by alate aphids to identify their host plants. The DVT-ols, *cis*-abienol, and SE are toxic to the tobacco aphid, and these compounds have potential as contact aphicides. Breeding of commercially acceptable *N. tabacum* cultivars with high levels of DVT-ols, *cis*-abienol and/or the SE would greatly decrease the level of tobacco aphid colonization and lessen the need to use conventional insecticides.

Table II. Laboratory Choice Experiments with Alate Aphids and No-Choice Experiments with Apterous Aphids on Leaf Discs of Susceptible and Resistant Nicotiana Tabacum Genotypes

	Alate Choice Experiments		Apterous No-Choice Experiments	
Tobacco Genotype	% Adults on treatment aftr 24 hr	Nymphs/ Adults on treatment after 72 hr	% Survival of adults after 72 hr	Nymphs/ Adults after 72 hr
NC 2326	50.0	6.0a[b]	97.0a	15.0ab
I-35	29.6**[a]	4.5ab	88.9a	11.7bcde
TI 1112	27.8*	4.3ab	97.0a	15.5a
TI 1024	16.5**	2.7ab	90.0a	13.3abcd
TI 1068	12.0**	1.1b	66.0b	7.4f
NFT	28.6*	3.4ab	87.8a	11.2cde
Red Rus	33.8*	4.5ab	85.6a	11.8bcde
TI 1123	18.2**	1.6b	68.9b	8.6ef

[a] Entries indicated by asterisks had significantly fewer aphids compared to NC 2326, (T-test significance, *, $P \leq 0.05$; **, $P \leq 0.01$).

[b] Means followed by same letter are not significantly different (Walter-Duncan k ratio procedure; k = 100, alpha = 0.05).

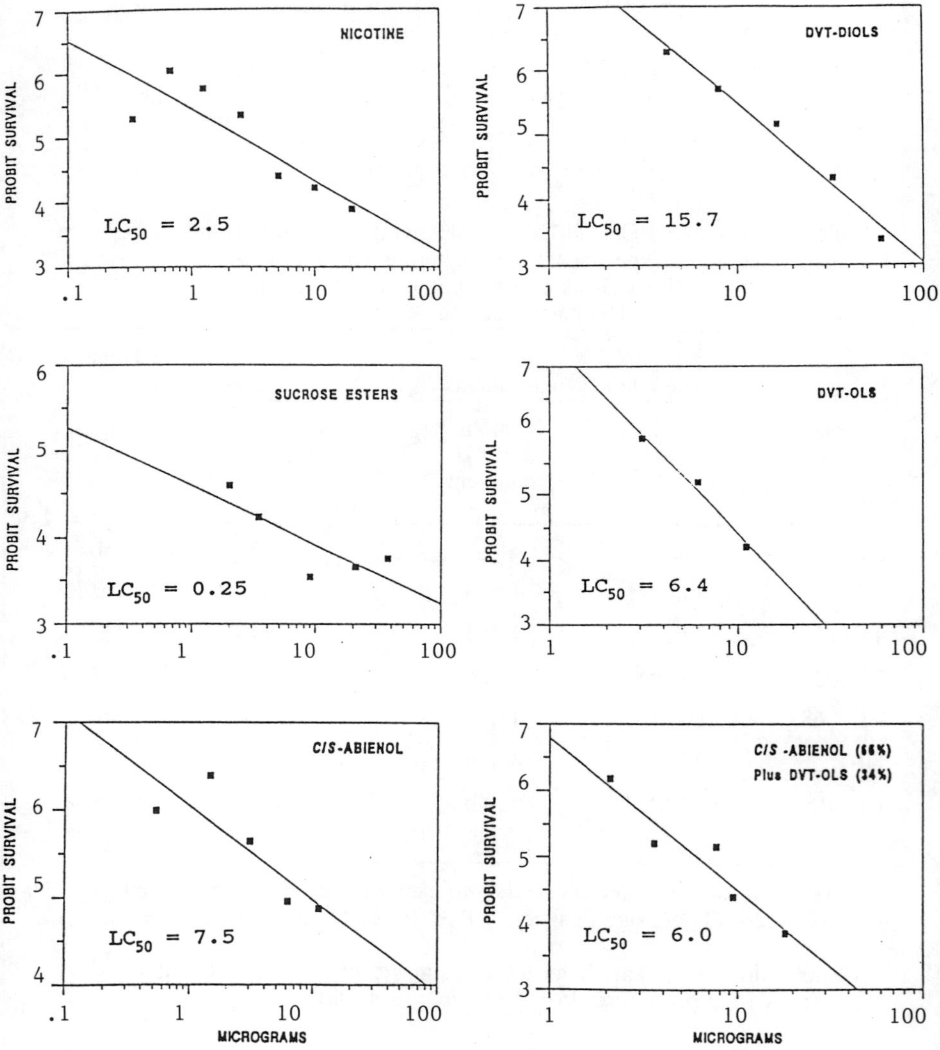

Figure 1. Probit analyses for survival rates of tobacco aphids that were topically dosed with components isolated from tobacco germplasm sources (N=60 aphids per dose; Regression R^2's: Nicotine, 0.75; Sucrose Esters, 0.66; *Cis*-Abienol, 0.73; α- and ß-4,8,13-duvatriene-1,3-diols (DVT-diols), 0.98; α- and ß-4,8,13-duvatrien-1-ols (DVT-ols), 0.98 DVT-ols + *Cis*-abienol, 0.90).

Literature Cited

1. Peters, D. C.; Webster, J. A; Chlouber, C. S. *Proc. Aphid-Plant Interactions: Populations to Molecules.* **1990**.
2. Blackman, R.L.; Eastop, V. E. *Aphids On the World Crops: An Identification and Informational Guide*. John Wiley and Sons, New York, NY. 1984; p 466.
3. Dixon, A.F.G. *Aphid Ecology*. Blackie, Glasgow, 1985, p 157.
4. Klingauf, F. . In: *Aphids: Their Biology, Natural Enemies and Control.* Minks, A. K and Harrewijn, P., Eds.; Host Plant Finding and Acceptance; Elsevier: New York, NY, 1987; Vol. 2A; p 209-233.
5. Lampert, E.P.; Dennis C. A. *Tob. Sci.* **1987**, *31*, 91-93.
6. Webster, J.A. *PROCEEDINGS Aphid-Plant Interactions: Populations to Molecules.* **1990**, pp. 87-111.
7. Thurston, R. *J. Econ. Entomol.* **1961**, *54*, 946-949.
8. Thurston, R.; Smith, W. T.; Cooper, B. P. *Entomol. Exp. Appl.* **1966**, *9*, 428-432.
9. Elsey, K.D.; Chaplin, J. F. *J. Econ. Entomol.* **1978**, *71*, 723-725.
10. Johnson, A.W. *Tob. Sci.* **1978**, *22*, 41-43.
11. Johnson, A. W.; Severson, R. F. *Tob. Sci.* **1982**, *26*, 98-102.
12. Severson, R. F.; Johnson, A. W.; Jackson, D. M. *Recent Advances in Tobacco Science.* **1985**, *11*, 105-174.
13. Jackson, D. M.; Sisson, V. A.; Severson, R. F. In: *Proc. Aphid-Plant Interactions: Populations to Molecules.* **1990**, p 281.
14. Severson, R. F.; Arrendale, R. F.; Chortyk, O. T.; Johnson, A. W.; Jackson, D. M.; Gwynn, G. R., Chaplin, J. F.; Stephenson, M. G. *J. Agric. Food Chem.* **1984**, *32*, 556-570.
15. Severson, R. F.; Arrendale, R. F.; Chortyk, O. T; Green, C. R.; Thome, F. A; Stewart, J. L; Johnson, A. W. *J. Agric. Food Chem.* **1985**, *33*, 870-875.
16. Severson, R. F.; Stephenson, M. G.; Johnson, A. W.; Jackson, D. M.; Chortyk, O. T. *Tob. Sci.* **1988**, *32*, 99-103.
17. Jackson, D. M.; Severson, R. F.; Sisson, V. A.; Stephenson, M. G. *J. Chem. Ecol.* **1991**, *17*, 2489-2515.
18. Jackson, D. M.; Severson, R. F.; Johnson, A. W. *Recent Advances in Tobacco Science* **1989**, *15*, 26-116.
19. Severson, R. F.; Jackson, D.M.; Johnson, A. W.; Sisson, V. A.; Stephenson, M. G. In: *Naturally Occurring Pest Bioregulators.* Hedin, P., Ed.; ACS Symposium Series No. 449; American Chemical Society; Washington, DC, 1991; pp 264-277.
20. SAS, *SAS/STAT Guide for Personal Computers*, Version 6 Edition, SAS Institute Inc., Cary, NC. 1987; p 1028.

RECEIVED April 30, 1993

PEPTIDES AND NEUROPEPTIDES

Chapter 13

Use of Peptides in the Pursuit of Novel Pest Control Agents

Edward P. Masler

Insect Neurobiology and Hormone Laboratory, Plant Sciences Institute, Agricultural Research Service, U.S. Department of Agriculture, Beltsville, MD 20705

Economic plant crops come under attack from a variety of natural enemies including fungi, insects and other plants. Development of chemical pesticides is expensive, and will become more restricted as environmental concerns are recognized. Natural products, specifically peptides and neuropeptides, offer vast potential for exploitation as leads to pest control agents. Benefits from research into biologically active peptides and related factors include the necessity to investigate and understand, in detail, the physiological context in which the peptide operates. In addition, research on peptide biosynthesis, mode of action, and catabolism reveals numerous biochemical steps which may prove to be targets for control agents. The variety of candidates for naturally-based control agent development is extensive, including microbial iron binding systems, fungal toxins, arthropod venom components, insect immune peptides, neuropeptides and others. In this symposium, these endogenous molecules are discussed with regard to the physiological processes they regulate and potential consequences of their manipulation, the need to explore leads to control agents in a wide variety of organisms, and the need to examine emerging technologies in an effort to bring laboratory successes to the field.

Biologically active peptides convey, within their structures, information derived from the genes coding for the peptides and the genes coding for the processing enzymes which are essential to the biogenesis of mature peptides. Peptides may be considered, then, as informational biomolecules which deliver messages to target receptors, resulting in defined physiological responses. The reality that peptides carry so much information, and are involved in regulating or affecting essentially all physiological activities, demands that they be explored as leads to pest control and crop protection agents. It would appear that, somehow, the information embedded within the peptide can be exploited, and turned against the crop pest. A second compelling reason for the interest in peptides is the real concern for the quality of the environment and the agents used to protect crops. Peptides are natural products, and there is a clear understanding that knowledge of the natural product and its action should lead to a valuable understanding of the physiological process in which the peptide is involved. A detailed study of the peptide and its action will lead to an understanding of the context in which the peptide functions, from gene activation and biogenesis, through interaction with the target and subsequent catabolism. This symposium sought to assemble participants who would describe widely varying approaches to the study and exploitation of peptides. As fields of study begin to mature, sub-fields emerge, and the tendency toward a provincial view of research, and its impact, must be resisted.

PLANT AND FUNGAL PEPTIDES

Plant crops can be threatened by insects, fungi or other plants. Control of these pests through the exploitation of peptides is being explored with a number of organisms. Natural herbicides, such as tentoxin (*1*), offer alternatives to traditional weed control. Their cyclic structure presents a challenge in the design of synthetic mimics, but the complexity of the structures may also offer possibilities for synthetic variants of higher toxicity. Similar possibilities exist for naturally occurring fungicides such as iturin (*2*). Fungal control is also being explored through the study of siderophores (*3*), cyclic peptides necessary for iron transport in microorganisms, including pathogenic fungi.

Other fungal peptides are toxic to insects and are being examined in detail (*4*). Insects present a distinct threat to a diversity of crops and are the focus of much research with regard to peptide biochemistry and physiology as they relate to leads in the development of crop protection agents.

NATURAL TOXINS

Anti-insect peptide and polyamine toxins are produced by a variety of organisms including fungi (*4*), scorpions (*5-9*), spiders (*10-13*), wasps (*14-16*), mites (*17,18*) and bacteria (*19-21*). In general, these toxins bind to gating receptors and disturb ion flow across the cell membrane. However, the molecular targets of the toxins vary. Delta-philanthotoxin from wasp venom (*16*) blocks ion channels in the membranes of muscle cells by binding to glutamate receptors. Toxic proteins from scorpion venom, such as charybdotoxin and AaHI toxin (*7,9*) block sodium and potassium channels. Members of the omega-agatoxin family in spider venom (*10-13*) block calcium channels. The effects of these arthropod toxins are on the neuromuscular system, usually resulting in paralysis. In contrast, the delta-endotoxins of the *Bacillus thuringiensis* bacterium bind to receptors in the midgut, disrupting potassium-regulated amino acid transport and causing cell lysis (*20*). Numerous investigations are targeted at understanding toxin pharmacology and exploiting toxins as insect control agents. Synthetic analogues of delta-philanthotoxin have been designed, prepared, and used to investigate the pharmacology of the natural toxin (*14,16*). Studies on the omega-agatoxins have revealed a family of toxins with varying pharmacological properties (*13*), suggesting a complex array of toxins acting to block a variety of different calcium ion channels. While *B. thuringiensis* provides a natural insecticide in the form of its endogenous toxin, genetic engineering of baculoviruses carrying selected foreign toxin genes is continuing. Genes coding for mite (*18*) and scorpion (*9*) neurotoxins have been expressed in a baculovirus system, and neurotoxic proteins obtained. Such successes indicate the feasibility of this molecular approach. However, various caveats need to be recognized. Cross-reactivity of the toxin with non-target species must be considered. The scorpion toxin had no effect in tests on mice (*9*), whereas the omega-agatoxins from spider venom affected a range of animals, including mammals (*13*). Close examination of the target specificity of toxins is an obvious essential component of any research on these compounds as potential pest control agents.

ENDOGENOUS INSECT PEPTIDES

The biochemistry and physiology of the insect itself are receiving attention as peptides and neuropeptides are investigated as leads to insect control. Three areas of particular interest include reproductive behavior, insect immunity and neuropeptide metabolism.

Reproductive Behavior. Male accessory glands contain peptides which affect female reproductive behavior. Female sexual receptivity is diminished after mating (*22*) and oviposition is stimulated (*22*). Peptides responsible for these behaviors have been isolated from *Drosophila* species (*22,23*) and sequenced. In addition, a peptide with similar behavioral influences in a lepidopteran has been isolated from *Helicoverpa zea (24)*. Miller (this volume) discusses the potential of these "sex peptides" in the development of unique insect control strategies. While males influence female reproductive behavior via the action of "sex peptides", females, especially in the *Lepidoptera*, influence male behavior through the release of pheromone. The production of pheromone, in turn, is controlled by the pheromone biosynthesis activating neuropeptide (PBAN; for a comprehensive review see *25*). PBAN-like activity has been observed in more than 20 insect species (*25*) and offers another avenue for exploitation of neuropeptides in insect control.

Insect Immunity. Insects defend themselves against bacterial attack through the production of anti-bacterial proteins (*26,27*). Such proteins have been identified in a number of insect species including coleopterans, dipterans and lepidopterans (*27,28*), and their genes have been characterized (*27,29*). This insect "immune response" is highly structured and exhibits complexity in the number of defensive proteins which can be induced by a bacterial challenge (*27*). An intriguing observation, which may have potential importance for pest control, is the appearance of anti-bacterial peptides in the midgut of non-infected, metamorphosing *Manduca sexta* larvae (P. E. Dunn, personal communication). This may represent a prophylactic action to protect the vulnerable metamorphosing animal. Clearly, a complex response to infection, manifested by the appearance of specific defensive proteins, affords the possibility of manipulating the immune response of insects and the development of very effective and unique control agents.

Neuropeptides. A third area of intense interest concerns the production and metabolic fate *per se* of insect neuropeptides. The journey of a bioactive peptide from gene activation through receptor binding, release and degradation offers numerous potential targets for the development of pest control agents (*30,31*). In addition, neuropeptides control nearly all physiological processes (*32-34*) in insects. Thus, exploitation of peptide processing and catabolic biochemistry (*31,33*) has the potential to lead to significant numbers of unique pest control agents. Essentially all of the events which occur between mRNA translation and mature peptide secretion (pre-secretion processing), and between secretion and final catabolism (post-secretion receptor binding and degradation), are dependent upon enzyme activities and various specific amino acid sequences (*31*). Knowledge of these enzymes and sequences becomes essential to exploitation. This is a relatively new field of exploration in insects, with the vertebrate literature providing much of the experimental precedent for processing (*35-37*), transport and secretion (*38,39*), receptor (*40*) and degradation (*41*) studies. However, exceptional progress has been made with insect systems (*31*). Molecular genetic studies have revealed precursor structures (*42-44*), and processing products have been identified (*45,46*). Structure-activity analyses (*47,48*) reveal the influence of amino acid sequence and residue structure on biological activity. The insight gained is used in the analysis of the three-dimensional requirements for peptide biological activity and the physical association of peptide ligand with receptor. Degradation mechanisms (*49-51*) appear to depend upon enzymes similar to those described in vertebrate systems (*41*). These successes suggest that the rate of progress in our understanding of insect neuropeptide biochemistry and molecular biology will increase rapidly. Continuing elucidation of peptide structures and enzyme activities will fuel this progress.

SUMMARY

Practical application of the discoveries made during studies on peptide and neuropeptide biochemistry and physiology is the ultimate goal of agricultural research. Practical aspects include environmental concerns, economics, efficacy, flexibility and diversity of approach. Clearly, the environment is a central issue in considering the development and application of alternative pest control agents. It is judicious, then, to examine natural products as leads to control. These products may already serve as natural control agents, such as the various toxins, or may be endogenous compounds, such as the neuropeptides, which are essential for life processes in the pest. An understanding of the toxin, its biochemistry, pharmacology and specificity can lead to the development of targeted analogues which may be highly specific and have low environmental persistence. Another approach is to incorporate the gene for the native toxins, or a modified gene, into a vector such as a baculovirus. Similar approaches can be taken with neuropeptides. Peptide mimetic analogues, antagonists or superagonists, can be directed toward specific processing enzymes or receptors, disrupting their normal activities. Such approaches have been successful in medicine for the control of hypertension (*52,53*). Certainly, the development of biologically-based pest control agents will not be straightforward. Significant technical obstacles need to be overcome. For example, often the engineering of a foreign gene (e.g. toxin) into a plant system for the control of insects confronts the problems of inadequate protein processing and secretion (*6*). Product may be expressed but be inactive or not secreted. Inhibitor design requires sophisticated chemical approaches, for example, the examination of substrate transition state chemistry (*53,54*). Bringing laboratory successes to the field is expensive and, as the research area of bioactive peptides in pest control matures, the economic realities of product development must be faced. The large investment of time, resources and money necessary to follow through on an idea and bring it to the field must be protected by the use of cooperative expertise and a broad outlook. The more knowledge and understanding we have of a physiological system and its biochemical components, the more efficient we will be in recognizing natural lead compounds and the more able we will be to respond to changing field and population conditions (e.g. resistance development; *55*), by modifying the active compound in a logical and rational way. Expertise from diverse areas of research, knowledge and exploitation of cutting-edge technology (e.g. methods for the production of peptide libraries; *56,57*) and constant monitoring of the goals of the control agent development program are essential for success.

LITERATURE CITED

1. Lax, A. R.; Shepherd, H. S. In *ACS Symposium Series 380*; Am. Chem. Society: Washington, D.C. **1988**; pp. 24-34.
2. Klich, M.A.; Lax, A.R.; Bland, J.M. *Mycopathologia* **1991**, *116*, 77-80.
3. Bakker, P.A.; VanPeer, R.; Schippers, B. In *Biological Control of Soil-Borne Pathogens*; Hornby, D. Ed; CAB Int'l.: Wallingford, England; **1990**; pp. 131-42.
4. Gupta, S.; Krasnoff, S. B.; Underwood, N. L.; Renwick, J. A.; Roberts, D. W. *Mycopathologia* **1991**, *115*, 185-89.
5. Darban, H.; Weber, C.; Braun, W. *Biochemistry* **1991**, *30*, 1836-45.
6. Pang, S. Z.; Oberhaus, S. M.; Rasmussen, J. L.; Knipple, D. C.; Bloomquist, J. R.; Dean, D. H.; Bowman, K. D.; Sanford, J. C. *Gene* **1992,** *116*, 165-72.

7. Bontems, F.; Roumestard, C.; Gilquin, B.; Menez, A.; Torna, F. *Science* **1991**, *254*, 1521-23.
8. Gurevitz, M.; Urbach, D.; Zlotkin, E.; Zilberberg, N. *Toxicon.* **1991**, *29*, 1270-72.
9. Stewart, L. M. D.; Hirst, M.; Ferber, M. L.; Merryweather, A. T.; Cayley, P. J.; Possee, R. D. *Nature* **1991**, *352*, 85-88.
10. Bindokas, V. P.; Venema, V. J.; Adams, M. E. *J. Neurophysiol.* **1991**, *66*, 590-601.
11. Olivera, B. M.; Imperial, J. S.; Cruz, L. J.; Bindokas, V. P.; Venema, V. J.; Adams, M. E. *Ann. NY Acad. Sci.* **1991**, *635*, 114-22.
12. Mintz, I. M.; Venema, V. J.; Swiderek, K. M.; Lee, T. D.; Bean, B. P.; Adams, M. E. *Nature* **1992**, *355*, 827-29.
13. Venema, V. J.; Swiderek, K. M.; Lee, T. D.; Hathaway, G. M.; Adams, M. E. *J. Biol. Chem.* **1992**, *267*, 2610-15.
14. Karst, H.; Piek, T.; vanMarle, J.; Lind, A.; vanWeeron Kramer, J.; *Comp. Biochem. Physiol. C.* **1991**, *98*, 471-77.
15. Blagbrough, I. S.; Bruce, M.; Bycroft, B. W.; Mather, A. J.; Usherwood, P. N. R. *Pesticide Sci.* **1991**, *30*, 397-403.
16. Brundell, P.; Goodnow, R.; Kerry, C. J.; Nakanishi, K., Sudan, H. L.; Usherwood, P. N. *Neurosci. Lett.* **1991**, *131*, 196-200.
17. Tomalski, M. D.; Kutney, R.; Bruce, W. A.; Brown, M. R.; Blum, M. S.; Travis, J. *Toxicon.* **1989**, *27*, 1151-67.
18. Tomalski, M. D.; Miller, L. K. *Nature* **1991**, *352*, 82-85.
19. Garczynski, S. F.; Crim, J. W.; Adang, M. J. *Appl. Environ. Microbiol.* **1991**, *57*, 2816-20.
20. Knowles, B. H.; Knight, P. J.; Ellar, D. J. *Proc. Roy. Soc. Lond. Biol.* **1991**, *245*, 31-35.
21. Wu, D.; Aronson, A. I. *J. Biol. Chem.* **1991**, *267*, 2311-17.
22. Chen, P. S.; Balmer, S. J. *J.Insect Physiol.* **1989**, *35*, 759-64.
23. Chen, P. S.; Stumm-Zollinger, E.; Aigaki, T.; Balmer, J; Bienz, M.; Bohlen, P. *Cell* **1988**, *54*, 291-98.
24. Kingan, T. G.; Raina, A. K.; Bodnor, W.; Hunt, D. F. , *Soc. Neurosci.* **1991**. *Abstr. 17*, 549.
25. Raina, A. K. *Annu. Rev. Entomol.* **1993**, *38*, 329-49.
26. Dunn, P. E. *Annu. Rev. Entomol.* **1986**, *31*, 321-39.
27. Kimbrell, D. A. *BioEssays* **1991**, *13*, 657-63.
28. Bulet, P.; Cociancich, S.; Dimarcq, J. L.; Lambert, J.; Reichart, J. M.; Hoffmann, D.; Hetru, C.; Hoffmann, J. A. *J. Biol Chem.* **1991**, *266*, 24520-25.
29. Tryselius, Y.; Samakovlis, C; Kimbrell, D. A.; Hultmark, D. *Eur. J. Biochem.* **1992**, *204*, 395-99.
30. Keeley, L. L.; Hayes, T. K.; Bradfield, J. Y. In *Insect Neurochemistry and Neurophysiology-1989*; Borkovec, A. B.; Masler, E. P. Eds.; Humana: Clifton, New Jersey, 1990; pp 163-203.
31. Masler, E. P.; Kelly, T. J.; Menn, J. J. *Arch. Insect Biochem. Physiol.* **1993**, *22*, 87-111.
32. Kelly, T. J.; Masler, E. P.; Menn, J. J. In *Pesticides and Alternatives*; Casida, J. E. Ed.; Elsevier Science Publ.: Amsterdam, 1990; pp. 283-97.
33. Masler, E. P.; Kelly, T.J.; Menn, J. J. In *ACS Symposium Series 453*; Am. Chem. Society: Washington, D.C. 1991; pp. 6-18.
34. Menn, J. J.; Borkovec, A. B. *J. Agri. Food Chem.* **1989**, *37*, 271-78.
35. Harris, R. B. *Arch. Biochem Biophys.* **1989**, *275*, 315-33.
36. Darby, N. J.; Smyth, D. G. *Bioscience Reports* **1990**, *10*, 1-13.
37. Schwartz, T. W. *FEBS Letters* **1986**, *200*, 1-10.
38. Fisher, J. M.; Sossin, W.; Newcomb, R.; Scheller, R. H. *Cell* **1988**, *54*, 813-22.

39. Sossin, W. S.; Fisher, J. M.; Scheller, R. H. *J. Cell Biology* **1990**, *110*, 1-12.
40. Hazum, E. In *Neuropeptides: A Methodology*; Fink, G.; Harmar, A. J. Eds.: Wiley, Chichester, England; 1989; pp. 225-43.
41. Turner, A. J. In *The Biology and Medicine of Signal Transduction;* Nishizuka, Y et al. Eds.: Raven, New York, 1990; pp. 467-71.
42. Davis, M. T.; Vakharia, V. N.; Henry, J.; Kempe, T. G.; Raina, A. K. *Proc. Nat'l. Acad. Sci. USA.* **1992**, *89*, 142-46.
43. Noyes, B. E.; Schaffer, M. H. *J. Biol. Chem.* **1990**, *265*, 483-89.
44. Nichols, R., Schneuwly, S. A..; Dixon, J. E. *J. Biol. Chem.* **1988**, *263*, 12167-70.
45. Hekimi, A.; Fischer-Lougheed, J.; O'Shea, M. *J. Neurosci.* **1991**, *11*, 3246-56.
46. O'Shea, M.; Rayne, R. C. *Experientia* **1992**, *48*, 430-38.
47. Raina, A. K.; Kempe, T. G. *Insect Biochem. Mol. Biol.* **1992**, *22*, 221-25.
48. Nachman, R. J.; Roberts, V. A.; Dyson, H. J.; Holman, G. M.; Tainer, J. A. *Proc. Nat'l. Acad. Sci. USA*. **1991**, *88*, 4518-22.
49. Isaac, R. E. *Biochem. J.* **1988**, *255*, 843-47.
50. Isaac, R. E.; *Biochem. J.* **1987**, *245*, 365-70.
51. Rayne, R. C.; O'Shea, M. *Insect Biochem. Mol. Biol.* **1992**, *22*, 25-34.
52. Greenlee, W. J. *Pharmaceutical Res.* **1987**, *4*, 364-74.
53. Lizuka, K.; Kamijo, T.; Harada, H.; Akahane, K.; Kubda, T.; Umeyama, H.; Ishida, T.; Kiso, Y. *J. Med. Chem.* **1990**, *30*, 2707-14.
54. Hammock, B. D.; Szekacs, A.; Hanzlik, T.; Maeda, S.; Philpott, M.; Bonning, B.; Possee, R. In *Recent Advances in the Chemistry of Insect Control, II;* Crombie, L. Ed.; The Royal Soc.: Cambridge, 1990, pp. 256-77.
55. Hochberg, M. E.; Waage, J. K. *Nature* **1991**, *352*, 16-17.
56. Foder, S. P.; Read, J. L.; Pirrung, M. C.; Stryer, L.; Lu, A. T., Solas, D. *Science* **1991**, *251*, 767-73.
57. Houghten, R. A.; Appel, J. R.; Blondelle, S. E.; Cuervo, J. H.; Dooley, C. T.; Pinilla, C. *BioTechniques* **1992**, *13*, 412-21.

RECEIVED September 27, 1993

Chapter 14

Sex Peptides

Potentially Important and Useful Regulators of Insect Reproduction

J. R. Miller[1,2], J. L. Spencer[1,2], A. J. Lentz[1,2], J. E. Keller[1,2], E. D. Walker[1], and J. F. Leykam[3]

[1]Department of Entomology, [2]Pesticide Research Center, and [3]Department of Biochemistry, Michigan State University, East Lansing, MI 48824

Male-produced peptides that, upon delivery to females at mating, turn off remating in the recipient and turn on oviposition (sex peptides) play a very important reproductive role in many true flies. These agents function in chemical mate guarding so as to allow males to gain numerous matings with high assurance of paternity. Although "sex peptide-like" effects have been demonstrated for about 30 fly species, only two *Drosophila* sex peptides have been chemically identified. They are 36mers unusually rich in hydroxyproline. Apparently, a single compound both turns off female remating and turns on oviposition. This chapter reviews and updates work on sex peptide biology and chemistry for *Drosophila*, *Aedes* mosquitoes, and vegetable flies (*Delia*). Sex peptides reveal a potentially important target for insect birth control, as activated virgin females readily lay unfertilized eggs. Moreover, a new type of biological control could be developed by capitalizing on sex peptide asymmetries across closely related and interbreeding pest and nonpest species. Nonpest males could sterilize virgin pest females by delivering active sex peptide but incompatible sperm. Sex peptides are likely to be critical factors in interspecific reproductive interference (satyrization).

This chapter is intended to help set the topic of peptides governing insect remating and oviposition (sex peptides) into current biological context and to report on the progress of sex peptide research for the true flies (Diptera). Additionally, we revisit and extend ideas about the potential impact of sex peptides on the reproductive biology of insects, as well as suggest how sex peptides might impact insect control.

0097–6156/94/0551–0189$06.25/0

Mammalian vs. Insectan Reproductive Strategies

Sexual reproduction in terrestrial animals is an admittedly clumsy business, apparently justified largely by the thwarting of natural enemies via genetic diversification of progeny (*1*). Spermatozoa must find the relatively few and ephemeral ova in both time and space, and zygotes must be provided an environment hospitable to embryogenesis for an appreciable time. Reproductive effort can reduce parental lifespan by competing with maintenance efforts like feeding, particularly when the two types of behaviors require different environments.

Although mammals and insects have both mastered the art of sexual reproduction in a heterogeneous environment, their approaches differ. In mammals this process is simplified by having females capable of making and sustaining a portable microhabitat (womb) favoring fertilization and embryogenesis. Wombs offer advantages equivalent to those that might accrue to an imaginary bird capable of carrying its nest along on foraging excursions. As an aid to synchronization of sperm presence and ovulation, many female mammals exhibit estrous, signaled by, e.g., changes in genitalia or their emissions, bouts of inviting vocalizations while assuming positions conducive to copulation, and emission of sex pheromones. Since males make themselves highly available to females in heat, and since unfertilized ova are present only infrequently, there is little reason for sperm to be long-lived in mammalian reproductive tracts. In fact, the effective lifespan of mammalian sperm averages only a few days (*2*, p. 38), with at least one notable exception; some bats (mostly hibernating species) can store viable sperm for up to 200 days (*3*).

Young mammals, by definition, are nourished with milk until they can forage. High parental investment, combined with relatively large size (which reduces niche availability (*4*)) render mammals quintessential k strategists.

Insects, on the other hand, are predominantly r strategists -- favoring high numbers of offspring, but with low parental investment in each. Except for the primitive orders (*5*), insects have simplified the process of reproducing in a heterogeneous environment by storing sperm in specialized organs (spermathecae) and dispensing them just a few at a time to fertilize individual eggs as they are being deposited. This advance allows female insects to produce and deposit many eggs over an extended period and without requiring the presence and participation of a male at each reproductive event. Compared to animals like fishes and frogs, which have achieved little decoupling of oviposition and mating, most insects have attained a high level of ovipositional liberation. Moreover, the typical gravid insect can balance maintenance vs. reproductive behaviors as conveniently as do pregnant mammals.

The Selfish Gene Perspective

The switch from "good-of-the-species" arguments to the "selfish gene" paradigm (*6*) has revolutionized research on animal behavior and reproduction. Now rather than expecting members of a species to cooperate so as to increase overall reproductive efficiency for the species, individuals are expected to do whatever increases their own reproductive success, particularly in relation to conspecifics. Some authors even argue that the fundamental units of selection are confederations of genes below the level of the individual (see (*6*) for review). Disconcerting as it may be to the kind-hearted, acts of altruism are now unexpected in the natural world, aside perhaps from some individuals among the "higher" life forms capable of rational and contemplative thought. The *law* of genetic selfishness is expected to manifest itself overtly in arms, antlers, mandibles, and claws, as well as subtly, e.g., in competition of sperm within a single ejaculate for the egg (*7*), in female choosiness among potential mates (*8*), or in certain cases, female promiscuity (*9*).

Challenge in Guarding Paternity

As selfish gene theory would predict, male animals act to enhance the probability that their deposited sperm yield offspring. For example, in addition to monopolizing preestrous female deer, a capable buck guards his mate closely after copulation and drives away other males who might deposit additional sperm that could share in or pre-empt paternity. Postcopulatory guarding in this case requires only a few days, as all available eggs are by then safely fertilized. There is no danger that a doe has stored and will fertilize her ova with sperm deposited during a previous estrous. When females cannot be monopolized successfully as in a harem, and instead, mate with multiple males during ovulation (e.g., various primates), selection has favored males having unusually large testes producing innumerable, vigorous sperm so as to win in a numbers game (*10*).

Guaranteeing paternity in the insect world can be considerably more complicated than for mammals. A female insect generally produces hundreds of eggs whose availability for fertilization can be spread over her lifetime. Thus, male insects using the strategy of physically guarding their mates may be forced to keep station for a much greater portion of a female's lifespan than is true for mammals. An effective way to physically guard a female insect is for the male to remain in copulatory position with his genitalia clasped over or very near the female's. Perhaps this insight somewhat deflates the reputation for stamina often accorded insects observed "to mate" for hours, days, or even ten continuous weeks in the case of certain walking sticks (*11*).

Much of the time dragonflies and damselflies fly in tandem is mate guarding rather than mating *per se*. In some of these Odonata, competition among males has become bizarre: male genitalia include a bristled horn that, like a bottle brush, scrubs sperm from previous matings out of the genital chamber of the female before the current male's sperm are transferred (*8*, p. 316). The only way such a male damselfly can guarantee paternity is to fly in tandem with his mate while she oviposits; eggs of insects are fertilized just before deposition. The male who uncouples to clasp a more attractive female nearby has no guarantee that his penultimate mate will bear him more offspring if males are abundant.

A second factor greatly complicating paternity in higher insect orders is the presence of the sperm-storing spermathecae. These often-sclerotized and glandular spheroids store ejaculated sperm in a densely packed and quiescent state (Figure 1). Not only do these organs liberate females to independently select where and when eggs are laid, they also can afford some choice of sires. A female finding a male more desirable than a previous mate(s), can accept new sperm into her spermathecae. For many insects, sperm are dispensed with a strong bias toward "last in-first out" (sperm precedence) (*2*, p. 11); thus, the new male will sire most of the eggs laid subsequently, or until his sperm are largely expended. In groups having multiple spermathecae, an interesting but open question is whether females exercise mate choice by segregating sperm from different matings for selective dispatch.

Chemical Mate Guarding in Insects

Given their genetic plasticity, diversity, vast numbers, and seniority on Earth, it is perhaps not surprising that some insects have evolved mechanisms of mate guarding more sophisticated than the prolonged physical guarding that severely reduces additional matings. The emerging picture for Diptera is that, in addition to sperm, males commonly deliver substances that: 1) turn off further mating receptivity in the female, and 2) turn on oviposition at least for a time. This *pair* of actions, mediated by one or more compounds delivered in the ejaculate, is offered as a tentative definition of chemical mate guarding in insects.

A synopsis of the developments leading to the study of sex peptides follows (see *12-14* for excellent reviews): As early as the 1950s, evidence was forthcoming that both testicular and paragonial materials influenced remating of *Drosophila* females. Maynard Smith (*15*) established that *D. subobscura* females became refractory to further matings for about one day when receiving only paragonial secretions (mating with testes-less males) and for about one week after a normal mating. Similar results were reported by Manning (*16*, *17*) for *D. melanogaster* and *D. simulans*; moreover, he suggested quite prophetically that remating might be best correlated with disappearance of chemicals associated with sperm that "operate via the hemolymph on the nervous system", and not directly by physical presence or absence of spermatozoa (a similar scenario was envisioned for grasshoppers (*18*)). Proof that paragonial secretions alone explained *D. melanogaster's* short-term mating refractivity was supplied by the transplantation experiments of Merle (*19*) and, most convincingly, by the chemical purification and injection bioassays of Chen and coworkers (*20*, *et ante*).

Evidence for the ovipositional component of chemical mate guarding also came in the late 1950s, when Kummer (*21*) found that injections of male paragonial gland extracts seemed to turn on oviposition in *D. melanogaster*. Garcia-Bellido (*22*) and Leahy (*23*) confirmed this by paragonial gland transplants. Moreover, upon extraction and partial purification, Leahy reported (*24*) and Chen and coworkers confirmed (*20*, *et ante*) the active factor was a peptide. It was unclear how Leahy's remating/ovipositional peptide was related to that of the male-specific, paper-chromatographic spot Fox *et al.* (25) suggested was responsible for *Drosophila* sex determination and thus named *sex peptide*. Although a biological role and identity were never convincingly assigned to Fox's peptide(s), the name "sex peptide" has transferred generically to peptides turning off mating receptivity and turning on oviposition.

These pioneering works generated a burst of research in the 1960s and early 1970s on RIS and FES -- Remating Inhibiting Substances and Fecundity Enhancing Substances, respectively, by the nomenclature of (*13*), of which sex peptides are a subset. In his 1988 review, Gillott (*13*) reported the following numbers of species for which RIS and FES, respectively, had been demonstrated: Orthoptera 0, 5; Hemiptera 0, 1; Coleoptera 0, 1; Lepidoptera 1, 2; and, prominently, Diptera 21, 9. The inequality between the incidences of RIS and FES, inconsistent with the hypothesis that most of these agents are involved in chemical mate guarding, appears to be related more to incomplete investigation than negative evidence.

A few additional examples have been reported since Gillott's 1988 review (*13*), mostly Diptera (*26-29*); however, Heady (*30*) has very recently confirmed a RIS in the planthopper, *Prokelisia dolus* (Homoptera: Delphacidae). Injection of a water-soluble factor from either the male accessory glands or the testes/seminal vesicles complex caused virgin females to cease acoustic calling of distant males and reject copulatory advances of nearby males. Again, oviposition was not investigated.

Not all RIS and FES are peptides. The well-known work of Destephano and Brady (*31*, *et ante*) and Stanley-Samuelson and Loher (*32*, *et ante*) on crickets established that male-produced esterified arachadonic acid and prostaglandin E_2 synthetase are transferred to females where the resulting prostaglandin facilitates egg output. A similar prostaglandin-synthesizing complex has been reported from males of the migratory grasshopper, *Locusta migratoria* (*33*). However, where chemical characterization has been accomplished, the preponderance of RIS and FES are peptides ranging widely in molecular weight estimates: about 500 in the common bean weevil (*34*), 750 in the house fly (*35*), 3,000 in *Culex* mosquitoes (*36*), 4,000 in *Drosophila* (*20*, *37*), 13,000 in *Locusta* grasshopper (*38*), 30,000 in *Melanoplus* grasshopper (*39*), and 30,000 - 60,000 in *Aedes* mosquitoes (*40*); the latter estimate is now suggested to be inflated (discussed below). More details on dipteran sex peptide chemistry and modes of action follow according to taxon.

Drosophila

Knowledge of dipteran sex peptides progressed immensely with the 1988 success of Chen and coworkers in isolating and identifying the sex peptide for *D. melanogaster* (*20*). Briefly, samples of several hundred paired paragonial glands and ejaculatory ducts were dissected into 80% methanol and sonicated. After centrifugation the supernatant was collected and lyophilized. The extract, dissolved in Bennett's solution (*20*), was fractionated on a Vydac C4 column eluted with gradients of 0.04% TFA in 95% acetonitrile, or 20 mM H_3PO_4 in 95% methanol. Repeated reverse-phase HPLC eventually resolved a single biologically active peak from a multitude of glandular constituents. The single material in this peak *both* suppressed remating *and* induced oviposition when injected into virgin females at a physiological dosage of 3 pmole/fly (*41*). Thus, in *D. melanogaster*, a single compound can conveniently effect mate guarding, at least for the short term.

Amino acid analysis and complete sequencing were performed on the purified peptide (*20*). The resulting 36mer, Drm-SP-1 according to the nomenclature of Raina and Gade (*42*), is detailed in Figure 2 along with the nearly identical sex peptide (Drs-SP-1) of *D. sechellia*, very closely related to *melanogaster* (*43*). No homology was found between these *Drosophila* sex peptides and peptides or proteins in sequence data banks (*20*). Drm-SP-1 and Drs-SP-1 contain numerous basic amino acids and an unusually high number of hydroxyprolines, suggestive of fibrous proteins (*44*). Perhaps sex peptides are derivatives of the proteinaceous packets (spermatophores) once used by most insects for transferring sperm, but secondarily lost in some advanced groups. In addition to the hydroxyprolines, these *Drosophila* sex peptides contain one additional posttranslationally modified residue at position 14; it is tentatively assigned as hydroxyisoleucine. Two cysteines at positions 24 and 36 (Figure 2) suggest an intrachain disulfide bridge to the C terminus. Deletions of residues 1-7 at the N terminus only slightly reduced Drm-SP-1 activity, whereas more extensive deletions or any involving the C terminal cysteine destroy biological activity (*41*). Drm-SP-1 is encoded by a short mRNA accumulating exclusively and at high levels in the male accessory gland (*20*). This peptide is thought to be synthesized as a 55 amino acid precursor (*41*) whose processing is not yet entirely elucidated. There is about 13 ng of Drm-SP-1 in a pair of fully charged accessory glands, about one-third of which is thought to be transmitted to the female per mating.

The gene for Drm-SP-1 has been localized by *in situ* hybridization to polytene chromosomes at 70A (*20*). This 266 bp single-copy gene with one intron has been cloned and expressed when linked to the heat shock promoter or the yolk protein 1 promoter (E. Kubli and coworkers as cited by (*41*)). Other aspects of Drm-SP-1 molecular biology are summarized by Chen (*41*).

Chen and coworkers also have identified several peptides from the paragonial glands of *D. funebris*. These are reviewed in (*41*); their effects on reproductive behavior are presently somewhat obscure.

Little is known about the mode of action of *Drosophila* sex peptides. The data of Bouletreau-Merle (*45*) suggest sex peptide is rapidly transported from the reproductive tract via the hemolymph to the pars intercerebralis (posterior midbrain); thus, Manning's 1962 insight (*16*) appears confirmed. Once in the brain, we suspect sex peptides modulate excitatory states in interneurons turning on or off the motor programs that drive mating and oviposition. Alternatively, they might turn on production of hormones known to activate target tissues like oviduct, e.g., ovulation in *Rhodnius prolixus* (Hemiptera) (*46*).

Despite the recent progress, it must be emphasized that these defined *D. melanogaster* and *D. sechellia* sex peptides reproduce only the one-day mating effect (*20*, *41*, *43*) previously shown in *D. melanogaster* by Maynard Smith (*15*) and Manning (*16*, *17*) to be attributable to the paragonial glands. Still a mystery is the *ca.* one-week mating effect attributed to the testes. By the combined use of testes-less (X/O) mutant males and consecutively mated normal males, Hihara (*47*) elegantly demonstrated that the prolonged reduction in propensity to remate and the pronounced increase in egg laying of normally mated *D. melanogaster* females is not due to sperm alone. Females mating with males that had just

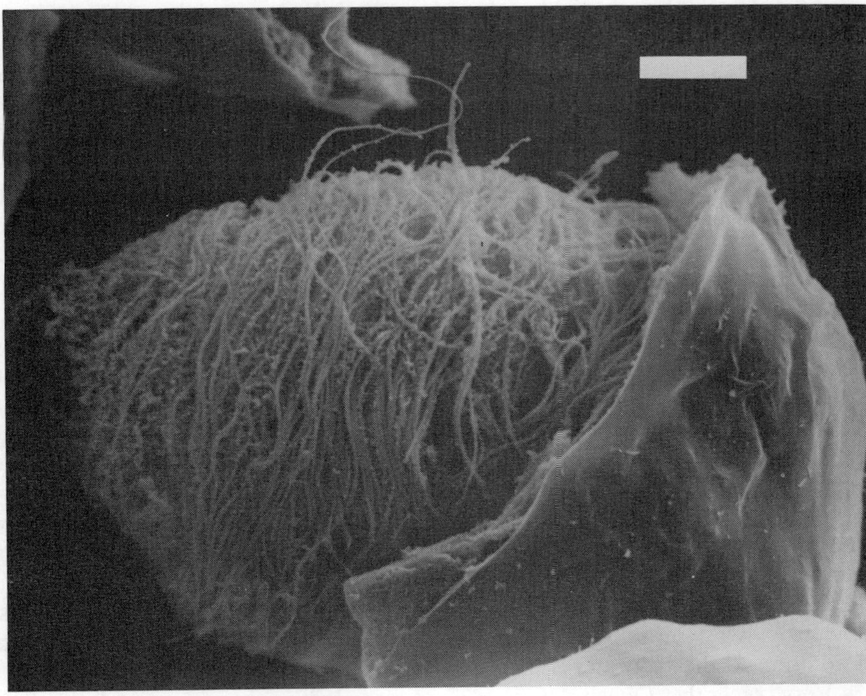

Figure 1. Scanning electron micrograph of an opened onion fly spermatheca revealing the orderly and dense packing of stored spermatozoa. The white bar in the upper right corner is 10 microns long.

	1				5					10			
Drm-SP-1	TRP	GLU	TRP	PRO	TRP	ASN	ARG	LYS	HPR	THR	LYS	PHE	HPR
Drs-SP-1	TRP	GLU	TRP	PRO	TRP	ASN	ARG	GLN	HPR	THR	ARG	TYR	HPR
		15					20					25	
	ILE*	HPR	SER	HPR	ASN	HPR	ARG	ASP	LYS	TRP	CYC	ARG	LEU
	ILE*	HPR	SER	HPR	ASN	HPR	ARG	ASP	LYS	TRP	CYC	ARG	LEU
				30					35				
	ASN	LEU	GLY	PRO	ALA	TRP	GLY	GLY	ARG	CYS			
	ASN	LEU	GLY	PRO	ALA	TRP	GLY	GLY	ARG	CYS			

Figure 2. Amino acid sequence of the sex peptides of *Drosophila melanogaster* (Drm-SP-1) and *D. sechellia* (Drs-SP-1) (adapted from 41). The isoleucines marked with an * are tentative assignments (20, 43).

previously mated with three other females in rapid succession received adequate sperm to fertilize an appreciable number of eggs, but behaved as if they were unmated, ostensibly because they received very little paragonial secretion. These females were induced to lay fertilized and fully viable eggs by additional matings with X/O males who could contribute only paragonial materials. The transience of such a remating (*47*) is explained by delivery only of Drm-SP-1. Hihara (*47*) suggests that the effect of full mating in *D. melanogaster* requires interaction between testis and paragonial gland products, possibly involving the action of sperm and the mysterious filamentous structures generated by the glandular cells in the apical half of the paragonial glands (*48*, *49*, *et ante*).

Despite these knowledge gaps, it seems apparent that the mating effect in *D. melanogaster* is a two-stage process: Drm-SP- 1 from the paragonial glands acts within minutes of delivery to suppress remating and activate oviposition. One day later, when this effect would have expired on its own, the longer-lasting mechanism involving testicular products is fully active. A tempting speculation is that the second mechanism might also involve Drm-SP-1 emanating from a precursor.

This two-stage mechanism makes sense from the perspective of chemical mate guarding by males: the quick-acting sex peptide effect may be critical in turning off female mating receptivity under heavy male pressure in aggregations where mating occurs (*50*). Perhaps the sex peptide causes females to depart mating aggregations (leks), while the longer-term effect might cause females to avoid aggregations of males and be resistive to more isolated mating attempts. Obviously, it will be important to test sex peptides in natural ecological contexts in addition to the laboratory.

Very recently, Ohashi *et al.* (*27*) succeeded in purifying and partially characterizing a sex peptide from male *D. suzukii* whole bodies. In this case, the bioassay was passage of eggs from the ovaries to the "uterus" (ovulation). The active substance is a peptide, Drsuz-SP-1 (*ca.* 40 ng per male), with chemical properties similar to the sex peptides of *D. melanogaster* and *D. sechellia*. The molecular weight is about 4,000, and it has about 36 amino acid residues, including some that are posttranslationally modified. This sex peptide proved unique in that the N terminus was blocked; hence, the full sequence is not yet known. In addition to stimulating ovulation, Drsuz-SP-1 also was highly effective in suppressing remating for one day after injection into virgin females. Again, a single chemical qualifies as a mate guarding agent. However, in contrast to *D. melanogaster*, one physiological dose of *D. suzukii* sex peptide stimulated ovulation for more than six days. Thus, not all sex peptide effects in *Drosophila* are of the transient type discussed above.

Mosquitoes

Chemical analyses of mosquito sex peptides have not progressed nearly as far as those on *Drosophila*. Work on mosquito sex peptides began shortly after such factors were demonstrated in *Drosophila*. In 1965, Leahy and Craig (*51*) reported that oviposition in blood-fed *Aedes aegypti* and *Ae. albopictus* could be stimulated by implants of male paragonial glands. Thereafter, these authors, respectively, showed that the same ovipositional effect could be generated by injection of aqueous extracts of these glands (*52*), and that such extracts also caused virgin females of this normally monogamous species to refuse mates for the remainder of their lives (*53*)! Moreover, female monogamy due to paragonial materials was demonstrated for 10 species of *Aedes*, as well as for *Anopheles quadrimaculatus* and *Culex pipiens* (*53*).

Fuchs *et al.* (*40*) chemically characterized the reproductively active paragonial material of *Ae. aegypti*. It was nondialysable, readily precipitated in 60% $(NH_4)_2SO_4$, and lost behavioral activity when exposed to 50° C for 5 min. Treatment with $MnCl_2$ or $MnSO_4$ to precipitate nucleic acids left the active material in the supernatant, but treatment with 5% trichloroacetic acid destroyed behavioral activity. Appreciable pH sensitivity was seen below 5 and above 9. Likewise, exposure to lipophilic solvents destroyed activity completely. The conclusion that the active substance(s) was proteinaceous was confirmed by loss of activity upon protease cleavage. Furthermore, activity was apparently lost due to aggregation and precipitation upon dialysis against distilled water; the interpretation (*40*) was that the active principle was a globular protein. After Sephadex gel filtration, it was reported that two fractions needed to be combined to achieve mating inhibition of virgin females.

Acetone-powder fractionation of whole-body *Ae. aegypti* male extracts resolved an alpha and beta fraction (*54*) having molecular weights between 30,000 and 60,000, but with differing activities: the former alone reportedly stimulated oviposition, but mating inhibition required both fractions (*55*). No further reports on the chemistry of *Ae. aegypti* have been forthcoming.

Using radiolabelled paragonial substances, Young and Downe (*56*) demonstrated that the heads (brains) of *Culex tarsalis* mosquitoes absorbed paragonial peptides. They also showed that the mating suppressor from carefully dissected paragonial glands of this mosquito falls in the molecular weight range of 2,000 - 5,000. Moreover, these authors showed that combining a relatively clean extract with material from whole bodies caused much of the active substance to bind to larger proteins. Therefore, much of the characterization of *Ae. aegypti* sex peptide(s) requires substantiation in light of the newer findings.

Finally it should be recognized for mosquitoes that paragonial substances may affect various behaviors and physiological functions in addition to mating suppression and activation of oviposition: e.g., increase in rate of digestion of blood meals (*57*), stimulation of vitellogenin synthesis (*58*), modification of circadian flight behaviors (*59*), and induction of preovipositional behaviors (*60*). Thus, the range of behaviors and reproductive functions performed by male accessory reproductive products may be broader than mate guarding, the emphasis of this chapter.

House Fly

Like the Diptera above, the male house fly (*Musca domestica*) delivers material to the female that suppresses remating (*61*). Because muscid flies do not possess paragonia, accessory reproductive secretions are produced by an expanded ejaculatory duct, shown to be the source of the behavioral effector (*61*). Activity was extractable in water, methanol, and ethanol, but not diethyl ether or chloroform. Extract injected into the hemocoel (*62*) was nearly as effective as normal matings. The active material proved heat stable, dialysable, and eluted from a sizing column at about M.W. 3,000 in one estimate (*63*) and about 800 in another (*63*).

Again, these accessory secretions also turn on oviposition (*64*). Interestingly, the threshold for ovipositional initiation in house fly appears to be more easily reached than that for mating refractivity. This stands in contrast to *D. melanogaster* where the dosage of synthetic sex peptide required to turn off mating in 50% of a sample of virgins is identical to that turning on oviposition (E. Kubli, personal communication). The latter case strongly suggests one receptor mediates both effects, while in the former there may be more than one receptor. Alternatively, unequal activation of different behavioral functions may be cued by a common receptor. As there is no reproductive reward for males to voluntarily deliver sex peptide to a female without accompanying sperm, it may be sensible and safe for a female receiving a tentative charge of sex peptide to lay some eggs but not pass up the opportunity to remate if it is presented.

After a hiatus of 20 years, sex peptides of house fly are again under investigation by the USDA (*65*). The *M. domestica* strain recently collected from the field responded more strongly than previously reported. Under similar bioassay conditions, crude aqueous extracts of ejaculatory ducts gave 100% inhibition of remating in the new work (*65*) compared to 70-80 % reduction seen for the Orlando house fly strain (*61*).

Strain effects could be very important in sex peptide research. Using the highly laboratory inbred Carolina Biological strain of *M. domestica*, we were unable to show any ovipositional stimulation whatsoever from injections of aqueous extracts of dissected ejaculatory ducts (A. J. Lentz and J. R. Miller, unpublished). A follow-up study is underway to determine if laboratory selection has suppressed sensitivity to sex peptides in this insect. Such shifts have not been noticed for, e.g., the highly domesticated drosophilids (*45*) or the screw-worm fly (*63*); but, perhaps this question was never under scrutiny.

Although modern separation techniques are now being applied to house fly sex peptide purification, a preliminary report suggests progress on isolation and identification remains very challenging (*65*). Upon HPLC fractionation, mating-suppressant activity was spread over various fractions and the recoverable behavioral activity was very transient compared to that in the starting material or the effect of natural matings. More work is needed on methodology.

Onion Fly

After a detailed quantitative analysis of the ovipositional behaviors of the onion fly, *Delia antiqua*, and associated external sensory cues (*66, 67*), our research group has moved to analyzing internal excitatory and inhibitory inputs, *sensu* the "rolling-fulcrum model" of resource acceptance (*68*). A long-range goal is to understand how external and internal behavioral effectors interact to shape insect foraging dynamics. We postulate sex peptides are a strong internal excitatory input, once absorbed by females

The paragonial glands of this anthomyiid fly produce a sex peptide that, when injected as an aqueous extract into the hemocoel of virgin females, turns off mating and activates oviposition identically to normal mating (*29*). This is a highly potent behavioral effector; 1/8 of a male equivalent of paragonial gland extract switches on the mated behavioral state for the entire *ca.* 6 wk lifetime of these normally monocoitic females (J. L. Spencer and J. R. Miller, unpublished). We have not tested whether other types of stimulation like mechanical manipulation of the female genitalia have any effect; however, they are not required once accessory gland materials are present in the hemolymph.

Many of the chemical properties of the active material from extracts of carefully dissected *D. antiqua* paragonia are similar to those summarized above for *Aedes* mosquitoes, but quite different from those of *Drosophila* sex peptides. This is curious given that dipteran phylogenies place the Anthomyiidae much closer to Drosophilidae than Culicidae (*69*). Activity is readily extracted into physiological saline or distilled water, but less readily into 20% aqueous methanol, a good solvent for *Drosophila* sex peptides (*20, 41*). Apparently onion fly sex peptide is more polar than Drm-SP-1. It does not pass through 100k MW cut-off dialysis tubing containing whole extract, but does pass through smaller pore-sized centrifuge microfilters when flushed with multiple washes of saline or water. Active material is precipitated from whole reproductive tract extracts by as little as 40% $(NH_4)_2SO_4$ and is lost upon 10 min exposure to temperatures > 60° C. It is, however, not lost upon months of storage as a crude aqueous extract at -20° C, or hours of exposure to room air as a dry sample on a centrifuge microfilter. As in *Aedes*, activity is lost when the pH drops to *ca.* 4 or below and upon sample exposure to lipophilic solvents, e.g., even 15% acetonitrile.

These data seem antithetical; most are supportive of a polypeptide or protein, while the centrifuge microfilter data suggest a small peptide. To date we have not been able to prove that there is more than one active substance, e.g., one large and one small. Perhaps a small active substance just happens to share a number of properties with polypeptides and proteins, or perhaps it binds strongly to a larger but inactive polypeptide or protein and masquerades as large. Such a problem has been demonstrated in *Culex* mosquitoes (*56*) and been identified as a pitfall in work on *Aedes* sex peptides.

As is apparent in many of the studies cited above, excluding *Drosophila*, we have had difficulty in recovering activity upon passing extracts through chromatographic systems, e.g., steric exclusion, ion exchange, or bonded hydrocarbon reverse-phase HPLC. This may be because of chemical fragility, loss by irreversible binding, or losses in working up fractions for bioassay. Substantial activity was recovered in one out of three runs using a microbore C18 column eluted with a linear gradient of ethanol and isopropanol (Figure 3). When lyophilized, redissolved in saline, and injected into virgin females, Peaks 1 and 2 both averaged 11 eggs/female/day, while the other chromatographic zones and the negative controls averaged only *ca.* three. Starting extract and normal mating yield *ca.* 12 eggs/female/day.

Peak 2 falls in the elution region expected for polar peptides of a few thousand MW. It is currently unknown whether the activity associated with Peak 1 is due to some unretained Peak 2 material traveling with other unretained substances. The number of paragonial gland equivalents per separation is being raised from a few hundred so that inevitable losses might be sustained without falling too near or below the threshold of sensitivity of the bioassay.

Ovipositional stimulation at about 4 times the negative control also has been electroeluted from a region near the front of a 10% native PAGE gel. The product was not visible when run on SDS-PAGE. Perhaps the quantity was undetectable, or the molecular weight was too low to detect by this analytical method. Capillary electrophoresis is a next step.

We have examined onion fly paragonial extract by matrix-assisted laser desorption-ionization time-of-flight (MALDI-TOF) mass spectrometry (*70*). Figure 4 A reveals the components in the molecular weight range suspected of containing onion fly sex peptide. We speculate that onion fly sex peptide will turn out to be a peak common to the onion fly and seedcorn fly extract (Figure 4 B), since extracts of paragonia from these two very closely related species are cross-reactive (J. L. Spencer and J. R. Miller, unpublished). Quantitative analyses of behavioral effects of these sex peptides await better qualitative and quantitative definition of stimuli.

Role of Sex Peptides in Reproductive Isolation

Provided their activities are not broadly overlapping within a major taxon, sex peptides could play a very important role as a reproductive isolating mechanism. In addition to finding, properly courting, and ejaculating into a receptive female, a male must be able to switch on oviposition if he is to sire offspring. Moreover, it is decidedly to his advantage to switch off female remating so that his sperm are not soon buried under those of another suitor.

From the perspective of a female sensitive to sex peptide, accepting an insemination does not irrevocably seal one's reproductive fate. If that mating does not prove to be a "mood-altering experience", eggs can be held and other mates accepted. Sex peptide communication may be the ultimate step in mate examination. Once oviposition is fully turned on and remating fully turned off, the reproductive die is cast, in some cases for life. Such females can do nothing about having received genetically incompatible sperm; no viable offspring will result.

For females, the advantage in getting to screen males via their sex peptides might be worth letting males play the guarding game. Alternatively, perhaps females are using these signals to lock out additional males once good sperm have been received. Importantly, this good sperm might become unavailable if over-wrapped by bad sperm. Rematings cannot continue indefinitely as there is finite storage space in spermathecae.

It is not yet clear just how taxon-specific sex peptides will turn out to be. Some of the early studies reported very wide heterospecificity -- all the way from interspecific (*41*, *71*) to interfamilial (*70*, *63*). However, scrutiny of the studies purporting widest heterospecificity reveals some deficits in controls, sample sizes, and statistical analyses that cloud interpretations and beg confirmations. Where paragonia were transplanted from one species to another, sex peptide titers might be much elevated over what mating would deliver, and release would be

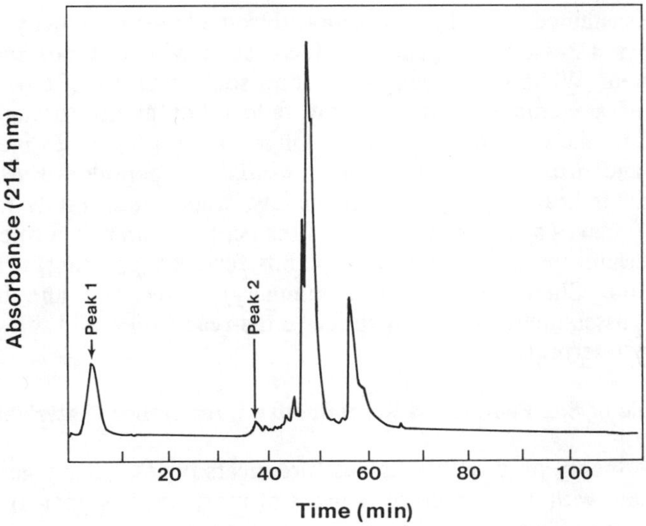

Figure 3. Elution profile of onion fly paragonial gland extract upon microbore C18 RP-HPLC fractionation using a linear gradient of ethanol/isopropanol in phosphate buffered saline. Ovipositional-stimulating activity was associated only with Peaks 1 and 2.

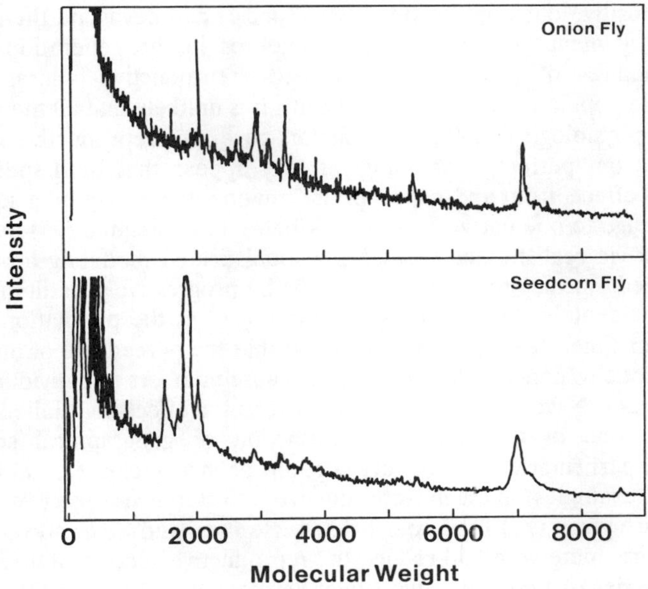

Figure 4. Matrix-assisted laser-desorption/ionization time-of-flight (MALDI-TOF) mass spectra for onion fly (*Delia antiqua*) and seedcorn fly (*Delia platura*) paragonial gland extracts. Sex peptides of other Diptera fall in the range of MW 3,000 - 5,000.

unnaturally sustained. Another problem with some heterospecificity data is that the bioassays assessed sex peptide effects at only one time, shortly after administration. While such a design can show some measure of cross-reactivity, the effect of a common treatment may fade differentially through time, as exemplified by the ovipositional data of Ohashi *et al.* (*71*) for *D. melanogaster*, *pulchrella*, and *suzukii* receiving purified *suzukii* sex peptide. Early bioassay results might indicate strong heterospecificity, while later results might not. Fuyama (*72*) makes a convincing case that sex peptides can play a role as species isolating mechanisms among some drosophilids, further supporting the substantial evidence from Chen (see *41* for a summary). Accurate and biologically meaningful assessments of sex peptide cross-reactivities will be crucial to evolutionary interpretations.

Possible Role of Sex Peptides in Reproductive Interference (Satyrization)

Selfish gene theory predicts that males (producers of less costly gametes) can afford to mate with any female (producers of more costly gametes) who *might* bear them some offspring (*8*, Chapter 3). If the cost of copulation is low, a few cross-species "mistakes" may be of little consequence to a male capable of many matings over a lifetime.

Why then should organisms like Diptera, whose copulations are relatively inexpensive, be respectful of species boundaries? In a landmark 1986 paper, Ribeiro and Spielman (*73*) demonstrated how interspecific matings, especially between closely related but genetically *incompatible* species, may have a startlingly positive outcome for one of the species and devastate the other.

Their argument (*73*) pivots on asymmetries in the generation of sterile hybrids, impaired offspring, or inviable eggs (reproductive failures). Where incompatible populations interact genetically, it is unlikely that all the ecological, behavioral, physiological, and genetic factors governing reproductive interaction will balance out perfectly. Although it may appear that both species suffer detrimental effects from an "inappropriate" mating, there may be a subtle long-term advantage to one party. However, Ribeiro and Spielman (*73*) and Ribeiro (*74*) explain via realistic and biologically sophisticated modeling how the "less disadvantaged" species may actually benefit by progressively outnumbering (at least proportionately) the one most disadvantaged as the population dynamics play out over time. It should be understood that the percentage of interspecific matings is not envisioned to be so high as to cause numbers of individuals in both species to crash. Nevertheless, even small differential effects multiplied over time could drive one of the species to extinction or geographical segregation (parapatry), particularly if the dietary portions of their niches (*73*) also overlap. Real-world examples of such reproductive interference (exact mechanisms unknown) are given (*73*) for *Ixodes* ticks, tsetse flies, and *Aedes* mosquitoes.

Males that mate with and reduce the reproductive success of a female of a different species or population have been termed *satyrs* (*73*, p. 527). In Greek mythology, satyrs were the horny, part-goat-part-man minor gods of the forest who attended Bachus' feasts and orgies (*75*) and who were reputed to have great sexual appetites and no hesitation about mating outside their species. *Satyrization* as used by Ribeiro and Spielman (*73*) is the act of mating with a female of an incompatible species [and reducing her reproductive success]. Ribeiro and Spielman (*73*) showed some ambivalence about insisting that a female who was satyrized was reproductively disadvantaged. We endorse this terminology with that inclusion.

Building upon the background developed in this chapter, sex peptides could have a major influence in shaping reproductive asymmetries, and hence, be prime candidates in any search for chemical satyrizing agents. For simplicity, consider the hypothetical case where two species are genetically incompatible and whose males achieve equal numbers of interspecific and intraspecific matings. If there were differences in sex peptides produced by the two types of males and/or differences in female reception of and responsiveness to them, pronounced asymmetries in sterilization could result (Figure 5). If virgin females of Species 1 were more responsive to conspecific than heterospecific sex peptide, they would ignore a heterospecific mating and have a good chance subsequently for a conspecific mating. With sperm precedence, this type of female would suffer little reproductively once she was properly mated. On the other hand, if females of Species 2 were nearly equally responsive to sex peptides from both types of males, they would be more heavily sterilized because of inability to discriminate more genetically compatible from less genetically compatible males. This differential sex peptide effect (particularly on remating receptivity), compounded over generations, could lead to displacement of one species by another.

We postulate that differential satyrization due to sex peptide asymmetry is a mechanism explaining the displacement in the Southern United States (*76*) of the yellow fever mosquito, *Ae. aegypti*, by the tiger mosquito, *Ae. albopictus*. Accumulating evidence, albeit preliminary, has not falsified this notion. It is known that interspecific matings of these mosquitoes, having strikingly similar biologies, yield no offspring in either direction. But, individuals of both species interact extensively and can intermate (J. Freier, CDC Division of Vector-Borne Infectious Diseases, personal communication to E. Walker). When we inject an aqueous extract of *albopictus* paragonial glands into *aegypti* virgins, cumulative egg output over the next week is virtually identical to that from conspecific sex peptide injections (26 vs. 28 eggs/female, respectively; normally mated positive controls and several negative control produced 15 and 5 eggs per female, respectively. In the reverse direction, *albopictus* females produced 9 vs 3 eggs/female for conspecific vs heterospecific injections, respectively, and 12 vs 1 for the positive and negative controls, respectively. *Ae. albopictus* always produces fewer eggs overall under our laboratory conditions; it has not been in culture as long as our Rockefeller strain of *Ae. aegypti*. However, the sample sizes in this experiment were large enough to support discrimination between sex peptides on the part of *albopictus* females but none by *aegypti*. Already in 1965, Leahy and Craig (*51*) reported the same pattern in ovipositional release by recently field-collected strains of these two mosquito species when paragonia were cross-transplanted. Collectively, these data are supportive of the *Ae. albopictus* satyrization hypothesis; but, much work must be done for a definitive answer to the very important question of whether sex peptide asymmetries are highly influential in satyrizations.

It is intriguing that in MALDI-TOF mass spectrometric analysis, *Ae. albopictus* contains many more ingredients than *aegypti* in the molecular size range typical of identified sex peptides (Figure 6). We wonder if *albopictus* might be armed with an arsenal of sex peptides capable of satyrizing various relatives. The tiger mosquito originates from the Pacific islands, which offer numerous possibilities for reproductive isolations and subsequent recontacts. Such conditions may have sharpened the sex peptide specificity of *Ae. albopictus* beyond that of mainland relatives.

Ways Sex Peptides Might be Used in Insect Control

It is but a small step from discussions of insect sterilizing agents to discussions of insect control. Immediately after discovery of their effects, it was recognized that sex peptides might be developed as pest control agents. For example, in 1969 Craig and Fuchs (*77*) successfully obtained a USA patent on a female mosquito sterilization agent from male mosquitoes.

The notion of exploiting sex peptides as insect birth control agents is appealing: they would be environmentally friendly, reasonably specific, and impact new physiological targets not preselected for resistance by past exposure to pesticides and phytochemicals. Thus, they would be a new tool for incorporation into integrated pest management schemes or pesticide rotations. The universal roadblock to the use of neuropeptides in insect control is getting active material inside insects (*78*). Peptides will not cross the integument and are likely to be degraded by proteases when administered through the gut. Moreover, conventional wisdom has held that the insect midgut cannot absorb intact peptides, and certainly not intact proteins (*79*). However, a few papers are starting to challenge the latter viewpoint. Allingham *et al.* (*80*, *et ante*) have found that in blood-feeding flies, some vertebrate immunoglobulins (MW > 100,000) appear in the insects' hemolymph totally intact and fully functional. This indicates that under conditions of bloodmeal digestion, intact proteins can traverse the midgut cells to the hemolymph. Perhaps insects are capable of some absorption of macromolecules by endocytosis, as is known in ticks (A. Raikhel, personal communication). Thus there is hope for absorption if peptides and proteins were somehow protected from proteases in the insect midgut. Even without protection, pheromone biosynthesis activating neuropeptide (PBAN, a 33mer) was reported to at least partially suppress pheromone titer of tobacco budworm females fed a sugar solution of the hormone (A. Raina, personal communication). We have been unable to demonstrate any measurable effects of feeding onion flies droplets of unprotected paragonial gland extracts.

Could special formulations aid peptide survival in the insect gut? Clinical Technology Associates (*81*) have recently announced development of "proteinoid" spheres that, upon pH manipulations, self assemble into hollow spheres so as to encapsulate whatever proteins are in solution. Upon shifting the pH upward, the spheres disintegrate to release their payloads. This product is exciting because it has successfully protected insulin (MW 6,000) from human stomach proteases and allowed uptake of the peptide in the small intestine. Moreover, this system has successfully delivered vaccines of over one million MW. If the technology

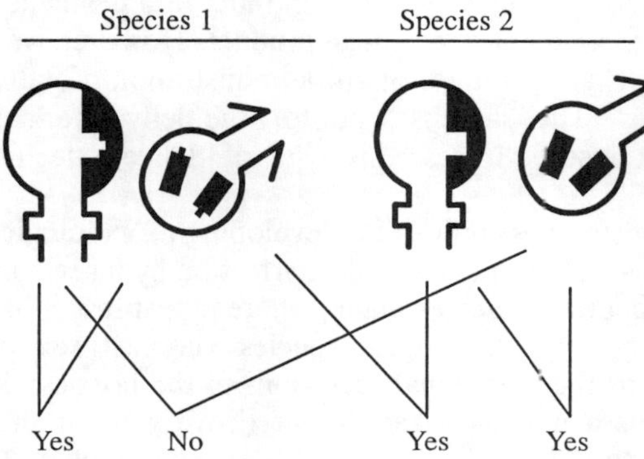

Figure 5. Schematic representation of how females of a species with a less discriminating sex peptide receptor (Species 2) can be activated by the sex peptide of a closely related species (Species 1). If these species were genetically incompatible, Species 2 could be sterilized (satyrized) while Species 1 would not.

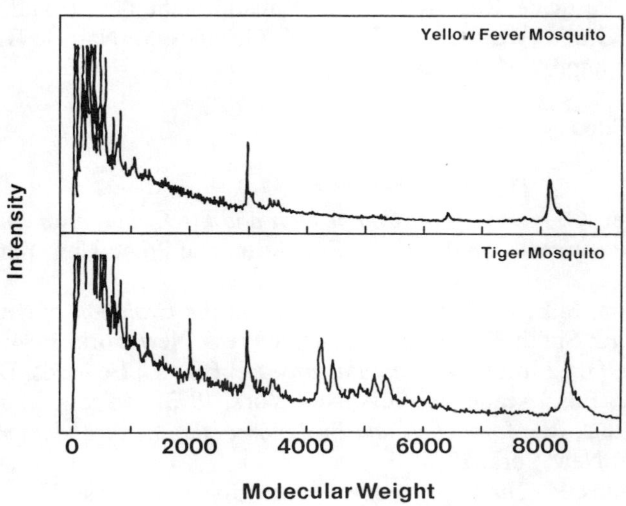

Figure 6. MALDI-TOF analyses of the paragonial gland extracts of the yellow fever mosquito (*Aedes aegypti*) and the tiger mosquito (*Ae. albopictus*). Males of the tiger mosquito contain more ingredients in the several thousand MW range than do those of the yellow fever mosquito.

can be developed for administering insulin to humans via pills containing microcapsules, similar technology should be tried to administer sex peptides to insects via their food. Admittedly, the economics of a pesticide product are less favorable than those for a pharmaceutical product. However, we believe attempts at "proof-of-concept" are justified for gut-administration of natural sex peptides or their analogues. The same is true for the delivery and expression of sex peptide genes via insect viruses; this type of delivery technology is rapidly developing (*82*).

Finally, we envision possibilities for developing new biological control agents whose effectiveness is due to enhanced satyrization by insects having genetically manipulated sex peptides. For example, where a pest insect had geographically displaced a closely related, nonpest species via past sex peptide-mediated satyrization, perhaps the paragonial secretions of the nonpest could be adjusted genetically to favor a new satyrization war ("love war" in the terminology of Ribeiro and Spielman (*74*)) with a humanly desired outcome. Compared to the Sterile Insect Release control technique, this new type of genetic control could have a great advantage; it could, like classical biological control, be self-perpetuating.

Acknowledgments

This work was supported by the USDA National Research Initiatives Grant Program, a Pesticide Research Center Enhancement Grant, and a Rackham Foundation Grant. We thank A. Peters, K. Grimnes, M. Nair, L. Besaw and M. Candolfi for improving the manuscript.

Literature Cited

1. Gutin, J. C. *Discover*. 1992, *13*(*6*): 32-9.
2. Parker, G. A. In *Sperm Competition and the Evolution of Animal Mating Mating Systems*; Smith, R. L., Ed.; Academic Press: New York, 1984; pp. 1-60.
3. Fenton, M.B. In *Sperm Competition and the Evolution of Animal Mating Systems*; Smith, R. L., Ed.; Academic Press: New York, 1984; pp. 573-87.
4. Bush, G. L. In *Evolutionary Patterns and Process*; Edwards, D., and Lees, D. R., Eds.; Academic Press: New York, 1993; In Press.
5. Matsuda, R. *Morphology and Evolution of the Insect Abdomen*. Pergamon Press: New York, 1976.
6. Dawkins, R. *The Selfish Gene*. Oxford University Press: New York, 1976.
7. Smith, R. L., Ed.; Academic Press: New York, 1984; pp. 85-115.
8. Thornhill, R.; Alcock, J. *The Evolution of Insect Mating Systems*. Harvard Univ. Press: Cambridge, MA, 1983.
9. Small, M. F. *Discover* 1992, *13*(*6*), 46-51.
10. Harvey, P. H.; Harcourt, A. H. *In Sperm Competition and the Evolution of Animal Mating Systems*; Smith, R. L., Ed.; Academic Press: New York, 1984; pp. 589-600.
11. Sivinski, J. *Natural History* 1992, *6*, 30-4.

12. Gillott, C.; Friedel, T. In *Advances in Invertebrate Reproduction*; Vol. *1*; Adiyodi, K, G.; Adiyodi, R. G., Eds; Peralam-Kenoth, Karivellur, 1976; pp. 199-218.
13. Gillott, C. *In Reproductive Biology of Invertebrates, Vol III, Accessory Sex Glands*; Adiyodi, K. G.; Adiyodi, R. G., Eds.; Johy Wiley & Sons: New York, 1988; pp. 319- 472.
14. Leopold, R. A. *Ann. Rev. Entomol.* 1976, *21*, 199-221.
15. Maynard Smith, J. *J. Genet.* 1956, *54*, 261-79.
16. Manning, A. *Nature*. 1962, *194*. 252-3.
17. Manning, A. *Animal Behav*. 1967, *15*, 239-50.
18. Haskell, P. T. *Animal Behav*. 1960, *8*, 76-81.
19. Merle, J. *J. Insect Physiol.* 1968, *14*, 1159-68.
20. Chen, P. S.; Stumm-Zollinger, E.; Aigaki, T.; Balmer, J.; Bienz, M.; Bohlen, P. *Cell* 1988, *54*, 291-8.
21. Kummer, H. Z. vergl. *Physiol.* 1960, *43*, 642-79.
22. Garcia-Bellido, A. *Z. Naturforsch*. 1964, *19b*, 491-5.
23. Leahy, M. G. *Drosoph. Info. Serv*. 1966, *41*, 145.
24. Leahy, M. G.; Lowe, M. L. *Life Sci*. 1967, *6*, 151-6.
25. Fox, A. S.; Mead, C. G.; Munyon, I. L. *Science* 1959, *129*, 1489-90.
26. Smith, P. H.; Gillott, C.; Barton Browne, L.; Gerwin, A. C. M. *Physiol. Entomol*. 1990, *15*, 469-81.
27. Ohashi, Y. Y.; Haino-Fukushima, K; Fuyama, Y. *Insect Biochem*. 1991, *21*, 413-19.
28. Bergh, J. C.; Harris, M. O.; Rose, S. *Ann. Entomol. Soc. Am*. 1992. *85*, 224-33.
29. Spencer, J. L.; Bush, G. L., Jr.; Keller, J. E.; Miller, J. R. *J. Insect Behav*. 1992, *5*, 689-697.
30. Heady, S. E. *Physiol. Entomol*. 1993. *In Press*.
31. Destephano, D. B.; Brady, U. E. *J. Insect Physiol*. 1977, *23*, 905-11.
32. Stanley-Samuleson, D. W.; Loher, W. *J. Insect Physiol*. 1983, *29*, 41-5.
33. Lange, A. B.; Loughton, B. G. Int. *J. Inv. Reprod. Dev*. 1984,
34. Das, A. K.; Barbier, M.; Quesneau-Thierry, A. *Experientia* 1980, *36*, 918-9.
35. Nelson, D. R.; Adams, T. S.; Pomonis, J. G. *J. Econ. Entomol*. 1969, *62*, 634-39.
36. Young, A. D. M.; Downe, A. E. R. *Physiol. Entomol*. 1987, *12*, 233-39.
37. Chen, P. S.; Balmer, J. *J. Insect Physiol*. 1989, *35*, 759-64.
38. Lange, A. B.; Loughton, B. G. *Gen. Comp. Endocr*. 1985, *57*, 208-15.
39. Friedel, T.; Gillott, C. *J. Insect Physiol*. 1976, *22*, 489-95.
40. Fuchs, M. S.; Craig, G. B., Jr.; Despommier, D. D. *J. Insect Physiol*. 1969, *15*, 701-9.
41. Chen, P. S. *Annls. Soc. ent. Fr.* 1991, *27*, 231-44.
42. Raina, A. K.; Gade, G. *Insect Biochem*. 1988, *18*, 785-7.
43. Chen, P. S.; Balmer, J. *J. Insect Physiol*. 1989, *35*, 759-64.
44. Voet, D.; Voet, J. G. *Biochemistry*. John Wiley & Sons: New York, 1990.
45. Bouletreau-Merle, J. *J. Insect Physiol*. 1976. *22*, 933-40.
46. Kriger, F. L.; Davey, K. G. *Can. J. Zool*. 1984, *61*, 684-86.
47. Hihara, F. *Zoological Magazine*. 1981, *90*, 307-16.

48. Tandler, B.; Williamson, D. L.; Ehrman, L. *J. Cell Biol.* 1968, *38*, 325-36.
49. Perrin-Waldemer, C. *Biol. Cell.* 1977, *30*, 171-6.
50. Droney, D. C. *Anim. Behav.* 1992, *44*, 1007-20.
51. Leahy, M. G.; Craig, G. B., Jr. *Mosquito News* 1965, *25*, 448-52.
52. Leahy, M. G. *J. Insect Physiol.* 1967, *13*, 1283-92.
53. Craig, G. B., Jr. *Science* 1967, *156*, 1499-501.
54. Fuchs, M. S.; Hiss, E. A. *J. Insect Physiol.* 1970, *16*, 931-39.
55. Hiss, E. A.; Fuchs, M. S. *J. Insect Physiol.* 1972, *18*, 2217-27.
56. Young, A. D. M.; Downe, A. E. R. *Physiol. Entomol.* 1987, *12*, 233-39.
57. Downe, A. E. R. J. *Insect Physiol.* 1975, *21*, 1835-9.
58. Borovsky, D. *Arch. Insect Biochem. Physiol.* 1985, *2*, 405-13.
59. Jones, M. D. R.; Gubbins, S. J. *Physiol. Entomol.* 1979, *4*, 345-51.
60. Klowden, M. J. *Experientia* 1990, *46*, 660-70.
61. Riemann, J. G.; Moen, D. J.; Throson, B. J. *J. Insect Physiol.* 1967, *13*, 407-18.
62. Adams, T. S.; Nelson, D. R. *Ann. Entomol. Soc. Am.* 1968, *61*, 112-6.
63. Nelson, D. R.; Adams, T. S.; Pomonis, J. G. *J. Econ. Entomol.* 1969, *62*, 634-9.
64. Riemann, J. G.; Thorson, B. J. *Ann. Entomol. Soc. Am.* 1969, *62*, 828-34.
65. Bird, T. G.; Raina, A. K.; Hayes, D. K. In *Naturally Occurring Pest Bioregulators*: Hedin, P. A., Ed.; American Chemical Society Press: Washington, DC, 1991; pp. 78-86.
66. Harris, M. O.; Miller, J. R. *J. Insect Behav.* 1991, *4*, 773-92.
67. Cowles, R. S.; Miller, J. R.; Hollingworth, R. M.; Abdel-aal, M. T.; Szurdoki, F.; Bauer, K.; Matolcsy, G. *J. Chem. Ecol.* 1991, *16*, 2401-28.
68. Miller, J. R.; Strickler, K. S. In *Chemical Ecology of Insects*: Bell, W. J.; Carde, R. T., Eds.; Chapman and Hall: London, 1984; pp.127-57.
69. McAlpine, J. F.; Wood, D. M., Eds.; *Manual of Nearctic Diptera*, Vol. 3; Canadian Government Publishing Centre Supply and Services Canada: Hull, Quebec, 1989.
70. Leahy, M. G. *J. Insect Physiol.* 1967, *13*, 1283-92.
71. Ohashi, Y. Y.; Haino-Fukushima, K.; Fuyama, Y. *Insect Biochem.* 1991, *21*, 413-9.
72. Fuyama, Y. *Experientia.* 1983, *39*, 190-2.
73. Ribeiro, J. M. C.; Spielman, A. *Am. Naturalist.* 1986, *128*, 513-28.
74. Ribeiro, J. M. C. *J. Med. Entomol.* 1988, *25*, 431-40.
75. Holme, B. *Bulfinch's Mythology*. Viking Penguin, Inc.: New York 1979.
76. Hobbs, J. H.; Hughes, E. A.; Eichold II, B. H. *J. Am. Mosq. Cont. Assoc.* 1991, 7, 488-9.
77. Craig, G. B., Jr.; Fuchs, M. *United States Patent*, 3, 450, 816, Sept. 17, 1969.
78. Menn, J. J.; Kelly, T. J.; Masler, E. P. *Insect Neuropeptides: Chemistry, Biology, and Action*. American Chemical Society Press: Washington, D.C., 1991.

79. Chapman, R. F. *The Insects: Structure and Function*. Harvard Univ. Press: Cambridge, MA, 1982.
80. Allingham, P.G.; Kerlin, R. L.; Tellam, R. L.; Briscoe, S. J.; Standfast, H. A. *J. Insect Physiol.* 1992, *38*, 9-17.
81. Ericson, D. *Scientific Amer.* 1992, *266*, 108-9.
82. Summers, M. D. In *Insect Neuropeptides: Chemistry, Biology, and Action*; Menn, J. J.; Kelly, T. J.; Masler, E. P., Eds.; ACS Symposium Series 453, Amer. Chem. Soc. Press, Washington, D.C., 1991; pp. 237-51.

RECEIVED June 28, 1993

Chapter 15

Pseudopeptide Mimetic Analogs of Insect Neuropeptides

Ronald J. Nachman[1], Jefferson W. Tilley[2], Timothy K. Hayes[3], G. Mark Holman[1], and Ross C. Beier[1]

[1]Food Animal Protection Research Laboratory, Agricultural Research Service, U.S. Department of Agriculture, College Station, TX 77845
[2]Roche Research Center, Hoffmann-LaRoche, Inc., Nutley, NJ 07110
[3]Department of Entomology, Texas A&M University, College Station, TX 77843

Pseudopeptide analogs that feature acyl group, reduced-bond, and/or stable chemical-construct replacements for amino acids or peptide linkages mimic the biological activity of the insect kinin, sulfakinin, and myosuppressin peptide families. The non-peptide mimetic CP-96,345-1 demonstrated both agonism and antagonism of the cockroach hindgut myostimulatory activity of substance P, which shows sequence homology with the insect locustatachykinins. Biological evaluation of a rigid cyclic analog reveals the conformational preference of the pyrokinin/PBAN active-core during interaction with hindgut/oviduct contractile and pheromone production receptors. The importance of information on the conformational requirements for neuropeptide activity for the development of peptidomimetics is also discussed.

Strategies for the development of future pest insect management agents from insect neuropeptides have been previously advocated and outlined because these chemical messengers regulate critical processes in insects (*1-3*). However, insect neuropeptides in and of themselves hold little promise as traditional insect control agents because of their susceptibility to enzymatic degradation in the target insect, lability under environmental conditions, and inability to pass through the hydrophobic insect cuticle. The removal of the peptide nature (i.e., the constituent peptide bonds) of insect neuropeptides leading to pseudopeptide and nonpeptide analogs represents a strategy that could overcome these limitations.

In this paper, we discuss pseudopeptide mimetic analogs that have been developed for selected insect neuropeptide families, including the insect kinins, sulfakinins, myosuppressins, and insect tachykinins. First members of these families were isolated from the cockroach or locust on the basis of their ability to either stimulate or inhibit contractions of the isolated cockroach/hindgut. Subsequently, these peptide families have been found in a range of insect species and associated with a variety of different physiological responses. The paper closes

0097–6156/94/0551–0210$06.00/0

with a discussion on the conformational preference of the pyrokinin active-core for interaction with hindgut/oviduct myostimulatory and pheromonotropic receptor sites, and on the utility of this information for the development of potent pseudopeptide and nonpeptide mimetic agonists and antagonists.

Insect Kinins

Members of the insect kinin peptide family have been isolated from such diverse sources as the cockroach (Leucophaea maderae), locust (Locusta migratoria), cricket (Acheta domesticus), and mosquito (Culex salinarius) (*1,4*) and share the conserved C-terminal pentapeptide sequence Phe-X^1-X^2-Trp-Gly-NH_2 [X^1 = Asn, His, Ser, or Tyr, X^2 = Pro or Ser]. All of the insect kinins demonstrate myostimulatory properties on the isolated cockroach hindgut preparation (*5*) at potent-thresholds of between 10^{-10} and 10^{-11} M. Achetakinins, leucokinins, and culekinin depolarizing peptide stimulate fluid secretion in Malpighian tubules of the cricket and mosquito (*5-7*). Therefore, the kinins may regulate water and ion balance in addition to hindgut motility in a variety of insects.

The C-terminal pentapeptide sequence common to the insect kinins is all that is required to elicit a physiological response in myotropic and diuretic assays. In particular, the active core sequence Phe-Tyr-Pro-Trp-Gly-NH_2 is equipotent with the parent nonapeptide in hindgut myotropic (*8*) and cricket Malpighian tubule secretion (*6*) assays. Within the active core pentapeptide, the aromatic residues Phe^1 and Trp^3 are of paramount importance for myotropic and diuretic activity (*6-8*), whereas position 2 tolerates wide variations in side chain character ranging from acidic to basic or hydrophobic to hydrophilic. Aromatic residues, such as Tyr or Phe, in the variable position 2 promote the highest potencies in myotropic and Malpighian tubule fluid secretion assays (*6-8*).

Superagonist activity has been observed for the unnatural hexapeptides Ala-Phe-Phe-Pro-Trp-Gly-NH_2 (*1*) and Arg-Phe-Phe-Pro-Trp-Gly-NH_2 (Nachman et al., unpublished data) in cockroach hindgut myotropic and cricket Malpighian tubule fluid secretion assays, with the latter analog demonstrating at least 550-fold greater myotropic potency than the pentapeptide Phe-Phe-Pro-Trp-Gly-NH_2 and/or natural leucokinins of longer chainlength. The leucokinin analog Ala-Tyr-Ser-Trp-Gly-NH_2, which replaces one of the two critical aromatic residues with an Ala, appears to antagonize the activity of the leucokinins/achetakinins in the cricket Malpighian fluid secretion assay at concentrations of between 0.1 and 1 μM or higher (Coast and Nachman, unpublished data). This result underscores the importance of the Trp residue to the receptor-binding affinity and biological activity of the insect kinins.

Pseudopeptide Analogs. Nonpeptide modification of the insect kinin pentapeptide core, leading to analogs with retention of biological activity, was first achieved by replacing the amide bond (-C(O)NH-) between Phe^1 and Phe^2 of the sequence Phe-Phe-Ser-Trp-Gly-NH_2 with a reduced-bond linkage (-CH_2NH-). The resulting pseudopeptides Phe ψ[CH_2-NH] Phe-Ser-Trp-Gly-NH_2 (Figure 1) (*9*) and Phe ψ[CH_2-NH] Phe-Pro-Trp-Gly-NH_2 demonstrated threshold values of about 3 nM, which represents a significant 1% of the activity of the parent pentapeptides. These pseudotripeptides were followed by the synthesis of a series of simple aryl, acyl pseudopeptides of the core pentapeptide Phe-Tyr-Pro-Trp-Gly-NH_2 (Figure 2). The amino acid or amino acid blocks Phe, Phe-Tyr, and Phe-Tyr-Pro were replaced with the aryl, acyl groups hydrocinnamic acid (Hca), 6-phenylhexanoic acid (6Pha), and 9-phenylnonanoic acid (9Pna), respectively. In each of the resulting pseudopeptide analogs Hca-Tyr-Pro-Trp-Gly-NH_2, 6Pha-Pro-Trp-Gly-NH_2, and 9Pna-Trp-Gly-NH_2, the distance between the critical N-terminal phenyl ring and the Trp residue was kept approximately constant via the introduction of methylene groups. An evaluation of the myotropic potency of the pseudopeptide 6Pha-Trp-Gly-NH_2 was also undertaken. Remarkably, the pseudotetrapeptide Hca-Tyr-Pro-Trp-Gly-NH_2 demonstrated a threshold concentration of 58 pM and an ED_{50} of 10 nM on the isolated cockroach hindgut preparation, retaining 70% of the potency of the parent pentapeptide. In addition, this pseudotetrapeptide analog retained 86% of the maximal contractile response of the core pentapeptide. This result demonstrated that the N-terminal amino group is of little consequence to the interaction between the insect kinin core peptide and the receptor site. With a threshold concentration of 8 nM and an ED_{50} of 2 μM, the pseudotripeptide proved to be about 2 orders of magnitude less potent than the pseudotetrapeptide, but retained 98% of the maximal myostimulatory activity of the parent pentapeptide. The two pseudodipeptide analogs 9Pna-Trp-Gly-NH_2 and 6Pha-Trp-Gly-NH_2 demonstrated a further drop in potency of about 2 orders of magnitude when compared with the pseudotripeptide (*10*). Nevertheless, despite major modifications, i.e. the replacement of three N-terminal core-amino acids with a simple aryl acyl group, these pseudodipeptides did not completely lose myostimulatory activity. As 9Pna-Trp-Gly-NH_2 contains 3 more methylene groups than does 6Pha-Trp-Gly-NH_2, the similar potencies of the two analogs suggests that the receptor exhibits tolerance towards modification of the distance between the phenyl ring of the acyl group and the Trp residue.

Taking the modification process one step further, the remaining peptide bond between Trp and Gly of 6Pha-Trp-Gly-NH_2 was replaced with a reduced-amide linkage (-CH_2NH-). The resulting non-peptide mimic 6Pha-Trp ψ[CH_2-NH]Gly-NH_2 (Figure 3) was observed to elicit low but discernable myostimulatory activity on the isolated cockroach hindgut assay and is expected to be resistant to exopeptidase degradation (Nachman, unpublished data). Using a similar strategy, a methyl group was placed on the amide bond between Trp and Gly of 6Pha-Trp-Gly-NH_2 to increase steric hindrance at this linkage. The resulting non-peptide mimic 6Pha-Trp-Sar-NH_2 (Figure 3) (Sar = sarcosine or N-methyl glycine) retained myostimulatory activity at a threshold concentration of 10 μM. The additional steric hindrance at the amide linkage of this pseudopeptide analog will increase resistance to peptidase degradation (Nachman, unpublished data).

Pseudopeptide analogs of the insect kinin family involving modifications of the C-terminus have also been reported. The C-terminal acid forms of the insect kinin sequences are inactive, perhaps as a result of unfavorable interaction between the negatively-charged carboxyl moiety and the myotropic receptor. However, the methyl ester and N-methyl amide analogs of the Ala-Phe-Phe-Pro-Trp-Gly kinin sequence proved active, although at a reduced level. These two analogs demonstrated reductions in myostimulatory potency over the parent peptide of about 2 orders of magnitude, with threshold concentrations in the 0.75 - 1.0 nM range. The N,N-dimethylated amide analog proved completely inactive in the cockroach

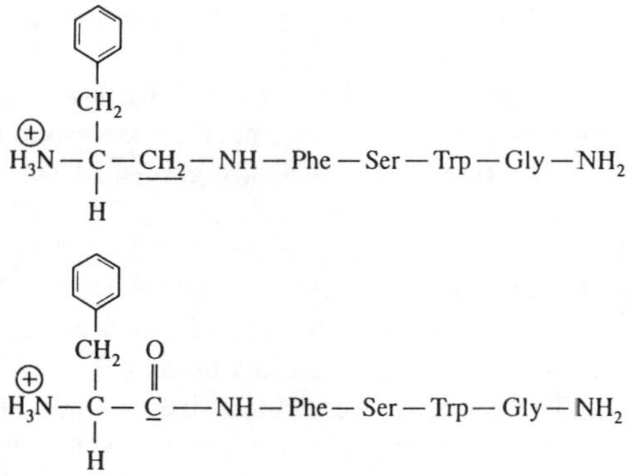

Figure 1. A pseudopeptide analog (top) of the leucokinin C-terminal pentapeptide Phe-Phe-Ser-Trp-Gly-NH_2 (bottom) in which the amide bond (-C(O)-NH-) between the two phenylalanine residues is replaced by a reduced bond linkage (-CH_2NH-) (*1,18*). (Reproduced with permission from ref. 1. Copyright 1993 Wiley Liss.)

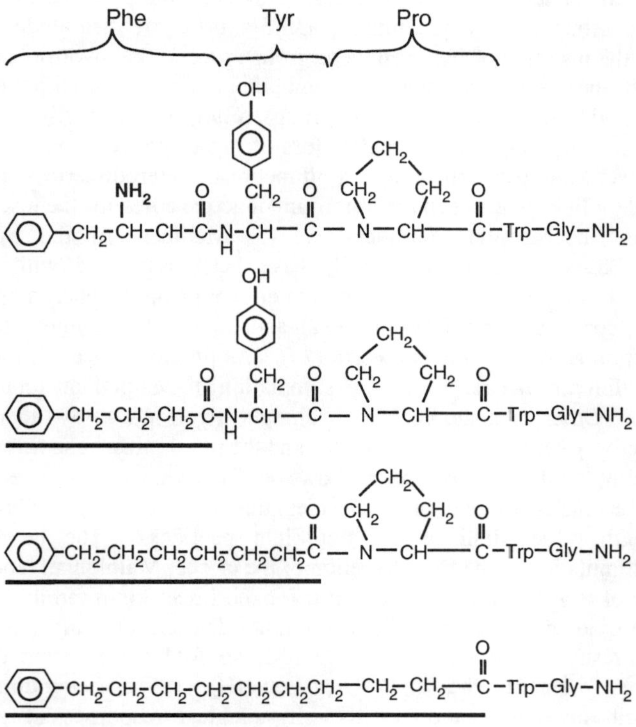

Figure 2. A series of pseudopeptides in which the amino acid and amino acid blocks Phe, Phe-Tyr, and Phe-Tyr-Pro of the achetakinin C-terminal pentapeptide Phe-Tyr-Pro-Trp-Gly-NH_2 (top) are replaced with hydrocinnamic acid, 6-phenylhexanoic acid, and 9-phenynonanoic acid, respectively, is illustrated (*10*). (Reproduced with permission from ref. 10. Copyright 1993 Muunksgaard.)

hindgut myotropic assay, perhaps as a result of greater steric hindrance at the C-terminal amide moiety.

Another series of C-terminally modified insect kinin analogs was synthesized in order to determine the effect of moving the C-terminal amide group away from the α-carbon of the terminal Gly residue. This was accomplished via the addition of methylene group spacers between the α-carbon center of Gly and the C-terminal amide group. The resulting analogs Ala-Phe-Phe-Pro-Trp-ßAla-NH_2, Ala-Phe-Phe-Pro-Trp-γAbu-NH_2 (γAbu $\equiv$ γ-aminobutyroyl), Ala-Phe-Phe-Pro-Trp-ϵAca-NH_2 (ϵAca $\equiv$ ϵ-aminocaproyl), Ala-Phe-Phe-Pro-Trp-Aha-NH_2(Aha $\equiv$ 7-aminoheptanoyl) (Figure 4) contain 1, 2, 4, and 5 methylene groups inserted between the α-carbon and the C-terminal amide. These analogs were synthesized via previously described conditions (*9*) and the structures confirmed by FAB mass spectrometry. Notably, all of these analogs demonstrated myotropic activity, with threshold concentration values of 1 $\pm$ 0.6 nM (*1*), 78 $\pm$ 15 nM, 20 $\pm$ 10 nM, and 2.1 $\pm$ 1.3 μM, respectively. The insertion of one methylene group results in a 2-order of magnitude drop in potency in relation to the parent superagonist peptide, similar to the potency drop observed for the methyl ester and N-methyl amide analogs of AFFPWG-NH_2 (*1*). Thus, some tolerance to movement of the C-terminal amide group is observed for the myotropic receptor. However, the presence of two and four methylene group spacers precipitates a 3.5-4 order of magnitude drop in potency. With the insertion of five methylene group spacers, the myotropic potency experiences a further drop of 5 orders of magnitude in comparison with the parent peptide. Thus, addition of the methylene groups results in a stepwise decline in potency but does not precipitate a complete loss of myotropic activity.

Another insect kinin pseudopeptide is a bifunctional, heterodimeric analog that features a covalent linkage between the N-termini of active cores for the insect kinin and "pyrokinin" neuropeptide families (*11*). Members of the pyrokinin (FXPRLamide) insect neuropeptide family have been associated with oviduct contractile activity in the cockroach, locust, and cricket (*1,8,11*); pheromonotropic activity in the corn earworm Heliothis (*12*) and silkworm Bombyx (*13*), and diapause induction activity in the silkworm (*14*). As previously stated, the insect kinins elicit a diuretic-like response, i.e. stimulation of Malpighian tubule fluid secretion (*6,7*). Other than the cockroach hindgut myotropic assay, there is no known cross-activity between the insect kinins and the pyrokinins. The heterodimer [aGWPFF-succinyl-FSPRLa] (Figure 5), however, has been shown to elicit two physiological responses in one insect species that are normally the exclusive province of either the kinin or the pyrokinin peptides. The heterodimer demonstrated stimulation of the fluid secretion of the cricket Malpighian tubule with a notable EC_{50} of 0.6 nM, a typical response for the insect kinin family. On the isolated cricket oviduct preparation, the heterodimer demonstrated myostimulatory activity at a threshold concentration of 7.7 nM, six-fold more potent than the pyrokinin core fragment Phe-Ser-Pro-Arg-Leu-NH_2. The heterodimer can potentially elicit any combination of the physiological responses of diuresis, pheromone biosynthesis, oviduct contraction, and diapause induction in suitable insect species. Non-peptide and pseudopeptide mimetics fashioned after this and other insect neuropeptide-heterodimers would have the potential to disrupt more

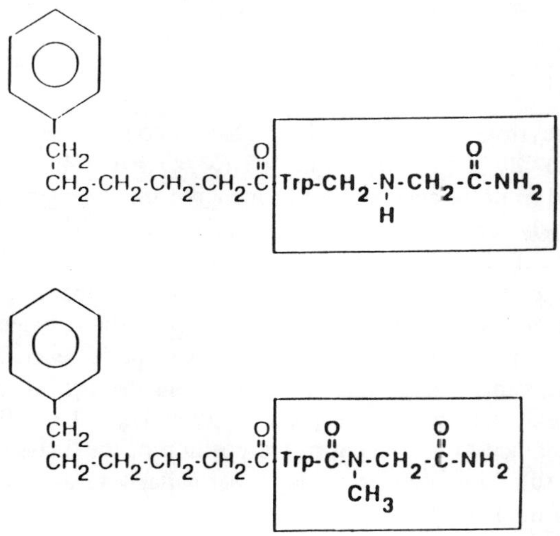

Figure 3. A non-peptide analog (top) of the C-terminal achetakinin pentapeptide Phe-Phe-Pro-Trp-Gly-NH_2 in which the amino acid block Phe-Phe-Pro is replaced by 6-phenylhexanoic acid and a reduced bond linkage (-CH_2-NH-) replaces the amide bond (-C(O)-NH). At bottom is a pseudodipeptide analog of the same pentapeptide in which the amino acid block Phe-Phe-Pro is replaced by 6-phenylhexanoic acid and a methyl group is inserted into the amide bond linking Trp and Gly (Nachman, unpublished data).

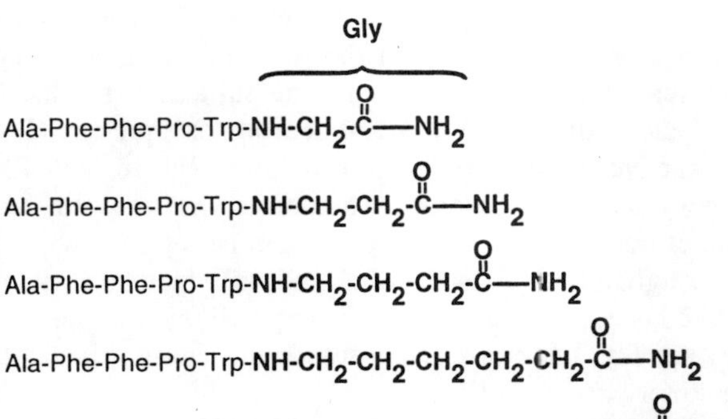

Figure 4. The achetakinin hexapeptide analog Ala-Phe-Phe-Pro-Trp-Gly-NH_2 (top) and C-terminally modified analogs containing, respectively, one, two, four, and five methylene groups between the α-carbon center and the terminal amide group (Nachman et al., unpublished data).

than one physiological process critical to insects and, therefore, promise to be more efficient and effective pest management agents.

Sulfakinins and Myosuppressins

The gastrin/cholecystokinin-like sulfakinins share a common octapeptide sequence X-Asp-Tyr(SO_3H)-Gly-His-Met-Arg-Phe-NH_2 (X = Asp or Glu). These myotropins have been isolated from the cockroaches L. maderae and Periplaneta americana, the locust L. migratoria, and the fleshfly N. bullata (*1,8,15,16*), suggesting that the sulfakinins are widely distributed among insects.

Only the C-terminal octapeptide fragment is required to elicit a full myostimulatory response on the isolated cockroach hindgut, indeed this fragment is seven-fold more potent than the parent undecapeptide leucosulfakinin (LSK) peptide. The C-terminal hexapeptide, which contains the critical sulfate moiety, is the active core sequence for myotropic activity (*17*). The sulfakinin receptor exhibits some tolerance to movement of the position of the Tyr-sulfate group in interactions with sulfakinin peptide analogs, particularly to a shift of one position towards the N-terminus (*18*).

Conversely, the myosuppressins are members of the FLRFamide peptide family, inhibit the spontaneous contractions of the isolated cockroach hindgut, and share the common C-terminal heptapeptide sequence Asp-His-Val-Phe-Leu-Arg-Phe-NH_2. They have been isolated from the cockroach L. maderae, locusts Schistocerca gregaria and L. migratoria, and flies Neobellieria bullata and Drosophila melanogaster (*1,8,19,20*). Leucomyosuppressin (LMS) inhibits evoked transmitter release at the neuromuscular junction of the mealworm Tenebrio molitor (*21*).

Despite eliciting contrasting biological responses, a comparison of the sulfakinin and myosuppressin C-terminal heptapeptide fragments (Figure 6) indicates that they share Asp, Arg, and Phe residues in analogous positions. Furthermore, the two aromatic residues in the myosuppressin fragment (His and Phe) have residue counterparts with aromatic character in the sulfakinin fragment (Tyr[SO_3H] and His). The myosuppressin residue Leu and the sulfakinin residue Met are both hydrophobic. Leucosulfakinin (LSK) and leucomyosuppressin (LMS) from the cockroach are inactive in the C-terminal acid form (*1*). In both LMS and LSK, partial retention of activity is observed when basic Lys replaces the Arg residue. The His residue of both of these neuropeptides can be replaced with other aromatic amino acids, though not with basic Lys, and still maintain biological activity (*8,17,22*). Of course, a sulfate group is required by the sulfakinins to elicit myostimulatory activity, but non-sulfated sulfakinins are inactive and, thus, do not display myoinhibitory activity (*17*).

In addition to the presence of the uncommon sulfate group in the sulfakinins, structure-activity studies suggest that the contrasting biological activity of the two peptide families result in large measure from either the presence or absence of branched-chain character in myosuppressin C-terminal heptapeptide positions 3 and 5 and the disparate importance of the shared Asp residue (Figure 6) to peptide-receptor interaction. For instance, replacement of the Met, or isosteric Nle, in the sulfakinin sequence with the myosuppressin branched-chain residue Leu (Figure 6) led to a loss of over 2 orders of magnitude in myostimulatory activity on the isolated cockroach hindgut preparation. Furthermore, replacement of the Gly of the sulfakinin sequence with the myosuppressin branched-chain residue Val (Figure 6) was accompanied by over a 4-order of magnitude decrease in myostimulatory activity for the resulting peptide analog. Truncation of the Asp residue in position 1 of the C-terminal myosuppressin heptapeptide fragment resulted in a 2-order of magnitude decrease in myoinhibitory potency on the cockroach hindgut bioassay. Conversely, truncation of the analogous Asp residue in the sulfakinin C-terminal heptapeptide fragment, led to a 3-fold increase in myostimulatory activity (*22*).

In summary, the structure-activity data explain the structural basis for the disparate biological responses elicited by the sulfakinins and myosuppressins. However, the structural similarities leave open the possibility that analogs of either family could demonstrate reversal of activity at a relatively weak potency level. For example, the sulfakinin pentapeptide deletion-analog Tyr(SO_3H)-His-Nle-Arg-Phe-NH_2 demonstrates weak myoinhibitory activity on the hindgut preparation at a threshold concentration of 240 nM (*17*). In this analog, the Gly is deleted and the Tyr(SO_3H) moiety shifts one position towards the C-terminus.

Sulfakinin Pseudopeptide Analogs. Pseudopeptide analog studies of the sulfakinins have focused on the replacement of the labile Tyr(SO_3H) moiety with a more stable chemical construct. The sulfate group is highly susceptible to hydrolytic cleavage under acidic conditions.

In a recent study, carboxylic acids of varying lengths were utilized to replace the acidic Tyr(SO_3H) group and/or another amino acid in the sulfakinin sequence. In one analog, the Tyr(SO_3) moiety was replaced by azelaic acid leading to the pseudopentapeptide Aze-Gly-His-Nle-Arg-Phe-NH_2 [Aze = azelayl group: $HO_2C(CH_2)_7C(O)$-] (Figure 7). Pseudotetrapeptide analog, Ddca-His-Nle-Arg-Phe-NH_2 [Ddca = dodecanedioyl: $HO_2C(CH_2)_{10}C(O)$-], was synthesized by using dodecanedioic acid as a replacement for both Tyr(SO_3H) and Gly (Figure 7). In the design of both pseudopeptides, an appropriate number of methylene groups was incorporated to enable the carboxyl group to reach the region of the sulfakinin receptor that interacts with the sulfate group. The pseudotetrapeptide analog retained myostimulatory activity at a threshold level of 820 nM, about 2 orders of magnitude less potent than its sulfated counterpart Tyr(SO_3)-Gly-His-Nle-Arg-Phe-NH_2. Conversely, the pseudopentapeptide analog demonstrated reversible, myoinhibitory activity at a threshold concentration of 120 nM, similar in potency with the previously mentioned [Tyr-SO_3H]-containing deletion analog (Tyr(SO_3H)-His-Nle-Arg-Phe-NH_2) and 570-fold less potent than the myosuppressin C-terminal heptapeptide fragment. This pseudopentapeptide analog provides another example of activity reversal (*10*).

A greater myostimulatory potency was achieved by replacement of the Tyr(SO_3H) residue with the isosteric, isoelectronic p-carboxymethylphenylalanine residue (*24*). The resulting analog Glu-Asp-[Phe-p-(CH_2CO_2H)]-Gly-His-Nle-Arg-Phe-NH_2 (Figure 8) demonstrated myostimulatory activity at 73 ± 28 nM, about 1-order of magnitude more potent than the acyl pseudotetrapeptide above, though 550-fold less potent than the sulfated octapeptide counterpart, Glu-Asp-Tyr(SO_3H)-Gly-His-Nle-Arg-Phe-NH_2. Conversely, the hexapeptide analog [Phe-p-(CH_2CO_2H)]-Gly-His-Nle-Arg-Phe-NH_2 (Figure 8) inhibited spontaneous

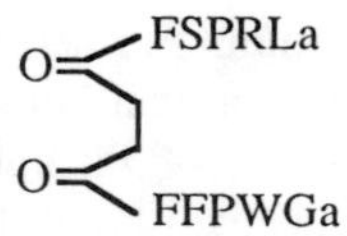

Figure 5. A bifunctional, heterodimeric neuropeptide analog linking the C-terminal pentapeptide core of the pyrokinin family (top) with the C-terminal pentapeptide core of the insect kinin family (bottom) (*11*). (Adapted from ref. 11.)

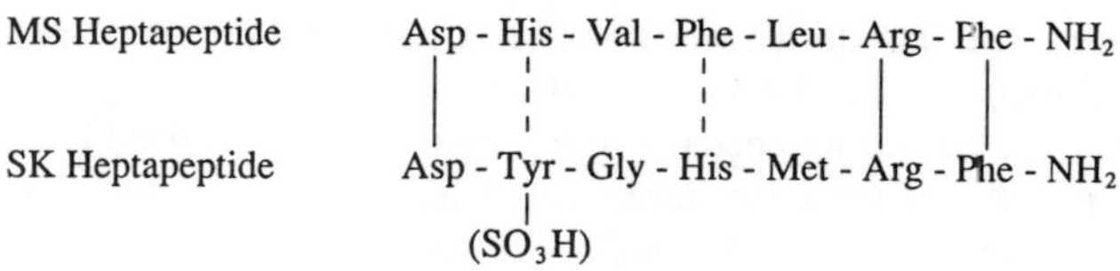

Figure 6. A comparison of the C-terminal heptapeptide sequences of the myoinhibitory myosuppressins (MS) and the myostimulatory gastrin/CCK-like sulfakinins (SK) (*22*). Solid lines denote residue matches and dotted lines indicate shared aromatic character. (Reproduced with permission from ref. 22. Copyright 1993 Pergamon Press.)

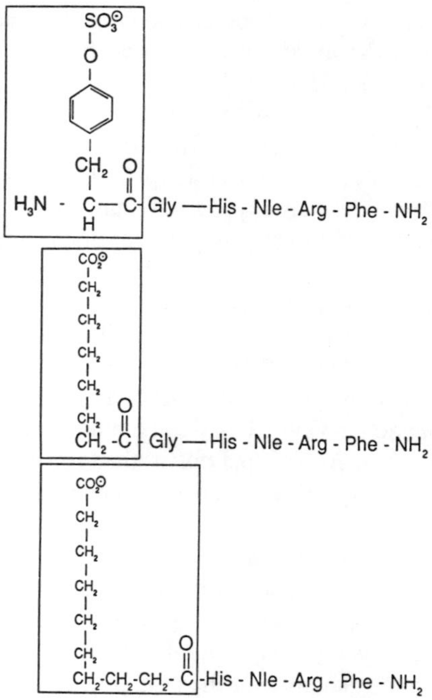

Figure 7. The Tyr(SO_3H) moiety of the C-terminal hexapeptide fragment of the sulfakinins (top) is replaced with azelaic acid in a sulfakinin pseudopentapeptide analog (middle). Dodecanedioic acid replaces both Tyr(SO_3H) and Gly in a sulfakinin pseudotetrapeptide analog (*10*).

a

b

c

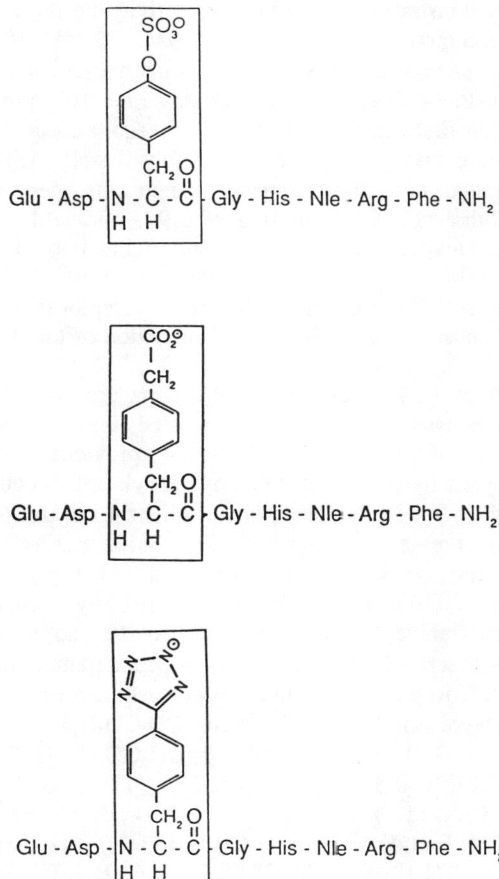

Figure 8. The acidic Tyr(SO_3H) residue of the C-terminal octapeptide sulfakinin fragment (a) is replaced with p-carboxymethylphenylanine (b) and p-tetrazoylphenylalanine (c) (Materials and Methods).

contractions of the isolated cockroach hindgut at 970 ± 700 nM, about 8-fold less potent than the above myoinhibitory acyl pseudopentapeptide analog.

A second acidic residue analog, D,L-p-tetrazolyphenylalanine [D,L-Phe-p-(Taz)] (*24*), was investigated as an isosteric, isoelectronic substitute for Tyr(SO_3H). It proved to be a poorer Tyr(SO_3H) replacement than p-carboxymethyl-phenylalanine. The octapeptide analog Glu-Asp-[L- or D-Phe-p-(Taz)]-Gly-His-Nle-Arg-Phe-NH_2 (Figure 8) demonstrated myostimulatory activity on the hindgut preparation at a threshold concentration of 5.8 ± 2.4 μM, 12-fold less active than the p-carboxymethylphenylalanine octapeptide counterpart. The second diasteromer Glu-Asp-[D or L-Phe-p-(Taz)]-Gly-His-Nle-Arg-Phe-NH_2, proved to be 5-fold less potent than the first diasteromer, with a threshold concentration of 28 ± 9 μM. The hexapeptide analog [D,L-Phe-p(Taz)]-Gly-His-Nle-Arg-Phe-NH_2, which could not be separated into diastereomeric components, demonstrated myoinhibitory activity at a threshold concentration of 0.9 ± 0.2 μM. The observed activity crossover from myostimulatory to myoinhibitory activity of the hexapeptide analogs may be a result of a preferential interaction of the acidic moiety in the myoinhibitory sulfakinin analogs with the Asp-acceptor region of the myosuppressin receptor as opposed to the sulfate acceptor region of the sulfakinin receptor.

Materials and Methods. Leucophaea maderae cockroach hindguts, free of central nervous system tissues, were dissected, immersed in saline, and prepared for recording of myogenic activity as previously described (5). The carboxymethylphenylanine and tetrazoylphenylalanine containing peptides were synthesized via t-Boc methodology on a Milligen-Biosearch 9600 automated peptide synthesizer as previously described (*9*). Sidechains of the amino acids were protected during synthesis as follows: Asp(O-benzyl), Glu(O-benzyl), His(N-dozyl), Arg(N-Tos), carboxymethylphenylalanine(O-benzyl), and tetrazoylphenylalanine(N-tBu). The latter two Boc amino acids were synthesized as previously described (*24*). The cleaved peptide amide mixture was purified on a Waters C18 Novapak column under previously described conditions (*9*). Amino acid analyses were as follows: E-D-[Phe-p-(CH_2CO_2H)]-G-H-Nle-R-F-NH_2, D (0.9), E (0.9), F (1.0), G (0.9), H (0.9); [Phe-p-(CH_2CO_2H)]-G-H-Nle-R-F-NH_2, F (1.0), G (0.8), H (0.8), R (0.8); E-D-[L- or D-Phe-p-(Taz)]-G-H-Nle-R-F-NH_2, D (0.9), E (1.1), F (1.0), G (1.1), H (1.1), R (1.1); E-D-[D- or L-Phe-p-(Taz)]-G-H-Nle-R-F-NH_2, D (1.0), E (1.1), F (1.0), G (1.1), H (1.1), R (1.1), [D,L-Phe-p-Taz]-G-H-Nle-R-F-NH_2, F (1.0), G (1.0), H (1.1), R (1.2). The more active of the two octapeptide tetrazoylphenylananine diastersomers probably contains the [L-Phe-p-(Taz)] enantiomer. Fast atom bombardment mass spectra were determined on a VG-70-250 EHF mass spectrometer (VG Analytical, Manchester, U.K.) according to previously described procedures (*10*). The structural identity of the synthetic peptide analogs were confirmed by the presence of the following molecular ions (MH^+) in the mass spectra: E-D-[Phe-p-(CH_2CO_2H)]-G-H-Nle-R-F-NH_2, [Calc. MH^+ for $C_{49}H_{68}N_{14}O_{14}$: 1077.51] 1077.77; [Phe-p-(CH_2CO_2H)]-G-H-Nle-R-F-NH_2, [Calc. MH^+ for $C_{40}H_{56}N_{12}O_8$: 833.44] 833.57; E-D-[L- or D-Phe-p-(Taz)]-G-H-Nle-R-F-NH_2, [Calc. MH^+ for $C_{48}H_{66}N_{18}O_{22}$: 1087.52] 1087.70; E-D-[D-or L-Phe-p-(Taz)]-G-H-Nle-R-F-NH_2, [Calc. MH^+ for $C_{48}H_{66}N_{18}O_{22}$: 1087.52] 1087.56; [D,L-Phe-p-(Taz)]G-H-Nle-R-F-NH_2, [Calc. MH^+ for $C_{39}H_{54}N_{16}O_6$: 843.45] 843.87. The peptide analogs had the following retention times (conditions described previously [9]): E-D-[Phe-p-(CH_2CO_2H)]-G-H-Nle-R-F-NH_2, 16.5 min; [Phe-p-(CH_2CO_2)]-G-H-Nle-R-F-NH_2, 12.0 min; E-D-[L- or D-Phe-p-(Taz)]-G-H-Nle-R-F-NH_2, 16.5 min; E-D-[D-or L-Phe-p-(Taz)]-G-H-Nle-R-F-NH_2, 16.0 min; [D,L-Phe-p-(Taz)]-G-H-Nle-R-F-NH_2, 18.0 min.

Myosuppressin Pseudopeptide Analogs. While the C-terminal pentapeptide represents the actual active-core sequence, it is markedly less potent than the parent leucomyosuppressin peptide. Therefore, development of pseudopeptide analogs of the inhibitory myosuppressin neuropeptide family focused on the C-terminal heptapeptide sequence (Asp-His-Val-Phe-Leu-Arg-Phe-NH_2), the smallest fragment capable of retaining the full myoinhibitory potency of the parent peptide. In order to retain the carboxylic acid side-chain of Asp, critical to full myosuppressin activity, diacids of varying lengths were utilized to replace amino acid residues in the N-terminal region of the heptapeptide fragment. A pseudohexapeptide analog with the structure Suc-His-Val-Phe-Leu-Arg-Phe-NH_2 (Suc = succinoyl: $HO_2C(CH_2)_2C(O)$-) (Figure 9) was found to retain myoinhibitory activity on the hindgut preparation at a threshold concentration of 0.15 nM, 2-orders of magnitude more potent than the myosuppressin C-terminal hexapeptide (threshold concentration (TC): 17 nM) and at least as potent as the myosuppressin heptapeptide fragment (TC: 0.21 nM). In this pseudohexapeptide analog, succinic acid replaces the Asp of the myosuppressin C-terminal heptapeptide fragment. The full potency demonstrated by the pseudohexapeptide analog suggests that the N-terminal amino group has little influence on the interaction between the myosuppressin hepapeptide and its receptor site. Pseudopentapeptide and pseudotetrapeptide myosuppressin analogs were also synthesized in which pimellic acid replaced the N-terminal residue block Asp-His and sebacic acid replaced the block Asp-His-Val (Figure 9). The pseudotetrapeptide analog Sba-Phe-Leu-Arg-Phe-NH_2 (Sba = sebacoyl: $HO_2C(CH_2)_8C(O)$-) retained myoinhibitory activity on the hindgut bioassay at a threshold of 2.7 μM, only about 75-fold less potent than the myosuppressin C-terminal pentapeptide fragment (TC: 35 nM). The myosuppressin C-terminal tetrapeptide fragment (i.e., FLRFamide) is inactive on the isolated cockroach hindgut bioassay. Therefore, the myosuppressin active core has been reduced from a pentapeptide to a pseudotetrapeptide.

However, the pseudopentapeptide analog Pim-Val-Phe-Leu-Arg-Phe-NH_2 (Pim = pimelyl: $HO_2C(CH_2)_5C(O)$-) demonstrated weak myostimulatory activity on the cockroach hindgut preparation at a threshold concentration of 8.6 μM, providing an example of a myosuppressin analog that exhibited activity-crossover (Nachman, unpublished data). The activity-reversal behavior of the pseudopentapeptide may be a result of a preferential interaction between the acyl carboxyl group with the sulfate-acceptor region of the sulfakinin receptor site as opposed to the Asp-acceptor region of the myosuppressin receptor.

Locustatachykinins

A family of four neuropeptides, containing the common C-terminal hexapeptide Gly-Phe-Tyr-Gly-Val-Arg-NH_2, was isolated from the locust L. migratoria based on their ability to stimulate contractions of the isolated cockroach hindgut bioassay (*25*). The family of peptides was named the locustatachykinins, due to sequence homology with the vertebrate physaelamin and substance P tachykinin subfamilies. For instance, 40-45% of the residues of the nonapeptide locustatachykinin I (Lom-TK-I) and decapeptides Lom-MT-II and Lom-MT-III are identical to analogous residues in physaelamin (*25*). Furthermore, Monnier and coworkers (*26*) cloned a developmentally regulated Drosophila receptor complex, present in the nervous system, which demonstrates 38% homology with the mammalian NK G-protein coupled tachykinin receptor complex. The Drosophila receptor complex was expressed in a mouse cell line and found to exhibit specific binding affinity with Lom-TK-II and not with substance P or physaelamin. The structural homology observed between the peptide families themselves and between two sets of receptor-complexes suggests that the tachykinins have been conserved during animal evolution. The locustatachykinins stimulate contraction of the isolated cockroach hindgut at threshold concentrations of between 0.1-0.3 nM, the locust foregut at threshold concentrations of between 0.3-5 nM, and the locust oviduct at threshold concentrations of between 2-40 nM (*25*).

Members of a number of vertebrate tachykinin subfamilies also elicit myostimulatory activity on the cockroach hindgut bioassay. The most potent of these were substance P and physaelamin with threshold concentrations of 8 nM and 7 nM, respectively. The potencies of the C-terminal pentapeptides of substance P (Phe-Phe-Gly-Leu-Met) and physaelamin (Phe-Tyr-Gly-Leu-Met) proved to be 288 nM and 147 nM, respectively (*27*).

Nonpeptide Tachykinin Mimic. Given the homology between the locustatachykinins and substance P and other vertebrate tachykinins along with the activity displayed by these vertebrate tachykinins in the cockroach hindgut bioassay, the substance P antagonist CP-96,345-1 (Figure 10) (*28*) was evaluated on the insect myotropic preparation. The substance P antagonist CP-96,345-1 was obtained from Pfizer, Inc. (Groton, CT) and the isolated cockroach hindgut bioassay was prepared and operated as previously described (*5*). Surprisingly, non-peptide CP-96,345-1 demonstrated myostimulatory activity on the L. maderae cockroach hindgut preparation at a threshold concentration of 6.5 $\pm$ 2.5 μM, behaving as an agonist of substance P. The non-peptide was only 20-fold less potent than the C-terminal pentapeptide of substance P.

At concentrations higher than 10 μM, CP-96,345-1 was found to prevent completely the myostimulatory response of substance P at a dose required to elicit a half maximal response, but could not block the half-maximal responses of other insect myotropic peptide families such as the leucokinins, pyrokinins, and sulfakinins. In this circumstance, the non-peptide was behaving as an antagonist. The sequence homologies between the insect and vertebrate tachykinin peptides and receptors suggests that the non-peptide CP-96,345-1 is interacting with the insect tachykinin ("locustatachykinin") receptor on the L. maderae hindgut.

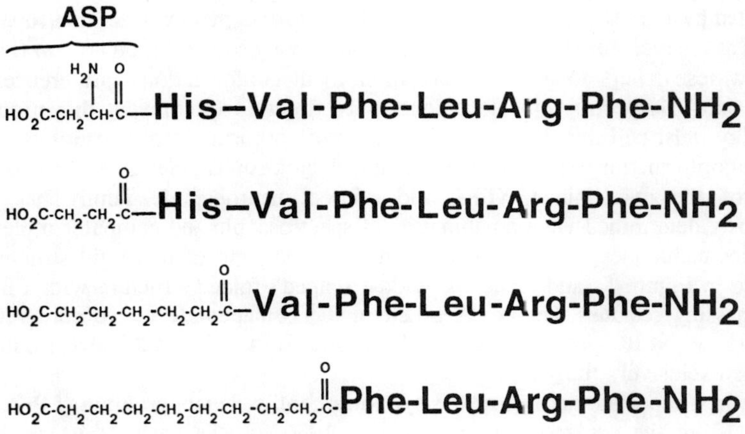

Figure 9. The amino acid and/or amino acid blocks Asp, Asp-His, and Asp-His-Val of the C-terminal heptapeptide fragment (top) of the myosuppressins is replaced with, respectively, succinic acid, pimellic acid, and sebacic acid (Nachman et al., unpublished data).

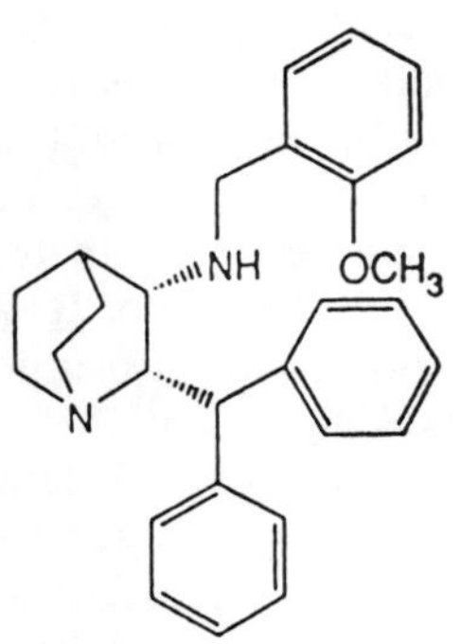

ARG–PRO–LYS–PRO–GLN–GLN–<u>PHE–PHE–GLY–LEU–MET–NH</u>$_2$

Substance P

Figure 10. The vertebrate tachykinin substance P is illustrated below with its C-terminal pentapeptide active core underlined. Above is the structure of the non-peptide substance P mimetic CP-96,345-1 (*28*).

Pyrokinin (FXPRLa) Active Conformation for Hindgut Contraction, Oviduct Contraction, and Pheromone Biosynthesis

The development of potent pseudopeptide and nonpeptide analogs will be greatly facilitated by a knowledge of the active conformation adopted by insect neuropeptide ligands at the receptor site. Most small peptides in aqueous solution are too flexible to allow researchers to gather information about conformational preferences via NMR and CD spectroscopic techniques. In order to circumvent this problem, flexibility must be limited by the introduction of conformational constraints.

The conformation of a rigid, cyclic analog (cyclo[Asn-Thr-Ser-Phe-Thr-Pro-Arg-Leu]) of the pyrokinin (FXPRLamide) insect neuropeptide family has been previously determined via a combination of spectroscopic and computer molecular dynamics techniques. A computer graphics presentation of this rigid structure is depicted in Figure 11 and reveals a "saddle shaped" folded structure with a ß-turn encompassing residues Thr-Pro-Arg-Leu in the pentapeptide active core region. Remarkably, NMR spectroscopic data demonstrated that the bond between Thr and Pro was exclusively trans (*23*).

The first reported member of the pyrokinin family, leucopyrokinin (LPK), was isolated from the cockroach L. maderae in 1986, on the basis of its ability to stimulate contractions of the isolated cockroach hindgut. Since that time a series of insect peptides containing the pyrokinin C-terminal pentapeptide Phe-X-Pro-Arg-Leu-NH_2 (X = Gly, Ser, Thr, or Val) have been isolated from various insects and have been associated with a number of biological activities in addition to cockroach hindgut contraction. For instance, six hindgut/oviduct myotropins of the pyrokinin family have also been isolated from the locust L. migratoria (*1,29*). In addition, the C-terminal pentapeptide Phe-X-Pro-Arg-Leu-NH_2 is also found in pheromone biosynthesis activating neuropeptides (PBANs) isolated from the corn earworm H. zea (*12*) and silkworm B. mori (*13*), the diapause hormone isolated from B. mori (*14*), and a peptide isolated from the armyworm P. separata that demonstrates pheromonotropic activity. Considerable cross-activity between these neuropeptides has been observed. For example, the PBANs of H. zea and B. mori demonstrate significant myostimulatory activity on the isolated cockroach hindgut (*8,30*). Conversely, the myotropic pyrokinins from the cockroach and locust show considerable pheromonotropic activity (*30,31*). Indeed, locust pyrokinin Lom-MT-II is as pheromonotropically active as oxidized Bom-PBAN-I in B. mori, yet contains only a quarter of the amino acid components (*30*). Orthopteran pyrokinins elicit significant diapause induction activity in the dipteran silkworm, with Lom-PK demonstrating 3-fold greater activity than the native Bom-DH-I[19-Cys] (*32*). The C-terminal pentapeptide common to the pyrokinins has proven to be the active core for the diverse biological activities of hindgut contraction (*23*), pheromone production (*30*), and diapause induction (*14*).

The rigid, cyclic pyrokinin analog demonstrates significant activity in the cockroach hindgut (*23*) and oviduct (Nachman et al., unpublished data) myotropic bioassays as well as a silkworm pheromonotropic assay (*32*). In the cockroach hindgut and oviduct myotropic assays, the cyclic analog demonstrated threshold concentrations of 35 nM and 3.1 μM, retaining 4% and 1%, respectively, of the myostimulatory activity of the linear counterpart (*23*; Nachman et al., unpublished data). With a threshold dose of 3 pM/female, the cyclic analog proved to be as potent as the C-terminal hexapeptide fragment of PBAN, or 10% of the pheromonotropic activity of natural Bom-PBAN-I in the silkworm (*30*).

Three linear fragments containing an intact pentapeptide active-core sequence could potentially arise from in situ endopeptidase cleavage of the cyclic analog. These are Thr-Ser-Phe-Thr-Pro-Arg-Leu-Asn-OH, Ser-Phe-Thr-Pro-Arg-Leu-Asn-Thr-OH, and Phe-Thr-Pro-Arg-Leu-Asn-Thr-Ser-OH. In both hindgut myotropic and pheromonotropic bioassays, all three of the approximately equipotent fragment-analogs proved to be at least 2 orders of magnitude less potent than the cyclic pyrokinin/PBAN analog. The results demonstrated that the activity observed for the cyclic analogs is not attributable to linear fragments generated in situ by enzymatic degradation. The results also suggest that much of the reduction in potency observed for the cyclic analog can be attributed to modification of the sensitive C-terminus resulting from addition of amino acid residues at this position. Thus, it is valid to compare the activity of the cyclic analog not only to PBAN, LPK, and other pyrokinins but to the C-terminally appended fragment-analogs. The negative charge of the carboxylic acid group is removed from the site of the pyrokinin active-core, C-terminal Leu by one, two, and three residues, respectively, in the appended fragment-analogs. The virtual equipotency observed for these three appended fragment-analogs suggests that the negatively charged carboxyl species in those positions has little influence on biological activity.

Despite the restraints of rigidity and C-terminal modification placed on the cyclic pyrokinin analog, it retains significant activity in three different bioassay systems. The retention of activity suggests that the rigidly held type I ß-turn (w/trans Pro) over active-core residues Thr-Pro-Arg-Leu represents the conformation necessary (Figure 11) to interact with the cockroach L. maderae hindgut and oviduct contractile, as well as the silkworm B. mori pheromonotropic receptors. The results further suggest that these receptor systems located on different tissues and in different insect species share similar binding requirements, particularly at the point of interaction between the C-terminal active-core region and the receptor site (Figure 12).

Knowledge of conformational or shape requirements for successful neuropeptide-receptor interaction can aid in the design of potent pseudopeptide and nonpeptide peptidomimetic analogs. Specific ß-turn mimetic systems (*1*) that replace the ß-turn hydrogen bond with stable covalent linkages can be adapted to synthesize pseudopeptide mimetics with the appropriate conformation and shape. Computer-aided methodologies can also be used to search for known organic compounds that share with the peptides certain chemical and shape characteristics that are required for biological activity. Alternatively, antibodies raised against the rigid, cyclic pyrokinin analog discussed in this chapter or other rigid, conformationally-restricted analogs could represent "receptor models" (Figure 13). With techniques such as ELISA, these receptor-model antibodies, containing a mirror-image receptor-like "imprint" of the active conformation, could be utilized to select for pseudopeptide and nonpeptide mimetic agonist and antagonist candidates based on immunochemical affinity.

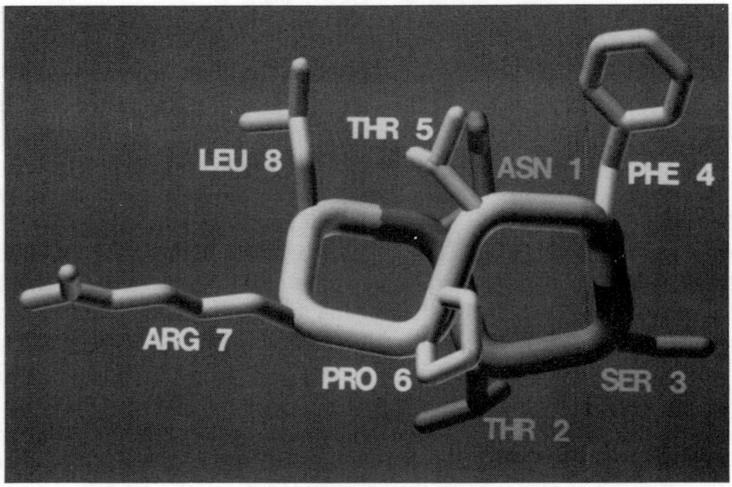

Figure 11. The backbone of the pyrokinin/PBAN analog *cyclo*[Asn-Thr-Ser-Phe-Thr-Pro-Arg-Leu] is in a conformationally constrained "saddle shaped" fold. Within the pentapeptide active core, residues Thr-Pro-Arg-Leu adopt a type I β-turn with an exclusively *trans* peptide bond between Thr and Pro (*23, 32*). (Reproduced with permission from ref. 32. Copyright 1993 Academic Press.)

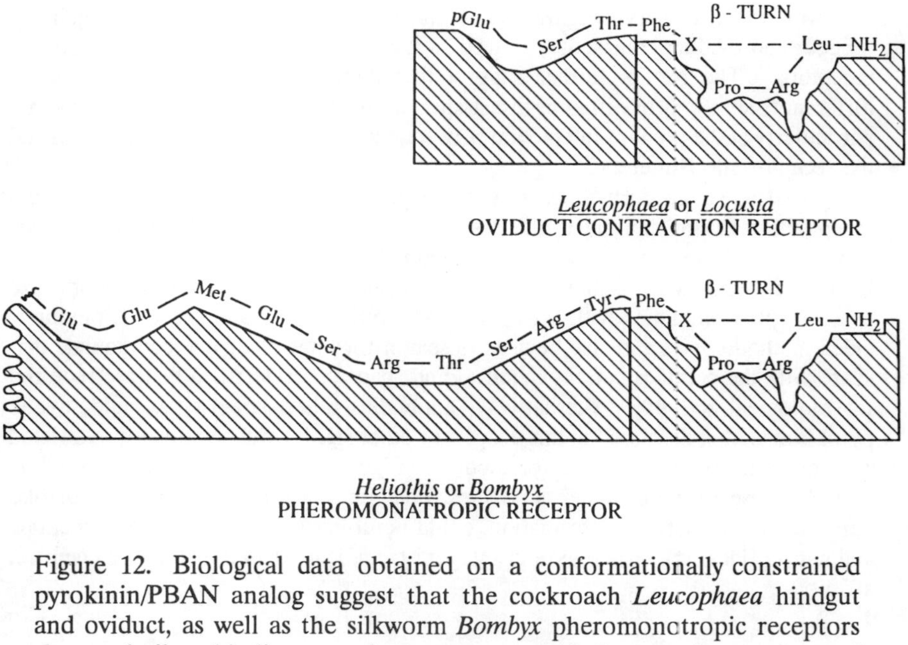

Figure 12. Biological data obtained on a conformationally constrained pyrokinin/PBAN analog suggest that the cockroach *Leucophaea* hindgut and oviduct, as well as the silkworm *Bombyx* pheromonotropic receptors share similar binding requirements, particularly at the region of interaction between the C-terminal active-core region and the receptor sites (*1, 23, 32*).

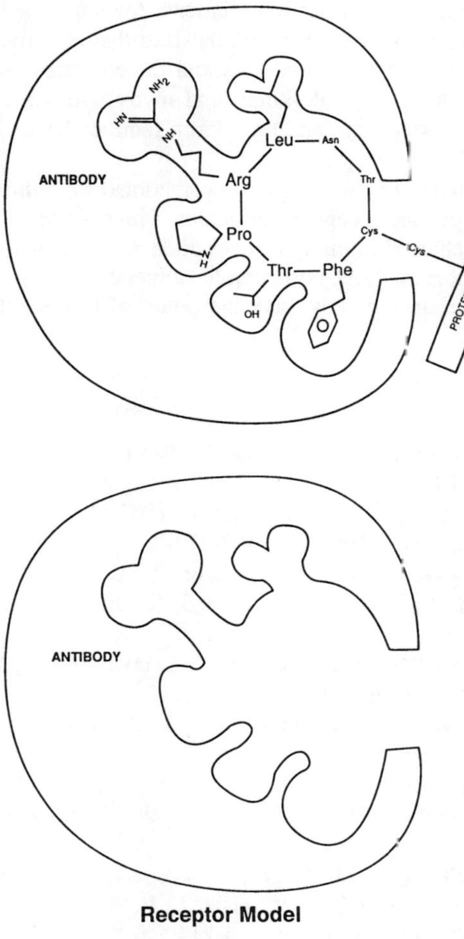

Figure 13. Antibodies raised against a rigid, cyclic pyrokinin analog (pictured) or other rigid, conformationally restricted insect neuropeptide analogs could represent "receptor models" that could be used to screen nonpeptide mimetic agonist/antagonist candidates on the basis of immunochemical affinity.

Summary

The preceding sections recount recent advances made in the development of pseudopeptide and nonpeptide analogs capable of mimicing the biological activity of selected insect neuropeptide families. Utilizing acyl-pseudopeptide, reduced-bond, and/or stable chemical-construct replacements for amino acids and peptide linkages, pseudopeptide analogs have been synthesized that effectively reduce both the amino acid chain length of the active core and the enzymatic susceptibility of insect neuropeptides of the kinin, sulfakinin, and myosuppressin families. Non-peptide mimetics of the insect kinin and tachykinin families have shown retention of myotropic activity.

In the future, knowledge of the active conformational or 3-dimensional shape requirements for neuropeptide-receptor interaction can help to expand upon the progress that has been achieved in insect neuropeptide analog development, and aid in the design of pseudopeptide and non-peptide mimetic analogs of even greater potency. Some of these analogs may have the potential to disrupt critical insect processes.

Literature Cited

1. Nachman, R. J.; Holman, G. M.; Haddon, W. F. *Arch. Insect Biochem. Physiol.* **1993,** *22*, 181-197.
2. Keeley, L. L.; Hayes, T. K. *Insect Biochem.* **1987,** *17*, 639-651.
3. Menn, J. J. *J. Pestic. Sci.* **1985,** *10*, 372-377.
4. Hayes, T. K.; Holman, G. M.; Pannabecker, T. L.; Wright, M. S.; Strey, A. A.; Nachman, R. J.; Hoel, D. F.; Olson, J. K.; Beyenbach, K. W. *Regul. Pept.*, in press.
5. Holman, G. M.; Nachman, R. J.; Schoofs, L.; Hayes, T. K.; Wright, M. S.; De Loof, A. *Insect Biochem.* **1991,** *21*, 107-112.
6. Coast, G. M.; Holman, G. M.; Nachman, R. J. *J. Insect Physiol.* **1990,** *36*, 481-488.
7. Hayes, T. K.; Pannabecker, T. L.; Hinckley, D. J.; Holman, G. M.; Nachman, R. J.; Petzel, D. H.; Beyenbach, K. W. *Life Sci.* **1989,** *44*, 1259-1266.
8. Nachman, R. J.; Holman, G. M. In *Insect Neuropeptides: Chemistry, Biology, and Action*; Menn, J. J.; Kelly, T. J.; Masler E. P., Eds.; American Chemical Society: Washington, D.C., 1991; pp 194-214.
9. Nachman, R. J.; Holman, G. M.; Haddon, W. F.; Vensel, W. H. *Int. J. Pept. Protein Res.* **1991,** *37*, 220-223.
10. Nachman, R. J.; Holman, G. M.; Hayes, T. K.; Beier, R. C. *Int. J. Pept. Protein Res.*, in press.
11. Nachman, R. J.; Coast, G. M.; Holman, G. M.; Haddon, W. F. *Int. J. Pept. Protein Res.* **1992,** *40*, 423-428.
12. Raina, A. K.; Jaffe, H.; Kempe, T. G.; Keim, P.; Blacher, R. W.; Fales, H. M.; Riley, C. T.; Klun, J. A. *Science* **1989,** *244*, 796-798.

13. Kitamura, A.; Nagasawa, H.; Kataoka, H.; Inoue, T.; Matsumoto, S.; Ando, T.; Suzuki, A. *Biochem. Biophys. Res. Commun.* **1989,** *163*, 520-526.
14. Nachman, R. J.; Holman, G. M.; Schoofs, L.; Yamashita, O. *Peptides*, in press.
15. Nachman, R. J.; Holman, G. M.; Haddon, W. F.; Ling, N. *Science* **1986,** *234*, 71-73.
16. Nachman, R. J.; Holman, G. M.; Cook, B. J.; Haddon, W. F.; Ling, N. *Biochem. Biophys. Res. Commun.* **1986,** *140*, 357-364.
17. Nachman, R. J.; Holman, G. M.; Haddon, W. F.; Hayes, T. K. *Pept. Res.* **1989,** *2*, 171-177.
18. Nachman, R. J.; Holman, G. M.; Haddon, W. F.; Vensel, W. H. *Int. J. Pept. Protein Res.* **1989,** *33*, 223-229.
19. Fonagy, A.; Schoofs, L.; Proost, P.; Van Damme, J.; Bueds, H.; De Loof, A. *Comp. Biochem. Physiol.* **1992,** *102C*, 239-245.
20. Nichols, R. *J. Mol. Neurosci.* **1992,** *3*, 213-218.
21. Yamamoto, D.; Ishikawa, S.; Holman, G. M.; Nachman, R. J. *Neurosci. Lett.* **1988,** *95*, 137-142.
22. Nachman, R. J.; Holman, G. M.; Hayes, T. K.; Beier, R. C. *Peptides* **1993,** *14*, 665-670.
23. Nachman, R. J.; Roberts, V. A.; Dyson, H. J.; Holman, G. M.; Tainer, J. A. *Proc. Natl. Acad. Sci. USA* **1991,** *88*, 4518-4522.
24. Tilley, J. W.; Danho, W.; Lovey, K.; Wagner, R.; Swistok, J.; Makofske, R.; Michaelewsky, J.; Triscari, J.; Nelson, D.; Weatherford, S. *J. Med. Chem.* **1991,** *34,* 1125-1136.
25. Schoofs, L.; Holman, G. M.; Hayes, T. K.; Kochansky, J. P.; Nachman, R. J.; De Loof, A. *Regul. Pept.* **1990,** *31*, 199-212.
26. Monnier, D.; Colas, J.-F.; Rosay, P.; Hen, R.; Borrelli, E.; Maroteaux, L. *J. Biol. Chem.* **1992,** *267*, 1298-1302.
27. Holman, G. M.; Nachman, R. J.; Wright, M. S. In *Progress in Comparative Endocrinology*; Epple, A.; Scanes, C. G.; Stetson, M. H., Eds.; Wiley-Liss: New York, 1990, Vol. 342; pp 35-39.
28. Snider, R. M.; Constantine, J. W.; Lowe, J. A.; Longo, K. P.; Lebel, W. S.; Woody, H. A.; Drozda, S. E.; Desai, M. C.; Vinick, F. J.; Spencer, R. W.; Hess, H. J. *Science* **1991,** *251*, 435-437.
29. Schoofs, L.; Holman, G. M.; Nachman, R.; Proost, P.; Van Damme, J.; De Loof, A. *Comp. Biochem. Physiol.*, in press.
30. Kuniyoshi, H.; Nagasawa, H.; Ando, T.; Suzuki, A.; Nachman, R. J.; Holman, G. M. *Biosci. Biotechnol. Biochem.* **1992,** *56*, 167-168.
31. Fonagy, A.; Schoofs, L.; Matsumoto, S.; De Loof, A.; Mitsui, T. *J. Insect Physiol.* **1992,** *38*, 651-657.
32. Nachman, R. J.; Kuniyoshi, H.; Roberts, V. A.; Holman, G. M.; Suzuki, A. *Biochem. Biophys. Res. Commun.* **1993,** *183*, 661-666.

RECEIVED August 19, 1993

Chapter 16

The Role of Propeptide Hormone Protein Conformation in Limited Endoproteolysis

Processing Mammalian Hypothalamic Progonadotropin-Releasing Hormone (GnRH) Precursor Protein

Robert B. Harris[1], Jun ling You[1], Saskia C. F. Milton[2], Raymond C. DeLisle Milton[2], N. S. Rangaraju[1], and Christopher P. Baker[1]

[1]Department of Biochemistry and Molecular Biophysics, Virginia Commonwealth University, Box 614, Richmond, VA 23298
[2]The Scripps Research Institute, 10666 North Torrey Pines Road, La Jolla, CA 92037

We propose that precursor proteins of bioactive peptide-hormones are conformationally dynamic and their response to changing intracellular conditions helps regulate the process of their own maturation by limited endoproteolysis. To do rigorous conformational studies, a strategy was developed to prepare bovine pro-GnRH precursor protein (69-residues) by condensation of long-chain peptide fragments. Synthetic bovine pre-pro-GnRH protein (92-residues) possesses a unique conformation that is modulated by changes in temperature and pH. Furthermore, both synthetic proteins are substrates for GAP-releasing enzyme, consistent with the idea the primary processing site sequence is located at the surface of the precursor protein and acts as a recognition element for the endoproteinase.

Most peptide-hormones are synthesized in secretory cells as precursor proteins of higher molecular weight whose sequences have often been revealed using molecular biology techniques. The pro-hormones are usually biologically inactive and it has been recognized for some time that they must undergo a variety of post-translational processing steps to yield the active peptides (for reviews, see *1-4*).

Post-translational processing occurs in a highly ordered, stepwise fashion; a common step essential for maturation of all precursor proteins is limited endoproteolysis at specific loci within the protein. It was originally proposed that the initial endoproteolytic event occured once the precursor was packaged into secretory granules (discrete, membrane bound, subcellular organelles from which the bioactive peptides are released)(*2-4,6,7*), and in neuronal cells, during axonal transport (*3,8*). More recent evidence, however, suggests that processing of the precursor may be initiated at the Golgi apparatus where sorting and packaging of the secretory components is thought to take place (*9-13*).

0097–6156/94/0551–0230$06.00/0

Limited Endoproteolysis of Precursor Proteins

The recognition signal for endoproteolysis of pro-hormone precursor proteins is accepted to be a pair of basic amino acid residues (Lys-Arg, Lys-Lys, or Arg-Arg) which bracket the peptide hormone (Figure 1). However, we (*5,14-18*) and others (19-24) are suggesting that the presence of doublets alone is not sufficient for processing specificity; more likely, the doublet is part of an extended recognition site which exists as a defined structural feature of the pro-protein. The structural integrity of the processsing site contributes to processing enzyme specificity. Limited proteolysis of a pro-protein is possible because the processing enzymes are specific for particular cleavage sites and specificity is derived in part from the conformation and primary sequence of the pro-hormone at the primary processing site (*5,17,18*). In addition to the doublet of basic amino acids, we (*5,18*) and now others (*24*), also suggest that the processing site sequence contains a strongly polar (cationic or anionic) residue in close sequence proximity to the doublet which may serve to bring the processing site to the surface of the protein. Furin, for example, requires both the doublet of basic amino acids and an Arg residue P_4 to the doublet for fidelity of cleavage (*24*).

In the absence of any extensive crystallographic information regarding any known pro-hormone precursor protein, predictive methods are helpful for initial examination of the potential secondary structure assumed by the processing site sequence. Based on experimental and predictive evidence, we proposed (*5*) that the processing site forms a turn at the surface of the pro-protein and that it functions as a recognition sequence for the endoproteinase. Bek and Berry (*23*) investigated the association of known cleavage sites with defined classes of secondary structural features of proteins and showed that an Ω-loop adequately describes the cleavage site sequence found in many of the pro-hormone sequences examined. Furthermore, they proposed that Ω loops reside at the surface of the protein and that processing endoproteinases interact with a binding site in the middle of the Ω loop; cleavage occurs at the basic residues located at the neck of the loop. Brakch et al. (*22*) proposed that the amino acid residues found at the primary cleavage site contribute to the formation of a ß-turn-α-helix structure. Recently, the structure of human pro-insulin was examined using high-field NMR protocols and a stable, localized structure (defined as a "knuckle") was found to exist at the processing site between the C- and A-peptides (*21*). This stable structure was suggested to form a "recognition element for the Type II proinsulin endopeptidase." Clearly, substrate recognition by the processing endoproteinase must involve several subsites in which structural determinants play a critical role.

A Family of Processing Endoproteinases?

Candidate enzymes that catalyze the endoproteolytic event for numerous pro-peptide hormones have been identified (see *5* and ref.cited therein) and other enzymes that catalyze post-translational modifications have also been reported. Secretory granule-associated carboxypeptidase E has been purified (*25-27*) and cloned (*28*) and peptidyl glycine monooxygenase α-amidating enzyme has been purified to homogeneity from pituitary tissue (*29,30*) and has also been identified in other regions of rat brain, including the hypothalamus (*31*). However, the only unambiguously identified processing endoproteinase previously characterized is the

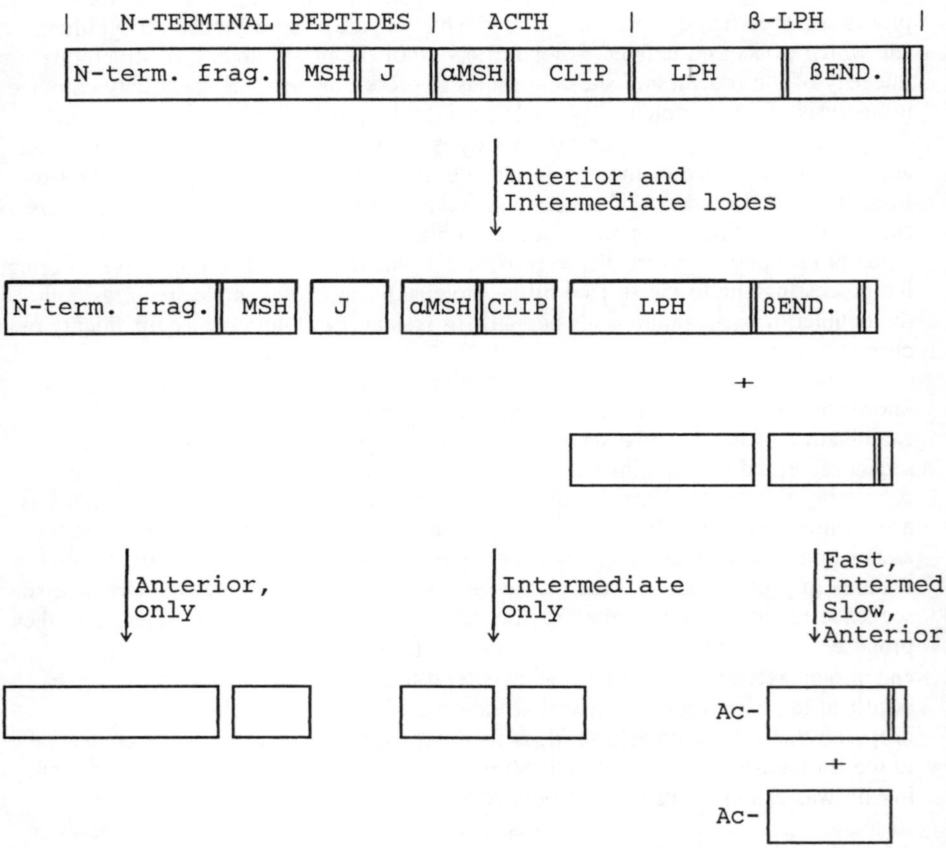

Figure 1. Endoproteolysis of pro-opiomelanocortin precursor protein in the pituitary gland. Processing takes place at specific loci, generally accepted to be doublets of basic amino acids (vertical lines) which bracket the sequences of the bioactive peptide-hormones.

Kex 2 gene product (now known as kexin) of the yeast Saccharomyces cerevisiae (*32,33*).

Kexin is a calcium dependent, serine proteinase with a bacterial subtilisin-like catalytic domain whose physiological substrate is the pro-α-mating factor precursor protein. Recently, mammalian homologs of kexin have been identified using various cDNA cloning techniques. Furin (*34,35*), prohormone convertase (PC) type 1/3 (*36-38*), PC2 (*36,39*), and one, if not two endoproteinases implicated in pro-insulin processing (*40,41*) are the first endocrine and neuroendocrine processing enzymes molecularly characterized in mammalian tissues. Since these initial discoveries, other kexin homolog enzymes have been found in other mammalian tissues (*24,42-44,45*) and it is now postulated that these enzymes form a distinct family of processing enzymes involved in the maturation of a wide variety of secretory and membrane proteins in endocrine and non-endocrine cells (Figure 2).

Pro-Gonadotropin Releasing Hormone Precursor Protein.

Hypothalamic gonadotropin releasing hormone (GnRH) regulates the secretion of follicle stimulating hormone and luteinizing hormone from anterior pituitary gonadotroph cells via receptor mediated activation of phosphoinositide hydrolysis (*46*). GnRH is released from neurosecretory granules; it is an amidated decapeptide which is also synthesized in gonads, placental tissue, mammary tissue and in other regions of the brain, but the function of GnRH in these other tissues is not understood.

The primary amino acid sequence of pre-pro-GnRH/GAP protein has been deduced from its cloned cDNA (*47,48*). The pro-hormone contains the sequence for GnRH and a sequence of 56 amino acids called gonadotropin releasing hormone associated peptide (GAP) which inhibits prolactin secretion and also stimulates the release of gonadotrophic peptides from rat pituitary cells in culture (*47*). The potency of GAP-induced prolactin inhibition is disputed and so the function of GAP is now considered controversial. However, GAP and GnRH are co-secreted into the hypothalamo-hypophyseal portal blood of ovariectomized ewes (*49*) and immunocytochemical evidence shows that GnRH and GAP are co-localized into hypothalamic secretory granules of the rat (*50,51*) and primate (*52*).

In addition, peptides encompassing the first 14 and 24 amino acids of GAP also stimulate the release of gonadotrophic hormones from human and baboon anterior pituitary cells in culture (*49,53*), and immunocytochemical staining of rat, ewe, and rhesus monkey brains suggests that further processing of GAP into shorter chain peptide fragments may take place (*54*). Pro-GnRH/GAP thus contains at least two (and possibly more) distinct biologically active peptides (Figure 3).

GAP-releasing enzyme. Bovine hypothalamic neurosecretory granules contain an enzyme that possesses the requisite specificity to be considered a processing enzyme of pro-GnRH/GAP protein (*5,14,17,18,45,55*). Thus, GAP-releasing enzyme displays the appropriate cleavage specificty with short-chain peptide substrates, longer chain peptide substrates, and with the synthetic proteins prepared for these studies. The proposed Enzyme Commission name for GAP-releasing enzyme is *pro-gonadoliberin proteinase*.

GAP-releasing enzyme is routinely purified to electrophoretic homogeneity

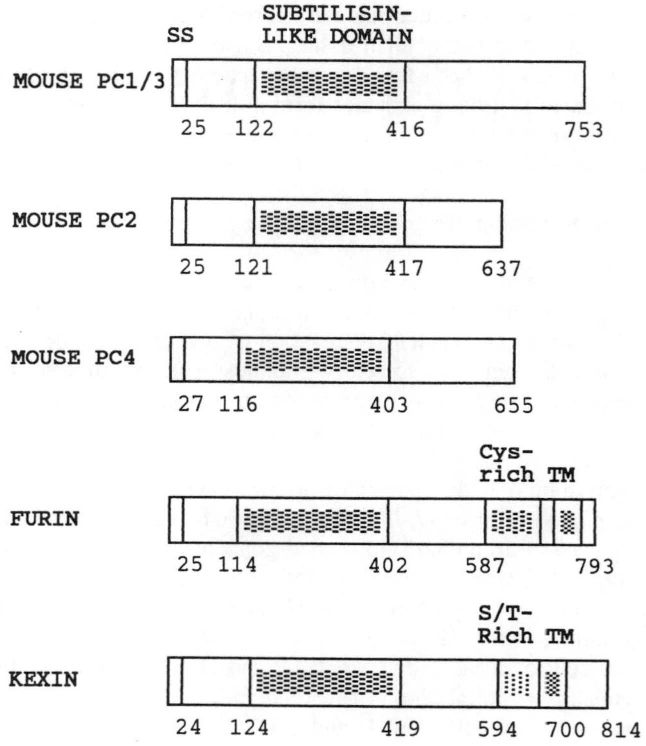

Figure 2. Schematic diagram of the family of kexin-like mammalian endoproteinases. Each protein is numbered beginning with the proposed signal sequence (SS) to show the conserved subtilisin-like catalytic domains, the Cys-rich and Ser/Thr rich regions (S/T), and proposed transmembrane spanning domains (TM). Adapted from ref. *24*.

GnRH GAP PEPTIDE

```
1         10        20        30
|         |         |         |
pQHWSYGLRPG GKR DAENLIDSFQEIVKEVG-

          40        50        60
          |         |         |
QLAETQRFECTTHQPRSPLRDLKGALESLIEEETGQKKI
```

Figure 3. Primary amino acid sequence of pro-GnRH/GAP. Gly^{10} becomes the amidated C-terminal residue of GnRH (Gln^{1}-Gly^{10}); Gly^{11} is the amide nitrogen donor. Asp^{14} is the N-terminal residue of GAP peptide (Asp^{14}-Ile^{69}) and the active 14- and 24-residue peptides. The processing recognition sequence for GAP-releasing enzyme (discussed below) is *underlined*. Other potential processing sites include Arg^{37}-Phe^{38}, Arg^{46}-Ser^{47}, and Lys^{67}Lys^{68}-Ile^{69}. See Wetsel et al.(*63*).

from isolated hypothalamic neurosecretory granules and is a neutral pH, calcium dependent serine proteinase (*17,18*). We now routinely obtain about 50μg of electrophoretically homogeneous enzyme from neurosecretory granules isolated from freshly obtained heifer hypothalamic tissue. Compositional analysis indicates the presence of one Cys and no Cystine residues per mol; hence there are no disulfide bonds. The protein is non-glycosylated and is a single polypeptide chain of molecular weight $\approx$65,000. The N-terminus of the protein appears compromised; repetitive sequencing attempts using solution-phase or blotted protein samples have been unsuccessful. We are currently sequencing proteolytically generated peptide fragments derived from GAP-releasing enzyme.

The recognition sequence for the enzyme within pro-GnRH/GAP has been mapped to G^{6}LRPGGKR^{13} using numerous synthetic peptide substrates (*17,18*). The recognition sequence thus encompasses both an "upstream" monobasic amino acid and the doublet of basic amino acids. The composition and sequence of the doublet of basic residues is important for substrate specificity, and extending the peptide sequence to more closely resemble the sequence in the precursor protein improves both kinetic constants (*14*).

GAP-releasing enzyme is likely to be the relevant physiological processing enzyme of pro-GnRH/GAP. It is located within the appropriate subcellular compartment(s), it possesses the appropriate physicochemical properties, and is immunologically related to murine pituitary PC1/3 proteinase (*45*).

Indirect ELISA and Western blot analyses were used to assess the extent of immunological reaction between GAP-releasing enzyme and anti-human PC2 and anti-murine PC1/3. Non-related proteins including bovine serum albumin, ovalbumin, trypsin, carboxypeptidases Y, and subtilisin Carlsberg do not react with either anti-PC2 or anti-PC1/3. and these same proteins and enzymes also do not react with non-immune serum. However, three different stage purity preparations of GAP-releasing enzyme showed significant cross reaction with anti-PC1/3. On western blot analysis, GAP-releasing enzyme reacted with anti-PC1/3 revealing a protein band with apparent molecular weight of 65,000, identical to the molecular weight of GAP-releasing enzyme visualized with silver stain and identical to the molecular weight of the enzyme as previously reported.

Thus, GAP-releasing enzyme likely shares sequence homology with PC1/3 and may represent bovine hypothalamic PC1/3. In any case, GAP-releasing enzyme should be considered a member of the subtilisin-like family of pro-hormone processing enzymes. PC1/3 and PC2 are found in various regions of the mouse hypothalamus (*36,37,44*). However, PC1/3 is the more abundant enzyme.

In addition to GAP-releasing enzyme, we have discovered a second endoproteinase, termed atrial granule serine proteinase, which is a candidate processing enzyme of pro-atrial natriuretic factor (pro-ANF) (*15,16*). The atrial natriuretic peptides are secreted from atrial granules which morphologically resemble neurosecretory granules and which are contained largely within atrial myocytes. The peptides possess potent natriuretic, diuretic and smooth muscle relaxant properties and because the target organ of these peptides are the kidneys, their existence has implicated the mammalian heart as an endocrine organ which participates in the hormonal regulation of fluid volume, electrolyte balance and blood pressure. This enzyme is distinct from other potential processing enzymes in that its specificity is directed towards monobasic amino acid cleavage sites. Nonetheless, the recognition sequence for this enzyme encompasses 7 residues including a doublet of basic amino acids (*5,15,16*). As an aside, if GAP peptide is further processed to lower molecular weight fragments at Arg 46-Ser 47 as has been suggested, then the hypothalamus also likely contains a monobasic amino acid cleaving enzyme. The PCs do not cleave at monobasic amino acid sites and GAP-releasing enzyme also does not cleave at the pro-ANF monobasic amino acid site (*18*).

Processing pathway of pro-GnRH/GAP. In addition to GAP-releasing enzyme, hypothalamic neurosecretory granules also contain peptidyl glycine α-amidating monooxygenase and carboxypeptidase E (*17*). Processing of pro-GnRH/GAP protein must occur in an ordered, step-wise fashion. The initial endoproteolytic event is catalyzed by GAP-releasing enzyme, which produces bioactive GAP and a form of GnRH extended at the C-terminus by the sequence Gly-Lys-Arg. Carboxypeptidase E then catalyzes the sequential removal of the cationic residues with equal facility, although the des-Arg product may accumulate until it reaches sufficient concentration to be a substrate for the enzyme. The α-amidating enzyme will only act on substrates in which the amide nitrogen donor is present as the free C-terminal residue and thus, will only act on the peptide product generated by carboxypeptidase E. The amidating enzyme will not act directly on the product produced by GAP releasing enzyme. Finally, the amino terminus of GnRH must cyclize to form pyro-Glu which may occur either spontaneously or perhaps through the action of a hypothalamic pyro-Glu forming enzyme.

Based on the pH optima, substrate specificities, and subcellular localization of the three processing enzymes, we can conclude that initial proteolysis of pro-GnRH/GAP occurs during formation of the secretory granule at the Golgi where the pH is thought to be near neutrality. Further processing likely occurs within the fully formed secretory granule. Thus, the three enzymes are enriched in isolated secretory granules and also in the microsmal subcellular fraction which contains elements of the Golgi membranes. Our results predict that the mature peptides are the storage form of pro-GnRH protein within the granules which is consistent with the immunocytochemical evidence.

Regulated Processing of Pro-hormone Precursor Proteins.

Precursor proteins which contain the sequences of numerous peptide-hormones, such as pro-opiomelanocortin or pro-GnRH/GAP, often are processed differently in different regions of the same tissue (eg., the anterior and posterior lobes of the pituitary gland). How can we account for this differential processing scheme?

The PC enzymes are widely distributed in numerous tissues and it is postulated that differential processing is possible because each PC has a different dibasic amino acid site preference (*36,37,44,56,57*). Thus, PC1/3 and PC2 acting alone or in combination could give rise to tissue specific maturation products (*44*) (Figure 4). We know, for instance, that GAP-releasing enzyme and atrial granule serine proteinase act with specificity only at their own recognition sequences regardless of whether the recognition sequence is placed within the peptide framework of its own precursor protein or in the peptide framework of the alternate precursor protein (*18*).

As discussed, there is little to distinquish the primary amino acid sequences found at each of the processing sites. It is likely, therefore, that "differences in dibasic amino acid site preference" are due either to the conformational uniqueness of the processing sites or to conformational constraints placed on the processing site by the remainder of the protein structure. The intra-cellular and/or intra-secretory granule environment may serve to modulate the conformation of the pro-hormone so that different processing sites become accessible during the course of maturation (*58*). In any case, it is clear that not all processing sites are treated equally by the processing endoproteinases (*59,60*).

Finally, it should be noted that intracellular conditions also may regulate processing by positively (or negatively) affecting the catalytic potential of the processing proteinases. For instance, atrial granule serine proteinase will be inactive (*15,16*) within the atrial myocyte secretory granule due to adverse conditions of low pH and high ionic strength (*61*). Intact pro-ANF is thus the storage form found in the myocyte (*62*) until it is processed during secretion, when changing cellular conditions are postulated to activate atrial granule serine proteinase.

We thus set out to examine the role of pro-hormone conformation in its own maturation by solving the solution phase conformation of pro- (and pre-pro) GnRH protein through the use of NMR and molecular modeling protocols.

Synthetic Pro-GnRH and Pre-Pro-GnRH Precursor Proteins.

A strategy of solid-phase peptide synthesis was chosen to prepare suitable amounts of the synthetic proteins needed for study. Pro-GnRH precursor protein was assembled following a strategy of condensation of individual long-chain peptide fragments. Fragment condensation has long been used for preparation of medium to long chain-length peptides and for preparation of elongated segments of proteins (cf., *64-67*). In this particular synthesis, Fmoc- and BOC chemistries were used to

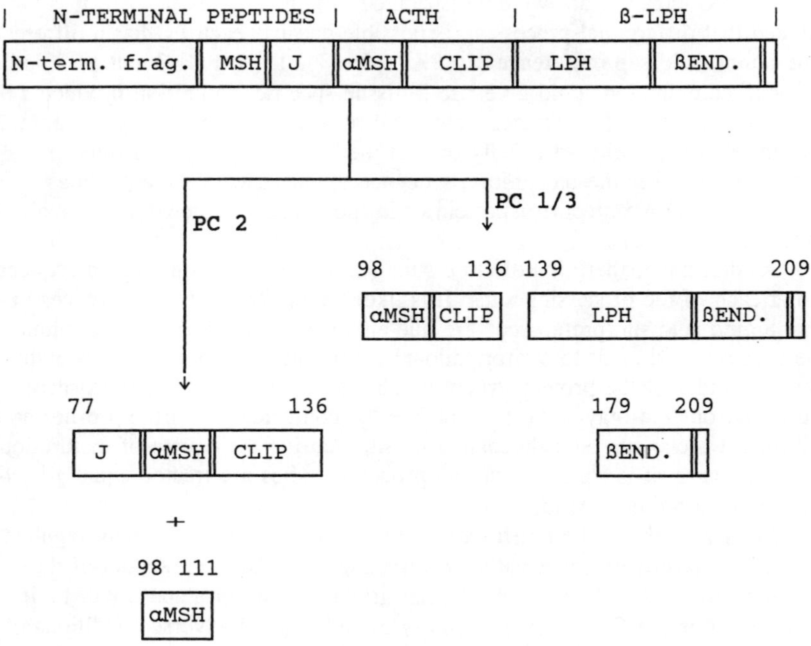

Figure 4. Cleavage products derived from pro-opiomelanocortin following processing by PC1/3 and/or PC2. Adapted from ref. *44*.

prepare the individual peptide fragments; the availability of Fmoc-acylated amino acids carrying strong acid labile side-chain protecting groups made the synthesis strategy possible. On the other hand, homogeneous pre-pro-GnRH was recently synthesized in hundred mg quantities (in the MRC Regulatory Peptides Research Group, University of Cape Town, South Africa) by an elaborate scheme of step-wise solid phase synthesis (*68,69*). No conformational analyses were done following the initial synthesis, but surprisingly, the synthetic pre-pro-protein, which normally does not escape the endoplasmic recticulum, was found to inhibit release of prolactin from cultured pituitary cells at concentrations comparable to those of bromocryptine. This is the first demonstration of significant biological activity associated with a precursor protein and provides the rationale for its further study.

Pro-GnRH is amenable to total synthesis by condensation of individual peptide fragments. To preclude the possibility of interchain disulfide bond formation, Ser was substituted for Cys at position 40 (Figure 3). Thus, three peptide fragments of pro-GnRH were synthesized corresponding to residues Q^1-I^{19}, D^{20}-F^{38}, and E^{39}-I^{69} (Figure 5). The C-terminal peptide (E^{39}-I^{69}) was synthesized by step-wise condensation of the appropriate protected Boc-amino acid on tBoc-Ile-PAM-resin. The N-terminal (Q^1-I^{19}) and middle fragments (D^{20}-F^{38}) were assembled using N(acyl)Fmoc-derivatized amino acids whose side chains, where appropriate, were blocked with strong acid-labile protecting groups (eg.,Fmoc-Ser(Bzl), Fmoc-Asp(OBzl), Fmoc-Thr(Bzl); Fmoc-Lys(2-Cl-Z), etc.). Thus, as the Fmoc group is repeatedly removed during synthesis, the side chain protecting groups are not cleaved. These fragments were synthesized on substituted Dasrin synthesis resins (Peptide Laboratories) which permit quantitative cleavage under relatively mild acid conditions of 25%(v/v) trifluoroacetic acid in dichloromethane (30 min, room temperature). These conditions do not cause deprotection of the amino acid side chains. The N-terminal Fmoc group is intentionally left on each fragment by escaping the normal synthesis protocol prior to the last deprotection step. Each fully protected Fmoc-peptide fragment was then purified separately by reverse-phase preparative HPLC, and its composition was determined following 36h acid hydrolysis.

Now, because the only functional group available on the middle fragment is its C-terminal carboxyl group, it was condensed in a single step with the deacylated C-terminal fragment (still carrying its side-chain protecting groups and still coupled to its synthesis resin). The fragment was coupled in a 4-fold molar excess over the C-terminal fragment using two, 2-hour coupling steps (diisopropylcarbodiimide was the coupling agent).

The Fmoc group was then removed and D^{20}-I^{69} was next coupled with the Fmoc-N-terminal fragment. A small portion of the completed resin-bound peptide (Q^1-I^{69}) was cleaved from the resin and deprotected under anhydrous HF conditions. However, the bulk of Q^1-I^{69} was acetylated directly on the synthesis resin prior to cleavage and deprotection in HF.

The protein was then purified yielding about 14 μmol of synthetic pro-GnRH protein [N(Ac)Q^1-I^{69}]. Based on the amount of C-terminal fragment-resin used to prepare the entire protein (0.3 g, equivalent to 40 μmol) the protein was obtained in about 35% yield.

The unacetylated peptide was sequenced through 43 cycles which

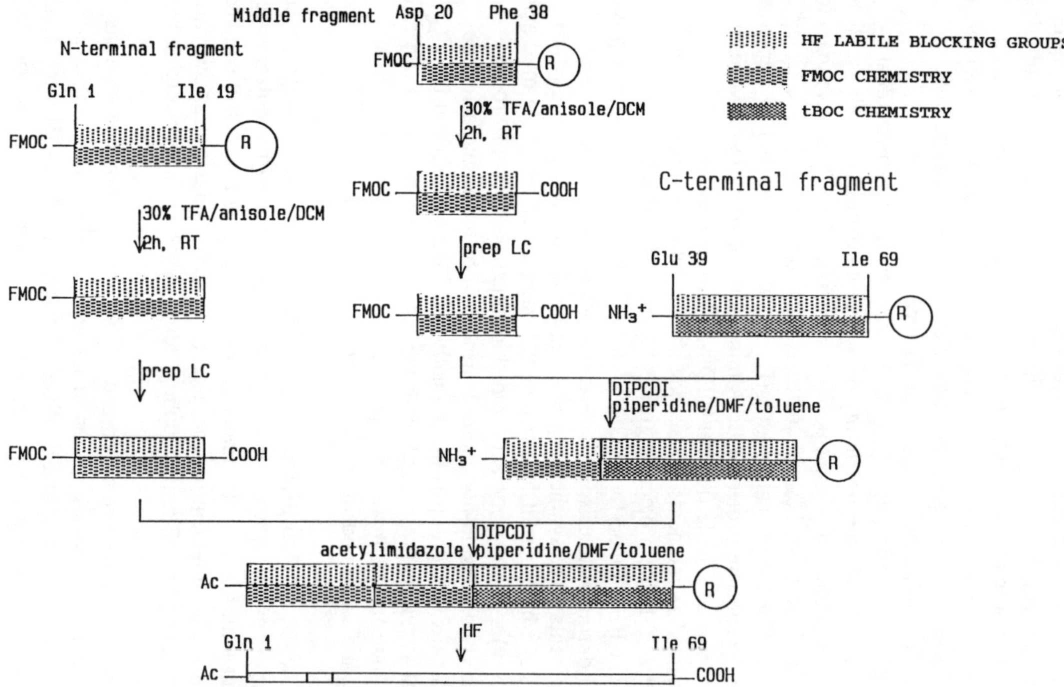

Figure 5. Strategy of fragment condensation followed for preparation of pro-GnRH protein. The C-terminal fragment (E^{39}–I^{69}) was synthesized following tBOC protocols while the middle (D^{20}–F^{38}) and N-terminal (Q^{1}–I^{19}) fragments were prepared following Fmoc protocols. Each fragment was synthesized using strong acid-labile side chain protecting groups. Following synthesis, the peptide was acetylated directly on the resin prior to cleavage and deprotection under anhydrous HF conditions.

encompasses both condensation junctions, and the expected residue was obtained at each position. The sequence results revealed only a single N-terminal residue, indicating virtual purity of the peptide. The amino acid composition was correct and the peptide showed a single band on SDS-PAGE in 15-25% linear gradient gels (molecular weight 8100). The protein thus appeared homogeneous on the basis of its sequence, amino acid composition, and molecular weight. However, low molecular weight impurities were revealed by mass spectral analysis, and for this reason, the protein is being repurified by a combination of ion exchange and reverse-phase HPLC chromatographies.

Proteolysis of Synthetic Pro-GnRH and Pre-Pro-GnRH Proteins. Each of the synthetic proteins is structurally similar (if not identical) to the native proteins as evidenced by the fact that both proteins are substrates for GAP-releasing enzyme. With pro-GnRH, GAP-releasing enzyme yields the two expected products, N(Ac)Q^{1}-R^{13}, and GAP peptide (D^{14}-I^{69}). With pre-pro-GnRH, GAP-releasing enzyme catalyzes the formation of GAP peptide (D^{14}-I^{69}) and presumably, M^{-23}-R^{13}, although this latter peptide has not been unambiguously identified (*70*). Of the serine proteinases tested (trypsin, kallikrein, plasmin), only GAP-enzyme shows this specificity of cleavage. The other enzymes cleave at multiple locations in the pro- and pre-pro-proteins (*70*), yielding numerous peptide products (Table I). Thus, it would appear that many of the potential cleavage sites for these various enzymes are accessible in the pro- and pre-pro-proteins, but only GAP-releasing enzyme displays limited specificity. The pre-pro-protein is also a substrate for endoproteinase Glu-C (V8 protease) which cleaves at E^{28}, E^{34}, E^{39}, and E^{63} (Table I).

Some of the hydrolysis products obtained with the non-specific proteinases are derived from longer chain peptides initially generated by cleavage at specific residues. A time course of hydrolysis of pre-pro-GnRH done with trypsin, for example, shows that after prolonged incubation, virtually all of the possible peptide products that could be generated from pre-pro-GnRH/GAP are detected. However, in short-time incubations, significant levels of cleavage occur only at R^{14} and R^{37} (to form D^{14}-R^{37}), and at R^{50} and K^{67} (to form F^{38}-R^{50} and S^{51}-K^{67}). These peptide then become substrates, yielding further breakdown products (*70*).

The initial products generated by each of the enzymes always represents cleavage at the primary processing site. So, although the hierarchal cleavage observed with trypsin supports the idea that peptide products other than GAP and GnRH might be formed by cleavage at monobasic amino acids within the pro-hormone sequence, the primary processing site is accessible in the pro- and pre-pro-proteins.

Conformational Analysis of Pre-Pro-GnRH. Because our preparation of pro-GnRH requires further purification, we have begun our structural studies with pre-pro-GnRH (*70*).

The microenvironment(s) containing the two Trp residues of pre-pro-GnRH can be studied by intrinsic Trp fluorescence emission spectroscopy. The emission wavelength is blue shifted relative to N(Ac)Trp-amide, indicating that the microenvironment(s) containing the Trp residues are buried within the protein interior. Although there is a monotonic decrease in fluorescence yield as a function

Table I. Hydrolytic products derived from synthetic pre-pro-GnRH catalyzed by GAP-releasing enzyme and other proteinases[a]

Enzyme	Peptide Product	Hydrolysis Reaction Time 30 min	18h
GAP releasing enz.	D14-I69	not detected (ND)	present
Trypsin	D14-R37	present	ND
	D51-K67	present	ND
	F38-R50	present	present
	K68-I69	present	present
	D14-K27	trace	present
	E28-R37	trace	present
	G54-K67	trace	present
	D51-K53	trace	present
	S47-R50	trace	present
	M-23-K-18	trace	present
plasmin[b]	D14-K27		
	R13-K27		
	S47-R50		
	S47-K53		
	E28-R46		
	G54-K67		
	G54-I69		
V8 Endoprot. Glu-C[b]	V29-E34		
	T35-E39		
	T64-I69		

SOURCE: Adapted from ref. *70*.

[a] The pre-signal sequence is M^{-23}KPIQKLLAGLILLTWCVEGCSS^{-1}. Only GAP-peptide and M^{-23}-K^{-18}, derived from hydrolysis at potential sites within the signal sequence, have so far been unambiguously identified.

[b] With plasmin, only 18 h digest samples were analyzed. With Endoproteinase Glu-C, only 30 minute digest samples were analyzed.

of temperature (Figure 6A), thermal induced conformational changes are at least 96% reversible as judged by repeated emission experiments. Fluorescence polarization measurements showed only modest changes in polarization values between 20° and 80°C. Clearly, the microenvironment(s) encompassing the Trp residues are resistant to thermal induced structural reorganization.

Thermal titration circular dichroism spectroscopy confirmed the results obtained by fluorescence spectroscopy. Only subtle changes were observed in the shape of the spectral envelope, and a monotonic decrease in helix content was observed (Figure 6B). Deconvolution of the spectra reveal that the protein is about 30% helix in the range of 2-40°C, but even at 90°C, the peptide retains nearly 50% of its initial helix character. Apparently, pre-pro-GnRH does not pass through a transition state as a function of temperature. Rather the protein appears to flex, resulting in minor and subtle (and probably localized) changes in secondary structure.

The marked stability of the protein with increasing temperature is contrasted with the marked instability of the protein as a consequence of pH changes. There are 3, perhaps 4, distinct pKas involved in maintaining structure as judged by fluorescence spectroscopy; pKa ≈ 6.5 is easily assigned as due to His, pKa ≈ 3.7 can be attributed to either Asp or Glu residues, while pKas in the alkaline range are assignable to both Lys and Arg residues (Figure 7A). On either side of the neutral pH range, substantial changes in structure are taking place which are not at all reversible in the acid range and only moderately reversible in the alkaline range.

When plotted as relative mean residue ellipticity at 222nm versus pH, the involvement of a His residue(s) in maintaining structure is clearly indicated (Figure 7B). The likely involvement of Lys residues (in preference to Arg residues) can be concluded from the titration break at pH $\approx$ 10. Thus, while Asp/Glu and Arg residues may contribute to structural stability, His and Lys residues predominate. Even in the pH range 10-13, the protein retains nearly half of its original helix content. Only in the pH range 7.1-7.9 does the protein display maximum structure. Interestingly, in the neutral pH range, there appears to be a marked absence of β-strand structural elements.

Concluding Remarks

Pre-pro- (and likely pro-) GnRH possesses sufficient structure at neutral pH to warrant further study of its solution phase conformation by high field NMR techniques. Spectra have already been acquired for synthetic GAP peptide (D^{14}-I^{69}; unpublished observations) which also appears to assume a relatively ordered structure. The primary processing site for GAP releasing enzyme is accessible in the pre-pro-hormone, consistent with our hypothesis that the recognition sequence is located at the surface of the protein and acts as a recognition element for the processing endoproteinase. These results show that precursor proteins are conformationally dynamic and their structures can be modulated by changes in pH or temperature. By extension, it is not unreasonable to postulate that intracellular (and intragranular) conditions may play a role in regulation of endoproteolyis. For instance, the pH at the trans face of the golgi apparatus is thought to be near neutrality, while the interior of secretory organelles is established to be in the acid

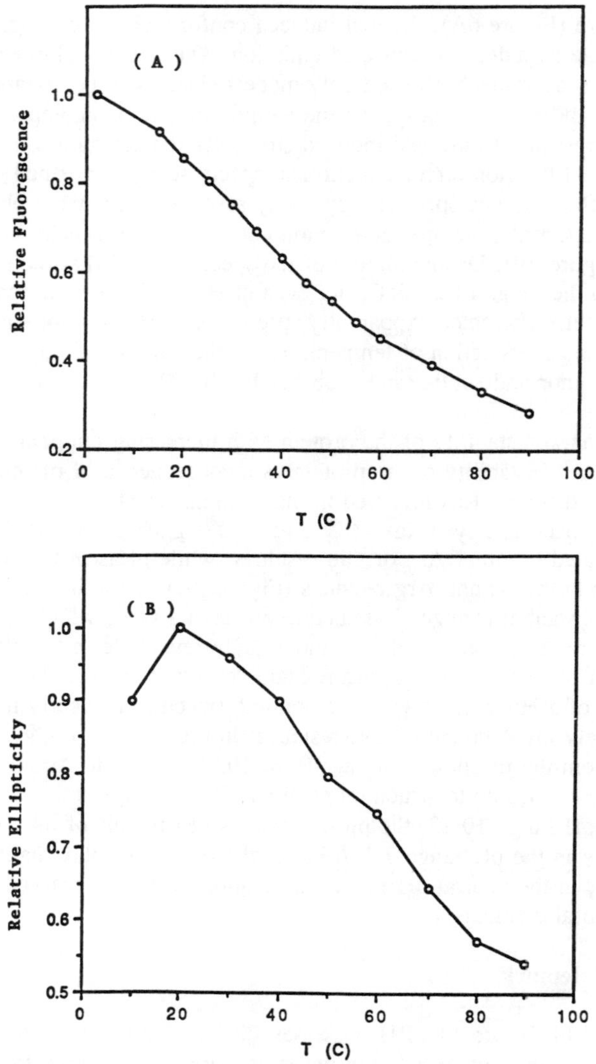

Figure 6. Thermal titration fluorescence and far ultraviolet circular dichroism (CD) spectroscopies of pre-pro-GnRH/GAP. (A) The fluorescence emission intensity of Trp (excitation 290 nm; emission 338 nm) of pre-pro-GnRH/GAP (32 nM) was determined with increasing temperature and the values obtained were normalized to the maximum observed value. (B) CD spectra of pre-pro-GnRH/GAP (50 μM) were acquired with increasing temperature; the mean residue ellipticities at 222 nm were normalized to the maximum observed value. Reproduced with permission from reference *70*.

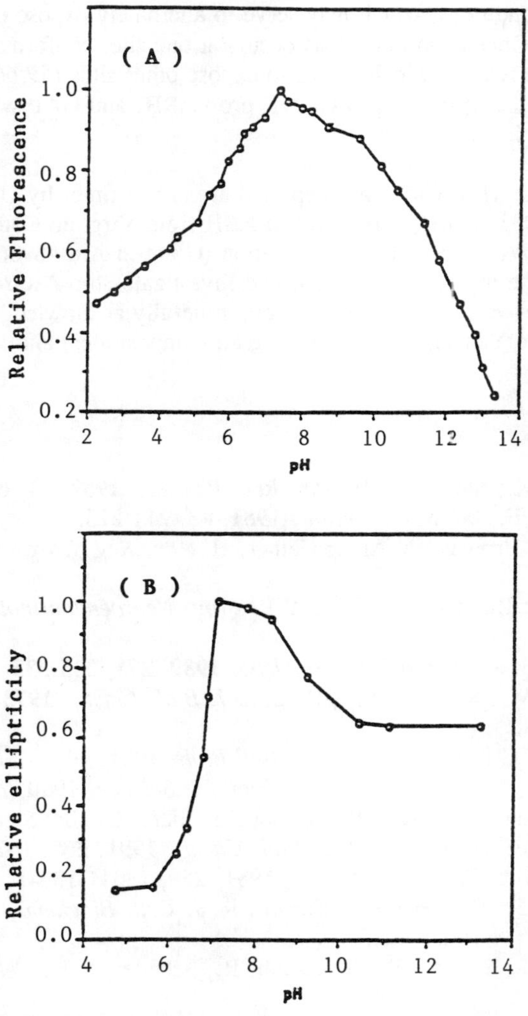

Figure 7. pH titration (A) fluorescence and (B) far ultraviolet CD spectroscopies of pre-pro-GnRH/GAP protein. The protein was dissolved in pH titration buffer containing 40mM Hepes, 25mM sodium acetate, 25mM glycine, and 1mM EDTA at 20°C, pH 7.4. Addition of small amounts of either acid or base to this solution results in a new pH determined by means of a pH probe. Reproduced with permission from reference *70*.

range due to very active proton pumps. Also, the interior of a secretory organelle is at very high ionic strength relative to cellular cytoplasm. Thus, pro-hormones may flex under these conditions, which may serve to alternately expose one processing site over another. Once endoproteolysis occurs at one site, conformational constraints may be relaxed which serves to expose other sites (*59,60*). Solving the solution phase structures of pre-pro-GnRH, pro-GnRH, and GAP will directly address these questions.

Acknowledgments. This work was supported at various times by USPHS grants HL 34578 and NS 24933 (RBH), and grants to RBH from Virginia Commonwealth University and the American Heart Association (Grant-in-Aid #890646). This work was done during the tenure of an Established Investigatorship Award to RBH from the American Heart Association. The authors gratefully acknowledge the technical skill of Ms. Bonnie Yanchisin in performing all compositional and sequence analyses.

Literature Cited

1. Docherty, K.; Steiner, D.F. *Ann. Rev. Physiol.*, **1982**, 44, 625-638.
2. Habener, J.F. *Ann Rev. Physiol.*, **1981**, 43, 211-223.
3. Loh, Y.P.; Brownstein, M.J.; Gainer, H. *Ann. Rev.Neurosci.*, **1984**, 7, 189-222.
4. Gainer, H.; Russell, J.T.; Loh, Y.P. *Prog. Neuroendocrinol.*, **1985**, 40, 171-184.
5. Harris, R.B. *Arch. Biochem. Biophys.*, **1989**, 275, 315-333.
6. Kemmler, W.; Steiner, D.F.; Borg, J. *J. Biol. Chem.*, **1973**, 248, 4544-4549.
7. Loh, Y.P.; Gritsch, H.A. *Eur. J. Cell Biol.*, **1981**, 26, 177-183.
8. Brownstein, M.J.; Russel, J.T.; Gainer, H. *Science*, **1980**, 207:373-378.
9. Lepage-Lezin, A.; Joseph-Bravo, P.; Devilliers, G.; Benedetti, L.; Launay, J.M.; Gomez, S.; Cohen, P. *J. Biol. Chem.*, **1991**, 265, 1679-1688.
10. Glembotski, C.C. *J. Biol. Chem.*, **1981**, 259, 13041-13048.
11. Steiner, D.F.; Docherty, K.; Carroll, R. *J. Cell. Biochem.*, **1984**, 24:121-130.
12. Fisher, J.M.; Sossin, W.; Newcomb, R.; Scheller, R.H. *Cell*, **1988**, 54, 813-822.
13. Davidson, H.W.; Rhodes, C.J.; Hutton, J.C. *Nature*, **1988**, 333, 93-96.
14. Palen, T.E.; Harris, R.B.; Wypij, D.M.; Wilson, I.B. *Peptides*, **1987**, 8, 21-24.
15. Wypij, D.M.; Harris, R.B. *J. Biol. Chem.*, **1988**, 263, 7079-7086.
16. Wypij, D.M.; Harris, R.B. *Life Sci.*,**1992**, 50, 523-531.
17. Rangaraju, N.S.; Xu, J.F.; Harris, R.B. *Neuroendocrinol.* **1991**, 53, 20-28.
18. Rangaraju, N.S.; Harris, R.B. *Arch. Biochem. Biophys.*, **1991**, 290, 418-426.
19. Clamagirand, C.; Creminon, C.; Fahy, C.; Boussetta, H.; Cohen, P. *Biochemistry*, **1987**, 26, 6018-6023.
20. Benoit, R.; Ling, N.; Esch, F. *Science*, **1987**, 238, 1126-1129.

21. Weiss, M.A.; Frank, B.H.; Khait, I.; Pekar, A.; Heiney, R.; Shoelson, S.E.; Neuringer, L.J. *Biochemistry*, **1990**, 29, 8389-8401.
22. Brakch, N.; Boussetta, H.; Rholam, M.; Cohen, P. *J. Biol. Chem.*, **1989**, 264, 15912-15916.
23. Bek, E.; Berry, R. *Biochemistry*, **1990**, 29,178-183.
24. Nakayama, K.; Kim, W.S.; Torii, S.; Hosaka, M.; Nakagawa, T.; Ikemizu, J.; Baba, T.; Murakami, K. *J. Biol. Chem.*, **1991**, 267, 5897-5900.
25. Fricker, L.D. *TINS*, **1985**, 25, 210-214.
26. Hook, V.Y.H.; Loh, Y.P. *Proc. Natl. Acad. Sci USA*,**1984**, 81, 2776-2780.
27. Supattapone, S.; Fricker, L.D.; Snyder, S.H. *J. Neurochem*, **1984**, 42, 1017-1022.
28. Fricker, L.D.; Evans, C.J.; Esch, F.S.; Herbert, E. *Nature*, **1986**, 323, 461-464.
29. Bradbury, A.F.; Finnie, M.D.A.; Smyth, D.G. *Nature*, **1982**, 298, 686-688.
30. Eipper, B.A.; Mains, R.E.; Glembotski, C.C. *Proc. Natl. Acad. Sci. USA*, **1983**, 80, 5144-5148.
31. Eipper, B.A.; Myers, A.C.; Mains, R.E. *Endocrinology*, **1985**, 116, 2497-2504.
32. Julius, D.; Brake, A.; Blair, L.; Kunisawa, R.; Thorner, J. *Cell*, **1984**, 37, 1075-1089.
33. Fuller, R.S.; Sterne,R.E.; Thorner, J. *Ann. Rev. Physiol.*, **1988**, 50, 345-362.
34. Fuller, R.S.; Brake, A.J.; Thorner, J. *Science*, **1989**, 246, 482-486.
35. Van Den Ouweland, A.M.W.; Van Duijnhoven, H.L.P.; Keizer, G.D.; Dorssers, L.C.J.; Van De Ven, W.J.M. *Nucl. Acids Res.*,**1990**, 18, 664.
36. Seidah, N.G.; Gaspar, L.; Mion, P.; Marcinkiewicz, M.; Mbikay, M.; Chrétien, M. *DNA and Cell Biol.*, **1990**, 9, 415-424.
37. Seidah, N.G.; Marcinkiewicz,M.; Benjannet, S.; Gaspar, L.; Beaubien, G.; Mattei, M.G.; Lazure, C.; Mbikay, M.; Chretien, M. *Molec. Endocrinol.*,**1991**, 5, 111-122.
38. Smeekens, S.P.; Avruch, A.S.; LaMendola, J.; Chan, S.J.; Steiner, D.F. *Proc. Natl. Acad. Sci. USA*, **1991**, 88, 340-344.
39. Smeekens, S.P.; Steiner, D.F. *J. Biol. Chem.*,**1990**, 265, 2997-3000.
40. Davidson, H.W.; Rhodes, C.J.; Hutton, J.C. *Nature*,**1988**, 333, 93-96.
41. Docherty, K.; Rhodes, C.J.; Taylor, N.AS.; Shennan, K.I.J.; Hutton, J.C. *J. Biol. Chem.*, **1989**, 264, 18335-18339.
42. Kiefer, M.C.; Tucker, J.E.; Joh, R.; Landsberg, K.E.; Saltman, D.; Barr, P.J. *DNA Cell Biol.*,**1991**, 10, 757-769.
43. Christie, D.L.; Batchelor, D.C.; Palmer, D.J. *J. Biol. Chem.*, **1991**, 266, 15679-15683.
44. Seidah, N.G.; Day, R.; Marcinkiewicz, M.; Benjannet, S.; Chretien, M. *The Proteins* (in press).
45. Rangaraju, N.S.; Harris, R.B. *Life Sci.*, **1992**, (in press).
46. Morgan, R.O.; Chang, J.P.; Catt, K.J. *J. Biol. Chem.*, **1987**, 262, 1166-1171.
47. Nikolics, K.; Mason, A.J.; Szonyi, E.; Ramachandran, J.; Seeburg, P.H. *Nature (London)*, **1985**, 316, 511-517.

48. Seeburg, P.H.; Adelman, J.P. *Nature (London)*, **1984**, 311, 666-668.
49. Milton, R.C. deL.; Wormald, P.J.; Brandt, W.; Millar, R.P. *J. Biol. Chem.*, **1986**, 261, 16990-16997.
50. King, J.A.; Anthony, E.L.P. *Peptides*, **1983**, 4, 963-967.
51. Phillips, H.S.; Nikolics, N.K.; Branton, D.; Seeburg, P.H. *Nature*, **1985**, 316, 542-546.
52. Song, T.; Nikolics, K.; Seeburg, P.H.; Goldsmith, P.C. *Peptides*, **1987**, 8, 335-339.
53. Millar, R.P.; Wormald, P.J.; DeLisle Milton, R.C. *Science*, **1986**, 232, 68-73.
54. Silverman, A.J.; Silverman, R.; Lehman, M.N.; Witken, J.W.; Millar, R.P. *Brain Res.*,**1987**, 402, 346-350.
55. Palen, T.E.; Wypij, D.M.; Wilson, I.B.; Harris, R.B. *Arch Biochem. Biophys.*, **1986**, 251, 543-550.
56. Benjannet, B.; Rondeau, N.; Chretien, M.; Seidah, N.G. *Proc. Natl. Acad. Sci. USA*, **1991**, 88, 3564-3568.
57. Thomas, L.; Leduc, R.; Thorner, B.A.; Smeekens, S.P.; Steiner, D.F. *Proc. Natl. Acad. Sci. USA*, **1991**, 88, 5297-5301.
58. Benjannet, S.; Reudelhuber, T.; Mercures, C.; Rondeau, N.; Chretien, M.; Seidah, N.G. *J. Biol. Chem.*, **1992**, 267, 11417-11423.
59. Berry, R.W. *Biochemistry*, **1981**, 20, 6200-6205.
60. Berry, R.W.; Yates, M.E. *Peptides*, **1986**, 7, 637-645.
61. Somlyo, A.P. *Proc. Natl. Acad. Sci. USA*, **1988**, 85,6222-6226.
62. Thibault, G; Garcia, R.; Gutkowska, J.; Bilodeau, J.; Lazure, C.; Seidah, N.G.; Chretien, M.; Genest, J.; Cantin, M. *Biochem. J.*, **1987**, 241, 265-272.
63. Wetsel, W.C.; Culler, M.D.; Johnston, C.A.; Negro-Vilar, A. *Molec. Endocrinol.*,**1988**, 2, 22-31.
64. Gunnarsson, K.; Grehn, K.; Ragnarsson, U. In: *Peptides 1986, 19th European Symposium*; Theodoropoulos, D., Ed. Gruyter: Berlin, Germany, **1986**, pp 123-126.
65. Gauthier, J.; Bruderlein, F.; Aubry, N.; Rakhit, S.; Valois, S.; Bousquet, Y. In: *Synthetic Peptides: Approaches to Biological Problems*. Tam, J.P., and Kaiser, E.T. Eds. UCLA Symposia on Molecular and Cellular Biology, **1989**, vol 86; pp 63-68.
66. Borin, G.; Marchiori, F.; Calderan, A.; Corradin, G.; Wallace, C.J.A. *Biopolymers*, **1986**, 25, 2271-2280.
67. Fujii, N.; Shimokura, M.; Yajima, H.; Shono, F.; Tsuda, M.; Yoshitake, A. *Chem. Pharm. Bull (Tokyo)*, **1984**, 32, 1193-1199.
68. Milton, S.C.F. *Ph.D. dissertation*, **1991**, University of Cape Town, Department of Chemical Pathology, Cape Town, South Africa.
69. Milton, C.S.F.; Brandt, W.F.; Schnolzer, M.; deLisle Milton, R.C. *Biochemistry*, **1992**, 31, 8799-8809.
70. You, J.L.; Milton, S.C.F.; Milton, R.C. deLisle; Rangaraju, N.S.; Harris, R.B. *J. Prot. Chem.* (in press).

RECEIVED April 30, 1993

Chapter 17

Venom Toxins Reveal Ion Channel Diversity

Michael E. Adams

Departments of Entomology and Neuroscience, University of California, Riverside, CA 92521

Naturally-occurring venoms contain toxins of extraordinary potency and selectivity. The properties of these toxins are proving useful in the discrimination of pharmacologically distinct receptor and ion channels types in nerve and muscle. Of relevance to pest control is the occurrence of phylogenetically selective toxins, some of which have already been shown to increase the efficacy of insecticidal baculovirus strains. Venoms may thus facilitatate introduction of engineered biological agents which may be considered fourth generation insecticides of the future.

Venoms are used by many animals for prey capture. Injected in extremely small amounts by wasps, spiders, scorpions, snails, or snakes, these complex biochemical cocktails are extraordinary in their ability to induce rapid paralysis. The biochemistry underlying this process involves mainly peptides and proteins, acting as potent and selective ligands for a variety of membrane-bound ion channels vital to nerve or muscle function.

A hallmark of venom toxins is their selectivity. For example, α- and β-scorpion toxins bind only to sodium channels in vertebrate animals, modifying their properties to produce excitation or block (1). All other ion channels (e.g., potassium, calcium, chloride) are unaffected by these toxins. Interestingly, while these toxins have no effect on insects, the natural prey of scorpions, other toxins occurring in the venom are equally as selective for insect sodium channels. Similarly, black widow spider venom contains a mixture of latrotoxins that are highly specific for either vertebrates or invertebrates (2-4). These examples of remarkable discrimination between phylogenetically distinct subtypes of toxin receptors serve as a starting point for considering novel strategies for pest control; i.e., the development of engineered biological insecticides that are insect-specific.

Peptides and proteins traditionally have not been considered to be candidate insecticides, having no topical or oral toxicity. As is obvious from observing a spider capture its prey, peptide toxins must be delivered by injection. However, genetic engineering technology offers the means for packaging and delivering them using engineered baculoviruses. Modified viruses expressing either a scorpion toxin (5,6) or an acarine toxin (7) have been shown to have increased efficacy over wild type strains. These demonstration experiments point to the potential for developing baculoviruses as "designer" insecticides, which could would be engineered to express combinations of insecticidal or insectostatic gene products appropriate to particular pest insect problems.

0097–6156/94/0551–0249$06.00/0

Various combinations of genes expressing toxins (or mixtures of insect specific enzymes or hormones) aimed at different molecular targets could be packaged as cocktails in analogous fashion to those that occur naturally in venoms. This feature, combined with the growing realization that many venom toxins have high insect selectivity, adds impetus for continued research on novel toxins that may serve as candidate gene products for engineered pesticides of the future.

Two major themes that will be emphasized in this article are toxin diversity and the ability of toxins discriminate differences in heterogeneous receptor populations. These attributes of toxins and their receptors hold great promise for future development of agricultural pest control agents as well as pharmaceutical drugs.

Toxin Diversity Reflects Receptor Heterogeneity

Studies of venom chemistry have led to some general insights about toxins and their receptors. First, venoms tend to be a complex mixture of toxins. Second, toxins fall into related classes, each of which targets a different receptor type. Hence, venoms cause paralysis through the combination of different physiological actions. One of the best known illustrations of this phenomenon is found in snake venom, which contains α- and β-bungarotoxins, the α- toxins binding to the postsynaptic receptor and the β-toxins to a presynaptic receptor. Another example is marine cone snail venom (8) in which groups of toxins occur, each with specificity for a distinct target. α-Conotoxins block the nicotinic acetylcholine receptor, μ-agatoxins block muscle sodium channels, and ω-conotoxins block neuronal calcium channels. All of these toxins act only on vertebrate ion channels; there is no report of these toxins affecting invertebrates. An analogous biochemical strategy directed at insects has been described for the funnel web spider, *Agelenopsis aperta*, which injects three ion channel specific classes of paralytic toxins (see below).

Toxin diversity may be advantageous in several respects. It serves the venomous predator to the extent that an advantageous mutation in one toxin receptor (i.e., one that confers resistance to the toxin) is not likely to result in resistance to the venom, since additional toxins acting on different receptors would be present in the venom. In other words, genes for several toxin receptors are not likely to be altered in a short length of evolutionary time. This minimizes the chance that individuals in the prey population will become resistant to the venom. Similarly, an engineered virus formulation which expresses several toxins aimed at different sites of action might also be less likely to select resistant insects.

Second, toxin diversity may be a dynamic interplay between receptor and toxin evolution. The multiplicity of toxins aimed at both sodium channels and calcium channels is possibly a strategy used by prey animals to contend with the constantly changing pattern of molecular targets in their environment. This is indicated by the fact that toxins are actually able to discriminate subtypes of particular ion channels (e.g., calcium).

Toxin Receptors in the Insect Nervous System

From a basic physiology standpoint, venom toxins demonstrate that insects can be incapacitated in many different ways. The drawing in Figure 1 shows four target sites at a generalized insect neuromuscular junction. Toxins that are insect selective and interact with these sites are listed in Table I. Three sites of action occur on the presynaptic nerve terminal. A common target for spider and scorpion toxins is the voltage-gated sodium channel ("1" in Figure 1), to which the toxins bind and modify inactivation kinetics and voltage-dependent activation. The insect-selective scorpion toxins AaIT (9,10) and $LqhIT_2$ (11), and the μ-agatoxins from funnel web spider venom (12,13) are selective for insect neuronal sodium channels (Table I). A second site on the presynaptic terminal is the voltage-activated calcium channel ("2" in Figure

1). Block of this channel in insects by ω-agatoxins (14-16) interrupts synaptic transmission by calcium-dependent transmitter release. A third site is a receptor protein for latrotoxins from black widow spider venom ("3" in Figure 1). The latrotoxin receptor appears to be part of a protein scaffold involved in the docking of synaptic vesicles to the inner leaflet of the nerve terminal plasma membrane (17). α-Latrotoxin binding to its receptor leads to massive exocytotic release, presumably due to disruption of vesicle docking protein interactions. It is noteworthy that α-latrotoxin action is vertebrate-specific, while a newly described variant of this toxin in the same venom, α-latroinsectotoxin, is highly selective for the insect receptor (4).

A postsynaptic target for venom toxins is the receptor channel activated by the neurotransmitter glutamate ("4" in Figure 1). In this case, aromatic polyamines from orb weaver venoms (argiotoxins; 18-20) or funnel web spider venoms (α-agatoxins; 12, 21-23) bind the channel in its open state, causing a block following activation by the transmitter.

Discovering Novel Ion Channel Toxins from Spider Venoms

Our research activities have gone through several phases of evolution, beginning with investigations into the biochemical mechanisms of venom-induced insect paralysis. We eventually focused on a heterogenous family of toxins that discriminate subtypes of voltage-activated calcium channels. Our characterization of unique venom toxins was driven by the use of sensitive and discriminating assays for detection of a particular type of biological activity.

We initially sought to explain the biochemical basis of venom-induced insect paralysis by spider venom components. Using an electrophysiological assay for synaptic transmission, we defined three classes of ion channel specific toxins, the α-, μ- and ω-agatoxins from *Agelenopsis aperta* venom. Two different types of ω-agatoxins (Types I and II) were found to block insect presynaptic calcium channels.

Table I. Sites of Action for Arthropod Venom Toxins Targeting Insects

Mechanism	*Toxin*
1 - Na channel inactivation	AaIT (*Androctonus*)[a] μ-agatoxins (*Agelenopsis*)[b]
1 - Na channel inact. & block	$LqhIT_2$ (*Leiurus*)[c]
2 - Ca channel block	ω-agatoxins (*Agelenopsis*)[d]
3 - Exocytosis: vesicle depletion	α-latroinsectotoxin (*Latrodectus*)[e]
4 - Ligand-gated channel block	α-agatoxins (*Agelenopsis*)[f] argiopines/argiotoxins (*Argiope*)[g]

a (9-10, 32)
b (12, 13)
c (11) Zlotkin et al., 1991
d (14-16)
e (2-4)
f (12, 21-23)
g (18-20)

Interest then focused on this group of toxins, which turned out to be a very heterogeneous group of calcium channel antagonists with different subtype selectivities.

A second phase of the work made use of a radioligand assay for N-type channel binding in the chick brain. Using radiolabelled ω-conotoxin GVIA (ω-CTX), a peptide calcium channel antagonis, we confirmed the diversity of ω-agatoxin in funnel web spider venom and described a new class of toxins (Type III) which have broad specificity for vertebrate high threshold calcium channels. Subsequently, we used potassium stimulated synaptosomal calcium flux to discover toxins that block P-type calcium channels in the mammalian brain.

We thus adopted biological assays for insect and vertebrate calcium channel function, each of which led to the discovery of a new class of toxins with pronounced differences in their calcium channel specificity. Experimental approaches to the purification of spider toxins and investigations into their mechanisms of action are described below.

HPLC Fractionation

We obtained large quantities of spider venoms from Spider Pharm (Feasterville, PA), a commercial supplier. Venoms were collected by an electrical milking technique devised by Mr. Charles Kristensen. Our approach to purification of toxins from the venom has been very simple: direct fractionation by reversed-phase liquid chromatography. By bypassing preliminary purifications using ion exchange or size exclusion methods, we found that yields of scarce toxins were optimized. It should be emphasized that this approach selects for relatively low molecular weight toxins, since large proteins are precipitated when taken up in aqueous TFA prior to application to the column.

For initial fractionations of crude venom (Figure 3), a Brownlee C_8 wide pore analytical cartridge column equipped with a guard column was used routinely. First, purifications were typically via a linear gradient of aqueous acetonitrile (2-45%, 0.5%/min) in constant 0.1% trifluoroacetic acid. Fractions were vacuum-centrifuged to dryness in polypropylene tubes and stored at -20°C. Fractions were re-dissolved in appropriate buffer or saline for subsequent biological assay. For subsequent purifications, different organic modifiers (acetonitrile, methanol, n-propanol, isopropanol), ion pairing agents (trifluoroacetic acid, heptafluorobutyric acid) and columns (C_4, C_8, C_{18}) were used to achieve different selectivities. Additional information regarding purification and bioassay procedures may be found in Adams et al., (12,15), Mintz et al., (24), and Venema et al., (25).

Electrophysiological Assay of Spider Venom Fractions

We used larval insect body wall muscles as an *in vitro* electrophysiological assay for block of synaptic transmission. The body wall muscles (longitudinal ventrolateral muscles 6 and 7) of pre-pupal house flies (*Musca domestica*) were used because, at this developmental stage, spontaneous and neurally-evoked contractions are minimal, allowing for greater ease of intracellular recordings. Individual muscle cells were impaled with a microelectrode to measure two responses: the excitatory junctional potential (EJP) in response to nerve root stimulation, and a depolarizing "ionophoretic potential" (IP) evoked by ionophoretic application of the transmitter glutamate close to a synaptic junction (Figure 2). Whereas the EJP response depends on a multi-step sequence of events leading to transmitter release and its binding to postsynaptic receptors, the IP is a direct assay of postsynaptic membrane response to applied transmitter applied directly. In this way, clues can be obtained regarding the likely pre-

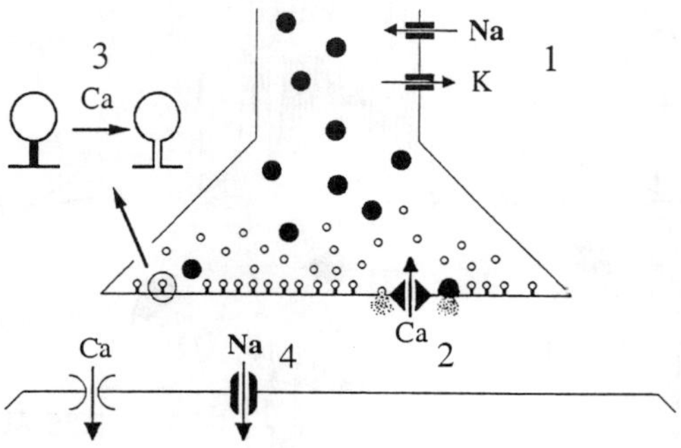

Figure 1. Diagram showing some ion channel targets for venom toxins a generalized insect synaptic junction. Sites 1-3 are presynaptic, 4 is postsynaptic. Site 1 is the insect voltage active sodium channel, targeted by scorpion toxins and spider toxins. Site 2 is the voltage-activated calcium channel. Block of this channel prevents calcium entry and transmitter release. Site 3 is the latrotoxin receptor, which is involved in the exocytotic release mechanism. Site 4 is the transmitter activated cation channel, which is blocked by acylpolyamine spider toxins. (See Table I for list of toxins acting at these sites).

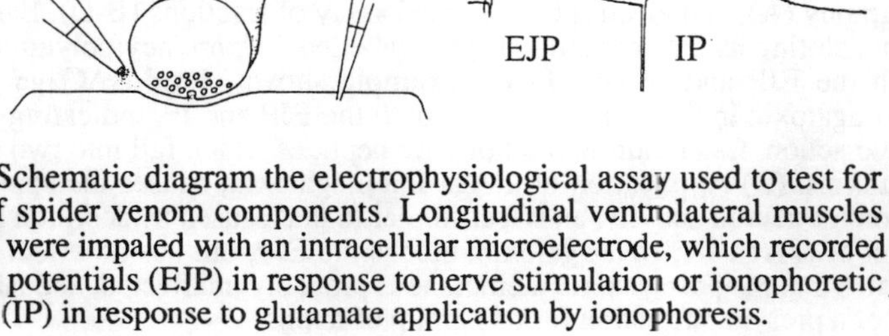

Figure 2. Schematic diagram the electrophysiological assay used to test for biological activity of spider venom components. Longitudinal ventrolateral muscles of *Musca domestica* were impaled with an intracellular microelectrode, which recorded excitatory junctional potentials (EJP) in response to nerve stimulation or ionophoretic junctional potentials (IP) in response to glutamate application by ionophoresis.

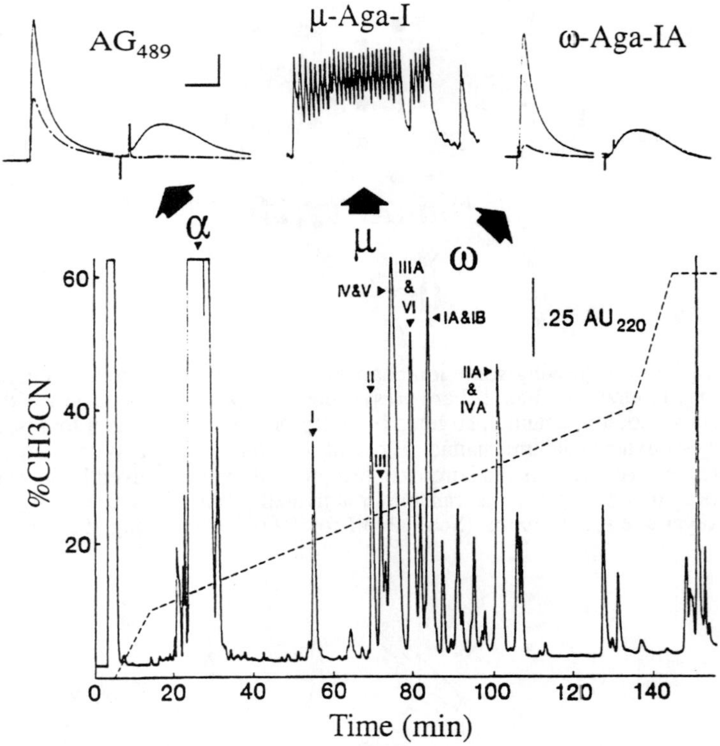

Figure 3. Fractionation of crude *Agelenopsis aperta* spider venom by reversed phase chromatography (A), and electrophysiological assay of fractions (B-D). Early eluting α-agatoxins, eluting as a mixture in the peak labelled "α", are acylpolyamines which block both the EJP and IP (B). In the example shown, 1 mM AG_{489} (the most abundant α-agatoxin in the venom), blocks both the EJP and IP, indicating a possible postsynaptic action. Later eluting fractions are peptides which fall into two classes: μ- and ω-agatoxins. (B) Application of μ-Aga-I, a μ-agatoxin, causes the appearance of repetitive motor neuron activity, an effect similar to that caused by scorpion toxins. (C) Application of ω-Aga-IA, an ω-agatoxin, blocks the EJP, but has no effect on the IP. This indicates that the postsynaptic membrane is probably unaffected, and suggests the likelihood of a presynaptic action for this group of toxins.

or postsynaptic action of a toxin. Whole venom or HPLC fractions diluted in insect saline were applied in the bath to the neuromuscular preparation.

This approach allowed us to distinguish three types of ion channel specific toxins in the venom of the funnel web spider, *Agelenopsis aperta* (12, 15). Early eluting fractions (Figure 3A, α-agatoxins) caused a highly reversible decrease in both the EJP and the IP, indicating that this type of block was accompanied by a change in the response of the postsynaptic membrane to the transmitter. This effect is illustrated in Figure 3B after application of the most abundant α-agatoxin in the venom, AG_{489}. The reversible action of AG_{489} correlated the reversible paralysis observed after its injection into whole insects. A group of fractions eluting later in the chromatogram (ω-agatoxins) caused a long-lasting reduction in the EJP, but had no effect on the IP (Figure 3D). The ω-agatoxins clearly produced synaptic block that was not associated with a change in postsynaptic receptor sensitivity to the neurotransmitter. We subsequently showed in a series of experiments that this effect resulted from block of presynaptic calcium channels (14, 16) by two types of ω-agatoxins (Types I and II). Finally, a third group of toxins, (the μ-agatoxins in Figure 3) led to repetitive activity originating in the distal branches of motor neurons. These toxins are highly selective for insect sodium channels, functioning in an analagous manner to the scorpion toxin AaIT.

The paralytic effects of the α- and μ-agatoxins are synergistic. The explanation for this phenomenon may lies in the use-dependent nature of α-agatoxin action. The α-agatoxins require activation of postsynaptic, glutamate-sensitive ion channels before their binding sites are revealed. Since the excitatory μ-agatoxins cause increased spontaneous neurotransmitter release, we expected that they may enhance the paralytic actions of the α-agatoxins. These expectations were demonstrated by co-injection of both α- and μ- agatoxins into adult insects, which showed higher frequency of paralysis than were observed following injection of either toxin alone (12).

We concluded that paralysis in insects caused by injection of funnel web spider venom was caused by the joint actions of actions of the α-, μ- and ω-agatoxins, each acting at a distinct ion channel target at the neuromuscular junction. The only toxins showing lethal actions when injected into insects were the μ-agatoxins, which also appear to be specific for insect sodium channels.

Radioligand Binding Assay for Toxin Discovery

Further studies showe that the ω-agatoxins are the most heterogeneous of the three classes of agatoxins. Having discovered Type I and Type II agatoxins which block insect presynaptic calcium channels, we proceeded to test them against vertebrate calcium channels in collaboration with Baldomero Olivera and colleagues at the University of Utah. Our approach was to test spider venom fractions for inhibition of radiolabelled ω-conotoxin GVIA (ω-CTX) binding. ω-CTX is a specific antagonist of N-type calcium channels in vertebrate neurons. The $[^{125}I]$-ω-CTX radioligand was prepared by standard direction iodination procedures using iodogen as a catalyst and chick synaptosomal membranes were prepared by the method of Cruz and Olivera, (26) 1986). Membranes were resuspended in buffer and incubated at $4^{o}C$ with various concentrations of unlabelled toxins for 30 min. Following the pre-incubation period, $[^{125}I]$ω-conotoxin GVIA was added to determine the number of unblocked binding sites remaining.

We found that Type II, but not Type I ω-agatoxins inhibited $[^{125}I]$ω-CTX binding to chick synaptosomal membranes and blocked ^{45}Ca entry through these channels at concentrations some 30x lower than ω-CTX. This finding demonstrated that the ω-agatoxins not only were structurally distinct, but that they also detected phylogenetic differences in calcium channels. These experiments also led us to re-test all fractions

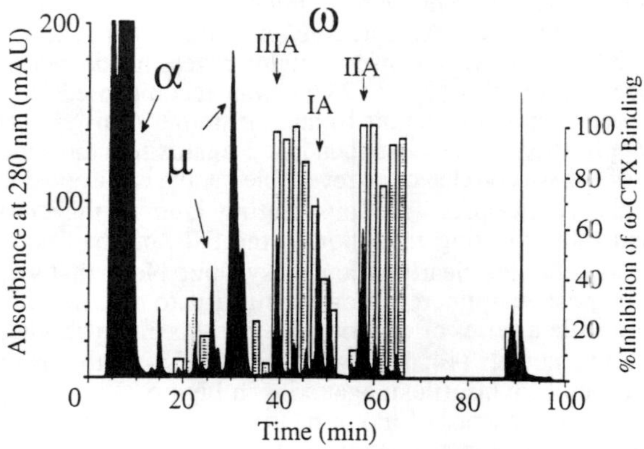

Figure 4. Fractions from *Agelenopsis aperta* spider venom block the binding of [^{125}I]ω-conotoxin to chick synaptosomal calcium channels. HPLC chromatogram of *Agelenopsis aperta* spider venom (shown in black), upon which a histogram is superimposed shows fractions from the venom that block radioligand binding. These fractions occur in the "ω" region of the chromatogram, which also showed antagonism of insect presynaptic calcium channels.

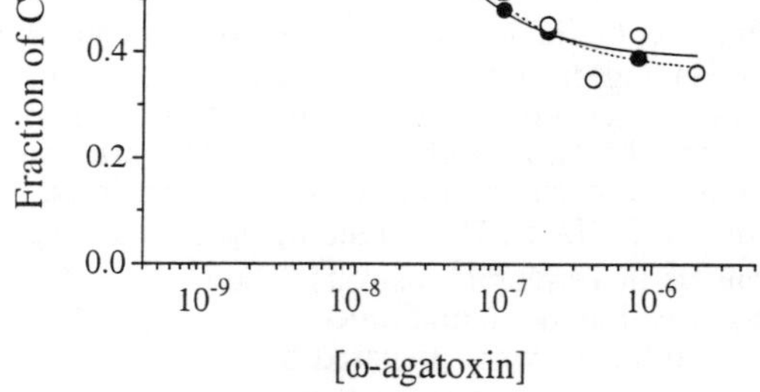

Figure 5. Block of potassium-stimulated ^{45}Ca entry into rat brain synaptosomes by two ω-agatoxins, ω-Aga-IVA and ω-Aga-IVB. These toxins selectively antagonize "P-type" calcium channels in the mammalian brain. This flux assay was used to screen for this type of biological activity in the crude venom.

from crude venom in the binding assay (Figure 4) as a means of screening for novel toxins. This resulted in the identification of a new peptide toxin, ω-Aga-IIIA, (25) Venema et al., 1992). A novel property of ω-Aga-IIIA is its ability to recognize a binding site common to all high-threshold calcium channels in vertebrates. Somewhat surprisingly, this toxin does not block calcium channels in *M. domestica*, the experimental animal originally used to discover the ω-agatoxins in electrophysiological assays. This again illustrates the high degree of phylogenetic selectivity found in many of these venom toxins.

Synaptosomes as a Screen for Calcium Channel Toxins

ω-Aga-IIIA and ω-CTX were found to be potent blockers of calcium channels in chick brain synaptosomes, but neither had any appreciable effect on rat brain synaptosomal calcium channels. This resistant calcium channel in rat brain nerve terminals was reminiscent of reports by Bean and colleagues (27, 28) showing that certain neurons in the rat brain express calcium channels that are insensitive to blockers of both N- and L-type calcium channels. Since this current is pronounced in cerebellar Purkinje neurons, it was dubbed "P-type" by Rodolfo Llinas and colleagues (29). We decided to employ the synaptosomal flux assay as a means of detecting toxins that might block resistant Ca channels in rat brain synaptosomes. Several fractions did in fact have this activity and we used the synaptosome assay to purify two toxins, ω-Aga-IVA (24) and ω-Aga-IVB (30). These toxins block about 60% of the potassium-stimulated calcium entry into rat brain synaptosomes, (Figure 5). Subsequently, it was shown using electrophysiological methods that ω-Aga-IVA and ω-Aga-IVB are highly selective for P-type calcium channels in mammalian central and peripheral neurons (31).

Implications of Toxin Diversity for Pest Control

Natural venoms are a vast reservoir of toxins with diverse pharmacological properties and phylogenetic specificities. It seems likely that toxin diversity is an evolutionary response of venomous predators to a dynamic heterogeneity of receptors in prey animals. Toxins are thus valuable tools for distinguishing differences in ion channel targets. The potent effects exerted by toxins on the nervous system has obvious applications for both pest control, drug development, and basic neuroscience.

Since peptide toxins are the direct products of gene expression, they can be incorporated and expressed by insect pathogens for increased insecticidal efficacy. This has already been demonstrated in several instances in which insect-selective toxins have been incorporated into baculoviruses, resulting in reduction of time to mortality (5-7). Given the relatively small number venoms that have been analyzed, it is probable that many more toxins with desirable properties remain to be discovered and considered as active ingredients in engineered pathogens with tailor-made selectivities. These highly potent and selective agents may be part of a fourth generation of insecticides, which combine the approaches of directed screening of natural products with gene cloning. The eventual uses of such insecticides may depend not so much on their efficacies as on the willingness of society to permit their use in the field.

Literature Cited

1. Catterall, W. A. *Science* **1988**, *242*, 50-61.
2. Frontali, N.; Grasso, A. *Arch. Biochem. Biophys.* **1964**, *106*, 213-218.

3. Fritz, L. C.; Tzen, M. C.; Mauro, A. *Nature* **1980**, *283*, 486-7.
4. Magazanik, L. G.; Fedorova, I. M.; Kovalevskaya, G. I.; Pashkov, V. N.; Bulgakov, O. V.; Grishin, E. V. *Neuroscience* **1992**, *46*, 181-8.
5. Tomalski, M. D.; Miller, L. K. *Nature* **1991**, *352*, 82-85.
6. Maeda, S.; Volrath, S. L.; Hanzlik, T. N.; Harper, S. A.; Majima, K.; Maddox, D. W.; Hammock, B. D.; Fowler, E. *Virology* **1991**, *184*, 777-80.
7. Stewart, L. M. D.; Hirst, M.; Ferber, M. L.; Merryweather, A. T.; Cayley, P. J.; Possee, R. D. *Nature* **1991**, *352*, 85-88.
8. Olivera, B. M.; Gray, W. R.; Zeikus, R.; McIntosh, J. M.; Varga, J.; Rivier, J.; de Santos, V.; Cruz, L. J. *Science* **1985**, *230*, 1338-1343.
9. Zlotkin, E.; Rochat, H.; Kopeyan, C.; Miranda, F.; Lissitzky, S. *Biochimie* **1971**, *53*, 1073-1078.
10. Walther, C.; Zlotkin, E.; Rathmayer, W. *J. Insect Physiol.* **1976**, *22*, 1187-1194.
11. Zlotkin, E.; Eitan, M.; Bindokas, V. P.; Adams, M. E.; Moyer, M.; Burkhart, W.; Fowler, E. *Biochemistry* **1991**, *30*, 4814-21.11.
12. Adams, M. E.; Herold, E. E.; Venema, V. J. *J. Comp. Physiol. [A]* **1989**, *164*, 333-42.
13. Skinner, W. S.; Adams, M. E.; Quistad, G. B.; Kataoka, H.; Cesarin, B. J.; Enderlin, F. E.; Schooley, D. A. *J. Biol. Chem.* **1989**, *264*, 2150-5.
14. Bindokas, V. P.; Adams, M. E. *J. Neurobiol.* **1989**, *20*, 171-188.
15. Adams, M. E.; Bindokas, V. P.; Hasegawa, L.; Venema, V. J. *J. Biol. Chem.* **1990**, *265*, 861-7.
16. Bindokas, V. P.; Venema, V. J.; Adams, M. E. *J. Neurophysiol.* **1991**, *66*, 590-601.
17. Petrenko, A. G.; Perin, M. S.; Davletov, B. A.; Ushkaryov, Y. A.; Geppert, M.; Sudhof, T. C. *Nature* **1991**, *353*, 65-8.
18. Grishin, E. V.; Volkova, T. M.; Arseniev, A. S.; Reshetova, O. S.; Onoprienko, V. V.; Magazanic, L. G.; Antonov, S. M.; Federova, I. M. *Bioorg. Khim.* **1986**, *12*, 110-112.
19. Bateman, A.; Boden, P.; Dell, A.; Duce, I. R.; Quicke, D. L. J.; Usherwood, P. N. R. *Brain Res.* **1985**, *339*, 237-244.
20. Adams, M. E.; Carney, R. L.; Enderlin, F. E.; Fu, E. T.; Jarema, M. A.; Li, J. P.; Miller, C. A.; Schooley, D. A.; Shapiro, M. J.; Venema, V. J. *Biochem. Biophys. Res. Commun.* **1987**, *148*, 678-83.
21. Jasys, V. J.; Kelbaugh, P. R.; Nason, D. M.; Philips, D.; Saccomano, N. A.; Stroh, J. G.; Volkmann, R. A. *J. Am. Chem. Soc.* **1990**, *112*, 6696-6704.
22. Jasys, V. J.; Kelbaugh, P. R.; Nason, D. M.; Phillips, D. *J. Org. Chem.* **1992**, *57*, 1814-1820.
23. Quistad, G. B.; Suwanrumpha, S.; Jarema, M. A.; Shapiro, M. J.; Skinner, W. S.; Jamieson, G. C.; Lui, A.; Fu, E. W. *Biochem. Biophys. Res. Commun.* **1990**, *169*, 51-6.
24. Mintz, I. M.; Venema, V. J.; Swiderek, K. M.; Lee, T. D.; Bean, B. P.; Adams, M. E. *Nature* **1992**, *355*, 827-829.
25. Venema, V. J.; Swiderek, K. M.; Lee, T. D.; Hathaway, G. M.; Adams, M. E. *J. Biol. Chem.* **1992**, *267*, 2610-5.
26. Cruz, L. J.; Olivera, B. M. *J. Biol. Chem.* **1986**, *261*, 6230-6233.
27. Regan, L. J. *J. Neurosci.* **1991**, *11*, 2259-69.
28. Regan, L. J.; Sah, D. W.; Bean, B. P. *Neuron* **1991**, *6*, 269-80.
29. Llinas, R.; Sugimori, M.; Lin, J.-W.; Cherksey, B. *Proc. Natl. Acad. Sci. USA* **1989**, *86*, 1689-1691.
30. Adams, M. E.; Mintz, I. M.; Reily, M. D.; Thanabal, V.; Bean, B. P. *Mol. Pharm.* **1993**, *In Press.*
31. Mintz, I. M.; Adams, M. E.; Bean, B. P. *Neuron* **1992**, *9*, 85-95.
32. Pelhate, M.;Zlotkin, E. *J. Exp. Biol.* **1982**, *97*, 67-77.

RECEIVED August 17, 1993

Chapter 18

Targeting Locust Muscle Glutamate Receptors with Polyamine-Containing Toxins

Possible Strategy for Insecticide Development

P. N. R. Usherwood

Department of Life Science, University of Nottingham, Nottingham NG7 2RD, United Kingdom

Glutamate receptors present in the surface membranes of excitable cells in insects represent an unexploited target for insecticide action. The polyamine amide toxins isolated from venoms of insects and spiders are potent antagonists of glutamate receptors, including those of insects, which gate cation-selective ion channels. These toxins may interact with glutamate receptors of insect muscle at 4 separate sites, one of which is intracellular. By appropriate structural modifications it may be possible to produce analogues of these toxins which are highly potent at their target sites and which can enter pest insects and reach their targets.

Receptors for the neurotransmitter L-glutamate have been found in the excitable tissues of many insect species (*1,2*). They appear to be less diverse than their known counterparts in vertebrate central nervous systems (*3*), although progress currently being made in cloning insect glutamate receptors (GluR) may lead to changes in this perception (*4*). One other perceived major difference between GluR of insects and vertebrates is that the former may be associated with either cation-selective or anion-selective ion channels (*5*), whereas the latter seem to be associated exclusively with cation-selective channels. The GluR of locust muscle, on which the majority of polyamine amide studies in insects have been made, comprise two pharmacologically and physiologically distinct populations; the one population gates cation-selective channels (*6*) and is sensitive to L-quisqualate (qGluR) (*7*); the other population gates chloride channels and is sensitive to L-ibotenate (*8*). A somewhat similar situation exists in the locust central nervous system (*9*).

Impairment of qGluR function would have a devastating, probably fatal impact on an insect. These receptors occupy a relatively exposed position on skeletal and

0097–6156/94/0551–0259$06.00/0

visceral muscle in terms of access to haemolymph-borne toxicants. It is surprising, therefore, that there are no insecticides which target these macromolecules. Rational strategies designed explicitly to exploit qGluR for insecticide purposes have so far been largely unsuccessful. However, the discoveries of polyamine amides (aryl amines) in venoms of certain species of spider and parasitic wasp which noncompetitively antagonise qGluR (*10*) (and also nicotinic acetylcholine receptors (nAChR) of insect central nervous systems (*11, 12*)) provide a new, and perhaps more realistic chance of exploiting these receptors.

This chapter reviews recently published data on the interactions of synthetic philanthotoxins with qGluR of locust (*Schistocerca gregaria*) muscle. It will be demonstrated that the effects of this toxin on qGluR are more complex than was first envisaged (*13, 14*), and that a case can now be made for multiple sites of action of these and other polyamine amides. Having established that there are potent interactions between insect qGluR and polyamine amides, is there now an opportunity to develop these compounds into a new class of insecticides?

The Philanthotoxins

δ-Philanthotoxin was discovered in the venom of a parasitic wasp (*Philanthus triangulum*) by Piek and colleagues (*15, 16*) and subsequently characterised as tyrosyl-butanoyl-thermospermine (PhTX-433; the numerals refer to the number of methylenes spanning the amines of the polyamine moiety) (*17, 18*). The close synthetic analogue PhTX-343 (*17*), which contains spermine rather than thermospermine, has formed the basis for extensive synthesis programmes leading to the production of over 100 analogues of this toxin (*19, 20*). Most of these compounds have been bioassayed on insect nerve-muscle preparations (obtained mainly from the locust *Schistocerca gregaria*) to provide a detailed structure-activity profile, but some have also been tested on vertebrate excitable tissues (reviewed in *21*).

Information on structure-activity relationships, the similar pharmacological properties of the philanthotoxins and other polyamine amide toxins (e.g. argiotoxins), and the interactions of philanthotoxins with invertebrate and vertebrate excitable systems has been extensively reviewed (*10, 21, 22*). A summary of the structure-activity relationships pertaining to the interactions of philanthotoxin with locust muscle qGluR is presented in Fig. 1.

qGluR of Locust Muscle

Skeletal muscle fibres of locusts and other insects are innervated by excitatory motoneurones which are glutamatergic (*23*). Some fibres may also receive innervation from one or more inhibitory motoneurones which release the neurotransmitter γ-aminobutyric acid (GABA) (*24*). Present evidence suggests that the philanthotoxins have no effect on peripheral GABA receptors in locust (*25*). qGluR are found postjunctionally at glutamate synapses on locust muscle. These receptors gate cation-selective (Na^+, K^+, Ca^{2+}) ion channels (*6, 8*). qGluR of this type are also found in low population density on the extrajunctional membrane of locust muscle (*25*).

Potentiation of qGluR by philanthotoxins

Like argiotoxin-636 (ArgTX-636) (*27, 28*), philanthotoxins increase the open probability of the channel gated by locust muscle qGluR (*29*). This potentiation of the response to agonist is reflected in an increase in the amplitude of the postsynaptic current (EPSC) seen following stimulation of the excitatory input of locust muscle (*24*). The potentiation is not voltage-sensitive and occurs at concentrations of PhTX-343 which are lower than those which antagonise qGluR (see below). It is proposed that a site (Site 1; Fig. 2) located extracellularly on qGluR is involved in this potentiation. It is further proposed that when toxin binds to Site 1 it allosterically increases the affinity of the agonist binding site(s) on qGluR. PhTX-343 binds to Site 1 even when qGluR is in its closed channel conformation, since the potentiation caused by this toxin is not dependent on prior activation of qGluR by agonist.

Antagonism of qGluR by Philanthotoxins

During it extracellular application to locust muscle either in a patch pipette (at ca 10^{-9}M), in single channel studies, or in the muscle bath (at $\geq 10^{-7}$M), for example in EPSC and muscle twitch contraction studies, PhTX-343 antagonises qGluR. This antagonism occurs in two stages. The first is voltage-independent and use-independent. This probably results from interaction of the toxin with the closed channel conformation of qGluR. Subsequently, a voltage-dependent antagonism occurs, during which the open channel gated by qGluR is blocked by toxin. It follows that two antagonist binding sites are involved in these reactions: the one located extracellularly on qGluR outside the membrane potential field of the receptor (Site 2; Fig. 2), the other located within the field, e.g. in the channel gated by this receptor (Site 3; Fig. 2). Site 2 could be the glutamate binding site, which implies competitive antagonism, but this remains uncertain. Antagonism at Site 3 is use-dependent, which confirms that this site is available only when qGluR occupies an open channel conformation. A third antagonist site (Site 4; Fig. 2) is indicated from studies of intracellularly-injected philanthotoxins (*30*). For example, when PhTX-343 is microinjected into locust muscle fibres by either current pulses or by pressure, the EPSC and qGluR channel openings are inhibited. This antagonism is neither use-dependent nor voltage-dependent. It is proposed that Site 4 is on the intracellular domain of qGluR and available when this receptor is in a closed channel conformation.

Philanthotoxins as tools in neuroscience research

Affinity columns designed from polyamine-containing synthetic toxins have already been used to isolate and purify GluR from mammalian brain. Hossain et al. (*31*) used a 1-naphthylacetyl spermine-formyl-cellulofine affinity column to isolate a ca

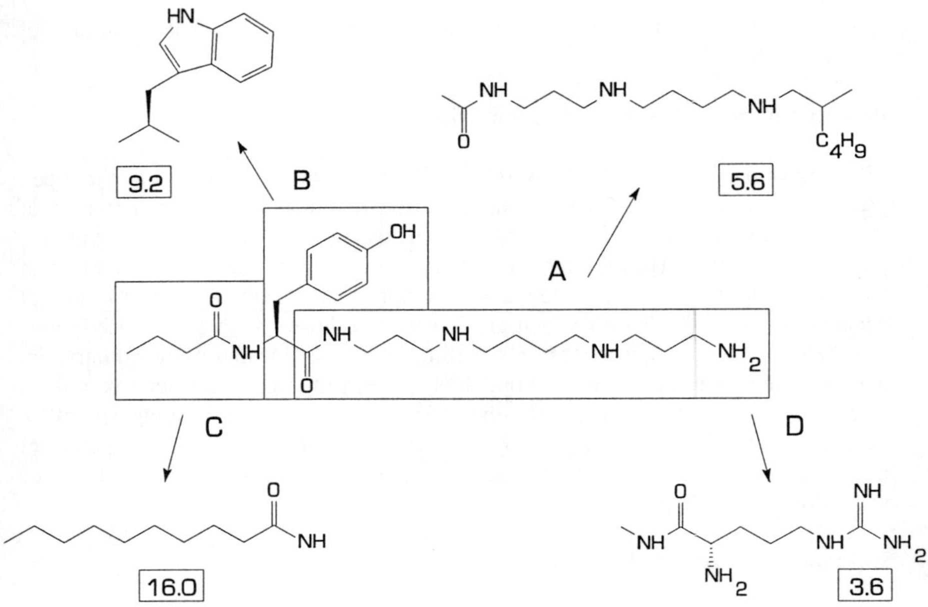

Figure 1. Summarised structure-activity properties of polyamine amides. Modifications to each of regions A-D of PhTX-343 lead to increases in potency, as assessed by inhibition of the locust retraction unguis muscle. (After Usherwood (*22*)).

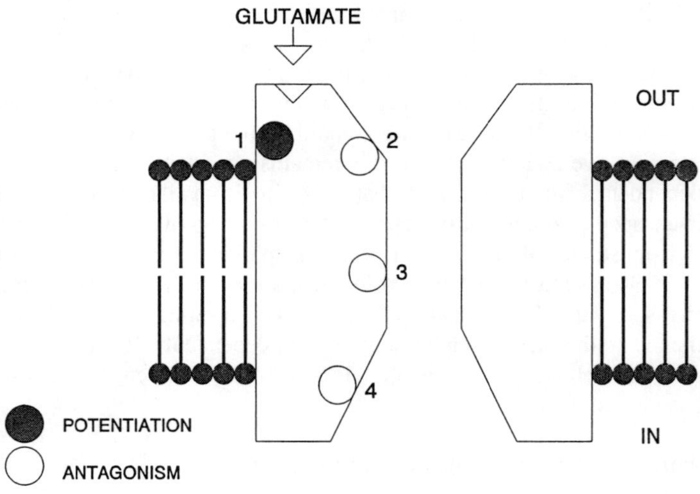

Figure 2. Sites (1-4) of interaction of polyamine amides with qGluR. When toxin binds to site 1 it potentiates the response of the receptor to agonists. Binding of toxin to sites 2-4 causes antagonism of qGluR. Site 3 is in the receptor channel and binding of toxin to this site covers open channel block. Access to site 4 could arise by transit of toxin through the open channel of qGluR (*35*).

60 kDa protein from bovine brain. The N-terminal region of this protein exhibits striking homology with a Ca^{2+}-binding protein, calreticulin, of muscle sarcoplasmic reticulum. Polyclonal antibodies raised in guinea pigs against this protein block transmission at a lobster glutamatergic neuromuscular junction and at glutamatergic synapses on CA1 neurones in guinea pig hippocampal slices. It is suggested (*32*) that the antibody interacts with postjunctional GluR in these preparations. However, there is an alternative possibility, namely that the antibody binds to Ca^{2+} channels on the presynaptic terminals of these synapses and, thereby, inhibits transmitter release. Affinity columns prepared from immobolized ArgTX-636 and an immobilised analogue of PhTX-343 are also being used to isolate GluR. In addition, photosensitive analogues of PhTX-343 are available for this purpose (*33*). When appropriately irradiated these photosensitive compounds irreversibly antagonise locust muscle qGluR (*34*). Photosensitive philanthotoxins are currently being employed in the labelling of GluR and nAChR, so that a clearer view of the sites of interactions of polyamine-containing toxins with ligand-gated membrane receptors can be established (*33*).

Commercial Applications of Philanthotoxins

A number of patents have been filed which cite natural and synthetic polyamine-containing toxins for use in agriculture. In some, it is claimed that these compounds are lethal either when they are applied topically to or when they are ingested by certain insects.

Some of the properties that would be required of a polyamine amide insecticide are an ability to gain entry into pest insects, resistance to metabolic degradation within these insects, and high potency at its site of action. There are, of course, other equally important considerations, such as persistence in the environment, cost of production, toxicity to man and other non-pest animals, and the impact on the ecosystem of the use of analogues of wasp and spider toxins as insecticides.

The spider and wasp toxins and their synthetic analogues are essentially hydrophilic whereas most insecticides are hydrophobic, a property which enables them to penetrate lipid-containing barriers. It may be possible to design hydrophobic analogues of polyamine amides which are transformed, by *in vivo* metabolism when they cross insect cuticle and reach the pest insect circulatory system, i.e. a "pro-insecticide" approach (Fig. 3). However, even if this objective is achieved, the toxins may have problems in reaching the qGluR (and nAChR) of central nervous system and muscle in the face of adsorption, absorption, and enzymic degradation. When cockroaches, for example, are injected with $10\mu g\ gm^{-1}$ ArgTX-636 and PhTX-343 they are initially paralysed but they recover after about 2h (*35*). It is not known what happens to these toxins when they have entered an insect, although the peptide bonds and terminal amines that they contain could make them subject to enzymic degradation in the haemolymph, in fat body and other tissues. Conversion of the terminal amines in these toxins to aldehydes by diamine oxidase would certainly cause a major reduction in their activity on qGluR. Removal of the positive charged moiety which characterises the terminal region of all polyamine amide toxins results in almost complete loss of activity (*19*). Also

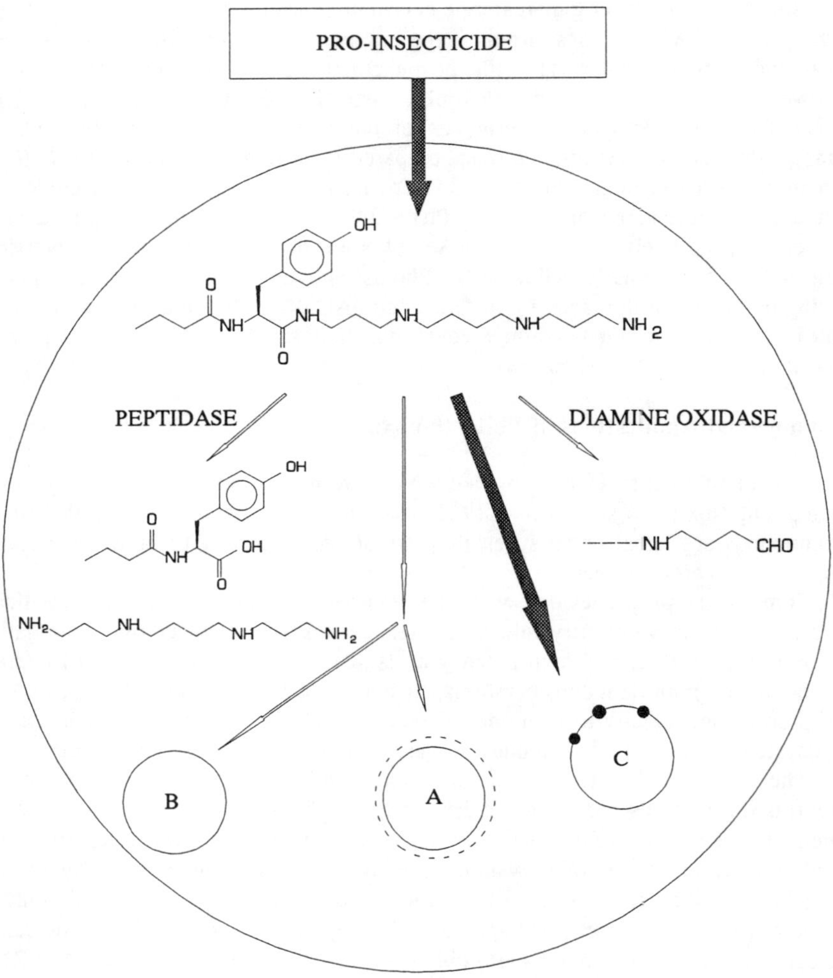

Figure 3. Requirements for the development of an insecticide using polyamine amides. Entry of a pro-insecticide into the pest insect would signal the release of the toxin in its active form. The toxin would then face enzymic degradation through, for example, the action of peptidase and diamine oxidase. Adsorption to cell surfaces (A) and absorbtion by cells (B) might further deplete it before it reaches its target site, i.e. qGluR (D, filled circles).

scission of the peptide bonds in these molecules would release metabolites (e.g. polyamines) with zero or much lower activity on qGluR (*19*).

Toxicity to Humans

It has been established that qGluR of insects and GluR of vertebrates (including man) have many pharmacological properties in common. In view of this commonality, it is hardly surprising that polyamine-containing toxins are potent antagonists of vertebrate GluR. Indeed, some of these toxins and their synthetic analogues have been the subject of patent applications with respect to their potential therapeutic utility. It is highly probable, therefore, that insecticides based upon polyamine amides will be toxic to man and other animals, which would could mean that their potential use in agriculture for pest control seems limited or even doomed. However, this is an argument that could be applied equally well to many, if not all, extant commercial insecticides.

Conclusions

The continuing interest in the polyamine amides as lead structures comes at a time of major changes in the insecticide industry, not least of which is a reappraisal of the so-called rational approach towards insecticide discovery. This may slow industrial research on these compounds. Also, developments involving natural products as starting points are being examined very critically by environmentalists and the polyamine amide toxins have not escaped their attention. It has been suggested that the use of polyamine amide insecticides could lead to resistance to the natural products present in the venoms of spider and wasp. However, the potential consequences of this for the ecosystem in general are difficult to identify, let alone to estimate.

Acknowledgments

Some of the work reported herein was supported by grants from the U.K. SERC and AFRC, NATO and the US Army European Office.

Literature Cited

1. Piek, T. In *Comparative Insect Physiology & Biochemistry*; Kerkut, G.A. & Gilbert, L.A., Eds.; Pergamon Press: Oxford, UK, 1985, Vol 11; pp 55-118.
2. Duce, I. R. In *Comparative Invertebrate Neurochemistry*; Lunt, G.G. & Olsen, R.W., Eds.; Croom-Helm, London, UK, 1988, pp 42-89.
3. Sommer, B.; Seeberg, P. H. *TINS*, 13, 291-296, 1992.
4. Schuster, C. M.; Ultsch, A.; Schloss, P.; Cox, J. A.; Schmitt, B.; Betz, H. *Science*, 1991, 254, 112-114.
5. Usherwood, P. N. R. *Adv. Comp. Physiol. Biochem.* 1978, *7*, 227-310.
6. Anwyl, R.; Usherwood, P. N. R. *Nature* 1974, *252*, 591-593.

7. Usherwood, P. N. R. In *Glutamate as a Neurotransmitter*; Di Chiari, G. & Gessa, G.L., Eds.; Raven Press New York, 1981, pp 183-193.
8. Lea, T. J.; Usherwood. P. N. R. *Comp. Gen. Pharmac.* 1973, *4*, 333-350.
9. Giles, D.; Usherwood, P. N. R. *Comp. Biochem. Physiol.* 1985, *89C*, 231-236.
10. Usherwood, P. N. R.; Blagbrough, I. S. *Pharm. Therap.* 1991, *52*, 245-268.
11. Rozental, R.; Scoble, G. T.; Albuqerque, E. X.; Idriss, M.; Sherby, S.; Sattelle, D. R.; Nakanishi, K.; Eldefrawi, A. T.; Eldefrawi, M. E. *J. Pharm. Exptl. Therap.*, 1989, *249*, 123-130.
12. Piek, T.; Hue, B.; Pelhate, M.; Spanjer, W.; Veldsema-Currie, R. D. *Arch. Insect Biochem. Physiol.* 1984, *1*, 297-306.
13. Usherwood, P. N. R.; Duce, I. R.; Boden, P. J. *Physiol. (Paris)*, 1984, *79*, 241-245.
14. Adams, M. E. In *Neurotox '88: Molecular Basis of Drug and Pesticide Action*; Lunt, G. G., Ed.; Excerpta Medica, Elsevier, Amsterdam, 1989. pp. 49-60.
15. Piek, T. *Comp. Biochem. Physiol.* 1982, *72C*, 311-315.
16. Clark, R. B.; Donaldson, P. L.; Gration, K. A. F.; Lambert, J. J.; Piek, T.; Ramsey, R. L.; Spanjer, W.; Usherwood, P. N. R. *Brain Res.* 1982, *241*, 105-114.
17. Eldefrawi, A. T.; Eldefrawi, M. E.; Konno, K.; Mansour, N. A.; Nakanishi, K.; Oltz, E.; Usherwood, P. N. R. *Proc. Natl. Acad. Sci. USA*, 1988, *85*, 4910-4913.
18. Blagrough, I. S.; Bruce, M.; Bycroft, B. W.; Mather, A. J.; Usherwood, P. N. R. *Pesticide. Sci.* 1990, *30*, 397-403.
19. Bruce, M.; Bukownik, R.; Eldefrawi, A. T.; Eldefrawi, M. E.; Goodnow, R., Jr.; Kallimopoulos, T.; Konno, K.; Naskanishi, K.; Niwa, M.; Usherwood, P. N. R. *Toxicon* 1990, *28*, 1333-1346.
20. Nakanishi, K.; Goodnow, R.; Kalivretenos, A.; Choi, S-K.; Fushiya, S.; Chiles, G.; Eldefrawi, A.; Eldefrawi, M.; Usherwood, P.N.R. In *New Leads for Targets in Drug Research*; Krogsgaard-Larsen, P., Christensen, S. Bregger, Kofod, H. Eds.; Alfred Benzon Symposium, Munksgaard, Copenhagen, 33, 1992, 138-148.
21. Anis, N.; Sherby, S.; Goodnow, R. Jr.; Niwa, M.; Konno, K.; Kallimopoulos, T.; Bukownik, R.; Nakanishi, K.; Usherwood, P. N. R.; Eldefrawi, A.; Eldefrawi, M. *J. Pharm. Exptl. Therap.* 1990, *254*, 764-773.
22. Usherwood, P.N.R. In *Excitatory Amino Acids*, Meldrum, B.S., Moroni, F., Simon, R.P., Woods, J.H. Eds.; Raven Press, New York, 1991, pp 379-395.
23. Blagbrough, I. S.; Brackley, P. T. H.; Bruce, M.; Bycroft, B. W.; Mather, A. J.; Millington, S.; Sudan, H. L.; Usherwood, P. N. R. *Toxicon*, 1992, *30*, 303-322.
24. Usherwood, P. N. R.; Grundfest, H. *J. Neurophysiol.* 1965, *28*, 497-518.
25. Sudan, H. I., Usherwood, P. N. R. (In preparation).
26. Cull-Candy, S. G. & Usherwood, P. N. R. *Nature* 1973, *246*, 62-64.

27. Usherwood, P. N. R. In *Sites of Action of Neurotoxic Pesticides*, Hollingworth, R. M., Green, M. B., Eds.; ACS Symp Ser. 1987, 356, 298-315.
28. Kerry, C. J.; Ramsey, R .L.; Sansom, M. S. P.; Usherwood, P. N. R. *Brain Res.* 1988, *459*, 312-327.
29. Karst, H. & Piek, T. *Comp. Biochem. Physiol.* 19 , *98C*, 479-489.
30. Brundell, P.; Goodnow, R, Jr.; Kerry, C. J.; Nakanishi, K.; Sudan, H. L.; Usherwood, P. N. R. *Neurosci. Letts.* 1991, *131*, 196-200.
31. Hossain, A.; Hagiwara, K.; Kawai, N.; Nakajima, T. *Chem. Pharm Bull.* 1991, *39*, 3079-3081.
32. Hossain, A.; Miwa, A.; Tagashira, M.; Hagiwara, K.; Kawai, N.; Nakajima, T. *Biomed Res.* 1991, *12*, 291-295.
33. Chiles, G. Choi, S-K.; Eldefrawi, A.; Eldefrawi, M.; Fushiya, S.; Goodnow, R. Jr.; Kalivretenos, A.; Nakanishi, K.; Usherwood, P. N. R. In *Neurotox '91: Molecular basis of Drug and Pesticide Action* Duce, I.R., Ed.; Elsevier, London, 1991, pp. 3-18.
34. Goodnow, R.A. Jr.; Nakanishi, K.; Sudan, H. L.; Usherwood, P. N. R. *Neurosci. Letts.* 1991, *125*, 62-64.
35. Usherwood, P. N. R.; Blagbrough, I. S. In *Insecticide Action: From Molecule to Organism*; Narahashi T., Chambers, J.E., Eds.; Plenum Press, New York, 1989, pp 13-31.

RECEIVED April 30, 1993

Chapter 19

Biorational Control of Weeds and Fungi with Peptides

Alan R. Lax, John M. Bland, and Hurley S. Shepherd

Southern Regional Research Center, Agricultural Research Service, U.S. Department of Agriculture, New Orleans, LA 70179

Tentoxin is a cyclic tetrapeptide produced by the fungus *Alternaria alternata* that disrupts chloroplast development in most of the major weed species in soybean and johnsongrass in corn while having no effect on either of these crop species. The major impediment to the development of tentoxin as a herbicide is its limited availability because of low biosynthetic yields. Iturins are a family of cyclic octapeptides having a unique ß-amino acid, iturinic acid, incorporated into the peptide backbone. Iturins have a broad antifungal activity, but low biosynthetic yields may limit their commercialization. Chemical synthesis of both molecules is possible and the production of analogues has led to meaningful structure/activity relationships. To date no analog has provided greater activity or specificity than the parent molecule for either of the two agents. Molecular genetic manipulation is considered the best strategy for the increase in production of both of these compounds for cost effective synthesis. Success of cloning of the biosynthetic genes may determine the eventual deployment of these biorational pesticides.

The cost of pest control in agriculture is staggering; direct costs of pesticides are estimated to be at least $20 billion annually, with unestimated indirect costs making the exact amount impossible to accurately state. Recent popular concern for the environment and legislation toward banning some of the more traditional pesticide compounds has led to the development of new strategies for the control of pests which include microbes as biological control organisms and the use of the biorational approach in which natural products provide a basis for the development of highly effective compounds having great specificity and inherently lower effect on non-target populations. Herein we will focus on the biorational approach to the development of two peptide pesticides, one a herbicide based upon the fungal product, tentoxin, and the second a fungicide based on the bacterial product, iturin. Both of these molecules are highly active with the major constraint to their use in agriculture their limited availability.

The commonly used herbicides today fall into relatively few chemical classes (*1,2*). The cost of the development of new chemistries is ever expanding in part because so many of the possible classes of chemicals have been exhausted and because of the increased cost for research and particularly registration for agricultural use. Secondary metabolites offer some particular advantage in this regard because many of the compounds produced in secondary metabolism are difficult syntheses which would likely not be tried in the more traditional "synthesize and spray" approach and also in part because of the potentially less stringent regulatory requirements facing biological and biorational compounds (*1,2*). Regulatory issues are increasingly a factor in the development of pesticides with future development likely to be more impacted because of public demand for low non-target toxicity, increased specificity and short environmental longevity.

Although the total usage of fungicides is significantly lower than that of either herbicides or insecticides the potential for human health risks is most significant with fungicides because ninety percent (by weight) of those fungicides used are known to cause tumors in laboratory animals (*3*). While one cannot directly extrapolate these findings to human health, the significance of these data on possible future regulatory actions by EPA or FDA is alarming, given the increased reliance on such fungicides in the marketplace (*3*). Agricultural scientists must be ready to offer alternatives to traditional pesticides when these more toxic traditional pesticides are replaced. The biorational approach to pesticide design will be but one of several techniques which must be intergarted for successful pest control.

Biologically active peptides represent tremendous potential as pesticides in the increasingly stringent development environment. Peptides are conspicuously under-represented as pesticides, while as a class they offer many advantages including the potential for exceptional specificity of effects, and low non-target activity (*1,2*). Because of their structure it is anticipated that their fate in the environment would be short lived, again environmentally advantageous, while reducing their potential for long term pest control via residual activity (*4*).

Work in our laboratories has centered upon two cyclic peptides of diverse origin as model systems for the development of both herbicides and fungicides. The two will be discussed separately below with the emphasis on the development of iturin as a fungicide because of the work previously presented on the potential of tentoxin as a herbicide (*1,2,5*). However, some of the limitations to the development of tentoxin will be pointed out as illustrative and some of the potential solutions to the problems will presented. These molecules are important not only because of their anticipated use as pesticides but also because of their potential to identify new target sites in the host organisms through mode of action studies.

Tentoxin

Tentoxin is a cyclic tetrapeptide produced by the fungus, *Alternaria alternata* (*6*), having the structure depicted in Figure 1. Tentoxin affects a large number of weeds in both soybean and corn while having no discernible effect on either of these species (see mode of action below)(*1,2,7*). While the synthesis of tentoxin has been achieved in the laboratory (*8-11*), it is a difficult synthesis, achieving a limited final yield because of the reaction at the cyclization step of the synthesis. While its fate in soil is as yet uncharacterized, tentoxin should be fairly readily broken down into its constituent amino acids by microbial action.(*10*)

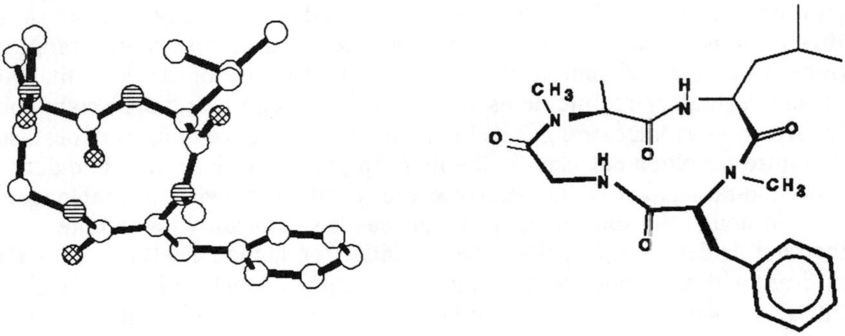

A B

Figure 1. Structure of tentoxin [cyclo(N-methylalanyl-leucyl-N-methyldehydrophenylalanyl-glycyl)] . A) Molecular model, B) chemical structure

The use of tentoxin as a herbicide has been hampered by its relatively low biosynthetic yield by even selected high producing strains of *Alternaria* (*1,2,5,12*), even with its potent biological activity in the micromolar range. This is probably the major obstacle to the development of tentoxin; because of their generally potent activity and consequently low biosynthesis this promises to be problematic with other, as yet undiscovered, microbial toxins as pesticides. Little is known concerning its biosynthesis, and it has yet to be determined whether tentoxin synthesis proceeds through thioester template typical of many cyclic peptides or through an as yet undiscovered mechanism. Data from our lab (*5*) and that of Liebermann and Ihn (*12*) indicate a mechanism different from the thioester pathway, however, this pathway has not been ruled out. Strain selection, media composition and fermentation parameters have not yielded significantly higher production of the toxin. Further discussion concerning improving biosynthesis follows in a later section. As an alternative to higher natural production, we have undertaken extensive structure activity relationship studies utilizing computer modelling to guide synthetic efforts at simpler more readily synthesized molecules which retain both the potent activity and the desired specificity (*13*). In numerous analogues synthesized and tested none of the compounds has demonstrated greater activity than the parent molecule (*8,9,13*).

Structure Activity Studies. Structure activity studies have led us to speculate that the composition and conformation of the leucyl-dehydophenylalanyl portion of tentoxin are essential for its activity (*8-11,13*). We have undertaken to produce simpler compounds whose conformations at this position match the parent compound. Using a computer modelling system we have identified several analogues, such as leucyl-dehydrophenylalanyl diketopiperazine, which have the predicted conformations of the dehydrophenylalanine, with the correct planar orientation of the phenyl ring. When tested in the lettuce seedling bioassay system we saw no chlorosis nor growth inhibitory effects. However, the products of this synthesis are extremely non-polar and have exceedingly low solubilities in water. We are attempting to formulate these compounds in such a way as to render them more available for uptake in this system and to evaluate the effects of sidechain modifications to improve solubility. One such analogue was the tripeptide diketopiperazine alanyl-leucyl-N-methyl-dehydrophenylalanyl diketopiperazine, containing the diketopiperazine of the analogue mentioned above, but with a third amino acid attached to the leucyl amine, thus adding a polar amino functionality. This compound, however, was also inactive. A large number of linear analogues have also been tested for chlorosis activity because of their increased water solubility, but with little if any positive results.

Because we have not been successful in the synthesis of compounds having the required activity and ease of synthesis necessary for pesticidal application it is likely that the biosynthesis of tentoxin itself will have to be improved in order to make this compound a viable product unless formulations provide the required solubility to allow activity of previously synthesized compounds.

Identification of Target Sites Through Mode of Action. Tentoxin is believed to affect the development of chloroplasts in one of two ways: 1) through disruption of import into the chloroplast of at least one protein, polyphenol oxidase (*14-17*), or 2) inhibition of chloroplast coupling factor 1 (CF1) (*18,19*). It has been demonstrated that tentoxin affects the relative quantities of several other proteins in addition to polyphenol oxidase in treated chloroplasts while these have not been identified (*5*). Tentoxin has also been shown to inhibit the activity of one of two chloroplast envelope ATPases presumably in a similar fashion to that of CF1 (*20*). The selective inhibition of one of the two envelope ATPases has been invoked to explain the more or less specific inhibition of polyphenol oxidase while other chloroplast proteins which are synthesized in the cytoplasm are unaffected by tentoxin treatment (*5,20*). Although tentoxin is known to inhibit chloroplast coupling factor 1, there are several compelling arguments to invoke another mechanism of chloroplast disruption (*15,16*). Recent work indicates that sensitive species have a CF1 having a β-subunit with a single amino acid substitution compared to resistant species. Moreover transformation of Chlamydomonas CF1 to the sensitive amino acid composition yields tentoxin sensitivity not seen in the wild type (*19*). Further research is need to resolve this intriguing problem and determine if the two effects are in some way connected. Resolution of this problem will allow further development of compounds having the desirable specificity shown by tentoxin and will allow the development of crops resistant (if they are not already immune) to the pesticide when it is introduced into the marketplace. Furthermore discovery of a target different from CF1 whose disruption has such effects on chloroplast development would provide a new molecular target for pesticide design having the specificity inherent to tentoxin.

Because tentoxin specifically inhibits chloroplast development while having no discernable effect on other plant organelles or on mitochondrial coupling factor, it can be assumed that tentoxin would have no non-target effects. Even within the same genus, different species have different sensitivities to tentoxin indicating a very specific site of action with limited effect. This is one of the more compelling reasons to develop tentoxin as a herbicide given the increasing public support for safer pesticides.

Iturin as a Fungicide

Bacillus subtilis, the organism which produces iturin(s) has been registered for use in the prevention of storage losses of peach to brown rot and has been suggested for the protection of rice from infection by *Aspergillus* sp and thus protection from aflatoxin (*21,22*). This organism moreover has been suggested for the protection of soil-borne diseases of greenhouse crops (*23*). Because of the often different environmental windows for growth of fungi and this potential biocontrol bacterium, we have begun investigations into the use of the antibiotic iturin when the producing organism is either unable to compete or when neither bacterial nor fungal growth is desirable (*24*). Little is known about the mode of action of the iturins. Unlike tentoxin's unique specificities iturin affects virtually all fungi tested and we will not discuss mode of action for this compound in detail.

Chemistry of the Iturins. The iturins are a family of cyclic lipopeptides having seven α-amino acids and an unusual ß-amino fatty acid, iturinic acid as depicted in Figure 2. Unlike tentoxin which is composed solely of L-amino acids the iturins contain both D-and L-forms. Members of the iturin family include the iturins, bacillomycins, and mycosubtilin. In all cases the LDDLLDL sequence of the α-amino acids is kept constant, as well as the ß-amino acid, D-Tyr^2, and D-Asn^3. Within the family of iturins there can be amino acid substitutions which change the character of the activity of the compound. Even within a single class of iturin the ß-amino fatty acid can vary in its length (13 to 17 carbons) and branching (*n*-, *iso*-, or *anteiso*-configuration) to provide a number of different compounds having the same amino acid complement but differing only in the aliphatic side chain (*25,26*). The structure-activity relationships of the various naturally occurring iturins and several chemically modified analogs have been partially investigated and yield some insight into the portions of the molecule which are required for fungicidal activity. Much research remains to fully elucidate the exact functional moieties and spatial relationships required for their antifungal properties. The naturally occurring iturins contain a D-tyrosyl moiety; chemical modification of the residue by methylation or acetylation completely abolishes the antifungal activity of the iturins. However, similar modification of the serine residue in iturin A only slightly diminishes the activity. Modification at both of the residues likewise destroys the activity. It can thus be concluded that the tyrosyl moiety is essential for fungal inhibition (*27*). Other naturally occurring modifications in the iturin structure are amino acid substitutions. These substitutions also provide insight into the conformational constraints on the cyclic peptide required for activity. Iturin C differs from iturin A only in the substitution of L-aspartic acid for L-asparagine at position 1, and is totally inactive against fungi. Because this residue is between the lipophilic ß-amino acid and the essential tyrosine (position 2), it might be assumed that ionic interference results in loss of activity. However, bacillomycin L also contains an analogous L-aspartyl residue at position 2 yet retains antifungal activity. It is therefore uncertain at this time which amino acid residue(s) are required for antifungal activity. Although naturally occurring iturins contain a mixture of eight methylene homologs and structural isomers of the iturinic acid, it had not been shown whether the aliphatic moiety is required for antifungal activity. We have previously reported the synthesis of iturin analogs with iturinic acid substituted by both ß-alanine and α-nonyl-ß-aspartate and demonstrated the loss of antifungal activity when the side chain was not present or included a polar functionality. HPLC separation of the natural mixture of iturin A, containing iturinic acid with *n*-C_{14}, *anteiso*-C_{15}, *iso*-C_{15}, *n*-C_{15}, *iso*-C_{16}, *n*-C_{16}, and *anteiso*-C_{17} configurations, enabled the effect of ß-amino acid side chain length and type of branching on antifungal activity to be determined. The relative activity of the homologues was determined on *Aspergillus* and *Penicillium sp.* showing an increase in activity proportional to chain length and for branching type giving an order of *iso* > *normal* > *anteiso*. Recently we have synthesized iturin A2 and shown that its activity is identical to that of the natural compound (*28*).

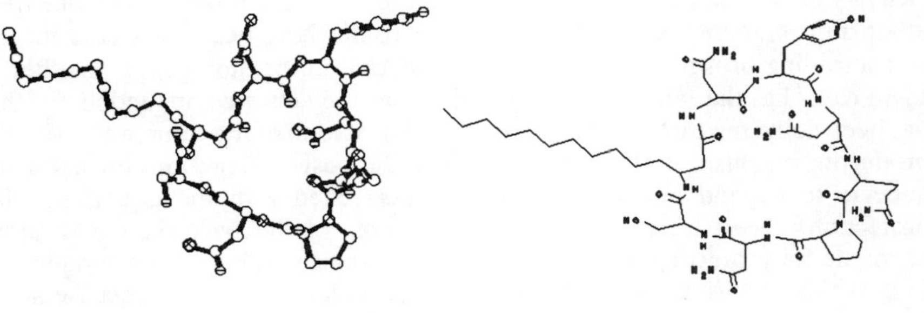

A B

Figure 2. Structure of iturin A2 [cyclo(ituryl-asparagyl-D-tyrosyl-D-asparagyl-glutamyl-prolyl-D-asparagyl-seryl)]. A) Molecular model, B) chemical structure.

Rationale for Development. A partial list of the fungi which are affected by iturin includes numerous species of *Aspergillus, Penicillium,* and *Fusarium*; among them are some of the most troublesome fungi from a food safety standpoint because of the potent mycotoxins which they produce (*24*). It was originally hoped that iturin may prove to protect stored grains from aflatoxin contamination caused by species of *Aspergillus flavus or parasiticus*. While some of the aflatoxin producing strains in our laboratory proved to be sensitive to high concentrations of iturin, we have found that some of the producing strains are not sensitive to the toxin and that in some cases the synthesis of the aflatoxin in liquid culture at least is increased by the iturin (*29*). It is questionable whether iturin can be used to control all species of *Aspergilli* and their toxins.

The most compelling arguments for using iturin *per se* for the protection of stored grain are ecological. Many of the fungal organisms which are associated with stored grains are more or less xerophytic organisms which thrive in dry conditions associated with stored agricultural commodities. *B. subtilis* however is not active under these conditions, so we felt that the compound responsible for the fungal inhibition may be better suited to the protection of such stored grains than the organism itself. We have not as yet explored the possibility that iturins, under these conditions, which may be less ideal than the liquid culture media employed, do in fact protect from infestations by the aflatoxin producing organisms. We are currently exploring this still intriguing possibility. Notwithstanding the possible failure of iturin at protecting against aflatoxin production by all isolates of the fungus, the potential for development of itruin as a fungicide in other agricultural settings is significant. Only five new fungicides introduced into the marketplace between 1975 and 1988 have gained greater than five percent of the market for any major crop (*3*). Increased presence of pesticides in groundwater (*3*) and potential health problems previously alluded to with continued use of current fungicides, create a potentially bright future for the use of iturin or its derivatives as a fungicide.

Delivery of Iturin. *Bacillus subtilis* has been used for the protection of rice from aflatoxin contamination and *B. subtilis* suspensions have been registered for use in controlling brown rot of stone fruits (peach) during storage. In neither of these cases has the iturin itself been isolated as the active compound, but rather has been delivered through the concomitant infestation of the commodity with the producing organism itself. Whether this is the best strategy for delivery again appears to depend on ecological factors associated with the organs needing protection. Seed treatments with living spores of the bacterium may prove adequate for protecting the rhizosphere of emerging seedlings from damping off fungi which require moist conditions; however, organisms which infect the aerial portions of the plant or phylloplane are often inhibited by just those conditions in which *Bacillus* would thrive, while the fungi are activated under conditions in which the bacterium would sporulate and go dormant. In these cases it may prove beneficial to have a ready source of the active compound to be applied in a more traditional manner.

Antibiotic Production

Production of iturin in culture has been more studied than that of tentoxin, while none of the biochemical steps *per se* has been identified. Optimal production of iturin is afforded by the inclusion of aspartic acid into the medium, but the highest reported synthesis of iturin in culture to date is reported to be around 60 mg per liter after 40 hours of fermentation (*30*). Tentoxin biosynthesis has been similarly shown to be affected by medium composition, and culture conditions but to date the highest expected biosynthetic yields are around 90 mg per liter (*31*). Tentoxin production is found to be maximal after 6 to 14 days depending on the isolate of the fungus used. Tentoxin biosynthesis on a large scale is also hampered by the fact that still culture of the fungus is required. Aeration provided by swirling or bubbling air inhibits the production of the toxin making commercial fermentations of this toxin seem prohibitive at this time (*31*).

The biosynthesis of tentoxin has been correlated with the presence of a fungal virus; any cultural manipulation which reduces the presence of the virus similarly affects the production of the toxin, while those cultural conditions which were reported to prevent the production of the toxin reduce the presence of the fungal viruses. Purification of the virus followed by electrophoretic separation of the coat proteins and western blotting using antibody produced against tentoxin showed antigenic similarities (*5*). We have been unable to show directly the relationship between the virus and tentoxin production since the virus has so far shown no infectivity toward non-virus containing strains, nor have we been able to cure the fungus of the virus; after removal of the conditions which reduce it the virus returns to previous levels (*5,31*). While there is precedence in the literature for the production of cyclic peptides from longer precursors which are subsequently cleaved and cyclized, the majority of cyclic peptides are synthesized using thioester templates such as that involved in the biosynthesis of tyrocidine and gramicidin by *Bacillus brevis* (*32*). The enzymes for a large number of these multifunctional peptide synthetase have been isolated and characterized. There are regions of homology in many of these peptide syntetases (*32*) which may make them amenable to molecular genetic probing for analogous enzymes in both tentoxin and itruin production. Because of the low biosynthetic potential of both of these important biologically active peptides we have begun an intensive search in our laboratories for analogous genes for their production with the hope of increasing the production of these compounds through genetic manipulation. Work has already begun on the identification of the genes for the production of iturin with promising results (*34*). Whether success in this area provides sufficient material for pesticide production remains to be seen.

Conclusions

Much research remains to be conducted prior to the release of either of these potential products into the marketplace. Because of their potent activity and specificity which should provide good control of the target pests with fewer side effects seen with more traditional pesticides there is justification for continued investigations. Research into the genetics of the production of these compounds may permit economical production while at the same time offers the possibility to determine the role of tentoxin in the phytopathogenicity of *Alternaria alternata* and the role of iturin in control of such diseases by *Bacillus* isolates. Development of these promising compounds would surely lead the way to discovery of other potentially useful peptide toxins and further biorational pest control research. It is our hope that this work stimulates such discovery and in some small way guides the development of such products in the near future.

Literature Cited

1. Duke, S.O. *Rev. Weed Sci.*, **1986**, *2*, 15-44.
2. Duke, S.O.; Lydon, J. *Weed Technology*, **1987**, *1*, 122-128.
3. Anonymous, In *Alternative Agriculture*, National Research Council, 1989. National Academy Press, Washington, DC. 448pp.
4. Edwards, J.V.; Dailey, O.D.; Bland, J.M.; Cutler, H.G. In *American Chemical Society-Agrochemicals, Proceedings*, 194th National Meeting, August 1987, New Orleans, Louisiana. Chapter 3, pp 35-56.
5. Lax, A.R.; Shepherd, H.S., In *American Chemical Society-Agrochemicals, Proceedings*, 194th National Meeting, August 1987, New Orleans, Louisiana. Chapter 2, pp24-34.
6. Templeton, G.E., In *Microbial Toxins*, Kadis, S.; Ciegler, A.; Ajl, S. J. eds.; Academic: New York, 1972; chapter 7, pp 169-192.
7. Rich, D.H., In *Toxins in Plant Disease. Mechanism of Action*; Durbin, R.D., Ed.; Academic: New York, 1981. Physiological Ecology Series, Chapter 8.
8. Rich, D. H.; Mathiaparanam, P., *Tetrahedron Lett.*, **1974**, *15*, 4037-40.
9. Rich, D. H.; Bhatnagar, P.; Mathiaparanam, P.; Tam, J. P. *J. Org. Chem.*, **1978**, *43*, 296-302.
10. Edwards, J.V.; Lax, A.R.; Lillehoj, E.B.; Boudreaux, G.J. *Int. J. Peptide Protein Research*, **1986**, *28*, 603-612.
11. Jacquier, R.; Verducci, J. *Tetrahedron Lett.*, **1984**, *25*, 2775-2778.
12. Liebermann, B.; Ihn, W. *J. Basic Microbiol.*, **1988**, *28*, 63-70.
13. Bland, J.M.; Edwards, J.V.; Eaton, S.R.; Lax, A.R. *Pesticide Science*, **1993**. in press.
14. Lax, A.R.; Vaughn, K.C., Sisson, V.A.; Templeton, G.E. *Photosynth. Res.*, **1985**, *6*, 113-120.
15. Lax, A.R.; Vaughn, K.C. *Physiol. Plant.*, **1986**, *68*, 384-391.
16. Vaughn, K.C.; Duke, S.O. *Physiol. Plant.*, **1981**, *53*, 421-428.
17. Vaughn, K.C.; Duke, S.O. *Physiol. Plant.*, **1984**, *60*, 257-261.
18. Steele, J.A.; Uchytil, T.F.; Durbin, R.D.; Bhatnagar, P.; Rich, D.H. *Proc. Nat. Acad. Sci. U.S.A.*, **1976**, *73*, 2245-2248.
19. Avni, A.; Anderson, J.D.; Holland, N.; Rochaix, J-D., Gromet-Elhanan, Z.; Edelman, M. *Science*, **1992**, *257*, 1245-1247.
20. Lax, A.R. *Plant Physiol.*, **1987**, *83S*, 53.
21. Pusey, P.L.; Wilson, C.L. 1988 United States Patent, 4,764,371.
22. Klich, M.A.; Lax, A.R.; Bland, J.M.; Scharfenstein, L.A. *Mycopathologia*, **1993**, in press.
23. Bochow, H. *Developments in Agricult. and Managed Forest Ecology*, **1990**, *23*, 236-240.
24. Peypoux, F.; Guinand, M.; Michel, G.; Delcambe, L.; Das, B.; Lederer, E. *Biochemistry*, **1978**, *17*, 3992-3996.

25. Isogai, A, Takayama, S.; Murakoshi, S.; Suzuki, A. *Tetrahedron Lett*, **1982**, *23*, 3065-3068.
26. Besson, F.; Peypoux, F.; Michel, G. *J. Antibiotics*, **1979**, *32*, 828-833.
27. Phae, C-G.; Shoda, M. *J. Fermentation Bioengineering*, **1991**, *71*, 118-121.
28. Matsuno, Y.; Hiraoka, H.; Ano, T.; Shoda, M. *FEMS Microbiol. Lett.*, **1990**, *67*, 227-230.
29. Matsuno, Y.; Ano, T; Shoda, M. *J. Gen. Appl. Microbiol.*, **1992**, *38*, 505-509.
30. Bland, J.M.; Lax, A.R.; Klich, M.A. *Peptides 1992: Proceedings, Twenty-Second European Peptide Symposium*, **1992**, 332-333.
31. Shepherd, H.S. *J. Virol.*, **1988**, *62*, 3888-389.
32. Krause, M.; Marahiel, M.A.; von Dohren, H.; Kleinkauf, H. *J. Bacteriol.*, **1985**, *162*, 1120-1125.
33. Mittenhuber, G.; Weckermann, R.; Marahiel, M.A. *J. Bacteriol.*, **1989**, *171*, 4881-4887.
34. Matsuno, Y.; Hitomi, T.; Ano, T.; Shoda, M. *J. Gen. Appl. Microbiol.*, **1992**, *38*, 13-21.

RECEIVED August 19, 1993

Chapter 20

Design of Antifungal Peptides for Agricultural Applications

Nicholas P. Everett

InterLink Associates Inc., Princeton Corporate Center, 5 Independence Way, Suite 300, Princeton, NJ 08540

The objective of this research is to identify antifungal peptides that have the appropriate properties to provide protection against plant diseases in major crops. The feasibility of designing antifungal peptides which combine stability against plant proteases, antibacterial activity and/or minimal phytotoxicity has been demonstrated. The peptides are sufficiently small to be amenable to chemical synthesis of both the peptide and the corresponding DNA sequence for a gene. This attribute permitted the research to start with naturally occurring peptides and then, using principles of rational design, move on to unique proprietary peptides that have improved qualities for agriculture. Thus it should be possible to design antifungal peptides which are suitable for their production in, and use on, crops.

A wide range of higher organisms make use of antimicrobial peptides as part of their defense system against infection. Examples of such peptides include the cecropins, attacins and diptericins involved in the cell-free immunity in insects (*1*), the apidaecins from honeybees (*2*), the defensins from mammalian phagocytes (*3*), the magainins from frog skin (*4*), and the thionins from plants (*5*). The high cost of chemical synthesis of peptides would preclude them from being considered for use in agriculture were it not for the possibility of using biotechnology to engineer their synthesis in crop plants or other recombinant production systems.

It is therefore not surprising that antimicrobial peptides have been considered as sources of disease resistance genes for transgenic plants (*6*). However, these naturally occurring peptides suffer from a number of problems which limit their usefulness against fungal diseases in plants. Many antimicrobial peptides are antibacterial rather than antifungal; the defensins and thionins have internal disulfide linkages (and recombinant proteins with internal disulfides often do not fold correctly in a foreign host); most antimicrobial peptides are membrane-active and have the potential for non-specific cytotoxicity; and many peptides are potentially unstable in

0097–6156/94/0551–0278$06.00/0

plants because of their short half-lives in their natural host and their exposure to different proteases in plants. For these reasons it was thought likely that the naturally occurring versions of many of these peptides would be unsuitable for use in plants. Accordingly, peptide design, synthesis and bioevaluation studies were initiated to determine the feasibility of designing synthetic antifungal peptides (mycopathogenins or MYPs) with improved properties for agricultural uses.

One class of peptides selected for initial studies was the antimicrobial frog skin peptide class known as magainins (*7*). These peptides combined a number of features that suggested their suitability as a starting point for peptide design:

Size. Magainin-like peptides of 20-25 amino acid residues in length are amenable to chemical synthesis of both the peptides and oligonucleotides that can be assembled into the coding sequence of a synthetic gene. Small peptides are also more amenable than large proteins, or more complex molecules, to computer modeling of structure-function relationships.

Structure. Magainins contain only normal amino acid residues found in proteins. Unlike many peptide antibiotics they do not contain D-amino acid residues, side-chain modifications or disulfide linkages, and they are not cyclic. They adopt an alpha-helical structure with amphiphilic character (*8*) that is important for biological activity. Knowledge of the likely three-dimensional structure allows modeling of the effects of changing individual amino acid residues; progress can be made using rational design instead of empirical testing of random changes.

Antimicrobial Activity. The natural magainins and some synthetic derivatives have been shown to have relatively broad spectrum activity against a wide range of pathogens (*9,10*) including important plant pathogens such as *Cercospora* spp., *Cochliobolus* spp., *Fusarium* spp., *Pythium* spp., *Trichoderma* spp., *Verticillium* spp., *Erwinia* spp., *Pseudomonas* spp. and *Xanthomonas* spp. (Everett et al., unpublished).

Low Cytotoxicity. Mature magainin peptides have been shown to have low hemolytic activity against human erythrocytes (*9*), a factor that has stimulated research on their potential use as pharmaceutical agents and should aid their treatment as "safer" fungicidal agents during registration. As small proteins, they would be expected to be degraded rapidly after ingestion or in the soil environment, leaving no toxic residues. Magainins do show potential toxicity against mitochondria (*11*) and isolated chloroplasts (Everett et al., unpublished data) *in vitro*. However, their synthesis in plants could be designed to involve the secretion of pre-pro peptides into the intercellular spaces in a way that is analogous to their production in frog skin and to the production of thionins in plants.

Information Base. Interest in the potential pharmaceutical use of magainin-like peptides has produced an explosion of publications on their structure, mode of action, synthesis, molecular biology and antimicrobial activity. This now provides a strong knowledge base upon which to build their adaptation for agricultural uses.

Novelty. The potential problems that peptides will encounter in plants are quite different from those in mammals. The solutions to such problems are likely to result in the design of peptides that are significantly different from both their natural progenitors and derivatives intended for human and animal healthcare.

Goal. Based on this background, the major goal was to design peptides that were stabilized against protease degradation, retained potent antifungal activity and exhibited minimal phytotoxicity.

Methods

Precise details of all the methods used in this study can be found in a European patent application (*12*).

Peptide Synthesis. Peptides were synthesized using standard solid-phase methodology on an ABI Model 430A peptide synthesizer. The peptides were cleaved from the resin using HF, converted to the acetate form, and purified by reverse phase HPLC using a TFA/acetonitrile gradient to elute the peptides from a Vydac C-4 column.

Bioassays. The biological activities of the synthetic peptides were determined using assay methods that were adapted to require only small amounts of peptide. Antifungal and antibacterial bioassays were performed in 96-well microtiter plates. Potential phytotoxicity was routinely evaluated using an oxygen electrode to measure inhibition of photosynthesis or respiration. All tests included magainin 1 or magainin 2 as a standard reference. Assay conditions were established to compare the relative activities of peptides, the absolute values cannot be extrapolated directly to activities in the field.

Antifungal Activity. Fungal spores isolated from cultures maintained on solid medium were used as a convenient and reproducible source of fungal inocula. A standard density of spores, or pre-germinated spores, in potato dextrose broth (total volume 100 uL) was incubated with a series of concentrations (usually in the range 10-150 ug/mL) of the test peptides for 48 hours. Spore germination and hyphal growth were scored by microscopic observation. The antifungal activity of each peptide was recorded as the MCIC (minimum complete inhibitory concentration), which corresponded to the lowest peptide concentration at which no spore germination or hyphal growth occurred. A *Fusarium* isolate (P3) from maize was used for primary screening.

Antibacterial Activity. Mid-logarithmic growth phase bacterial cultures were diluted to about 10^5 colony forming units per mL and incubated with a series of concentrations (usually in the range 10-150 ug/mL) of the test peptides for 44 hours at 28 C. The optical density (at 630 nm) of each culture well after 20 and 44 hours was used to determine the lowest peptide concentration at which no bacterial growth occurred (MCIC). Primary screening used *Erwinia carotovora* strain SR319.

Potential Phytotoxicity. Tests for potential phytotoxicity included inhibition of photosynthesis in isolated chloroplasts, inhibition of respiration in suspension culture cells, and protoplast lysis. Routine testing used the inhibition of oxygen evolution from isolated chloroplasts in an illuminated oxygen electrode (Hansatech Instruments Ltd., England). Chloroplasts were isolated mechanically from spinach leaves and purified on a Percoll gradient. Freshly isolated chloroplasts (equivalent to 26 ug chlorophyll) were added to a temperature-equilibrated, illuminated oxygen electrode. Once a stable rate of oxygen evolution had been established, peptide was added and the resulting reduction in rate of oxygen evolution determined.

Resistance to Proteolysis. As the peptides were expected to be either secreted from plant cells in transgenic plants or applied to the external surfaces of plant tissues, the extracellular proteases present in the intercellular spaces of plant tissues were considered to be most important. Accordingly, peptides were tested against crude preparations of enzymes from the extracellular fluid (ECF) of plant leaves. Extracellular fluids from leaves of maize, tobacco and potato, and spent medium from cultured plant cells produced similar results; tobacco ECF was used routinely.

Peptide solutions (1 mg/mL) were incubated with diluted ECF (1-10% v/v) at 37 C for various times (e.g. 1 hour). The reaction was stopped by adding trifluoroacetic acid (TFA, to 1% final volume) and the resulting peptide fragments analyzed by reverse phase HPLC on a Vydac C-4 column eluted with a gradient (0-60%) of acetonitrile in TFA. Resistance to proteolysis was determined as the relative rate of disappearance of the parent peptide.

Results

The Importance of Alpha-Helix Formation. Early results obtained with magainins 1 and 2, which differ only by the amino acid residues at positions 10 and 22, showed that relatively small changes in amino acid sequence could have profound effects on antibacterial activity and potential phytotoxicity without altering the fungicidal activity against a plant pathogen (Table I). The importance of the potential to form alpha-helical structure was tested by the incorporation of proline, a well-known alpha-helix-breaker, and the substitution of alanine for glycine at positions 13 and 18, changes which would be expected to increase the propensity to form alpha-helices. Proline substitutions almost completely eliminated antifungal and antibacterial activity (Table I). Alanine substitutions in magainin 1 restored antibacterial activity to that seen with magainin 2 without significantly affecting antifungal activity. However, this increase in antibacterial activity also caused a concomitant increase in potential phytotoxicity, negating an otherwise potentially useful advantage of magainin 1 for use in plants.

Broad Spectrum Antifungal Activity. The natural magainins 1 and 2 exhibited similar activities against a number of important plant pathogens (Table II). *Colletotrichum* and *Phytophthera* represented two fungal species that appeared to be resistant to magainins 1 and 2, but which were weakly sensitive to two synthetic magainin derivatives (Table II; MYPs 20 and 21). The results obtained with

Table I. Effects of Proline and Alanine Substitutions on Magainins

Peptide Structure	*Relative Antifungal Activity*	*Relative Antibacterial Activity*	*Relative Potential Phytotoxicity*
Magainin 2	120	110	71
Magainin 1	120	17	15
[Pro^{10}]-Mag 2	0*	0	8
[Pro^{11}]-Mag 2	12	0	62
[Ala^{18}]-Mag 1	113	103	75
[Ala^{13,18}]-Mag 1	117	130	100

*Zero relative activity corresponds to a MCIC greater than 150 ug/mL

peptides representing four different classes of modification (Table II; MYPs 26, 30, 31, 33 and 34) showed that this general antifungal activity was relatively tolerant of a variety of structural modifications.

Associations Between Antibacterial Activity and Potential Phytotoxicity. Analysis of approximately fifty magainin-related peptides showed that peptides with high antibacterial activity also tended to have high potential toxicity against chloroplasts (Figure 1). This association does not represent an absolute correlation because peptides lacking antibacterial activity showed a full range of potential phytotoxicities. Five peptides with significant antibacterial activity (>50) showed modest potential phytotoxicity (<50%).

Associations Between Antifungal Activity and Potential Phytotoxicity. A similar analysis of antifungal activity showed less of an association with phytotoxicity (Figure 1). Unlike antibacterial activity, almost all of the peptides studied had significant antifungal activity against a *Fusarium* isolate. Nineteen peptides combined useful antifungal activity (>50) with modest (<50%) potential phytoxicity (Figure 1; bottom, right quadrant).

Reduction of Proteolysis in Plants. Both magainin 1 and magainin 2 showed at least 90% degradation after a one-hour exposure to diluted protease preparations from the extracellular compartment of plant leaves. HPLC purification of peptide fragments from short exposures to these proteases revealed two primary fragments that, when analyzed by FAB-MS and amino acid analysis of hydrolysates, identified a primary cleavage site between residues 7 and 8 (data not shown). Other minor fragments were suggestive of the progressive action of peptidases on the C- and N-termini of the peptides. Accordingly, modifications to positions 7, 8 and the C-

Table II. Broad Spectrum Antifungal Activity of Magainins and Synthetic Peptides (MYPs)

	Relative Antifungal Activity*						
Peptide Identity	*Fusarium sp. P3 isolate*	*Trichoderma reesei*	*Cercospora sp.*	*Helminthosporium carbonum*	*Fusarium moniliforme*	*Colletotrichum graminicola*	*Phytophthora parasitica*
Magainin 1	130*	105	125	123	103	0	0
Magainin 2	130	102	115	103	107	0	0
MYP 20	130	100	137	110	87	0	53
MYP 21	130	113	130	90	130	60	13
MYP 26	125	85	105	112			
MYP 30	130	100	110	130			
MYP 31	110	90	28	47			
MYP 33	130	115	82	73			
MYP 34	135	130	112	97			

*A relative activity of 130 is equivalent to MCIC = 20 ug/mL

Table III. Effects of Modifications Designed to Reduce Proteolysis in Plants

Peptide Identity	*Amino Acid Position 7*	*Amino Acid Position 8*	*C-Terminal Amino Acid*	*Percentage Proteolysis*	*Antifungal Activity*	*Antibacterial Activity*	*Potential Phytotoxicity*
Magainin 1	His	Ser	Ser	90	130	60	16
Magainin 2	His	Ser	Ser	90	130	105	72
MYP 4	His	Thre	Ser	92	110	0	35
MYP 5	His	Ala	Ser	98	130	80	90
MYP 6	His	Glu	Ser	66	110	0	35
MYP 7	Lys	Ser	Ser	78	120	115	56
MYP 8	Arg	Ser	Ser	90	120	120	84
MYP 9	Phe	Ser	Ser	100	100	20	41
MYP 24	His	Ser	Met	80	130	0	18
MYP 27	Arg	Glu	Lys	30	130	80	30
MYP 30	Arg	Glu	Pro	0	130	70	59

terminus were tested for their ability to reduce susceptibility to proteolysis, and for any unintended effects on antifungal activity, antibacterial activity and potential phytotoxicity (Table III).

The single amino acid residue change that produced the greatest reduction in protease susceptibility was Ser^8 to Glu^8 (MYP 6). Changing the C-terminus to Met from Ser (MYP 24) produced a 10% reduction in proteolysis. Most successful were combinations of changes at positions 7, 8 and the C-terminus. For example, the sensitivity of MYP 27 was reduced to one-third of that of magainins 1 and 2, and MYP 30 showed no degradation under the same conditions.

The above-mentioned modifications to reduce proteolysis had little effect on antifungal activity. In some cases, antibacterial activity was eliminated (MYPs 4, 6, and 24). Potential phytotoxicity showed both increases and decreases as a result of these modifications. MYP 27 and MYP 30 are examples of synthetic peptides that combine low susceptibility to proteolysis with potent antifungal activity, significant antibacterial activity and modest potential phytotoxicity.

Interactions with Hydrolytic Enzymes from Plants. Peptides used to protect plants against fungal pathogens will also encounter other plant-derived potentially antifungal agents such as chitinases and glucanases. Antifungal bioassays, of mixtures of MYP 20 and a chitinase purified from maize seeds, showed that the presence of chitinase activity could reduce the concentration of peptide required for complete inhibition of a fungal pathogen (Table IV).

Table IV. Interaction Between an Antifungal Peptide and a Hydrolytic Enzyme

Combinations causing complete fungal* inhibition	
MYP 20	Maize chitinase 1
15 ug/mL	0 ug/mL
10 ug/mL	1 ug/mL
5 ug/mL	8 ug/mL
0 ug/mL	10 ug/mL

**Fusarium* sp. isolate P3

Discussion

Structure-Function Relationships. The foundation for designing improved peptides is an understanding of the role of structural features in determining biological activity. With such information the number of peptide modifications to be tested can be reduced significantly from those dictated by empirical testing of random modifications.

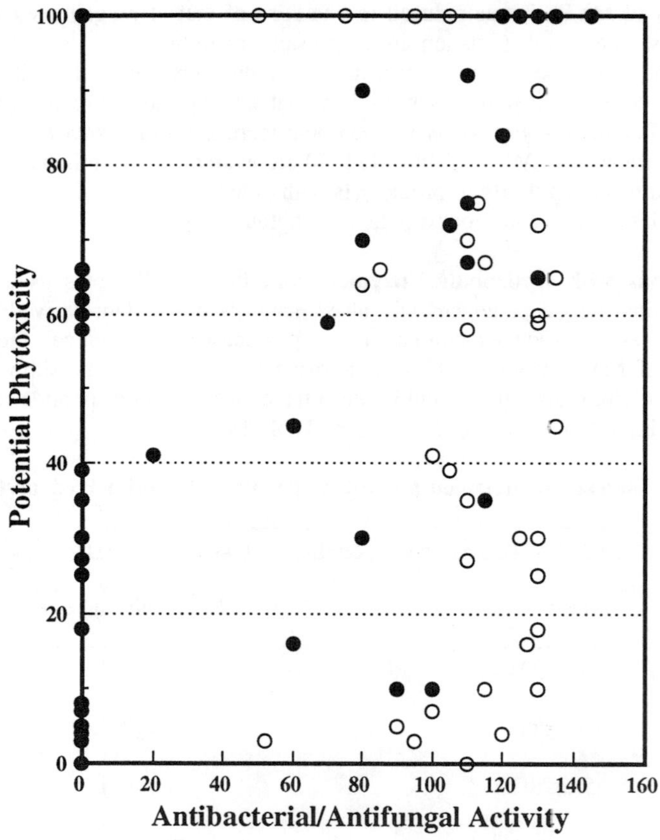

Figure 1. Scattergram Analysis of Relationship Between (a) Antibacterial Activity (closed circles) or (b) Antifungal Activity (open circles) and Phytotoxicity *in vitro*.

A number of design principles are not specific to magainin-related peptides, but are shared with many peptides, or domains of proteins, which disrupt membrane function. Common features of working models include an abundance of basic amino acids to provide an electrostatic attraction to negatively-charged membranes; the adoption of an alpha-helical structure in hydrophobic environments; and the assembly of peptide monomers into a multimeric structure that can span a membrane and form a hydrophilic pore (e.g. *13-16*).

Magainins are basic peptides which have been shown to adopt an alpha-helical structure with amphiphilic character (*8*), to be capable of spontaneous assembly into multimers (*17*), and to cause disruption of membrane potentials (*18*) and leakage from negatively-charged lipid vesicles (*19*). However, recent results from NMR analysis of magainin peptides in lipid bilayers (*20*) suggest that, instead of forming membrane-spanning pores like most lytic peptides, magainins may remain bound with the alpha-helix orientated parallel to the membrane surface.

Disruption of alpha-helix potential, by the substitution of helix-breaking proline residues, almost completely eliminated the antifungal and antibacterial activity of magainin 2 (Table I). Conversely, the substitution of alanine residues, to increase alpha-helix potential, increased the antibacterial activity of magainin 1 (Table I and ref. *21*). Chen et al. (*21*) concluded that substitution of helix-promoting residues at Gly^{13} and Gly^{18} of magainins seemed to be the most important modification for increase in antimicrobial activity. They also showed that these changes caused no appreciable increase in hemolytic activity and only a small effect on the reduction of membrane potential and uncoupling of respiration (*22*). It was therefore rather surprising that these same changes were found to increase dramatically the ability of magainin 1 to inhibit photosynthesis in isolated chloroplasts (Table I). As the rate of photosynthesis was altered within seconds of adding peptide, it is unlikely that this increased activity against chloroplasts is due to reduced susceptibility to proteolysis in the presence of membranes (*22*).

It appears that retention of the ability to adopt a helical conformation is important, but that changes in individual amino acid residues may have different effects in different biological situations. Comparison of magainins 1 and 2 alone support this view because they showed similar antifungal activity but very different activities against bacteria and chloroplasts. Interestingly, one of the amino acid residues (#10) that distinguishes magainin 1 from magainin 2 lies within a cationic region that has been proposed as a common sequence feature of many lytic peptides (*23*).

Specificity. Ideally, an antifungal peptide for agricultural applications should be active against a wide range of fungal pathogens. Magainins, as a class, have now been shown to have significant activity against a number of pathogenic fungal species, an activity that seems to be relatively resilient to many different classes of modifications (Figure 1, Table 2).

These modifications included certain additions, deletions or substitutions at the amino- and carboxy-termini and at internal sites, and combinations thereof. *Phytophthora* and *Colletotrichum* showed low sensitivity to magainins 1 and 2. It is not known whether this represents low inherent sensitivity to magainins or possibly rapid degradation or inactivation of magainins. The low but significant activity of MYPs 20 and 21 (Table 2) suggests that there is an opportunity to design peptides that are more active against these two important pathogens.

Increasing broad-spectrum antifungal activity of peptides that have already shown activity against fungi, bacteria, protozoa (*10*), mitochondria (*11*), chloroplasts (Table 1), rat mast cells (*24*) and malaria parasites (*25*), might be expected to increase non-specific cytotoxicity. However, it has been possible to identify peptides with high antifungal activity and low activity against isolated chloroplasts even though it proved to be more difficult to separate antibacterial activity from this potential phytotoxicity (Figure 1). Only in a few specific cases have modifications to magainin-like peptides produced increased hemolytic activity (data not shown). Such evidence of differential effects on target membranes from various sources calls into question the general model of biological activity originating from the formation of pores in membranes. Combined with NMR evidence that magainins may remain on the membrane surface, this may suggest that some of the biological activity is mediated by interaction with membrane proteins instead of the physical penetration of the lipid bilayer to create a pore. For example, Ser^8 of magainin 2 has been shown to be capable of phosphorylation by protein kinase C (PKC), and certain magainin analogs are potent inhibitors of PKC (*26*). This protein kinase has been implicated in transmembrane signaling to regulate the functions of cell surface receptors, transporters, ion pumps and ion channels (*27*).

Such interactions with other membrane components could certainly provide an attractive basis for the specificity of some peptides that appear to have a similar general structure. This situation might then be analogous to that of a group of structurally related peptides from snake venom which have separate defined regions that are responsible for either cytotoxicity or neurotoxin activity (*28*).

Stability. A common problem in the biotechnology industry is that proteins and peptides produced in hosts and/or cellular compartments different from their natural occurrence suffer from problems of degradation by proteases. If antifungal peptides are to be used to protect plants from invading fungal pathogens, the peptides will certainly encounter whatever proteases are present in the extracellular environment of plant tissues. Magainins 1 and 2 have been shown to be degraded rapidly by crude mixtures of such extracellular plant proteases (Table III). Thus, to test the potential of antifungal peptides in or on plants, it was considered critical to address this problem of stability against protease degradation.

In their natural host, magainins undergo a proteolytic cleavage after they are discharged from the granular glands. This cleavage occurs between residues 10 and 11 to produce two half-peptide fragments that no longer exhibit antibiotic activity (*29*). The endopeptidase responsible for this inactivating cleavage recognizes not only the required basic amino acid residue(s) at the cleavage site but also alpha-helical secondary structure (*29*). HPLC, FAB-MS and amino acid analyses of peptide fragments from the action of plant extracellular fluids on magainins revealed two primary degradation products resulting from an internal cleavage between His^7 and Ser^8 instead of Lys^{10} and Lys^{11}. Additional degradation products were suggestive of sequential exo-peptidase activity at the amino- or carboxy-termini. Certain substitutions at positions 7 and 8 significantly reduced susceptibility to proteolysis and, when combined with a modified carboxy-terminus, essentially eliminated degradation under the *in vitro* conditions. This peptide, MYP 30, is considered to be a good candidate for further research and development since it combines resistance to extracellular plant proteases with potent antifungal activity and modest potential phytotoxicity and antibacterial activity.

Additional stabilizing modifications may be desirable if the peptide is to be produced in a microbial host, but other methods to control *in vivo* proteolysis are also available for this situation (*30*). If chemical synthesis becomes economical, the use of all-D enantiomers can be considered because they have increased resistance to enzymatic degradation and retain the same activity as the all-L enantiomers (*31-34*). This activity of D-enantiomers also indicates that a close molecular contact with chiral components of the cell membrane is not required for biological activity (*31*).

Peptide Behavior *in vivo*. *In vitro* tests are useful for the comparison of peptides during the design phase, but they do not represent the conditions under which the peptides are expected to perform in the field. A number of additional factors need to be considered to address the question of whether membrane-active peptides can be used to inhibit pathogens without being inhibitory to the crop plant.

The positive charge of basic peptides provides the basis for an electrostatic interaction with negatively-charged membranes that is thought to be the first stage in membrane binding (*32*). Similar electrostatic attractions are also likely to cause binding to cell wall components. Peptide molecules are thus rapidly removed from solution (data not shown) and become bound to cellular components. Biological responses are then a reflection of the number of peptide molecules bound to sensitive sites (i.e. dose-response) rather than a response to their original concentration in solution.

Antifungal peptides for agricultural applications will encounter plant-derived products that may either decrease or increase their efficacy against plant pathogens. Plant proteases represent one example of a deleterious interaction that has already been addressed. A positive interaction exists between magainins and hydrolytic enzymes such as chitinases and glucanases. This interaction is reported to be even more significant when using crude preparations of chitinases and glucanases (*35*) rather than the purified maize chitinase reported here (Table IV). The effect of this interaction may be particularly important in intact plants because most plant tissues contain extracellular chitinases and glucanases that accumulate in response to fungal pathogens or stress.

Potential phytotoxicity is a concern that may or may not be justified. Currently, some antibacterial activity is being sacrificed to design peptides with modest potential phytotoxicity *in vitro*. The rapid binding of peptides applied to plant surfaces, or their secretion in transgenic plants, may result in organelles such as chloroplasts and mitochondria never experiencing a significant dose of peptide. Also thionins, which are antimicrobial peptides produced by plants, show toxicity to plant protoplasts *in vitro* (*36*). Thus plants are already accomplished at managing peptides that have potential phytotoxicity. If phytotoxicity proves not to be a problem, the scope for peptide design and providing simultaneous protection against fungal and bacterial pathogens will be significantly enhanced.

Clearly, experiments with whole plants are required to address some of these issues. Now that a class of antifungal peptides with enhanced stability against plant proteases is available, such experiments can be performed with transgenic plants and the external application of peptides to intact plant tissues.

Acknowledgments

This research was initiated at the Biotechnology Laboratories of EniChem Americas Inc., Princeton, New Jersey. In addition to the author, members of the research team included J. Williams, M. Swerdloff, C. Mapelli, N. Bascomb, C. DeRobertis, J. Stahl, B.-F. Huang and J. Einset. The technology is now being developed further by InterLink Associates Inc., under license from EniChem.

Literature Cited

1. Boman, H. G. and Hultmark, D. Ann. Rev. Microbiol. 1987, 41, 103.
2. Casteels, P.; Ampe, C.; Jacobs, F.; Vaeck, M. and Tempst, P. The EMBO Journal 1989, 8, 2387.
3. Ganz, T.; Selsted, M. E. and Lehrer, R. I. Eur. J. Haematol. 1990, 44, 1.
4. Bevins, C. L. and Zasloff, M. Ann. Rev. Biochem. 1990, 59, 1559.
5. Bohlmann, H.; Clausen, S.; Behnke, S.; Giese, H.; Hiller, C.; Reimann-Philipp, U.; Schrader, G.; Barkholt, V. and Apel, K. The EMBO Journal, 1988, 7, 1559.
6. Jaynes, J. M.; Xanthopoulos, K. G.; Destefano-Beltran, L. and Dodds, J. H. BioEssays 1987, 6, 263.
7. Zasloff, M. Proc. Natl. Acad. Sci. USA 1987, 84, 5449.
8. Marion, D.; Zasloff, M. and Bax, A. FEBS Lett. 1988, 227, 21.
9. Soravia, E. S.; Martini, G. and Zasloff, M. FEBS Lett. 1988, 228, 337.
10. Zasloff, M.; Martin, B. and Chen, H.-C. Proc. Natl. Acad. Sci. USA 1988, 85, 910.
11. Westerhoff, H. V.; Hendler, R. W.; Zasloff, M. and Juretie, D. Biochim. Biophys. Acta 1989, 975, 361.
12. Bascomb, N. F.; Mapelli, C.; Swerdloff, M. D.; Williams, J. I. and Everett, N. P. European Patent Application 1992, EP 0472987 A.
13. Dawson, C. R.; Drake, A. F.; Helliwell, J. and Hider, R. C. Biochim. Biophys. Acta 1978, 510, 75.
14. Terwilliger, T. C. and Eisenberg, D. J. Biol. Chem. 1982, 257, 6016.
15. Vogel, H. and Jahnig, F. Biophys. J. 1986, 50, 573.
16. Fink, J.; Boman, A.; Boman, H. G. and Merrifield, R. B. Int. J. Peptide Protein Res. 1989, 33, 412.
17. Urrutia, R.; Cruciani, R. A.; Barker, J. L. and Kachar, B. FEBS Lett. 1989, 247, 17.
18. Westerhoff, H. V.; Juretic, D.; Hendler, R. W. and Zasloff, M. Proc. Natl. Acad. Sci. USA. 1989, 86, 6597.
19. Matsuzaki, K.; Harada, M.; Handa, T.; Funakoshi, S.; Fujii, N.; Yajima, H. and Miyajima, K. Biochim. Biophys. Acta. 1989, 981, 130.
20. Bechinger, B.; Zasloff, M. and Opella, S. J. Biophys. J. 1992, 62, 12.
21. Chen, H.-C.; Brown, J. H.; Morell, J. L. and Huang, C. M. FEBS Lett. 1988, 236, 462.
22. Juretic, D.; Chen, H.-C.; Brown, J. H.; Morell, J. L.; Hendler, R. W. and Westerhoff, H. V. FEBS Lett. 1989, 249, 219.

23. Kini, R. M. and Evans, H. J. Int. J. Peptide Protein Res. 1989, 34, 277.
24. Hook, W. A.; Tsuji, S. and Siraganian, R. P. Proc. Soc. Exp. Biol. Med. 1990, 193, 50.
25. Gwadz, R. W.; Kaslow, D.; Lee, J.-Y.; Maloy, W. L.; Zasloff, M. and Miller, L. H. Infect. Immun. 1989, 57, 2628.
26. Nakabayashi, H.; Brown, J. H.; Morell, J. L.; Chen, H.-C. and Huang, K.-P. FEBS Lett. 1990, 267, 135.
27. Nishizuka, Y. Science 1986, 233, 305.
28. Klopman, G. and Venegas, R. E. In *Biocatalysis and Biomimetics;* Burrington, J. D. and Clark, D. S., Eds.; ACS Symposium Series No. 392; American Chemical Society: Washington, D.C., 1989; pp 52-64.
29. Resnick, N. M.; Maloy, W. L.; Guy, H. R. and Zasloff, M. Cell 1991, 66, 541.
30. Enfors, S.-O. TIBTECH 1992, 10, 310.
31. Wade, D.; Boman, A.; Wahlin, B.; Drain, C. M.; Andreu, D.; Boman, H. G. and Merrifield, R. B. Proc. Natl. Acad. Sci. USA 1990, 87, 4761.
32. Merrifield, R. B.; Boman, H. G.; Andreu, D.; Li, Z.-q.; Fink, J. and Wade, D. In *Peptides 1990;* Giralt, E. and Andreu, D.,Eds.; ESCOM Science Publishers B. V., 1991; pp 3-16.
33. Bessalle, R.; Kapitkovsky, A.; Gorea, A.; Shalit, I. and Fridkin, M. FEBS Lett. 1990, 274, 151.
34. Maloy, W. L.; MacDonald, D. and Brasseur, M. In *Peptides 1990;* Giralt, E. and Andreu D., Eds.; ESCOM Science Publishers B.V.: 1991, pp 743-745.
35. Ryals, J. A.; Gay, P. B.; Ahl Goy, P. A. and Garcia-Olmedo, F. European Patent Application 1991, EP 0448511 A1.
36. Reimann-Philipp, U.; Schrader, G.; Martinoia, E.; Barkholt, V. and Apel, K. J. Biol. Chem. 1989, 264, 8978.

RECEIVED June 28, 1993

Chapter 21

Insect Neuropeptides

Current Status and Avenues for Pest Control

Thomas J. Kelly, Edward P. Masler, and Julius J. Menn

Plant Sciences Institute, Agricultural Research Service, U.S. Department of Agriculture, Beltsville, MD 20705

The physiological processes regulated by insect neuropeptides are extensive and include growth, development, molting, reproduction, diapause, behavior, color change, ion and water balance, and muscle contraction. Nearly 80 novel insect neuropeptides have been sequenced to date, 10 or so gene sequences have been determined, numerous analogs have been synthesized, and neuropeptide genes have been expressed in vector systems. Investigations into sites of synthesis and release, and tissue specificity and action, continue to reveal complexity in the (classically simple) insect neuroendocrine system. The rapidly increasing knowledge in this area suggests that useful prototypes for the design of selective pest control agents will emerge in the forseeable future.

The field of insect neuroendocrinology has expanded rapidly, fostering numerous reviews within the past five years (*1-21*), including one book entirely devoted to insect neurohormones (*22*) and one to insect neuropeptides (*23*). These publications discuss the discovery, physiology, history and chemistry of insect neuropeptides and present potential applications to pest control. These applications may include: blocking neuropeptide synthesis through expression of antisense RNA or inhibition of specific processing enzymes; inactivation of circulating neuropeptides by binding with specific antibodies, antipeptides or by degradation with specific peptidases; disruption of target binding or signal transduction by peptidomimetics; and disruptive expression of natural or modified neuropeptides through baculoviruses or other expression vectors. As suspected in 1988 (*4*), the number of purified and sequenced insect neuropeptides has continued to increase, and many of the corresponding nucleotide sequences have followed in rapid succession. Nearly eighty novel amino acid sequences for insect neuropeptides have been identified, and novel nucleotide sequences approach ten. This review updates the recent progress in identification of insect neuropeptide structures and their genes and other areas relevant to the development of pest control agents, including insect neuropeptide synthesis, metabolism, pharmacology, sites of release and receptors.

Although a wide variety of neurohormone-regulated processes has been identified in insects (see reviews, *1,8,14,24*; and books, *22,23,26*), the major ones fall into two broad categories: 1. homeostatic and behavioral neuropeptides, and 2. developmental and reproductive neuropeptides (*23*). The former includes the myotropins, AKH/RPCH family, diuretic hormones, eclosion hormone, and PBAN and the latter includes the allatotropins/allatostatins, ecdysteroidogenins and oostatic hormone (see Table I for a list of functions and sequences). Classification into the various groups is based on currently available information. However, considerable overlap in structure and function is already apparent for certain groups, and as more is known about the complete functions, cellular localization and preprohormone structure of these various neuropeptides, more accurate classification may be possible. Each of these groups is discussed according to the aspects outlined above.

Myotropins

Following the identification of proctolin (Arg-Tyr-Leu-Pro-Thr-OH) from the cockroach *Periplaneta americana* by Starratt and Brown (*28*), a neuropeptide which acts as a neurotransmitter in visceral and somatic muscle (*29*) and effects neurons in the central nervous system (CNS) (*17,18*), considerable effort has been invested in the isolation and identification of myotropic peptides. Several recent reviews document this progress, including assessments of their mode of action, localization and pharmacology (*17,18,30-34*). The major effort has been in the identification of orthopteran myotropins other than proctolin, now classified into five groups and first identified in the cockroach, *Leucophaea maderae* (Table I). Numerous studies with proctolin and its analogs have examined both its physiological and pharmalogical effects (see reviews, *29,30,34*; also *40-43*), and a recent study has shown proctolin binding to hindgut membrane preparations from the African migratory locust, *Locusta migratoria* (*44*). The work initiated by the *L. maderae* studies has resulted in the identification of 30 or so novel peptides (*17,18,33*) including the recent identification of locustakinin (Ala-Phe-Ser-Ser-Try-Gly-NH_2)(*45*), the locustamyotropins III (Arg-Gln-Gln-Pro-Phe-Val-Pro-Arg-Leu-NH_2) and IV (Arg-Leu-His-Gln-Asn-Gly-Met-Pro-Phe-Ser-Pro-Arg-Leu-NH_2) (*46*), the myotropic inhibiting peptide (Ala-Trp-Gln-Asp-Leu-Asn-Ala-Gly-Trp-NH_2) (*47*), and the locustatachykinins (I-IV; Table I)(*38,48*) from *L. migratoria*. The locustatachykinins are 30% homologous with the substance P subfamily of vertebrates, and their action on the motility of the alimentary canal is similar in insects and vertebrates (*49*). Also recently identified is the oviduct myotropin (Ile-Ala-Tyr-Lys-Pro-Glu-NH_2) from the Colorado potato beetle, *Leptinotarsa decemlineata* (*50*). Myotropic neuropeptides range in size from 3-14 amino acids and have been isolated from 8 species in 3 insect orders, Orthoptera, Diptera, and Coleoptera. They are present at pmol/head or less in orthopterans, evoke stimulatory or inhibitory responses at 10^{-9} M or less, and are commonly assayed on the gut, heart or oviduct (*32,33*). The *L. maderae* hindgut preparation has been instrumental in the isolation of these neuropeptides, as recently documented by Holman *et al.* (*51*).

Recent studies on the distribution and mode of action of the myotropins have revealed tissue specificity in distribution and activity. Furthermore, some of them occur in the intestine as well as the brain, similar to the brain-gut peptides of vertebrates (*52*). Studies on locustamyotropin immunoreactive cells, using highly specific antibodies to the carboxy-terminus (*53*) revealed about 100 immunoreactive cells in the head, metathoracic and five abdominal ganglia of *L. migratoria*, including intensive labelling of the cell bodies in the subesophageal

TABLE I. REPRESENTATIVES FROM CLASSES OF INSECT NEUROPEPTIDES

General Class	Function	Examples[1]	Source	Structure[2]	Reference
		HOMEOSTATIC AND BEHAVIORAL NEUROPEPTIDES			
Myotropins[3]	stimulate muscle contraction	Leucokinin I (Lem-K-I)	*Leucophaea maderae*	DPAFNSWG-NH2	35
	"	Leucopyrokinin (Lem-PK-I)	"	pQTSFTPRL-NH2	36
	"	Leucosulfakinin (Lem-SK-I)	"	EQFEDY(SO3)GHMRF-NH2	37
	"	Locustatachykinin (Lom-TK-I)	*Locusta migratoria*	GPSGFWGVR-NH2	38
	suppress muscle contraction	Leucomyosupressin (Lem-MS)	*Leucophaea maderae*	pQDVDHVFLRF-NH2	39
AKH/RPCH[4] Family	stimulate sugar and lipid metabolism	Adipokinetic Hormone I (Lom-AKH-I)	*Locusta migratoria*	pQLNFTPNWGT-NH2	75

Diuretic Hormones	regulate water retention	(AVP-like-Lom-DH)	*Locusta migratoria*	CLITNCPRG-NH_2 CLITNCPRG-NH_2	107
	"	(Mas-DH)	*Manduca sexta*	RMPSLSIDLPMSVLRQKLSLE-KERKVHALRAAANRNFLNDI-NH_2	113
Eclosion Hormones	initiate the ecdysial process	(Mas-EH)	*Manduca sexta*	NPAIATGYDPMEICIENCAQ-CKKMLGAWFEGPLCAESCIKF-KGKLIPECEDFASIAPFLNKL-OH	128,129
Pheromone Biosynthesis Activating Neuropeptides	stimulate sex pheromone synthesis	(Hez-PBAN)	*Helicoverpa (Heliothis)zea*	LSDDMPATPADQEMYRQDPE-QIDSRTKYFSPRL-NH_2	145
DEVELOPMENTAL AND REPRODUCTIVE NEUROPEPTIDES					
Allatotropins	stimulate juvenile hormone synthesis	(Mas-AT)	*Manduca sexta*	GFKNVEMMTARGF-NH_2	173
Allatostatins	inhibit juvenile hormone synthesis	(Dip-AS-I)	*Diploptera punctata*	APSGAQRLYGFGL-NH_2	178,179

Continued on next page

TABLE I.—*Continued*

Ecdysteroido-genins	initiate the molting and/or egg development process	Bom-bombyxin	*Bombyx mori*	A chain:[5] GIVDECCLRPCSVDVLLSYC-OH	209
				B chain: pQQPQAVHTYCGRHLARTLA-DLCWEAGVD-OH	210
	"	(Bom-PTTH)[6]	"	GNIQVENQAIPDPPCTCKYK-KEIEDLGENSVPRFIETRNC-NKTQQPTCRPPYICKESLYS-ITILKRRETKSQESLEIPNE-LKYRWVAESHPVSVACLCTR-DYQLRYNNN-OH	227,228
Oostatic Hormones	inhibit ovarian maturation	(Aea-TMOF)[7]	*Aedes aegypti*	YDPAPPPPP-OH	239

1 See Raina and Gade (*27*) for definition of nomenclature.
2 Underlined letters indicate conserved amino acids for that group.
3 This class also includes proctolin (RYLPT-OH), and FMRFamide-like peptides.
4 AKH/RPCH is the abbreviation for adipokinetic hormone/red pigment concentrating hormone.
5 The sequence for 4K-PTTH-II is given. The A and B chains are linked by disulfide bridges.
6 PTTH is the abbreviation for "prothoracicotropic hormone". It is a homodimer linked by disulfide bridges. The monomer sequence is given.
7 TMOF is the abbreviation for "trypsin modulating oostatic factor" which appears to be the primary action of TMOF, inhibiting ovarian maturation through inhibition of midgut trypsin activity and thus blood meal digestion (*240*).

ganglion of the silkworm, *Bombyx mori*, where pheromone biosynthesis activating neuropeptide (PBAN) is produced. These antibodies apparently recognize the carboxy-terminal pentapeptide sequence, Phe-X-Pro-Arg-Leu-NH_2 of the pyrokinins, a class of myotropic neuropeptides which now include PBAN (*54*). Recent studies with polyclonal antibodies to leucomyosuppressin revealed variation in the distribution of immunoreactive cells in *L. maderae* and the stable fly, *Stomoxys calcitrans*, with staining in the brain and thoracic ganglia of both species, and staining in the subesophageal and abdominal ganglia of *S. calcitrans*, only (*55,56*). Immunocytochemical studies with antibodies to leucokinin I revealed immunoreactive cells and neurons in the brains of cockroach and blowfly species (*57*) as well as the abdominal ganglia that innervate the perisympathetic neurohemal organs, including moth species (*58*). A study with antibodies to proctolin and FMRFamide revealed two separate peptidergic systems derived from the abdominal ganglia and supplying structures in the hindgut (*59*).

The selectivity of action and variation in tissue distribution of the myotropins has also been suggested by HPLC-separated factors from *L. migratoria* foreguts, midguts, hindguts and Malpighian tubules (*52*). Similar results on the selectivity of action have been demonstrated in more detailed kinetic studies with *L. maderae* by Cook and co-workers (*60-62*) who showed that leucokinins I-VIII gave a threshold contractile response on the hindgut at 10^{-10} M, whereas the foregut and oviduct were 100- to 1000-fold less sensitive, and the heart showed no response. Leucomyosuppressin, on the other hand, suppressed spontaneous contractions of the foregut and hindgut at 10^{-11} M, and the heart and oviduct were 100- to 1000-fold less sensitive.

A few recent studies have examined the mechanism by which the myotropins and various neurotransmitters initiate, modulate or suppress contraction. Following the lead of *Meola et al.* (*55*), who demonstrated leucomyosuppressin antibody staining of *S. calcitrans* tissues, Cook *et al.*. (*63*) and Cook and Wagner (*64*) showed excitation thresholds for proctolin on the hindgut and oviduct of *S. calcitrans* at less than 10^{-10} M, but were unable to evoke any substantial pharmacological response to leucomyosuppressin, suggesting that the previously described antibodies recognized only partial sequences within the leucomyosuppressin molecule. Similar studies by Lange *et al.* (*65*) with the locust myosuppressin (SchistoFLRFamide) have shown on the oviduct that SchistoFLRFamide inhibits or reduces the amplitude and frequency of spontaneous contractions, relaxing basal tonus, and reducing the amplitude of neurally-evoked, proctolin-induced, glutamate-induced and high potassium-induced contractions. The sulfakinins and myosuppressins have C-terminal homology with the FMRFamides which are widely distributed throughout the metazoa (*66*) and which sulfakinin and myosuppressin polyclonal antibodies may recognize. Examination of the structural requirements for peptide activity at the neuromuscular junction suggests the presence of at least four different receptors, two presynaptically and two postsynaptically (*34*). Much of the work in designing modified myotropins and pseudopeptides necessary to understand the structural and chemical basis for normal and enhanced activity has been performed by Nachman and co-workers (see current volume). No myotropic neuropeptide receptors have been isolated, and only the myotropin genes for drosulfakinin (*67*) and the *Drosophila* FMRFamide-related neuropeptide (*68-71*) have been sequenced, the latter leading to studies on its organization (*72*) and expression in various tissues during development (*73,74*).

AKH/RPCH Family

The first insect neuropeptide structurally identified in this class was adipokinetic hormone I (see Table I), isolated from the corpus cardiacum (CC) of *L. migratoria* and the desert locust, *Schistocerca gregaria* and structurally identified by Stone *et al.* (*75*) in 1976. Since that time, 20 or so novel neuropeptides have been isolated from 30 or so species in 5 orders of insects including the Orthoptera, Lepidoptera, Coleoptera, Diptera and Odonata (see *17,18,76-81* for a complete list of sequenced neuropeptides and current reviews of this area). Recent identification of novel neuropeptides in this family include *Locusta* adipokinetic hormone III (*82*), containing a novel Trp at position 7, *Polyphaga aegyptiaca* hypertrehalosemic hormone (*83*), containing a novel Ile at position 2, *Phormia terraenovae* hypertrehalosemic hormone (*84*; Schaffer *et al.* (*85*) determined the same sequence for *D. melanogaster*), containing a novel Asp at position 7 and providing an unusual positive charge, and *Melolontha melolontha* CC peptide (*86*), containing the novel Asp at position 7 plus an unusual Tyr at position 4. Except for the latter substitution at position 4 which affected both efficacy and potency, these substitutions affected only AKH potency (*87*). Keeley *et al.* (*78,79*) and Gade (*87*) provide an explanation of terminology and a summary of structure/activity studies with *Blaberus discoidalis* and *Periplaneta americana* hypertrehalosemic activities. In general, positions 1, 4 and 8, the conserved positions for this family (see Table I), are also required for maximum activity and play a decisive role in receptor binding. However, this is clearly dependent on the activity investigated since some members of this family also have cardioacceleratory activity (eg., Pea-CAH-I or neurohormone D) where positions 2, 6, and 7 appear most important for receptor recognition (*88*). A β-turn (positions 5-8) is also required (*79,87*) for activity and binding in lipolemic and hyperglycemic assays, with the recent exception of *Manduca sexta* (*89*), where the β-turn appears unimportant for binding of peptides. Fox and Reynolds (*90*) have recently shown, by co-injecting various AKH members along with the native *M. sexta* AKH, that AKH antagonists exist and suggest that a single receptor population may mediate the AKH response in *M. sexta*. Also, the AKH of *B. mori* was recently sequenced and shown to be identical to that of *M. sexta* and *Helicoverpa (Heliothis) zea* (*91*).

The term "adipokinetic hormone" was coined from the lipid-mobilizing activity of this neuropeptide in *L. migratoria* (*75*), and the term "red pigment concentrating hormone" was coined from its chromatophorotropic activity in the prawn, *Pandulus borealis* (*92*), although no chromatophorotropic activity has been demonstrated in insects. AKHs range in size from octa- to decapeptides and function in lipid mobilization in some species, hyper- and hypotrehalosemia in others, and protein-synthesis inhibition and stimulation in others (*79*). In addition to characteristically causing metabolite mobilization in the fat body, several other actions were identified for the AKH/RPCH family peptides in *B. discoidalis* (*93*), including elevating carbohydrate metabolism 4-fold and fat body mitochondrial cytochrome heme synthesis 3-fold, although carbohydrate synthesis was 10 times more sensitive to hypertrehalosemic hormone than heme synthesis. *Acheta domestica* AKH, on the other hand, inhibited ^{3}H-leucine incorporation at 10^{-11} M in crickets and locusts but was two orders of magnitude less sensitive (10^{-9} M) in stimulating an adipokinetic response (*94*). AKH also had a negative control over vitellogenesis in adult *L. migratoria* (*95*), and, in *M. sexta*, controlled lipid metabolism in adults and carbohydrate metabolism in larvae (*96*). Hypertrehalosemic hormone regulation of a *B. discoidalis* fat body cytochrome P450 gene was recently demonstrated (*97*).

PreproAKH structures have been determined for *M. sexta* (*98*), *L. migratoria* (*99*), and *Schistocerca nitans* (*100*), showing similar features, i.e., signal peptide, AKH, processing site, C-terminal peptide (*79*), the latter having no known function and showing little homology between species. AKH represents the most thoroughly studied insect neuropeptide with regard to neuropeptide processing (*76,81*). It has the unprecedented feature that the immediate precursors to AKH-I and II are not linear prohormones, but dimeric constructs. *In vitro* systems are being developed with *L. migratoria* extracts for purification of the converting enzymes, none of which have been identified in insects (*81*). Studies on AKH inactivation in *L. migratoria* reveal that degradative activity does not exist in hemolymph but is membrane-bound in specific tissues such as fat body, muscle and the Malpighian tubules and that the specific endopeptidase resembles mammalian endopeptidase 24.11 and may be effective against a variety of N- and C-terminally blocked Phe-containing peptides (*81,101*). However, in *M. sexta* a hemolymph endopeptidase exists with a capacity equal to that of the Malpighian tubules (*102*).

Diuretic Hormones

This class of homeostatic and behavioral neuropeptides has been recently reviewed with regard to function, chemistry and assay (*18,103-106*). Diuretic activity has been found throughout the insect neuroendocrine system (*104*), and a diuretic neuropeptide was first isolated and sequenced from *L. migratoria* extracts of subesophageal and thoracic ganglia using antibodies to vertebrate arginine vasopressin to follow purification (*107*). This molecule, AVP-like diuretic hormone, is a nonapeptide homodimer linked by disulfide bridges in antiparallel fashion (see Table I) with 67% homology to arginine vasopressin and 78% homology to arginine vasotocin. It stimulates water transport by increasing the cAMP level in Malpighian tubules (*108*) and stimulates urine secretion by Malpighian tubules *in vitro* at 10^{-10} M (*107*). AVP-like molecules with similar immunological and chromatographic properties have been shown in the cockroach, stick-insect and bumblebee CNSs (*109*). Immunocytochemistry, Lucifer Yellow or cobalt intercellular staining, and intracellular recording have been used to characterize the pair of vasopressin-like immunoreactive (VPLI) neurons with cell bodies in the subesophageal ganglion of *L. migratoria* (*110,111*). Synaptic input to the VPLI neurons comes from a pair of descending brain interneurons with photosensitive cell bodies apparently located in the pars intercerebralis (*111*). Synthetic studies by Picquot and Proux (*112*) revealed in *L. migratoria* enzymatically controlled synthesis of the monomer of AVP-like diuretic hormone in the subesophageal ganglion where initiation of dimer conversion begins. This is followed by release of the monomer and dimer into the hemolymph with continued dimer formation, and enzymatic degradation of the dimer by the Malpighian tubules.

An unrelated 41-residue diuretic peptide (Table I) has been isolated and sequenced from *M. sexta* (*113*). It stimulates fluid excretion *in vivo* following injection into larval *M. sexta* at 0.1 µg/larva and has 37% homology with the family of vertebrate corticotropin releasing factors. Two 46-residue diuretic peptides with 56% and 41% homology to Mas-DH were recently isolated from *L. migratoria* brains and CC (*114*), and from *L. migratoria* (*115*) and *A. domestica* (*116*) whole heads. The homologous amino acids are spread throughout the molecule, and consideration is given for additional N-terminal and internal sequences. A 30-residue diuretic peptide with 25% homology to Mas-DH, primarily in the C-terminal region, was also isolated from the median neurosecretory cells (MNCs) and CC/corpus allatum (CA) complexes of *M. sexta* (*117*). These peptides apparently function through a cAMP second messenger cascade (*114-116,118*). Immunocytochemical studies with antisera to the N- and C-terminal ends of Mas-DH revealed staining in two median cells on each side of the brain protocerebral groove in larvae and staining of a group of 80 median cells as well as their axons and terminals in the CC of adults (*119*). A synthetic gene for Mas-DH has been expressed by BmNPV in fifth instar *B. mori* showing a slight increase in mortality over animals infected with wild-type baculovirus (*120*). Injections of high doses (500 pmol) of Mas-DH into last-instar larvae of the cotton budworm, *Heliothis virescens* resulted in depressed water and food consumption, suggesting that hormonal suppression of feeding by DH could be a useful pest-control strategy (*121*).

Recently, a novel function was shown for the leuco- and achetakinins, which, as mentioned previously, stimulate contraction in *L. maderae* hindguts. Leucokinins modulate the transepithelial potential and rate of fluid secretion in isolated Malpighian tubules of *Aedes aegypti* (*122*). Achetakinins double the rate of fluid secretion in isolated Malpighian tubules of *A. domestica* (*123*). The achetakinins were active at 10^{-9} M, acted independently of cAMP, and required the core, C-terminal sequence of the leucokinins Phe-X-Ser/Pro-Trp-Gly-NH_2 (*123*).

Eclosion Hormones

The majority of the research on this class of neurohormones has been carried out by Truman and co-workers, primarily with *M. sexta* (see recent reviews, *124-127*). Mas-EH is a 62-residue peptide (*128,129*) with 80% homology to Bom-EH (*130*; Table I). It is released from the ventral nervous system in larvae and pupae of *M. sexta* and the CC/CA complex in adults (*131*). It has numerous effects on the developing insect, acting as a trigger for larval, pupal and adult ecdyses, inducing cuticle plasticization during the molt, and activating a stereotyped program of intersegmental muscle cell death during pharate adult development (*128*). Progress in the past few years has been primarily in two areas: determination of the site of release of Mas-EH in larvae and pupae, and isolation, sequencing and expression of the Mas-EH gene.

In the adult *M. sexta*, EH is produced in 5 pairs of group Ia cells in the brain protocerebrum that project ipsilaterally into the CC/CA complex (*132*). The release of adult EH is regulated both by the developmental state of the insect and a circadian clock (*124*). In larvae and pupae, on the other hand, no EH activity is present in the CC/CA complex, and the timing of ecdysis is not circadian-regulated (*124*). These facts suggested an alternate site of EH synthesis and release in larvae and pupae. By means of retrograde and anterograde cobalt fills and EH immunocytochemistry, Truman and Copenhaver (*131*) localized the larval and pupal source of EH to two pairs of ventromedial (VM) cells in the ventromedial region of the brain. Axons from these cells run ipsilaterally along the ventral nerve cord into the proctodeal nerve, a novel neuroendocrine pathway in insects. Hewes and Truman (*133*) further showed in prepupal animals that removal of the proctodeal nerve eliminated the appearance of blood-born EH, although ecdysis behavior occurred on schedule. It now appears that centrally released EH (i.e. within the CNS) acts to stimulate the ecdysis motor program, while peripherally released EH acts on peripheral targets, such as the dermal glands, which do not respond in animals with the proctodeal nerve ablated (*127,133*). Furthermore, EH appears to activate the nervous system through an ecdysteroid-permissive cGMP pathway, followed by a novel ecdysteroid-inhibited phosphorylation of two membrane-associated proteins (EGPs, the acronym for EH and cyclic GMP regulated phosphoproteins) of unknown function (*125,126*).

In 1989, Horodyski *et al.* (*134*) isolated and sequenced the Mas-EH gene. It is present as a single copy per haploid genome and distributed in 3 exons over 7.8 kb of DNA. Unlike the genes for most biologically active peptides, which are processed from precursors containing multiple copies of active peptides, the EH gene produces a single copy of EH from 0.8 kb of mature mRNA that results in a single translation product, pre-EH, that is 88 amino acids in length (i.e., signal peptide plus a single copy of EH). *In situ* hybridization analyses have consistently revealed EH mRNA in the VM cells of larvae, pupae and adult *M. sexta*, whereas it has never been detected in the 5 pairs of group Ia cells (*134*; Horodyski *et al.*, personal communication). Furthermore, the VM cells make a contribution to the EH released at adult eclosion, following growth of a collateral axon connecting the VM cells to the CC/CA complex during metamorphosis (*127*). Thus, the relative role of the group Ia cells in adult ecdysis is still unresolved.

Eldridge *et al.* (*135*) have expressed the *M. sexta* gene in an *Autographa californica* multiple nuclear polyhedrosis virus (AcMNPV) cloning-expression vector, obtaining 10 µg of active peptide per 10^6 cells. The secreted peptide was precipitated by EH antisera and eluted with the natural hormone on reverse-phase HPLC. However, determination of identity with the natural hormone awaits further analysis by mass spectrometry and other physico-chemical methods. Eldridge *et al.* (*136*) recently examined the efficacy of this EH-expressing baculovirus as a pesticide by infecting the fall armyworm, *Spodoptera frugiperda*, where Mas-EH was confirmed to be biologically active. Expression of Mas-EH in fifth or sixth (last) instars had no effect on the ability of AcMNPV to limit insect feeding or hasten death. In neonates, it slightly reduced the infectivity of the virus. This lack of enhanced efficacy may not be surprising considering that, at least in later stages, wild-type, virus-infected insects normally die before ecdysis when EH would be expected to act. Clearly, more research is required to elucidate the physiological and behavioral effects of natural and engineered baculoviruses on insect development.

In *B. mori*, Kono *et al.* (*137*) recently localized Bom-EH by immunocytochemistry to two pairs of median neurosecretory cells in the day-0 pupal brain. These cells may correspond to the VM cells of *M. sexta* (*131*), as later suggested by Kamito *et al.* (*138*) where similar cells in 5th instar *B. mori* larvae were localized by *in situ* hybridization to cDNA probes. The latter authors also isolated and sequenced the cDNA for Bom-EH. Bom-EH has been expressed in yeast (*139*) and *Escherichia coli* (*140*), the latter used to establish the location of the 3 disulfide bonds in Bom-EH and to show that the 6 residues at the N-terminal end are dispensable whereas the 4 residues at the C-terminal end are required for activity. An identical location for the disulfide bonds of Mas-EH (Cys^{14}-Cys^{38}, Cys^{18}-Cys^{34}, Cys^{21}-Cys^{49}) was recently determined (*141*). Using a nucleotide probe from Mas-EH and PCR methods, Horodyski and co-workers determined a 73-amino-acid sequence for *D. melanogaster* EH which has ca. 60% homology to Mas-EH (*127*; Horodyski *et al.*, personal communication). Kobayashi *et al.* (*142*) suggested that cGMP may act as the second messenger for Bom-EH, as it does for Mas-EH. Efforts by Sakakibara and Fugo (*143*) and Sakakibara *et al.* (*144*) to establish an *in vitro* assay for release of Bom-EH should be useful in examining control of release.

Pheromone Biosynthesis Activating Neuropeptide

PBAN was first isolated and sequenced from the corn earworm, *Helicoverpa (Heliothis) zea* (*145*; Table I). It is a 33-residue peptide with the C-terminal pentapeptide representing the minimum sequence necessary for activity in *H. zea* (*146*), although Gazit *et al.* (*147*) indicate that the C-terminal hexapeptide has very little activity in *Heliothis peltigera*. This sequence, Phe-Ser-Pro-Arg-Leu-NH_2, also represents the core C-terminal sequence necessary for activity of the leucopyrokinins (i.e., Phe-X-Pro-Arg-Leu-NH_2; see previous section on myotropins and *54,148*). Although PBAN was isolated and sequenced only 3 years ago, recent progress has been significant and results of structure/activity studies, sites of synthesis and release, stability, mode of action, and species distribution have been documented in several species (see reviews, *149-153*).

Hez-PBAN was isolated from 5000 brain-subesophageal-ganglion complexes (*145*) and has 79% homology with PBAN-I from *B. mori* (*154*). The sequence of Bom-PBAN-I is identical to that of Bom-PBAN-II with the addition of an arginine at the N-terminus (*155*). The sequence of Bom-PBAN-I is also identical to that of *B. mori* melanization and reddish coloration hormone (MRCH) (*156*), which induces cuticular melanization in the common armyworm, *Leucania separata*, but has no activity in *B. mori* (*157*). A novel, 18-residue, pheromonotropic peptide (Lys-Leu-Ser-Tyr-Asp-Asp-Lys-Val-Phe-Glu-Asn-Val-Glu-Phe-Thr-Pro-Arg-Leu-NH_2) has recently been isolated and sequenced from the armyworm, *Pseudaletia separata* (*158*). The only homology with Hez-PBAN is in the myotropic C-terminal pentapeptide which is responsible for biological activity. Structure/activity studies with Bom-PBAN-I gave similar results to studies with Hez-PBAN (*159,160*), and recent studies with C-terminal fragments of Bom-PBAN-I modified at the N-terminal amino group showed enhanced activity over the unoxidized Bom-PBAN-I, suggesting enhanced stablity in the hemolymph (*161*). The role of the central versus the C-terminal region of the PBAN molecule has yet to be determined (*146,147*), but the presence of two separate receptors is suggested (*162*) with one possibly acting at low PBAN concentrations (high affinity receptor) and another at higher PBAN concentrations (low affinity receptor) (*163*).

PBAN acts through cAMP (*164*) and calcium (*165*). However, there remains considerable controversy over the target organ of PBAN and its route of transport from the subesophageal ganglion to the pheromone gland in the tip of the abdomen, and the sites of action may vary between species (see previous reviews, *150-153*). Suggested sites of action include the corpus bursae, the terminal abdominal ganglion and direct action on the pheromone gland. Possible routes of transport include the ventral nerve cord and the hemolymph. It appears that, at least for photoperiodic control of pheromone production, the controlling factors operate at the level of neuropeptide release, since synthesis is continuous (A. K. Raina, personal communication), and not at the level of the target tissue (*163*). Studies on PBAN degradation in *B. mori* suggest the importance of a hemolymph aminopeptidase (*161*). Inhibitor studies showed it to be a metalloprotease with no participation by cysteine or serine proteases.

The gene for Hez-PBAN has recently been isolated and sequenced (*166*). It was isolated from a lambda genomic library, prepared from adult bodies, and contains a 0.63-kb intron within the region coding for amino acid 14. Two regions, 5′ and 3′ to the PBAN gene, contain sequences encoding Phe-X-Pro-Arg-Leu followed by Gly-Arg, allowing cleavage from the PBAN sequence and C-terminal amidation. This amidated sequence represents the core sequence for leucopyrokinin activity (see previous myotropin section). Active PBAN has been expressed in AcMNPV (A. K. Raina, personal communication) and also in *E. coli* using a synthetic gene with a C-terminal-Gly followed by enzymatic amidation (*167*).

PBAN serves as the primary regulator of pheromone production in many Lepidoptera and some species from other orders (*151-153*). However, there are some species such as the cabbage looper, *Trichoplusia ni*, where pheromone production is linked to changes in ecdysteroid titer (*168*). Pheromone production is also regulated by inhibitory factors found in the bursa, ovaries and hemolymph of senescing and mated females and other inhibitory factors passed from the accessory glands of males during mating (*149, 169-171*). One of these factors, a peptide, has recently been purified and sequenced from *H. zea* (*172*; A. K. Raina, personal communication).

Allatotropins/Allatostatins

The members in this class of reproductive and developmental neuropeptides have been isolated and sequenced from *M. sexta* and the cockroach, *Diploptera punctata*. The allatotropin (AT) from *M. sexta* (see Table I) is the only AT to be isolated and sequenced (*173*). This 13-residue peptide was isolated from 10,000 trimmed heads of pharate adults and stimulated juvenile hormone (JH) synthesis *in vitro* by CA from newly emerged adults. It was active at 10^{-10} M on adult CA only, including *M. sexta* and *H. virescens*, but not *P. americana*, *S. nitans* or the beetle, *Tenebrio molitor*. Full activity was elicited with the truncated fragment 6-13, and at the time of publication, Mas-AT had no sequence homology with any known peptide. Unni *et al.* (*174*), using modified analogs of the core sequence of Mas-AT, confirmed its inactivity on larval CA and inferred from *in vitro* experiments that Mas-AT stimulated a rate-limiting enzyme in farnesoate synthesis. Recent work by Reagan *et al.* (*175*) showed that Mas-AT activated an inositol 1,4,5-triphosphate pathway in the CA and that increased free intracellular calcium stimulated JH synthesis.

Several allatostatins (AS) have been isolated and sequenced from *D. punctata* and this work has been reviewed (*176,177*). Four allatostatins (Dip-AS-I-IV) were isolated and sequenced by Woodhead *et al.* (*178*) and one of these, Dip-AS-I, by Pratt *et al.* (*179*) (see Table I). A fifth, Dip-AS-V, was recently isolated and sequenced (*180*). Dip-AS's range in size from 8-18 amino acids and have a consensus C-terminal sequence Leu/Val-Tyr-X-Phe-Gly-Leu-NH_2. They reversibly inhibit synthesis and release of JH by *D. punctata* adult CA *in vitro* at 10^{-7} to 10^{-9} M (*178*), show similar sensitivities on penultimate- and final-instar larvae (*181*) and are active on *P. americana* CA, as well (*178,182*). Structure/activity studies with Dip-AS-I by Pratt *et al.* (*176*) showed that the N-terminal region was important for receptor binding (i.e., the address) whereas the C-terminal region provided for signal transduction (i.e., the message). Later studies with Dip-AS-IV by Guan *et al.* (*183*) suggested, however, that the C-terminal region was important for receptor binding. The ASs show some similarity in amino acid sequences to the vertebrate enkephalin-related peptide (Met-8) and may be sensitive to degradation by enkephalinases (*184*). Photoaffinity labeling of AS-receptor proteins demonstrated specific binding proteins (38-60 kD) in the CA, brain and fat body (*185*). Cusson *et al.* (*186*) have suggested that the sensitivity of the CA to AS may be dependent on the quantity of receptor protein.

Using polyclonal (*177*) and monoclonal (*187*) antibodies to Dip-AS-I, immunoreactivity was demonstrated in the NCC II that traverse the CC and branch extensively in the CA. Stay *et al.* (*187*) showed prominent staining of the lateral neurosecretory cells that exit the brain as NCC II, but no staining of NCC I. Similar pathways were demonstrated in *D. punctata* by antisera to Met-enkephalin-Arg^6-Gly^7-Leu^8 (Met-8), which has C-terminal homology with Dip-AS (*188*), although the Met-enkephalin antisera stained, in addition, cells in the pars intercerebralis that send axons to the NCC I. Furthermore, the amidated met-8 peptide was required at 10^5 M greater concentration than Dip-AS-I to achieve a similar level of inhibition of the *D. punctata* CA *in vitro* (*189*). Prominent staining was also demonstrated with the Dip-AS-I monoclonal antibodies in NCC III (*187*) and its varicosities along the antennal pulsatile organ muscle, a myogenic muscle that functions in circulation of hemolymph through the antennae (*190*). These observations and the fact that specific low-level immunostaining was found throughout many areas of the brain including interneurons (*187*) suggested that Dip-AS may have multiple functions (i.e., neuromodulator, neurotransmitter) like proctolin and the FMRFamides. An AS was recently isolated and sequenced from *M. sexta* (pGlu-Val-Arg-Phe-Arg-Gln-Cys-Tyr-Phe-Asn-Pro-Ile-Ser-Cys-Phe-OH) (*191*). It has no sequence homology with any known neuropeptides, is active *in vitro* at 10^{-9} M, and its effect is reversible. Mas-AS is also active on *H. virescens* CA, but not *P. americana*, *T. molitor* or the grasshopper, *Melanoplus sanguinipes*.

Ecdysteroidogenins

This class of developmental and reproductive neuropeptides includes the egg development neurosecretory hormone (EDNH) and the pre-adult ecdysteroidogenins, bombyxin and prothoracicotropic hormone (PTTH; see Table I). EDNH stimulates egg maturation and ecdysteroid production by adult ovaries in Anophiline mosquitos (see review, *192*; also *193*). PTTH stimulates ecdysteroid production by the prothoracic gland (PG) and thus initiates the cascade of events involved in molting and metamorphosis. Mosquito EDNH (*Aedes aegypti*) has been purified (*192,194*) and a sequence obtained (*192*), although the synthetic free amide of this sequence was inactive in EDNH assays. Results suggest a peptide family for the EDNHs much like that of the pre-adult ecdysteroidogenins (*192*) with at least one member possibly related to bombyxin (see below) (*195*).

Studies on the pre-adult ecdysteroidogenins are extensive but include few species. Much of the work was recently reviewed, including studies on the biosynthesis and secretion of ecdysteroids by the PGs (*196*). More specific reviews on *B. mori* have dealt primarily with peptide and gene isolation, cellular localization and titer (*197-201*), specific reviews on *M. sexta* have dealt primarily with cellular localization, mode of action and titer (*202-205*), and those on the gypsy moth, *Lymantria dispar* primarily with isolation and titer (*206,207*).

The term "bombyxin" was coined by Mizoguchi *et al.* (*208*) to refer to the smaller forms of pre-adult ecdysteroidogenins in *B. mori* (ca. 4 kD), previously known as 4K-PTTHs or small PTTHs. Members of this group were first isolated and sequenced by Nagasawa *et al.* (*209,210*) (Table I) and stimulated ecdysteroid production by *B. mori* PGs *in vitro* at 10^{-11} M and adult development by debrained *Samia cynthia ricini* pupae at 0.1 ng per animal (*211*). Five classes of bombyxins (I-V) were identified based on amino acid sequence variations in the A and B chains (*212*). The A and B chains are connected by 2 disulfide bridges with an additional internal disulfide bridge in the A chain. Three bombyxin gene families (A,B,C) were identified in *B. mori* (*213-215*) with the A and B families recently identified in *S. cynthia ricini* (*216*). Initial cDNA isolation revealed a preprohormone structure comparable to insulin and the insulin-like growth factors (*213,217*). Lagueux *et al.* (*218*) found a similar structure for the preprohormone of *L. migratoria* insulin-related peptide. Recently, bombyxins II and IV were synthesized at greater than 25% yield and the natural disulfide bond arrangement determined (A6-A11, A7-B10, A20-B22-cystine) (*219-222*). Bombyxin is produced in four pairs of large dorsomedial neurosecretory cells in the protocerebrum and is present at high levels during the first half of each instar (i.e., the feeding period) and low levels during the second half (i.e., the molting period) (*212*). These observations agree with similar observations in the wax moth, *Galleria mellonella*, using bombyxin antibodies (*223*), except that 4 medial and 4 lateral immunopositive MNCs were found along with one pair of cells in the thoracic ganglia. Ecdysteroidogenic activity was recently demonstrated in the gut proctodeum of the European corn borer, *Ostrinia nubilalis* and *L. dispar* (*224*) and appears to be of low molecular weight (*225*). It is now believed, however, that bombyxin cannot be regarded as a physiological PTTH because of its *in vitro* activity at exceedingly high doses (*226*).

The sequence for large PTTH (22K-PTTH) has been determined only from *B. mori* (*201,227,228*) (Table I). It is a 109-residue peptide with a consensus sequence for glycosylation (Asn-X-Thr) at residue 41 and exists as a dimeric structure of identical, or nearly identical subunits held together by disulfide bridges (*228*). The preprohormone structure determined from cDNA isolation and sequencing revealed that it is synthesized as a 224-amino acid precursor (*227*). It contains three proteolytic cleavage sites that would result in two polypeptides cleaved from the 5′ end of the preprohormone (designated by the authors as p2K and p6K and of unknown function) and one PTTH subunit cleaved from the 3′ end of the preprohormone. The Bom-PTTH gene is present as a single copy per haploid genome (*229*) and has recently been expressed in *E. coli* (*230*) and AcMNPV (L. K. Miller, personal communication). Bom-PTTH was active at 10^{-11} M on *B. mori* PGs *in vitro* and at 0.1 ng per animal in brainless *B. mori* pupae (*201*). Immunocytochemical and *in situ* hybridization studies demonstrated that Bom-PTTH is produced by two pairs of dorso-lateral neurosecretory cells in the brain protocerebrum of *B. mori* (*227,231*). In contrast, a recent study with *L. migratoria* brains, using the Bom-PTTH antibodies, showed immunoreactivity in a large number of median neurosecretory cells of the pars intercerebralis (*232*).

Recent studies with the PTTH from *M. sexta* showed a dimeric structure similar to Bom-PTTH (*233,234*). However, partial sequencing revealed a similarity of Mas-PTTH to cellular retinol binding protein (CRBP), although it possessed an epitope similar to Bom-PTTH since Mas-PTTH could be detected with anti-Bom-PTTH antibody following SDS-PAGE/electroblotting (*234*). Davis *et al.* (*235*) used the polymerase chain reaction to amplify a 221 base-pair fragment from *L. dispar* brain cDNA that represents 68% of the PTTH gene and has 96% homology to Bom-PTTH. Immunocytochemical studies with monoclonal antibodies to Mas-PTTH recently examined the developmental expression of Mas-PTTH in developing embryos (*236*) and the three-dimensional architecture of Mas-PTTH-expressing neurons in the larval CNS (*237*). These studies suggested that PTTH may be one of the earliest neuropeptides expressed in insect embryos (immunoreactivity appearing at 24-30% of embryonic development) and that along with its previously demonstrated expression in the group III lateral neurosecretory cells, larval PTTH may be co-expressed in the ventral medial (VM) cells with eclosion hormone (see previous section). Furthermore, the axons of the VM cells traverse the length of the ventral nerve cord to release sites in the hindgut proctodeum. Mas-PTTH was recently shown to stimulate phosphorylation of a 34-kD protein in intact PGs. The time-course of phosphorylation suggests a role for this protein as an intermediate between PTTH stimulation of adenylate cyclase and ecdysteroid synthesis (*238*).

Oostatic Hormone

The only member of this class of reproductive and developmental hormones to be isolated and sequenced is mosquito Aea-OH by Borovsky *et al.* (*239*) (Table I). It is an unusual decapeptide containing six C-terminal prolines. Evidence of oostatic or antigonadotropic hormones has accumulated for a number of insect species, as recently reviewed (*240,241*). In the Diptera, these factors may affect oogenesis through modulation of ovarian ecdysteroid synthesis, gut trypsin synthesis, or EDNH release. In the blood-sucking hemipteran, *Rhodnius prolixus*, on the other hand, JH action on ovarian yolk uptake seems to be the major point of action (see reviews). Aea-OH modulates gut trypsin synthesis, preventing the breakdown of the bloodmeal into amino acid components necessary for vitellogenesis (*239,240*) and has been termed trypsin-modulating oostatic factor (TMOF). Aea-OH was purified from the ovaries of *A. aegypti* (*239*) and is synthesized in the follicular epithelium of the ovary 24-48 hours after a blood meal (*242*). It is active at ca. 1 µg/insect when injected (*239*), has activity in a number of dipteran species, and may act on similar trypsin and chymotrypsin enzymes in larval stages, suggesting potential for development as a unique larval and adult pesticide (*240,243*). A synthetic gene for Aea-OH was recently expressed in *E. coli* (D. A. Carlson, personal communication).

Novel Insect Neurohormones

In addition to the previously discussed insect neuropeptides, there are some whose functions do not strictly fit into these classes (e.g., neuroparsins, diapause hormone). Furthermore, there are some chemically-identified insect neuropeptides whose functions in insects are not yet known (e.g., pigment-dispersing factors (*244*) and the *L. migratoria* 5-6 kD CC peptides (*245*)), and many more whose functions are known, at least in part, but are not completely chemically identified (e.g., *246*; see reviews, 1,8,14,24; and books, 22,26).

Neuroparsins A and B were isolated from the nervous part of the *L. migratoria* CC by Girardie *et al.* (*247,248*). Neuroparsin B is a homodimer of a 78-residue peptide (*247*). The neuroparsins are made in the MNCs, released from the CC, and have multiple actions, including a brief elevation (15 min) of hemolymph trehalose without decreasing fat body glycogen, inhibition of the CA-JH-system, antidiuresis and hyperlipemia. Heitter *et al.* (*249*) recently obtained sequences superimposable to those of Girardie *et al.* (*247,248*) and proposed that the neuroparsins exist as monomers rather than dimers. Lagueux *et al.* (*250*) obtained a cDNA sequence for neuroparsin A and showed that neuroparsin exists as a single peptide sequence in the neuroparsin precursor.

The diapause hormone, isolated from 55,000 complexes of the subesophageal and first thoracic ganglion of day-1, *B. mori* pupae, was recently sequenced (*251*). It is a 24-residue peptide (Thr-Asp-Met-Lys-Asp-Glu-Ser-Asp-Arg-Gly-Ala-His-Ser-Glu-Arg-Gly-Ala-Leu-Cys-Phe-Gly-Pro-Arg-Leu-NH_2) containing the C-terminal pentapeptide sequence (Phe-X-Pro-Arg-Leu-NH_2) necessary for activity (*252*) and homologous to the C-terminal sequences of the PBANs and the myotropic pyrokinins (see previous sections). Diapause hormone is present in the subesophageal ganglion throughout pupal-adult development in *B. mori* and stimulates embryonic diapause through its action on the developing ovary to stimulate trehalase activity through protein synthesis (*251,252*).

Conclusions

Significant progress in the isolation and sequencing of insect neuropeptides and their genes has occurred within the past five years. Perhaps, several hundred insect neuropeptides (*49*) will eventually be isolated and sequenced. With the addition of invertebrate peptide growth factors (*253*), to which insulin and neuroparsin can be added (*254*), many more bioactive peptides are expected to be isolated and sequenced. However, many obstacles and considerations need to be addressed before insect neuropeptides and their analogs, genes, etc. can be used in insect control (see reviews in the Introduction). Paramount among these may be the existence of significant homology between many vertebrate and invertebrate peptides (*49*). Furthermore, some of the insect neuropeptides have shown activity in vertebrate (*255,256*) and non-insect, invertebrate systems (*257,258*).

Although substantial progress has been made in the past five years in the isolation and chemical identification of insect neuropeptides and their genes, resulting in baculovirus expression of Mas-DH, Mas-EH, Hez-PBAN and Bom-PTTH (see previous sections), considerable effort remains to develop peptidomimetic insect neuropeptides comparable to the peptidomimetic drugs of humans (*9*). Furthermore, considerable effort remains in the isolation of insect neuropeptide processing enzymes (see Harris *et al.*, this volume for a comparison to vertebrate systems) and degrading enzymes. No insect neuropeptide receptors have been isolated and sequenced with the possible exception of a partial sequence for the AKH receptor from the face fly, *Musca autumnalis* (*259*; personal communication) and a nucleotide sequence for a tachykinin receptor from *D. melanogaster* (260). The photoaffinity labelling of insect neuropeptides (*261*) should facilitate receptor isolation. Recent advances in the rapid synthesis and screening of peptide analogs will enhance the isolation of potent agonists and antagonists (*262-265*). Furthermore, rational design is beginning to play a role in the design of antagonists for human neurohormones (*266*), and problems of peptide transport across the human (*267*) and insect (*268*) intestine are being addressed.

Finally, it is useful to consider some of the mechanisms by which host neuroendocrine systems may be naturally regulated by parasites or disease-causing agents. Insect baculoviruses may regulate host development by altering ecdysteroid metabolism through synthesis of ecdysteroid UDP-glucosyltransferase (*269*). Similarly, the parasitic infection of the armyworm, *Pseudaletia separata*, by the wasp, *Apantales kariyai*, results in synthesis of a 4.5-kD peptide that causes 50% repression of hemolymph juvenile hormone esterase when 6.5 pmol are injected into day-3, last-instar larvae (*270*). At the neurohormone level, parasite infections of insects can inhibit the synthesis and release of PTTH (*271*). In fresh water snails, parasites can stimulate the synthesis and/or release of a 79-amino-acid gonadotropin antagonist, schistosomin (*272,273*). It therefore seems reasonable to assume that some natural interactions between organisms, where hormone/neurohormone regulation is modified, may involve interorganismal transmission of neuropeptides or their agonists/antagonists. Such interactions would include mating, where the transfer of peptide and proteinaceous material from male to female occurs, and sex peptides from the male can be transported to the brain of the female causing an all-or-nothing switch from virgin to mated behavior (see Miller, this volume). This process often includes stimulation of vitellogenesis, usually JH-regulated in adult insects. In this regard, careful analysis of available insect neuropeptide sequences reveals an unusually high 77% homology between the male adult accessory gland myotropin of *L. migratoria*, Lom-AG-MT-I (*274*) and Mas-AT (Fig. 1), suggesting that Lom-AG-MT-I may act as an allatotropin in *L. migratoria* females following its transmission from the male.

Mas-AT	Gly-Phe-Lys-Asn-Val	Glu-Met-Met	Thr-Ala-Arg-Gly-Phe-NH_2
Lom-AG-MT-I	Gly-Phe-Lys-Asn-Val	Ala-Leu- Ser	Thr-Ala-Arg-Gly-Phe-NH_2

Figure 1. Homology between Mas-AT (*173*) and Lom-AG-MT-I (*274*).

This close homology may further explain the diverse range of immunostaining from brain to ventral nerve cord with antibodies to Mas-AT in *D. melanogaster* (*275*) and antibodies to Lom-AG-MT-I in *L. migratoria* (*276*) As yet, neither a Lom-AT nor a Mas-AG-MT has been identified for comparison to the structures found in Figure 1. Further studies may elucidate the physiological significance, if any, of this homology and the role of another accessory gland peptide, Lom-AG-MT-II (Ala-His-Arg-Phe-Ala-Ala-Glu-Asp-Phe-Gly-Ala-Leu-Asp-Thr-Ala-OH) (*277*), in reproductive development. Certainly, numerous neuropeptides and their functions remain to be discovered in the class Insecta.

Acknowledgments

Thanks to Philip Copenhaver, Mark Holman, Frank Horodyski, Larry Keeley, Ashok Raina and Barbara Stay for critical reading of the manuscript and to numerous other individuals for providing preprints and unpublished information. Thanks also to Carol Masler and Carol Robinson for editorial assistance and Sherry Cohen for typing the manuscript.

Literature Cited

1. Keeley, L. L.; Hayes, T. K. *Insect Biochem.* **1987**, *17*, 639-51.
2. Thorpe, A.; Duve, H. *NATO Ser. A, Life Sciences* **1987**, *141*, 133-69.
3. Duve, H.; Thorpe, A. In *Neurohormones in Invertebrates*; Thorndyke, M. C.; Goldsworthy, G. J., Eds.; Cambridge University: Cambridge, **1988**; pp 115-39.
4. Holman, G. M.; Wright, M. S.; Nachman, R. J. *ISI Atlas of Science: Plants and Animals* **1988**, *1*, 129-36.
5. Keeley, L. L. In *Biotechnology for Crop Protection*; Hedin, P.A.; Menn, J. J.; Hollingworth, R. M., Eds.; American Chemical Society: Washington, D.C., **1988**; pp 147-59.
6. Thorpe, A.; Duve, H. In *Stimulus-Secretion Coupling in Neuroendocrine Systems, Current Topics in Neuroendocrinology*; Ganten, D.; Pfaff, D., Eds.; Springer-Verlag: New York, **1988**; Vol. 9, pp 185-230.
7. Mordue, W.; Siegert, K. J. In *Neurohormones in Invertebrates*; Thorndyke, M. C.; Goldsworthy, G. J., Eds.; Cambridge University: Cambridge, **1988**; pp 99-114.
8. Wheeler, C. H.; Gade, G.; Goldsworthy, G. J. In *Neurohormones in Invertebrates*; Thorndyke, M. C.; Goldsworthy, G. J., Eds.; Cambridge University: Cambridge, 1988; pp 142-57.
9. Menn, J. J.; Borkovec, A. B. *J. Agr. Food Chem.* **1989**, *37*, 271-78.
10. Menn, J. J.; Raina, A. K.; Edwards, J. P. In *Progress and Prospects in Insect Control*, McFarlane, H. R., Ed.; BCPC Mono. No. 43: Farnham, Surrey, United Kingdom, **1989**; pp 89-106.
11. Penzlin, H. *Naturwissenschaften* **1989**, 76, 243-52.
12. Adams, M. E. In *Cockroaches as Models for Neurobiology: Applications in Biomedical Research*; Huber, I.; Masler, E. P.; Rao, B. R., Eds.; CRC: Boca Raton, **1990**; Vol. 2, pp 3-33.
13. Holman, G. M.; Nachman, R. J.; Wright, M. S. *Ann. Rev. Entomol.* **1990**, 35, 201-17.
14. Keeley, L. L.; Hayes, T. K.; Bradfield, J. Y. In *Insect Neurochemistry and Neurophysiology-1989*; Borkovec, A. B.; Masler, E. P., Eds.; Humana: Clifton, New Jersey, **1990**; pp 163-203.
15. Kelly, T. J.; Masler, E. P.; Menn, J. J. In *Pesticides and Alternatives: Innovative Chemical and Biological Approaches to Pest Control*; Casida, J., Ed.; Elsevier: Amsterdam, **1990**; pp 283-97.
16. Schooley, D. A.; Carney, R. L.; Kataoka, H.; Kramer, S. J.; Li, J. P.; Toschi, A.; Troetschler, R. In *Molecular Insect Science*; Hagedorn, H. H.; Hildebrand, J. G.; Kidwell, M. G.; Law, J. H., Eds.; Plenum: New York, **1990**; pp 199-212.
17. Gade, G. In *Proceedings of a Conference on Insect Chemical Ecology*; Hrdy, I., Ed.; Academia Prague and SPB Acad. Publ.: The Hague, **1991**; pp 407-27.
18. Gade, G. In *Advances in Regulation of Insect Reproduction*; Bennettova, B.; Gelbic, I.; Soldan, T., Eds.; Czech Acad. Sci.: Ceske Budejovice, **1992**; pp 67-80.
19. Masler, E. P.; Kelly, T. J.; Menn, J. J. In *Insect Neuropeptides: Chemistry, Biology and Action*; Menn, J. J.; Kelly, T. J.; Masler, E. P., Eds.; American Chemical Society: Washington, D.C., **1991**; pp 6-18.
20. Masler, E. P.; Kelly, T. J.; Menn, J. J. *Arch. Insect Biochem. Physiol.* **1993**, *22*, 87-111.
21. Nagasawa, H. *Experientia* **1992**, *48*, 425-30.
22. Raabe, M. *Recent Developments in Insect Neurohormones*; Plenum: New York, **1989**; 503 pp.

23. Menn, J. J.; Kelly, T. J.; Masler, E. P., Eds. *Insect Neuropeptides: Chemistry, Biology and Action*; American Chemical Society: Washington, D.C., **1991**; 260 pp.
24. Cook, B. J.; Holman, G. M. In *Comprehensive Insect Physiology, Biochemistry and Pharmacology*; Kerkut, G. A.; Gilbert, L. I., Eds.; Pergamon: Oxford, **1985**; Vol. 11, pp 531-93.
25. Wheeler, C. H.; Gade, G.; Goldsworthy, G. J. In *Neurohormones in Invertebrates*; Thorndyke, M. C.; Goldsworthy, G. J., Eds.; Cambridge Univ.: Cambridge, **1988**; pp 142-57.
26. Raabe, M. *Insect Neurohormones*; Plenum: New York, **1982**; 352 pp.
27. Raina, A. K.; Gade, G. *Insect Biochem.* **1988**, *18*, 785-87.
28. Starratt, A. N.; Brown, B. E. *Life Sci.* **1975**, *17*, 1253-56.
29. O'Shea, M.; Adams, M. *Adv. Insect Physiol.* **1986**, *19*, 1-28.
30. Bishop, C. A.; Witten, J. L.; O'Shea, M. In *Cockroaches as Models for Neurobiology: Applications in Biomedical Research*; Huber, I.; Masler, E. P.; Rao, B. R., Eds.; CRC: Boca Raton, **1990**; Vol. 2, 35-51.
31. Cook, B. J.; Wagner, R. M. In *Cockroaches as Models for Neurobiology: Applications in Biomedical Research*; Huber, I.; Masler, E. P.; Rao, B. R., Eds; CRC: Boca Raton, **1990**; Vol. 2, 53-83.
32. Cook, B. J.; Wagner, R. M. In *Insect Neuropeptides: Chemistry, Biology and Action*; Menn, J. J.; Kelly, T. J.; Masler, E. P., Eds.; American Chemical Society: Washington, D.C. **1991**; pp 51-64.
33. Holman, G. M.; Nachman, R. J.; Wright, M. S.; Schoofs, L.; Hayes, T. K.; De Loof, A. In *Insect Neuropeptides: Chemistry, Biology and Action*; Menn, J. J.; Kelly, T. J.; Masler, E. P., Eds.; American Chemical Society: Washington, D.C., **1991**; pp 40-49.
34. Walther, C.; Zittlau, K. E.; Murck, H.; Nachman, R. J. In *Comparative Aspects of Neuropeptide Function*; Florey, E.; Stefano, G. B., Eds.; Manchester University: Manchester, United Kingdom, **1991**; pp 175-86.
35. Holman, G. M.; Cook, B. J.; Nachman, R. J. *Comp. Biochem. Physiol., Part C* **1986**, *84*, 205-11.
36. Holman, G. M.; Cook, B. J.; Nachman, R. J. *Comp. Biochem. Physiol., Part C* **1986**, *85*, 219-24.
37. Nachman, R. J., Holman, G. M.; Haddon, W. F.; Ling, N. *Science* **1986,** *234*, 71-73.
38. Schoofs, L.; Holman, G. M.; Hayes, T. K.; Nachman, R. J.; De Loof, A. *FEBS Lett.* **1990,** *261*, 397-401.
39. Holman, G. M.; Cook, B. J.; Nachman, R. J. *Comp. Biochem. Physiol., Part C* **1986,** *85*, 329-33.
40. Bartosz-Bechowski, H.; Rosinski, G.; Konopinska, D.; Sujak, P.; Sobotka, W. *Intl. J. Pep. Prot. Res.* **1990**, *36*: 450-56.
41. Konopinska, D.; Rosinski, G.; Bartosz-Bechowski, H.; Lesicki, A.; Sujak, P.; Sobotka, W. *Intl. J. Pep. Prot. Res.* **1990**, *35*, 12-16.
42. Nawrot, J.; Konopinska, D.; Sobotka, W.; Bartosz-Bechowski, H. *J. Appl. Ent.* **1991**, *111*, 104-108.
43. Paemen, L.; Amelinckx, M.; De Loof, A. In *Advances in Regulation of Insect Reproduction*; Bennettova, B.; Gelbic, I.; Soldan, T., Eds.; Czech Acad. Sci.: Prague, **1992**; pp 185-91.
44. Puiroux, J.; Pedelaborde, A.; Loughton, B. G. *Insect Biochem. Molec. Biol.* **1992**, *22*, 547-51.
45. Schoofs, L.; Holman, G. M.; Proost, P.; Van Damme, J.; Hayes, T. K.; De Loof, A. *Reg. Peptides* **1992**, *37*, 49-57.
46. Schoofs, L.; Holman, G. M.; Hayes, T. K.; Nachman, R. J.; Kochansky, J. P.; De Loof, A. *Insect Biochem. Molec. Biol.* **1992**, 22, 447-52.
47. Schoofs, L.; Holman, G. M.; Hayes, T. K.; Nachman, R. J.; De Loof, A. *Reg. Peptides* **1991**, *36*, 111-19.

48. Schoofs, L.; Holman, G. M.; Hayes, T. K.; Kochansky, J. P.; Nachman, R. J.; De Loof, A. *Reg. Peptides* **1990,** *31*, 199-212.
49. De Loof, A.; Schoofs, L. *Comp. Biochem. Physiol., Part B* **1990,** *95*, 459-68.
50. Spittaels, K.; Schoofs, L.; Grauwels, L.; Smet, H.; Van Damme, J.; Proost, P.; De Loof, A. *Peptides* **1991**, *12*, 31-36.
51. Holman, G. M.; Nachman, R. J.; Schoofs, L.; Hayes, T. K.; Wright, M. S.; De Loof, A. *Insect Biochem.* **1991**, *21*, 107-12.
52. Schoofs, L.; Schroven, W.; De Loof, A. *Comp. Biochem. Physiol., Part C,* **1990**, 97, 195-99.
53. Schoofs, L.; Tips, A.; Holman, G. M.; Nachman, R. J.; De Loof, A. *Reg. Peptides* **1992**, *37*, 237-54.
54. Kuniyoshi, H.; Nagasawa, H.; Ando, T.; Suzuki, A.; Nachman, R. J.; Holman, G. M. *Biosci. Biotech. Biochem.* **1992**, *56*, 167-68.
55. Meola, S. M.; Wright, M. S.; Holman, G. M.; Thompson, J. M. *J. Med. Ent.* **1991**, *28*, 712-18.
56. Meola, S. M.; Wright, M. S.; Holman, G. M.; Thompson, J. M. *Neurochem. Res.* **1991**, *16*, 543-49.
57. Nassel, D. R.; Lundquist, C. T. *Neurosci. Lett.* **1991,** *130*, 225-28.
58. Cantera, R.; Hansson, B. S.; Hallberg, E.; Nassel, D. R. *Cell Tiss. Res.* **1992,** *269*, 65-77.
59. Cantera, R.; Nassel, D. R. *Comp. Biochem. Physiol., Part C,* **1991,** *99*, 517-25.
60. Cook, B. J.; Holman, G. M.; Wagner, R. M.; Nachman, R. J. *Comp. Biochem. Physiol., Part C*, **1989**, *93*, 257-62.
61. Cook, B. J.; Holman, G. M.; Wagner, R. M.; Nachman, R. J. *Comp. Biochem. Physiol., Part C*, **1990**, *95*, 19-24.
62. Cook, B. J.; Wagner, R. M. *Comp. Biochem. Physiol., Part C*, **1991**, *99*, 95-99.
63. Cook, B. J.; Wagner, R. M.; Peterson, T. L. *J. Insect Physiol.* **1991**, *37*, 635-45.
64. Cook, B. J.; Wagner, R. M. *Comp. Biochem. Physiol., Part C*, **1992**, *102*, 273-80.
65. Lange, A. B.; Orchard, I.; Te Brugge, V. A. *J. Comp. Physiol. A* **1991**, *168*, 383-91.
66. Price, D. A.; Greenberg, M. J. *Biol. Bull. Woods Hole* **1989**, *177*, 198-205.
67. Nichols, R.; Schneuwly, S. A.; Dixon, J. E. *J. Biol. Chem.* **1988**, *263*, 12167-70.
68. Nambu, J. R.; Murphy-Erdosh, C.; Andrews, P. C.; Feistner, G. J.; Scheller, R. H. *Neuron* **1988**, *1*, 55-61.
69. Schneider, L. E.; Taghert, P. H. *Proc. Natl. Acad. Sci. USA* **1988**, *85*, 1993-97.
70. Taghert, P. H.; Schneider, L. E. *J. Neurosci.* **1990**, *10*, 1929-42.
71. Taghert, P. H.; Schneider, L. E.; O'Brien, M. A. In *Molecular Insect Science*; Hagedorn, H. H.; Hildebrand, J. G.; Kidwell, M. G.; Law, J. H., Eds.; Plenum: New York, **1990**; pp 213-221.
72. Schneider, L. E.; Taghert, P. H. *J. Biol. Chem.* **1990**, *265*, 6890-95.
73. Schneider, L. E.; O'Brien, M. A.; Taghert, P. H. *J. Comp. Neurol.* **1991**, *304*, 608-22.
74. O'Brien, M. A.; Schneider, L. E.; Taghert, P. H. *J. Comp. Neurol.* **1991**, *304*, 623-38.
75. Stone, J. V.; Mordue, W.; Batley, K. E.; Morris, H. R. *Nature* **1976,** *263*, 207-11.
76. O'Shea, M.; Hekimi, S.; Schulz-Aellen, M.-F. In *Molecular Insect Science*;

Hagedorn, H. H.; Hildebrand, J. G.; Kidwell, M. G.; Law, J. H., Eds.; Plenum: New York, **1990**; pp 189-97.
77. Gade, G. *Zool. Jb. Physiol.* **1992**, *96*, 211-25.
78. Keeley, L. L. In *Cockroaches as Models for Neurobiology: Applications in Biomedical Research*; Huber, I.; Masler, E. P.; Rao, B. R., Eds.; CRC: Boca Raton, **1990**; Vol. 2, pp 149-58.
79. Keeley, L. L.; Hayes, T. K.; Bradfield, J. Y.; Lee, Y.-H. In *Insect Neuropeptides: Chemistry, Biology and Action*; Menn, J. J.; Kelly, T. J.; Masler, E. P., Eds.; American Chemical Society: Washington, D.C., **1991**; pp 65-82.
80. Penzlin, H.; Baumann, E. In *Comparative Aspects of Neuropeptide Function*; Florey, E.; Stefano, G. B., Eds.; Manchester University: Manchester, United Kingdom, **1991**; pp 222-32.
81. O'Shea, M.; Rayne, R. C. *Experientia* **1992**, *48*, 430-38.
82. Oudejans, R. C. H. M.; Kooiman, F. P.; Heerma, W.; Versluis, C.; Slotboom, A. J.; Beenakkers, A. M. T. *Eur. J. Biochem.* **1991**, *195*, 351-59.
83. Gade, G.; Kellner, R. *Gen. Comp. Endocr.* **1992**, *86*, 119-27.
84. Gade, G.; Wilps, H.; Kellner, R. *Biochem. J.* **1990**, *269*, 309-13.
85. Schaffer, M. H.; Noyes, B. E.; Slaughter, C. A.; Thorne, G. C.; Gaskell, S. J. *Biochem. J.* **1990**, *269*, 315-20.
86. Gade, G. *Biochem. J.* **1991**, *275*, 671-77.
87. Gade, G. *J. Insect Physiol.* **1992**, *38*, 259-66.
88. Baumann, E.; Gade, G.; Penzlin, H. *J. Comp. Physiol. B* **1990**, *160*, 423-29.
89. Ziegler, R.; Eckart, K.; Jasensky, R. D.; Law, J. H. *Arch. Insect Biochem. Physiol.* **1991**, *18*, 229-37.
90. Fox, A. M.; Reynolds, S. E. *J. Insect Physiol.* **1991**, *37, 373-81.*
91. Ishibashi, J.; Kataoka, H.; Nagasawa, H.; Isogai, A.; Suzuki, A. *Biosci. Biotech. Biochem.* **1992**, *56*, 66-70.
92. Fernlund, P.; Josefsson, L. *Science* **1972**, *177*, 173-75.
93. Keeley, L. L.; Hayes, T. K.; Bradfield, J. Y.; Sowa, S. M. *Insect Biochem.* **1991**, *21*, 121-29.
94. Cusinato, O.; Wheeler, C. H.; Goldsworthy, G. J. *J. Insect Physiol.* **1991**, *37*, 461-69.
95. Moshitzky, P.; Applebaum, S. W. *Insect Biochem.* **1990**, *20*, 319-23.
96. Ziegler, R.; Eckart, K.; Law, J. H. *Peptides* **1990**, *11*, 1037-40.
97. Bradfield, J. Y.; Lee, Y.-H.; Keeley, L. L. *Proc. Natl. Acad. Sci. USA* **1991**, *88*, 4558-62.
98. Bradfield, J. Y.; Keeley, L. L. *J. Biol. Chem.* **1989**, *264*, 12791-93.
99. Schulz-Aellen, M.-F.; Roulet, E.; Fischer-Lougheed, J.; O'Shea, M. *Neuron* **1989**, *2*, 1369-73.
100. Noyes, B. E.; Schaffer, M. H. *J. Biol. Chem.* **1990**, *265*, 483-89.
101. Rayne, R. C.; O'Shea, M. *Insect Biochem. Molec. Biol.* **1992**, *22*, 25-34.
102. Fox, A. M.; Reynolds, S. E. *Peptides* **1991**, *12*, 937-44.
103. Spring, J. H. *J. Insect Physiol.* **1990**, *36*, 13-22.
104. Wheeler, C. H.; Coast, G. M. *J. Insect Physiol.* **1990**, *36*, 23-34.
105. Proux, J.; Fournier, B.; Girardie, J.; Picquot, M. In *Comparative Aspects of Neuropeptide Function*; Florey, E.; Stefano, G. B., Eds.; Manchester University: Manchester, United Kingdom, **1991**; pp 233-45.
106. Schooley, D. A. In *Insect Neuropeptides: Chemistry, Biology, and Action*; Menn, J. J.; Kelly, T. J.; Masler, E. P., Eds.; American Chemical Society: Washington, D.C., **1991**, pp 83-94.
107. Proux, J. P.; Miller, C. A.; Li, J. P.; Carney, R. L.; Girardie, A.; Delaage, M.; Schooley, D. A. *Biochem. Biophys. Res. Comm.* **1987**, *149*, 180-86.
108. Proux, J. P.; Herault, J.-P. *Neuropeptides* **1988**, *12*, 7-12.
109. Proux, J.; Beydon, P. *Insect Biochem. Molec. Biol.* **1992**, *22*, 185-92.

110. Thompson, K. S. J.; Tyrer, N. M.; May, S. T.; Bacon, J. P. *J. Comp. Physiol. A* **1991**, *168*, 605-17.
111. Thompson, K. S. J.; Bacon, J. P. *J. Comp. Physiol. A* **1991**, *168*, 619-30.
112. Picquot, M.; Proux, J. *Reg. Peptides* **1990**, *31*, 139-56.
113. Kataoka, H.; Troetschler, R. G.; Li, J. P.; Kramer, S. J.; Carney, R. L.; Schooley, D. A. *Proc. Natl. Acad. Sci. U.S.A.* **1989**, *86*, 2976-80.
114. Lehmberg, E.; Ota, R. B.; Furuya, K.; King, D. S.; Applebaum, S. W.; Ferenz, H.-J.; Schooley, D. A. *Biochem. Biophys. Res. Comm.* **1991**, *179*, 1036-41.
115. Kay, I.; Wheeler, C. H.; Coast, G. M.; Totty, N. F.; Cusinato, O.; Patel, M.; Goldsworthy, G. J. *Biol. Chem. Hoppe-Seyler* **1991,** *372*, 929-34.
116. Kay, I.; Coast, G. M.; Cusinato, O.; Wheeler, C. H.; Totty, N. F.; Goldsworthy, G. J. *Biol. Chem. Hoppe-Seyler* **1991,** *372*, 505-12.
117. Blackburn, M. B.; Kingan, T. G.; Bodnar, W.; Shabanowitz, J.; Hunt, D. F.; Kempe, T.; Wagner, R. M.; Raina, A. K.; Schnee, M. E.; Ma, M. C. *Biochem. Biophys. Res. Comm.* **1991**, *181*, 927-32.
118. Troetschler, R. G.; Kramer, S. J. *Arch. Insect Biochem. Physiol.* **1992**, *20*, 35-47.
119. Veenstra, J. A.; Hagedorn, H. H. *Cell Tissue Res.* **1991**, *266*, 359-64.
120. Maeda, S. *Biochem. Biophys. Res. Comm.* **1989**, *165*, 1177-83.
121. Keeley, L. L.; Chung, J. S.; Hayes, T. K. *Experientia* **1992**, *48*, 1145-48.
122. Hayes, T. K.; Pannebecker, T.; Hinckley, D.; Holman, G.; Nachman, R.; Petzel, D.; Beyenbach, K. *Life Sci.* **1989**, *44*, 1259-66.
123. Coast, G. M.; Holman, G. M.; Nachman, R. J. *J. Insect Physiol.* **1990**, *36*, 481-88.
124. Truman, J. W. In *Molting and Metamorphosis*; Ohnishi, E.; Ishizaki, H., Eds.; Japan Sci. Soc.: Tokyo, **1990**; pp 67-82.
125. Truman, J. W.; Horodyski, F. M.; Morton, D. B.; Riddiford, L. M. In *Molecular Insect Science*; Hagedorn, H. H.; Hildebrand, J. G.; Kidwell, M. G.; Law, J. H., Eds.; Plenum: New York, **1990**; pp 223-29.
126. Morton, D. B.; Truman, J. W. In *Comparative Aspects of Neuropeptide Function*; Florey, E.; Stefano, G. B. Eds.; Manchester University: Manchester, United Kingdom, **1991**; pp 187-97.
127. Truman, J. W.; Horodyski, F. M.; Hewes, R. S.; Riddiford, L. M. In *Insect Neuropeptides: Chemistry, Biology and Action*; Menn, J. J.; Kelly, T. J.; Masler, E. P., Eds.; American Chemical Society: Washington, D.C., **1991**; pp 95-99.
128. Kataoka, H.; Troetschler, R. G.; Kramer, S. J.; Cesarin, B. J.; Schooley, D. A. *Biochem. Biophys. Res. Comm.* **1987**, *146*, 746-50.
129. Marti, T.; Takio, K.; Walsh, K. A.; Terzi, G.; Truman, J. W. *FEBS Lett.* **1987**, *219*, 415-18.
130. Kono, T.; Nagasawa, H.; Isogai, A.; Fugo, H.; Suzuki, A. *Agric. Biol. Chem.* **1987**, *51*, 2307-8.
131. Truman, J. W.; Copenhaver, P. F. *J. Exp. Biol.* **1989**, *147*, 457-70.
132. Copenhaver, P. F., Truman, J. W. *J. Neurosci.* **1986**, *6*, 1738-47.
133. Hewes, R. S.; Truman, J. W. *J. Comp. Physiol. A* **1991**, *168*, 697-707.
134. Horodyski, F. M.; Riddiford, L. M.; Truman, J. W. *Proc. Natl. Acad. Sci. U.S.A.* **1989**, *86*, 8123-27.
135. Eldridge, R.; Horodyski, F. M.; Morton, D. B.; O'Reilly, D. R.; Truman, J. W.; Riddiford, L. M.; Miller, L. K. *Insect Biochem.* **1991**, *21*, 341-51.
136. Eldridge, R.; O'Reilly, D. R.; Miller, L. K. *Biological Control* **1992**, *2*, 104-10.
137. Kono, T.; Mizoguchi, A.; Nagasawa, H.; Ishizaki, H.; Fugo, H.; Suzuki, A. *Zool. Sci.* **1990**, *7*, 47-54.
138. Kamito, T.; Tanaka, H.; Sato, B.; Nagasawa, H.; Suzuki, A. *Biochem. Biophys. Res. Comm.* **1992**, *182*, 514-19.

139. Hayashi, H.; Nakano, M.; Shibanaka, Y.; Fujita, N. *Biochem. Biophys. Res. Comm.* **1990**, *173*, 1065-71.
140. Kono, T.; Nagasawa, H.; Kataoka, H.; Isogai, A.; Fugo, H.; Suzuki, A. *FEBS Lett.* **1990**, *263*, 358-60.
141. Kataoka, H.; Li, J. P.; Lui, A. S. T.; Kramer, S. J.; Schooley, D. A. *Int. J. Pept. Prot. Res.* **1992**, *39*, 29-35.
142. Kobayashi, K.; Higuchi, K.; Kurioka, A.; Kono, T.; Hugo, H.; Nagasawa, H.; Suzuki, A. *J. Seric. Sci. Jpn.* **1991**, *60*, 257-62.
143. Sakakibara, M.; Fugo, H. *J. Insect Physiol.* **1990**, *36*, 489-93.
144. Sakakibara, M.; Orikasa, C.; Fugo, H. *Jap. J. Appl. Ent. Zool.* **1990**, *34*, 63-70.
145. Raina, A. K.; Jaffe, H.; Kempe, T. G.; Keim, P.; Blacher, R. W.; Fales, H. M.; Riley, C. T.; Klun, J. A.; Ridgway, R. L.; Hayes, D. K. *Science* **1989**, *244*, 796-98.
146. Raina, A. K.; Kempe, T. G. *Insect Biochem.* **1990**, *20*, 849-51.
147. Gazit, Y.; Dunkelblum, E.; Benichis, M.; Altstein, M. *Insect Biochem.* **1990**, *20*, 853-58.
148. Fonagy, A.; Schoofs, L.; Matsumoto, S.; De Loof, A.; Mitsui, T. *J. Insect Physiol.* **1992**, *38*, 651-57.
149. Tumlinson, J. H.; Teal, P. E. A. In *Molecular Insect Science*; Hagedorn, H. H.; Hildebrand, J. G.; Kidwell, M. G.; Law, J. H., Eds.; Plenum: New York, **1990**; pp 243-55.
150. Raina, A. K.; Kempe, T. G.; Jaffe, H. In *Insect Neuropeptides: Chemistry, Biology and Action*; Menn, J. J.; Kelly, T. J.; Masler, E. P., Eds.; American Chemical Society: Washington, D.C., **1991**; pp 100-109.
151. Altstein, M.; Gazit, Y.; Dunkelblum, E. *Arch. Insect Biochem. Physiol.*, **1993**, *22*, 153-68.
152. Raina, A. K.; Menn, J. J. *Arch. Insect Biochem. Physiol.* **1993**, *22*, 141-51.
153. Raina, A. K. *Ann. Rev. Entomol.* **1993**, *38*, 329-49.
154. Kitamura, A.; Nagasawa, H.; Kataoka, H.; Inoue, T.; Matsumoto, S.; Ando, T.; Suzuki, A. *Biochem. Biophys. Res. Comm.* **1989**, *163*, 520-26.
155. Kitamura, A.; Nagasawa, H.; Kataoka, H.; Ando, T.; Suzuki, A. *Agric. Biol. Chem.* **1990**, *54*, 2495-97.
156. Matsumoto, S.; Kitamura, A.; Nagasawa, H.; Kataoka, H.; Orikasa, C.; Mitsui, T.; Suzuki, A. *J. Insect Physiol.* **1990**, *36*, 427-32.
157. Mase, K.; Matsumoto, S.; Kiguchi, K. *J. Seric. Sci. Jpn.* **1991,** *60*, 186-91.
158. Matsumoto, S.; Fonagy, A.; Kurihara, M.; Uchiumi, K.; Nagamine, T.; Chijimatsu, M., Mitsui, T. *Biochem. Biophys. Res. Comm.* **1992,** *182*, 534-39.
159. Kuniyoshi, H.; Kitamura, A.; Nagasawa, H.; Chuman, T.; Shimazaki, K.; Ando, T.; Suzuki, A. In *Peptide Chemistry 1990*; Shimonishi, Y., Ed.; Protein Research Foundation: Osaka, **1991**; pp 251-54.
160. Kuniyoshi, H.; Nagasawa, H.; Ando, T.; Suzuki, A. In *Peptide Chemistry 1991*; Suzuki, A., Ed.; Protein Research Foundation: Osaka, **1992**; pp 221-26.
161. Kuniyoshi, H.; Nagasawa, H.; Ando, T.; Suzuki, A. *Insect Biochem. Molec. Biol.* **1992,** *22*, 399-403.
162. Raina, A. K.; Kempe, T. G. *Insect Biochem. Molec. Biol.* **1992,** *22*, 221-25.
163. Rafaeli, A. *Proc. XIXth Intl. Congr. Entomol.* **1992**, Abstract no. 93A.
164. Rafaeli, A.; Soroker, V.; Kamensky, B.; Raina, A. K. *J. Insect Physiol.* **1990,** *36*, 641-46.
165. Jurenka, R. A.; Jacquin, E.; Roelofs, W. L. *Proc. Natl. Acad. Sci. U.S.A.* **1991,** *88*, 8621-25.
166. Davis, M.-T. B.; Vakharia, V. N.; Henry, J.; Kempe, T. G.; Raina, A. K. *Proc. Natl. Acad. Sci. U.S.A.* **1992,** *89*, 142-46.

167. Kawano, T.; Kuniyoshi, H.; Nakayama, J.; Nagasawa, H.; Suzuki, A. *Agric. Biol. Chem.* **1991,** *55*, 1943-45.
168. Tang, J. D.; Wolf, W. A.; Roelofs, W. L.; Knipple, D. C. *Insect Biochem.* **1991,** *21*, 573-81.
169. Teal, P. E. A.; Tumlinson, J. H.; Oberlander, H. *Experientia* **1990**, *46*, 1047-50.
170. Bird, T. G.; Raina, A. K.; Hayes, D. K. In *Naturally Occurring Pest Bioregulators*; Hedin, P. A., Ed.; American Chemical Society: Washington, D.C., **1991**; pp 78-86.
171. Teal, P. E. A.; Tumlinson, J. H. In *Naturally Occurring Pest Bioregulators*; Hedin, P. A., Ed.; American Chemical Society, Washington, D.C., **1991**; pp 66-77.
172. Raina, A. K.; Kingan, T. G.; Giebultowicz, J. M. *Proc. XIXth Intl. Congr. Entomol.* **1992**, Abstract no. 93B.
173. Kataoka, H.; Toschi, A.; Li, J. P.; Carney, R. L.; Schooley, D. A.; Kramer, S. J. *Science* **1989**, *243*, 1481-83.
174. Unni, B. G.; Bhaskaran, G.; Dahm, K. H.; Hayes, T. K. *Arch. Insect Biochem. Physiol.* **1991**, *17*, 129-42.
175. Reagan, J. D.; Miller, W. H.; Kramer, S. J. *Arch. Insect Biochem. Physiol.* **1992**, *20*, 145-55.
176. Pratt, G. E.; Farnsworth, D. E.; Siegel, N. R.; Fok, K. F.; Feyereisen, R. In *Insect Neuropeptides: Chemistry, Biology and Action*; Menn, J. J.; Kelly, T. J.; Masler, E. P., Eds.; American Chemical Society: Washington, D.C., **1991**, pp 177-92.
177. Stay, B.; Woodhead, A. P.; Joshi, S.; Tobe, S. S. In *Insect Neuropeptides: Chemistry, Biology and Action*; Menn, J. J.; Kelly, T. J.; Masler, E. P., Eds.; American Chemical Society: Washington, D.C., **1991**; pp 164-76.
178. Woodhead, A. P.; Stay, B.; Seidel, S. L.; Khan, M. A.; Tobe, S. S. *Proc. Natl. Acad. Sci. U. S. A.* **1989**, *86*, 5997-6001.
179. Pratt, G. E.; Farnsworth, D. E.; Siegel, N. R.; Fok, K. F.; Feyereisen, R. *Biochem. Biophys. Res. Comm.* **1989**, *163*, 1243-47.
180. Pratt, G. E.; Farnsworth, D. E.; Fok, K. F.; Siegel, N. R.; McCormack, A. L.; Shabanowitz, J.; Hunt, D. F.; Feyereisen, R. *Proc. Natl. Acad. Sci. U. S. A.* **1991**, *88*, 2412-16.
181. Stay, B.; Joshi, S.; Woodhead, A. P. *J. Insect Physiol.* **1991**, *37*, 63-70.
182. Weaver, R. J. *J. Insect Physiol.* **1991**, *37*, 111-18.
183. Guan, X.-C.; Hayes, T. K.; Tobe, S. S. *Proc. XIXth Intl. Congr. Entomol.* **1992**, Abstract no. 116C.
184. Yagi, K. J.; Yu, C. G.; Tobe, S. S. *Experientia* **1992,** *48*, 758-61.
185. Cusson, M.; Prestwich, G. D.; Stay, B.; Tobe, S. S. *Biochem. Biophys. Res. Comm.* **1991**, *181*, 736-42.
186. Cusson, M.; Prestwich, G. D.; Stay, B.; Tobe, S. S. In *Insect Juvenile Hormone Research: Fundamental and Applied Approaches - Chemistry, Biochemistry and Mode of Action*; Mauchamp, B.; Couillaud, F.; Baehr, J. C., Eds.; INRA: Paris, **1992**; pp 183-191.
187. Stay, B.; Chan, K. K.; Woodhead, A. P. *Cell Tiss. Res.* **1992,** *270*, 15-23.
188. Duve, H.; Thorpe, A.; Tobe, S. S. *Cell Tiss. Res.* **1991**, 263, 285-91.
189. Stay, B.; Joshi, S.; Tobe, S. S. In *Advances in Regulation of Insect Reproduction*; Bennettova, B.; Gelbic, I.; Soldan, T., Eds.; Czech Acad. Sci.: Prague, **1992**; pp 281-84.
190. Woodhead, A. P.; Stoltzman, C. A.; Stay, B. *Arch. Insect Biochem. Physiol.* **1992**, *20*, 253-63.
191. Kramer, S. J.; Toschi, A.; Miller, C. A.; Kataoka, H.; Quistad, G. B.; Li, J. P.; Carney, R. L.; Schooley, D. A. *Proc. Natl. Acad. Sci. U. S. A.* **1991**, *88*, 9458-62.

192. Masler, E. P.; Kelly, T. J.; Kochansky, J. P.; Woods, C. W.; Borkovec, A. B. In *Insect Neuropeptides: Chemistry, Biology and Action*; Menn, J. J.; Kelly, T. J.; Masler, E. P., Eds.; American Chemical Society, Washington, D.C., **1991**; pp 124-32.
193. Birnbaum, M. J.; Kelly, T. J.; Woods, C. W.; Imberski, R. B. *Gen. Comp. Endocr.* **1984**, *56*, 9-18.
194. Matsumoto, S.; Brown, M. R.; Suzuki, A.; Lea, A. O. *Insect Biochem.* **1989**, *19*, 651-56.
195. Lea, A. O.; Brown, M. R. In *Molecular Insect Science*; Hagedorn, H. H.; Hildebrand, J. G.; Kidwell, M. G.; Law, J. H., Eds.; Plenum: New York, **1990**; pp 181-88.
196. Sakurai, S.; Gilbert, L. I. In *Molting and Metamorphosis*; Ohnishi, E.; Ishizaki, H., Eds.; Japan Sci. Soc.: Tokyo, **1990**; pp 83-106.
197. Ishizaki, H. In *Ecdysone: From Chemistry to Mode of Action*; Koolman, J., Ed.; Thieme: Stuttgart, **1989**; pp 204-10.
198. Iwami, M. In *Molting and Metamorphosis*; Ohnishi, E.; Ishizaki, H., Eds.; Japan Sci. Soc.: Tokyo, **1990**; pp 49-66.
199. Mizoguchi, A. In *Molting and Metamorphosis*; Ohnishi, E.; Ishizaki, H., Eds.; Japan Sci. Soc.: Tokyo, **1990**; pp 17-32.
200. Nagasawa, H.; Kataoka, H.; Suzuki, A. In *Molting and Metamorphosis*; Ohnishi, E.; Ishizaki, H., Eds.; Japan Sci. Soc.: Tokyo, **1990**; pp 33-48.
201. Kataoka, H.; Nagasawa, H.; Kawakami, A.; Oka, T.; Mizoguchi, A.; Iwami, M.; Ishizaki, H.; Suzuki, A. In *Insect Neuropeptides: Chemistry, Biology and Action*; Menn, J. J.; Kelly, T. J.; Masler, E. P., Eds.; American Chemical Society: Washington, D.C., **1991**; pp 20-26.
202. Gilbert, L. I. In *Ecdysone: From Chemistry to Mode of Action*; Koolman, J., Ed.; Thieme: Stuttgart, **1989**; pp 448-71.
203. Watson, R. D.; Spaziani, E.; Bollenbacher, W. E. In *Ecdysone: From Chemistry to Mode of Action*; Koolman, J., Ed.; Thieme: Stuttgart, **1989**; pp 188-203.
204. Smith, W. A.; Sedlmeier, D. *Invert. Reprod. Dev.* **1990**, *18*, 77-89.
205. Westbrook, A. L.; Bollenbacher, W. E. In *Molting and Metamorphosis*; Ohnishi, E.; Ishizaki, H., Eds.; Japan Sci. Soc.: Tokyo, **1990**; pp 3-16.
206. Masler, E. P.; Kelly, T. J.; Bell, R. A.; Thyagaraja, B. S.; Borkovec, A. B. *Ad. Bios.* **1989**; *8*, 101-28.
207. Kelly, T. J.; Masler, E. P.; Bell, R. A.; Thyagaraja, B. S.; Davis, R. E.; Fescemyer, H. W.; Borkovec, A. B. In *Insect Neuropeptides: Chemistry, Biology and Action*; Menn, J. J.; Kelly, T. J.; Masler, E. P., Eds.; American Chemical Society: Washington, D.C., **1991**; pp 27-37.
208. Mizoguchi, A.; Ishizaki, H., Nagasawa, H.; Kataoka, H.; Isogai, A.; Tamura, S.; Suzuki, A.; Fujino, M.; Kitada, C. *Molec. Cell. Endocr.* **1987,** *51*, 227-35.
209. Nagasawa, H.; Kataoka, H.; Isogai, A.; Tamura, S.; Suzuki, Ak.; Ishizaki, H.; Mizoguchi, A.; Fujiwara, Y.; Suzuki, At. *Science* **1984**, *226*, 1344-45.
210. Nagasawa, H.; Kataoka, H.; Isogai, A.; Tamura, S.; Suzuki, Ak.; Mizoguchi, A.; Fujiwara, Y.; Suzuki, At.; Takahashi, S. Y.; Ishizaki, H. *Proc. Natl. Acad. Sci. U. S. A.* **1986**, 83, 5840-43.
211. Nagasawa, H.; Kataoka, H.; Hori, Y.; Isogai, A.; Tamura, S.; Suzuki, A.; Guo, F.; Zhong, X.; Mizoguchi, A.; Fujishita, M.; Takahashi, S. Y.; Ohnishi, E.; Ishizaki, H. *Gen. Comp. Endocr.* **1984**, *53*, 143-52.
212. Mizoguchi, A.; Hatta, M.; Sato, S.; Nagasawa, H.; Suzuki, A.; Ishizaki, H. *J. Insect Physiol.* **1990**, *36*, 655-64.
213. Iwami, M.; Kawakami, A.; Ishizaki, H.; Takahashi, S. Y.; Adachi, T.; Suzuki, Y.; Nagasawa, H.; Suzuki, A. *Dev. Growth Differ.* **1989**, *31*, 31-37.

214. Iwami, M.; Adachi, T.; Kondo, H.; Kawakami, A.; Suzuki, Y.; Nagasawa, H.; Suzuki, A.; Ishizaki, H. *Insect Biochem.* **1990**, *20*, 295-303.
215. Kawakami, A.; Iwami, M.; Nagasawa, H.; Suzuki, A.; Ishizaki, H. *Proc. Natl. Acad. Sci. U. S. A.* **1989**, *86*, 6843-47.
216. Kimura-Kawakami, M.; Iwami, M.; Kawakami, A.; Nagasawa, H.; Suzuki, A.; Ishizaki, H. *Gen. Comp. Endocr.* **1992**, *86*, 257-68.
217. Adachi, T.; Takiya, S.; Suzuki, Y.; Iwami, M.; Kawakami, A.; Takahashi, S. Y.; Ishizaki, H.; Nagasawa, H.; Suzuki, A. *J. Biol. Chem.* **1989**, *264*, 7681-85.
218. Lagueux, M.; Lwoff, L.; Meister, M.; Goltzene, F.; Hoffmann, J.A. *Eur. J. Biochem.* **1990**, *187*, 249-54.
219. Maruyama, K.; Nagata, K.; Tanaka, M.; Nagasawa, H.; Isogai, A.; Ishizaki, H.; Suzuki, A. *J. Prot. Chem.* **1992**, *11*, 1-12.
220. Maruyama, K.; Nagasawa, H.; Isogai, A.; Ishizaki, H.; Suzuki, A. *J. Prot. Chem.* **1992**, *11*, 13-20.
221. Nagata, K.; Maruyama, K.; Nagasawa, H.; Tanaka, M.; Isogai, A.; Ishizaki, H.; Suzuki, A. In *Peptide Chemistry 1991*; Suzuki, A., Ed.; Protein Research Foundation: Osaka, **1992**; pp 1-6.
222. Nagata, K.; Maruyama, K.; Nagasawa, H.; Urushibata, I.; Isogai, A.; Ishizaki, H.; Suzuki, A. *Peptides* **1992**, *13*, 653-62.
223. Zitnan, D.; Sehnal, F.; Mizoguchi, A.; Ishizaki, H.; Nagasawa, H.; Suzuki, A. *Dev. Growth Differ.* **1990**, *32*, 637-45.
224. Gelman, D. B.; Thyagaraja, B. S.; Kelly, T. J.; Masler, E. P.; Bell, R. A.; Borkovec, A. B. *Experientia* **1991**, *47*, 77-80.
225. Gelman, D. B.; Thyagaraja, B. S.; Bell, R. A. *Proc. Xth Ecdysone Workshop*, **1992**, p 39.
226. Kiriishi, S.; Nagasawa, H.; Kataoka, H.; Suzuki, A.; Sakurai, S. *Zool. Sci.* **1992**, *9*, 149-55.
227. Kawakami, A.; Kataoka, H.; Oka, T.; Mizoguchi, A.; Kimura-Kawakami, M.; Adachi, T.; Iwami, M.; Nagasawa, H., Suzuki, A.; Ishizaki, H. *Science* **1990**, *247*, 1333-35.
228. Kataoka, H.; Nagasawa, H.; Isogai, A.; Ishizaki, H.; Suzuki, A. *Agric. Biol. Chem.* **1991**, *55*, 73-86.
229. Ishizaki, H.; Suzuki, A. *Proc. XIXth Intl. Congr. Entomol.*, **1992**. Abstract no. 95C.
230. Kataoka, H.; Ishibashi, J.; Suzuki, A.; Kawakami, A.; Ishizaki, H. *Proc. XIXth Intl. Congr. Entomol.*, **1992**, Abstract no. 95D.
231. Mizoguchi, A.; Oka, T.; Kataoka, H.; Nagasawa, H.; Suzuki, A.; Ishizaki, H. *Dev. Growth Differ.* **1990**, *32*, 591-98.
232. Goltzene, F.; Holder, F.; Charlet, M.; Meister, M.; Oka, T. *Cell Tiss. Res.* **1992**, *269*, 133-40.
233. Gray, R. S.; Muehleisen, D. P.; Bollenbacher, W. E. *Amer. Zool.* **1991**, *31*, 86A.
234. Muehleisen, D. P.; Gray, R. S.; Katahira, E. J.; Bollenbacher, W. E. *Amer. Zool.* **1991**, *31*, 86A.
235. Davis, R. E.; Kelly, T. J.; Masler, E. P.; Thyagaraja, B. S.; Borkovec, A. B. *49th Annual Symposium of the Society for Developmental Biology*, **1990**, Abstract no. 67.
236. Westbrook, A. L.; Bollenbacher, W. E. *Dev. Biol.* **1990**, *140*, 291-99.
237. Westbrook. A. L.; Haire, M. E.; Kier, W. M.; Bollenbacher, W. E. *J. Morph.* **1991**, *208*, 161-74.
238. Rountree, D. B.; Combest, W. L.; Gilbert, L. I. *Insect Biochem. Molec. Biol.* **1992**, *22*, 353-62.
239. Borovsky, D.; Carlson, D. A.; Griffin, P. R.; Shabanowitz, J.; Hunt, D. F. *FASEB J.* **1990**, *4*, 3015-20.

240. Borovsky, D.; Carlson, D. A.; Hunt, D. F. In *Insect Neuropeptides: Chemistry, Biology and Action*; Menn, J. J.; Kelly, T. J.; Masler, E. P., Eds.; American Chemical Society: Washington, D.C., **1991**; pp 133-42.
241. DeMilo, A. B.; Kelly, T. J.; Masler, E. P.; Bell, R. A. In *Insect Neuropeptides: Chemistry, Biology and Action*; Menn, J. J.; Kelly, T. J.; Masler, E. P., Eds.; American Chemical Society: Washington, D.C., **1991**; pp 143-51.
242. Borovsky, D.; Carlson, D. A.; Ma, M.; Song, Q. *Proc. XIXth Intl. Congr. Entomol.* **1992**, Abstract no. 91D.
243. Borovsky, D.; Mahmood, F.; Carlson, D. A. *J. Fla. Anti-Mosq. Assoc.* **1989,** *60*, 66-70.
244. Rao, K. R.;. Mohrhrerr, C. J.; Bonomelli, S. L.; Riehm, J. P.; Kingan, T. G. In *Insect Neuropeptides: Chemistry, Biology and Action*; Menn, J. J.; Kelly, T. J.; Masler, E. P., Eds.; American Chemical Society: Washington, D.C., **1991**; pp 110-122.
245. Hietter, H.; Luu, B.; Goltzene, F.; Zachary, D.; Hoffmann, J.; Van Dorsselaer, A. *Eur. J. Biochem.* **1989,** *182*, 77-84
246. Girardie, J.; Richard, O.; Girardie, A. *J. Insect Physiol.* **1992,** *38*, 215-21.
247. Girardie, J.; Girardie, A.; Huet, J.-C.; Pernollet, J.-C. *FEBS Lett.* **1989,** *245*, 4-8.
248. Girardie, J.; Huet, J.-C.; Pernollet, J.-C. *Insect Biochem.* **1990,** *20*, 659-66.
249. Hietter, H.; Van Dorsselaer, A.; Luu, B. *Insect Biochem.* **1991,** *21*, 259-64.
250. Lagueux, M.; Kromer, E.; Girardie, J. *Insect Biochem. Molec. Biol.* **1992,** *22*, 511-16.
251. Imai, K.; Konno, T.; Nakazawa, Y.; Komiya, T.; Isobe, M.; Koga, K.; Goto, T.; Yaginuma, T.; Sakakibara, K.; Hasegawa, K.; Yamashita, O. *Proc. Japan Acad., Ser. B.* **1991,** *67*, 98-101.
252. Yamashita, O.; Ikeda, M.; Sato, Y.; Isobe, M.; Imai, K. *Proc. XIXth Intl. Congr. Entomol.* **1992**, Abstract no. 627A.
253. Ferkovich, S. M.; Oberlander, H. *In Vitro Cell. Dev. Biol.* **1991,** *27A*, 483-86.
254. Vanhems, E.; Delbos, M.; Girardie, J. *Eur. J. Neurosci.* **1990,** *2*, 776-82.
255. Muramatsu, H. *J. Jap. Diab. Soc.* **1990,** *33*, 743-49.
256. Fone, K. C. F.; Johnson, J. V.; Putland, A. P.; Bennett, G. W. *J. Neurochem.* **1991,** *57*, 845-51.
257. Groome, J. R.; Townley, M. A.; De Tschaschell, M.; Tillinghast, E. K. *J. Insect Physiol.* **1991,** *37*, 9-19.
258. Wenning, A.; Greisinger, U.; Proux, J. P. *J. Exp. Biol.* **1991,** *158*, 165-80.
259. Minnifield, N. M.; Hayes, D. K. *Prep. Biochem.* **1992,** *22*, 215-28.
260. Monnier, D.; Colas, J.-F.; Rosay, P.; Hen, R.; Borrelli, E.; Maroteaux, L. *J. Biol. Chem.* **1992**, *267*, 1298-302.
261. Prestwich, G. D. *Insect Biochem.* **1991,** *21*, 27-40.
262. Scott, J. K.; Smith, G. P. *Science* **1990,** *249*, 386-90.
263. Devlin, J. J.; Panganiban, L. C.; Devlin, P. E. *Science* **1990,** *249*, 404-6.
264. Fodor, S. P. A.; Read, J. L.; Pirrung, M. C.; Stryer, L.; Lu, A. T.; Solas, D. *Science* **1991,** *251*, 767-73.
265. Brenner, S.; Lerner, R. A. *Proc. Natl. Acad. Sci. U.S.A.* **1992,** *89*, 5381-83.
266. Fuh, G.; Cunningham, B. C.; Fukunaga, R.; Nagata, S.; Goeddel, D. V.; Wells, J. A. *Science* **1992,** *256*, 1677-80.
267. Kleinert, H. D.; Rosenberg, S. H.; Baker, W. R.; Stein, H. H.; Klinghofer, V.; Barlow, J.; Spina, K.; Polakowski, J.; Kovar, P.; Cohen, J.; Denissen, J. *Science* **1992,** *257*, 1940-43.
268. Allingham, P. G.; Kerlin, R. L.; Tellam, R. L.; Briscoe, S. J.; Standfast, H. A. *J. Insect Physiol.* **1992,** *38*, 9-17.
269. O'Reilly, D. R.; Brown, M. R.; Miller, L. K. *Insect Biochem. Molec. Biol.* **1992,** *22*, 313-20.

270. Hayakawa, Y. *J. Biol. Chem.* **1990,** *265*, 10813-16.
271. Tanaka, T.; Agui, N.; Hiruma, K. *Gen. Comp. Endocr.* **1987,** *67*, 364-74.
272. Hordijk, P. L.; Schallig, H. D. F. H.; Ebberink, R. H. M.; De Jong-Brink, M.; Joosse, J. *Biochem. J.* **1991,** *279*, 837-42.
273. De Jong-Brink, M.; Elsaadany, M.; Soto, M. S. *Parasitology* **1991,** *103*, 371-78.
274. Paemen, L.; Tips, A.; Schoofs, L.; Proost, P.; Van Damme, J.; De Loof, A. *Peptides* **1991,** *12*, 7-10.
275. Zitnan, D.; Sehnal, F.; Bryant, P. J. *Dev. Biol.***1993**, *156*, 117-35.
276. Paemen, L.; Schoofs, L.; De Loof, A. *Cell Tiss. Res.* **1992,** *268*, 91-97.
277. Paemen, L.; Schoofs, L.; Proost, P.; Decock, B.; De Loof, A. *Insect Biochem.* **1991,** *21*, 243-48.

RECEIVED August 28, 1993

Natural and Engineered Viral Agents

Chapter 22

Natural and Engineered Viral Agents

James L. Vaughn

Insect Biocontrol Laboratory, Plant Sciences Institute, Agricultural Research Service, U.S. Department of Agriculture, BARC-West, Beltsville, MD 20705–2350

Of the many viruses infecting insects, the nuclear polyhedrosis viruses and the granulosis viruses, both members of the family Baculoviridae, are the most promising candidates for use in insect pest management because they have many of the desired characteristics of the ideal microbial agent. They are found only in Arthropods; infecting the immature, feeding stage of the insect and causing devastating epizootics in natural populations. The range of susceptible insects is limited to closely related species, often only within the same genera. In addition, the infectious virion is embedded in a protein body that protects it from inactivation in storage and in the field. Increased levels of resistance to chemical insecticides, the public's concern with environmental contamination and the possible health effects of chemicals in their food, have focused attention on the use of alternative methods for pest management. The papers in this session review the current research on these viruses, including altering the viruses with recombinant DNA techniques to develop more effective pest management tools.

More than 460 years ago an Italian poet, Vida, in a poem about silkworms, described a disease that was probably caused by a virus. Presumably the same disease, called "grasserie" in French, was described in the scientific literature by Nysten in 1908 (*1*). In 1890-1892 a disease was described that killed enormous numbers of the nun moth, *Lymantria monacha*, in the spruce forests of central Europe (*2*). Wahl in 1909 (*3*) identified the causative agent of the disease as a virus. In 1921 Allen reported an outbreak of a polyhedrosis disease in the fall armyworm, *Spodoptera frugiperda*, that left dead larvae hanging from the tips of blades of grass in considerable numbers (*4*). These and other similar outbreaks of virus diseases in a number of insect pests often were quite spectacular and they attracted the attention of entomologists working on pest management, who sought to use them to control pest insects.

The first recorded attempt to use a virus in insect control was by German foresters who tried to use the pathogen of the nun moth to control later infestations of this pest. However, since viruses had not been described at that time, the foresters did not know how to produce the pathogen and the results were not consistently favorable. In 1938 a heavy infestation of the European spruce sawfly, *Gilpinia hercyniae*, caused extensive damage in Quebec, Canada. By 1940 the insect population had declined to the point where no great damage was caused. A virus of unknown origin was eventually credited with causing the decline and viral epizootics followed the spreading insect population into Northeastern United States where the virus continued to suppress the population (*1*). These and many subsequent epizootics have resulted in a continuing interest in the viruses as control agents.

More than 1600 virus isolates from about 1100 species of insects and mites have been recorded (*5*). These viruses are classified into seven families (*6*). Members of three families, Poxviridae, Reoviridae, and Baculoviridae, have shown the most potential as biocontrol agents.

Viruses belonging to the family Poxviridae cause disease in grasshoppers, forest lepidoptera, and turf beetles (As examples see ref. *7-9*). However, because of their similarity to the pox viruses from vertebrates they were not considered good candidates for insect control until detailed study of the viral genome by restriction enzyme nucleotide fragment patterns demonstrated that they were quite different. There is no record of the entomopox viruses causing disease in vertebrates (*10*). Therefore it is likely that these viruses will become useful control agents in the future.

Reoviruses from insects, called "cytoplasmic polyhedrosis viruses," are unique among the members of this family in that they are the only ones in the family that have the infectious virions enclosed in a protein crystal. Only the cytoplasmic polyhedrosis virus from the pine caterpillar, *Dendrolimus spectabilis*, has been widely used as a control agent (*11*). These viruses are not highly lethal, but rather persist in the infected population causing lethargy in infected larvae and lower fecundity in adults. They are most suitable for use in situations, such as the forest, where some insect damage can be tolerated.

Most of the viruses thus far considered for biocontrol of insects belong to the family Baculoviridae. This family contains three genera: Nuclear polyhedrosis viruses, Granulosis viruses, and Non-occluded viruses. This session will deal primarily with the Nuclear Polyhedrosis viruses and the Granulosis viruses. Members of these genera occur exclusively in arthropods, primarily in insects. This provides a basis for the presumed safety of these viruses. All the viruses in these two genera have the infectious virion enclosed in a protein matrix that provides protection and stability during storage, shipping and following application. The viruses infect the larval stage of the insects and cause an acute disease most often resulting in death of the infected insect. Thus, they have most of the characteristics desired of a microbial pest management agent; lethality, safety, and stability.

Most of the pest management agents discussed during this Conference were similar to synthetic chemicals in the form in which they are applied in the field and the way in which they affect the pest. Viruses are different in significant ways: They are applied as living organisms, and as such, can multiply in the field after application. They are obligate pathogens, meaning that they can multiply only in the tissues and cells of a susceptible host. The effect on the insect is the result of a series of physiological processes - a disease, not a poison. These differences give the viruses some of their advantages over the other control agents and at the same time result in some restraints on their use and effectiveness as pest management agents.

The specificity of the viruses can be either an advantage or a disadvantage. Because the viruses are host specific, they will not kill beneficial insects such as honey bees or parasitic insects. Therefore, using viruses does not remove the natural pressures on the target insect or on other potential pests controlled by parasites or predators and lead to increased need for chemical control. They will not harm plants or vertebrates; nor will they accumulate in other organisms in the food chain as do some chemicals. Because of their specificity and the fact that they can multiply in the field after application, no tolerance levels are established for viruses; thus, there is no residue problem.

Specificity is also a disadvantage in some situations. If a crop is attacked by more than one pest, a single virus may not prevent economic loss. From the commercial viewpoint, the specificity limits the potential market size for a product and the earnings from sales in that market may not be sufficient to recover the costs of development. If independent applications of different viruses are required to control several pests, the cost to the grower may also be a disadvantage.

Since no effect on the pest is obtained until there has been some replication of the applied virus, there is a time lag between the application and the appearance of signs of the disease. During this incubation period the insect continues to feed and may cause economic losses even though the pest population is eventually reduced. This must be considered when a decision whether or not to use viruses is being made.

Developing a virus for pest management is influenced greatly by the production methods available. Because viruses must be produced in the tissues of their hosts, the only means of producing them generally is to rear and infect large numbers of the host insect. In developed countries this has discouraged companies from producing viruses for a commercial market. This is particularly true if there is no artificial diet for the insect and it must be grown on foliage. The recent developments in large-volume cell culture indicate that this will soon be a more attractive alternative production system for some viruses.

Until recently patent protection for viral pest management products was not considered possible. Thus, companies that did make the considerable investment required to develop a viral product had no protection from competitors while they recovered this cost. However, recently patents have been issued for two virus strains obtained from nature. Certainly patent protection will be available for genetically altered viruses such as will be described in following chapters. This should serve as encouragement for further development by commercial firms.

In this brief introduction I have attempted to provide a broad overview of the reasons why certain viral pathogens of pest insects have held the attention of entomologists for at least a century. I have also introduced some of the problems that have prevented or delayed the fulfillment of the promise that these organisms showed 100 years ago in Germany. In the first three chapters prepared by the speakers in this session the details of this history will be presented. In the final chapter one of the exciting new areas in the field of insect virology, the genetic manipulation of the nuclear polyhedrosis viruses, will be discussed. This new technology holds great promise for overcoming some of the restraints to the use of viruses in pest management. These recent advances and the increased pressure to reduce the use of chemical insecticides have provided a new opportunity for the incorporation of viruses into pest management and are the reasons for including this topic in the Conference.

Literature Cited

1. Steinhaus, E. A.; *Principles of Insect Pathology*; McGraw-Hill: New York NY, **1949**; pp 417-545.
2. Hofmann, O.; *Die Schlaffsucht (Flacherie) der Nonne (Liparis monacha) nebst einem Anhang. Insektentotende Pilze mit besonderlr. Berucksichtigung der Nonne.* P. Weber: Frankfurt a. M., **1891**.
3. Wahl, B.; *Zentr. Gesell. Forstw.* **1909**; 35, 164.
4. Allen, H. W.; *J. Econ. Entomol.*; **1921**; 14, 510-511.
5. *A Catalog of Viral Diseases of Insects, Mites, and Ticks*; Martignoni, M. E.; Iwai, P. J. Eds.; General Technical Report PNW-195; Pacific Northwest Research Station, United States Department of Agriculture: Washington, D. C., **1986**; 4th Edition Revised.
6. Classification and Nomenclature of Viruses, Fifth Report of the International Committee on Taxonomy of Viruses; Francki, R. I. B.; Fauquet, C. M.; Knudson, D. L.; Brown, F. Eds.; *Archives of Virology Suppl. 2*; Springer - Verlag: Wein, Austria, **1991**.
7. Bird, F. T.; Sanders, C. J.; Burke, J. M. *J. Invertebr. Pathol.* **1971**, 18, 159-161.
8. Milner, R. J.; Lutton, G. G. *Entomophaga* **1975**, 20, 213-220.
9. Henry, J. E.; Nelson, B. P.; Jutila, J. W. *J. Virol.* **1969**. 3, 605-610.
10. Langridge, W. H. R.; Oma, E.; Henry, J. E. *J. Invertebr. Pathol.* **1983**, 42, 327-333.
11. Katagiri, K.; In *Microbial Control of Pests and Plant Diseases 1970-1980*; Burges, H. D. Ed.; Academic Press: New York, NY, **1981,** pp 433-440.

RECEIVED April 30, 1993

Chapter 23

Nuclear Polyhedrosis Viruses for Insect Control

John J. Hamm

Insect Biology and Population Management Research Laboratory, Agricultural Research Service, U.S. Department of Agriculture, P.O. Box 748, Tifton, GA 31793–0748

Nuclear polyhedrosis viruses have been extensively researched and are used more often for microbial control of insect pests than any of the other viruses. Their generally high degree of host specificity and ability to produce epizootics are among their most favorable attributes. The broader host ranges of a few NPVs offer hope for using a single virus to control several key pests and expands the potential market for these viruses. The rather rapid inactivation of NPVs by sunlight and the longer time required to kill pests, compared to chemical pesticides, are unfavorable attributes. In the future, the addition of feeding stimulants, UV protectants, and natural or chemical enhancers to increase infectivity of viruses may make viruses more competitive with chemicals for insect control.

There have been some very good reviews of baculoviruses, and nuclear polyhedrosis viruses in particular, in recent years (*1,11,40*). The process of registration of baculoviruses as pesticides was reviewed by Betz (*5*) and some practical factors influencing the utilization of baculoviruses as pesticides was discussed by Bohmfalk (*6*). Since this conference deals with pest management agents, I will concentrate on the characteristics of nuclear polyhedrosis viruses (NPVs) as pest management agents, and how these characteristics affect the strategies for their use.

Characteristics of Nuclear Polyhedrosis Viruses

Ingestion and Infection. The primary route of virus infection is via the alimentary tract during larval feeding. The polyhedral occlusion bodies dissolve in the alkaline midgut, releasing the virions, or enveloped nucleo-

Published 1994 American Chemical Society

capsids. Once released into the midgut, the virions are subject to degradation by the gut juices. There is a dynamic balance between the destruction of virions in the midgut and passage of virions through the peritrophic membrane and entry into the midgut epithelial cells. The envelope of the virion is believed to fuse with the cell membrane of the gut microvilli. Once inside the midgut epithelial cell, nucleocapsids migrate to the nuclear membrane where they apparently enter the nucleus through nuclear pores. The virus replicates in the nucleus. In the case of sawflies (Hymenoptera), the polyhedral occlusion bodies are produced in the midgut epithelium as that is the only tissue infected (*10*). In Lepidoptera and other orders, nucleocapsids or virions are not occluded in the nuclei of midgut cells but leave the nuclei and pass to the basal membrane. Nucleocapsids acquire a host-mediated envelope with a peplomere structure at one end as they bud through the basal membrane into the hemocoel. These virions infect various other tissues, including fat body, epidermis, and tracheal epithelium, where the majority of virions are occluded in the polyhedra. When larvae die, the cadavers tend to liquify or "melt," releasing the occlusion bodies into the environment. This results in a tremendous increase in the pathogen population. Obviously, from the standpoint of insect control, two of the most critical points in this cycle are the times when the occluded virus is exposed to the environment and when the virions are exposed to the contents of the midgut.

Host Range. Adams et al. (*1*) list 523 species of insects from 52 families and 8 orders as hosts of NPVs. The majority, 455 species, are in the order Lepidoptera and 107 species are in the family Noctuidae, which contains many agricultural pests of economic importance. The order Hymenoptera has 31 species listed as hosts of NPVs, 19 of which are in the family Diprionidae and are pests of forest and shade trees. The order Diptera contains 27 species listed as hosts of NPVs, 20 of which are in the family Culicidae which are important pest mosquitoes. Thus, a great number of economically important insects are potential targets for commercially produced NPVs if the NPVs can compete with other control agents.

Specificity. The specificity and safety of baculoviruses were reviewed by Doller (*11*) and Groner (*23*). All tests indicate that the NPVs do not infect man and other vertebrates or, for the most part, other orders of insects. Most NPVs are pathogenic to only a few closely related species of insects. Some NPVs have been demonstrated to infect only a single species of insect. A few broader spectrum NPVs, such as the *Autographa californica* NPV, the *Anagrapha falcifera* NPV, and the *Mamestra brassicae* NPV (*26,12*), infect many species of Lepidoptera in the family Noctuidae and several species in other families of Lepidoptera. There are a few reports of NPVs from Lepidoptera infecting insects in other orders. A low level of replication of the *Autographa californica* NPV was reported in a mosquito cell line, order Diptera (*36*). *Spodoptera littoralis* NPV was reported to cause a lethal disease in two species of locusts, order Orthoptera (*4*) and also to infect a species of termites, order Isoptera (*17*). But in general it appears that NPVs can be used without endangering nontarget species, and especially beneficial species such as parasitoids and predators that help keep the pest populations down.

NPVs have few deleterious effects on other biological control agents. NPVs do not directly affect predators and adult parasitoids as chemical insecticides often do, and NPVs do not infect parasitoid larvae in virus-infected hosts. However, the early mortality of host larvae, due to virus, may result in death of parasitoid larvae if the parasitoids do not have time to complete development.

Parasitoids and predators can play an important role in the dissemination of viruses (*25,31*). Female parasitoids which sting virus-infected larvae can transmit the virus when they sting subsequent larvae. This can be particularly important in moving virus from tree to tree in the forest or from row to row in agricultural crops (*25*). Because the polyhedral occlusion bodies, which dissolve in the basic midgut of host larvae, are resistant to the more acid digestive tracts of predacious insects, birds, and small mammals, they remain infectious when passed out in the feces of most predators. This helps to keep the virus circulating through the environment and provides for both short and long range dissemination of virus by birds (*25*).

Epizootic Potential. One of the main advantages of viruses as microbial control agents for insects is their epizootic potential. Fuxa and Tanada (*19*) define epizootic as an unusually large number of cases of disease in a host population. Epizootics are sporadic and limited in duration being characterized by a sudden change in prevalence and incidence of the disease. Two important factors are the ability of the pathogen to increase and its ability to spread though the host population (*15*). While only one, or a few, viral occlusion bodies may infect an insect larva, the infected larva may release billions of occlusion bodies into the environment when it dies (one large corn earworm larvae can produce approximately 6×10^9 occlusion bodies). When these occlusion bodies contaminate feeding sites of susceptible larvae, the disease can spread rapidly through the host population (*13*).

Environmental Stability

Importance of Occlusion Body. Another important factor is the stability of the pathogen in the environment. The occlusion body provides a significant degree of protection to the virus. NPVs stored as intact occlusion bodies may retain activity for several years when stored under cool dark conditions, whereas free virions lose their infectivity within weeks or months, even when stored at 4°C (*28*). An envelope, containing both protein and polysaccharides, surrounds the occlusion body and adds to its stability. When the occlusion body envelope is ruptured, cracks may form in the occlusion body, resulting in degradation of the virions (*1*). Thus, when producing virus it is important to allow the maximum number of occlusion bodies to complete development and be enveloped in order to maximize stability and infectivity of the viral preparation (*32,33*). Infected larvae may be e collected before they die in order to minimize loss of recoverable virus due to melting of the larvae and to minimize the level of bacterial contamination. However, the activity of the virus may be significantly greater when larvae are collected after they die.

UV Inactivation. Most NPVs lose 50% or more of their original activity in one or two days when sprayed on plants in the field. This inactivation is associated with exposure to sunlight and more specifically, the UV portion of sunlight (*28*). This has led to a search for materials that could be added to viral formulations to protect them from UV inactivation.

Persistence in Soil. NPVs can persist for years in the soil where the virus is protected from UV irradiation. Thus the soil acts as a reservoir of virus and provides inoculum to initiate new infections each year.

Comparison of NPVs in Lepidoptera and Hymenoptera

Lepidoptera. There are some important differences between NPVs of Lepidoptera and Hymenoptera. In Lepidoptera, the virus generally replicates in the midgut without producing occlusion bodies and then passes into the hemocoel where it infects a variety of tissues and produces the occlusion bodies. There is very little release of virus into the environment from the midgut, and most lateral transmission occurs after the infected larvae die and release the occlusion bodies as the cadaver melts or liquifies.

Probably the most successful use of an NPV to control an agricultural crop pest is the use of NPV to control velvetbean caterpillar in Brazil (*30*). In this case, the government was involved in the production of limited supplies of the virus, which was distributed to extension service personnel and farmer cooperatives. The supply of virus was increased either by treating naturally occurring larvae or by releasing large numbers of larvae into soybean fields treated with the virus and collecting the dead larvae. The dead larvae were collected 7 to 10 days after treatment and frozen for further use or processing. In the early years, the virus was used by the farmers as a crude preparation. More recently, the government has been processing the virus into a kaolin-based wettable powder formulation for distribution to farmers. They still recommend that farmers collect virus-killed larvae to apply the virus to larger areas, or store frozen for use in the subsequent season. The combination of laboratory production with other less costly methods, taking advantage of low labor costs in the country, has allowed application of formulated velvetbean caterpillar NPV at a cost of US $2.00 per ha compared to US $5.00 per ha for chemical insecticides. Also, only one application of the virus is required compared to several applications of chemical insecticide. The estimated area treated with velvetbean caterpillar NPV in Brazil increased form 2,000 ha in the 1982-83 season to 1,000,000 ha in the 1989-90 season.

Hymenoptera. In Hymenoptera the development of NPVs is restricted to the midgut. The occlusion bodies are produced in the infected midgut nuclei and are released into the lumen of the midgut as the infected cells rupture. Thus, great numbers of occlusion bodies are released into the environment in the fecal material of infected larvae. Some infected larvae survive to the adult stage. Because the adult midgut is also infected, infected adult sawflies can aid in dispersal of the virus (*10*).

The most virulent entomopathogenic viruses described to date are NPVs infecting several species of sawflies (*9,31*). Currently, sawflies are relatively minor forest pests. There are no artificial diets for sawflies and it is rather expensive to rear larvae on foliage in the laboratory. Therefore, viruses are produced by spraying heavily infested plantations and harvesting the virus-killed larvae. This is relatively inexpensive, especially because only 50 virus-killed larvae are required to treat a hectare.

Strategies for Using Nuclear Polyhedrosis Viruses.

Introduction. Harper (*24*) described the goals of applied epizootiology related to different strategies for using pathogens for biological control. Permanent reduction of the general level of pest population density is the goal of classical biological control through introduction of natural enemies which persist in the new environment. While there are examples of NPVs being introduced into a new area where they have persisted in the pest population, in most cases they do not keep the pest population below the economic threshold. The European spruce sawfly, *Gilpinia hercyniae*, is an example of an introduced pest being brought under control by the introduction of a NPV (*14*). Apparently the NPV was accidentally introduced into Canada with the importation of parasites of European spruce sawfly (*8*).

Augmentation. More commonly, NPVs are used to initiate short-term epizootics in which the pathogen may cycle and cause mortality in one to several generations (*18,31*). This is accomplished by augmenting the number of pathogens over those already present in the habitat. This increases disease prevalence above natural levels and can result in temporary lowering of pest population density. This differs from the temporary suppression of pest populations through application of rapidly acting entomopathogens which do not recycle, such as *Bacillus thuringiensis*. *Bacillus thuringiensis* is used in a "microbial insecticide" strategy where multiple applications of the pathogen or its toxic product are applied to compensate for the lack of epizootic potential, i.e. ability to replicate and/or survive in the field. Multiple applications of virus are often made to compensate for the time between application of the virus and recycling of the virus. However, once there is significant release of new virus additional applications may not be required.

Integration with Other Methods of Control

IPM. Integrated pest management, or IPM, has been defined as "a system that, in the context of the associated environment and the population dynamics of the pest species, utilizes all suitable techniques and methods in as ecologically compatible a manner as possible and maintains pest population levels below those causing economic injury" (*7*). In an editorial in a recent Society for Invertebrate Pathology Newsletter, Mark Goettel asked the question "Whatever Happened to the 'I' in IPM?" He points out that the development of *Bacillus thuringiensis* as a microbial control agent was expected to strengthen Integrated Pest Management but in most cases it was used as a more selective insecticide in a chemical insecticide approach to control, with the same resulting problems of resistance development.

Yearian and Young (*41*) point out that in short term comparisons most viruses do not compare favorably with chemical insecticides, from either an economic or effectiveness standpoint, in terms of crop protection or productivity. However, insect viruses are ideally suited for IPM systems. Their host specificity limits their direct effects to target pests or closely related species. This minimizes the chances of pest population resurgence or release of secondary pests as a result of reduction in parasitoids and predators, as often occurs with the use of broad-spectrum insecticides. In IPM systems, where the objective is maintenance of pest populations below the economic threshold, not total elimination of the pest, the efficacy of many NPVs may be quite adequate. The biggest impediment to incorporation of insect viruses into IPM systems is the lack of an adequate supply of formulated virus (*1,16,41*).

Population Suppression. The *Heliothis* NPV, Elcar, was originally registered for use only on cotton (*27*). From a population management standpoint, the time to get control of cotton bollworm and tobacco budworm is during the first two generations, before they are serious pests on cotton. The populations of bollworm and budworm are lowest in the first two generations when the moths emerge from over-wintering pupae in the soil and oviposit on wild flowers and weeds, before crop plants are attractive. On these wild host plants the larvae are exposed to parasitoids and predators, and thus the population does not increase significantly between the over-wintering generation and the first generation produced in the spring. The rationale for area-wide management of bollworm and budworm, with emphasis on reducing survivors in the first two generations, was presented by Knipling and Stadelbacher (*29*). This led to preliminary tests for area-wide treatment of wild host plants with NPV to reduce the populations of bollworm and budworm in the Delta area of Mississippi (*2,3*).

A single aerial application of the *Heliothis* NPV (Elcar) to wild geranium, at a rate of 6 X 10^{11} occlusion bodies per ha resulted in an 88% reduction in tobacco budworm and a 100% reduction in bollworm adult emergence in June (*2*). A single application of 3 X 10^{11} occlusion bodies per ha resulted in reductions of 65% in budworms and 57% in bollworms.

We need a better understanding of potential interactions between NPVs and host plant resistance. In general, smaller, less mature larvae are more susceptible to infection by NPVs than are larger, more mature larvae. Some host plants exhibit resistance through antibiosis, resulting in a reduced growth rate of insects feeding on them compared to more susceptible plants. This reduced growth rate of the insect can result in increased susceptibility of the insect to virus infection and longer exposure to parasitoids and predators.

Some varieties of plants may alter the feeding sites of the insect pest or allow better coverage of virus sprays, resulting in more effective control of insects. The presence of salt glands on some plants such as cotton may increase the pH of dew on the leaves and contribute to inactivation of NPVs (*13*). To my knowledge, no effort has been made to screen varieties of cotton for their effects on the stability of insect viruses.

In programs involving the release of sterile or substerile insects for suppression of pest populations, the released insects could be contaminated with NPV and allowed to disseminate the virus to the natural population. This would reduce the natural population, and thus the number of released insects required to obtain the desired ratio of sterile or substerile insects to wild insects.

Area-wide suppression of pest populations is the goal of many control programs. This involves treating large areas that could not be treated with broad-spectrum chemical insecticides. Even a pesticide as broad-spectrum as *Bacillus thuringiensis* might not be as desirable as a more host-specific NPV for application to large areas if there is a single key pest to be targeted.

If NPVs were used as one of a series of pest control agents it might slow the development of resistance to the various control agents. For example, the *Heliothis* NPV (Elcar) could be used against the first generation of bollworm and tobacco budworm on wild host plants, and possibly on corn, where bollworm populations expand rapidly. In subsequent generations, when these pests attack cotton, *Bacillus thuringiensis* could be used. If the pest populations reach too high a threshold, chemical insecticides could be used. This should reduce the pressure for the pest populations to develop resistance to any one of the control measures (chemical, *Bacillus thuringiensis*, or virus) compared to using the same control agent against all generations on all hosts. Perhaps, rather than considering NPVs as direct competitors with *Bacillus thuringiensis,* we should consider NPVs as tools for sustaining the usefulness of *Bacillus thuringiensis* and some of the chemical insecticides.

Recent Advances

Fluorescent Brighteners. Shapiro (*J. Econ. Entomol.*, in press) demonstrated the UV protectant properties of a series of optical or fluorescent brighteners. These are chemical agents added to detergents to make your clothes "whiter than white and brighter than bright." More importantly, five of the optical brighteners, including Tinopal LPW, enhanced the infectivity of the gypsy moth NPV for gypsy moth larvae even when the virus was not exposed to UV irradiation (*35*). Tinopal LPW (also know as Calcofluor white M2R and Fluorescent Brightener 28) is a stilbene and is believed to be a chitin synthetase inhibitor. Tinopal LPW at 0.1% enhanced the activity of gypsy moth NPV by 118-fold and at 1% it enhanced the activity by 184- to 1,670-fold. Tinopal LPW, at 0.1%, enhanced infectivity of the fall armyworm NPV for fall armyworm larvae by 164- to 303,000-fold (Hamm, J. J.; Shapiro, M., *J. Econ. Entomol.* in press). Thus the optical brightener Tinopal LPW has been shown to greatly enhance the infectivity of NPVs in two important lepidopteran pests: the gypsy moth, a forest pest in the family Lymantriidae, and the fall armyworm, an agricultural pest in the family Noctuidae. Due to the unique level of enhancement of viral infectivity for lepidopteran larvae produced by this fluorescent brightener, a patent for the use of fluorescent brighteners in
biological control was awarded 23 June 1992 (*34*).

Factors Associated with Granulosis Viruses that Enhance Infectivity of NPVs.

Tanada (*37*) reported that the Hawaiian strain of the armyworm, *Pseudaletia unipuncta*, granulosis virus (GV) enhanced infectivity of the armyworm NPV when they were fed simultaneously. Since then, a viral lipoprotein in the GV occlusion body was identified as the synergistic factor (SF). Preliminary tests indicated that the SF enhanced infectivity of GVs and NPVs of the armyworm, the cabbage looper (*Trichoplusia ni*), and the beet armyworm (*Spodoptera exigua*) in their respective hosts (*38*). Uchima et al. (*39*) demonstrated that the SF binds to midgut membranes and may serve as attachment sites for the enveloped virions.

A different viral enhancing factor (VEF) was found associated with the GV of cabbage looper (*21*). This VEF acts primarily by disrupting the structural integrity of the peritrophic membrane, allowing more easy passage of virions to the midgut, and apparently also aiding in attachment of virions to the midgut epithelial cells. In addition to enhancing infectivity of cabbage looper NPV for cabbage looper, it also enhances infectivity of the *Autographa californica* NPV and the *Anticarsia gemmatalis* NPV for cabbage looper. The VEF also decreased the time to mortality (*20*). The potency ratios, comparing virus alone and virus with 40 ng of VEF, were 16 for TnSNPV and 10 for AgMNPV.

The VEF is 10 times more stable than the virions to UV inactivation and is heat stable; this, plus the ability to enhance infectivity in later instars make it an attractive additive for field applications of virus (*20*). Granados and Carsaro (*21*) suggested that the genome for the VEF might be incorporated into the genome of other baculoviruses to enhance the infectivity of the baculoviruses or even into transgenic plants to increase the infectivity of naturally occurring baculoviruses.

The incorporation of these enhancers, either a fluorescent brightener or one of GV products, into NPV formulations could make the NPVs more competitive with other control agents. The enhancers decrease, slightly, the time to mortality and decrease the amount of virus necessary to produce mortality. They may effectively increase the host range of some of the broad spectrum NPVs by making them more effective against pests which they can infect at high doses but against which they are only marginally effective without an enhancer.

Summary

In summary I would like to paraphrase from Jim Harper's message from the president in a recent newsletter of the Southeastern Biological Control Working Group: Sustainable agriculture is today's buzzword and represents a concept that is crucial for the long-term conservation of resources that we will need to feed and clothe ourselves for decades and centuries to come. Integrated pest management is one of the most technologically advanced of the many components of sustainable agriculture. Central to IPM are biological control techniques that naturally promote sustainability because they are either self-perpetuating (and thus sustainable) or are generally much less disruptive to non-target organisms in the environment than are synthetic chemical pesticides. This specificity promotes maintenance of diversity in the environment being manipulated, resulting in greater stability and ultimately greater sustainability. Nuclear polyhedrosis viruses, with their specificity and epizootic potential, fit well into the IPM approach to sustainable agriculture. But these viruses must be available as commercial products that are economically feasible for relatively small markets. NPVs should not be used as chemical insecticides have been, as the only means of control for all generations of pests on all crops. This approach leads to development of resistance to the control agent and loss of sustainability. However, if NPVs are made available for use in conjunction with other control agents (other biocontrol agents, host plant resistance, *Bacillus thuringiensis*, and chemical insecticides) in ways that deter development of resistance to any of these agents, then NPVs can play an important role in pest management for many years to come. A major problem is how to make profitable the registration and production of viruses for relatively small markets. If the cost of registration could be reduced significantly, that could help solve the problem.

Literature Cited

1. Adams, J. R.; McClintock, J. T.; Couch, J. A. In *Atlas of Invertebrate Viruses;* Adams, J. R.; Bonami, J. R., Eds.; CRC Press: Boca Raton, FL., 1991, pp 87-204.
2. Bell, M. R. In *Vth International Colloquium on Invertebrate Pathology and Microbial Control*; Proceedings and Abstracts, The XXIII Annual Meeting of the Society for Invertebrate Pathology: Adelaide, Australia, 1990, pp 486-490.
3. Bell, M. R. *J. Econ. Entomol.* 1991, *84*, 851-854.
4. Bensimon, A.; Zinger, S.; Gerassi, E.; Hauschner, A.; Harpaz, I.; Sela, I. *J. Invertebr. Pathol.* 1987, *50*, 254-260.
5. Betz, F. S. In *The Biology of Baculoviruses Volume II Practical Application for Insect Control*; Granados, R. R.; Federici, B. A., Eds.; CRC Press: Boca Raton, FL., 1986, pp 203-222.
6. Bohmfalk, G. T. In *The Biology of Baculoviruses Volume II Practical Application for Insect Control*; Granados, R. R.; Federici, B. A., Eds.; CRC Press: Boca Raton, FL., 1986, pp 223-235.
7. Croft, B. A. In *Integrated Management of Insect Pests of Pome and Stone Fruits*; Croft B. A.; Hoyt, S. C., Eds.; John Wiley & Sons: New York, NY., 1983, pp 1-18.
8. Cunningham, J. C. In *Microbial and Viral Pesticides;* Kurstak, E., Ed.; Marcel Dekker, Inc.: New York, NY, 1982, pp 335-386.
9. Cunningham, J. C. In *Vth International Colloquium on Invertebrate Pathology and Microbial Control*; Proceedings and Abstracts, XXIII Annual Meeting of the Society for Invertebrate Pathology: Adelaide, Australia, 1990, pp 164-168.
10. Cunningham, J. C.; Entwistle, P. F. In *Microbial Control of Pests and Plant Diseases 1970-1980*; Burges, H.D., Ed.; Academic Press, Inc.: New York, NY., 1981, pp 378-407.
11. Doller, G. In *Viral Insecticides for Biological Control*; Maramorosch, K.; Sherman, K. E., Eds.; Academic Press, Inc.: Orlando, FL., 1985. pp 399-439.
12. Doyle, C. J.; Hirst, M. L.; Cory, J. S.; Entwistle, P. F.; *Apl. Environ. Microbiol.* 1990, 56, 2704-2710.
13. Evans. H. F. In *The Biology of Baculoviruses Volume II Practical Application for Insect Control*; Granados, R. R.; Federici, B. A., Eds.; CRC Press: Boca Raton, FL., 1986, pp 89-132.
14. Evans, H. F.; Entwistle, P. F. In *Microbial and Viral Pesticides*; Kurstak, E., Ed.; Marcel Dekker, Inc.: New York, NY., 1982, pp 449-461.
15. Evans, H. F.; Entwistle, P. F. In *Epizootiology of Insect Diseases*; Fuxa, J. R.; Tanada, Y., Eds.; John Wiley & Sons, Inc.: New York, NY., 1987, pp 257-322.
16. Falcon, L. A. In *Biological control in Agricultural IPM Systems*; Hoy, M. A.; Herzog, D. C., Eds.; Academic Press, Inc.: Orlando, FL., 1985, pp 229-242.

17. Fazairy, A. A. Al; Hassan, F. A.; *Insect Sci. Applic.* 1988, *9*, 37-39.
18. Fuxa, J. R. In *Annual Review of Entomology;* Mittler, T. E.; Radovsky, F. J.; Resh, V. H., Eds. Annual Reviews, Inc.; Palo Alto, CA, 1987, Vol. 32; pp 225-251.
19. Fuxa, J. R.; Tanada, Y. In *Epizootiology of Insect Diseases*; Fuxa, J. R.; Tanada, Y., Eds.; John Wiley & Sons, Inc.: New York, NY., 1987, pp 3-21.
20. Gallo, L. G.; Corsaro, B. G.; Hughes, P. R.; Granados, R. R. *J. Invertebr. Pathol.* 1991, *58*, 203-210.
21. Granados, R. R.; Corsaro, B. G. In *Vth International Colloquium on Invertebrate Pathology and Microbial Control;* Proceedings and Abstracts, XXIII Annual Meeting of the Society for Invertebrate Pathology: Adelaide, Australia, 1990, pp 174-178.
22. Granados, R. R.; Williams, K. A. In *The Biology of Baculoviruses Volume I Biological Properties and Molecular Biology*; Granados, R. R.; Federici, B. A., Eds.; CRC Press: Boca Raton, FL., 1986, pp 89-108.
23. Groner, A. In *The Biology of Baculoviruses Volume I Biological Properties and Molecular Biology*; Granados, R. R.; Federici, B. A., Eds.; CRC Press: Boca Raton, FL., 1986, pp 177-202.
24. Harper, J. D. In *Epizootiology of Insect Diseases*; Fuxa, J. R.; Tanada, Y., Eds.; John Wiley & Sons, Inc.: New York, NY., 1987, pp 473-496.
25. Hostetter, D. L.; Bell, M. R. In *Viral Insecticides for Biological Control*; Maramorosch, K.; Sherman, K. E., Eds.; Academic Press, Inc.: Orlando, FL., 1985, pp 249-284.
26. Hostetter, D. L.; Puttler, B.; *Environ. Entomol.* 1991, *20,* 1480-1488.
27. Ignoffo, C. M.; Couch, T. L. In *Microbial Control of Pests and Plant Diseases 1970-1980*; Burges, H. D., Ed.; Academic Press, Inc.: New York, NY., 1981, pp 329-362.
28. Jaques, R. P. In *Viral Insecticides for Biological Control*; Maramorosch, K.; Sherman, K. E., Eds.; Academic Press, Inc.: Orlando, FL., 1985, pp 285-360.
29. Knipling, E. F.; Stadelbacher, E. A. *Bull. Entomol. Soc. Am.* 1983, *29* (4) 29-37.
30. Moscardi, F.In *Vth International Colloquium on Invertebrate Pathology and Microbial Control*; Proceedings and Abstracts, XXIII Annual Meeting of the Society for Invertebrate Pathology: Adelaide, Australia, 1990, pp 184-187.
31. Podgwaite, J. D. In *Viral Insecticides for Biological Control*; Maramorosch, K.; Sherman, K. E., Eds.; Academic Press, Inc.: Orlando, FL., 1985, pp 775-797.
32. Shapiro, M. In *Microbial and Viral Pesticides*; Kurstak, E., Ed.; Marcel Dekker, Inc.: New York, NY., 1982, pp 463-492.
33. Shapiro, M. In *The Biology of Baculoviruses Volume II Practical Application for Insect Control*; Granados, R. R.; Federici, B. A., Eds.; CRC Press: Boca Raton, FL., 1986, pp 31-61.

34. Shapiro, M.; Dougherty, E. M.; Hamm, J. J. 1992, U.S.Patent No. 5, 124,149.
35. Shapiro, M.; Robertson, J. L. *J. Econ. Entomol.* 1992, *85*, 1120-1124.
36. Sherman, K. E.; McIntosh, A. H. *Inf. Immunity* 1979, *26*, 227-234.
37. Tanada, Y. *J. Insect Pathol.* 1959, *1*, 215-231.
38. Tanada, Y. *J. Invertebr. Pathol.* 1985, *45*, 125-138.
39. Uchima, K.; Harvey, J. P.; Omi, E. M.; Tanada, Y. *Insect Biochem.* 1988, 18, 645-650.
40. Vaughn, J. L.; Dougherty, E. M.; In *Viral Insecticides for Biological Control*; Maramorosch, K.; Sherman, K. E., Eds.; Academic Press, Inc.: Orlando, FL., 1985, pp 569-633.
41. Yearian, W. C.; Young, S. Y. In *Microbial and Viral Pesticides*; Kurstak, E., Ed.; Marcel Dekker, Inc.: New York, NY., 1982, pp 387-423.

RECEIVED April 30, 1993

Chapter 24

Granulosis Viruses in Insect Control

Louis Falcon

Department of Entomological Sciences, University of California, Berkeley, CA 94720

Over 150 insect species from the order Lepidoptera have been reported to be susceptible to one or more granulosis virus (GV) isolates. Many of the species are important pests (e.g. codling moth, *Cydia pomonella*, imported cabbageworm, *Artogeia* (=Pieris) *rapae*, greasy cutworm, *Agrotis segetum*, potato tuber moth, *Phthorimaea operculella*, Indianmeal moth, *Plodia interpunctella*, etc.). While the GVs have been little used for insect control since being discovered in the 1920s, research is continuing around the globe. In 1981, codling moth GV became the first and, so far, the only candidate GV for USEPA registration. In Europe, two GVs are registered for use and a third is in the Registration Phase. The GVs are attractive for development as microbial insecticides mainly because of their specificity and safety. Many have been successfully multiplied in vivo, a few in vitro, and, it appears, they may be genetically engineered.

The first publication reporting a GV is said to have appeared in 1926 *(1)* when a Frenchman named Paillot described a disease which he called "pseudo-grasserie", from larvae of the European cabbageworm, *Pieris brassicae*. Twenty two years later, Bergold *(2)* determined that such diseases were caused by viruses when he examined GV infected larvae of the European fir budworm, *Choristoneura murinana*. Steinhaus *(3)* was the first to use the term "granulosis" in reference to the minute granules of the virus he saw in the hemocoel of an infected insect larva of the variegated cutworm, *Peridroma margaritosa*. Both Paillot and Steinhaus experimented with, and wrote about the use of these naturally occurring viral agents for insect control. The use of GVs in insect control has been reviewed by several authors *(4-11)*. The classification, structure, composition, gross pathology, replication, cytopathology and pathophysiology of GVs was recently reviewed by Crook *(12)* and Tanada and Hess *(13)*. Approximately 150 GV isolates had been reported by the mid-1980s *(14)*.

0097–6156/94/0551–0336$06.00/0

Current Activities with Granulosis Viruses for Insect Control

A review of the scientific literature produced the last 10 years (1982-1992) provides a panorama of the state of affairs with GVs for insect control. The research and development of GVs, like that of other pest control agents, follows a pattern and a progression. In this chapter, I have chosen to organize the current state of research and development of GVs into three phases: (1) Exploratory Phase; (2) Pre-registration Phase; and (3) Registration Phase.

The Exploratory Phase. This phase encompasses those GVs that have only recently been discovered or rediscovered. For the most part, the research activities are in the descriptive and characterization stages. Most GVs have been identified only on the basis of their morphology and the host species from which they were isolated. They are named on the basis of this information (e.g. *Agrotis segetum* GV = AsGV). As discussed by Crook *(12)*, the system while informative, suffers from a number of disadvantages. For example: (1) the views of insect taxonomists in classifying the host species can change from time to time and be different from one country to another, thus insect names may be different from one country to another [e.g. *Pieris* (Artogeia) *rapae*]; (2) the same GV may be isolated from more than one host species and thus have more than one name [e.g. GVs isolated from *P. brassicae*, *A. rapae* and *A. napi (15, 16)*]; and (3) more than one GV may be isolated from the same species with all having the same name [e.g. 2 isolates from *Pseudaletia unipuncta* (PuGV) *(17)*]. While the host species may be different, the morphology of most all GVs is very similar. Gradually, as more sophisticated methods are developed, they are being employed in the characterization of GVs. These methods include serology, polypeptide analysis, DNA hybridization and restriction enzyme analysis. Thus far, these studies have not uncovered any particular similarities and dissimilarities within the GVs. Crook *(15)* arranged several isolates of PrGV into subtypes on the basis of their restriction profiles. The subtypes corresponded closely to the geographic origins of the isolates depending on where they came from (Asia, Australia, Europe or North America).

The Pre-registration Phase. This phase consists of GVs that are undergoing development and field testing. The purpose of these activities is to determine the insect control potential of the GV. The activities should include sufficient safety testing to ascertain that the candidate GV does not pose a hazard to mammals and other nontarget species. Adequate GV multiplication methods must be in place to provide sufficient quantities of the GV to conduct the desired studies. Quality control and standardization procedures should be employed to ensure the integrity, purity and potency of the GV isolate throughout the testing. In the USA and other countries, an Experimental Use Permit (EUP) must be obtained from the responsible government regulatory bodies in order to conduct field testing on more than 10 acres. During this phase, it may be useful to distribute the candidate GV to other workers so that testing can be done under a variety of conditions and situations.

The Registration Phase. This phase encompasses candidate GVs that have demonstrated sufficient promise to merit consideration for registration and commercialization. It is the final and necessary phase that all microbial agents which are to be sold as microbial insecticides must pass through before they can be commercialized in the USA and other countries of the world. This means, that all of the safety tests deemed necessary to convince the regulatory authorities that a candidate GV does not pose undue hazards to nontarget organisms must be completed satisfactorily. In most situations, it also means that the GV has been shown to be effective and useful. As this can be an expensive and lengthy process, only the most promising and useful GVs may go through this phase.

Current State of Research & Development of GVs

1.**GVs in the Exploratory Phase** (see Note after 3.3)
1.1. *Cacoecia occidentalis*/Tortricidae/ citrus leafroller/CoGV: morphology *(18)* South Africa
1.2. *Chilo sacchariphagus indicus*/Pyralidae/sugarcane/ CsGV: survey *(19)* cross infectivity *(20)* and DNA characterization *(21)* India
1.3. *Cryptophlebia leucotreta*/Tortricidae/false codling moth (citrus, cotton, maize--Africa)/ClGV: biochemical *(22)* physical map *(23)* Germany
1.4. *Epinotia pygmaena*/Tortricidae/ needleminer/EpGV: morphology *(24)* Germany
1.5. *Euxoa auxiliaris*/Noctuidae/army cutworm (polyphagous)/ EaGV: pathology *(25)* USA
1.6. *Hyphantria cunea*/Arctiidae/fall webworm (apple, ash, oak--USA, Europe, Japan)/HcGV: morphology *(26)* Netherlands
1.7. *Iragoides conjuncta*/Cochlidiidae/saddle-back caterpillars/IcGV: morphology *(27)* PRC (=People's Republic of China)
1.8. *Latoia lepida*/Limacodidae/saddleback caterpillars/ (pineapple, coconut, persimmon)/LlGV: morphology *(28)* PRC
1.9. *Lecanobia oleracea*/Noctuidae/tomato moth/LoGV: biochemical *(29)* England
1.10. *Spodoptera litura*/Noctuidae/tobacco leaf caterpillar (cabbage, etc.)/SlGV: morphology *(30)* India
1.11. *Xestia c-nigrum*/Noctuidae/spotted cutworm (grasses)/Xc-nGV: biochemical *(31)* Japan
2.**GVs in the Pre-registration Phase**
2.1. *Adoxophyes orana fasciata*/Tortricidae/fruit tortrix, smaller tea tortrix--cosmopolitan/AoGV: descriptive *(32)* Switzerland; production *(33)* simulated field *(34)* field *(35)* Japan
2.2. *Agrotis segetum*/Noctuidae/greasy cutworm (corn, okra, potato, sugar beet, tobacco, dwarf asters, etc.--cosmopolitan) /AsGV: production *(36)* specificity *(37)* Denmark; field *(38)* Pakistan; production *(39)* field *(40)* Spain; field *(41)* Germany;
2.3. *Artogeia* (=Pieris) *rapae*/Pieridae/imported cabbageworm (crucifers--cosmopolitan)/ArGV: all field-- *(42)* PRC; *(43)* England; *(44)* Canada; *(45)(46)* ROC (=Republic of China); *(47-49)* USA
2.4. *Chilo infuscatellus*/Pyralidae/sugarcane borer/CiGV: field *(19,50-52)* India
2.5. *Choristoneura occidentalis*/Tortricidae/western spruce budworm/ CoGV: field *(53)* Canada
2.6. *Cydia pomonella*/Tortricidae/codling moth (pome fruit, walnut--cosmopolitan)/CpGV: all field-- *(54)* England; *(55,56)* Europe; *(57)* Canada; *(58)* USA; *(59)* Pacific Rim

2.7. *Harrisina brillans*/Zygeaenidae/western grapeleaf skeletonizer/HbGV: field *(60)* USA
2.8. *Homona magnanima*/Tortricidae/oriental tea tortrix/HmGV: field *(35)* Japan
2.9. *Phthorimaea operculella*/Gelechiidae/potato tuber moth/PoGV: all field-- *(61)* Australia; *(62)* Peru; *(63,64)* Tunisia
2.10. *Plodia interpunctella*/Pyralidae/Indianmeal moth/PiGV: all field--*(65-67)* USA
2.11. *Plutella xylostella*/Plutellidae/diamondback moth (crucifers--cosmopolitan)/PxGV: all field-- *(68)* Malaysia; *(45,46)* ROC
2.12. *Pseudaletia unipuncta*/Noctuidae/armyworm (polyphagous--cosmopolitan)/PuGV: synergistic factor, *(69)* Japan
2.13. *Trichoplusia ni*/Noctuidae/cabbage looper (cotton, crucifers, etc.--North America)/TnGV: enhancement factor *(70)* field *(47,71)* USA

3.**GVs in the Registration Phase**

3.1. *Adoxophyes orana*: CAPEX, Switzerland, EUP 12/88 *(56)*
3.2. *Agrotis segetum*: AGROVIR Denmark, 1/90 (notification only, no registration needed) *(56)*
3.3. *Cydia pomonella*: SAN 406, USA, EUP 6/81; DECYDE, USA, EUP 6/84; UCB-87, USA, EUP 6/87; MADEX, Switzerland, Registration 12/87; GRANUPOM, Germany, Registration 3/89 *(56)*; SPECIFIC-T-1, USA, Under review for registration 10/92

-- (Note: cosmopolitan = widely distributed in world; polyphagous = wide host range; field = evaluation of control potential) --

Discussion. Of the twenty four GVs listed above, 11 are classified as in the Exploratory Phase, 13 are placed in the Pre-registration Phase and 3 of the 13 also appear in the Registration Phase. The three GVs listed in the Registration Phase are also listed in the Pre-registration Phase because they are not undergoing registration in all areas where they may be used. It is possible other GV candidates should be listed in the Registration Phase. One is ArGV which may be registered for use in PRC and ROC. When I visited PRC in 1982, I was handed a package of ArGV and told it was being used for insect control by farmers in local communes.

Nine families are represented with the Tortricidae and the Noctuidae each having 7 species, the Pyralidae recorded 3, and there are 6 families with 1 species each. Most, if not all species, are pests of some economic importance. At least 10 are notorious pests including 1.10, 2.1, 2.2, 2.3, 2.5, 2.6, 2.10, 2.11, 2.12 and 2.13. The information sources are from many parts of the world with all continents except Antarctica represented. While not all of the potentially useful GVs are found in the list above, the information demonstrates where the GV research and development activities appear to be at this time. The results of this survey suggest there is continued, and possibly increased interest and activity in the research and development of GVs for insect control around the world. This certainly appears true for GVs in the Registration Phase, because GVs only reached this phase for the first time beginning with CpGV in 1981. Since then, 2 more GVs (AoGV and AsGV) were added.

The first GV to be registered was CpGV in Switzerland in 1987, 24 years after its discovery in 1963. The AsGV product (AGROVIR), which may be used by growers for insect control in Denmark, became available in 1990, 55 years after the GV was discovered by Paillot in 1935 (72). The only other GV to be listed in the Registration Phase is AoGV which was issued an EUP in 1988 by the regulatory authorities in Switzerland. AoGV was discovered in 1974, so it took 14 years to obtain an EUP. The research and development process to get other microbial agents into the registration mode has been done in a lot less time. For example, the bacterium, *Bacillus thuringiensis israeliensis* was discovered in 1977 and about five years later it was registered for insect control in the USA. *Heliothis zea* NPV was registered in 1975, about 10 years after the process began. The US Forest Service followed a similar course with the three NPVs it registered during the 1970s. The factors that determine the progress to registration of any pest control agent as a pesticide are many including: need, efficacy, usefulness, market potential, ease of production, packaging and storage requirements, and the execution of the research needed to provide the answers to these questions.

The Development of CpGV for Insect Control

CpGV is the most advanced of the GVs and it appears useful to examine the research and development of this GV in more detail.

The Pre-registration Phase. CpGV was discovered, described and characterized in the mid-1960s at the University of California @ Berkeley (UCB) *(73)*. Preliminary field tests were conducted in California in the late 1960s *(74)*. The field results were sufficiently encouraging and CpGV samples were made available to other researchers around the world. In the beginning, each researcher produced the CpGV needed for field tests in their own laboratory *(75-77)*. Beginning in 1973, an intensive and continuous CpGV research program was initiated at the Institute for Biological Pest Control, Darmstadt, Germany. Methods for the mass production of CpGV were developed. Subsequently, most of the CpGV used in field trials conducted in many European countries under the auspices of the IOBC/WPRS working group during 1976 to 1977 was produced at Darmstadt *(78)*. Based on the methods developed at Darmstadt, additional CpGV production was established at the Glasshouse Crop Research Institute in Littlehampton, UK, and at INRA in Montfavet-Avignon, France. A new phase started in 1979 with a program of the European Communities (CEC) on "Biological Control in Apple Orchards". During 4 years CEC supported CpGV research projects in France, Germany, Greece, the Netherlands, and the United Kingdom. A vast amount of information on production, standardization, formulation and application of CpGV was gathered *(79,80)*.

The Registration Phase in the USA. In 1980, the UCB-CpGV Program headed by L. Falcon, provided Sandoz, Inc., USA, with CpGV and start-up technology. Sandoz developed an experimental product labelled "SAN 406" and USEPA issued an EUP with an exemption from a tolerance for it in 1981. The product received wide distribution and was tested by several researchers in more than two dozen orchards in the USA *(59)* and Canada *(57)* and 10 European countries *(56)* from 1981 through 1986. The results were useful and encouraging and set the stage for the commercialization of CpGV. Although Sandoz, Inc. withdrew from the insect virus insecticide business in 1982, the UCB-CpGV Project renewed the CpGV EUP. In 1984 the UCB-CpGV Project provided CpGV and start-up technology to MicroGeneSys, Inc., a new biotechnology company. Their product was named DECYDE and it was tested under the extended EUP originally issued in 1981. Again, the product received wide distribution and testing and much useful information was developed. Two years after they began, MicroGeneSys, Inc. stopped development of DECYDE to concentrate fully on AIDS research. Having no other potential private companies interested in registering and commercializing CpGV, the UCB-CpGV Project, with encouragement and support from the IR-4 Project, decided to proceed with the registration of CpGV. In 1987, it was encouraged to go forth with the registration effort by a group of interested fruit growers who organized a nonprofit company named the Association for Sensible Pest Control (ASPC) to help finance the UCB-CpGV Project. The experimental CpGV product was designated UCB-87, the registered product will be called SPECIFIC-T-1.

The Registration Phase In Europe. In France, INRA began a multi-institute program for IPM in apple orchards based on control of codling moth with CpGV. A standardized product named CARPOVIRUSINE was developed between INRA and Solvay, S.A. This preparation was tested in several orchards in France, Israel and Italy *(81)*. When Solvay, S.A. lost interest in the project a new contract was issued in 1989 between INRA and Calliope, S.A. Calliope, a small company located in southern France, already had other insect virus insecticides in its sales program. At about the same time in the United Kingdom, Agricultural Genetics Company developed a CpGV product designated AGC 200 which was short-lived *(82)*. In Switzerland, a small company, Andermatt-Biocontrol AG, developed a CpGV preparation called MADEX which was registered in 1987. Starting in 1984, Hoechst AG in Germany began production of CpGV for field trials in different climatic areas of Germany. A CpGV product named GRANUPOM was registered in Germany in 1989. The commercial product is produced by a small company, the "Institute fur Umweltanalytik" in Pforzheim, Germany *(83)*.

The Registration Process in the USA. The CpGV registration process in California was started in 1987 by the UCB-CpGV Program. Testing protocols were developed using the then current *USEPA Pesticide Testing Guidelines, Subdivision M, Microbial and Biochemical Pest Control Agents (Part A Microbial)* which was in draft form at the time. The 1981 version was used until the 1987 draft edition became available. In 1987, it was determined there were no commercial laboratories in the USA with relevant experience available to perform the Tier 1 safety testing required for the registration program. Therefore, the animal tests were performed at the University of California @ Davis, Animal Resources Service (ARS) by qualified veterinarians under strictly controlled experimental conditions. The following toxicity/pathogenicity tests were conducted between April 1987 and January 1989: acute oral; acute dermal; dermal irritation; primary eye; acute intravenous; and acute pulmonary. Tissues from all tests were stored at -70 C. and storage stability testing was performed. When qualified commercial laboratories became available in 1989, the dermal toxicity and the acute pulmonary tests were repeated to reflect changes in the *Subdivision M Guidelines*. These were revised several times between 1986 and March 1989, when the final version was produced. Qualified commercial laboratories were used to conduct the wildlife and non-target insect tests from April 1989 through February 1991. For the wildlife studies, toxicity/pathogenicity studies were performed on 3 species: bobwhite quail; rainbow trout; and *Daphnia magna*. The insect studies utilized 4 species: green lacewing, *Chrysopa carnea*; ladybird beetle, *Hippodamia convergens;* the parasitic hymenopteran, *Uga menoni*; and the honey bee, *Apis mellifera*. Good Laboratory Practices were followed throughout all the studies *(Pesticide Programs; Good Laboratory Practice Standards; Final Rule; Part IV; November 29,1983*; Environmental Protection Agency).

As a part of the safety testing protocols, the fate of all CpGV in animal tissues had to be determined, and the potency of all CpGV inocula verified. In 1987, bioassay (Falcon & Berlowitz, unpublished) and Enzyme Linked Immunosorbent Assay (ELISA) were compared to assess their sensitivity in the detection of CpGV in animal tissue (Berlowitz and Falcon, unpublished). Also, the use of DNA probes was explored (before Polymerase Chain Reaction technology was available). The bioassay method was chosen because 1) it is the most sensitive for detecting infectious viral DNA, and 2) it detects viable virus, not just viral protein. About 25,000 neonate codling moth larvae were used in 160 animal tissue and 87 concurrent dose bioassays to develop the data needed for the CpGV registration package submitted to USEPA in 1992 (Berlowitz and Falcon, unpublished data). The cost of the UCB-CpGV registration program (salaries+depreciation+overhead +supplies+equipment) was ca. US$2.5M for the 6 year period that ended in mid-1992 (Falcon, unpublished).

The Commercialization and Use of CpGV for Insect Control. Field experiments with SPECIFIC-T-1 have demonstrated it can be used effectively to control the codling moth in a variety of ways and situations (Falcon and Berlowitz, unpublished). The product can be used throughout the season as part of a certified organic crop production program. It may be integrated with all other pest control agents and methods throughout the growing season. It may be used together with pheromone confusion to enhance the overall effectiveness of both control methods. Where codling moth resistance to azinphosmethyl or other chemical insecticides is a problem, the addition of SPECIFIC-T-1 may ameliorate the resistance problem. CpGV products may be expanded to control insects other than the codling moth. The name SPECIFIC-T-1 emphasizes that CpGV is specific to the Tortricidae (=T) and it is the first (=1) of a product line of GVs to be registered by USEPA. The specificity of CpGV which, like GVs in general, is infective only for members of its own taxonomic family, the Tortricidae {*Grapholitha molesta (74)*, *Cydia nigricana (84)*, *Rhyacionia buoliana*, *G. funebrana*, *Lathronympha strigana (85)*, *R. frustrana (86)* and *Cryptophlebia leucotreta (22)*}. All of these species can be pests and may need to be controlled at one time or another. SPECIFIC-T-1 may be used or new GV products derived from CpGv may be developed for each species. This would broaden the market potential for CpGV and enhance its economic position.

The University of California and ASPC will be the registrants of the CpGV product SPECIFIC-T-1 when it is registered by USEPA. The research and development of CpGV will be continued in an effort to improve formulations, extend effective field life, keep costs and market prices reasonable and competitive and develop new uses for CpGV products. Every effort will be made to establish CpGV products as readily available, useful and reliable. Private industry will be invited to participate as licensees, in the research, development and marketing of CpGV products.

CpGV in the Future. Initial success multiplying CpGV in vitro was reported by Naser et al (87) and Miltenburger et al (88). More recently, Winstanley and Crooke (89) selected a cloned cell line from embryonic cell cultures which supported the replication of CpGV. This information will be used when they explore the feasibility of producing a genetically-modified CpGV which, when ingested by the codling moth larva, would result in reduced feeding damage (90). A model of the effect of CpGV on larvae was developed by Brain & Glen (91). Research is also continuing to investigate the encapsulation of CpGV to extend its residual field life (Falcon & Berlowitz unpublished). Combining CpGV with other pest control agents has resulted in improved performance of the virus (Gallo & Falcon, unpublished). New strains of CpGV may be found which may allow for the establishment of patents such as was done recently by the USDA with new strains of gypsy moth and celery looper nuclear polyhedrosis viruses. A less costly CpGV production method for CpGV was reported by Reiser & Groner (92) using larvae of *C. leucotreta*.

Suggestions for Accelerating the Development of GVs for Insect Control

As discussed in this paper, of the more than 150 insect species reported to be susceptible to one or more GV isolates, at least 24 insect species and their respective GV isolates, have received attention from scientists in various parts of the globe during the past 10 years. Of the 24 GVs listed, three are in the Registration Phase with the first one appearing in 1981. While several other GVs are good candidates for the Registration Phase, it has taken an inordinate amount of time after discovery for a GV isolate to reach the Registration Phase. If the potential demonstrated by GVs for insect control is to be realized, something must be done to accelerate the development and registration of all promising GV candidates. By so doing, GVs can assume their rightful place in the arsenal of weapons available to farmers and others needing to control pests around the world. The urgency has never been greater because this arsenal of pest control weapons is diminishing rapidly as more and more pest species develop resistance to once effective chemical insecticides, and to the microbial insecticides based on *B. thuringiensis*; as more and more useful chemical insecticides are being removed from the market for regulatory and environmental reasons; as fewer and fewer new chemical insecticides are being developed. A product vacuum is forming between the loss of once useful insecticides and the new pest control solutions that people hope/expect will be provided by molecular biology and genetic engineering. The most immediate way to fill the product vacuum is to make available in a useful form, the many products offered by nature for combatting insect pests. The GVs, as well as the NPVs and other microbial agents, represent a tremendous arsenal of raw materials to be exploited for insect control. Many can be converted into useful pest control agents rather quickly and relatively easily. In addition to providing insect control, the products derived from GVs and many other microbial agents will not pollute the environment and will not injure or destroy nontarget species.

In order to speed up the Registration Process, of GV-based microbial insecticides, regulatory agencies must be encouraged to accept the facts. The facts are that, of all the safety tests conducted with GVs and NPVs (=Baculoviruses) in the world, I am unaware of evidence showing that the GVs or NPVs examined are a hazard to living things other then one or a few insect species. Thus, I believe it is reasonable that regulatory agencies could permit a "generic registration" whereby all GVs and NPVs would not need to undergo the costly and time consuming animal and wildlife testing currently required in the USA and other countries of the world. What I believe needs to be regulated, is GV/NPV insecticide product purity, integrity and quality. That is, the producer/vendor of a GV product insecticide to be sold for insect control is responsible under penalty of the law, that the virus product contains only the microbial agents declared on the label. GV product integrity requires that the GV isolate in the product be maintained, i.e. not allowed to degrade, and not be lost or replaced by another strain during production and storage. Quality control necessitates maintaining a minimum potency of GV in the product, and uniformity of the GV product so that it behaves the same way each time it is used.

Organization and funding is required to research, develop, register and commercialize GVs for insect control. With registration requirements reduced to a minimum, private firms are more apt to invest in GV product development. While markets for individual GV products may be small, companies may increase income by producing a wide selection of GVs. Some GVs such as those needed only sporadically, could be produced in relatively small quantities and stored until needed. Other GVs could be produced on a seasonal basis. Once GVs are made available in quantities, researchers and users will have greater opportunities to refine and enhance the ways the GVs may be used for insect control. Most of the biological problems that interfere with commercialization (e.g. in vivo multiplication, bioassay methods, formulation, application, etc.) have been dealt with in various ways and much information is available in the scientific literature. The available information can be modified for specific situations, and new information developed where needed. This is possible within the framework of the industries that may become involved. In my opinion, it is this type of action that is needed now if GVs are ever to be developed as useful insect control agents.

Acknowledgment

The author thanks Staff Research Associate Arthur Berlowitz for his assistance in preparing this chapter.

Literature Cited

1. Paillot, A. *C. R. Acad. Sci.* **1926**, *182*, pp. 180-182.
2. Bergold, G. H. *Z. Naturforsch.* **1948**, *3b*, pp. 338-342.
3. Steinhaus, E. A. *Science* **1947**, *106*, pp. 323-327.

4. Huger, A. In *Insect Pathology, An Advanced Treatise*; Steinhaus, E. A. Ed.; Academic Press, New York, **1963**, Vol. 1 pp. 531-575.
5. World Health Organization. *Tech. Rep. Ser.*, **1973**, *531*, 150 pp.
6. David, W. A. L. *Ann. Rev. Entomol.* **1975.** *20*, pp. 97-117.
7. Tinsley, T. W. *Ann. Rev. Entomol.* **1979**, *24*, pp. 63-87.
8. Payne, C. C. *Parasitology*, **1982**, *84*, pp. 35-77.
9. Payne, C. C. *Fortschr. Zool.* **1986**, *32*, pp. 183-225.
10. Huber, J. In *The Biology of Baculoviruses*; Granados, R. R. & Federici, B. A. Eds.; CRC Press, Boca Raton, **1986**, *2*, pp. 181-221.
11. Cunningham, J. C. *Outlook Agric.* **1988**, *17*, pp. 10-15.
12. Crook, N. E. In *Viruses of Invertebrates*; Kurstak, E. Ed,; Marcel Dekker, Inc., New York, **1991**, pp. 73-110.
13. Tanada, Y.; Hess, R. T. In *Atlas of Invertebrate Viruses*, Adams, J. R.; Bonami, J. R. Eds. **1992**, CRC Press, Boca Raton, FL, pp.227-257.
14. Martignoni, M. E.; Iwai, P. J. *A Catalog of Viral Diseases of Insects, Mites, and Ticks*, 4th ed., Gen. Tech. Rep. PNW-195, U. S. Department of Agriculture, Forest Service, Pacific Northwest Research Station, Portland, OR, **1986**, 51 pp.
15. Crook, N. E., *J. Gen. Virol.* **1986**, *67*, pp. 781-785.
16. Crook, N. E., *Virology* **1981**, *115*, pp. 173-177.
17. Harvey, J. and Tanada, Y. *J. Invertebr. Pathol.* **1985**, *46*, pp. 174-178.
18. Smith, D. H.; DaGraca, J.V.; Whitlock, V. H. *J. Invertebr. Pathol.* **1990**, *55*, pp. 319-324.
19. Easwaramoorthy, S.; Jayataj, S. *Trop. Pest Manage.*, **1987**, 33, pp. 200-201, 249-250, 254.
20. Easwaramoorthy, S.; Jayataj, S. *J. Entomol. Res.* **1987**, *11*, pp. 170-174.
21. Easwaramoorthy, S.; Cory, J. S. *Archives of Virology* **1990**, *110*, pp. 113-119.
22. Fritsch, E.; Huber, J.; Samson, R. A.; Vlak, J. M.; Peters, D. Eds., Foundation 4th International Colloq. Invertebr. Pathol., Wageningen, The Netherlands, **1986**, pp. 112.
23. Jehle, J. A.; Backhaus, H.; Fritsch, E.; Huber, J. *J. Gen. Virol.* **1992**, *73*, pp. 1621-1626.
24. Maksymov, J. K.; Bachler, M.; Wehrli, E; *Anz. Schad. Pflan. Umwelt.* **1989**, *62*, pp. 123-127.
25. Jackson, J. J.; Sutter, G. R. *J. Kans. Entomol. Soc.* **1985**, *58*, pp. 353-355.
26. Chukhrii, M. G.; Bondar, Y. S.; Bondar, E. S. *Biological and Chemical Methods of Plant Protection*; Popushoi, I. S. Ed., Balkema, A. A., Rotterdam, **1988**, pp. 48-52.
27. Sun, F. R.; *Ping Hsueh Chih Virol. Sin.* **1988**, *1*, pp. 59-63.
28. Wu, J. F.; Tai, G. Q.; Shi, M. B.; Huang, Z. H.; Zian, B. C. *Phytophylacica Sin.* **1983**, *10*, pp. 69-70.
29. Foster, G. N.; Crook, N. E. *Bull. IOBC, Wageningen*, **1982**, *6*, pp. 163-166.
30. Narayanan, K. *Current Sci*, **1985**, *54*, pp.128 8-1289.
31. Goto, C.; Minobe, Y.; Iizuka, T. *J. Gen. Virol.* **1985**, *73*, pp. 1491-1497.
32. Schmid, A.; Cazelles, O.; Benz, G. *Bull. Soc. Entomol. Suisse* **1983**, *56*, pp. 225-235.
33. Sato, T. *Shokubutsu-boeki* (Plant Protection), **1984**, *38*, pp. 366-369.
34. Sekita, N.; Kawashima, K.; Aizu, H.; Shirasaki, S.; Yamada, M. *Japanese Society of Applied Entomology and Zoology* **1984**, *19*, pp. 498-508.

35. Sato, T.; Oho, N.; Kodomari, S. *Jap. Agric. Res. Q.* **1986**, *19*, pp. 271-275.
36. Bolet, B. *Dan. J. Plant Soil Sci.* **1983**, *87*, pp. 417-424.
37. Zethner, O.; Ogaard, L. *Dan. J. Plant & Soil Sci.* **1982**, *86*, pp. 471-475.
38. Zethner, O.; Khan, B. M.; Chaudry, M. I.; Bolet, B.; Khan, S.; Khan, H.; Gul, H.; Ogaard, L.; Zaman, M.; Nawaz, G. *Entomophaga*, **1987**, *32*, pp. 449-455.
39. Carranza, P. *Universidad de Cordoba*, **1989**, 107pp.
40. Caballero, P.; Vargas-Osuna, E.; Santiago-Alvarez, C. *J. Appl. Entomol.* **1991**, *112*, pp. 59-64.
41. Geissler, K.; Schliephake, E. *Arch. Phytopath. and Plt. Prot.* **1991**, *27*, pp. 79-80.
42. Lui, N.; Liang, D. *Proc. Chinese Aca. Sci. -U. S. Nat. Aca. Sci. Joint Symp. Biol. Cntrl. Insects*, Adkisson, P. L.; Shijun, Ma. Eds. **1982**, Science Press, Beijing, China, pp. 401-407.
43. Tatchell, G. M.; Payne, C. C. *Entomophaga*, **1984**, *29*, pp. 133-144.
44. Jaques, R. P. *Can. Entomol.* **1988**, 120, pp. 575-580.
45. Su, C. Y. *Plt. Protect. Soc. Rep. China*, **1987**, *29*, pp. 293-296.
46. Su, C. Y. *Plt. Protect. Soc. Rep. China*, **1987**, *29*, pp. 387-399.
47. Tompkins, G. J.; Linduska, J. J.; Young, J. M.; Dougherty, E. M. *J. Econ. Entomol.* **1986**, *79*, pp. 497-501.
48. Webb, S. E.; Shelton, A. M. *Environ. Entomol.* **1991**, *19*, pp. 111-116.
49. Webb, S. E.; Shelton, A. M. *Entomophaga* **1991**, *36*, pp. 379-389.
50. Easwaramoorthy, S.; Jayataj, S. *Trop. Pest Manage.*, **1988**, *34*, pp. 426-428.
51. Easwaramoorthy, S.; Jayataj, S. *Trop. Pest Manage.*, **1989**, *35*, pp. 352-353.
52. Easwaramoorthy, S.; Jayataj, S. *Trop. Pest Manage.*, **1991**, *37*, pp. 134-137.
53. Otvos, I. S.; Cunningham, J. C.; Kaupp, W. J. *Can. Entomol.* **1989**, *121*, pp. 209-217.
54. Richards, M. G. *Proc. 4th International Colloq. Invertebr. Pathol.*, Wageningen, The Netherlands, **1986**, pp. 95-98.
55. Burgerjon, A. *Proc. 4th International Colloq. Invertebr. Pathol.*, Wageningen, The Netherlands, **1986**, pp. 102-105.
56. Huber, J.; *Brighton Crop Protection Conference*, Brighton, England, **1990**, pp. 601-606.
57. Jaques, R. P.; Laing, J. E.; Laing, D. R.; Yu, D. S. K. *Can. Entomol.* **1987**, *119*, pp. 1063-1067.
58. Vail, P. V.; Barnett, W.; Cowan, D. C.; Sibbett, S. *J. Econ. Entomol.* **1991**, *84*, pp. 1448-1453.
59. Falcon, L. A.; Berlowitz, A. *Proc. 4th International Colloq. Invertebr. Pathol., Wageningen, The Netherlands,* **1986**, 99-101.
60. Stern, V. M.; Federici, B. A. *Calif. Agric.* **1990**, *44*, pp. 21-22.
61. Briese D. T.; Mende, H. A. *Bull. Entomol. Res.* **1981**, *71*, pp. 11-18.
62. Raman, K. V.; Alcazar, J. *Asian Potato Assoc. (APA) Proc.* **1988**, 173-174.
63. von Arx, R.; Gebhardt, F. *Entomophaga.* **1990**, *35*, 151-159.
64. Bensalah, H.; Aalbu, R. In *Agriculture, Ecosystems and Environment.* Elsevier Science Publishers B. V., Amsterdam, **1992**, pp. 119-126.
65. Kellen, W. R.; Hoffman, D. F. *Environ. Entomol.*, **1987** *16*, pp.1231-1234.
66. Cowan, D. K.; Vail, P. V.; Kok-Yokomi, M. L.; Schreiber, F. E. *J. Econ. Entomol.* **1986**, *79*, pp. 1085-1090.

67. Vail, P. V., *J. Stored Products*, **1991**, *27*, pp. 103-107.
68. Abd Kadir, H. In *First Regional Symposium on Biological Control*; Hussein, M. Y.; Ibrahim, A. G. Eds. Universiti Pertanian Malasia, Serdang, **1985**, pp. 261-273.
69. Hukuhara, T.; Zhu, Y. *J. Invertebr. Pathol.* **1989**, *54*, pp. 71-78.
70. Gallo, L. G.; Corsaro, B. G.; Hughes, P. R.; Granados, R. R. *J. Invertebr. Pathol.* **1991**, *58*, pp. 203-210.
71. Dougherty, E. M.; Kelly, T. J.; Rochford, R.; Forney, J. A.; Adams, J. R. *Physiol. Entomol.* **1987**, *12*, pp. 23-30.
72. Paillot, A. *C. R. Acad. Sci. Paris*, **1935**, *201*, pp. 1062-1066.
73. Tanada, Y. *J. Invertebr. Pathol.* **1964**, *6*, pp. 378-380.
74. Falcon, L. A.; Kane, W. R.; Bethell, R. S. J. Econ. Entomol. **1968**, *61*, pp. 1208-1213.
75. Morris, D. S. *Abstracts 14th International Congress of Entomology*, Canberra, Australia, **1974**, pp. 238.
76. Keller, S. *Zeitschrift fur Angewandte Entomologie,* **1973**, *73*, pp. 137-181.
77. Dickler, E. *Mitteilungen der Biologischen Bundesanstalt fur Land- und Fortwirtschaft* (Berlin-Dahlem), **1978**, *180*, 120pp.
78. Huber, J. *Mitteilungen der Deutschen Gesellschaft fur Allegmeine und Angewandte Entomologie,* **1981**, 2, pp. 141-145.
79. Cavalloro, R.; Piavaux, A. *C. E. C. Programme on Integrated and Biological Control*, EUR 8273 EN, Office for European Communities, Luxembourg, **1983**, 349pp.
80. Cavalloro, R.; Piavaux, A. *C. E. C. Programme on Integrated and Biological Control*, EUR 8689, Office for European Communities, Luxembourg, **1983**, 476pp.
81. Burgerjon, A.; Sureau, F. *Proc. 5e Colloq. sur les recherches fruitieres-Bordeaux*, **1985**, pp. 53-66.
82. Ballard, J. *Grower*, pp. **1988**, pp. 14-15.
83. Huber, J. *Proc. 5th International Colloquium on Invertebrate Pathology and Microbial Control*, Adelaide, Australia, **1990**, pp. 424-427.
84. Payne, C. C. *J. Invertebr. Pathol.* **1981**, *38*, pp. 71-77.
85. Huber, J. *Proceedings Third International Colloquium on Invertebrate Pathology*, **1982**, Brighton, U. K., 119.
86. Stephen, F. M.; Yearian, W. C.; Young, S. Y.; Wallis, G. W. *J. Georgia Entomol.* **1982**, *17*, pp. 398-402.
87. Naser, W. L.; Miltenburger, H. G.; Harvey, J. P.; Huber, J.; Huger, A. M. *Microbiological Letters*, **1984**, *24*, pp. 117-121.
88. Miltenburger, H. G.; Naser, W. L.; Harvey, J. P.; Huber, J.; Huger, A. M. *Zeitschr. Naturfor. Reihe C,* **1984**, *39*, pp. 993-1002.
89. Winstanley, D.; Crook, N. E.; *Proc. 5th International Colloquium on Invertebrate Pathology and Microbial Control*, Adelaide, Australia, **1990**, pp. 431-433.
90. Winstanley, D.; Crook, N. E.; *Aspects of Applied Biology*, Monograph Series, **1990**, *24*, pp. 11-15.
91. Brain, P.; Glen, D. M. *Ann. Appl. Biol.*, **1989**, *11*, pp. 129-140.
92. Reiser, M.; Groner, A. *Proc. 5th International Colloquium on Invertebrate Pathology and Microbial Control*, Adelaide, Australia, **1990**, pp. 444.

RECEIVED April 30, 1993

Chapter 25

Recombinant Baculovirus Expressing an Insect-Selective Neurotoxin

Characterization, Strategies for Improvement, and Risk Assessment

B. F. McCutchen and Bruce D. Hammock

Departments of Entomology and Environmental Toxicology, University of California, Davis, CA 95616

With the advent of recombinant DNA technology, we now possess the ability to improve the insecticidal properties of naturally occurring microorganisms. We have engineered a recombinant nuclear polyhedrosis virus to express an insect-selective neurotoxin, AaIT, which significantly increases its insecticidal value. Toxicological studies indicate a 30 to 40% reduction in the time to kill host insects as well as a dramatic reduction in feeding. We propose several strategies to further enhance the insecticidal properties of these viruses, including pharmacological, chemical, molecular, and entomological approaches. Furthermore, we assess the risks associated with developing these recombinant viruses for use as pest control agents.

Synthetic chemicals are an integral component of modern agriculture, and these chemical agents have provided an efficient and effective means to greatly increase the yields of many crops. It is generally accepted that artificial pest control is necessary to maintain our current level of agricultural productivity (*1-3*). However, over reliance on non-selective pesticides has lead to resistance, pest resurgence, destruction of non-target organisms, contamination of the environment, and a reduction in profitability. As a result of these problems alternative control strategies have been explored, including one of the most promising, biological control. Agricultural industry now perceives biological control agents as attractive supplements or replacements for synthetic pesticides in integrated pest management programs.

Biological control, by definition, is the action of predators, parasites, or pathogens in maintaining another organism's population density at a lower level than would occur in their absence (*4*). Pathogens can be used to regulate pest populations through the intentional creation of epizootics or by the utilization of naturally occurring epizootics. This practice is commonly referred to as microbial

0097–6156/94/0551–0348$06.00/0

control (*5*). One particularly promising practice of microbial control is the inundative release of microorganisms to achieve relatively rapid suppression of a pest population. However, many of the possible entomopathogens that might be considered, lack the quick killing action that we have grown accustomed to with the classical synthetic insecticides. This killing property and/or the "quick kill" mentality may eventually be addressed through research leading to agricultural practices which are conducive for epizootiological control of pests. However, this framework is not currently available for most row-crop agriculture (*5,6*). Alternatively, we now possess the ability to change the properties of some of these entomopathogens. One approach to microbial insect control is genetic engineering, because biotechnology offers tools to enhance the insecticidal properties of naturally occurring microorganisms. For instance, the genes that encode toxins, hormones, or enzymes can be cloned and inserted into appropriate microorganisms to produce recombinant insecticides. Potential vectors for such genetic manipulation include insect-specific viruses from the family Baculoviridae.

Baculovirus Characteristics

Most baculoviruses isolated to date have been from the insect order Lepidoptera; however, these viruses have been isolated from other orders including Coleoptera, Hymenoptera, Diptera, Trichoptera and from the non-insect order of decapod crustaceans, Decapoda. Baculoviruses are pathogenic to invertebrates and are characterized by having double-stranded, circular DNA genomes ranging from 80 to 200 kilobases in size. The family Baculoviridae has a single genus, Baculovirus, which is divided into three subfamilies including the non-occluded baculoviruses (NOVs), the granulosis viruses (GVs), and the nuclear polyhedrosis viruses (NPVs). The NOVs are characterized by the absence of an occlusion body surrounding the virion. The GVs are comprised of a single virion encased in the occlusion body, while the NPVs usually have several virions surrounded by an occlusion body.

For the NPVs the occlusion body is a protein matrix referred to as a polyhedron. The polyhedra range from 1-10 μm in size and are comprised mainly of a single protein, polyhedrin. In the infection process the polyhedra are ingested by the insect and passed to the midgut. The alkaline conditions in the midgut solubilize the proteinacious coat releasing the virions. These virions, referred to as occluded virions, are nucleocapsids containing the viral DNA genome which attach to and infect the epithelial cells of the midgut (*7,8*). These occluded virions are responsible for the horizontal transmission of the virus between insects.

Upon infection of cells a second virion phenotype is produced which is referred to as a non-occluded virion or budded virion. These budded virions are produced when nucleocapsids move to the plasma membrane, shed the nuclear derived coat in the cytoplasm, and bud through the cytoplasmic membrane into the hemocoel. This process is repeated in subsequent cells leading to a systemic infection within the insect. In most cases the infected insect will continue to develop, sometimes for weeks, allowing the virus to propagate and exponentially increase in number over this period of time. Late in the infection process of most cell types, virions become occluded in order to facilitate horizontal transmission of the virus. It has been estimated that up to 10^{10} polyhedra can be produced and can account for 30% of the dry weight of the insect at death (*9*). Characteristically, an infected larva will disintegrate following death facilitating dissemination of the polyhedra.

Although certain GVs and NOVs have been studied, NPVs are the most thoroughly characterized due in part to their pharmaceutical value and due to the availability of cell culture and cloning systems. The NPVs have been successfully employed as highly efficient eukaryotic expression vectors (*10-14*). Efforts have been focused on expression of foreign proteins utilizing NPVs, especially *Autographa californica* NPV. These efforts have resulted in the rapid development of established insect cell lines, transfer vectors, and techniques for viral isolation and purification. Not only have these advances resulted in an expression system that provides high yield and biologically active, recombinant proteins, but indirectly this technology has resulted in the ability to modify these insect-specific viruses to have increased insecticidal properties.

NPV Characteristics. Not only does the encapsulation of the occluded virions facilitate horizontal transmission, but it dramatically enhances stability of the virus by providing protection from environmental factors such as ultraviolet light. The NPVs have long been considered as potential biopesticides due to their many advantageous characteristics including their environmental compatibility. In fact, NPVs and other baculoviruses are often important in the natural regulation of pest insects. Several viruses have been registered as biological insecticides, although their use has been limited (*15*).

Another attribute of the NPVs is their host specificity. Viruses from the family Baculoviridae are found only in arthropods and generally possess narrow host ranges even within an insect order. Host specificity studies utilizing electron microscopy, DNA hybridization, and recombinant DNA technology have demonstrated that the baculovirus lacks the ability to transfer viral DNA into the mammalian cell nucleus even when an established cell line is used for infection (*16-18*). In addition, extensive studies have shown that baculovirus insecticides are safe for non-target species and the environment (*19,20*).

In spite of these advantages, a variety of disadvantages have limited their use in row-crop agriculture including field instability, problems associated with mass production and a limited host range. Probably the single biggest obstacle curtailing the use of baculoviruses in row-crop agriculture is the virus' inability to act quickly, thereby allowing an infected larva to continue to feed and develop. As a natural control agent this 'slow kill' characteristic allows the virus to replicate to tremendous numbers. Although this would be considered an attribute in a natural control strategy, this trait is a severe limitation in row-crop agriculture, and as a result baculoviruses can not compete with the synthetic insecticides.

Genetic Modification of Nuclear Polyhedrosis Viruses

In order to increase the speed of kill, it has been suggested that NPVs could be engineered to express genes encoding proteins such as toxins, hormones, or enzymes to enhance the insecticidal properties of the NPV (*8,11,21-25*). NPVs possess several characteristics that make them favorable candidates for recombinant modification. The production of NPVs both *in vitro* and *in vivo* has been well studied with respect to extraction and harvest methods, quality control, and environmental impacts.

An ideal insecticide should be tailored to physiological targets specific to insects. We propose utilization of insect-selective NPVs to express an insecticidal protein which will accelerate the killing action of the virus. Actually, the NPV is used as a highly selective mediator to introduce an insecticidal protein by overcoming the natural defense mechanisms of the insect. Presently, it is not possible to introduce these insect-selective proteins to insects in the field in sufficient amounts due to the instability of the protein in the environment, degradation in the host gut, and inability to penetrate the insect cuticle. These problems, which are not as prevalent with classical insecticides, may be circumvented by utilizing the special properties of the NPVs. The development of such recombinant NPVs should be very compatible with any integrated pest management program. Furthermore, we can minimize the risk to non-target organisms by introducing proteins which we have thoroughly characterized and which are known to be insect selective. In fact, a number of research teams have embarked on closely related studies in an attempt to enhance the rate of kill of host insects through genetic modification of the NPV genome (Table I: *22,23,26-40*).

Recombinant Baculoviruses Affecting the Insect Endocrine System Several recombinant viruses have been constructed to express insect neurohormones, although only one factor has been reported to produce an increased insecticidal effect. Maeda *et al.* (*22*) incorporated a synthetic gene that encodes for the diuretic hormone from *Manduca sexta* into the *Bombyx mori* NPV (BmNPV) genome in an attempt to disrupt the osmoregulation of infected insects. Over-expression of the peptide resulted in an approximate 20% reduction in the time to kill hosts when compared to wild-type BmNPV. Recently, Eldridge *et al.* (*26*) have incorporated the gene that encodes for eclosion hormone from *M. sexta* into AcNPV. Although expression of eclosion hormone was confirmed and the hormone was biologically active, the rate of kill by the recombinant virus was not enhanced.

The first attempt to express an insect enzyme in NPV was reported by Hammock *et al.* (*23*) in which a gene encoding for juvenile hormone esterase (JHE) was inserted into AcNPV. The expressed enzyme significantly reduced the amount of diet consumed in first instar larvae of *Trichoplusia ni*, and ultimately inhibited growth of these insects. However, this effect could not be duplicated in other instars of *T. ni*. Recently, site-directed mutagenesis has been used to modify the cDNA of the JHE gene in an attempt to stabilize the enzyme *in vivo*. Expression of the modified JHE has resulted in an esterase with increased insecticidal activity, and as a result the recombinant AcNPV expressing this modified JHE significantly reduces the rate of kill of host insects (*29*).

A novel approach to increase the insecticidal properties of NPV was demonstrated by O'Reilly and Miller (*30*). The researchers deleted an indigenous gene that encodes for ecdysteroid UDP-glucosyltransferase (EGT) from the AcNPV genome. EGT transfers the sugar moiety from a UDP-sugar to ecdysone, effectively

Table I. Recombinant Nuclear Polyhedrosis Viruses Constructed for Increased Insecticidal Activity

Expressed Protein or Peptide	Virus[a]	References
A. Endocrine System		
Diuretic Hormone	BmNPV	*(22)*
Eclosion Hormone	AcNPV	*(26)*
Juvenile Hormone Esterase	AcNPV / BmNPV[b]	*(23,27,28)*
Modified Juvenile Hormone Esterase	AcNPV	*(29)*
Ecdysteroid UDP-Glucosyltransferase[c]	AcNPV	*(30-32)*
B. Toxins		
Buthus eupus (BeIT, scorpion)	AcNPV	*(33)*
Bacillus thuringiensis (B.t., bacteria)	AcNPV	*(34,35)*
Pyemotes tritici (TxP1, mite)	AcNPV	*(36,37)*
Androctonus australis (AaIT, scorpion)	AcNPV / BmNPV	*(38-40)*

[a] Ac = *Autographa californica*; Bm = *Bombyx mori*; NPV = nuclear polyhedrosis virus.
[b] Hanzlik, unpublished data.
[c] Deletion of viral gene ecdysteroid UDP-glucosyltransferase.

blocking the activity of this molting hormone (*31,32*). As a result, the larval molt is delayed, allowing the virus to propagate producing even greater numbers. However, deletion of this gene from the AcNPV genome prevents the delay in molting leading to accelerated death of the larva.

Recombinant Baculoviruses Expressing Insect Toxins. The first attempt to modify a NPV for increased insecticidal activity was conducted by Carbonell *et al.* (*33*) in which an insect-specific toxin (BeIT) from the scorpion *Buthus eupus* was incorporated into the AcNPV genome. Although low levels of toxin expression were confirmed, the recombinant NPV did not result in quicker killing rates of host insects. Failure to improve killing activity may be due to an insufficient level of toxin expression by AcNPV, an incorrect sequence of toxin DNA, and/or insufficient or incorrect folding of the toxin following expression.

Recently, a number of insect toxins have been incorporated into NPVs. Merryweather *et al.* (*34*) incorporated the δ endotoxin from *Bacillus thuringiensis* (*B.t.*) subsp. *kurstaki* into polyhedrin-negative and -positive constructs of AcNPV. Expression of the *B.t.* toxin was confirmed although a reduction in the killing rate of host insects was not observed. While these results were discouraging, they were not unexpected since the site of action for the toxin is the larval midgut, and expression of *B.t.* toxin by the recombinant NPV would not occur in the midgut. Martens *et al.* (*35*) incorporated the Cry 1A(b) toxin of *B. thuringiensis* subsp. *aizawai* into AcNPV. Although lethal times of the virus were not reduced, expression of the toxin was confirmed, and the *B. t.* crystals produced *in vitro* showed insecticidal activity when fed to host larvae.

Tomalski and Miller (*36*) have successfully expressed a mite toxin, TxP1, isolated from *Pyemotes tritici* in AcNPV. The first recombinant virus encoding this toxin was a polyhedrin-negative construct, and injection of host larvae resulted in a 30 to 40% reduction in the time to kill when compared to wild-type AcNPV. The mite protein which is expressed is actually a protoxin which is cleaved by proteases to produce an active toxin. Additionally, new polyhedrin-positive AcNPV constructs encoding the mite toxin under the control of a variety of promoters have been engineered (*37*). This study emphasized the importance of the promoter-type in order to maximize the timing and amounts of expression necessary to provide quicker killing action. It was shown that a hybrid promoter of late and very late origin ($P_{cap/polh}$) resulted in the highest expression of toxin and quickest kill of host insects.

Recently three studies have described the successful expression of a scorpion toxin, AaIT (*38-40*). Maeda *et al.* (*38*) used a polyhedrin-negative BmNPV construct to assess the insecticidal properties produced by the expression of AaIT in larvae of *B. mori.* Injection of the virus resulted in a 30 to 40% reduction in the killing time of larvae as well as dramatic reductions in larval weight gain. Stewart *et al.* (*39*) and McCutchen *et al.* (*40*) developed polyhedrin-positive constructs of AcNPV using a viral gp67 signal sequence and the bombyxin signal sequence, respectively. Direct comparison of the two engineered AcNPV constructs indicate similarity in their ability to reduce time of kill (30 to 40%) as well as reduction in feeding damage (R. Possee, personal communication).

In the remainder of this chapter we will describe the construction and characterization of AcAaIT by McCutchen *et al.* (*40*), assess its potential as an insecticidal agent, provide insight for possible strategies to improve the insecticidal properties of NPVs, and assess risks associated with genetically modified NPVs.

Development and Characterization of a Recombinant Nuclear Polyhedrosis Virus, AcAaIT

It is well documented that certain arthropods produce neurotoxins, which are highly specific for insects. We selected the insect-selective neurotoxin, AaIT, for expression in AcNPV primarily because AaIT was the most thoroughly characterized of the insect-selective toxins isolated to date. This resource of information was instrumental in augmenting our toxicological characterization of the recombinant baculovirus expressing AaIT.

The Insect-Selective Neurotoxin, AaIT. The North African scorpion, *Androctonus australis* Hector, possesses three classes of toxins in its venom. One of these classes is specific for insects although one anti-insect toxin has been shown to induce a response in both insect and mammalian systems (*41*). *A. australis* possess an insect toxin (AaIT) which is a single chained polypeptide composed of seventy amino acids (M_r = 8kDa) cross linked by four disulfide bridges (*42*). The insect selectivity of AaIT has been documented by toxicity, binding and electrophysiological assays (*43*), and studies with mice have shown that 50 mg/kg (subcutaneous) produced no adverse physiological symptoms (*44*).

AaIT binds to the sodium channel of the insect nervous system, affecting the gating properties of the channels. It is believed that this excitatory toxin causes the following two actions on sodium conductance: 1) a small depolarizing action resulting from increased sodium permeability at resting membrane potential without suppression of activatable peak sodium conductance, resulting in increased peak sodium conductance and 2) voltage-dependent slowing of sodium current inactivation (*45,46*). These presynaptic effects induce repetitive firing of the insect's motor neurons resulting in massive and uncoordinated stimulation of the respective skeletal muscles.

Although the AaIT molecule has several characteristics in common with other toxins isolated from *A. australis*, such as a conserved hydrophobic surface conducive to binding to its receptor, AaIT is unique in that it possesses an atypical disulfide bridge and an unusually long C-terminus (*42*). It has also been demonstrated that chemical modification of amino acid residues of the AaIT molecule results in a decreased toxicity level to insects (*47*).

Construction of the Recombinant Nuclear Polyhedrosis Virus, AcAaIT. A detailed description of the construction of AcAaIT can be found in the manuscript by McCutchen *et al.* (*40*). The following is a condensed version of the construction of AcAaIT accompanied by a flow diagram (Figure 1). The AaIT gene including a signal sequence of bombyxin (*48*) required for secretion was inserted into the pTZ18R plasmid between the Sac I and Xba I sites. The resulting plasmid contained

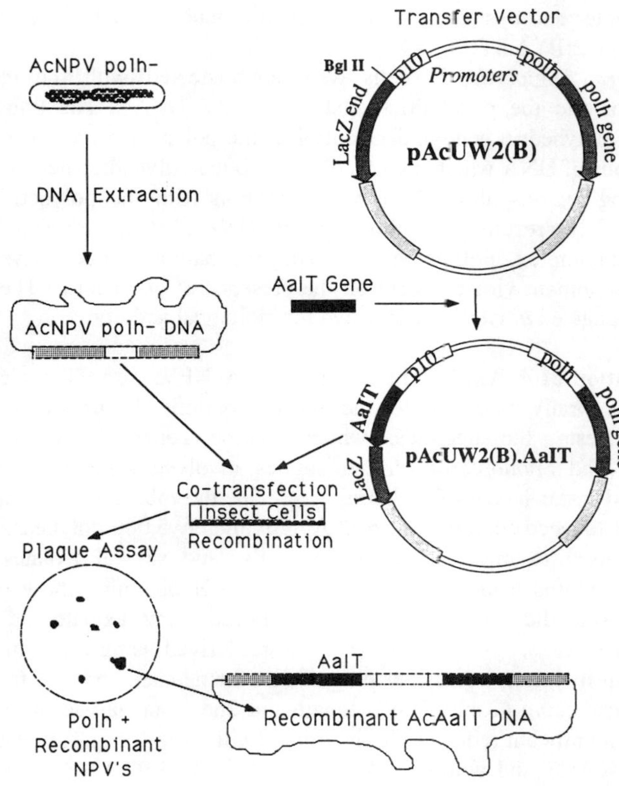

Figure 1. Flow diagram detailing the construction and isolation of a recombinant nuclear polyhedrosis virus expressing the insect-selective neurotoxin, AaIT. The AaIT gene was introduced at a Bgl II site in a pUC18 derived transfer vector, pAcUW2(B), under the control of the p10 promoter and upstream of the polyhedrin gene and promoter. Following cotransfection of the transfer vector (polyhedrin positive) with AcNPV (polyhedrin negative) in *Spodoptera frugiperda* cells, recombination between the homologous fractions of viral and vector DNA resulted in recombinant AcNPV encoding the AaIT gene. Polyhedrin positive isolates were plaque purified for further characterization. polh = polyhedrin.

two BamH I sites, one at each end of the AaIT gene. The gene was excised at the BamH I sites and the AaIT gene fragment was ligated into the Bgl II cloning site of the plasmid vector pAcUW2(B) (*49*), and then transformed into the JM 101 strain of *E. coli*. Construction resulted in the AaIT cDNA sequence inserted downstream of the p10 promoter and upstream of the polyhedrin gene producing the recombinant plasmid pAcUW2(B).AaIT.

Spodoptera frugiperda 21 cells were cotransfected with the recombinant transfer vector and the polyhedrin-negative AcNPV DNA. The transfer vector contained the polyhedrin gene under control of the polyhedrin promoter as well as flanking regions of DNA which are homologous to the polyhedrin-negative AcNPV. These flanking regions allowed for recombination with the polyhedrin-negative AcNPV DNA. The resulting recombinant viral DNA contained the AaIT gene and the polyhedrin complex. Following a 4 to 6 day incubation, plaque assays were used to isolate recombinant viruses based on the presence of polyhedra. These isolates were then propagated *in vitro* and screened for biological activity.

Characterization of AcAaIT. The recombinant AcNPV, AcAaIT, which we have constructed is orally infective to the most economically important group of lepidopteran pests, the noctuids, which include genera such as *Heliothis*, *Trichoplusia*, and *Spodoptera*. Initial studies involved screening our selected isolates in 2nd instar larvae of *Heliothis virescens,* the tobacco budworm (a major pest of cotton, tobacco and other crops). Approximately 5,000 polyhedrin inclusion bodies were used to infect these larvae orally, and several isolates produced neurotoxic symptoms in the larvae (Figure 2). After obtaining these preliminary results, detailed studies of mortality were conducted using neonate and 2nd instar larvae of *H. virescens*. Lethal times (LTs) were derived using a common dose of 250 polyhedrin inclusion bodies per plug of diet for neonate and 2nd instar larvae. A dose-response curve (Table II) of neonate and 2nd instar larvae of *H. virescens* illustrates the significant reduction in the time of kill. The LT_{50}s for neonate larvae treated with AcAaIT and wild-type AcNPV are 67.3 and 97.9 h p. i., respectively (Betana and McCutchen, unpublished data). Bioassays with 2nd instar larvae resulted in LT_{50}s (Table II) of 88.0 and 125 h p.i., respectively (*40*). A likelihood-ratio test (equal slopes and intercepts, C.I. = 0.95) showed a significant difference among treatments for both stages of larvae. Specifically, the larvae treated with AcAaIT had significantly lower LT values than larvae infected with wild-type AcNPV. These data represent over a 30% reduction in the rate of kill by AcAaIT when compared with wild-type AcNPV.

Additional observations made during these studies showed that larvae infected with AcAaIT typically started showing symptoms of paralysis and stopped feeding 10 to 15 hours prior to death. As a result, these larvae can functionally be considered dead, and treatment with AcAaIT represents an approximately 40-45% reduction in the time required to kill host larvae when compared to wild-type AcNPV. With the reduced killing time and early immobilization of host larvae by paralysis, AcAaIT may effectively curtail food consumption which may translate into reduced feeding damage.

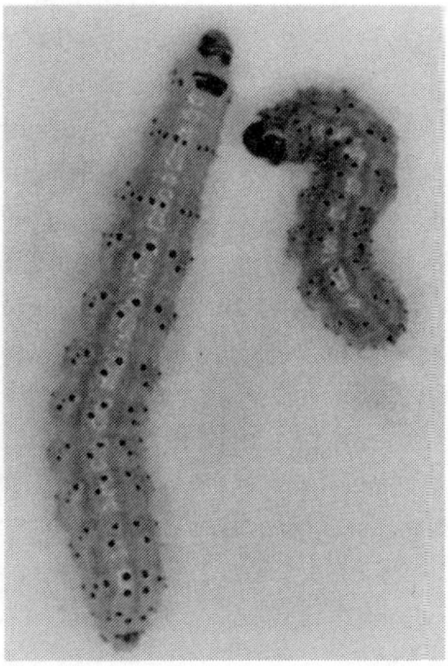

Figure 2. Symptoms of 2nd instar larva of *Heliothis virescens* treated with AcAaIT. The larva on the right shows typical contraction of the larval segments associated with the paralyzing effects after exposure with the recombinant virus AcAaIT. The larva on the left was treated at the same time with wild-type AcNPV and shows no signs of infection. (Picture by Dr. Michael P. Parrella, Department of Entomology, UC Davis).

Table II. The Lethal Times of Larvae of *H. virescens* with Treatments of AcAaIT and Wild-type AcNPV

Treatment[a]	Lethal Times (hours)[b]		
	10	50	90
1st Instar Larvae[c]			
AcAaIT	46.2 (42.2 - 49.6)	67.3 (64.3 - 70.4)	98.1 (92.2 - 106)
Wild-type AcNPV	57.1 (50.8 - 62.3)	97.9 (91.6 - 106)	167 (147 - 201)
2nd Instar Larvae[d]			
AcAaIT	59.8 (49.8 - 67.3)	88.0 (80.7 - 94.7)	129 (118 - 146)
Wild-type AcNPV	91.4 (78.1 - 101)	125 (117 - 135)	172 (156 - 203)

[a] 250 polyhedrin inclusion bodies per plug of diet.
[b] POLO probit analysis program with lower and upper limits (C.I. 0.95).
[c] Betana and McCutchen unpublished data.
[d] McCutchen et al. 1991 (*40*).

Confirmation of Toxin Expression by AcAaIT. Using an established bioassay with the dipteran larvae of *Sarcophaga falculata,* the blow fly (*50*), we made estimations of the concentration of AaIT expressed by the recombinant virus. A range of AaIT doses was administered to larvae by subcutaneous injection. A hypodermic needle was laterally inserted to an approximate depth of 5 mm through an abdominal intersegmental membrane. Paralysis was qualified as complete immobilization of the blow fly larvae and was monitored at intervals up to 24 h post injection. Contraction paralysis, which is the immediate and fast contraction of the larvae, was assessed as described by Zlotkin *et al.* (*50*). A dose-response curve was constructed with blow fly larvae injected with purified AaIT collected from scorpion venom. Using these results, we estimated that 4th instar larvae (showing paralytic symptoms) contained 1-2 ng of AaIT per μl of hemolymph (*40*). In addition, these hemolymph samples were shown to produce an immunoreactive band which co-migrated with purified AaIT.

Possible Strategies for Improving AcAaIT and other Recombinant NPVs

Several approaches can be taken to improve the insecticidal properties of AcAaIT, and these approaches should be applicable to most other genetically engineered viruses. Figure 3 shows a flow chart indicating four routes of research which may ultimately lead to more effective baculovirus insecticides.

Pharmacological Approaches. In a recent study by Herrmann *et al.* (*51*) it was shown that lepidopteran larvae were significantly more tolerant than dipteran larvae when similar doses of AaIT were injected into insects. In fact, the PD_{50} (paralytic dose) for larvae of *Spodoptera littoralis* was 2.4 μg/100 mg compared to a PD_{50} or CPU (contraction paralysis unit) of 0.003 ug/100 mg for larvae of *S. falculata.* These data indicate that blowfly larvae were approximately 800 times more sensitive to AaIT than lepidopteran larvae. This study suggests that AaIT may not be a favorable candidate for achieving the dramatic enhancement in the killing rate of NPVs associated with lepidopteran insects. However, our data indicate that introduction of AaIT into the baculovirus genome significantly reduces the rate of kill. Assuming the levels of AaIT found in the hemolymph (1-2 ng/μl) of larvae infected by AcAaIT are representative of AaIT concentrations found throughout the body, a 100 mg larvae would harbor 100 - 200 ng of AaIT at peak levels (when the insect is showing clear contraction paralysis; Fig 2). According to Herrmann *et al.* (*51*), injected doses of over 1.0 μg of AaIT were needed to produce even slight paralysis in similar larvae of this size (100 mg). In other words, the levels of AaIT expressed *in vivo* by AcAaIT are approximately ten times lower than the levels needed to produce visible paralytic symptoms much less produce mortality.

Our results may represent a classical example in drug delivery. Caterpillars have a protective sheath surrounding their nervous system. The neuromuscular synapses are completely covered by glial cells which invaginate and strongly intertwine with muscular processes (*52*). It is likely that this protective sheath provides a physiological barrier preventing injected AaIT from reaching its site of action. However, it is known that NPVs infect the nervous tissue of insects. As a result, AaIT may be expressed in nervous tissue resulting in a relatively high and localized concentration of toxin at its site of action, the insect neuronal sodium channel. Removal of the sheath around the central nervous system dramatically increases the response of the motor nerves in *in situ* binding assays (*51*). Perhaps this precise delivery, along with the continual expression of AaIT, explains the significant decrease in the rate of kill associated with AcAaIT. Certainly, pharmacodynamic studies are needed to investigate this phenomenon which may lead to novel approaches in delivery of insecticides.

Herrmann *et al.* (*51*) have provided an initial understanding of the pharmacodynamics of AaIT in lepidopteran larvae. It was shown that injections of purified AaIT produced paralysis in lepidopteran larvae; however, the larvae recovered within 48 hours. Although the degradative mechanism(s) of AaIT in the insect body were not determined, it was demonstrated that the toxin is completely stable *in vitro* in insect hemolymph and is rapidly eliminated *in vivo* by the excretory system with 20% of the original activity recovered in the feces. These results suggest that a majority of AaIT may be absorbed in some tissue(s), and degraded by proteolytic cleavage, in addition to rapid elimination.

Assuming proteolytic degradation occurs, AaIT could be radiolabelled using the AcNPV expression system with ^{3}H or ^{35}S and purified by reverse-phase high performance liquid chromatography (HPLC). Then physiological and pharmacological aliquots of labeled AaIT could be injected into larvae, and hemolymph and tissue samples could be monitored for activity. After localization of tissue(s) suspected of toxin uptake, degradative products could be isolated by thin layer chromatography, reverse-phase HPLC, and/or immunoaffinity chromatography. These metabolic fragments could be analyzed by N-terminal sequencing and/or mass spectrometry. These data might provide insight into discovery of amino acid residues susceptible to proteolytic hydrolysis. Subsequently, site-directed mutagenesis of the toxin DNA could be used to modify these residues leading to a more stable toxin in lepidopteran larvae.

Chemical and Modeling Approaches. Another approach to improving potency or speed of kill by AcAaIT would employ analytical chemistry and protein modeling. The foremost objective would be to scale-up and improve *in vitro* NPV expression of AaIT. Then using a combination of purification techniques such as antibody columns and reverse phase HPLC, purified AaIT could be obtained for X-ray crystallography, nuclear magnetic resonance (NMR) imaging and circular dichroism analyses. Based on these results, predictions might be made concerning the structure of AaIT. These models could then be used to predict modifications which may increase the toxicity or specificity of the toxin by increasing receptor binding or stability.

Modified toxins could be produced with site-specific mutations to produce toxins with amino acid residue substitutions, additions or deletions. These modified toxins could then be characterized by a variety of techniques including bioassays, binding studies, X-ray crystallography and NMR analyses. Additionally, these studies will lead to a better understanding of the insect sodium channel. In fact, results gathered from X-ray analysis, binding assays and NMR studies may ultimately lead to the development of a new class of synthetic chemical insecticides which may have a high degree of specificity for insects.

Molecular Approaches. As described earlier, one molecular approach which has been successful was the deletion of a viral gene encoding ecdysteroid UDP-glucosyltransferase (EGT) from the AcNPV genome (*31*). This simple modification prevents disruption of the molting process and allows the larva to continue its development resulting in quicker kill. Most viruses probably contain genes like EGT which either hamper, or at least are not essential, for use of the virus as an insecticidal agent. Deletion of such non-essential DNA could increase effectiveness, improve production characteristics, and/or reduce the ability of the recombinant virus to recycle in the field.

Another approach to improve the efficacy of recombinant baculoviruses includes the use of different promoters for expression of the foreign protein. Alternate promoters could lead to earlier or improved transcription, more stable messages, and/or improved translation. Tomalski and Miller (*37*) have recently reported the expression of the *Pyemotes* mite toxin by several different promoters. These included an early promoter, a late promoter, and two hybrid promoters derived of a late and a very late promoter sequence. The study indicated that expression of the mite toxin with one of the hybrid promoters resulted in slightly quicker kill and a reduction in feeding. Interestingly, the use of an early promoter did not result in larval paralysis suggesting that the levels of expressed toxin were not adequate to overcome the insect. However, this study should not discourage the use of earlier promoters, since a more potent toxin may provide more dramatic results (*53*).

Other molecular approaches to improve the insecticidal properties of recombinant baculoviruses include methods to stabilize RNA's or screening signal sequences for improved expression of AaIT. Changes leading to improved stability of the protein or peptide could include C-terminal amidation and/or N-terminal pyroglutaminylation.

Other modifications such as the addition of genes from other baculoviruses or deletion of indigenous genes may provide for tailor-made viruses with increased insecticidal activities. Possible modifications include engineering viruses with quicker replication rates or with modified host ranges. A perfect candidate for such genetic modification is AcNPV, since this virus has a comparably slow replication rate and a broad host range which might be narrowed or broadened based on circumstance. Maeda and coworkers (*54*) have made substantial discoveries regarding the relative similarities in host ranges of BmNPVs and AcNPVs. Studies characterizing host range may prove significant in the further development of NPVs as effective insecticidal agents, since we may be able to tailor host specificity through genetic modification.

Other viruses such as the GV's are much more infective than the NPVs. The success demonstrated by the engineering of AcNPV to increase its effectiveness should provide the impetus to collect and characterize viruses from the field. Hopefully, cloning systems will be developed for viruses which demonstrate promising infectivity and host range. In fact, the modification of other baculoviruses for control of other insect pests is already underway. However, AcNPV remains the most thoroughly characterized baculovirus to date, and as a result further improvements in its insecticidal properties may initially be realized in AcNPV. The improvements discussed above provide guidelines for similar modifications in other insect viruses.

Entomological Approaches. Another strategy for improving the insecticidal properties of both the recombinant and wild-type baculoviruses could involve classical selection. One approach is to select isolates from cell culture with increased rates of toxin expression or quicker replication rates which might result in enhanced killing rates of host insects. Another strategy to improve lethal times is to select viral stocks from host insects which die at quicker rates, possibly as part of a program of classical breeding based on natural or enhanced recombination of viruses. By repeatedly passing these viruses through host insects and isolating virus from insects which die at quicker rates, one should be able to effectively improve the killing rates by the virus. Similar approaches may provide significant reductions in the lethal doses of virus required to produce mortality in host insects. Additionally, for NPVs with broad host ranges such as AcNPV, improved insecticidal characteristics might be obtained by isolating and propagating viral stocks in different species of insects targeted for pest control tactics.

A major effort should be taken to isolate new baculoviruses in other insects especially those insects considered major pests. However, entomologists must consider several factors including the host's life cycle, before attempting to isolate new baculoviruses. For example, one might consider the cotton bollweevil as a perspective candidate; however, consideration of the life cycle of the bollweevil suggests that the probability of effectively controlling this pest with a baculovirus is low. Not only does the larval stage develop solely within the confines of the cotton fruit, but the adult primarily feeds on cotton squares enclosed in brachts. As a result, conventional insecticide application technology would not provide sufficient insecticide coverage to effectively infect the bollweevil. However, other insect pest such as mosquitoes or the corn rootworm might be ideal candidates since the larval stages of these insects are accessible to insecticide application.

Risk Assessment of Genetically Modified Baculoviruses

As with any new technology, there are always concerns, both social and scientific, about potential risks to mankind and the environment. A particularly controversial topic is the public's reaction to the release of genetically engineered organisms - organisms with altered genetic material that make them capable of producing new substances or performing new functions - into the environment. Undoubtedly, the release of modified baculoviruses is an area in which careful examination, planning, and risk assessment are imperative to ensure safe and effective application of this new technology. The potential benefits of genetic engineering in agriculture are many and diverse but do not come without possible risks. Deliberate release of genetically engineered baculoviruses poses potential risks associated with replication of released organisms, physical and genetic stability of these recombinant viruses, potential for gene exchange, and effects of these microbes on ecological processes.

Regulations and Monitoring Methodologies. The time has come for policymakers to evaluate possible risks along with the potential benefits and make regulatory decisions concerning the control of this technology. Regulatory oversight of agricultural biotechnology products is administered at the federal level through statutory authorities such as the Environmental Protection Agency. These agencies will have to make appropriate decisions regarding the release of genetically engineered baculoviruses. The regulations adopted must be sufficient to ensure the safety of non-target organisms and the environment, but can not be so stringent as to deter industry from developing this technology.

One of the most pertinent technologies to develop and streamline prior to any large scale release of recombinant baculoviruses will be the ability to monitor these genetically modified viruses throughout the environment. According to Kearney and Tiedji (*55*), there are three major categories of detection methods including culture and metabolic techniques, genetic techniques, and immunological techniques. Immunological and genetic techniques will probably be used for the initial screening of samples from the field since these techniques are amenable to large sample sizes. Genetic techniques rely on the specificity of gene probes which are able to select for sequences of nucleic acids. This methodology can be used to detect organisms without culturing and without the necessity of having a specific selectable marker (*55*). Gene probes also allow monitoring of the foreign gene even if it has been transfered to another organism. Certainly technologies based on the polymerase chain reaction appear promising for monitoring of both wild-type and recombinant viruses. Immunochemical techniques for monitoring genetically engineered microorganisms involve the use of an antibody specific for the gene product or microorganism of interest. Immunoassays are rapid, simple, cost effective, and adaptable to laboratory and field conditions. Culture and metabolic techniques are traditional methods used by microbiologists for decades and will probably be used for confirmation. In fact, all of these techniques should be used collectively to effectively monitor the movement of these recombinant viruses through the environment.

Risks to Non-target Organisms. Concerns have been raised over the competitiveness of the recombinant NPVs compared to the wild-type NPVs. Williamson (*56*) has suggested that recombinant viruses may have a detrimental effect on the natural fauna. Specifically, Williamson has expressed concern that recombinant NPVs may be able to infect non-target insects and become established in the environment. This is a legitimate concern and should be addressed accordingly. It is of utmost importance is to test these genetically modified viruses in an array of different non-target organisms, especially closely related Lepidoptera and those organisms which are most likely to be exposed to these recombinant insecticidal agents. For example, predators and parasitoids of target pests such as *H. virescens* should be screened thoroughly for any adverse effects associated with the recombinant baculoviruses (Figure 4).

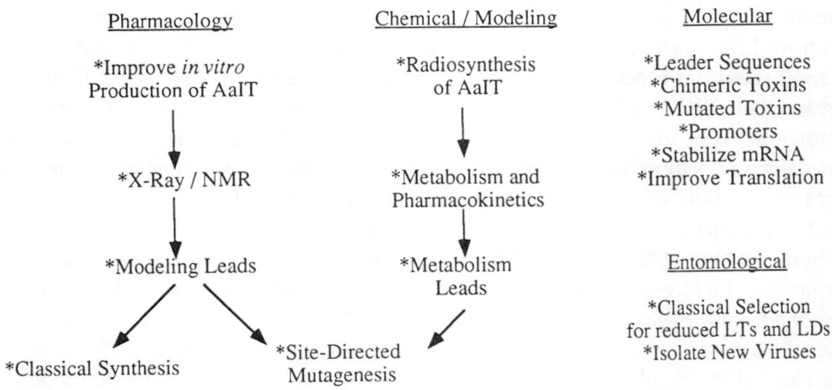

Figure 3. Flow chart indicating possible routes to further improve the efficacy of the insecticidal properties of NPVs, including: pharmacological, chemical, molecular and entomological approaches. LT = lethal time; LD = lethal dose.

Figure 4. A predaceous insect, *Chrysopa carnea* (green lacewing larva) feeding on a 2nd instar larva of *Heliothis virescens*. (Picture by Jack K. Clark, UC IPM).

It should be emphasized that recombinant baculoviruses are being courted as insecticides, not as biological control agents (*57,58*). Therefore, industry will develop these insecticides so that they only moderately persist in the environment thereby requiring repeated applications during the season. Vlak *et al.* (*59*) have shown that NPVs with a p10 gene deletion do not induce the characteristic disintegration of the host larva. As a result the larva is not prone to rupture, thus preventing dissemination of polyhedra. However, genetic modifications to prevent dispersal of polyhedra from recombinant viruses may not be necessary. In our laboratory, larvae infected with either AcAaIT or AcJHE (modified JHE) die without disintegrating since it is the insecticidal protein and not the virus that kills the larvae. Identical results were found with the recombinant viruses developed by Tomalski and Miller (*35,36*). Additionally, the quick killing action of these recombinant viruses reduced the number of progeny virus produced (McCutchen and Herrmann, unpublished data). Interestingly, we do not see a dramatic reduction in the number of polyhedra produced that would be expected given the reduced killing times of the recombinant viruses. In fact, infected larvae are paralyzed for a period before actual death allowing the virus to continue propagating. At least one other factor contributes to the inability of recombinant viruses to spread. The ability of viruses expressing paralyzing agents to immobilize host larvae should result in insects falling from the foliage, further reducing the possibility that viruses will be disseminated by predators. These three factors place recombinant viruses at a severe competitive disadvantage and should facilitate the registration process. We feel that, once well documented, these three factors are sufficient to insure that recombinant viruses will not become established in the environment. However, if it is considered important by regulatory agencies, there are a variety of genetic approaches that could be used to further reduce the ability of recombinant viruses to compete or even block recycling of the recombinant virus altogether. For instance, one could employ a virus which produces polyhedra under laboratory but not field conditions.

Closing Remarks.

Although this work represents a novel approach to controlling insects, it is critical to understand the capabilities and limitations of recombinant baculoviruses at this early stage in their development. The development of recombinant NPVs will provide a basis for further enhancement of the insecticidal activity of baculoviruses and other agents. Currently, chemical pesticides cannot be replaced by NPVs in row-crop agriculture with the expectation that they will provide the same timely reduction in even the most susceptible pest populations. Depending upon the particular baculovirus and dosage used, it may take 5-16 days after application before insect pests die. During this time the pests may continue to feed resulting in crop damage and economic loss. Our studies suggest that these engineered viruses could compete with classical insecticides in limited markets, and that these viruses may become part of integrated pest management in the near future.

It is critical that these recombinant NPVs not be over exploited in order to curtail the development of resistance. In fact, several cases of insect tolerance to baculoviruses have been reported (*60*). Recently, Fuxa and Richter (*61*) noted decreased infectivity of AcNPV to *S. frugiperda* involving a factor at the gut level. Realistically, this initial infection process can be considered the weakest link of viral insecticides and may contribute to the bulk of resistance problems associated with wild-type and recombinant viruses. With the recombinant NPVs reported to date the foreign protein expressed for insecticidal activity will not be selected against, since the virus will ultimately kill even those insects which escape the effects of the toxin. On the positive side, most known mechanisms for cross resistance should not be effective with viruses. In addition, the short generation time of viruses should help them to overcome the potential for insect resistance.

The construction and apparent success of these genetically engineered baculoviruses represent a considerable advance in the continuing effort to develop effective baculoviruses for use as pest control agents. The development of a more effective, faster-acting baculovirus will provide a novel agent in the pesticide arsenal including a unique mode of action for combatting resistant or specific pest species. The development of selective recombinant insecticides will augment any Integrated Pest Management program by reducing the impact on non-target species, including beneficial insects.

Acknowledgments

We would like to thank Jack K. Clark of UC IPM and Dr. Michael P. Parrella of UC Davis for taking photographs. This work was supported by NIEHS grant ES02710 and USDA grant 90-37250-5239. BFM is a NIEHS fellow and B.D.H. is a Burroughs Wellcome Scholar in Toxicology.

Literature Cited

1. Boyce, A. M. In *The Future for Insecticides: Needs and Prospects;* Metcalf, R. L.; McKelvey, J. J., Jr., Eds.; Advances in Environmental Science and Technology; John Wiley and Sons, Inc.: New York, NY, **1974**, Vol. 6; pp 469.
2. Croft, B. A., 1990. In *Arthropod Biological Control Agents and Pesticides*; R. L. Metcalf & W. Stumm, Eds.; Environmental Science and Technology; John Wiley & Sons, Inc.: New York, NY, **1990**, pp 194-287.
3. Mellor, J. W.; Adams, R. H., Jr. *Chem. Eng. News* **1984**, *62*, 32-39.
4. Debach, P. In *Biological control of insect pests and weeds;* DeBach, P.; Schlinger, E. I. Eds.; Reinhold, NY. **1964**, pp. 429-458.
5. Harper, J. D. In *The epizootiology of insect diseases;* Fuxa, J. R.; Tanada, Y. Eds.; John Wiley and Sons, Inc.: New York. **1987**.
6. Evans, H. F. In *The biology of baculoviruses*; Granados, R. R.; Federici, B. A., Eds.; CRC Press: Boca Raton, FL. **1986**, pp. 89-132.
7. Granados, R. R.; Lawler, K. A. *Virology* **1981**, *108*, 297-308.
8. Vlak, J. M.; Rohrmann, G. F. In *Viral Insecticides for Biological Control*, Maramorosch, K.; Sherman, K. E., Eds.; Academic Press, New York, NY. **1985**, pp 489-542.
9. Miller, L. K.; Lingg, A. J.; Bulla, L. A. Science **1983**, *219*, 715-721.

10. Miller, L. K. *Ann. Rev. Microbiol.* **1988**, *42*, 177-199.
11. Maeda, S. *Ann. Rev. Entomol.* **1989**, *34*, 351-372.
12. Luckow, V. A. In *Recombinant DNA Technology and Application*; Prokop, A.; Bajpai, R. K.; Ho, C.S., Eds.; McGraw-Hill, Inc.: New York, NY, **1991**, pp 97-152.
13. O'Reilly, D. R.; Miller, L. K.; Luckow, V. A. *Baculovirus Expression Vectors: A Laboratory Manual*, W. H. Freeman and Company, New York, NY, **1992**; 347.
14. King, L. A.; Possee, R. D. *The Baculovirus Expression System*, Chapman and Hall, London, New York, **1992**; 229.
15. Betz, F. S. In *The Biology of Baculoviruses*; Granados, R. R.; Federici, B. D. A. Eds.; CRC Press: Boca Raton, FL. **1986**, Vol. I. pp 1-36.
16. Carbonell, L. F.; Klowden, M. J; Miller, L. K. *J. Virol.* **1985**, *56*, 153-160.
17. Carbonell, L. F.; Miller, L. K. *Appl. Environ. Microbiol.* **1987**, *53*, 1412-1417.
18. Brusca, J.; Summers, M.; Couch, J.; Courtney, L. *Intervirol.* **1986**, *26*, 207-222.
19. Summers, M. D.; Engler, L.; Falcon, A.; Vail, P. In *Baculoviruses for Insect Control: Safety Considerations;* American Society for Microbiology, Washington D.C. **1975;** pp. 186.
20. Groner, A. In *Safety of Microbial Insecticides;* Laird, M.; Lacey, L. A.; Davidson, E. W.; CRC Press, Boca Raton, FL, **1989**, pp. 135-147.
21. Kirshbaum, J. B. *Ann. Rev. Entomol.* **1985**, *30*, 51-70.
22. Maeda, S. *Biochem. Biophys. Res. Commun.* **1989**, *165*, 1177-1183.
23. Hammock, B. D.; Bonning, B.; Possee, R. D.; Hanzlik, T. N.; Maeda, S. *Nature* **1990**, *344*, 458-461.
24. Keeley, L. L.; Hayes, T. K. *Ins. Biochem.* **1987**, *17*, 639-651.
25. Leisy, D. J.; Van Beek, N. *Chemistry and Industry* **1992**, *251*, 250-254.
26. Eldridge, R.; Horodyski, F. M.; Morton, D. B.; O'Reilly, D. R.; Truman, J. W.; Riddiford, L. M.; Miller, L. K. *Insect Biochemistry* **1991**, *21*, 341-351.
27. Bonning, B. C.; Hammock, B. D. Biotechnol. Gen. Eng. Rev. **1992**, 10, 453-487.
28. Eldridge, R.; O'Reilly, D. R.; Hammock, B. D.; Miller, L. K. *Appl Env. Microbiol*. **1992**, *58*, 1583-1591.
29. Bonning, B. C.; Hammock, B. D. In *Natural and Derived Pest Management Agents* Hedin, P., Hollingworth, R.; Menn, J. Eds.; ACS Symposium Series, American Chemical Society, Washington, DC, **1993**.
30. O'Reilly, D. R.; Miller, L. K. *Biotechnol.* **1991**, *9*, 1086-1089.
31. O'Reilly, D. R.; Miller, L. K. *Science*. **1989**, *245*, 1110-1112.
32. O'Reilly, D. R.; Miller, L. K. *J. Virol*. **1990**, *64*, 1321-1328.
33. Carbonell, L.F.; Hodge, M.R.; Tomalski, M.D.; Miller, L.K. *Gene* **1988**, *73*, 409-418.
34. Merryweather, A. T.; Weyer, U.; Harris, M. P. G.; Hirst, M.; Booth, T.; and Possee, R. D. *J. Gen. Virol.* **1990**,*71*, 1535-1544.
35. Martens, J. W. M.; Honee, G.; Zuidema, D.; Van Lent, J. W. M.; Visser, B; Vlak, J. M. *Appl. Environ. Microbiol.* **1990**, *56*, 2764-2770.
36. Tomalski, M. D.; Miller, L. K. *Nature* **1991**, *352*, 82-85.
37. Tomalski, M. D.; Miller, L. K. *Biotechnol.* **1992**, *10*, 545-549.

38. Maeda, S.; Volrath, S. L.; Hanzlik, T. N.; Harper, S. A.; Maddox, D. W.; Hammock, B. D.; Fowler, E. *Virol.* **1991**, *184*, 777-780.
39. Stewart, L. M. D.; Hirst, M.; Ferber, M. L.; Merryweather, A. T.; Cayley, P. J.; Possee, R. D. *Nature* **1991**, *352*, 85-88.
40. McCutchen, B. F.; Choudary, P. V.; Crenshaw, R.; Maddox, D.; Kamita, S. G.; Palekar, N.; Volrath, S.; Fowler, E.; Hammock, B. D.; Maeda, S. *Biotechnol.* **1991**, *9*, 848-852.
41. Loret, E. P., Martin-Eauclaire, M. F., Mansuelle, P., Sampieri, F., Granier, C., Rochat, C. *Biochem.* **1990**, *30*, 633-640.
42. Darbon, H.; Zlotkin, E.; Kopeyan, C.; van Rietschoten, J.; Rochat, H. *Int. J. Pesticide Protein Res.* **1982**, *20*, 320-330.
43. Zlotkin, E. In *Neuropharmacology and Pesticide Action.* Ford, M. G. *et al.*, Horwood, Chichester, **1986** pp. 352-383.
44. De Dianous, S., Hoarau, F., Rochat, H. T. *Toxicon.* **1987**, *25*, 411-417.
45. Zlotkin, E.; Eitan, M.; Bindokas, V. P.; Adams, M. E ; Moyer, M.; Burkhart, W.; Fowler, E. *Biochem.* **1991**, *30*, 4814-4821.
46. Pelhate, M.; Zlotkin, E. *J. Exp. Biol.* **1982**, *97*, 67-77.
47. Loret, E.; Mansuelle, P. P.; Rochat, H.; Granier, C. *Biochem.* **1990**, *29*, 1492-1501.
48. Adachi, T.; Takiya, S.; Suzuki, Y.; Iwami, M.; Kawakami, A.; Takahashi, S. Y.; Ishizaki, H.; Nagasawa, H.; Suzuki, A. *J. Biol. Chem.* **1989**, *264*, 7681-7685.
49. Weyer, U.; Knight, S.; Possee, R. D. *J. Gen Virol.* **1990**, *71*, 1525-1534.
50. Zlotkin E.; Fraenkel, G.; Miranda, F.; Lissitzky, S. *Toxicon.* **1971**, *9*, 1-8.
51. Herrmann, R.; Fishman, L.; Zlotkin, E. *Insect Biochem.* **1990**, *20,* 625-637.
52. Rheuben, M. B.; Reese, T. S. *J. Ultrastr. Res.* **1978**, *65*, 95-111.
53. Maeda, S.; Hammock, B. D. In *Newer Pest Control Agents and Technology with Reduced Environmental Impact;* Duke, S. O.; Menn, J. J.; Plimmer, J. R.; Eds.; ACS Symposium Series, American Chemical Society, Washington, DC, **1993**; pp 281-297.
54. Kondo, A.; Maeda, S. *J. Virol.* **1991**, *65*, 3625-3632.
55. Kearney, P. C.; Tiedje, J. M. In *Biotechnology For Crop Protection* Hedin, P. A.; Menn, J. J.; Hollingsworth, R. M., Eds., **1988**, pp 352-358. ACS Symposium Series.
56. Williamson *Nature*, **1991**, *353*, 394.
57. Hammock, B. D. *Nature*, **1992**, *355*, 119.
58. Hammock, B. D.; McCutchen, B. F.; Beetham, J. K ; Choudary, P.; Ichinose, R.; Ward, V. K.; Vickers, J.; Bonning, B. C.; Harshman, L. G.; Grant, D.; Uematsu, T.; Lanzrein, B.; Maeda, S. *Arch. Insect Biochem. Phys.* **1992**, *22*, 315-344
59. Vlak, J. M.; Klinkenberg, F. A.; Zaal, K. J. M.; Usmany, M.; Klinge-Roode, E. C.; Geervliet, J. B. F.; Roosien, J.; Van Lent, J. W. M. *J. Gen. Vir.* **1988**, *69*, 765-767.
60. Briese, D. T. In *The Biology of Baculoviruses.* Granados, R. R.; Federici, B. D. A. Eds., Vol. II, CRC Press, Boca Raton, FL, **1986**, pp 237-263.
61. Fuxa, J. R.; Richter, A. R. J. Inv. Path. **1990**, *55*, 272-277.

RECEIVED August 28, 1993

Chapter 26

Insect Control by Use of Recombinant Baculoviruses Expressing Juvenile Hormone Esterase

Bryony C. Bonning and Bruce D. Hammock

Departments of Entomology and Environmental Toxicology, University of California, Davis, CA 95616

Anti-juvenile hormone activity, and the resulting cessation of feeding, has long been considered by the agricultural chemical industry as a means for insect control. Juvenile hormone esterase (JHE) contributes to the decline in JH titer at certain times during larval development. This chapter reviews the efforts made to develop a fast acting recombinant baculovirus insecticide by creating a virus which will express artificially high levels of JHE in the insect at inappropriate times. The findings are discussed in the light of our current understanding of the regulation of larval development and the role of JHE. Prospects for the newly developed, highly effective baculoviruses expressing modified forms of JHE are also considered.

Regulation of Larval Development in Lepidoptera

The regulation of larval development in Lepidoptera is complex, finely controlled, and still incompletely understood. There is also great interspecies variation, and data have been acquired from numerous species. Juvenile hormone (JH) has a role central to the course of larval growth and, in simplistic terms, may be considered to keep the larva in its feeding stage. A drop in the titer of JH signals the insect to stop feeding in readiness for molting. JH also prevents metamorphosis to the adult insect (*1*). The titer of JH remains high during early instars. A drop in the titer during the last larval instar is followed by a further transitory peak, which is crucial for normal development of the pupa. Molting is initiated and coordinated by 20-hydroxyecdysone, and the nature of each molt is determined by the titer of JH (Figure 1a). Ecdysone is released by the prothoracic glands of the insect in response to the action of prothoracicotropic hormone (PTTH) released by the brain. A low initial titer of ecdysteroid during each larval instar is followed by a molt-inducing surge of ecdysteroids (*2*). The absence of JH during the last larval instar triggers release of PTTH and a small surge of ecdysteroids which result in the loss of larval

0097–6156/94/0551–0368$06.00/0

commitment of the tissues (*3*,*4*). The larval tissues reprogram to become pupal at the molt, in response to the action of 20-hyroxyecdysone (*2*; Figure 1a).

The titer of JH in the hemolymph is controlled by the rate of biosynthesis in the *corpora allata* (*5*), and by the rate of degradation. Both epoxide hydrolases and esterases act on JH. However, in the hemolymph of Lepidoptera just before wandering, JH degradation is mainly due to the action of juvenile hormone esterase (JHE) (*6*). This enzyme hydrolyzes JH to the biologically inactive JH acid (*6*; Figure 1b). JHE is a scavenger enzyme with a high k_{cat}/K_m ratio. All JH's found in Lepidoptera are susceptible to breakdown by JHE (*7*). There is some variation between lepidopteran species in the profile of JHE titers during the course of development (*6*). However, in general, titers of JHE are low during the early larval instars. A brief period of increased JHE activity approximately 24 hours prior to each ecdysis is seen in *Trichoplusia ni* (*8*). JHE contributes to the drop in the level of JH which results in cessation of feeding prior to the molt. Inhibition of JHE during the last larval instar of *Manduca sexta* results in a titer of JH in the hemolymph which is sufficient to keep the insect artificially in the feeding stage, and giant insects result (*9*). The JHE titer peaks twice during the final larval instar. The first peak of JHE coincides with the initial drop in JH, and seems to be controlled by a head factor (*9*,*10*). The precipitous drop in JH is followed by secretion of PTTH which initiates events leading to pupation. Release of PTTH occurs within a critical window of time determined in part by the size of the larva (*11*). The second peak of JHE is thought to be induced by the prepupal peak of JH, which must be degraded for normal development. JHE appears in many tissues with high levels in fat body and epidermal cells (*12*). The enzyme is likely to be produced by the fat body although the source is not proven (*13*,*14*). During the last instar of *Heliothis virescens*, an increase in JHE mRNA in the fat body corresponds to an increase in JHE activity in the fat body and hemolymph (*12*).

The Role of Juvenile Hormone Binding Proteins. All JH in the hemolymph associates with juvenile hormone binding protein/s (JHBP). These binding proteins are also found in the fat body, hemolymph and epidermal cells in Lepidoptera. Some workers have claimed that such binding proteins protect JH from degradation by non-specific hemolymph esterases; however, hard data to support this hypothesis are lacking. They may also protect JH from degradation in cells. There is better evidence that binding proteins prevent non-specific adsorption of JH which is lipophilic (*15*). They also facilitate transport and distribution of JH from the site of synthesis to the target site. The titer of JHBP in the hemolymph of *T. ni* and *M. sexta* remains at a relatively high level which varies roughly with protein content, throughout the larval instars (*16*,*17*). Regulation of JHBP is independent of the regulation of JH (*15*). The titer of JHBP does not influence larval development *per se*, but the presence of JHBP is necessary for transport and action of JH.

The association of JH with JHBP is such that JHBP facilitates rapid distribution of JH only in the absence of JHE. In the presence of JHE, which has a higher affinity for the hormone, the JH released by the JHBP is rapidly hydrolyzed (*7*).

In its morphogenetic action, JH is thought to act at the transcriptional level through an unknown signal transduction pathway. One theory is that the JH-JHBP complex is recognized (*15*), although again there is little evidence. The alternative

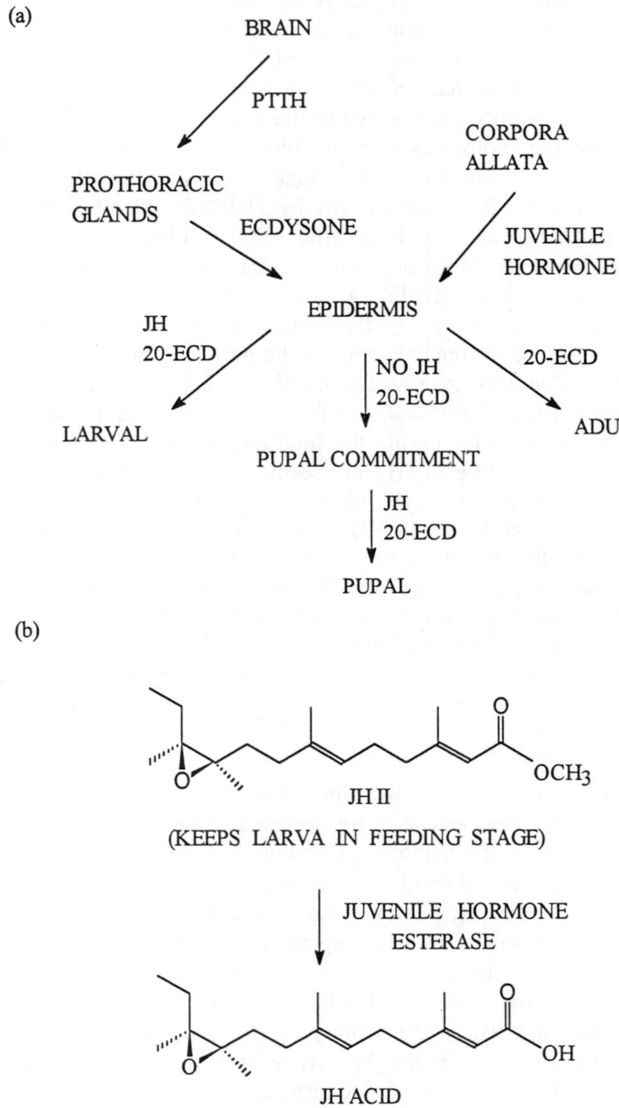

Figure 1
(a) Simplified scheme for determination of the course of Lepidopteran development by 20-hydroxyecdysone (20-ECD) and juvenile hormone (JH).
(b) JHE degrades JH to JH acid which is inactive. JH II (the major JH of the Lepidoptera examined) is depicted in this case.

theory for JH action is akin to steroid hormone action with specific cellular receptors (*18*).

Juvenile Hormone Analogs and Anti-Juvenile Hormone Agents for Insect Control

Larval development is so finely controlled, that the general belief has been that a minor disruption may result in a major effect for the insect. Disruption of the endocrine system for insect control was first proposed by Williams (*19,20*). Considerable attention has been paid to JH analogs (JHAs) for insecticidal use in the field. The JHAs have been exploited for control of a number of pest insects including stored product pests, fleas and mosquito larvae (*21,22*).

Anti-JH activity may potentially result in precocious development, cessation of feeding, and an associated reduction in crop damage. There are a number of compounds with widely different structures which have anti-JH activity, and their potential exploitation for insect control purposes has received considerable attention (*23,24*).

JHE is highly voracious, but is not an abundant enzyme. The development of an efficient affinity purification system (*25*) permitted isolation of sufficient quantities for generation of antibodies. In addition, partial amino acid sequencing was carried out on JHE from a number of different insect species, most notably from the tobacco budworm, *H. virescens,* which is a major pest of cotton. A cDNA library derived from the fat body of fifth instar (day 2) larvae of *H. virescens* was screened, and a clone coding for JHE was identified (*26*).

Baculoviruses: Protein Expression Vectors and Insect Control Agents

With the advent of recombinant DNA technology, baculoviruses have been engineered for production of a large number of both prokaryotic and eukaryotic proteins (*27,28*). The cDNA sequence for the protein concerned is introduced into the viral genome under control of a viral promoter. These recombinant viruses then prove to be powerful tools for high level production of the foreign protein (*29*). As a result of thorough analysis of the viral genome for exploitation by this technology, a considerable amount has been learned of viral gene expression, and baculovirus expression vectors have undergone a number of significant improvements over the years (*28,30,31*).

Prior to the interest shown in baculoviruses as protein expression vectors, augmentation of natural populations of these insect-specific viruses had been in use for decades, as an insect control strategy (*32*). Due to the nature of baculovirus infection of lepidopteran larvae, it takes several rounds of viral replication within the larva before the larva succumbs to the infection. This may take from days to weeks according to environmental conditions. Baculoviruses are used therefore mainly against pests of crops which can sustain some degree of damage, without significant economic loss (*32*).

The development of novel pest control agents is receiving renewed attention in response to increased concern over the impact of chemical insecticides on health and the environment. Genetic engineering of baculoviruses to introduce cDNA sequences coding for insecticidal proteins represents such an alternative approach to pest control.

The logical extension of the interest in anti-JH action, and in use of baculoviruses for insect control, in combination with the recently acquired knowledge of baculovirus genetics, led to the question of whether incorporation of the cDNA sequence for JHE into the baculovirus genome, might result in a quick-kill insecticidal agent. JHE is not orally active, and the baculovirus presents a means of introducing the enzyme into insect larvae. A number of other candidate sequences for other potentially insecticidal agents have also been considered, and expressed in recombinant baculoviruses with varying degrees of success (*33*). These include agents derived from the insects own system, such as diuretic hormone (*34*), eclosion hormone (*35*), and the insect specific toxins derived from a scorpion (*36-38*) or mite (*39-41*).

The baculoviruses which have received the majority of attention to date are two of the nuclear polyhedrosis viruses (NPV). One was originally derived from the alfalfa looper *Autographa californica*, and the other from the silkworm *Bombyx mori*. These viruses are referred to as AcNPV and BmNPV respectively. The engineering of baculoviruses to express JHE was undertaken in both of these systems by a number of different groups. Details of methodology for construction and purification of recombinant baculoviruses are given elsewhere (*28*,*31*).

Expression of JHE by AcNPV and BmNPV

The first work on baculovirus expression of JHE was done in AcNPV. A 1.7 kbp fragment from the 3hv16 clone was used (*26*,*42*). JHE was expressed in a non-occluded virus, AcRP23.JHE, under control of the viral promoter for the polyhedrin protein (Table I; *43-53*). It was shown that the baculovirus-expressed JHE was biologically active, by carrying out the black *Manduca* larval assay (*54*). This commonly used assay for anti-juvenile hormone agents utilizes the fact that *M. sexta* deposits melanin in the cuticle in response to a reduced titer of JH (*24*,*55*). The anti-juvenile hormone agent is injected into larvae of the tobacco hornworm, *M. sexta*. Careful timing of injection results in the larva turning black at the ensuing molt. This effect is reversible by application of a juvenile hormone analog to the injected larva. Of note is the fact that it takes less JHE to induce blackening with increasing instar possibly due to reduced biosynthesis of JH, or lower sensitivity of tissues to JH in the later stadia (*54*).

Despite high level expression of JHE by the recombinant virus in cell culture, titers of JHE in the hemolymph of infected larvae of the cabbage looper, *T. ni*, were low (*47*). Injection of polyhedrin-negative virus is preferable to infection by feeding for such a study, because infectivity of the non-occluded (polyhedrin-negative) viruses is largely uncharacterized. The viral polyhedrin coat provides protection for the enclosed infectious nucleocapsids, until it is dissolved in the alkaline environment of the larval gut. It is likely that non-occluded virus may be inactivated before infecting larval tissues. This may explain the low JHE titers seen *in vivo* on infection with AcRP23.JHE (Table II). For field use occluded viruses are preferable, due to their significantly greater stability.

Table I. Characteristics of Recombinant Baculoviruses Expressing JHE

Virus (*reference*)	promoter (*vector reference*)	Sequences at p10 locus	Sequences at polyhedrin locus	p10	polyhedrin
AcUW2(B).JHE (*43*)	p10 (*44*)	p10	polyhedrin, JHE	+	+
AcPR1 (*45*)	p10 (*46*)	JHE	polyhedrin	-	+
AcRP23.JHE (*47*)	polyhedrin (*48*)	p10	JHE	+	-
AcPR3 (*45*)	polyhedrin (*49*)	p10	JHE	+	-
AcMP1.JHE[a]	basic protein (*50*)	polyhedrin	JHE	-	+
vJHEwt (*51*)	modified polyhedrin (*52*)	p10	JHE	+	-
vJHEEGTZ[b] (*51*)	modified polyhedrin (*52*)	p10	JHE	+	-
vJHEEGTD[c] (*51*)	modified tandem[d] (*52*)	p10	polyhedrin, JHE	+	+
BmNPV.JHE[e]	polyhedrin (*53*)	p10	JHE	+	-

a Bonning (unpublished).
b *egt* disrupted by lacZ.
c non-functional egt gene.
d modified polyhedrin and synthetic promoter placed in tandem. Synthetic promoter derived from concensus sequences of late and very late promoters.
e Hanzlik, Kamita, Maeda and Hammock (unpublished).

Table II. JHE Activity in Hemolymph of Larvae Infected with Recombinant Baculoviruses

virus	Insect species (*reference*)	Instar infected	JHE [a]	JHE [b]	days post infection
			nmol JHIII/min/ml		
Nonoccluded:					
AcRP23JHE	Trichoplusia ni[c] (*47*)	L1	93	10	3
	Trichoplusia ni[d,e]	L5D1	93	200	2.5
				700	3.3
				500	4.4
	Heliothis virescens[e,f]	L5D1	65	1200	2.5
				4100	3.3
				1600	4.4
vJHEEGTD	Trichoplusia ni[f] (*51*)	L4D1	93	39	1
				3040	2
				1210	3
BmNPVJHE	Bombyx mori[e,g]	L2D1	31	250	3
				500	4
		L3D1		1000	3
				1200	4
		L4D1		1500	3
				2600	4
		L5D1		50	2
				700	3
				1800	4
Occluded:					
AcUW2BJHE	Trichoplusia ni (*43*)	L2D2	93	105	2.6
				100	3.6
				115	4.6
				101	5.6
AcPR1	Spodoptera exigua[i]	L4D1	91	9	3
				8	4

a maximum JHE activity seen during normal development.
b JHE activity in hemolymph of infected larvae.
c virus ingested.
d injected with 2 x 10^6 pfu.
e Hanzlik *et al*, unpublished
f injected with 5 x 10^6 pfu.
g injected with 5 x 10^4 pfu.
h injected with 1 x 10^7 pfu.
i Roelvink *et al*, unpublished

Despite the low titers of JHE detected *in vivo*, some feeding inhibition was noted on infection of first instar larvae of the cabbage looper *T. ni*, with the recombinant baculovirus. This inhibition, and subsequent decline in growth, was partially reversed by dorsal application of the juvenile hormone analog, epofenonane, to larvae previously infected with the recombinant virus (*47*).

Juvenile hormone esterase was also expressed under control of the polyhedrin promoter in BmNPV (Hanzlik, Kamita, Hammock and Maeda, unpublished). A 3kbp insert from the clone 3hv1, was introduced into BmNPV (*42*). The *in vivo* expression of AcNPV and BmNPV constructs were compared side by side on injection of *B. mori*, *T. ni* and *H. virescens* (Table II).

Larvae were injected early during the fifth instar with between 2 and 10 $\times 10^6$ plaque forming units of recombinant virus. In all cases, titers of JHE produced by the recombinant viruses were higher than the maximum titers of JHE seen during development (31, 52 and 65 nm JH III hydrolyzed/min/ml hemolymph for *B. mori*, *T. ni* and *H. virescens* respectively; Table II). Following viral expression, the specific activity of JHE was high and similar for both the AcNPV construct in *H. virescens*, and the BmNPV construct in *B. mori*. The total amount of JHE expressed was highest for the recombinant BmNPV in *B. mori*, which can be attributed to the larger size of the insect. However, no morphological changes were seen in *B. mori* in response to overexpression of JHE. In this, and the previous study (*47*), the extent of glycosylation of JHE produced in cell culture was different from that produced by the insect. It is not known whether there is any biological significance associated with these differences in glycosylation of the enzyme.

AcNPV expression of JHE *in vivo* was enhanced by incorporating the cDNA sequence for JHE under control of the viral promoter for the p10 protein in a polyhedrin-positive (occluded) virus AcUW2(B).JHE (*43*). JHE was also expressed in AcNPV under control of the viral basic protein promoter (AcMP1.JHE; *50*), which gave a further improvement in expression relative to the p10 construct (Table I; Bonning, unpublished).

Bioassay data for the time taken for the recombinant baculoviruses to kill, and for the effective doses of the recombinant viruses in *T. ni*, showed no improvement relative to the wild type, non-engineered virus (*38,43*; Table III). Similar results were acquired for bioassays, carried out in larvae of the beet armyworm *Spodoptera exigua* (Roelvink, van Meer, Hammock and Vlak, unpublished; *45*; Table III). Weight gain monitored during the development of infected larvae also indicated that feeding behavior was not altered by the high levels of JHE expressed by the recombinant baculoviruses (*33*).

The Role of Ecdysteroid UDP-Glucosyltransferase (EGT). Identification of a virally encoded ecdysteroid UDP-glucosyltransferase (EGT) (*56,57*) raised the possibility that the action of this enzyme could be blocking the biological effect of the baculovirus expressed JHE. This enzyme inactivates ecdysone by conjugating the C-22 hydroxyl group of ecdysone with a sugar (*58*). The effect of EGT in larvae infected with AcNPV expressing this enzyme, is to block molting, which is normally initiated by ecdysone. It was also noted that larvae infected with viruses lacking the gene for EGT (termed *egt*), feed less during the course of viral infection (*59*). This

Table III. Bioassay Data for Recombinant Baculoviruses Expressing JHE

Insect species	Virus	Instar	LT50 (hours)	LRT50[a]	Instar	LD50 (pibs)	LRD50[a]
Spodoptera exigua	AcPR1	2	117.6	1.065	2	145	1.460
Trichoplusia ni	AcPR1	-	-	-	2	32	1.140[d]
	vJHEEGTD	1	78	0.950[b]	1	(8910)[c]	1.071
	AcUW2BJHE	1	118	1.017	2	74	4.625
	AcmJHE[e]	1	84	0.790	-	-	-
Heliothis virescens	AcUW2BJHE	1	126	1.105	-	-	-
	AcUW2BJHE[f]	2	143	1.144	2	27.2	1.242
	AcmJHE[e]	1	80	0.702	-	-	-

a Lethal ratio for time (LRT) or dose (LRD) calculated as: $\frac{\text{LD50 or LT50 for test virus}}{\text{LD50 or LT50 for control virus}}$ (*33*).
b LRT50 for vJHEEGTD calculated with reference to control virus with non-functional egt gene (vEGTDEL; *51*).
c Expressed as pibs/ml diet. Neonate larvae fed for 24h (*51*).
d LRD50 calculated with reference to control virus lacking p10 gene (Roelvink et al., unpublished).
e Virus expressing JHE modified by site-directed mutagenesis (see text).
f (*38*)

question of the masking of possible effects of JHE by EGT was addressed by Eldridge *et al* (*51*).

JHE was expressed under control of a modified viral promoter in AcNPV constructs with, and without, expression of EGT (Table I). *egt*-minus viruses (without JHE) are slightly more effective in terms of speed of kill, than *egt*-positive viruses (*59*). Analysis of weight gain of larvae infected with the test viruses, and of bioassay data for lethal time and viral dose, indicated that there was no difference in the activity of expressed JHE in the presence or absence of EGT. The reason for this is unclear. However, the bioassays reported were carried out on fourth and fifth instar larvae of *T. ni*; Precocious development would not be detectable at this late stage, as the larvae would normally pupate very shortly after the time chosen for infection. Bioassays on early instar larvae are also preferable in terms of field applicability, as early instars will be targeted for optimal pest control.

Possible Reasons for the Lack of Biological Effect of Wild Type JHE. During the normal decline in JH titer early in the last larval instar, the rate of JH biosynthesis declines dramatically while JHE levels in both hemolymph and tissues increase. Levels of epoxide hydrolase acting on JH may be high in some tissues, especially key target tissues such as the imaginal discs. The level of JH binding protein in the hemolymph and presumably in most tissues remains high, while one can only speculate on changes in receptor type and quantity. The precipitous decline in both hemolymph and total JH titers caused by these events leads to a cessation of feeding and an onset of wandering behavior.

The failure of the artificially high titers of JHE to disrupt larval development may be due to several factors. Since the reduction in JHE is mediated by a variety of events, one can predict that for recombinant JHE to be effective, levels in the hemolymph must greatly exceed those normally present in the last larval instar. In normal larvae JH metabolism by esterase and epoxide hydrolase is high in most tissues while in infected larvae JHE levels increase in the hemolymph and infected, but presumably not uninfected tissues. Thus tissue levels of JH may not decrease sufficiently to induce metamorphosis. Since epoxide hydrolase levels are not affected there is also the possibility that JH acid may be remethylated to an active form in some tissues. The JH binding protein is hypothesized to increase the degradation of JH in the presence of JHE by keeping the JH in a pool accessible to the enzyme. However, the titers of JH binding protein in infected insects are not known.

Removal of expressed JHE may occur before anti-JH effects are induced, or JH biosynthesis may be increased in response to the artificially high titers of JHE. As shown in Table II, levels of JHE in late larval instars of most species tested can reach quite high levels. Although these levels of JHE activity far exceed the normal concentrations leading to pupation, demonstrating precocious development in such late instars is unlikely. In earlier instars, the levels of expression are quite low. It is unlikely that they will overcome the endogenous rates of JH biosynthesis sufficiently to result in precocious development. Even very low concentrations of JH in *T. ni* (about 30 pg/g tissue) are capable of maintaining active feeding behavior by the larva and preventing onset of wandering behavior as a prelude to pupation (*60*), and the rate of clearance of JH will also decrease with decreasing JH titer.

The plethora of literature that deals with hormonal regulation of insect development gives some indication of the complexities involved. The role of JHE varies between species during the early larval instars, and the reduced sensitivity of tissues to anti-JH effects via JHE in later instars (*54*) may well be limiting the insecticidal efficacy of the recombinant baculoviruses expressing JHE.

Coexpression of JHE with JHBP (to facilitate JH degradation by JHE), or epoxide hydrolase (for enhanced JH degradation within the cells) may enhance the insecticidal potential of baculovirus-expressed wild type JHE. Other factors which reduce JH biosynthesis, or affect JH-receptor type and density may also have potential for a multi-component approach.

Pharmacokinetics of JHE. When high titers of purified recombinant JHE from *H. virescens* were injected into *M. sexta*, a rapid decline of this JHE in the hemolymph was noted (*61*). JHE was removed by a saturable process, possibly by receptor mediated endocytosis, and there was no difference in clearance rate between different larval stadia of *M. sexta* and *H. virescens*. As noted above, it is possible that the recombinant virus-expressed JHE may be removed before having a significant biological effect. However, baculovirus expressed JHE reaches very high activity levels *in vivo*, up to 20 or 40 times the maximum titers seen during the course of normal larval development in some cases (Table II; *51*). Despite this fact, precocious development of larvae infected with the various recombinant viruses is not seen.

Further investigation by electron microscopy brought to light that JHE is removed from the hemolymph by the pericardial cells of the insect (*62,63*). JHE marked by immunogold labeling can be seen concentrated in the lysosome-like granules of the pericardial cells where the enzyme presumably is degraded (Figure 2).

Modification of JHE. Analysis of the cDNA sequence encoding JHE showed various characteristic sequences believed to be involved in the degradation of proteins (*64-67*). A potential lysosome-targeting sequence was also identified. Site-directed mutagenesis was carried out within these areas to alter single amino acid residues within these regions of the sequence. This was done in order to stabilize the enzyme against proteolytic attack. Expression of the modified forms of JHE in AcNPV (using the AcUW2(B) construct; *44*) and subsequent bioassay, showed that several highly effective viral insecticides had been generated (Bonning and Ward, unpublished). The time taken by these recombinant viruses to kill larvae of *H. virescens* and *T.ni* was reduced by about 30% (Figure 3). This represents the same degree of efficacy attained by the highly effective recombinant baculoviruses expressing an insect-specific toxin derived from a scorpion (*37,41,68*). The reduction in the time taken to kill represents a significant decrease in feeding. Further analysis of the viruses expressing modified JHE is underway in order to confirm the insecticidal mode of action.

Prospects for an Effective Insecticidal Baculovirus Expressing Modified JHE

The enhanced speed of kill of baculoviruses expressing modified JHE is very similar to the best reported enhancement with insect-specific toxins. Thus, the main advantage of these viruses expressing modified JHE over those expressing insect-

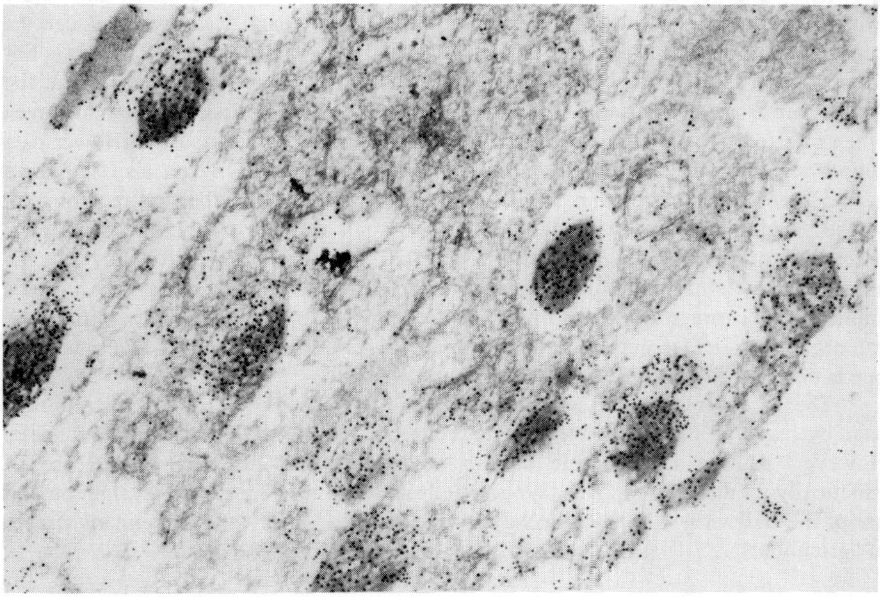

Figure 2
Electron micrograph of pericardial cells isolated from larvae of *Trichoplusia ni*. The section is immunogold labelled for JHE (*62*). The pericardial cells have a characteristic vacuolated appearance, with the JHE concentrated in lysosome-like granules. (EM courtesy of T.F. Booth).

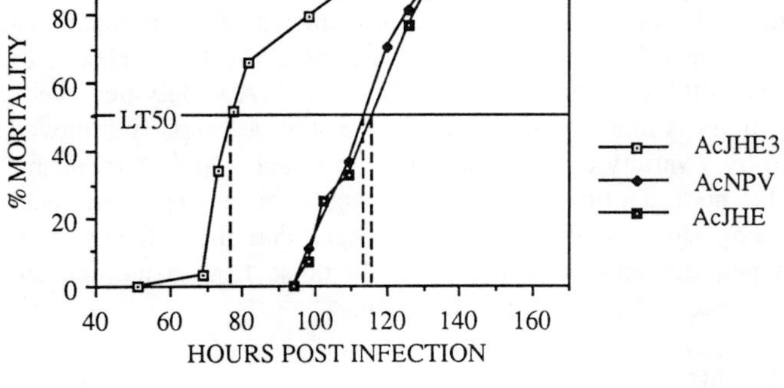

Figure 3
LT_{50} data for AcNPV expressing JHE (AcJHE is AcUW2(B)JHE; *43*), or modified JHE (AcJHE3; see text) compared to the wild type, non-engineered virus (AcNPV).

specific toxins, is one of public acceptability. Extensive risk assessment and safety testing is required before field trials of any recombinant baculoviruses can be considered (*69,70*), and keeping the public informed of events is crucial (*71*). For instance, there is extensive literature offering *in vivo* and *in vitro* evidence, that the AaIT toxin from the scorpion *Androctonus australis* offers no risk to mammals, even when injected directly into the central nervous system (*41*). There are far fewer data supporting the safety of the *Pyemotes*, or mite toxin, but these data also suggest a high degree of safety to mammals. However, it is important that the first recombinant viruses used have not only unequivocal efficacy and absolute safety, but also have a perception of safety. The modified JHE not only has demonstrated efficacy, but the absence of the term toxin in its name may be perceived as being less of a threat than a similar virus expressing even a highly insect-specific, toxic peptide. In addition one can argue that JHE is present naturally in all stages of the pest insect examined and one is only increasing these levels in the pest.

The extent of the influence of such emotive issues on use of recombinant baculoviruses in the field, remains to be seen, and is essentially unpredictable. However, caution is advisable as the regulations and protocol for release of genetically modified organisms take shape. It is hoped that the careful evaluation and use of the JHE virus will pave the way for the use of a variety of recombinant viruses in agriculture.

Improvements in Recombinant Baculoviruses.

Possibly the greatest impact of these highly effective viruses expressing JHE is the demonstration that the activity of the original construct can be dramatically improved. When a lead is found in the screening of classical insecticides, this information is taken as a basis upon which to modify structures to enhance a variety of properties including action at the target site, distribution, and metabolic stability. When similar activity is observed with recombinant organisms resulting from much smaller research efforts, there seems to be a tendency to evaluate the organisms as practical control agents immediately, rather than as leads for improved products. Numerous methods for the systematic improvements of recombinant baculoviruses have been outlined (*68,72*). One of these methods is illustrated clearly by these data on the enhanced biological activity of modified JHE. A widely observed principle in regulatory biology is that potent chemical mediators are rapidly removed from their site of action by a variety of processes including uptake and metabolism. Given this knowledge, the poor activity observed with baculoviruses expressing neurohormones is not surprising. Just as with JHE, it is likely that the efficacy of a variety of recombinant peptides and proteins, especially those from a natural origin, can be dramatically increased by simple modification.

Acknowledgments

We thank P.W. Roelvink and T.N. Hanzlik for permission to present unpublished data. Research carried out by P.W. Roelvink was supported by BAP-0415-UK and BAP-0416-NL of the risk assessment program of the Commission of the European Communities.

Preparation of this manuscript was supported in part by grants from the National Science Foundation (DCB-8818875), the US Department of Agriculture (91-37302-6186), the North Atlantic Treaty Organization (CG 900955) and the USDA Forest Service (23-696). B.D. Hammock is a Burroughs Wellcome Scholar in Toxicology.

Literature Cited

(*1*) Riddiford, L.M.; Palli, S.R. In *Molecular Insect Science;* Hagedorn, H.H.; (Hildebrand, J.G.; Kidwell M.G.; Law J.H. Eds.; Plenum Press: New York, 1990; pp 77-81.
(*2*) Sehnal, F. In *Ecdysone, from Chemistry to Mode of Action;* Koolman, J. Ed.; Thieme: New York, 1989; pp 271-278.
(*3*) Riddiford, L.M. In *Progress in Ecdysone Research;* Hoffman, J.A. Ed.; Elsevier: Amsterdam, 1980; pp 409-430.
(*4*) Sehnal, F. In *Comprehensive Insect Physiology, Biochemistry and Pharmacology*; Kerkut, G.A.; Gilbert, L.I. Eds.; Pergamon Press: New York, 1985, Vol. 2; pp 17-102.
(*5*) Feyereisen, R. In *Comprehensive Insect Physiology, Biochemistry and Pharmacology;* Kerkut, G.A.; Gilbert, L.I. Eds.; Pergamon Press: New York, 1985; pp. 391-430.
(*6*) Hammock, B.D. In *Comprehensive Insect Physiology, Biochemistry and Pharmacology;* Kerkut, G.A.; Gilbert, L.I. Eds.; Pergamon Press: New York, 1985, pp. 431-472.
(*7*) Abdel-Aal, Y.A.I.; Hammock, B.D. *Insect Biochem.* **1988,** *18*, 743-750.
(*8*) Hanzlik, T.N.; Hammock, B.D. *Arch. Insect Biochem. Physiol.* **1988,** *9*, 135-156.
(*9*) Sparks, T.C.; Hammock, B.D. *J. Insect Physiol.* **1979,** *25*, 551-560.
(*10*) Jones, G.; Wing, K.D.; Jones, D.; Hammock, B.D. *J. Insect Physiol.* **1981,** *27*, 85-91.
(*11*) Jones, D.; Jones, G.; Hammock, B.D. *J. Insect Physiol.* **1981**, *27*, 779-788.
(*12*) Wroblewski, V.J.; Harshman, L.G.; Hanzlik, T.N.; Hammock, B.D. *Arch. Biochem. Biophys.* **1990**, *278*, 461-466.
(*13*) Jones, G.; Jones, D.; Hiremath, S. *Insect Biochem.* **1987**, *17*, 897-904.
(*14*) Newitt, R.; Hammock, B.D. *Insect Biochem.* **1989**, *19*, 557-571.
(*15*) Goodman, W.G.; Chang, E.S. In *Comprehensive Insect Physiology, Biochemistry and Pharmacology;* Kerkut, G.A.; Gilbert, L.I. Eds.; Pergamon Press: New York, 1985, Vol. 7; pp 491-510.
(*16*) Wing, K.D.; Rudnicka, M.; Jones, G.; Jones, D.; Hammock, B.D. *J. Comp Physiol.* **1984,** *154*, 213-223.
(*17*) Goodman, W.G. *Insect Biochem.* **1985,** *15*, 557-564.
(*18*) Yamamoto, K.R. *Ann. Rev. Gen.* **1985**, *19*, 209-252.
(*19*) Williams, C.M. *Nature* **1956**, *178*, 212-213.
(*20*) Williams, C.M. *Science* **1967**, *217*, 13-17.
(*21*) Henrick, C.A.; Staal, G.B.; Siddall, J.B. In *The Juvenile Hormones;* Gilbert, L.I. Ed.; Plenum Press: New York 1976; pp 48-60.
(*22*) Sehnal, F. In *The Juvenile Hormones;* Gilbert, L.I. Ed ; Plenum Press: New York, 1976; pp 301-322.
(*23*) Bowers, W.S. In *The Juvenile Hormones* Gilbert, L.I. Ed.; Plenum Press: New York, 1976; pp 394-408.

(*24*) Staal, G.B. *Ann. Rev. Entomol.* **1986**, *31*, 391-429.
(*25*) Abdel-Aal, Y.A.I.; Hammock, B.D. *Science* **1986**, *233*, 1009-1124.
(*26*) Hanzlik, T.N.; Abdel-Aal, Y.A.I.; Harshman, L.G.; Hammock, B.D. *J. Biol. Chem.* **1989**, *264* (21), 12419-12425.
(*27*) Luckow, V.A.; Summers, M.D. *Biotechnol.* **1988**, *6*, 47-55.
(*28*) O'Reilly, D.R.; Miller, L.K.; Luckow, V.A. *Baculovirus expression vectors: a laboratory manual.* W.H. Freeman and Company: New York, NY, 1992; 347 pp.
(*29*) Miller, L.K. *Ann. Rev. Microbiol.* **1988**, *42*, 177-99.
(*30*) Bishop, D.H.L. *Sem. Virol.* **1992**, *3*, 253-264.
(*31*) King, L.A.; Possee, R.D. *The baculovirus expression system, a laboratory guide.* Chapman and Hall: London, New York, 1992; 229 pp.
(*32*) Entwistle, P.F.; Evans, H.F. In *Comprehensive Insect Physiology, Biochemistry and Pharmacology;* Kerkut, G.A.; Gilbert, L.I. Eds.; Pergamon Press: New York, 1985, Vol. 12; pp. 347-412.
(*33*) Bonning, B.C.; Hammock, B.D. *Biotechnol. Gen. Eng.Rev.* **1992**, *10*, 453-487.
(*34*) Maeda, S. *Biochem. Biophys. Res. Comm.* **1989**, *165*, 1177-1183.
(*35*) Eldridge, R.; Horodyski, F.M.; Morton, D.B.; O'Reilly, D.R.; Truman, J.W.; Riddiford, L.M.; Miller, L.K. *Insect Biochem.* **1991**, *21*, 341-351.
(*36*) Maeda, S.; Volrath, S.L.; Hanzlik, T.N.; Harper, S.A.; Maddox, D.W.; Hammock, B.D.; Fowler, E. *Virol.* **1991**, *184*, 777-780.
(*37*) Stewart, L.M.D.; Hirst, M.; Ferber, M.L.; Merryweather, A.T.; Cayley, P.J.; Possee, R.D. *Nature* **1991**, *352*, 85-88.
(*38*) McCutchen, B.F.; Choudary, P.V.; Crenshaw, R.; Maddox, D.; Kamita, S.G.; Palekar, N.; Volrath, S.; Fowler, E.; Hammock, B.D.; Maeda, S. *Biotechnol.* **1991**, *9*, 848-52.
(*39*) Tomalski, M.D.; Miller, L.K. *Nature* **1991**, *352*, 82-85.
(*40*) Tomalski, M.D.; Miller, L.K. *Biotechnol.* **1992**, *10*, 545-549.
(*41*) McCutchen, B.F.; Hammock, B.D. In *Natural and Derived Pest Management Agents* Hedin, P.; Hollingworth, R.; Menn, J. Eds.; ACS Symposium Series, American Chemical Society, Washington DC, 1993.
(*42*) Hammock, B.D.; Wroblewski, V.; Harshman, L.; Hanzlik, T.; Maeda, S.; Philpott, M.; Bonning, B.C.; Possee, R. In *Molecular Insect Science*; Hagedorn, H.H, Ed.; Plenum Press, New York, 1990; pp 49-56.
(*43*) Bonning, B.C.; Hirst, M.; Possee, R.D.; Hammock, B.D. *Insect Biochem. Molec. Biol.* **1992**, *22*, 453-458.
(*44*) Weyer, U.; Knight, S.; Possee, R.D. *Journal of General Virology* **1990**, *71*, 1525-1534.
(*45*) Roelvink, P.W.; Van Meer, M.M.M.; De Kort, C.A.D.; Possee, R.D.; Hammock, B.D.; Vlak, J.M. *J. Gen. Virol.* **1992**, *73*, 1481-1489.
(*46*) Vlak, J.M.; Schouten, A.; Usmany, M.; Belsham, G.J.; Klinge-Roode, E.C.; Maule, A.; Van Lent, J.W.M.; Zuidema, D. *Virol.* **1990**, *178*, 312-320.
(*47*) Hammock, B.D.; Bonning, B.C.; Possee, R.D.; Hanzlik, T.N.; Maeda, S. *Nature* **1990**, *344*, 458-61.
(*48*) Possee, R.D.; Howard, S.C. *Nucleic Acids Res.* **1987**, *15*, 10233-10248.
(*49*) Zuidema, D.; Schouten, A.; Usmany, M.; Maule, A.J.; Belsham, G.J.; Roosien, J.; Klinge-Roode, E.C.; Van Lent, J.W.M.; Vlak, J.M. *J. Gen. Virol.* **1990**, *71*, 2201-2209.

(*50*) Hill-Perkins, M.S.; Possee, R.D. *J. Gen. Virol.* **1990**, *71*, 971-976.
(*51*) Eldridge, R.; O'Reilly, D.R.; Hammock, B.D.; Miller, L.K. *Appl. Env. Microbiol.* **1992**, *58*, 1583-1591.
(*52*) Wang, W.; Ooi, B.G.; Miller, L.K. *Gene* **1991**, *100*, 131-137.
(*53*) Maeda, S. In *Invertebrate cell system applications;* Mitsuhashi, J. Ed.; CRC Press: Boca Raton, Florida, 1989; Vol. 1, pp 167-181.
(*54*) Philpott, M.L.; Hammock, B.D. *Insect Biochem.* **1990**, *20*, 451-459.
(*55*) Truman, J.W.; Riddiford, L.M.; Sefranek, L. *J. Insect Physiol.* **1973**, *19*, 195-205.
(*56*) O'Reilly, D.R.; Miller, L.K. *Science* **1989**, *245*, 1110-12.
(*57*) O'Reilly, D.R.; Miller, L.K. *J. Virol.* **1990**, *64*, 1321-1328.
(*58*) O'Reilly, D.R.; Brown, M.R.; Miller, L.K. *Insect Biochem. Molec. Biol.* **1992**, *22*, 313-320.
(*59*) O'Reilly, D.R.; Miller, L.K. *Biotechnol.* **1991**, *9*, 1086-1089.
(*60*) Jones, G.; Hanzlik, T.; Hammock, B.D.; Schooley, D.A.; Miller, C.A.; Tsai, L.W.; Baker, F.C. *J. Insect Physiol.* **1990**, *36*, 77-83.
(*61*) Ichinose, R.; Kamita, S.G.; Maeda, S.; Hammock, B.D. *Pestic. Biochem. Physiol.* **1992**, *42*, 13-23.
(*62*) Booth, T.F.; Bonning, B.C.; Hammock, B.D. *Tissue and Cell* **1992**, *24*, 267-282.
(*63*) Ichinose, R.; Nakamura, A.; Yamoto, T.; Booth, T.F.; Maeda, S.; Hammock, B.D. *Insect Biochem. Molec. Biol.* **1992**, *22*, 893-904.
(*64*) Rogers, S.; Wells, R; Rechsteiner, M. *Science* **1986**, *234*, 364-368.
(*65*) Rechsteiner, M. *Ann. Rev. Cell Biol.* **1987**, *3*, 1-30.
(*66*) Bachmair, A.; Finley, A.; Varshavsky, A. *Science* **1986**, *234*, 179-186.
(*67*) Bachmair, A.; Varshavsky, A. *Cell* **1989**, *56*, 1019-1032.
(*68*) Hammock, B.D.; McCutchen, B.F.; Beetham, J.K.; Choudary, P.; Ichinose, R.; Ward, V.K.; Vickers, J.; Bonning, B.C.; Harshman, L.G.; Grant, D.; Uematsu, T.; Lanzrein, B.; Maeda, S. *Arch. Insect Biochem. Physiol.* **1992**, *22*, 315 -344.
(*69*) Summers, M.; Engler, R.; Falcon, L.A.; Vail, P. *Baculoviruses for insect pest control: safety considerations*. American Society for Microbiology, Washington; 1975.
(*70*) Cory, J.S.; Entwistle, P.F. *Asp. Appl. Biol.* **1990**, *24*, 187-194.
(*71*) Bishop, D.H.L.; Entwistle, P.F.; Cameron, I.R.; Allen, C.J.; Possee, R.D. In *The Release of Genetically Engineered Micro-organisms*; Sussman, M.; Collins, C.H.; Skinner, F.A.; Stewart-Tull, D.E., Eds.; Academic Press: New York, 1988; pp 143-179.
(*72*) Maeda, S.; Hammock, B.D. In *Newer Pest Control Agents and Technology with Reduced Environmental Impact*; Duke, S.O.; Menn, J.J.; Plimmer, J.R. Eds.; ACS Symposium Series, American Chemical Society, Washington DC, 1993; pp 281-297.

RECEIVED August 28, 1993

Chapter 27

Formulation and Application of Viral Insecticides

S. Y. Young

Department of Entomology, University of Arkansas, Fayetteville, AR 72701

Methodologies most often used for formulation and application of viral insecticides are those developed for conventional chemical insecticides. Viral insecticides are most effectively formulated as wettable powders by lyophilization or spray dry methods. These formulations are best standardized using both counts of occluded virus particle concentration and bioassay activity. Viral insecticides are typically applied as sprays against larval pests of Lepidoptera and Hymenoptera (sawfly) using both aerial and ground equipment. Spray parameters for viral insecticides are not well understood and available equipment is not suitable for their most efficaceous use. Much of the research on virus application has been on development of adjuvants for tank mixtures to overcome problems with plant coverage and sunlight inactivation.

Viral diseases have been isolated from several hundred insect species (*1*). Although these viruses represent several families, those studied for use as control agents are almost exclusively limited to occluded viruses (nuclear polyhedrosis [NPV] and granulosis [GV]) of Baculoviridae. Known occluded baculoviruses are mostly restricted to larvae of Lepidoptera and Hymenoptera (sawflies), thus limiting viral insecticide development primarily to these two orders. These viruses are safe (*2, 3*), virulent, efficaceous (*4*) and can be produced in quantity in insect hosts. Virus efficacy is typically assessed in terms relative to the efficacy of standard chemical insecticides, and they seldom fare well in this comparison. Although numerous viruses have been tested in the field as insecticides, few have been developed as viable commercial products. Pesticide formulation and application technology has paralleled the development of synthetic pesticides, accordingly virus formulation and application methodology was copied from this.

0097–6156/94/0551–0384$06.00/0

The formulation and application of viral products using methodology for chemical insecticides may be expedient. However, the applicability of formulation criteria and application technology developed for the fast acting, contact poisons of low molecular weight may be limited for the relatively unstable, slow-acting viral pathogens that must be consumed to be infectious and effective. Application of viral products is in effect an inundative release of a biological control agent. They can be very efficacious when applied in IPM systems where their benign effects on beneficial insects (*5*) and disease producing potential are desirable and advantageous (*6, 7*).

The numerous research programs around the world on development of viral insecticides have resulted in a wide array of technical reports on their formulation and application. There is little current research in this area, and available information has been reviewed in depth in numerous reviews (*4, 8-14*). This report will summarize progress in these areas and elaborate on areas where the need for additional progress is most critical.

FORMULATION

Formulation requirements for viral insecticides are similar to those of conventional chemical insecticides (*10*). The formulation must be a standardized product with little loss of activity during the formulation process, it must be stable in storage, and have good mixing and spraying qualities in the field. The additives must be safe to animals (*2, 3*) and plants.

Methods of Formulation

Formulation as Sprays. Virus formulations have usually been for application as sprays using conventional application equipment. Most viruses tested by researchers in field trials have been aqueous suspensions of filtrates or precipitates from macerated virus-killed cadavers. These are stable for long periods under refrigeration or when frozen and usually are easily tank-mixed with water and other adjuvants (*12, 13*). Performance of these simple formulations in field trials usually is equal or superior to more elaborate formulations. They are seldom practical for commercialization, however. Commercial formulations are concentrated precipitates because stable and concentrated product is needed for storage and shipping.

Attempts to improve virus formulations have been primarily by the product registrant, especially private industry as this is where much of the formulation expertise exists. Commercial formulations have often been wettable powder, because the particulate nature of virus makes formulation of flowables difficult. Virus in dry formulations typically exhibits better shelf life than in liquid formulations (*12*). It must be noted, however, that not all dry formulations are stable, particularly those formulated as acetone precipitates. When formulated as an acetone extract, the early experimental formulations of *Helicoverpa zea* NPV, including Viron-H (International Minerals and Chemical Corp., Libertyville, IL) lost activity between shipment in the spring to the scientists studying their efficacy against cotton insects and the end of the cotton producing season in the summer (*10*). Although acetone precipitates mix well for use as sprays, this procedure is seldom desirable.

Lyophilization is often used effectively in formulation of viruses, but it may be expensive. A common problem with lyophilized products is clumping of virus in the lyophilization step. Although milling can powder the lyophilate, some clumping of polyhedra remains, limiting dispersal in the tank mixture (*15*). The U. S. Forest Service uses lyophilization followed by milling to formulate both *Lymantria dispar* NPV (Gypchek) and *Orgyia pseudotsugata* NPV (TM-Biocontrol-1).

Formulation as Dusts or Granules. Viruses have been occasionally formulated as dusts or granules, but these have usually been less effective than sprays. The *H. zea* NPV has been formulated as a dust using attaclay (*16*) and with cellulose plus cottonseed meal, wheat or corn (*17*). *Plodia interpunctella* GV has been formulated as a dust using wheat flour (*18*).

Formulation using Microencapsulation Techniques. Viruses have been formulated as wettable powders using microencapsulation techniques. The *H. zea* NPV was formulated on two occasions by microencapsulation using cellulose, gelatin or polymer formed from styrene maleic anhydride (*19, 20*). The greatest advance in formulation was made by Sandoz, Inc. (Palo Alto, CA) using spray-dry techniques. In this process a microencapsulated wettable powder is produced by spray-drying the virus with clay and other diluents. This procedure was used to produce several experimental formulations of *H. zea* NPV which were more stable, exhibited greater activity per unit of product and resulted in desirable spray qualities (*20, 21*). Sandoz 240-070-2WP, one of the most effective of these, was registered as Elcar and briefly marketed for suppression of heliothine species in cotton. Sandoz Agro, Inc. has since used this process to successfully formulate several other experimental viral products including *Syngrapha falcifera* NPV and *Autographa californica* NPV (Sandoz 404). This procedure is apparently not applicable to all NPV's, however, as the *Choristonura* NPV lost much activity when formulated by this process (*10*). Also, the spray dry formulation of *Laspeyresia pomonella* GV (Sandoz 406) was not stable as the loss of active product during storage was excessive.

Stability in Storage. Stability in storage is particularly important for viral insecticides. Most chemical insecticides have an array of potential usages and can be stored for lengthy periods at room temperature without loss of activity. The baculovirus host ranges are often limited to closely related species within an order and their efficaceous use is confined to one or a few pest species in appropriate IPM systems. Therefore, long periods may be required between opportunities for use. At present, it is recommended that virus products be stored refrigerated or frozen. These storage conditions are not readily available, thus putting virus products at a competitive disadvantage with chemical insecticides.

Standardization

Standardization of virus formulations is more difficult than for chemical insecticides. Although viruses have been standardized on a weight basis in dry formulations, this is unsatisfactory (*22*) as viruses may lose activity in storage. In addition, the quantity of virus produced in cadavers may be influenced by a variety of factors and comparable procedures of preparation may result in considerable differences in virus yield per unit weight. Viruses have also been standardized by counts of occluded virus under a light microscope. However, this method may provide erratic results as activity per polyhedra can differ and can be reduced during storage without discernable breakdown under the microscope.

Standardization of virus activity by bioassay against larvae of a susceptible host is a more successful method of standardizing virus formulations. However, they are time consuming, expensive and data are often less precise than desired (*22, 23*). The most commonly used bioassay is diet surface treatment (*24*). A recently developed bioassay is a virus suspension feeding method (*25*). For some situations this method may be preferred to the diet method, but although larvae older than neonates need to be starved prior to exposure for consistent results. The most successful standardization protocols have employed both polyhedral counts and bioassay.

Spray Adjuvants

Viral formulations may contain ingredients to increase insecticidal activity of the applied product such as surface active agents, ultraviolet light screens or gustatory stimulants. These materials may present problems when included in the primary formulation, and are often not added until the product is mixed in the spray tank (tank mixtures). Adjuvants added to the primary formulation often significantly dilute the product and may increase the expense of shipping and storage. Furthermore, adjuvants may not be compatible with the virus formulation, reducing activity during formulation and (or) decreasing stability during storage.

Acetone precipitates of virus have been produced with sugar (*26*) to obtain a product that is easily resuspended for spraying. Lyophilization of virus in sugar can also reduce the clumping of polyhedra encountered in this process. Several experimental formulations have been developed that retard sunlight inactivation and extend field persistence. Microencapsulated *H. zea* NPV products of National Cash Register Company (Dayton, OH) and Southwest Research Institute (San Santonio, TX) contained UV screens that provided a high level of persistence for several days in the field (*19, 27*). Both formulation encapsulation procedures result in unacceptable losses in virus activity during the process, however. Recently sunlight protectants were included in formulations produced by starch encapsulation technology. The addition of activated carbon, dye, stelbene fluorescent brightners (*28*) or polyflavanoid significantly increase persistence of the virus in formulations exposed to UV light. Although the process is simple, inexpensive and increases virus persistence, the particle size produced is considered too large for formulation of viruses (*29*). Sandoz, Inc. also developed spray-dry products of *H. zea* NPV, Sandoz-240-070-3 and Sandoz-240-070-5, that contain UV screen which extends virus activity in cotton fields (*21*).

Formulation of viral products has been more difficult than that of chemical insecticides. No single method has been successful for all viruses. Although aqueous preparations of filtrates of virus-killed cadavers are feasible for small quantities of product and inexpensive, they are not desirable for formulations of large quantities of virus due to the dilute nature of the preparations and instability from activity loss and bacterial contamination during long term storage at room temperature. The more successful formulations have been wettable powders from spray-dry methods, provided activity is not lost during the formulation or lyophilization processes or storage. Chemicals to improve performance of the virus in the field should seldom be included in the primary formulation if they can be effectively added to the tank mixture at application.

APPLICATION

Viral insecticides are typically used in situations where they must compete with chemical insecticides for a market share. Further, they are applied with equipment that was developed for dispensing chemical insecticides. Although they are seldom as efficaceous (i.e. less mortality and slower-acting) as most chemical insecticides, production practices and available equipment require that they be used in this way. Ideally viruses should be used in a manner that facilitates their dispersal potential and epizootic development within and between generations of the insect host. Where cropping practices and pest population size allow this, viral pesticides have been more successful (*30, 31*).

Timing of Application

Viral insecticides are slow-acting agents that are not very effective against older, more damaging larvae. Applications must be carefully timed so that mortality occurs while larvae are small. Timing may differ, however, if older larvae are more exposed during feeding than small larvae, as is the case with eastern spruce budworm, *Choristoneura fumiferana* in forests of North America (*31*). Foliage feeding larvae are the most likely candidates for control with viral insecticides (*10, 31*). *Helicoverpa zea* NPV has been used with some success against heliothine species, foliage and fruit feeders, on cotton. This is because application can be timed against small larvae that often feed on the upper terminals (i.e. foliage where they are exposed to spray deposits).

When a viral insecticide is applied against a subeconomic population for control of succeeding generations, it may be desirable to direct the application to larger larvae. The quantity of polyhedra produced in virus-killed cadavers will increase with their size at death. Timing the application so that treated larvae are large at death will result in higher deposits of viral inoculum against the target generation of the pest population. Timing virus applications to control future generations or in succeeding years is more practical against foliage feeders on perennial crops such as in forest situations (*31*).

Proper timing of viral insecticides is critical for their efficaceous use on crops. Much of the variation in efficacy in field tests can be attributed to improper timing (usually late) of the application. In many instances proper timing could mean treating when the target population is predominantly eggs and newly hatched larvae. Effective application timing requires a more extensive (and expensive) scouting program than is typically used for chemical insecticides. On some crops with high damage thresholds, where thresholds do not exist, or in areas where pest populations seldom reach economical levels, systematic survey programs rarely exist. Furthermore, insect control literature may recommend that insecticidal application be applied against large larvae. Viral insecticides are not likely to be effective in these systems where early population detection and application cannot be assured.

Viral insecticides kill slowly and result in lower mortalities than chemical insecticides in most instances. This often dictates that virus application be directed against pest populations of lower density than would be necessary for chemical insecticides. The lower levels of control obtained and the short residual period of viral insecticides may also require that virus treatments be timed closer together and applied more often than for chemical insecticides. Because viral insecticides are less effective than chemicals in producing quick results, they are less likely to control high population densities of the target population. Because producers are accustomed to evaluating the effectiveness of the fast-acting chemical insecticides, they find it difficult to delay their evaluation of virus application for the several days or even weeks often required for treated larvae to die (*31*). If not properly trained, users may switch to chemical insecticides prematurely and never realize that the virus would have been efficaceous.

Methods of Application

Application Equipment. For the most part equipment cost and availability dictate that viral insecticides be applied with equipment designed for chemical insecticides. Accordingly, most research on application of viral insecticides has been with this equipment. In earlier years some dust equipment was available, but at present viruses are applied as concentrated sprays from ground or air application equipment. Whether applied as sprays, dusts or granules, the use of equipment developed for contact chemical insecticides does not appear capable of providing the desired coverage needed for the viruses. Virus must be deposited at feeding sites of the target pest so that it can be ingested. This may require a method of application that directs the spray to areas of the plant canopy that can not be reached using conventional equipment. The parameters necessary for application of viral insecticides are not well understood (*12*).

Field trials in which viruses are formulated as dusts or granules have been limited. Results of these tests have often been discouraging as efficacy of viruses formulated as dusts or granules is less than that of spray formulations (*10*).

Air vs. Ground Application Equipment. Ground spray equipment is the method of choice for testing viral insecticides in field crops. It is readily available and lends itself well to use in small plots. With both air and ground equipment, a boom-type sprayer with flat fan or hollow-cone nozzles is often used. For control of forest pests, viral insecticides are typically applied by air, although control is generally superior when ground equipment is used in young plantations, parks, beside roads, etc. (*31*). High pressure spray equipment is more likely to be used to apply virus in fruit and vegetable IPM programs. Application volume is normally similar to that used for chemical insecticides, 10 to 47 l/ha and 19 to 114 l/ha for air and ground, respectively. Higher spray volumes generally increase coverage and efficacy, although there are exceptions (*10, 30, 32*). Use of high spray volume is limited primarily by the expense and time required to cover a given area (*4*).

There is often a poor correlation between virus rate and efficacy in tests in the field (*4, 31*), suggesting that available application methods are not effective in delivering the virus to feeding sites of the target pest. The desired spray parameters may differ with target insect and crop. In situations where many individuals feed secluded, the concept of "good" coverage may differ from that commonly accepted for conventional chemical insecticides. It may be desirable to maximize virus deposits to control small larvae in more exposed sites while ignoring the remaining larvae, as any gain obtained from attempts to direct the spray to less exposed larvae could be small and not compensate for reduced control of exposed larvae. Existing application methods have been used most successfully with viral insecticides against larval pests that feed on foliage in the outer canopy (*4, 30, 31*).

Unconventional Methods of Application. In addition to the conventional methods of insecticide application, several novel methods of virus delivery have been tested. Most of these were designed for long term population suppression by utilizing the disease producing features of dispersal and epizootic development. Autodissemination of the virus by male moths contaminated with virus in baited light or pheromone traps has been demonstrated (*33*). The males in turn contaminate the female genitalia during mating and eggs are contaminated externally when layed. Hatching larvae eat the virus deposits on the eggshell and become infected. Honey bee, *Aphis melliphora*, pollinators have been contaminated with *H. zea* NPV formulated as a dust while entering and leaving their hive (*33*). This virus was successfully dispersed into nearby clover fields and epizootics developed in *H. zea* populations were subsequently observed. Chemigation application with irrigation water has been used to apply the *Spodoptera frugiperda* NPV and *H. zea* NPV against their respective hosts on corn (*35*). Although offering potential for efficaceous use in some systems the expense of this type of application is prohibitive in many cropping systems unless irrigation is needed at the time of application.

Tank Mixtures

Much of the effort on viral insecticide research has been aimed at the development of adjuvants to improve efficacy. Although some effort has been directed at incorporation of adjuvants into the basic formulation, a majority of effort has been with spray tank adjuvants. Materials to overcome the potential efficacy limiting characteristics of the virus (e.g. stability and coverage) have been targets of research.

Surface Active Agents. Surface-active agents are most often added to tank mixtures. These may be evaporation retardants, spreader-stickers and/or wetting agents. A wide variety of available commercial materials are compatible with viral insecticides, although performance of these agents is not consistent (*9*, *12*).

Sunlight Screens. The rapid inactivation of viral insecticides by the UV spectrum of sunlight has led to the development of sunlight screens for use in tank mixtures that increase virus persistence on the plant. Shade™ (Sandoz, Inc.) and lignin sulfate are two effective sunlight protectants most commonly used with applications in forests and agricultural crops (*10*, *31*). Many other materials compatible with viral insecticides are effective sunlight screens (*13*, *28*). Although effective as sunlight screens, these materials often do not significantly increase virus efficacy (*4*, *31*). The increase in effectiveness of L. dispar NPV against L. dispar larvae observed with stilbene fluorescent brighteners does not appear to result from their activity as sunlight screens, since these chemicals are also effective synergists for the NPV in laboratory assays (36). These findings suggest that against many insects, virus persistence is adequate at the feeding sites of the target insect, such as on undersurface of leaves, terminals, etc. Furthermore, since sunlight screens add expense to the cost of application and at higher concentrations impart undesirable spray qualities (difficult to mix and spray), it may not be practical in many instances to use them at the concentrations needed to provide efficacious sunlight protection.

Buffers: Because solutions of high ionic concentrations will inactivate virus, it may be necessary to buffer the tank-mixture. Some buffers may also be necessary with some adjuvants and where local water has an extreme pH. In addition, leaf surface chemicals may result in a high pH leaf surface. Cotton leaves have a surface pH between 9 and 10 and have been shown to inactivate viral insecticides applied to the surface. Inactivation is slow (in comparison to sunlight inactivation) and does not appear to be important under field conditions where occasional rainfall occurs (*37*).

Gustatory Stimulants. The addition of gustatory stimulants to the tank mixture has been found to increase consumption of virus by some lepidopterous larval pests. These materials appear to offer potential to increase efficacy of viral insecticides on crops, particularly in situations where larvae feed at sites where coverage is not adequate. Most studies have been conducted with gustatory stimulants of heliothine species on cotton. Host plant extracts and low concentrations of sugars used at rates which increase virus consumption in the laboratory have seldom been efficacious in field tests (*4*). Sugars at higher concentrations (often molasses) and more complex adjuvants that have gustatory stimulant properties have sometimes proven to add to efficacy (*10, 38*). However, improvements may be caused by several other desirable characteristics of these adjuvants (e.g. weak sunlight screens, evaporation retardants, increased droplet density, improve coverage, etc.) (20). When an increase in efficacy occurs it is difficult to determine which desirable property(ies) is responsible. A complex adjuvant, Coax (CCT Corp., Litchfield Park, AZ), with gustatory stimulant and other desirable properties is available commercially but expense prohibits its use at effective rates in many IPM systems.

Mixtures with other Pesticides

An array of pesticides may be applied to most crops during the growing season. Compatibility among pesticides is important as it is often desirable and/or necessary to tank-mix products simultaneously. This is particularly true for virus insecticides because of their extremely limited host range, especially when a pest complex is present. Viral insecticides have been shown to be compatible with most conventional chemical insecticides such as pyrethroids, organophosphates and carbamates (*10, 39*). Viruses are typically compatible with organic pesticides with low ionic concentration (near neutral pH). Viruses have been tested in combination with a variety of insecticides for synergistic activity. In the laboratory, activity has been additive in most instances, with synergistic activity in some instances. In field trials, however, efficacy has seldom been other than additive. Mixtures of virus and *Bacillus thuringiensis* have been tested against several pests in the field with similar results (*40*).

SUMMARY

Production practices and availability of equipment require that viral insecticides be formulated and applied using technology designed for application of contact insecticides. Development of formulation and application methodology for viral insecticides has not proceeded as rapidly as anticipated. Many viral insecticides are efficaceous, but research has slowed because few are marketed. Although viral insecticide use is limited, their potential for use in IPM systems where other pesticides are not available or desirable provides a window of opportunity for greater use in the future. Improvements in formulation and application technology designed for viral insecticides are needed to help viruses reach their potential as commercial insecticides.

LITERATURE CITED

1. Payne, C. C. *Parasitology*. **1982**, *84*; pp 35-77.
2. Saik, J.E.; Lacey, L. A.; Lacey, C. M. In *Safety of Microbial Insecticides*; Laird, M; Lacey, L. A.; Davidson, E. W., Eds; CRC Press: Boca Raton, FL, 1989; pp 115-134.
3. Doller, G. In *Viral Insecticides for Biological Control*. Maramorosch, K.; Sherman, K. E., Eds; Academic Press: New York, NY, 1985; pp 399-440.
4. Yearian, W. C.; Young, S. Y. In *Microbial and Viral Pesticides*; Kurstak, E., Ed; Marcel Dekker: New York, NY, 1982; pp 387-424.
5. Groner, A. In *Safety of Microbial Insecticides*; Laid, M.; Lacey, L. A.; Davidson, E. W., Eds; CRC Press: Boca Raton, FL, 1989; pp 135-148.
6. Harper, J. D. In *Epizootiology of Insect Diseases*; Fuxa, J. R. and Tanada, Y., Ed; John Wiley & Sons: New York, NY, 1987; pp 473-496.
7. Podgwaite, J. D. In *Viral Insecticides for Biological Control*. Maramorosch, K.; Sherman, K. E., Eds.; Academic Press, Inc.: New York, NY, 1985; pp 775-797.
8. Couch, T. L.; Ignoffo, C. M. In *Microbial Control of Pests and Plant Diseases, 1970-1980*; Burges, H. D., Ed; Academic Press: New York, NY. 1981; pp 621-634.
9. Yearian, W. C.; Young, S. Y. *Misc. Publ., Entomol. Soc. Am*, **1978**, *10*; pp 21-26.
10. Young, S. Y.; W. C. Yearian. In *The Biology of Baculoviruses Vol. II. Practical Application for Insect Control*; Granados, R. R.; Federici, B. A., Ed.; CRC Press, Inc: Boca Raton, FL, 1986; pp 157-179.
11. Falcon, L. A. In *Proc. NSF-USDA-Univ. Fla. Workshop; Microbial Control of Insect Pests: Future Strategies in Pest Management Systems*; Allen, G. E.; Ignoffo, C. M.; Jaques, R. P., Eds; University of Florida: Gainesville, FL, 1978; pp 113-117.
12. Smith, D. B.; Bouse, L. F. In *Microbial Control of Pests and Plant Diseases, 1970-1980*; Burges, H. D., Ed; Academic Press, New York, NY, 1981; pp 635-654.
13. Jaques, R. P. *Misc. Publ. Entomol. Sci. Am.* **1977**, *10*; pp 99-116.
14. Hostetter, D. L.; Bell, M. R. In *Viral Insecticides for Biological Control*. Maramorosch, K.; Sherman, K. E., Eds; Academic Press: New York, NY, 1985; pp 249-284.
15. Ignoffo, C. M. *J. Insect Pathol.* **1964**, *6*; pp 318-326.
16. Montoya, E. L.; Ignoffo, C. M. *J. Invertebr. Pathol.*; **1966**, *8*; pp 251-254.
17. Stacey, A. L.; Yearian, W. C.; Young, S. Y. *Ark. Farm Res*; **1977**, *2b*; pp 3-4.
18. McGaughey, W. H. *J. Econ. Entomol.* **1975**, *68*; pp 346-348.
19. Ignoffo, C. M.; Batzer, O. F. *J. Econ. Entomol.*; **1971**, *64*; pp 850-853.
20. Ignoffo, C. M.; Hostetter, D. L.; Smith, D. B. *J. Econ. Entomol.*, **1976**, *69*; pp 207-210.

21. Ignoffo, C. M.; Yearian, W. C.; Young, S. Y.; Hostetter, D. L.; Bull, D. L. *J. Econ. Entomol.*, **1976**, *69*; pp 233-236.
22. Dulmage, H.; Burgerjon, A. *Entomophaga*, **1977**, *22*; pp 131-133.
23. Hughes, P. R.; Wood, H. A. In *The Biology of Baculoviruses. Vol. II. Practical Application for Insect Control*; Granados, R. R.; Federici, B. A., Eds; CRC Press: Boca Raton, FL, 1986; pp 1-30.
24. Ignoffo, C. M. *J. Invertebr Pathol.*, **1965**, *7*; pp 315-319.
25. Hughes, P. R.; Wood, H. A. *J. Invertebr. Pathol.* **1981**, *37*; pp. 154-159.
26. Dulmage, H. T.; Martinez, A. J.; Correa, J. A. *J. Invertebr. Pathol.* **1970**, *16*; pp 80-83.
27. Bull, D. L. *Misc. Pub. Entomol. Soc. Am.* **1978**, *10*; pp 11-20.
28. Shapiro, M. J. Econ. Entomol. **1992**. 85; pp. 1682-1686.
29. Ignoffo, C. M.; Shasha, B. S.; Shapiro, M. J. Invertebr. Pathol. **1991**, *57*; pp 134-136.
30. Huber, J. In *The Biology of Baculoviruses*. Vol. II. Practical Application for Insect Control; Granados, R. R.; Federici, B. A., Ed; CRC Press: Boca Raton, FL, 1986; pp 181-202.
31. Cunningham, J. C. In *Microbial and Viral Pesticides*; Kurstak, E., Ed; Marcel Dekker: New York, New York, 1982; pp 335-386.
32. Smits, P. H.; Rietstra, I. P.; Vlak, J. M. J. Econ. Entomol. **1988**, 81; pp 470-475.
33. Jackson, D. M.; Brown, G. C.; Nordin, G. L.; Johnson, D. W. *J. Econ. Entomol.* **1993**, *85*; pp 710-719.
34. Gross, H. R.; Hamm, J. J.; Carpenter, J. E. Environ. Entomol. **1993**, in press.
35. Hamm, J. J.; Hare, W. W.. J. Econ. Entomol. **1982**, *75*; pp 1074-79.
36. Shapiro, M.; Robertson, J. L. J. Econ. Entomol. **1992**, 85: pp 1120-1124.
37. Young, S. Y.; Yearian, W. C. J. Georgia Entomol. Soc. **1976**, *11*; pp 277-282.
38. Richter, A. R.; Fuxa, J. R. J. Georgia Entomol. Soc. **1984**, *19*; pp 383-387.
39. Benz, G. In *Microbial Control of Insects and Mites*; Burges, H. D.; Hussey, N. W., Eds; Academic Press: New York, NY, 1971; pp 327-356.
40. Harper, J. D. In *The Biology of Baculoviruses. Vol. II. Practical Application for Insect Control.* CRC Press: Boca Raton, FL, 1986; pp 133-156.

RECEIVED April 30, 1993

Evolving Approaches to Pesticide Discovery: Biochemistry and Computer-Aided Design

Chapter 28

EMIL, a System for Computer-Aided Structure Transformation of Bioactive Compounds

Application to Synthetic Pyrethroid Series

Toshio Fujita[1,4], Keiichiro Nishimura[1], Zhi-Ming Cheng[1], Hirosuke Yoshioka[2], Yoshihiro Minamite[3], and Yoshio Katsuda[3]

[1]Department of Agricultural Chemistry, Kyoto University, Kyoto 606–01, Japan

[2]Bioregulator Design and Synthesis Laboratory, Institute of Physical and Chemical Research, Wako, Saitama 351–01, Japan

[3]Central Research Laboratories, Dainippon Jochugiku Company, Ltd., Toyonaka, Osaka 561, Japan

Numerous structural "evolution" processes observed in a number of agrochemical and medicinal series of compounds were collected and arranged so that the substructural modification patterns involved in each process could be used as possible "rules" for bioisosteric transformations to be attempted in drug design studies. EMIL is a system that incorporates a database for these structural evolution examples and a data-processing engine constructed to release "higher-ordered" candidate structures from a "lower-ordered" input structure "automatically" with the aid of the database. The system was applied to series of synthetic pyrethroids. Some compounds belonging to a candidate structural class selected from structures "generated" from fenvalerate as the input structure were synthesized and bioassayed. Certain compounds possessing the isobutyranilidoxime *O*-phenoxybenzyl ether structure were shown to be highly active as house-dust miticides and patented.

Numerous structural series of bioactive compounds exhibiting specific biological effects have been known in agrochemicals as well as in medicines. Among examples in agrochemicals are carbamate (*1*) and phosphate insecticides (*2*), nitromethylene heterocycles (*3*), pyrethroids (*4*), azole fungicides (*5*), and diphenyl ether (*6*) and sulfonylurea herbicides (*7*). In each series, the ultimate prototype compound has been identified or disclosed first either as a bioactive principle contained in natural products or among organic compounds synthesized intentionally or unintentionally.

For synthetic chemists working in the area to develop new agrochemicals and medicines, the most challenging target is probably to identify or to generate the prototype compounds having novel lead structures. Regardless of whether the prototype compounds are from synthetic products or not, however, synthetic chemists are supposed to modify the prototype lead structure not only to optimize activity by systematic replacements of substituents and substructures, but also to explore even

[4]Current address: EMIL Project, Fujitsu Kansai Systems Laboratory, 2–2–6 Shiromi, Chuoku, Osaka 540, Japan

0097–6156/94/0551–0396$06.00/0

newer lead structures with more or less drastic structural changes. The explorations are usually continued one after another sometimes in different institutions independently to evolve "higher-ordered" or new generation structures from more "primitive" or "lower-ordered" lead compounds.

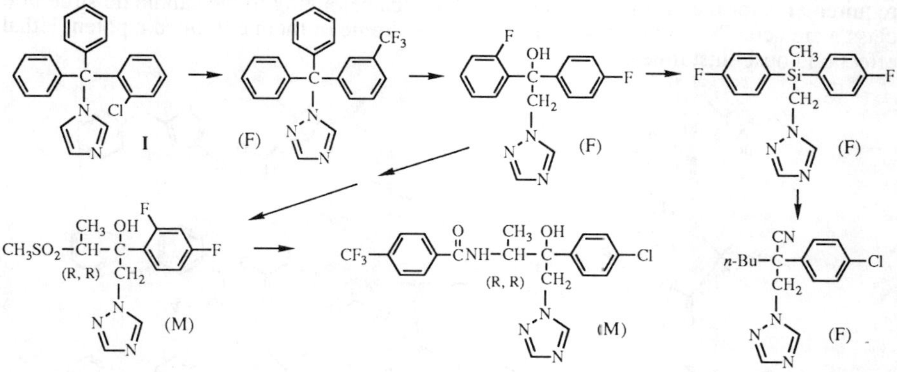

Fig. 1. Structural Evolution from Clotrimazole (**I**, topical antimycotic) to Agricultural Fungicides (F) and Oral Antimycotics (M) (Reproduced with permission from ref. 26. The preparation was assisted by Masanori Yoshida).

These consecutive structural transformations towards newer generation compounds could be called "lead evolution" (*8*). Examples of the simplified lead evolution "tree" for azole fungicides starting from clotrimazole (**I**) (*5,9*), nitromethylene heterocycles leading to imidacloprid (**II**) and analogs (*3,10*, Iwataki, I., Nippon Soda Co., personal communication, 1992), and synthetic pyrethroids (*4,11*) are shown in Figs. 1-3, respectively. The prototype structure has been transformed enormously with improvements in the activity profile through the lead evolution processes in these examples.

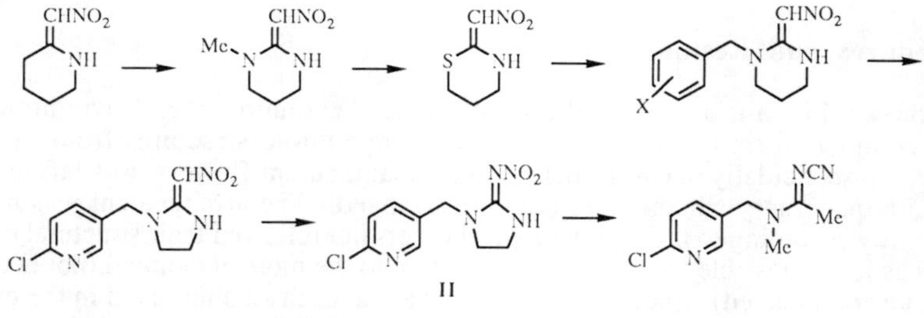

Fig. 2. Structural Evolution Leading to Imidacloprid (**II**) and Analogs.

A majority of individual steps in the structural evolution series might originally be attempted on trial-and-error bases. However, structural transformation patterns included in these steps, which have eventually been "utilized" to improve or at least to retain the bioactivity profile, are well recognized as being invaluable knowledge sources for "bioisosteric molecular transformation" in a broader sense. If numerous structural transformation examples are collected and organized as a database for possible bioisosteric transformation "rules" and the database is incorporated into a system so that any lower-ordered or primary lead structures introduced into the system are processed with the rules to release elaborated or higher-ordered candidate structures as the output "automatically", the system could be a great benefit for synthetic chemists.

We have been working on a project to construct such a system named EMIL: **E**xample-**M**ediated-**I**nnovation-for-**L**ead-Evolution (Fujita, T. In *Trends in QSAR and Molecular Modeling '92*; Wermuth, C. G., Ed.; ESCOM Science Publishers B. V. :

Leiden, The Netherlands, in press). In this article, we will describe an application of this system to synthetic pyrethroid series. A candidate structural class was selected from among a number of output structures generated from an input structure. The selection was made so that the structural features are "novel" but still in accord with structural requirements for the past pyrethroids. Compounds belonging to the candidate structural class were actually synthesized and bioassayed. Some of them exhibited a potent lethal effect on house-dust mites.

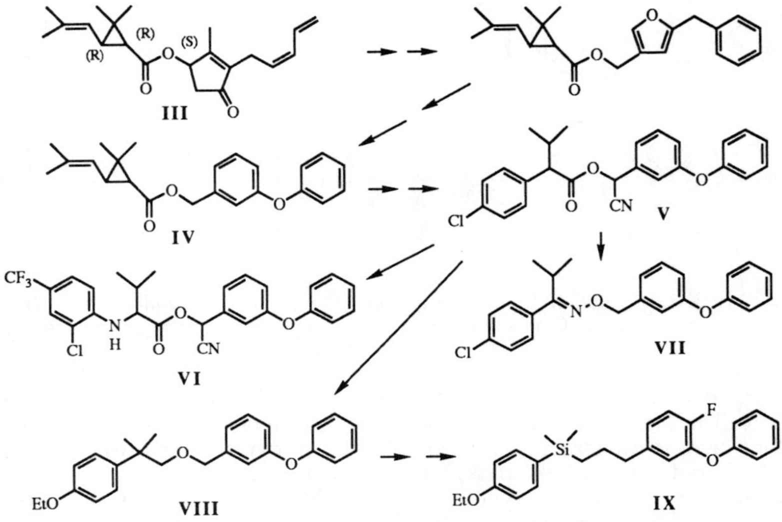

Fig. 3. Structural Evolution of Pyrethroid Insecticides; Compound (**III**) is pyrethrin-I and Compound (**IX**) is silafluofen.

Procedures and Results

Database : Fig. 3 is a simplified form of the lead evolution tree of pyrethroids. In a more complete form of the evolution tree, 33 pyrethroid structures from pyrethrin-I (**III**), an insecticidally active principle in chrysanthemum flowers, to silafluofen (**IX**) having a quaternary silicon atom (*11*) are arranged. The arrangement was made not necessarily according to their chronological orders but followed their structural evolution orders as far as possible from the more primitive to the more elaborated (not necessarily the more complicated) structure. Information about each step involved in the evolution tree was organized in such a data sheet as shown in Fig. 4. The most distinct feature of the EMIL database from conventional ones for drugs and chemicals is that each unit includes the data of the lower-ordered compound of Structure A as well as of the higher ordered compound of Structure B. Data for each compound consist of the use and effect, the mode of action, structural requirements, the potency and features of the activity, physicochemical data and others.

In each data sheet, the core of information is the identification of substructural modification patterns which could be utilized as possible "rules" for the bioisosteric transformation in the lead evolution phase of the drug design research not only within a single series, but also extended into other categories of bioactive compounds. There are numerous examples in which the structural evolution has occurred from agrochemicals to medicines (an example is the azole-type antimycotics shown in Fig. 1), from herbicides to fungicides as well as to insecticides and *vice versa* (Fujita, T. In *Trends in QSAR and*

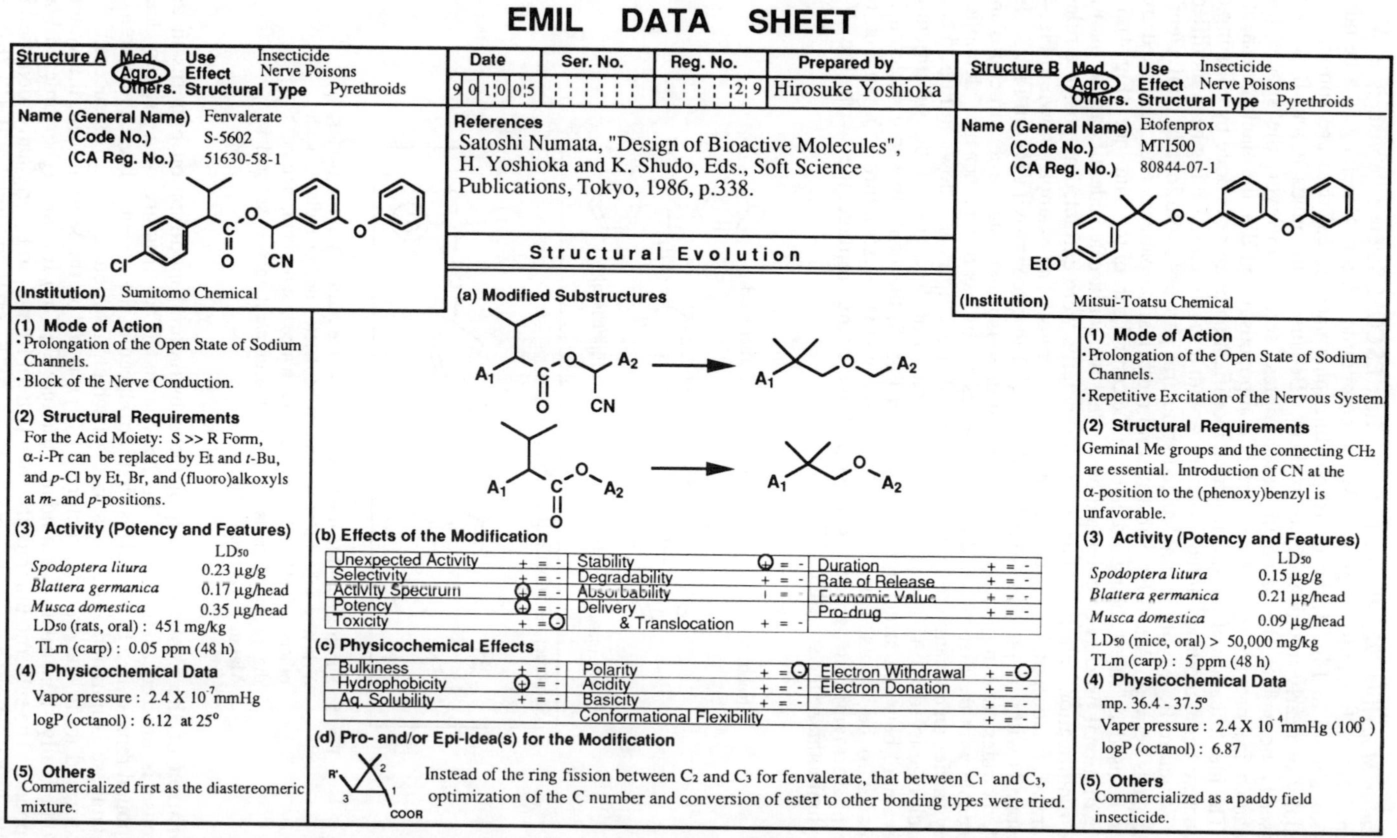

EMIL DATA SHEET

Structure A Med. (Agro) Others. **Use** Insecticide **Effect** Nerve Poisons **Structural Type** Pyrethroids

Date	Ser. No.	Reg. No.	Prepared by
9 0 1 0 0 5		2 9	Hirosuke Yoshioka

Structure B Med. (Agro) Others. **Use** Insecticide **Effect** Nerve Poisons **Structural Type** Pyrethroids

Name (General Name) Fenvalerate
(Code No.) S-5602
(CA Reg. No.) 51630-58-1
(Institution) Sumitomo Chemical

References
Satoshi Numata, "Design of Bioactive Molecules", H. Yoshioka and K. Shudo, Eds., Soft Science Publications, Tokyo, 1986, p.338.

Name (General Name) Etofenprox
(Code No.) MTI500
(CA Reg. No.) 80844-07-1
(Institution) Mitsui-Toatsu Chemical

Structural Evolution

(1) Mode of Action (Structure A)
- Prolongation of the Open State of Sodium Channels.
- Block of the Nerve Conduction.

(2) Structural Requirements
For the Acid Moiety: S >> R Form, α-*i*-Pr can be replaced by Et and *t*-Bu, and *p*-Cl by Et, Br, and (fluoro)alkoxyls at *m*- and *p*-positions.

(3) Activity (Potency and Features)

	LD_{50}
Spodoptera litura	0.23 μg/g
Blattera germanica	0.17 μg/head
Musca domestica	0.35 μg/head

LD_{50} (rats, oral) : 451 mg/kg
TLm (carp) : 0.05 ppm (48 h)

(4) Physicochemical Data
Vapor pressure : 2.4 X 10^{-7}mmHg
logP (octanol) : 6.12 at 25°

(5) Others
Commercialized first as the diastereomeric mixture.

(a) Modified Substructures

(b) Effects of the Modification

Unexpected Activity	+ = -	Stability	(+) = -	Duration	+ = -
Selectivity	+ = -	Degradability	+ = -	Rate of Release	+ = -
Activity Spectrum	(+) = -	Absorbability	+ = -	Economic Value	+ = -
Potency	(+) = -	Delivery & Translocation	+ = -	Pro-drug	+ = -
Toxicity	+ = (-)				

(c) Physicochemical Effects

Bulkiness	+ = -	Polarity	+ = (-)	Electron Withdrawal	+ = (-)
Hydrophobicity	(+) = -	Acidity	+ = -	Electron Donation	+ = -
Aq. Solubility	+ = -	Basicity	+ = -		+ = -
		Conformational Flexibility			+ = -

(d) Pro- and/or Epi-Idea(s) for the Modification
Instead of the ring fission between C_2 and C_3 for fenvalerate, that between C_1 and C_3, optimization of the C number and conversion of ester to other bonding types were tried.

(1) Mode of Action (Structure B)
- Prolongation of the Open State of Sodium Channels.
- Repetitive Excitation of the Nervous System.

(2) Structural Requirements
Geminal Me groups and the connecting CH_2 are essential. Introduction of CN at the α-position to the (phenoxy)benzyl is unfavorable.

(3) Activity (Potency and Features)

	LD_{50}
Spodoptera litura	0.15 μg/g
Blattera germanica	0.21 μg/head
Musca domestica	0.09 μg/head

LD_{50} (mice, oral) > 50,000 mg/kg
TLm (carp) : 5 ppm (48 h)

(4) Physicochemical Data
mp. 36.4 - 37.5°
Vaper pressure : 2.4 X 10^{-4}mmHg (100°)
logP (octanol) : 6.87

(5) Others
Commercialized as a paddy field insecticide.

Fig. 4. An Example of the EMIL Data Sheet (the preparation was assisted by Ryo Shimizu).

Molecular Modeling '92; Wermuth, C. G., Ed.; ESCOM Science Publishers B. V. : Leiden, The Netherlands, in press). When compounds A and B belong to a certain series as shown in Fig. 4, the corresponding pharmacological data are sometimes essentially or nearly identical between them. Because compounds A and B do not necessarily belong to the same pharmacological series, however, the spaces for pharmacological data are allocated to each of compounds A and B in the blank data sheet.

The identification of the substructural modification patterns is done by collating a substructure being modified in the Structure A with a substructure having been modified in the Structure B, leaving an unchanged structural part or "evolutionally equivalent" counterparts between A and B. Substructural modification patterns thus identified are what to be used as "rules", being indicated in the central part of the data sheet (Fig. 4). In each evolutional process of collected examples, we could often identify more than a single modification pattern. Thus, some 45 patterns were extracted as possible rules from transformation processes including 33 pyrethroids. The example shown in Fig. 4 corresponds with the process from fenvalerate (**V**) to etofenprox (**VIII**) in Fig. 3. In Fig. 4, the rules are extracted so that the α-substituted isovaleric acid ester structures are replaceable with the β-substituted isobutyl ethers regardless of whether the "alcoholic" moiety carries an α-cyano group or not. Some other examples are shown in Figs. 5 and 6. Each of circled A, A_1 and A_2 in the presentation of these rules is the substructure "equivalent" between structures A and B. In Fig. 5, the benzyltriazole skeleton is suggested to be "bioisosteric" with the triazolylethanol structure (*5*). In Fig. 6, dimethylformamidino group is regarded as being replaceable with the *N*-methyliminothiazoline structure (*12*).

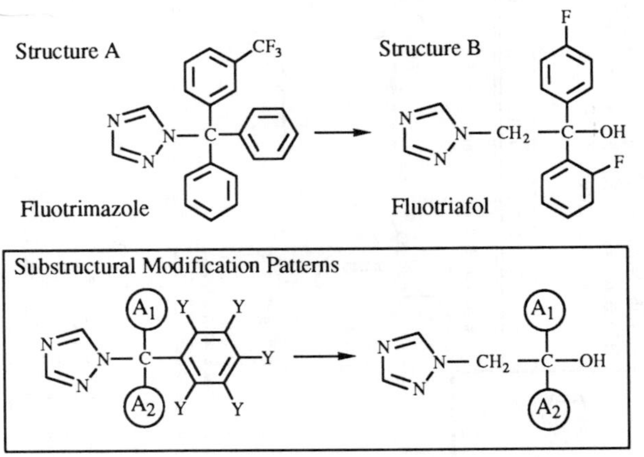

Fig. 5. Substructural Modification Pattern in the Azole-Type Fungicides; Y is any substituent which can be deleted and/or replaced with others, and A_1 and A_2 are the "unchanged" substructures through the structural transformation (the preparation was assisted by Masanori Yoshida).

Operation of the System : Each of the substructural modification patterns in the database is utilized as the "rule" in the system. The structure of a primary lead compound represented by R_1-(a_1) is first introduced in the system. Then, the search of the database is initiated. If an example, in which a structure S_1-(a_1) is successfully transformed to an elaborated structure S_1-(b_1), is hit by the search, then, the system "automatically" constructs a candidate structure R_1-(b_1) for the higher-ordered lead compound. The substructural modification pattern from a_1 to b_1 originally identified in the structural evolution example from Structure A, [S_1-(a_1)], to Structure B, [S_1-(b_1)], is used here as the rule for the substructural modification of R_1-(a_1) to R_1-(b_1). Usually, more than a single pattern are hit leading to quite a few "brother" structures. Of course, a_1 and b_1 are not necessarily simple substituents, but substructures of chemical significance. The operational functions of the system can be simplified as shown in Fig. 7. The cycle of the operation can be repeated as far as the output structure R_1-(b_1), which could be rewritten as R_2-(a_2), is able to match a rule with which S_n-(a_2) is converted into S_n-(b_2) in the database. Depending upon the judgement how many cycles are sufficient to yield a reasonable number of output structures, the operation can be terminated.

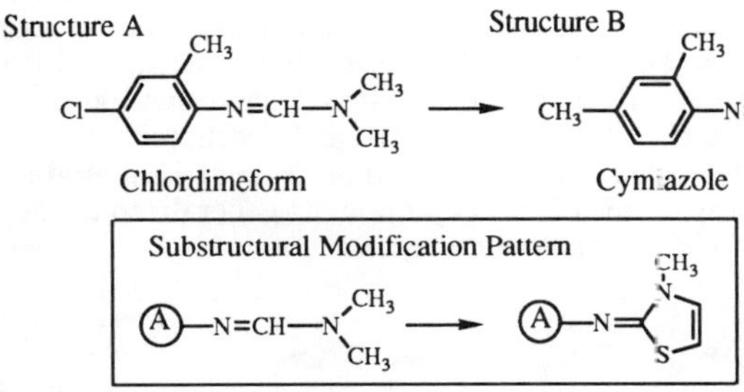

Fig. 6. Substructural Modification Pattern in the Chlordimeform-type Acaricides (the preparation was assisted by Isao Iwataki).

Results of the Data Processing : We selected fenvalerate (**V** in Fig. 3) as the input structure. Seventeen brother structures were released as those of the "second generation". Because some structural evolution processes starting from fenvalerate were collected in the database, structures released with the use of the corresponding rules were inevitably the same as structures registered in the database (n = 3 out of 17). The cycle of substructure-matching operation was repeated by applying the modification patterns to all of the output structures, releasing 127 "third generation" structures without duplication among them. The CPU time for the first and second cycles was 20 and 290 sec, respectively, with the use of a SPARCstation 2GX processing the database consisting of 45 unit rules.

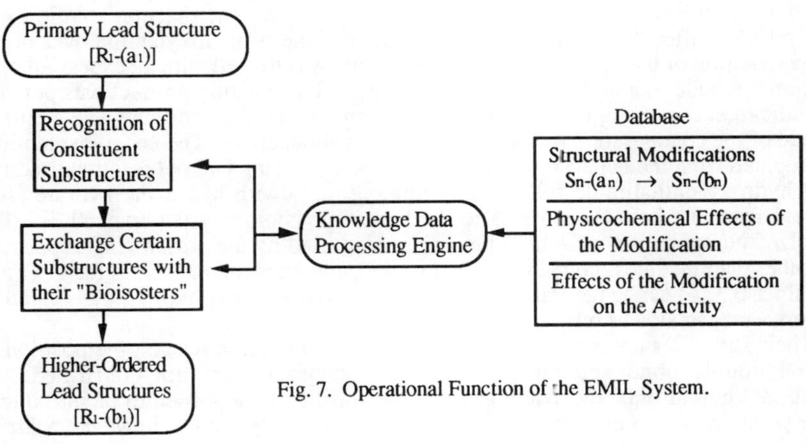

Fig. 7. Operational Function of the EMIL System.

Some output structures are shown in Fig. 8. Structural evolution series were limited within pyrethroids, so were the substructural modification rules utilized here. Therefore, variations of the released structures from that of fenvalerate were, in a way, not so "drastic", but mostly "systematic", being made by every possible combinations of the substructural modification rules applied to the structure of fenvalerate and its second generation outputs. Among these output structures, we were interested in the structure of such a substituted *N*-phenyl-*O*-benzyl isobutyramidoxime (**X** in Fig. 8). Two substructural modification rules were applied to generate this structure: first, the pattern from fenvalerate to fluvalinate (**VI** in Fig. 3); and second, the pattern from fenvalerate to SD 47443 (**VII** in Fig. 3). The transformation from SD 47443 to the isobutyranilidoxime ether corresponds with that from fenvalerate to fluvalinate. So, the substituted phenyl structure of SD 47443 was "suggested" to be replaceable bioisosterically with substituted anilino group by the system. Because the structural transformation from fenvalerate to fluvalinate is "evolutionary" (*13*), we considered the structural transformation from fenvalerate generating the isobutyranilidoxime ether structure is more drastic than merely systematic. The syntheses of this class of compounds generalized as the structure **XI** in Fig. 9 were considered to be worth attempting.

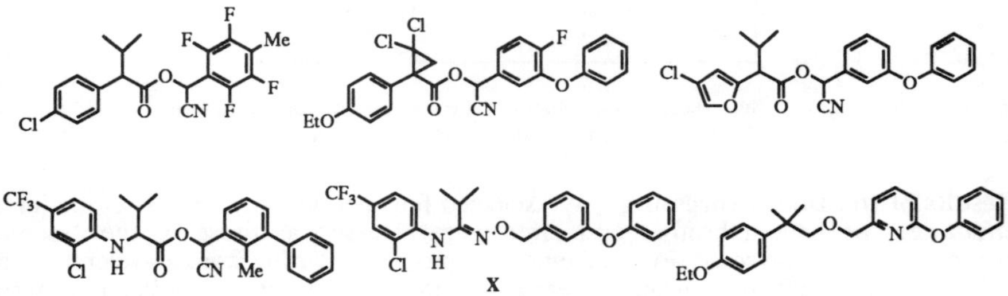

Fig. 8. Output Structures Generated from Fenvalerate as the Input Structure. Structure (**X**): Substituted *N*-Phenyl-*O*-benzyl Isobutyramidoxime.

Syntheses : In principle, the syntheses were made by condensation of the substituted isobutyranilides with the 3-phenoxybenzyloxylamine moiety as shown in Fig. 9. The isobutyranilides were reacted with an equivalent of phosphorus pentachloride in anhydrous chloroform under refluxing conditions for 3 hr. The reaction solution was further refluxed after the addition of an equivalent of the benzyloxylamine over night. After evaporation of the solvent *in vacuo*, the residue was treated with an excess amount of sodium ethoxide in anhydrous ethanol for 5 min. The reaction product was purified with a silica-gel chromatography with a chloroform-hexane (3:1) mixture as the eluent. The yield of the condensation reaction products was about 30 %. The benzyloxylamines were prepared from the benzyl chlorides via the corresponding Gabriel reaction products with *N*-hydroxy-phthalimide followed by treating them with hydrazine hydrate (*14*). The yield from the benzyl chlorides to the benzyloxylamines was about 60 %. The synthesized compounds shown in Table I were identified by the proton NMR spectrum peaks other than that attributable to the NH proton, and elemental analyses for C, H and N (within ± 0.3 %) as the neutral form. Except for compound **2**, m.p. 73 - 74 °C, all of the others were isolated as oil.

Their structures are represented as **XI** in Fig. 9. In fact, however, the position of the C=N double bond and the geometrical isomerism were not confirmed. In conventional pyrethroids, the cyano group is the substituent supposed to be introduced into the position of R in the structure **XI**. In this class of compounds, however, a strong electron-withdrawing effect of the α-cyano group would make the molecule unstable towards attacks of nucleophiles or basic biocomponents. In fact, no description of the α-cyano-benzyl analogs is found in the paper reporting syntheses and insecticidal activities of SD 47443 (**VII**) and its relatives (*15*), suggesting that the oxime *O*-(α-cyanobenzyl) ether compounds could not be isolated. Thus, instead of the cyano group, we introduced the electron-donating methyl group as the R substituent in a couple of compounds.

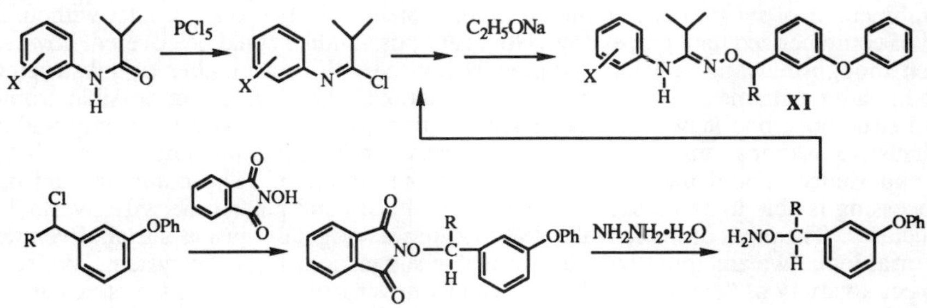

Fig. 9. Synthetic Route of Substituted *N*-Phenyl-*O*-benzyloxy Isobutyramidoximes (**XI**).

Biological Activity : All compounds in Table I were only slightly insecticidal against house flies and American cockroaches in preliminary tests, so further measurements were not made. Since some pyrethroids are effective miticides, we also examined miticidal activity against three species of house-dust mites as shown in Table I. Each compound was dissolved in acetone to make the concentration approximately 0.5 or 1 %. Then, a filter paper (5×10 cm^2) was soaked well in the acetone solution and air-dried. About 150 heads of mites (mixed ages) were released on the filter paper which was then folded and fixed. The miticidal activity was observed after 24 hr as the percentage of the killed mites. All miticidal operations were done under conditions of 24 °C and 75 % relative humidity.

As understood from data in Table I, compounds **1** and **6** exhibited house-dust miticidal activities similar to those of phenothrin (**IV** in Fig. 3) which is currently used as a domestic sanitary insecticide/miticide. Based on only six compounds, it was almost impossible to draw any clear picture about structure-activity relationships except for a similar trend existing among potency variations against three species of mites.

Table I. Compounds **XI** and Their Miticidal Activities

Compounds			Miticidal Percentage[a]		
Nr.	R	X	*D.f.*[b]	*D.p.*[c]	*T.p.*[d]
1	H	4-OEt	68	37	25
2	H	4-Cl	36[e]	17[e]	7[e]
3	H	2-Cl-4-CF_3	17	41	-[f]
4	H	2-Cl-4-CF_3-6-aza[g]	4	14	10
5	Me	4-OEt	0	3	13
6	Me	2-Cl-4-CF_3	74	63	29
Phenothrin (**IV** in Fig. 3)[h]			64 ± 12[i]	33 ± 10[i]	50 ± 15[i]

a) Percentage of the killed mites with the dose of 0.5 % solution in acetone, unless noted. b) *Dermatophagoides farinae*. c) *Dermatophagoides pteronyssinus*. d) *Tyrophagus putrescentiae*. e) With 1% solution in acetone. f) Not measured. g) 3-Cl-5-CF_3-2-pyridinyl. h) Control. i) Standard deviation for three repeats.

Discussion

In cases such as the present study, in which the input structure and the range of data sources are limited within a single series of compounds, some output structures have been registerd in the database. Many structures could be synthesized with various combinations of substructures required for the potent activity sooner or later without the aid of computerized data processing. Not every possibilities could be covered, however, even though the range of data sources is limited. With a number of substructural modification patterns and their possible combinations with a number of possible primary lead structures, one may overlook invaluable transformation processes easily without exhaustive examinations. Moreover, prejudices owing to personal preferences or biases in experiences could unconsciously worsen the situation. The computerized data processing is able to avoid such prejudices and biases and glean otherwise overlooked structures. The input of the fenvalerate structure was originally just as a trial. Even with the trial input, we are able to obtain invaluable suggestions from the system. We could expect a variety of "exotic" but "novel" candidate structures, if input structures are not included in the range of data sources and "rules" are expanded so that the "bioisosteric" molecular transformation examples in other categories of bioactive compounds are applicable.

One may consider that the structures and bioactivity profiles of compounds **1** and **6** are not spectacular. They were new, however, and patented (*16*). The reason why this class of compounds have not been disclosed might be due to some "implicit" preconceptions that the oxime *O*-ether structures are not very stable and/or that the structural transformations to increase polarity of the molecule are not favorable to the insecticidal activity. At the moment, it is uncertain whether further trials towards their development for the practical use are worth making. The hydrophobicity of compound **1** in terms of log P (P is the partition coefficient measured with the 1-octanol/water system) was approximately estimated as being about 6.0 (Leo, A. Pomona College, Claremont, CA, personal communication, 1993) which is lower than that of phenothrin, 7.56 (*17*). This could be an advantage to the formulation of water-based aerosols for domestic uses. The less hydrophobic compounds are believed to be safer in terms of undesirable toxic side effects (*18*). Because compound **1** shows an almost equivalent miticidal activity with phenothrin, the use of compound **1** could be more favorable for domestic use without loss of efficacy.

The present study is regarded as having disclosed an additional example of the bioisosteric relationship between substituted phenyl and substituted phenylamino groups. While the bioactivity profile was "slightly" changed from insecticidal to miticidal, the bioisosteric relationship between fenvalerate and fluvalinate was retained between SD 47443 and the isobutyranilidoxime *O*-ethers. Fig. 10 shows some past examples of bioisosteric relationships not only between phenyl and phenylamino groups but also including substituted phenoxy. In fact, we had applied bioisosteric relationships observed in synthetic auxins (*19*) and DDT analogs (*20*) to fenvalerate (*21*, *22*). The substituted phenylamino ("fluvalinate") (*21*) as well as the substituted phenoxy analogs (*22*) of fenvalerate had been synthesized and tested by our group to exhibit potent insecticidal activity before the paper "disclosing" fluvalinate and analogs was published by the Zoecon group (*13*). Our experience of successful applications of the bioisosteric substructural modification patterns to different category of bioactive compounds was one of the greatest motivations for constructing the system EMIL to process bioisosteric transformation data comprehensively without bias and omission in designing new compounds.

The enantiomerism and diastereomerism were not strictly considered in the extraction of the rule. This is because this system is primarily to present candidate skeletal structures to synthetic chemists. The relationship between stereochemistry and activity could be a problem challenged in the lead optimization phase. The performance of the system is heavily dependent on the quality and size of the database as well as the skill of the data-processing algorithm. These should be done by cooperation of scientists free from prejudices. The most recent version of the system has some additional functions to utilize information about the hierarchic classification and the physicochemical and pharmacological effects of the substructural modification patterns in sorting out the applicable rules. This system can also be combined with such software systems as that to calculate the log P value (*23*) and/or those to "predict" possible toxicities and pharmacokinetic (*24*) and environmental behaviors (*25*). Thus, we could confine candidate structures more "accurately" among a number of outputs in terms of their hydrophobicity as well as side-effect indices. The more complete article describing the concept of the EMIL will be published elsewhere (Fujita, T. In *Trends in QSAR and Molecular Modeling '92;* Wermuth, C. G., Ed.; ESCOM Science Publishers B. V. : Leiden, The Netherlands, in press).

Synthetic Auxins (R : Me, Et)

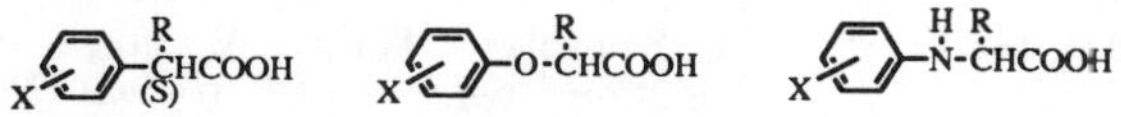

DDT Analogs (X, X' : Me, MeO, EtO, Cl)

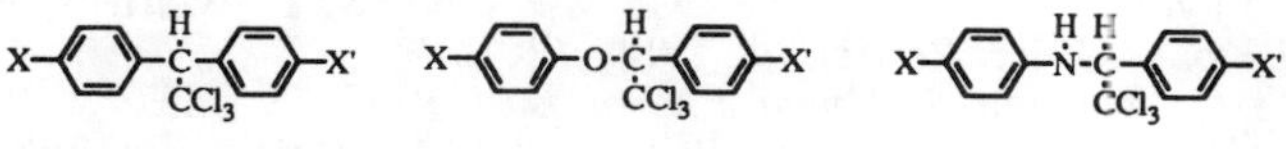

Synthetic Pyrethroids (X : halogen, CF_3, EtO)

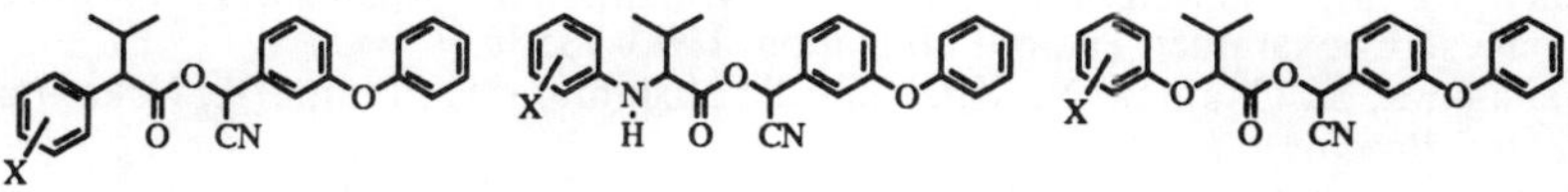

Fig. 10. Bioisosterism of Ph, PhO, and PhNH Structures

Acknowledgments : We wish to thank the members of the EMIL working group who have been devoting their effort to collect structural transformation examples and arrange them as database. We also extend our gratitudes to Mr. Masahiro Baba of Fujitsu Nagano System Engineering Ltd. for his assistance of the data processing. The artwork was made by a skilled assistance of Dr. Yoshiaki Nakagawa, Kyoto University.

Literature Cited.

1. Kuhr, R. J.; Dorough, H. W. *Carbamate Insecticides: Chemistry, Biochemistry and Toxicology*; CRC Press : Cleveland, OH, **1976**.
2. Eto, M. *Organophosphorus Pesticides : Organic and Biological Chemistry*; CRC Press: Cleveland, OH, **1974**.
3. Soloway, S. B.; Henry, A. C.; Kollmeyer, W. D.; Padgett, W. M., Powel, J. E.; Roman, S. A.; Tieman, C. H.; Corey, R. A.; Horne, C. A. In *Advances in Pesticide Science*; Geissbühler, H., Ed.; Pergamon Press: Oxford, **1979**, Part 2; pp. 206 - 217.
4. Yoshioka, H. In *Rational Approaches to Structure, Activity, and Ecotoxicology of Agrochemicals*; Draber, W.; Fujita, T., Eds.; CRC Press: Boca Raton, FL, **1992**; pp. 185 - 217.
5. Krämer, W. In *Chemistry of Plant Protection*; Haug, G.; Hoffmann, H., Eds.; Springer Verlag: Heidelberg, **1986**, Vol. 1; pp. 25 - 64.

6. Matsunaka, S. In *Herbicides: Chemistry, Degradation and Mode of Action*; Kearney, P. C.; Kaufman, D. D., Eds.; Mercel Dekker, Inc.: New York, NY, **1975**; pp. 709 - 739.
7. Levitt, G. In *Pesticide Chemistry: Human Welfare and the Environment*; Miyamoto, J.; Kearney, P. C., Eds.; Pergamon Press: Oxford, **1983**, Vol. 1; pp. 243 - 250.
8. Fujita, T. In *QSAR in Drug Design of Bioactive Compounds*; Kuchar, M., Ed.; J. R. Prous Science Publishers: Barcelona, **1992**; pp. 3 - 22.
9. Konosu, T.; Tajima, Y.; Takeda, N.; Miyaoka, T.; Kasahara, N.; Yasuda, H.; Oida, S. *Chem. Pharm. Bull.* **1991**, *39*, 2581.
10. a) Kagabu, S; Moriya, K.; Shibuya, K.; Hattori, Y.; Tsuboi, S.; Shiokawa, K. *Biosci. Biotech. Biochem.* **1992**, *56*, 362, b) Moriya, K.; Shibuya, K.; Hattori, Y.; Tsuboi, S.;Shiokawa, K.; Kagabu, S. *Biosci. Biotech. Biochem.* **1992**, *56*, 364.
11. Sieburth, S. M.; Manly, C. J.; Gammon, D. W. *Pestic. Sci.* **1990**, *18*, 289.
12. Immler, R. M.; Bouvard, H.; Ernst, G. H.; Knüsel, F.; Traber, W.; von Orelli, M. In *Proceedings of 1977 British Crop Protection Conference: Pests and Diseases*; British Crop Protection Council: London, **1977**, Vol. 2; pp. 383 - 396.
13. Henrick, C. A.; Garcia, B. A.; Staal, G. B.; Cerf, D. C.; Anderson, R. J.; Gill, K.; Chinn, H. R.; Lavovitz, J. N.; Leippe, M. M.; Woo, S. L.; Carney, R. L.; Gordon, D. C.; Kohn, G. K. *Pestic. Sci.* **1980**, *11*, 224.
14. Fujii, T.; Wu, C. C.; Yamada, S. *Chem. Pharm. Bull.* 1967, **15**, 345.
15. Bull, M. J.; Davies, J. H.; Searle, R. J. G.; Henry, A. C. *Pestic. Sci.* **1980**, *11*, 249.
16. Cheng, Z. M.; Nishimura, K.; Fujita, T.; Minamite, Y.; Katsuda, Y; Yoshioka, H. Japanese Unexamined Patent Publication, **1993**-43536.
17. Nakagawa, S.; Okajima, N.; Kitahaba, T.; Nishimura, K.; Fujita, T.; Nakajima, M. *Pestic. Biochem. Physiol.* **1982**, *17*, 243.
18. Hansch, C.; Björkroth, J. P.; Leo, A. *J. Pharm. Sci.* **1987**, *76*, 663.
19. Veldstra, H.; Booij, H. L.; *Biochim. Biophys. Acta* **1949**, *3*, 278.
20. Hirwe, A. S.; Metcalf, R. L.; Kapoor, I. P. *J. Agr. Food Chem.* **1972**, *20*, 818.
21. Katsuda, Y.; Nakajima, M.; Fujita, T. Japanese Unexamined Patent Publication, **1977**-82724; Japanese Examined Patent Publication, **1982**-48522.
22. Katsuda, Y.; Nakajima, M. Japanese Unexamined Patent Publication, **1977**-70021; Japanese Examined Patent Publication, **1983**-2202.
23. Leo, A. J. In *Comprehensive Medicinal Chemistry*; Hansch, C., Ed.; Pergamon Press: Oxford, **1990**, Vol. 4; pp. 295 - 319.
24. Moriguchi, I.; Komatsu, K.; Matsushita, Y. *Anal. Chim. Acta* **1981**, *133*, 625.
25. Liu, Q.; Hirono, S.; Matsushita, Y.; Moriguchi, I. *Environ. Toxicol. Chem.* **1992**, *11*, 953.
26. Fujita, T. In *Trends in QSAR and Molecular Modeling '92*; Wermuth, C. G., Ed.; ESCOM Science Publishers : Leiden, in press.

RECEIVED September 21, 1993

Chapter 29

Biochemical Approaches to Herbicide Discovery

Enzyme Target Selection and Inhibitor Design

Alan R. Rendina and Lynn M. Abell

Agricultural Products, Stine-Haskell Research Center, DuPont, Newark, DE 19714

Successful design of novel herbicides based on specific inhibition of selected enzyme targets requires careful consideration of the choice of the target, mechanism of the enzyme, design of potent inhibitors, delivery of the inhibitor to the target and metabolic fate of the inhibitor. Validated targets, those that produce phytotoxic effects upon partial inhibition, can be identified by genetic methods or by obtaining chemical leads. Genetic approaches include studies of conditionally lethal bacterial and plant mutants and use of antisense technology. In the absence of chemical leads with known sites of action, targets for validation may be selected by the following criteria: the target is essential to plants and, preferably, inhibition leads to multiple deleterious effects; the target is not present in mammals; the target has low intracellular concentration, i.e., has potential for low use rates; and the proposed inhibitors of the target are synthetically accessible. Enzyme inhibitors can be categorized into six major groups: ground-state analogues, group specific reagents, affinity labels, suicide substrates, reaction intermediate analogues and extraneous site inhibitors. Examples of each category, and their advantages and disadvantages as potential agrochemicals, will be discussed. Potent inhibition of the selected target may still not produce an effective herbicide. Studies of the uptake, translocation and metabolism of the inhibitor are needed to determine if the cause of poor *in vivo* performance is due to these factors or to an intrinsically poor target. Without full appreciation of each of these aspects of herbicide design, the chances for success with the target-site directed approach are reduced.

This paper will discuss our views on the critical components needed for the successful design of new herbicides using biochemical approaches. This topic has been the subject of a number of recent reports (*1-5*). While there have been no commercial successes to date using target-site based approaches, there **have** been a number of near misses, some of which will be described here, and there is tremendous potential for success in this approach. At Du Pont, we are firmly committed to biochemical approaches to the discovery of novel pesticides and believe that it will ultimately be successful from a commercial standpoint.

0097–6156/94/0551–0407$06.00/0

As the title indicates, the main focus of this report is to examine some of the key factors involved in a target-site based approach to inhibitor design. Most of the discussion is based on examples from the literature, and whenever possible, will use examples having agrochemical utility. Most of these studies are on established herbicide targets with empirically discovered herbicides. However, in many cases, especially those involving active site directed inhibitors, we believe modern synthetic design strategies could have produced the same compounds (for an example see (*4*)).

One of the key aspects to using enzyme targets to generate new leads for novel herbicides is that it requires very close interaction of a multidisciplinary team. The input of plant physiologists, enzymologists, molecular biologists, synthesis chemists, protein modelers and crystallographers is crucial to the overall success of this approach.

The biochemical approach can be divided into three main processes, target selection, inhibitor design and translation of *in vitro* activity into *in vivo* efficacy. Selection of a truly lethal target site is the key initial component and both genetic and chemical methods for validating potential target enzymes will be discussed. This discussion is primarily concerned with finding new targets that are not associated with a chemical lead. After genetic validation of the chosen target, or as part of the chemical validation process, the next step is to use detailed knowledge of the enzyme's chemical and kinetic mechanism to design potent inhibitors. The various categories of enzyme inhibitors will be discussed along with their advantages and disadvantages as potential herbicides. The last process, translation of *in vitro* potency into *in vivo* efficacy, will not be addressed here. Most of the discussion will focus on choosing targets and inhibitor design strategies. The last aspect may not require extraordinary effort if one is lucky, that is, a potent inhibitor will also be a good herbicide.

Criteria for Target Selection

Over the years, a number of enzymes have been examined as potential herbicide targets. Many interesting and potent enzyme inhibitors have been prepared, but often these are not active *in vivo* or are weakly active at best. Some of the lack of *in vivo* activity may be due to the inherent weakness of the chosen target enzyme as a lethal site. That is, greater than 95% inhibition may be required before toxic effects are manifested. In practice, this is a level of inhibition that is difficult to achieve even with very potent inhibitors. Therefore, it is very important to chose the right enzyme target. To help define what constitutes a lethal target site, additional studies are needed on the known enzyme targets to elucidate the sequence of events that ultimately lead to cell death. Such studies would help establish better criteria for testing unproven target sites. To narrow down the number of potential targets and to increase chances of success, the following criteria have been applied to the selection of potential targets by us and others, dating back to the mid 1970's at Fisons (now Schering) (*6*).

The target should be essential to plants and inhibition should result in multiple deleterious effects. For example, inhibition should deprive the plant of an essential metabolic intermediate, consume a high energy intermediate in a futile cycle and/or cause a toxic intermediate to accumulate. Simply blocking the production of the end product of a biosynthetic pathway may not produce cytotoxic results, but may only result in cell stasis or growth inhibition. A good target will probably require less than 80% inhibition to produce lethal effects.

To minimize the potential for mammalian or non pest toxicity the target should be present in plants, but not mammals. However, toxicity of xenobiotics is difficult to predict since it is often due to effects that are unrelated to inhibition of the target site. Most herbicidal inhibitors of glutamine synthetase (glufosinate) and acetyl-CoA carboxylase (diclofop and sethoxydim) have very low mammalian toxicity (*2*), yet both enzymes are found in animal tissues.

Potential targets should also be present at low intracellular concentrations so that inhibitors will be more likely to have lower use rates and reduced environmental impact. Ribulose-1,5-bisphosphate carboxylase/oxygenase (Rubisco) serves as a negative example of this point. While it plays a key role in plant specific metabolism, it is far too abundant to inhibit in a practical sense. Rubisco is thought to comprise up to 50% of the soluble leaf protein (*7*) and we estimate that even an ideal inhibitor, one that completely inhibits all of the enzyme in a 1:1 ratio of inhibitor to enzyme, would require a use rate of approximately 50 kg/ha to completely tie up all the enzyme molecules. Lethal effects would probably be observed with less than 100% inhibition, but the use rates would still be impractical.

The target enzyme should also offer practical opportunities for inhibitor design, synthesis and delivery to the target site. For example, if inhibitors are based on reaction intermediates, then enzymes using complex, highly charged, or highly lipophilic substrates should be avoided. Also, fewer inhibitor strategies are available for enzymes that use simple acid-base chemistry to achieve catalysis (kinases for example). For convenience, the target enzyme should also be readily obtainable. With the mechanism based inhibitor approach, the bacterial enzyme may be used as a model system since the mechanism of the bacterial and plant enzyme is likely to be the same.

Examination of the properties of known targets also suggests that in many cases the target for herbicides is, or is close to being, the rate limiting step in the biosynthetic pathway. Assuming that most biosynthetic enzymes are functioning at subsaturating concentrations of the metabolic intermediate, then comparison of the V_{max}/K_m for those intermediates with the corresponding enzymes in the pathway can be used to gauge which enzyme(s) is limiting the overall flux through the pathway (assuming product release is not rate limiting). In the absence of knowledge about other criteria, such as toxic intermediate buildup, this information can be used to select the most promising target.

Target Validation/Invalidation

Target selection based solely on the above criteria does not guarantee that the enzyme is a potent herbicide site. Genetic or chemical validation may provide sufficient confidence in a target to commit full resources (i.e., detailed mechanistic studies, crystal structure determination, and high through put screening) to thoroughly exploit the chosen target. A practical definition of validation is that suppression of the enzyme's activity by 80% or less, by either genetic or chemical means, produces lethal effects. In choosing the best target in a biosynthetic pathway, it is important to recognize that not all enzymes in an essential pathway are equally good targets. Each individual target enzyme must be validated or invalidated rather than the entire pathway. For example, in the branched chain amino acid biosynthetic pathway the first enzyme, acetolactate synthase (ALS), is the known target for several classes of commercial herbicides (*8, 9*), but other enzymes in the pathway are proving to be weaker targets. Inhibitors have been designed or discovered for the second and third enzymes in the valine/isoleucine pathway, ketol acid reductoisomerase (KARI)(*10, 11*) and dihydroxy acid dehydratase (DHAD)(*12-14*), respectively, and for the second and third enzymes specific to leucine biosynthesis, isopropylmalate isomerase (IMI)(*15*) and isopropylmalate dehydrogenase (IPMDH)(*16*), respectively. Although the inhibitors were not of equal potency, inhibitors of DHAD were not active *in vivo* while inhibitors of KARI, IMI and IPMDH showed weak to modest herbicidal activity. None of the inhibitors of these alternate targets in the pathway have yielded whole plant effects that approach the potency achieved with the inhibitors of targets of the commercial herbicides. With the exception of KARI and IPMDH, the reasons for this lack of *in vivo* potency have yet to be fully elucidated.

Genetic Validation. The most expedient way to validate potential targets may be to use genetic approaches. The genetic approaches to target validation should be designed to locate the lethal sites in a metabolic pathway and to estimate the level of inhibition required to produce toxic effects. Genetic approaches include characterizing plant mutants, using bacteria as models to look for lethal consequences of blocking individual steps in a biosynthetic pathway (*17*) and using antisense or other genetic constructs to vary the level of expressed activity of specific enzymes. The second approach is suspect because there are many uncertainties about the differences between bacterial and plant systems, especially in the regulation of biosynthetic pathways.

In the first approach, higher plant mutants are isolated and their phenotypes characterized under stringent conditions such as starvation for the end product of the pathway. To have the greatest utility, it is desirable to have a series of mutants expressing different levels of the targeted enzyme. In practice, there are only a limited number of examples of fully characterized plant auxotrophs. In most cases, they have lesions that completely block the pathway of interest which tends to validate the entire pathway rather than a specific target (*18-20*). A careful study of mutants in barley having different levels of expression of chloroplastic glutamine synthetase (GS) provides a good example of the potential of this approach (*21, 22*). GS is a key enzyme in the recycling of carbon lost during photorespiration and in the assimilation of ammonia in plants. Conditionally lethal plant mutants, defective in GS and other photorespiratory enzymes, are isolated under high CO_2 levels where photorespiration is suppressed. In air and in the dark these plants grow normally, but when exposed to light they exhibit varying degrees of chlorosis depending on the levels of GS activity. At ambient levels of CO_2 and O_2, mutant plants with less than 40% of the GS activity found in wild type plants show reduced rates of CO_2 fixation and begin to accumulate ammonia. Since only partial blockage of GS (60%) produces phytotoxic symptoms, this enzyme would be a good target for herbicide design. While GS was first validated chemically by the empirically discovered natural product phosphinothricin (*23, 24*), the active ingredient in the commercial herbicide glufosinate, the agreement between the chemical and genetic validation results suggests that targets may be validated or invalidated on genetic evidence alone.

The application of antisense technology (reviewed in *25, 26*) to quantitatively reduce the levels of expression of specific genes is being increasingly used in plant biochemistry to study gene function and has tremendous potential for elucidating novel herbicide target sites. The biotechnology industry is also exploiting this technique for a variety of commercial uses such as production of tomato lines with delayed fruit ripening (*27*). Antisense RNA is messenger RNA (mRNA) transcribed from the opposite strand of DNA than is normally used for biosynthesis of a protein. In most cases, the presence of both sense and antisense RNA reduces the expression of that protein. Pairing of the two RNA species is probably involved in the inhibition process, but the exact mechanism is unknown. Validation of herbicide targets using antisense involves cloning the gene or cDNA from a plant species, selecting a suitable promoter for the antisense construct (one that produces an expression pattern similar to the native gene or one that can be induced at the appropriate stage in development), preparing the antisense constructs, transforming and regenerating plants and correlating the phenotypes of transformants with the activity levels of the enzyme target, or the levels of its mRNA or protein. The relative strengths of the promoters for the native and antisense genes and the number of copies of antisense genes introduced determine the level of expression of the targeted enzyme.

In plants, antisense RNA or other gene disruption methods have been used to reduce the activity of nopaline synthase (*28,29*), chloramphenicol acetyltransferase (*30, 31*), chalcone synthase (*32-34*), polygalacturonase (*35,36*), phosphinothricin acetyltransferase (*37*), β-glucuronidase (*38, 39*), granule-bound starch synthase (*40*), Rubisco (*7,41,42*), 1-aminocyclopropane-1-carboxylate synthase (*43*) and oxidase (*44-46*), phenylalanine ammonia lyase (PAL) (*47*), stearoyl-acyl carrier protein desaturase (*48*), highly anionic peroxidase (*49*), tonoplast H^+ ATPase (*50*) and GS (*51*). In addition, there have been 15 posters on antisense in plants presented at the last two annual meetings of the American Society of Plant Physiologists indicating the growing interest in applications of this technology. While these studies were not conducted to validate these enzymes as potential herbicide targets, the results can be interpreted from that perspective. With the exception of Rubisco, the tonoplast ATPase and GS, noticeable symptoms or effects on the whole plant were observed only when the enzyme activity was reduced by greater than 90% compared to the wild type levels. Therefore, these enzymes are not good targets for herbicide design. The results also demonstrate the utility of the method in eliminating potential targets. In transgenic tobacco, where Rubisco levels were reduced by 82% with antisense RNA, significant growth reduction was observed (*7*); however, Rubisco is not a good herbicide target site due to its high concentration in plants (see earlier discussion). Transformed, regenerated carrots having 80 to 85% lower levels of the tonoplast ATPase activity than the wild type exhibited altered leaf morphologies, reduced cell expansion and shorter tap roots suggesting that this might be a potential target (*50*). More detailed studies are needed to determine how lethal this site is at intermediate levels of expression and to obtain more consistent phenotypes. The tonoplast ATPase would not be an ideal target because it is found in all eukaryotic cells and is probably relatively abundant. GS activity levels in tobacco were reduced by only 40% by introduction of an antisense alfalfa GS gene compared to wild type activity levels. Nevertheless, the decrease in GS activity corresponded with a 40% decrease in total protein and visible symptoms of nitrogen deficiency (*51*), again confirming that this is a very lethal site.

Chemical Validation. Most target site design research currently involves a chemical approach to validation of the target. Sites of action associated with a chemical lead having *in vivo* activity would be considered to be validated. In the absence of a such a lead, potent and selective inhibitors of the target enzyme are designed based on knowledge of the enzyme mechanism, synthesized and tested both *in vitro* and *in vivo*. Extremely potent enzyme inhibitors have been discovered, but these often have little or no effect on the whole plant. It is not clear in most of these cases if these compounds are weak herbicides due to poor uptake and/or translocation, metabolic detoxification or because the target is not a particularly lethal site. It is also possible that the lower herbicidal effect is due to the presence of an alternate enzyme activity that is not sensitive to the inhibitor. Ideally, when weak herbicidal activity is observed with a potent inhibitor, the next step should be to prepare radiolabeled inhibitor and study its uptake, translocation and metabolism to be sure that the compound reaches the target enzyme without being detoxified. Whenever possible, it is important to measure the target enzyme activity, *in vivo*, in the presence of inhibitor and compare it with the uninhibited control to ensure that the compound is selective for the chosen target (most easily measured with tight-binding or irreversible inhibitors). Slight modifications in the inhibitor structure can result in changes in the site of action. Alternatively, either accumulation of metabolic intermediates or their precursors, or decreases in the end products of the pathway can be monitored. Reversal of the phytotoxic symptoms produced by the inhibitor by biosynthetic intermediates that are downstream from the target site can also verify that the compound is truly affecting the intended target. If any of these factors limit the *in vivo* activity of a compound, then additional analogues or proherbicides can be designed and tested to overcome these barriers. On the other hand, if the obvious reasons for lack of efficacy of the inhibitor (uptake, translocation, detoxification and level and selectivity of inhibition) do not account for its lower *in vivo* potency, then inhibition of this enzyme may be intrinsically less phytotoxic than inhibition of known herbicide target sites. This enzyme may need to be inhibited to a greater extent, and/or for a longer period of time than can be achieved chemically, to result in plant death.

Studies conducted on inhibitors of KARI and IPMDH serve to illustrate these points. With KARI, all of the appropriate physiological studies have been conducted to demonstrate that the enzyme was a relatively poor target (*52*). KARI catalyses the reversible conversion of 2-aceto-2-hydroxybutyrate to 2,3-dihydroxy-3-methylpantanoate or 2-acetolactate to 2,3-dihydroxy-3-methylbutyrate. Along with detailed mechanistic studies, Aulabaugh and Schloss (*10*) were able to design and prepare potent inhibitors of KARI such as N-isopropyloxalylhydroxamate (IpOHA) (Figure 1, top). The iminol tautomer of IpOHA (shown in brackets) bears a strong resemblance to the proposed rearrangement transition state (also in brackets) which accounts for its potent inhibition of KARI. Similar potent inhibition was obtained with HOE 704, a phosphine oxide inhibitor of KARI, discovered via screening rather than by design (*11*). Although the oxalylhydroxamates are better inhibitors of KARI (dissociation constant of 22 pM with a half time for release of 6.2 days) than the most potent herbicidal inhibitors of ALS, they are only weak to moderate herbicides. Studies with radiolabeled inhibitor ruled out uptake, translocation and metabolism as significant factors lowering the herbicidal efficacy of these compounds in susceptible plants. Due to the extreme potency and slow binding properties of the oxalylhydroxamates, residual KARI activity could be measured *in vivo* (corn tissue culture) at different inhibitor levels. When KARI activity was inhibited by more than 90% *in vivo*, no cytotoxic effects (such as reduction in cell volume) were observed. Only when KARI activity was reduced by 95-98%, was accumulation of acetoin (formed from the substrate) observed along with reduction in cell volume. These results suggest that KARI is not as good a target enzyme as ALS, probably because the plant has more enzyme than it needs to survive.

The oxalyhydroxamate inhibitors are also an interesting example of how subtle changes in inhibitor structure can result in dramatic changes in target selectivity. In addition to inhibiting KARI, these compounds can also selectively inhibit IPMDH (*16*). IPMDH catalyzes the NAD^+-linked oxidative decarboxylation of β-isopropyl malate to α-ketoisocaproate (Figure 1, bottom). Oxalylhydroxamates substituted on the nitrogen are highly selective for KARI (such as IpOHA), while oxalyhydroxamates substituted on the oxygen are highly selective for IPMDH (such as *O*-isobutenyl oxalylhydroxamate, *O*-IbuHA). IpOHA has a K_i of 22 pM for KARI and a K_i of 5.9 μM for IPMDH, whereas *O*-IbuHA has a K_i of about 100 μM for KARI and a K_i of 31 nM for IPMDH. With a bulky substituent on the nitrogen (isopropyl), the oxalylhydroxamates are 268,000 fold better inhibitors of KARI than of IPMDH. In contrast, oxalylhydroxamates with a bulky substituent on oxygen (isobutenyl) are 3,200 fold better inhibitors of IPMDH than of KARI. In the case of IPMDH, the inhibitor is thought to mimic the enolate reaction intermediate (bracketed) formed upon decarboxylation of the β-keto acid intermediate. For both KARI and IPMDH, the coordination of magnesium to either inhibitor or the corresponding intermediate is thought to be important. The first indication of the switch in enzyme selectivity was obtained by observing that inhibition of pea root growth by IpOHA could be prevented or reversed by all three branched chain amino acids (valine, leucine, and isoleucine), but the same inhibition by *O*-IbuHA could be prevented by leucine alone. In the later case, reversal of inhibition by *O*-IbuHA methyl ester by α-ketoisocaproate (but not β-isopropylmalate) helped establish IPMDH as the new site of action. While the inhibitors of IPMDH (active at *ca.* 400 g/ha) are better herbicides than the KARI inhibitors (active at *ca.* 2000 g/ha), neither approaches the activity of ALS inhibitors (*ca.* 2 g/ha). With IPMDH inhibitors, the herbicidal activity is correlated with the accumulation of β-isopropylmalate in the plants and there is some indication that the metabolic degradation of both this intermediate and the inhibitor contribute to the reduced effectiveness of these compounds in certain plant species (*53*).

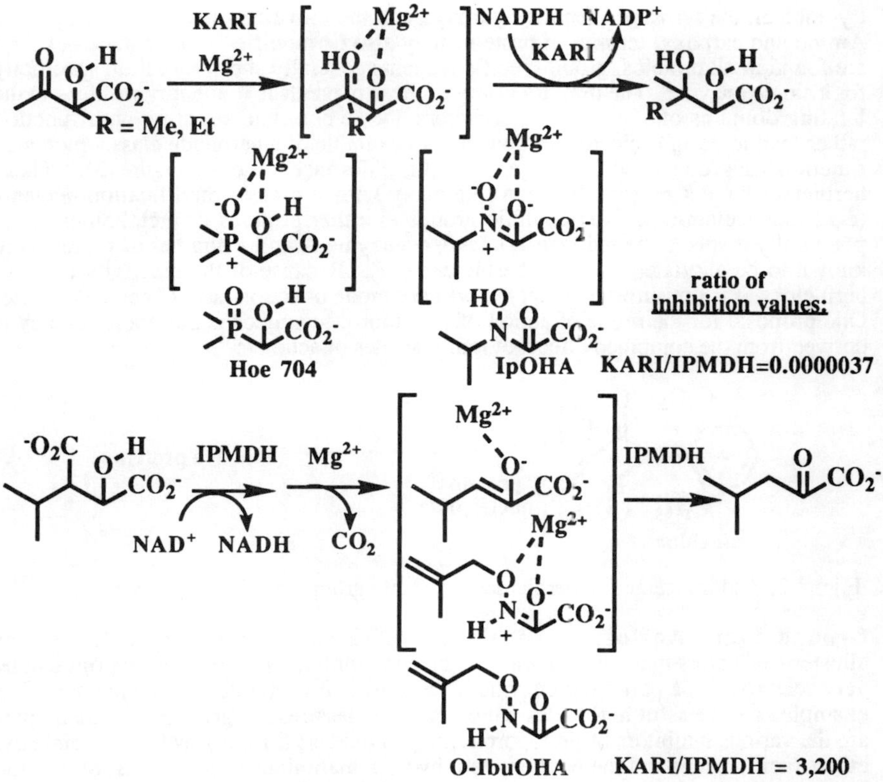

Figure 1. Reaction intermediate analogue inhibitors of ketol acid reductoisomerase (top) and isopropylmalate dehydrogenase (bottom).

Enzyme Inhibitor Design

Chemical validation depends on quickly obtaining an inhibitor of the target enzyme. This can be achieved by screening the enzyme for lead structures, using structure based approaches if a crystal structure with appropriate bound substrate analogs is available or by mechanism based design. In our opinion, the latter is the most expedient. If genetic methods have successfully validated a target, then additional resources can be invested more confidently in all three strategies to discover herbicidal inhibitors of the target. If the genetic approach involves cloning the target (antisense for example), then it may facilitate all three approaches by providing a ready means for overexpression to obtain large quantities of the target for detailed mechanism studies, high through put screening and for crystal structure elucidation. The following sections will briefly summarize six categories of enzyme inhibitors having agrochemical utility. For more examples and for an analysis of the kinetic characteristics of each type of inhibitor, see recent reviews (*54-56*).

Group specific Reagents. These inhibitors contain reactive functional groups which modify amino acid residues in proteins. Examples of amino acid side chains which can be modified are the ε-amino groups of lysine and arginine, cysteine sulfhydryl groups, carboxyl groups of glutamate or aspartate, the indole moiety of tryptophan, the tyrosine phenolic hydroxyl, and the thio ether linkage of methionine. Amino and carboxyl termini of proteins may also be modified. Since these residues are found in all proteins, group specific reagents generally show very little specificity for a single enzyme. The primary utility of these reagents is to identify residues in the binding domains of substrates or cofactors and to correlate structure with function rather than to design selective inhibitors. An example of a herbicide class which may function as group specific reagents is the chloroacetamides (Figure 2). These herbicides have a reactive halogen like many known protein modification reagents (e.g., iodoacetamide). Nucleophilic groups in either proteins or metabolites could potentially displace the halogen. *In vivo*, coenzyme A and a number of proteins are known to be alkylated by these herbicides (*57*). Because of the reactivity of these herbicides, it is not surprising that the primary mode of action has not been identified. One proposal for the mode of action of the chloroacetamides is that their potency is derived from the combined effects of multiple sites of action (*58*).

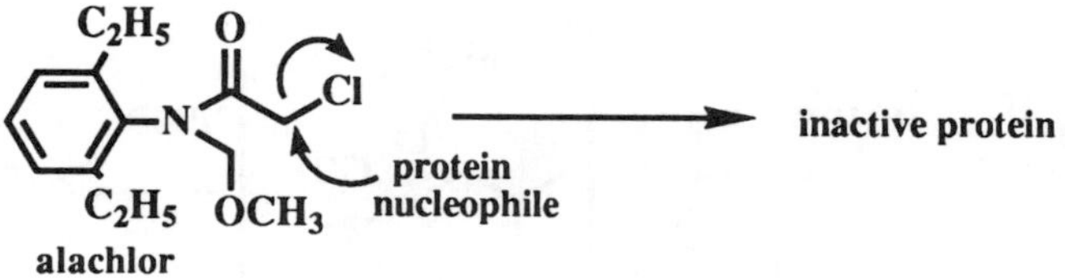

Figure 2. Chloroacetamide herbicides - possible group specific reagents.

Ground State Analogues. Analogues of substrates, products, cofactors, or allosteric effectors may also be practical enzyme inhibitors. Substrate analogs can be very selective for a particular enzyme if the substrate is unique to that enzyme. One example of successful herbicides which may be classified as ground state analogues are the various inhibitors of protoporphyrinogen oxidase (Protox) such as the diphenyl ethers (acifluorfen) and the N-phenyltetrahydrophthalimides (Figure 3). Protox is the last common enzyme in the synthesis of both chlorophyll and heme and is a membrane bound enzyme associated with both the chloroplast and mitochondrial envelopes. Recent studies on the mode of action of acifluorfen and other porphyrin biosynthesis inhibitors (recently reviewed (*59*)) suggest that inhibition of Protox results in the accumulation of the **product**, protoporphyrin IX, rather than the substrate, in the plasma membrane where it participates in rapid photodegradative damage to the membrane. In the presence of the Protox inhibiting herbicides, there is increasing evidence that protoporphyrinogen IX accumulates in the plastid envelope to only a limited extent, instead it diffuses out into the cytoplasm where it is rapidly oxidized to protoporphyrin IX both spontaneously and enzymically (not by Protox). Both compartmentalization of Protox and the proposed oxidizing enzyme, and the physical properties of inhibitors and substrates are thought to contribute to the phytotoxic consequences of Protox inhibition. Acifluorfen is a potent competitive inhibitor with respect to protoporphyrinogen IX (K_i = 6 to 120 nM)(*60, 61*). In contrast, the product, protoporphyrin IX, is unable to prevent binding of [^{3}H]acifluorfen (*61*). The I_{50} values for Protox inhibition by the diphenyl ether herbicides range from 0.008 to 420 μM (*59, 62*). Unlike most ground state analogue inhibitors, Protox inhibitors can be bound more tightly than the substrate, protoporphyrinogen IX (K_m = 0.4 to 5 μM)(*60, 63*). It has been proposed that the Protox inhibiting herbicides compete with protoporphyrinogen IX by matching one half of its structure (*59, 62*). In general, inhibitors that are analogues of substrates or products have only limited potency since their affinity for the enzyme is usually in the same range as that for substrates or products (1 μM to 1 mM). The inhibition may frequently be overcome by the build up of the natural substrate, especially if it is a primary metabolite (except in the unusual case of the Protox inhibitors where the accumulated substrate diffuses out of the compartment where the enzyme is found).

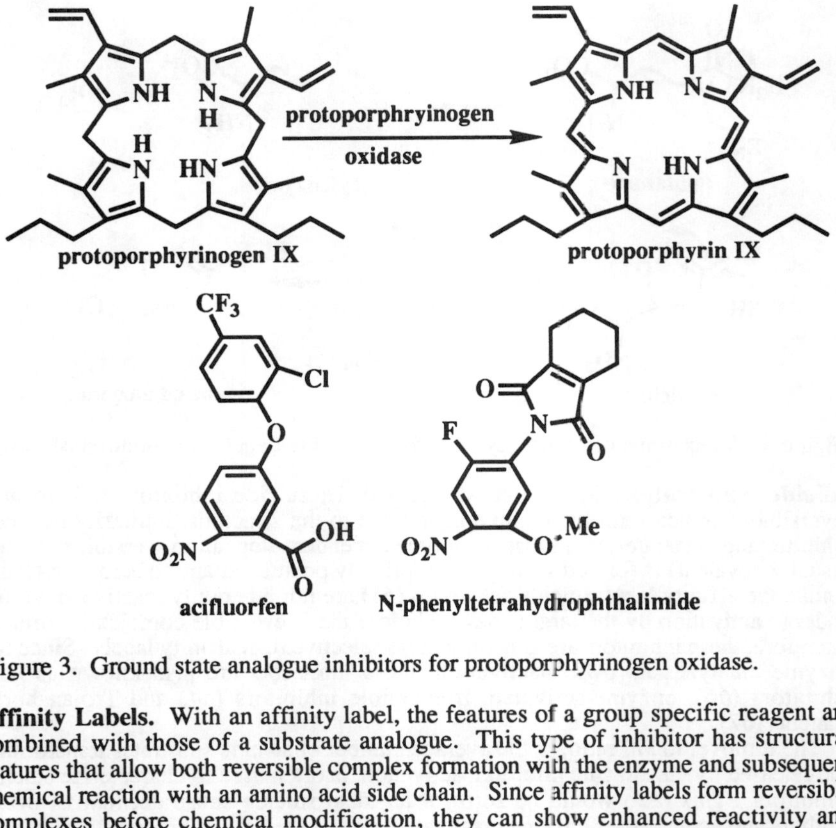

Figure 3. Ground state analogue inhibitors for protoporphyrinogen oxidase.

Affinity Labels. With an affinity label, the features of a group specific reagent are combined with those of a substrate analogue. This type of inhibitor has structural features that allow both reversible complex formation with the enzyme and subsequent chemical reaction with an amino acid side chain. Since affinity labels form reversible complexes before chemical modification, they can show enhanced reactivity and specificity compared to group specific reagents. However, nonspecific modification of proteins is inevitable due to the inherently reactive nature of affinity labels. Ideally, the target should have an appropriately positioned, modifiable amino acid side chain.

An analogue of glutamine, L-(αS,5S)-α-amino-3-chloro-4,5-dihydro-5-isoxazoleacetic acid (acivicin, AT-125) (Figure 4), has been found to inactivate glutamine-dependent amidotransferases including anthranilate synthase and glutamate synthase by alkylating an active site cysteine residue (*64*). Other examples of glutamine affinity label analogues are cycloserine and diazonorleucine. These enzymes catalyze the formation of an acyl enzyme intermediate with release of ammonia, which is transferred to a second substrate. Organisms can avoid the potential toxic consequences of ammonia by generating it in controlled amounts "*in situ*" by this mechanism. These types of amino transfer reactions are found in numerous biosynthetic pathways including purine, pyrimidine, amino acid, and cofactor biosynthesis. The amino donor is usually the glutamine amide. The modest herbicidal activity observed with acivicin probably results from inactivation of a number of enzymes. One explanation for the relatively weak *in vivo* activity is that acivicin may react with nonessential sulfhydryl groups in the plant, either enzyme-bound or protein-free. This inhibitor is not expected to be very specific since a large number of enzymes, including mammalian enzymes, use glutamine as an amino donor and appear to have similar chemical mechanisms involving a covalent enzyme-thiol-substrate complex. In this approach, as in the case of the substrate analogues, it is desirable for the substrate to be unique to the target enzyme.

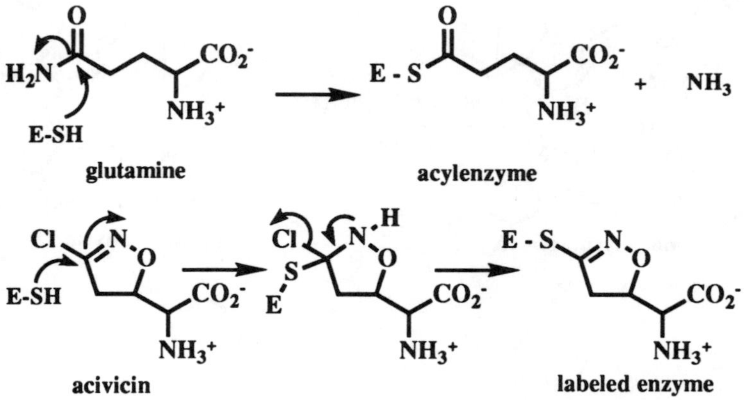

Figure 4. An example of an affinity label for glutamine-dependent amidotransferases.

Suicide Substrates. Much like affinity labels, suicide inhibitors first form a reversible complex with the target enzyme due to the structural similarity between inhibitor and substrate. In a subsequent time dependent step, an irreversible complex (usually covalent) is formed with an appropriately positioned amino acid side chain. Unlike the affinity label, suicide substrates (*65*) are not inherently reactive and must undergo activation by the target enzyme before the irreversible complex is formed. Therefore, these inhibitors are generally more selective than affinity labels. Since the enzyme catalyzes its own inactivation, these inhibitors are also known as kcat inhibitors (*66*), enzyme-activated, irreversible inhibitors (*54*) and Trojan horse reagents (*67*).

The turnover to kill ratio, or the average number of suicide substrates transformed and released by the enzyme prior to inactivation, is often used to characterize suicide inhibitors. This ratio would be zero for an ideal suicide substrate, that is, every inhibitor molecule causes enzyme inactivation with no release of product. The higher this ratio, the greater the potential for loss of specificity upon release of the transformed, reactive inhibitor, which may then modify and inactivate other, non targeted enzymes.

One example of a suicide substrate with herbicidal activity (*68, 69*) is the naturally occurring neurotoxin, 5-amino-1,3-cyclohexadienylcarboxylate (gabaculine) (Figure 5). Gabaculine is a potent, irreversible inhibitor of γ-aminobutyrate: α-ketoglutarate aminotransaminases (*70*). In plants, it produces herbicidal effects by inhibition of glutamate-1-semialdehyde aminotransaminase (GSA-AT), the last of three enzymes involved in converting glutamate to δ-aminolevulinate, the branchpoint intermediate to heme, chlorophyll, and phytochrome biosynthesis (*69*). The effect of gabaculine on chlorophyll biosynthesis is reversed by δ-aminolevulinate (*68*) and its effect on phytochrome biosynthesis is reversed by both δ-aminolevulinate and biliverdin (*71*). GSA-AT is a pyridoxal-5'-phosphate (PLP) dependent enzyme and catalyzes the transfer of the C-2 amino group of GSA to the C-1 position using a typical ping pong Bi Bi mechanism in which 4,5-diaminovalerate is a likely intermediate (*72*). The reaction mechanism, structural analysis and spectral changes of the bound cofactor suggest the GSA-AT is closely related to other aminotransferases. The mechanism proposed for the inactivation involves formation of the external aldimine of the pyridoxal cofactor and gabaculine, presumably via a tetrahedral intermediate. The second step, rearrangement to the ketamine of gabaculine and the pyridoxamine cofactor, requires the initial formation of an α-carbanion which is also a step in the normal reaction with GSA and is common to most PLP dependent enzymes. In the last step, a β-proton is abstracted forming a β-carbanion which leads to irreversible aromatization and formation of the stable anthranilate derivative (*73* and references therein). This covalent and irreversible adduct of the cofactor is bound extremely tightly to the active site of the enzyme and can be liberated under denaturing conditions (*70*). Gabaculine is a particularly effective suicide inactivator with inactivation rate constants close to those of substrate turnover ($0.1\ s^{-1}$) and, based upon spectral changes of the cofactor, is initially bound 10 times tighter than the substrate (*73*). The selectivity of gabaculine truly lies in the mechanism carried out by the enzyme and not in its structural similarity to normal substrates. A majority of PLP enzymes form α-carbanions, however, they are not inactivated by gabaculine because formation of the α-carbanion leaves the molecule unreactive. Only those PLP enzymes which can form β-carbanions will be inactivated by gabaculine (*74*).

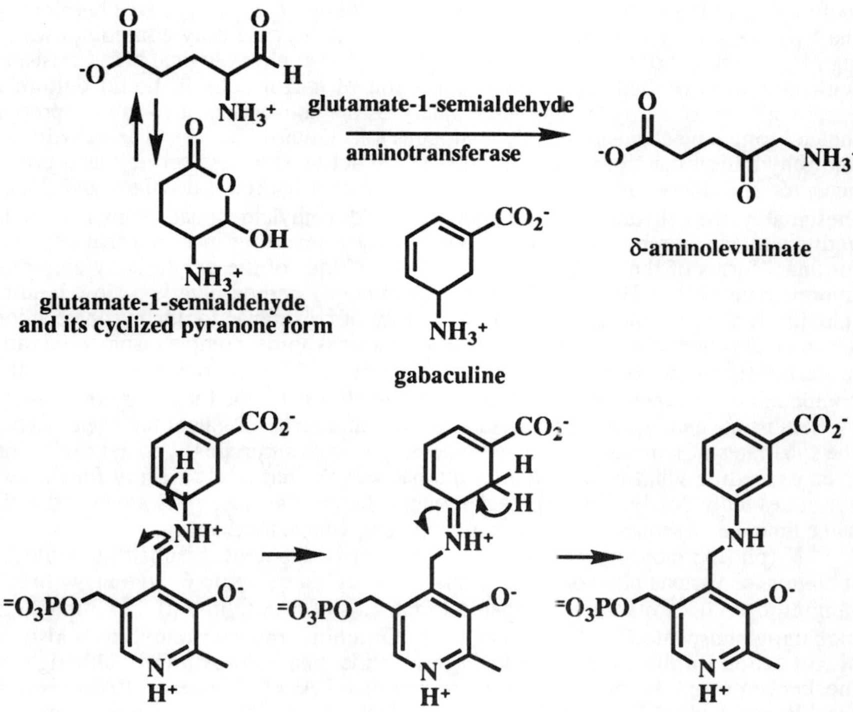

Figure 5. An example of a suicide substrate: gabaculine and the postulated mechanism of inactivation of glutamate-1-semialdehyde aminotransferase.

Reaction Intermediate Analogues. Reaction intermediate or transition state analogues are stable mimics of labile reaction intermediates (for thorough reviews see (*55, 56*). This type of inhibitor can be quite selective if the intermediate is unique to the targeted enzyme. These inhibitors should be much more potent than simple substrate analogues since enzymes bind reaction intermediates more tightly than either substrates or products during the course of a chemical transformation. In general, reaction intermediate analogues are not chemically reactive and, unlike affinity labels, there is no indiscriminate labeling of non targeted proteins. In contrast to affinity labels and suicide substrates, inhibition does not involve covalent bond formation with an appropriately placed nucleophile and is reversible. The resulting inhibition is a balance of the association rate and the dissociation rate for the analogue. By having slow dissociation rates, the effective reaction intermediate analogues often display time dependent inhibition kinetics. The $t_{1/2}$ for dissociation can be as slow as 1-14 years (*75*). While slow dissociation rates are desirable, slow association rates are undesirable since they increase the amount of time required before the onset of inhibition *in vivo* and may allow the plant to metabolize the inhibitor.

There are a number of literature examples of potent reaction intermediate analogues having herbicidal properties including the oxalylhydroxamate inhibitors for KARI and IPMDH and the successful herbicide, phosphinothricin, which inhibits GS (see earlier discussions). Recently, potent transition state inhibitors with herbicidal activity have been described for isopropylmalate isomerase (IMI), the second committed step in leucine biosynthesis (*15*). The nitronic acid forms of nitroisopropylmalate (*76*), 1-hydroxy-2-nitrocyclopentane-1-carboxylate, and 1-hydroxy-2-nitrocyclohexane-1-carboxylate (*15*) are all potent inhibitors of IMI (Figure 6, top). The best herbicide is the hydroxynitrocyclopentane carboxylate because it is the only compound with a significant amount of the nitronate form (pKa of 7.3) at physiological pH. Consistent with inhibition of IMI, the growth inhibition of carrot cells in liquid culture is specifically reversed by leucine. IMI catalyzes the conversion of α- to β-isopropylmalate using a mechanism directly analogous to aconitase (*77, 78*). The dehydration and rehydration reactions are facilitated by an active site base serving as a proton acceptor and donor, and a Fe_4S_4 cluster acting as a hydroxyl acceptor and donor. Presumably, the dehydrated alkene intermediate (dimethylcitraconate) flips 180° in the active site and is rehydrated on the opposite face resulting in isomerization. The nitronate forms of the inhibitors are excellent mimics of the *aci*-carboxylate anion intermediate (*91*). 1-Hydroxy-2-nitrocyclopentane-1-carboxylate is a slow binding inhibitor with an estimated dissociation constant of *ca.* 0.6 nM for the nitronate anion. The time dependent inhibition is biphasic with a weak initial complex isomerizing to a more tightly bound one with a rate of $\geq 5\ min^{-1}$. The dissociation rate for this compound is approximately $0.07\ min^{-1}$ with a modest half time for release of 10 min.

Bisubstrate analogues are a variation on this approach. Covalent linkage of two of the substrates of a multisubstrate enzyme may lead to surprisingly potent inhibition, even exceeding what is expected on the basis of the enzyme's affinity for the two substrates individually. This approach requires that the two substrates are bound at the same time, i.e., a sequential rather than ping pong kinetic mechanism.

N-(phosphonoacetyl)-L-ornithine (PALO) is a potent, bisubstrate analogue inhibitor for various plant ornithine transcarbamoylases (Figure 6, bottom) with K_i's ranging from 0.2 to 0.4 μM compared to the K_m's for ornithine (0.3 to 3 mM) and carbamoylphosphate (0.2 to 0.6 mM)(*79*). Ornithine transcarbamoylase is also the site of action for the chlorosis inducing tripeptide, phaseolotoxin (*80*), although the mechanism of inhibition is less well studied than PALO. However, PALO is only weakly herbicidal (*79*). The reasons for its lack of efficacy are unknown, since no metabolic studies have been reported for this compound. Ornithine transcarbamoylase catalyzes the condensation of carbamoylphosphate with ornithine to produce citrulline. The proposed chemical mechanism for this enzyme involves initial formation of a tetrahedral carbinolamine intermediate between ornithine and carbamoylphosphate which then dehydrates with loss of phosphate to yield citrulline. PALO resembles the tetrahedral intermediate that would be formed by attack of the amino group of ornithine on the carbonyl group of carbamoylphosphate. Unlike reaction intermediate analogues of tetrahedral adducts (such as octicidin), PALO maintains a trigonal carbon at the position corresponding to the tetrahedral carbon of the carbamoylphosphate-ornithine adduct.

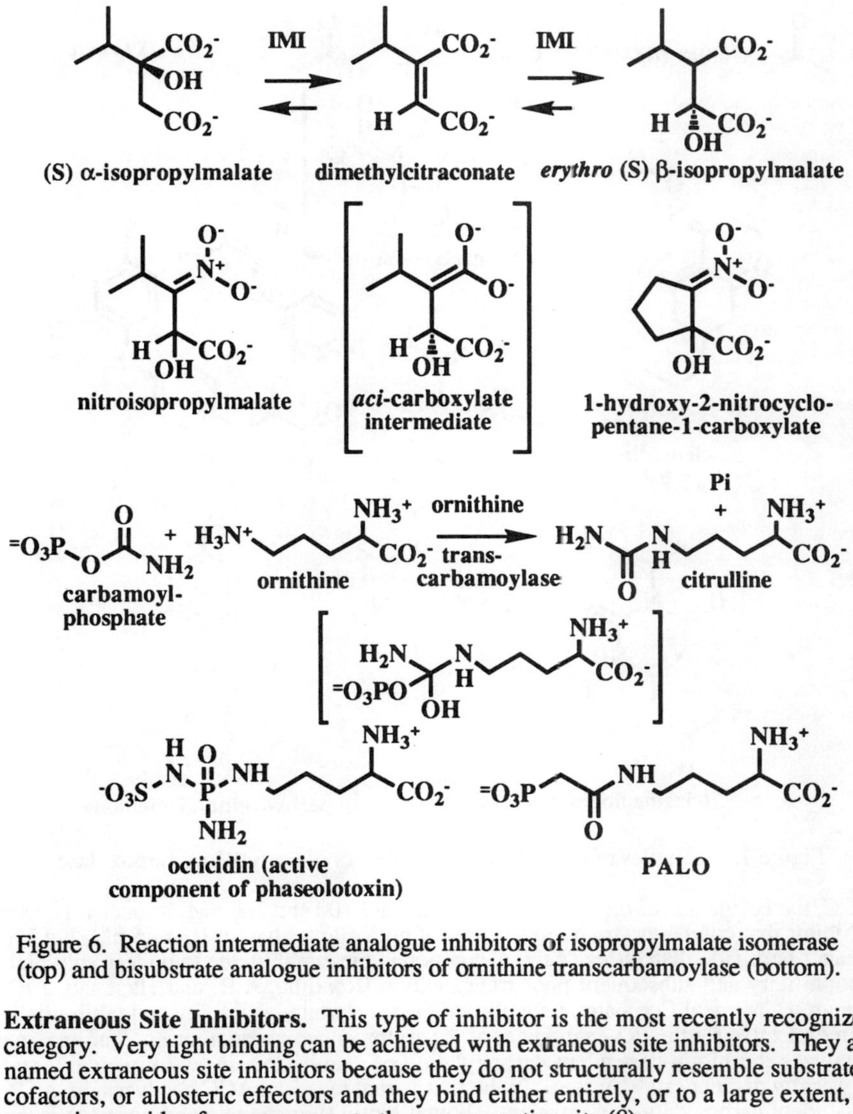

Figure 6. Reaction intermediate analogue inhibitors of isopropylmalate isomerase (top) and bisubstrate analogue inhibitors of ornithine transcarbamoylase (bottom).

Extraneous Site Inhibitors. This type of inhibitor is the most recently recognized category. Very tight binding can be achieved with extraneous site inhibitors. They are named extraneous site inhibitors because they do not structurally resemble substrates, cofactors, or allosteric effectors and they bind either entirely, or to a large extent, to some site outside of or extraneous to the enzyme active site (*9*).

The best characterized extraneous site inhibitors are the many chemical classes of herbicidal inhibitors of acetolactate synthase. These have been extensively reviewed elsewhere (*1, 8, 9, 52, 56, 81, 92*). Extraneous site inhibitors are also exemplified by four classes of grass selective herbicides; the aryloxyphenoxypropionic acids represented by diclofop, the cyclohexanediones typified by clethodim (for a recent review see (*57*)), the triazinediones (*82*) and the perhydroindolizinediones (*83*) (Figure 7). The following discussion is limited to the first two classes. Both classes of herbicides are potent, reversible inhibitors of acetyl-CoA carboxylase (ACC) from susceptible plants, the putative rate limiting enzyme in lipid biosynthesis. With wheat

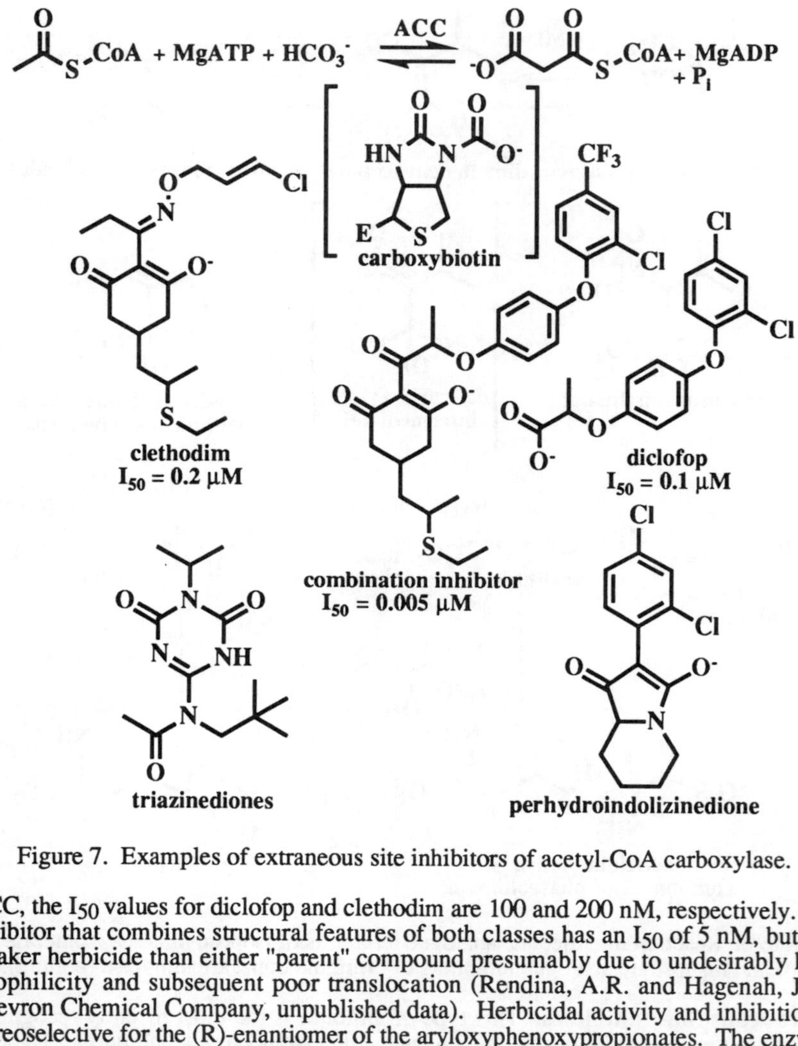

Figure 7. Examples of extraneous site inhibitors of acetyl-CoA carboxylase.

ACC, the I_{50} values for diclofop and clethodim are 100 and 200 nM, respectively. An inhibitor that combines structural features of both classes has an I_{50} of 5 nM, but is a weaker herbicide than either "parent" compound presumably due to undesirably high lipophilicity and subsequent poor translocation (Rendina, A.R. and Hagenah, J.A., Chevron Chemical Company, unpublished data). Herbicidal activity and inhibition is stereoselective for the (R)-enantiomer of the aryloxyphenoxypropionates. The enzyme catalyzes the biotin-dependent carboxylation of acetyl-CoA in two kinetically and physically distinct catalytic sites. In the first partial reaction, ACC catalyzes the ATP-dependent carboxylation of covalently bound biotin (bicarbonate and not CO_2 is the substrate). At the second catalytic site, the enzyme abstracts a proton from acetyl-CoA generating a carbanion that attacks the carboxybiotin intermediate releasing malonyl-CoA and recycling the biotin cofactor. Although inhibition of wheat ACC by clethodim and diclofop is noncompetitive versus MgATP, bicarbonate, and acetyl-CoA, they are nearly competitive with acetyl-CoA since the level of inhibition is most sensitive to the acetyl-CoA concentration (*84, 85*). Only isotope exchange and partial reactions catalyzed by the carboxytransfer site are inhibited by both classes of herbicides suggesting that the herbicide site is near this site and not the biotin carboxylation site (*86, 87*). Inhibition studies with pairs of inhibitors suggest that the two herbicide classes are mutually exclusive and that only the thioester region of malonyl- and acetyl-CoA overlaps with some structural feature of the herbicides, presumably the small alkyl side chains that are required for good herbicidal properties (CoA and the herbicides are not mutually exclusive while malonyl-CoA and the herbicides are) (*86, 87*). The enhanced binding of the combination inhibitor shown in Figure 7 suggests one way these two classes may be linked together and how they might exclude each other from a common portion of the binding site. CoA esters of diclofop and fluazifop were from 20 to 425-fold more potent than the corresponding unconjugated herbicides as inhibitors of rat liver (*88*) and wheat ACC (Rendina, A.R.; Taylor, W.S.; Hixon, M.S, E.I. Du Pont de Nemours, unpublished data) confirming that the CoA pocket is distinct from the herbicide pocket. These observations strongly suggest that the cyclohexanedione and aryloxyphenoxypropionate herbicide binding regions only partially overlap with the active site of ACC and that most of the herbicide structure occupies an extraneous site.

The major potential disadvantage of extraneous site inhibitors is that resistance can develop readily. Since the extraneous site is usually not essential for the function of the enzyme, mutations in that region are more likely to eliminate the inhibition without disrupting catalytic efficiency. Design of extraneous site inhibitors based on knowledge of the enzyme mechanism is nearly impossible due to the nature of these inhibitors. They are most readily discovered through both *in vitro* and *in vivo* screening. Optimization of activity can be greatly facilitated if the target is known. Knowledge of the crystal structure of the enzyme-inhibitor complex and the application of molecular modeling can also facilitate optimization of activity (for recent successes in structure based drug design see (*89, 90*)).

Summary of Inhibitor Types. The reaction intermediate analogues are the most promising inhibitors for herbicide design. They have both great selectivity and potency for the target enzyme. The same general strategy for design of a reaction intermediate analogue with one target enzyme can frequently be applied to other enzymes with similar reaction mechanisms. However, due to the specificity of reaction intermediate analogues for the target, there may be only a few structures that produce potent inhibition (e.g., glyphosate and phosphinothricin). As a result, this design approach will probably lead to non selective, broad spectrum herbicides. It is less likely that crop safety will occur by differential detoxification with this type of inhibitor because the scope of chemistry is very narrow. Instead, plant breeding or genetic engineering approaches will be used to introduce herbicide resistance to the inhibitor early in the discovery process in order to achieve crop tolerance (*93, 94*). In contrast to extraneous site inhibitors, development of natural resistance to this type of inhibition should be much less frequent. Mutations at the active site which might effectively overcome this type of inhibition are more likely to adversely affect the catalytic efficiency of the enzyme since these inhibitors are bound in the active site and are designed based on knowledge of the enzyme's function. While these inhibitors are not difficult to conceive, their successful design depends on a detailed knowledge of the enzyme's mechanism.

Concluding Remarks

We are confident that target-site directed research will ultimately succeed in producing novel pesticides. The approach has the potential to produce extremely low use rate compounds that are non toxic and environmentally friendly . Target-site design efforts should complement traditional screening approaches by providing new lead areas with novel modes of action in areas of chemistry that have previously been under explored, namely water-soluble chemicals. Molecular techniques such as antisense validation of new target sites will greatly facilitate this approach.

Acknowledgments

We thank John Schloss, Vern Wittenbach, Wendy Taylor and Jim Kerschen for many helpful discussions in the preparation of this manuscript.

Literature Cited

1. Abell, L.M.; Schloss, J.V.; Rendina, A.R. *Amer. Chem. Soc. Symp. Ser.* **1993**, *524*, 16-37.
2. Siehl, D.L. *Cur. Topics Plant Physiol.* **1992**, *7*, 146-162.
3. Goosey, M.W. *Pestic. Sci.* **1992**, *34*, 313-320.
4. Pillmoor, J.B.; Wright, K.; Lindell, S.D. *Proc. Brighton Crop Protection Conf.-Weeds* **1991**, *2*, 857-866.
5. Haworth, P.; Siehl, D.L. *J. Agric. Food Chem.* **1990**, *38*, 1271-1273.
6. Corbett, J.R. *Proc. British Insecticide Fungicide Conf.* **1975**, *8*, 981-993.
7. Hudson, G.S.; Evans, J.R.; von Caemmerer, S.; Arvidsson, Y.B.C.; Andrews, T.J. *Plant Physiol.* **1992**, *98*, 294-302.
8. LaRossa, R.A.; Schloss, J.V. *J. Biol. Chem.* **1984**, *259*, 8753-8757.
9. Schloss, J.V. In *Prospects for Amino Acid Biosynthesis Inhibitors Crop Protection and Pharmaceutical Chemistry* ; Copping, L.G.; Dalziel, J.; Dodge, A.D. Eds.; The British Crop Protection Council: Farnham, 1989, BCPC Monograph No. 42; pp 147-152.
10. Aulabaugh, A.; Schloss, J.V. *Biochemistry* **1990**, *29*, 2824-2830.
11. Schulz, A.; Sponemann, P.; Kocher, H.; Wengenmayer, F. *FEBS Lett.* **1988**, *238*, 375-378.
12. Westkaemper, R.B.; Zenk, P.C. *Eur. J. Med. Chem.* **1988**, *23*, 233-236.
13. Pirrung, M.C.; Ha, H.-J.; Holmes, C.P. *J. Org. Chem.* **1989**, *54*, 1543-1548.
14. Pirrung, M.C.; Holmes, C.P.; Horowitz, D.M.; Nunn, D.S. *J. Am. Chem. Soc.* **1991**, *113*, 1020-1025.
15. Hawkes, T.R.; Cox, J.M.; Fraser, T.E.M.; Lewis, T. *Z. Naturforsch.* **1993**, *48c*, 364-368.
16. Wittenbach, V.A.; Rayner, D.R.; Schloss, J.V. *Cur. Topics Plant Physiol.* **1992**, *7*, 69-88.
17. LaRossa, R.A.; Van Dyk, T.K. *BioEssays* **1987**, *7*, 125-129.
18. Schneider, T.; Dinkins, R.; Robinson, K.; Shellhammer, J.; Meinke, D.W. *Develop. Biol.* **1989**, *131*, 161-167.
19. Wallsgrove, R.M.; Risiott, R.; King, J.; Bright, S.W.J. *Plant Sci.* **1986**, *43*, 109-114.
20. Horsch, R.B.; King, J. *Planta* **1983**, *159*, 12-17.
21. Lea, P.J.; Blackwell, R.D.; Murray, A.J.S.; Joy, K.W. *Recent Adv. Phytochem.* **1989**, *23*, 157-189.
22. Blackwell, R.D.; Murray, A.J.S.; Lea, P.J. *J. Exp. Botany* **1987**, *38*, 1799-1809.
23. Rupp, W.; Finke, M.; Bieringer, H.;Langeleuddeke, P. *Patent: Ger. Offen. DE 2717440* **1977**.
24. Bayer, E.; Gugel, K.H.; Hagele, H.; Hagenmaier, H.; Jessipow, S.; Konig, W.A.; Zahner, H. *Helv. Chim. Acta* **1972**, *55*, 224-239.
25. Mol, J.N.M.; van der Krol, A.R.; van Tunen, A.J.; van Blokland, R.; de Lange, P.; Stuitje, A.R. *FEBS Lett.* **1990**, *268*, 427-430.
26. Mol, J.N.M.; van der Krol, A.R. *Antisense Nucleic Acids and Proteins;* Marcel Dekker, New York, 1991.
27. Moffat, A.S. *Science* **1991**, *253*, 510-511.
28. Rothstein, S.J.; DiMaio, J.; Strand, M.; Rice, D. *Proc. Natl. Acad. Sci. USA* **1987**, *84*, 8439-8443.

29. Sandler, S.J.; Stayton, M.; Townsend, J.A.; Ralston, M.L.; Bedrock, J.R.; Dunsmuir, P. *Plant Mol. Biol.* **1988**, *11*, 301-310.
30. Ecker, J.R.; Davis, R.W. *Proc. Natl. Acad. Sci. USA* **1986**, *83*, 5372-5376.
31. Delauney, A.J.; Tabaeizadeh, Z.; Verma, D.P.S. *Proc. Natl. Acad. Sci. USA* **1988**, *85*, 4300-4304.
32. van der Krol, A.R.; Mur, L.A.; de Lange, P.; Gerats, A.G.M.; Mol, J.N.M.; Stuitje, A.R. *Mol. Gen. Genet.* **1990**, *220*, 204-212.
33. van der Krol, A.R.; Stuitje, A.R.; Mol, J.N.M. In *Antisense Nucleic Acids and Proteins*, Mol, J.N.M.; van der Krol, A.R., Eds.; Marcel Dekker: New York, 1991; pp 125-140.
34. van der Meer, I.M.; Stam, M.E.; van Tunen, A.J.; Mol, J.N.M.; Stuitje, A.R. *Plant Cell* **1992**, *4*, 253-262.
35. Smith, C.J.S.; Watson, C.F.; Ray, J.; Bird, C.R.; Morris, P.C.; Schuch, W.; Grierson, D. *Nature* **1988**, *334*, 724-726.
36. Schuch, W., Bird, C.R.; Ray, J.; Smith, C.J.S.; Watson, C.F.; Morris, P.C.; Gray, J.E.; Arnold, C.; Seymour, G.B.; Tucker, G.A.; Grierson, D. *Plant Mol. Biol.* **1989**, *13*, 303-311.
37. Cornelissen, M.; Vandewiele, M. *Nuc. Acids Res.* **1989**, *17*, 833-843.
38. Robert, L.S.; Donaldson, P.A.; Ladaique, C.; Altosaar, I.; Arnison, P.G.; Fabijanski, S.F. *Plant Mol. Biol.* **1989**, *13*, 399-409.
39. Cannon, M.; Platz, J.; O'Leary, M.; Sookdeo, C.;Cannon, F. *Plant Mol. Biol.* **1990**, *15*, 39-47.
40. Visser, R.G.F.; Somhorst, I.; Kuipers, G.J.; Ruys, N.J.; Fenstra, W.J.; Jacobsen, E. *Mol. Gen. Genet.* **1991**, *225*, 289-296.
41. Rodermel, S.R.; Abbott, M.S.; Bogorad, L. *Cell* **1988**, *55*, 673-681.
42. Quick, W.P.; Fichtner, K.;Schulze, E.-D.; Wendler, R.; Leegood, R.C.; Mooney,H.;Rodermel, S.R.;Bogorad,L.;Stitt, M. *Planta* **1992**,*188*, 522-531.
43. Oeller, P.W.; Min-Wong, L.; Taylor, L.P.; Pike, D.A.; Theologis, A. *Science* **1991**, *254*, 437-439.
44. Hamilton, A.J.; Lycett, G.W.; Grierson, D. *Nature* **1990**, *346*, 284-287.
45. Bouzayen, M.; Hamilton, A.; Picton, S.; Barton, S.; Grierson, D. *Biochem. Soc. Trans.* **1992**, *20*, 76-79.
46. Penarrubia, L.; Aguilar, M.; Margossian, L.; Fischer, R.L. *Plant Cell* **1992**, *4*, 681-687.
47. Elkind, Y.; Edwards, R.; Mavandad, M.; Hedrick, S.A.; Ribak, O.; Dixon, R.A.;Lamb, C.J. *Proc. Natl. Acad. Sci. U.S.A.* **1990**, *87*, 9057-9061.
48. Knutzon, D.S.; Thompson, G.A.; Radke, S.E.; Johnson, W.B.; Knauf, V.C.; Kridl, J.C. *Proc. Natl. Acad. Sci. USA* **1992**, *89*, 2624-2628.
49. Sherf,B.A.;Bajar,A.M.;Kolattukudy,P.E. *Plant Physiol.* **1993**,*101*,201-208.
50. Gogarten, J.P.; Fichmann, J.; Braun, Y.; Morgan, L.; Styles, P.; Taiz, S.L.; DeLapp, K.; Taiz, L. *Plant Cell* **1992**, *4*, 851-864.
51. Temple, S.J.; Knight, T.J.; Unkefer, P.J.; Sengupta-Gopalan, C. *Mol. Gen. Genet.* **1993**, *236*, 315-325.
52. Wittenbach, V.A.; Aulabaugh, A.;Schloss, J.V. In *Pesticide Chemistry: Advances in International Research, Development, and Registration*, Frehse, H., Ed.; Pesticide Chemistry; VCH: Weinheim, 1991; pp 151-160.
53. Wittenbach, V.A.; Teaney, P.W.; Rayner, D.R.; Schloss, J.V. *Plant Physiol.* **1993**, *102*, 50.
54. Silverman, R.B. *Mechanism-Based Enzyme Inactivation : Chemistry and Enzymology*; CRC Press: Boca Raton, Florida, 1988, Vol. I & II.
55. Schloss, J.V. *Accounts Chem. Res.* **1988**, *21*, 348-353.
56. Schloss, J.V. In *Target Sites of Herbicide Action*; Boger, P.H.; Sandmann, G. Eds.; CRC Press: Boca Raton, Florida, 1989; pp 165-245.
57. Gronwald, J.W. *Weed Sci.* **1991**, *39*, 435-449.
58. Jaworski, E.G. *J. Agric. Food Chem.* **1969**, *17*, 165-170.

59. Nandihalli, U.B.; Duke, S.O. *Amer. Chem. Soc. Symp. Ser.* **1993**, *524*, 62-78.
60. Camadro, J.-M.; Matringe, M.; Scalla, R.; Labbe, P. *Biochem. J.* **1991**, *277*, 17-21.
61. Matringe, M.; Mornet, R.; Scalla, R. *Eur. J. Biochem.* **1992**, *209*, 861-868.
62. Nandihalli, U.B.; Duke, M.V.; Duke, S.O. *Pestic. Biochem. Physiol.* **1992**, *43*, 193-211.
63. Jacobs, J.M.; Jacobs, N.J. *Biochem. J.* **1987**, *244*, 219-224.
64. Tso, J.Y.; Bower, S.G.; Zalkin, H. *J. Biol. Chem.* **1980**, *255*, 6734-6738.
65. Walsh, C.T. *Horiz. Biochem. Biophys.* **1977**, *3*, 36-81.
66. Rando, R.R. *Science* **1974**, *185*, 320-324.
67. Walsh, C. *Tetrahedron* **1982**, *38*, 871-909.
68. Flint, D. *Plant Physiol.* **1984**, *75*, S-965.
69. Kannangara, C.;Schouboe, A. *Carlsberg Res. Commun.* **1985**, *50*, 179-191.
70. Rando, R.R. *Biochemistry* **1977**, *16*, 4604-4610.
71. Elich, T.D.; Lagarias, J.C. *Plant Physiol.* **1987**, *84*, 304-310.
72. Smith, M.A.; Kannangara, C.G.; Grimm, B.; von Wettstein, D. *Eur. J. Biochem.* **1991**, *202*, 749-757.
73. Smith, M.A.; Grimm, B. *Biochemistry* **1992**, *31*, 4122-4127.
74. Soper, T.S.; Manning, J.M. *J. Biol. Chem.* **1982**, *257*, 13930-13936.
75. Kaplan, A.P.; Bartlett, P.A. *Biochemistry* **1991**, *30*, 8165-8170.
76. Emptage, M.H.; Schloss, J.V. *Fed. Proc.* **1986**, *45*, 1536.
77. Emptage, M.H. In *Biosynthesis of Branched Chain Amino Acids*; Baraak, Z.; Chipman, D.M.; Schloss, J.V., Eds.; VCH: Weinheim, 1991; pp 315-328.
78. Lauble, H.; Kennedy, M.C.; Beinert, H.; Stout, C.D. *Biochemistry* **1992**, *31*, 2735-2748.
79. Acaster, M.A.; Scott-White, S.;Weitzman, P.D. *J. Exp. Bot.* **1989**, *40*, 1121-1125.
80. Patil, S.S.; Kwok, O.C.H.; Moore, R.E. *Toxicon* **1983**, *Suppl. 3*, 341-344.
81. Schloss, J.V.; Ciskanik, L.M.; Van Dyk, D.E. *Nature* **1988**, *331*, 360-362.
82. Walker, K.A.; Ridley, S.M.; Lewis, T.; Harwood, J.L. *Phytochem.* **1990**, *29*, 3743-3747.
83. Babczinski, P.; Fischer, R.; *Pestic. Sci.* **1991**, *33*, 455-466.
84. Rendina, A.R.; Felts, J.M. *Plant Physiol.* **1988**, *86*, 983-986.
85. Rendina, A.R.; Felts, J.M.; Beaudoin, J.D.; Craig-Kennard, A.C.; Look, L.L.; Paraskos, S.L.; Hagenah, J.A. *Arch. Biochem. Biophys.* **1988**, *265*, 219-225.
86. Rendina, A.R.; Beaudoin, J.D.; Craig-Kennard, A.C.; Breen, M.K. *Proc. Brighton Crop Protection Conf.-Weeds* **1989**, *1*, 163-172.
87. Rendina, A.R.; Craig-Kennard, A.C.; Beaudoin, J.D.; Breen, M.K. *J. Agric. Food Chem.* **1990**, *38*, 1282-1287.
88. Kemal, C.; Casida, J.E. *Life Sciences* **1992**, *50*, 533-540.
89. Ealick, S.E.;Babu, Y.S.;Bugg, C.E.;Erion, M.D.;Guida, W.C.;Montgomery, J.A.; Secrist III, J.A. *Proc. Natl. Acad. Sci. USA* **1991**, *88*, 11540-11544.
90. Kuntz, I.D. *Science* **1992**, *257*, 1078-1082.
91. Schloss, J.V.; Porter, D.J.T.; Bright, H.J.; Cleland, W.W. *Biochemistry* **1980**, *19*, 2358-2362.
92. Schloss, J.V. In *Chemistry and Biochemistry of Flavoenzymes*; Muller, F., Ed.; CRC Press: Boca Raton, Florida, 1992; Vol. III; pp 531-542.
93. Mullner, H.; Eckes, P.; Donn, G. *Amer. Chem. Soc. Symp. Ser.* **1993**, *524*, 38-47.
94. Gerard, B.; Kishore, G.; Padgette, S.; Taylor, M.; Kolacz, K.; Weldon, M.; Re, D.; Eichholtz, D.; Fincher, K.; Hallas, L. *Cur. Topics Plant Physiol.* **1992**, *7*, 139-145.

RECEIVED August 19, 1993

Chapter 30

Investigation of the Origins of Specificity and Reactivity in N-Linked Protein Glycosylation

B. Imperiali and K. W. Rickert

Department of Chemistry, California Institute of Technology, Pasadena, CA 91125

Post and co-translational protein processing reactions show a broad range of both specificity and reactivity. One such reaction, asparagine linked glycosylation, is specific for Asn-Xaa-Thr/Ser sequences and is interesting for the remarkable reactivity shown by the asparagine sidechain nitrogen. The effects of the substrate conformation are examined, and it is found that the ability to form an Asx-turn is important for the peptide to undergo glycosylation. The mechanism by which the reactivity of the asparagine is increased is investigated by examining the binding properties of a series of peptides containing asparagine analogs. A model mechanism, consistent with the observed results, is proposed.

The formation of peptide bonds between amino acid residues, which is directed by mRNA and catalyzed by ribosomes, is at the heart of protein biosynthesis. However, a wide variety of other processes are also necessary for proteins to achieve their biological function(*1,4*). All polypeptides must undergo noncovalent changes, such as folding of the polypeptide chain, association with other subunits, and translocation across membranes. Many also undergo covalent modifications of both the peptide backbone and amino acid side chains. These covalent modifications can drastically affect both protein structure and function.

Covalent modifications of proteins serve many purposes (*1-4*). Some are structural and affect the three dimensional structure of proteins, such as disulfide bonds or cross linking of collagen chains via allysine side chains. There are a many different modifications that allow for the attachment of a variety of non-peptide prosthetic groups to the protein. The attachment of the heme group to cysteine in c type cytochromes and that of biotin or pyridoxal phosphate to lysine are but a few examples. Some processes, such as cysteine isoprenylation or N myristoylation, allow proteins to become tightly associated with membranes. In other situations, a protein may be regulated by a reversible reaction, such as phosphorylation. The best know examples of this are serine, threonine, and tyrosine phosphate esters. In many other cases, the function of a particular modification is less evident.

0097–6156/94/0551–0425$06.00/0

Most protein processing reactions are highly specific, and the origin of specificity is of great interest. There are several mechanisms through which specificity can arise (*1,2,4*). The most obvious is compartmental specificity. If the enzymes that catalyze a specific modification are localized to a specific subcellular component, then only proteins that pass through that component can undergo modification. Of course, the specificity here is a result of the mechanisms that determine how proteins are targeted to various destinations. In many cases, specificity is also related to the amino acid sequence. For example, the protein N-glycosyl transferase, isoprenyl transferases, and certain protein kinases only modify amino acids that meet certain minimal sequence requirements. Other modifications which affect only a small number of proteins may have much more demanding specificity requirements.

Protein Glycosylation

Protein glycosylation is a widespread modification of eukaryotic proteins. Most protein glycosylation takes place at either serine or threonine residues (O-glycosylation) or at asparagine (N-glycosylation). In either case, a wide variety of oligosaccharides can be found attached to the protein. Glycosylation is known to be able to affect protein conformation, protein targeting, and recognition. In other circumstances, its role is less clear.

Protein O-Linked Glycosylation. Some of the roles of protein O-glycosylation are well established. Most proteins that undergo O-glycosylation contain regions that are heavily glycosylated interspersed with non glycosylated regions. In the heavily glycosylated regions, up to 40% of the residues may be glycosylated serine or threonine, drastically affecting conformational properties and yielding a much more rigid and extended peptide (*5*). For membrane proteins, the glycosylated region may form a stalk, protruding from the membrane surface. For other proteins, such as mucins, the rigidity allows interacting mucin molecules to form a gel. The high degree of glycosylation also confers a high degree of protease resistance to these regions. Protein O-glycosylation can also play a role in cell-cell recognition and may serve other roles (*6*).

The specificity of O-glycosylation, however, is very poorly understood. There is no required sequence for O-glycosylation, beyond the serine or threonine residue itself. A study of amino acid distributions reveals that proline occurs somewhat more frequently around O-glycosylated sequences than similar non-glycosylated sequences (*7*). This distribution may simply reflect the tendency of prolines to form reverse turns, which are often found near the protein surface. Since O-glycosylation appears to be a post-translational event (*3,7*), the accessibility of the site to the various glycosyltransferases involved is an important issue.

Protein N-Linked Glycosylation. Protein N-glycosylation is better understood than O-glycosylation. N-glycosylation generally takes place in three distinct stages (*8, 9*). First, a large core oligosaccharide ($Glc_3Man_9GlcNAc_2$) is assembled one sugar at a time upon a lipid donor, dolichol phosphate which is located in the membrane of the endoplasmic reticulum(*10*). Next, this core oligosaccharide is transferred to the asparagine from the dolichol donor (Figure 1). This takes place as the peptide is translocated into the endoplasmic reticulum. Finally, the oligosaccharide is modified by the removal and/or addition of various sugar residues. This further processing is known to take place in the endoplasmic reticulum, where the glucose and some mannose residues are removed, and the Golgi apparatus, where additional mannose residues can be removed, and other monosaccharides can be transferred to the oligosaccharide. Commonly found oligosaccharides include high-mannose oligosaccharides containing only N-acetylglucosamine and mannose, or complex oligosaccharides, which may contain a variety of sugars including fucose, sialic acid, galactose, and N-acetylgalactosamine (*1,3,9*). Almost all share a core pentasaccharide, $Man_3GlcNAc_2$. The exact nature of the final oligosaccharide depends not only on the identity of the protein and the site being glycosylated, but also the tissue where it is synthesized.

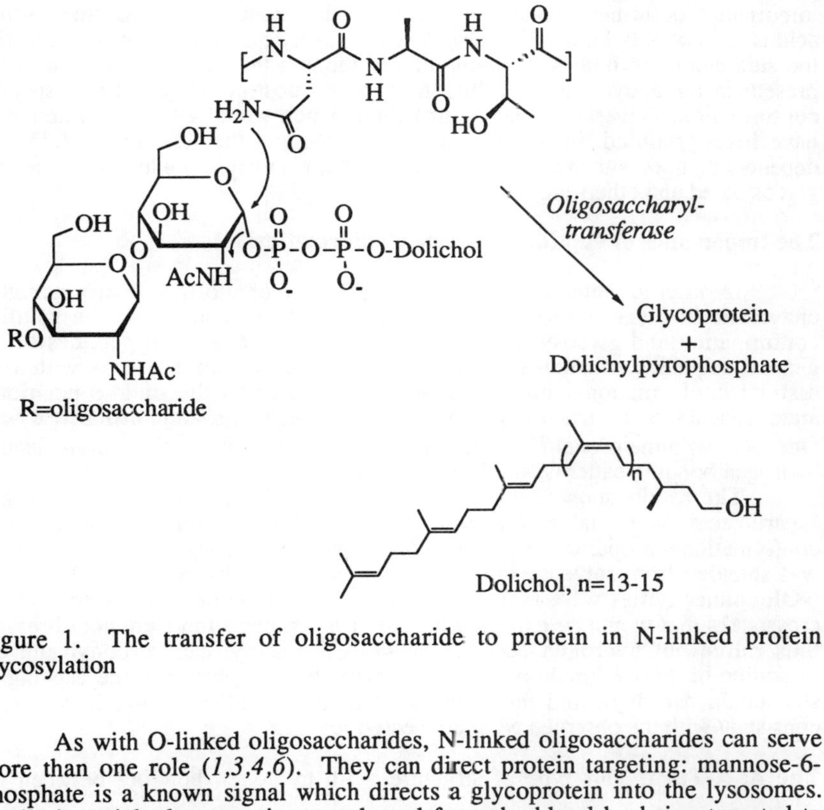

Figure 1. The transfer of oligosaccharide to protein in N-linked protein glycosylation

As with O-linked oligosaccharides, N-linked oligosaccharides can serve more than one role (*1,3,4,6*). They can direct protein targeting: mannose-6-phosphate is a known signal which directs a glycoprotein into the lysosomes. Similarly, asialoglycoproteins are cleared from the blood by being targeted to liver or macrophage cells. N-linked oligosaccharides may also mediate intracellular interactions. The oligosaccharide structure of one glycoprotein, the neural cell adhesion molecule N-CAM, varies extensively throughout development, and it has been suggested that the changes affect the self affinity of this molecule, reducing binding and increasing cell mobility during certain developmental stages. Since N-linked oligosaccharides are added shortly after translocation of a polypeptide into the ER, it has been hypothesized that they can affect the course of protein folding. Inhibition of N-linked glycosylation *in vivo* causes some proteins to aggregate, while others show impaired function, and some are not affected at all.

In sharp contrast to O-glycosylation, N-glycosylation has a clear sequence dependence. Asparagine must occur in the sequence Asn-Xaa-Ser/Thr to be glycosylated, where Xaa is any amino acid but proline. This is a minimum requirement as not all such sites are glycosylated (*7*). Factors that are known to affect the efficiency of glycosylation include the folding of the protein substrate and the availability of oligosaccharyl donor (*9,11*). Threonine is found roughly twice as often in glycosylated sites as serine (*16*). N -glycosylation sites are not heavily clustered as are O-glycosylation sites, although they are slightly more common in the N-terminal regions of proteins than the C-terminal regions.

The only naturally occurring amino acid not tolerated within the Asn-Xaa-Thr/Ser triplet is proline (*12,13*). Proline is unique in its conformational properties, implying that the conformational properties of the peptide substrate are important. As further evidence, variations in the structure of the hydroxy amino acid affect not only binding but also catalysis (*14*). This observation suggests that the side chains of both the asparagine residue and the hydroxy amino acid are present in the active site, and thus the peptide substrate is bound in a specific conformation. Several secondary structures which would allow this interaction have been proposed, including a β turn, loops, and the Asx-turn (*12,15*). A dependence upon substrate conformation might also explain why some sites are glycosylated and others are not.

The Importance of Peptide Substrate Conformation

In order to better understand the importance of substrate conformation to enzymatic N-glycosylation, a study was carried out comparing the solution conformation and glycosyl acceptor properties of a series of tripeptides of the general sequence Ac-Asn-Xaa-Thr-NH_2 (*16*). Several amino acids with very distinct conformational properties were substituted in the middle position : alanine, leucine, aspartic acid, proline, D-alanine, and α-aminoisobutyric acid. One and two dimensional ^{1}H NMR studies were used to determine intramolecular hydrogen bonding patterns, and NOE connectivities.

The results showed that those peptides containing alanine, leucine, and aspartic acid, were viable glycosyl acceptors. All three peptides showed similar conformational properties in solution (*16*). The threonine backbone amide proton was shielded from solvent exchange by intramolecular hydrogen bonding. The NOE connectivities were indicative of an extended peptide backbone, and the crosspeaks that would be expected for a β turn conformation were not observed, thus ruling out hydrogen bonding between the acetyl cap carbonyl and the threonine backbone amide proton. Hydrogen bonding between the asparagine side chain carbonyl, and the threonine backbone amide proton, however, is consistent with the observed NOE connectivities.

The ASX-Turn. This type of structure, with hydrogen bonding between the

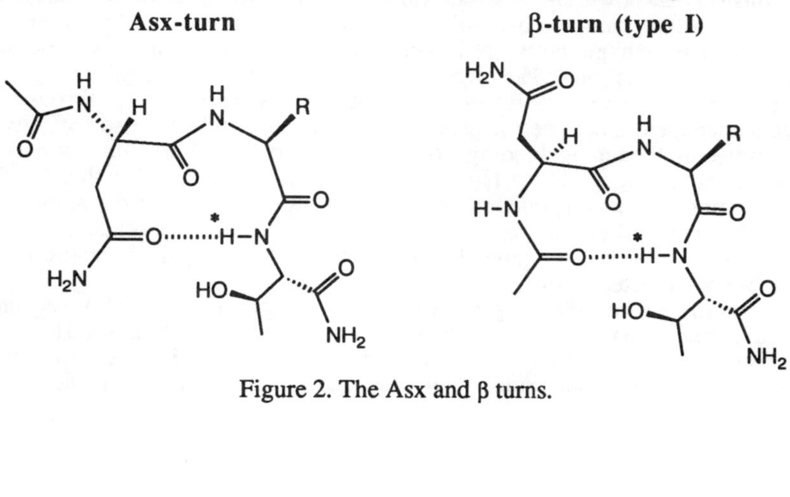

Figure 2. The Asx and β turns.

peptide backbone and the asparagine sidechain, is known as an Asx-turn (Figure 2) (*15*). The Asx-turn has been observed in several protein crystal structures, and occurs with nearly equal frequency to β turns for asparagine and aspartic acid residues(*17*). It is also observed in the crystal structures of small peptides (*18*). The topology of this motif is identical to the better known β turn; both have a hydrogen bond closing a 10 membered cycle.

The nonacceptor peptides, however, had very different conformational properties. Both the NOE connectivities and the intramolecular hydrogen bonding demonstrate that these peptides prefer very different conformations from the acceptor peptides (*16*).

This evidence demonstrates that the conformation of the peptide substrate is important for its recognition by the glycosyltransferase and that the acceptor peptides prefer the Asx-turn conformation. However, the solution conformation of a peptide may not reflect the conformation in which it actually binds to the enzyme. One approach to determining the bound conformation is to impose conformational constraints upon the peptide substrates, restricting them to a limited number of conformations or perhaps even a single conformation.

Probing the Conformation of a Bound Substrate

While the previous studies showed that substrate peptides were not found in β turn conformations in solution, they cannot rule out binding of the peptide to the enzyme in this conformation. In order to determine whether this was the case, peptides were synthesized which were constrained into a β turn (*19*). In the cyclic hexapeptides **2a** and **2b**, the proline-D-amino acid dipeptide forms a type II β turn. This locks the remaining residues into a type I β turn with the asparagine fixed into either the (i) or (i+1) position of this turn. One and two dimensional ^{1}H NMR confirmed that these peptides were in the predicted conformations with all the predicted NOE connectivities and with the expected intramolecular hydrogen bonding (*19*).

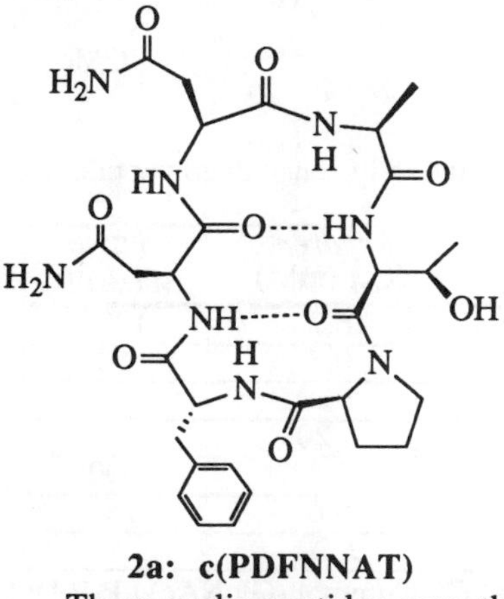

2a: c(PDFNNAT)

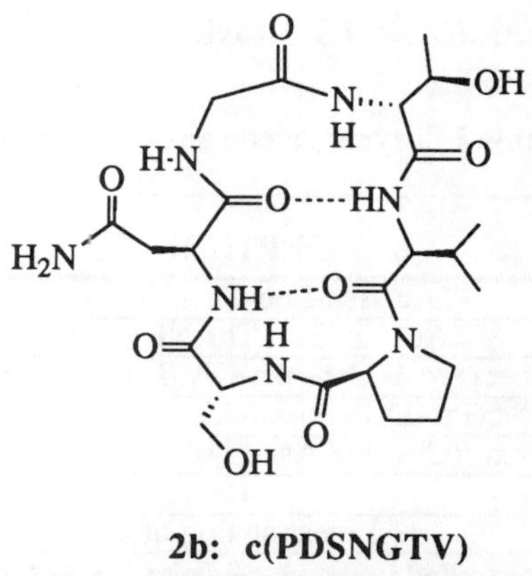

2b: c(PDSNGTV)

These cyclic peptides were then evaluated for their ability to undergo glycosylation and compared with the linear parent peptides (Table I). In both cases the linear peptides showed ordinary behavior as glycosyl acceptors. In contrast, the cyclized peptides **2a** and **2b** with the asparagine in either position of the β turn, did not undergo any significant degree of glycosylation (*19*). This demonstrates that glycosyl transferase is incapable of binding the Asn-Xaa-Thr motif in a β turn conformation, and that some other conformation must be recognized.

Applying other conformational constraints to the peptide can yield quite different results. Cyclization of a peptide by forming intramolecular disulfide bonds is a different method for imposing a conformational constraint and has previously been applied to this problem (*20*). The cyclic disulfide **3** retains reasonable properties as a glycosyl acceptor following cyclization (Table I) (*19*). While the solution conformation is not constrained to the Asx-turn motif, this peptide seems to be flexible enough to achieve it. It is, however, unable to adopt a β turn conformation. For other peptides constrained through disulfide bonds, cyclization causes a large decrease in the rate of glycosylation; many of these peptides are unable to form any contacts between the asparagine and hydroxy amino acid sidechains when cyclized (*20*).

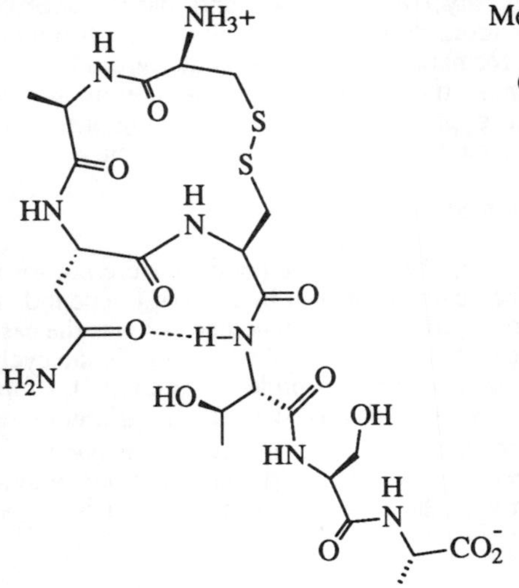

3: CANCTSA (oxid.)

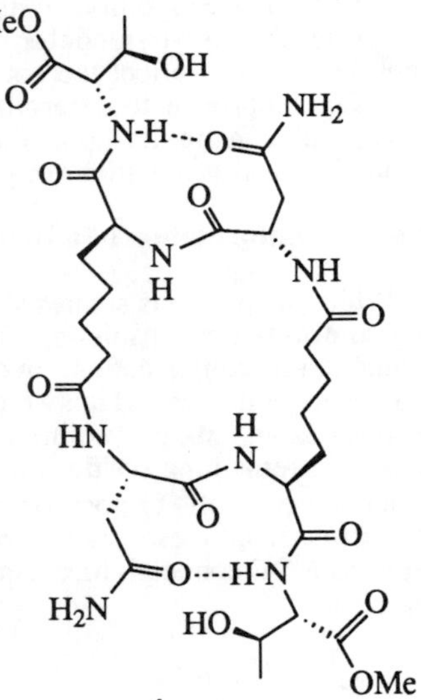

4

Table I Enzyme kinetic analysis of conformationally constrained peptides[a]

PEPTIDE	*Apparent* K_M (mM)	*relative* *V* (%)[b],
Ac-Asn-Leu-Thr-OMe **1a**	6.7	100
Ac-Asn-Leu-Thr-NHMe **1b**	0.3	150
c(Pro-D-Ser-Asn-Gly-Thr-Val) **2a**	> 20	-
c(Pro-D-Phe-Asn-Asn-Ala-Thr) **2b**	> 20	-
oxid(Cys-Ala-Asn-Cys)-Thr-Ser-Ala **3**	2.8	1600
4	0.6	75
Linear analog of **4**	> 20	-

[a]Determined in the presence of 200,000 dpm, 6.1 Ci/mmol $(GlcNAc)_2$-P-P-Dol ($4x10^{-9}$ M). (*16*)

[b]Peptide **1a** as standard

In order to constrain a peptide into an Asx-turn another approach must be used. The peptide analog **4** is constrained by a sidechain-backbone cyclization into the desired conformation. One and two dimensional ^1H NMR confirm that the cyclized peptide has adopted the desired conformation. The linear analog of the peptide **4**, is a poor substrate for glycosyl transfer. In contrast, the cyclized peptide shows much tighter binding and increased turnover with glycosyl transferase (Table I) (*19*). Since constraining a peptide into an Asx-turn enhances binding and turnover while constraining to a β turn decreases binding and turnover, this strongly suggests that the enzyme recognizes and binds the peptide in an Asx-turn conformation.

Understanding the Mechanism of N-Glycosylation

A second intriguing issue in protein N-glycosylation is that of mechanism. The carboxyamido nitrogen of the asparagine sidechain is not normally reactive and is a poor nucleophile. However, the displacement of the lipid donor from the oligosaccharyl group would appear to take place by a nucleophilic attack from the carboxyamido nitrogen. Furthermore, glutamine carboxyamido nitrogens have the same functionality, but are not reactive. The glycosyltransferase must therefore not only enhance the nucleophilicity of this nitrogen, but also do it with a high degree of specificity.

Recognition of the Asx-turn provides a great deal of specificity. Only asparagine and aspartic acid residues are able to form this structure: glutamine residues are unable to adopt a similar structure, and cannot be glycosylated. The conformation may also be important for catalysis of the reaction. It has previously been shown that the nature of the side chain on the hydroxy amino acid affects both binding and catalysis (*14*). Based on this observation, mechanistic models have proposed that intramolecular hydrogen bonding between the side chains of the asparagine and the hydroxy amino acid plays an important role in catalyzing glycosyl transfer.

Proposed Models for The Mechanism of N-Glycosylation. The first such model was that of Marshall, who suggested that the hydroxy amino acid sidechain donated a hydrogen bond to the asparagine carbonyl oxygen (Figure 3a) (*21*). This would in turn increase the acidity of the carboxyamido nitrogen, facilitating removal of a proton and its replacement with the oligosaccharide. Bause later presented a model with the opposite hydrogen bonding scheme, where the

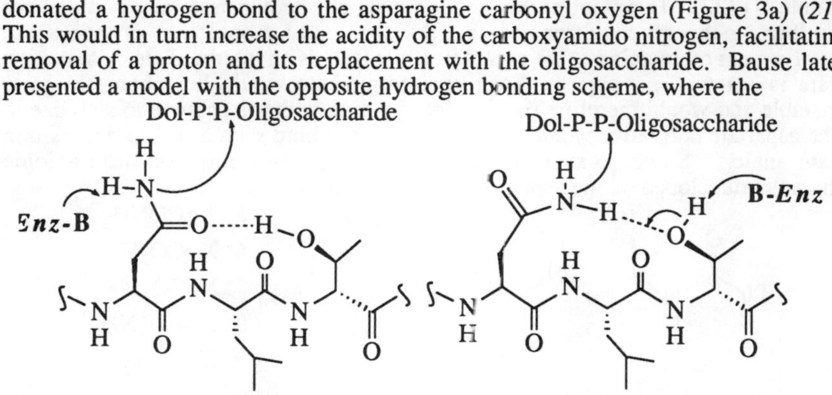

Figure 3a Marshall **Figure 3b Bause**

Figure 3. The models of Marshall and Bause for the enhancement of asparagine reactivity

asparagine sidechain nitrogen donates a hydrogen bond to the hydroxy amino acid, increasing the nucleophilicity of that nitrogen and allowing it to displace dolichol pyrophosphate from the oligosaccharide (Figure 3b) (*14*).

Neither of these models considered the effects of peptide conformation upon the intramolecular hydrogen bonding. Given that the ability to form an Asx-turn is important for a peptide to be a viable glycosyl acceptor, the hydrogen bonding within this motif might be of great importance. Implicit in the Asx-turn is the hydrogen bond between the asparagine side chain carbonyl and the backbone amide nitrogen of the hydroxy amino acid (*15*). This motif is also capable of forming another hydrogen bond, between the asparagine side chain carbonyl and the hydroxy amino acid sidechain. Such a hydrogen bond has been observed in a short peptide (*18*).

Probing the Mechanism of N-Glycosylation with Asparagine Analogs

All of the proposed mechanisms rely on the hydrogen bonding properties of the asparagine side chain and will be sensitive to changes in these properties. Since these mechanisms involve different hydrogen bonding patterns, they will predict different responses. Therefore, these models can be tested by synthesizing peptides where the asparagine side chain has been replaced with groups with differing ionization and hydrogen bonding properties. The substitutions should result in significant effects on the ability of the peptides to serve as glycosyl acceptors and to bind to the enzyme. Comparing these effects with those predicted by the various models for the enhancement of asparagine nucleophilicity should then allow a good evaluation of the merits of each model.

Several such peptides were synthesized: two contain the normal carbonyl but replace the asparagine nitrogen with groups with different hydrogen binding properties. The other two contain a hydrogen bond donor at the asparagine nitrogen but have changes in the carbonyl functionality (*22*). These peptides were evaluated both for the ability to undergo glycosylation and the ability to competitively inhibit glycosylation of a standard substrate.

In nature, Asp-Xaa-Thr/Ser sites are never glycosylated at Asp. The peptide **6**, containing such a site, is equally unable to undergo glycosylation (Table II) (*22*). In addition, this peptide does not demonstrate any competitive binding to the glycosyltransferase. As such, any model that proposes the involvement of an anion at the asparagine nitrogen as an intermediate or transition state is not well supported. Such an intermediate would be inherently quite unstable and would therefore require tight binding with the enzyme to stabilize it. The aspartate derivative would also be expected to bind such a site, as a transition state analog. Since no such binding is observed, there must be little anionic character developed on the asparagine side chain nitrogen.

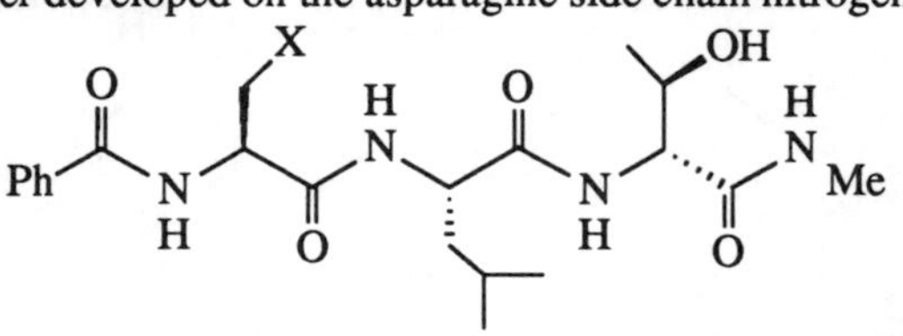

5: X=$CONH_2$
6: X=CO_2H
7: X=CO_2Me
8: X=$CH_2NH_3^+$
9: X=$CSNH_2$

The aspartate γ methyl ester derivative **7** does not bear any anionic character. It also does not undergo glycosylation, nor does it inhibit glycosylation of a normal substrate (Table II) (*22*). In this case, the hydrogen bond donors of the asparagine sidechain nitrogen have been removed, and the additional steric bulk of a methyl group has been introduced. It has previously been observed that a peptide containing γ N-methyl asparagine was neither a substrate for glycosylation nor a competitive inhibitor of glycosylation (*23*). Similar to the methyl ester, in this peptide one hydrogen bond donor from the asparagine sidechain nitrogen has been removed and replaced with the additional steric bulk of a methyl group. Considering that the methyl ester can freely rotate and that the active site can accommodate the bulk of a oligosaccharide adjacent to the asparagine, it seems likely that the removal of the hydrogen bond donor is an important factor in determining binding to the active site.

In sharp contrast, the peptide **8**, while not active as a glycosyl acceptor, does competitively inhibit glycosylation of a standard substrate such as Bz-Asn-Leu-Thr-NHMe (Table II) (*22*). This peptide does not contain a carbonyl group and cannot form the hydrogen bonding of the Asx-turn, although it can still adopt a similar geometry with the amine nitrogen assuming the position of the asparagine sidechain nitrogen. The peptide **9** containing a thioamide in place of the asparagine carboxyamide is a substrate for glycosylation with binding to the enzyme similar to that seen for a similar asparagine containing peptide. It is turned over at only a fraction of the rate.

Table II Enzyme kinetic analysis of tripeptides [a]

PEPTIDE	*Apparent* K_M (mM)	relative V^b	K_i (mM)
Bz-Asn-Leu-Thr-NHMe **5**	0.24	100	-
Bz-Asp-Leu-Thr-NHMe **6**	-	-	>10[c]
Bz-Asp(OγMe)-Leu-Thr-NHMe **7**	-	-	>10[c]
Bz-Amb-Leu-Thr-NHMe **8**	-	-	1.0
Bz-Asn(γS)-Leu-Thr-NHMe **9**	0.26	8.4	-

[a]Determined in the presence of 200,000 dpm (6.1 Ci/mmol; 4×10^{-9} M) $GlcNAc_2$-P-P-Dol.(*16*)

[b]Peptide **1** as standard

[c]No inhibition was observable at concentrations below 5 mM.

Evaluation of Previous Proposals. The original Marshall model does not adequately model the observed behavior. This model predicts deprotonation of the asparagine carboxyamido nitrogen to yield a potent nucleophile which can attack the oligosaccharide (*21*). Such a deprotonation would place anionic character on the carboxyamido nitrogen, which has already been ruled out. This model also does not reflect the hydrogen bonding interactions within the Asx-turn.

Bause and Legler proposed that the hydroxy amino acid might serve as a hydrogen bond acceptor from the asparagine carboxyamido nitrogen, enhancing the nucleophilicity of that site (*14*). This model was based on their observation that increased acidity of the hydroxy amino acid side chain lead to a decrease in the rate of glycosylation. This excluded any other effects, such as effects on peptide conformation which we now know to be important, and the hydrogen bonding within this model is not compatible with that in an Asx-turn.

The Bause model also fails to adequately predict the behavior of these substitutions. Their model would seem to predict enhanced turnover from the thioamide, given the increased acidity of the thioamide protons (*22*). This would also predict that the peptide **8** could be a viable substrate for glycosylation since it could readily transfer a proton to the hydroxy amino acid and form a potent nucleophile. While this peptide binds to the enzyme, no glycosylation or increased hydrolysis of dolichol-oligosaccharide is observed (*22*).

A Mechanistic Proposal for Protein N Glycosylation

Figure 4 depicts a mechanism which account for these observations (*22*). An essential feature is the intramolecular hydrogen bonding to the asparagine sidechain carbonyl from both the Asx-turn hydrogen bond and the essential hydroxy amino acid. This interaction allows protonation of the carbonyl oxygen, while simultaneously the carboxyamido nitrogen is deprotonated to form the imidate tautomer. This tautomer is then a competent nucleophile and able to displace dolichol pyrophosphate from the oligosaccharide. This neighboring group assistance accounts for both the sequence and conformational requirements observed for protein N-linked glycosylation. In this model, the enzyme provides catalysis of the rate determining tautomerization, as well as stabilization of the imidate tautomer.

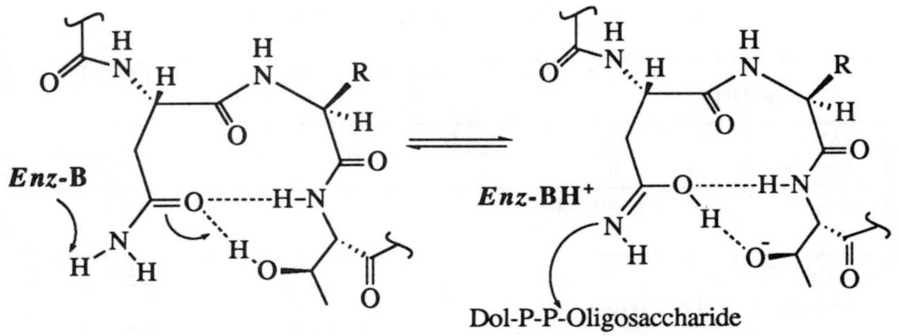

Figure 4. Proposed model for the mechanism of N-linked protein glycosylation.

This model can account for many of the observations made. For example, the tautomerization means that little negative charge is ever developed on the asparagine side chain nitrogen. Both the hydrogen bond donor of the asparagine side chain nitrogen and the hydrogen bond acceptor, the carbonyl oxygen, are required. Peptides that are lacking either, such as **8** or the aspartate methyl ester **7**, would therefore be unable to undergo glycosylation. The results observed for the thioamide **9** are also consistent. While the ability of the sulfur atom to support a negative charge increases acidity of the amide nitrogens, it also decreases the nucleophilicity of the corresponding imidate tautomer. This could result in the observed decrease rate of turnover.

This proposal demonstrates that the sequence and conformation of the peptidal substrate can both play a critical role in protein N glycosylation. These factors combine to form a unique environment, providing the specificity and reactivity required to link an oligosaccharide to an asparagine carboxyamido nitrogen. Similar factors may play an important role in other protein processing reactions.

Acknowledgements

This work was supported by the NIH (GM39334) and Sterling Drug Inc. In addition we gratefully acknowledge the award of a Fannie and John Hertz Foundation Fellowship to K.W.R.

Literature Cited

1. Freedman, R. B. *Biochem. Soc. Trans.* **1989**, *17*, 331-335.
2. Han, K.; Martinage, A. *Int. J. Biochem.* **1992**, *24*, 19-28.
3. Moore, B. R.; Free, S. J. *Int. J. Biochem.* **1985**, *17*, 283-289.

4. Wold, F. *Ann. Rev. Biochem.* **1981**, *50*, 783-814.
5. Jentoft, N. *Trends Biochem. Sci.* **1990**, *15*, 291-294.
6. Paulson, J. C. *Trends Biochem. Sci.* **1989**, *14*, 272-276.
7. Wilson, I. B. H.; Gavel, Y.; von Heijne, G. *Biochem. J* **1991**, *275*, 529-534.
8. Hubbard, S. C.; Ivatt, R. J. *Ann. Rev. Biochem.* **1981**, *50*, 555-583.
9. Kornfeld, R.; Kornfeld, S. *Ann. Rev. Biochem.* **1985**, *54*, 631-664.
10. Lennarz, W. J. *Biochemistry* **1987**, *26*, 7205-7210.
11. Kaplan, H. A.; Welply, J. K.; Lennarz, W. J. *Biochemica et Biophysica Acta* **1987**, *906*, 161-173.
12. Bause, E. *Biochem. J.* **1983**, *209*, 331-336.
13. Gavel, Y.; von Heijne, G. *Protein Engineering* **1990**, *3*, 433-442.
14. Bause, E.; Legler, G. *Biochem. J.* **1981**, *195*, 639-644.
15. Abbadi, A.; Mcharfi, M.; Aubry, A.; Prémilat, S.; Boussard, G.; Marraud, M. *J. Am. Chem. Soc.* **1991**, *113*, 2729-2735.
16. Imperiali, B.; Shannon, K. L. *Biochemistry* **1991**, *30*, 4374-4380.
17. Baker, E. N.; Hubbard, R. E. *Prog. Biophys. Mol. Biol.* **1984**, *44*, 97-180.
18. Pichon-Pesme, V.; Aubry, A.; Abbadi, A.; Boussard, G.; Marraud, M. *Int. J. Pept. Protein Res.* **1988**, *32*, 175-182.
19. Imperiali, B.; Shannon, K. L.; Rickert, K. W. *J. Am. Chem. Soc.* **1992**, *114*, 7942-7944
20. Bause, E.; Hettkamp, H.; Legler, G. *Biochem. J.* **1982**, *203*, 761-768.
21. Marshall, R. D. *Ann. Rev. Biochem.* **1972**, *41*, 673-702.
22. Imperiali, B.; Shannon, K. L.; Unno, M.; Rickert, K. W. *J. Am. Chem. Soc.* **1992**, *114*, 7944-7945
23. Welply, J. K.; Shenbagamurthi, P.; Lennarz, W. J.; Naider, F. *J. Biol. Chem.* **1983**, *258*, 11856-11863.

RECEIVED April 30, 1993

Chapter 31

Ethylene Biosynthesis from 1-Aminocyclopropanecarboxylic Acid

Interplay of Molecular Genetics and Organic Mechanism

Michael C. Pirrung

P. M. Gross Chemical Laboratory, Department of Chemistry, Duke University, Durham, NC 27708–0346

Advances in the molecular biology of ethylene formation in plants have provided essential knowledge for the isolation and purification of the ethylene-forming enzyme from apple fruit. Comparisons with related enzymes have provided guidance concerning the mechanism of ethylene biosynthesis, and the history of mechanistic postulates for this process has been reviewed in this light. The biosynthesis of ethylene is of interest because of the role that it plays in catabolic proceses in plant physiology: ripening, senescence, abscission, germination, and response to environmental stress.

Ethylene is an endogenous plant hormone that controls a number of processes in growth and development. Since the original report (*1*) that 1-aminocyclopropanecarboxylic acid (ACC) is the immediate precursor of ethylene (eq 1), the macromolecule that conducts the oxidative conversion of ACC to ethylene has been an important topic in biochemistry, and plant physiology. However, because over a decade of effort in protein purification had failed to provide authentic cell-free ethylene-forming enzyme (EFE) activity, the lore developed that the enzyme is membrane-bound and does not survive breakage of the cell wall (*2*). Consequently, the many mechanistic studies of ethylene production (*3-8*) have been limited to biosynthesis experiments in which only substrates and products can be inferred. Recent advances in molecular genetics, including the determination of the DNA sequence of the EFE gene, have provided information concerning the class of proteins to which EFE belongs but cannot address key issues concerning the mechanism. This chapter reviews the genetics and mechanisms of ethylene biosynthesis and connects them to our recent purification of the ethylene-forming enzyme.

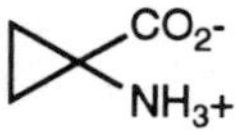

$$\longrightarrow H_2C{=}CH_2 + HCN + CO_2 \qquad (1)$$

0097–6156/94/0551–0436$06.00/0

MICHIGAN STATE UNIVERSITY TRANSMITTAL

TO Dr. Dave Dilley

PLEASE

☐ ANSWER ☐ copy to me ☐ CIRCULATE ☐ SEE ME

☐ for my signature ☐ COMMENT ☐ SIGN

☐ APPROVE ☐ FILE ☐ TAKE ACTION ☐ advise

☐ CALL ME ☐ FOR YOUR INFORMATION ☐ RETURN

RESPONSE TO YOUR REQUEST

☐ APPROVED ☐ ACTION TAKEN ☐ NOTED

REMARKS

Dave,

please see Chapter 31. When you are finished please return the book to me.

Thanks.

Murialee

FROM	DATE
DEPT.	PHONE NO.

MSU is an Affirmative Action/Equal Opportunity Institution O-12513 **Stock Order # 140-**

Genetics

Using tomato ripening as a model system for the developmental changes induced by ethylene, the lab of Donald Grierson at the University of Nottingham made key initial contributions to obtaining the EFE gene. This group had isolated a number of ripening-related cDNAs from tomato fruit that were of unknown function. Smith *et al.* and Holdsworth *et al.* showed that transcription and translation of one of these clones, pTOM13 (*9*), is induced during fruit ripening (*9,10*) and wounding. Hamilton *et al.* showed that pTOM13, when introduced into transgenic tomato plants in an antisense orientation, leads to reduced fruit softening and reduced ethylene biosynthesis in the presence of ACC (*11*). While one obvious explanation for these observations was that pTOM13 directly encodes EFE, it was also possible that it had some positive regulatory action on ethylene biosynthesis. Attempts to express this cDNA did not lead to enzymatic activity (*10*), further clouding the issue. Meanwhile, studies in the laboratory of Thomas Boller (Botanisches Institut of the University of Basel and the Friedrich Miescher-Institut) on the response of cultured tomato cells to fungal infection showed that ethylene synthesis is induced on exposure to fungal elicitor. Boller's student, Pietro Spanu, isolated total mRNA from the induced cell line and injected it into *Xenopus* oocytes in the presence of ACC (*12*). *These toad eggs produced ethylene!* This result indicated that the isolated mRNA contained all the information required to express EFE activity. A cDNA library in phage lambda was consequently prepared from induced tomato cells and, given its possible connection to ethylene biosynthesis, pTOM13 was used to probe the library. A full length EFE cDNA clone, pHTOM5, was obtained. This novel example of reverse biochemistry for obtaining the EFE gene is summarized in Figure 1 at the top of the following page.

The validation of pHTOM5 as the ethylene-forming enzyme used a classical test for enzymatic action and was necessary because of the great sensitivity of ACC. It is rapidly converted to ethylene by such oxidants as Fenton's reagent and bleach (*vide infra*). The crucial test is stereoselectivity with chiral alkyl-substituted ACC analogues. The logic behind such experiments is that, though itself achiral, ACC must interact with a chiral protein binding site. ACC analogues that possess chirality should interact with the binding site diastereomerically, with consequent distinguishable kinetic constants. A collaboration between Yang and Ichihara had resulted in the first example of stereoselectivity in the processing of ethyl ACC analogues to 1-butene in plant tissue, a criterion that has become the *sine qua non* of a valid ethylene-forming enzyme (*13*). Subsequent work from our laboratory established the absolute configuration of methyl and ethyl ACC analogues accepted by the enzyme (*14*). In the case of the protein expressed from pHTOM5, stereoselectivity for coronamic and allocoronamic acids (*cis*- and *trans*-2-ethyl ACC) was identical to the *in vivo* system, establishing that it is indeed the authentic EFE.

Comparison of the sequence of the EFE gene to pTOM13 demonstrated the reason for the failure of Grierson's laboratory to observe enzymatic activity on expression of pTOM13: there were two bases missing close to the 5' end of the cDNA. Hamilton directly sequenced an RNA homologous to pTOM13 and discovered these same cloning artifacts. A valid full-length EFE coding sequence was then constructed and expressed in yeast (*15*). The similarity of the pTOM13 DNA sequence to a flavanone-3β-hydroxylase, believed to be a non-heme iron containing dioxygenase (*16*), had been noted in the original report on the pTOM13

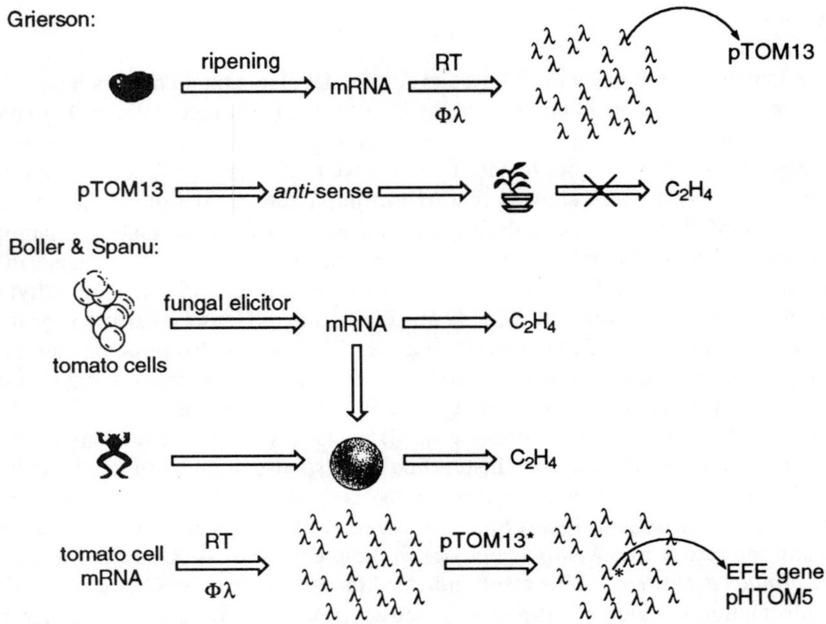

Figure 1. Cloning of the ethylene-forming enzyme through transgenic plants and *in vitro* expression.

antisense tomatoes and suggested that the EFE was a member of this family. The lability of non-heme iron enzymes makes their classification as members of this class far easier by "reverse biochemistry" than by classical protein purification. It also suggests that the earlier difficulties in obtaining valid cell-free EFE activity were due to this trait and not to membrane association of the protein. This observation prompted our laboratory to conduct a search of GenBank for other homologous proteins. As summarized in Figure 2 on the following page, there are bacterial proteins involved in the biosynthesis of penicillin and cephalosporin antibiotics and plant proteins involved in several biosynthetic pathways that have homology to EFE. A common trait of this family of non-heme iron-containing proteins is several conserved histidine residues (indicated by arrows) which may contribute to ligation of the iron, and indeed iron has been shown to be an essential cofactor for ethylene biosynthesis in tomato cells.

Mechanism of Non-Heme Iron Proteins

The enzymes of this non-heme iron family use the 4 oxidizing equivalents of molecular oxygen in three different ways as summarized in Figure 3 at the top of the following page. IPNS removes 4 hydrogen atoms from the ACV tripeptide to directly produce isopenicillin N and two water molecules (*17*). Studies of the mechanism of oxygen activation by this enzyme are meager; it is believed that the iron is ligated by three histidine residues and that the thiol of the tripeptide is bound to the iron in the catalytically-significant complex (*18-20*). It has been suggested in the extensive studies of modified substrates for IPNS by Baldwin that the β-lactam ring is first closed to generate a thiol-ligated iron-oxo intermediate that activates the methine of

F3H	178	HTDPGTITLLLQD...257	VVNLGDHGHFLSNGRFKNADHQA
H6H	236	HIDIGFVTILLQD...254	VVNLGLTLKVITNEKFEGSIHRV
pHTOM5	157	HTDAGGIILLFQD...194	VVNLGDQLEVITNGKYKSVLHRV
pAE12	177	HSDAGGIILLFQD...214	VINLGDQIEVITNGKYKSVMHRV
DAOCS	184	HYDLSTLTLVHQT...224	VVFCGAVGTLAIGGKVKAPKHRV
IPNS	214	HEDVSLITVLYQS...252	LINCGSYMAHITDDYYPAPIHRV

Figure 2. Deduced protein sequence comparison between the ethylene-forming enzyme and other known non-heme iron proteins.

the C-terminal valine residue for closure of the thiazolidine ring. He has shown that the intermediate, in which the enzyme lies two oxidation equivalents higher than the resting state, can promote oxidation of alkenes in unsaturated tripeptide analogues (*21,22*). Molecular rearrangements observed in the processing of cyclopropyl-containing tripeptides are consistent with the idea that the enzyme generates a carbon-based radical.

Many non-heme iron dioxygenase enzymes use α-ketoglutarate as co-substrate, converting it to succinate and CO_2 while introducing an oxygen atom into substrate (*16*). Consequently, they possess only two oxidizing equivalents for substrate. The well-known prolyl hydroxylase involved in collagen biosynthesis in mammals is one example of an α-ketoglutarate-dependent dioxygenase enzyme, but the chicken prolyl hydroxylase has no homology to any of those enzymes in Figure 2. Flavanone-3β-hydroxylase (F3H) (*23*) and hyoscyamine-6β-hydroxylase (H6H) (*24*) both add an oxygen atom to their substrates. An intermediate commonly invoked in these hydroxylations is the iron-oxo, which could be generated via the mechanism originally proposed by Siegel (*25*) (eq 2). It could presumably introduce its oxygen atom into substrate in a pathway analogous to that proposed for cytochrome P450. There is strong support for the presence of the high-valent iron-oxo in these heme proteins (*26*). Clearly, no intermediates have been established in the family of non-heme iron proteins to which EFE belongs, but neither does a superior working hypothesis exist. A considerable body of work has suggested that oxidation of C-H bonds by iron-oxos proceeds by a H-abstraction, OH-rebound mechanism (eq 3) (*27*). Using radical clocks, the rate of rebound when cytochrome P450 hydroxylates methyl groups has been estimated to be $\sim 2 \times 10^{10}$ s^{-1} (*28*). This value is relevant to evaluation of mechanistic pathways for the EFE in later sections of this chapter.

Deacetoxycephalosporin C synthetase (DAOCS) (*29*) is not an overt oxidase; no oxygen atoms are introduced into the substrate. However, the ring expansion introduces an unsaturation, so the product lies one oxidation state higher than starting material. Furthermore, the *C. acremonium* enzyme is bifunctional, further converting deacetoxycephalosporin C to deacetylcephalosporin C, a hydroxylation. DAOCS decarboxylates two moles of α-ketoglutarate in conducting the ring expansion and hydroxylation, thereby accounting for the oxidizing equivalents of the two moles of oxygen consumed. It is therefore reasonable that the enzyme uses an iron-oxo for both steps. A variety of experiments conducted by Baldwin have suggested that the enzyme generates a substrate alkyl radical at the methyl group (*30*). This could occur via hydrogen atom abstraction by the iron-oxo, notably the first step in the postulated

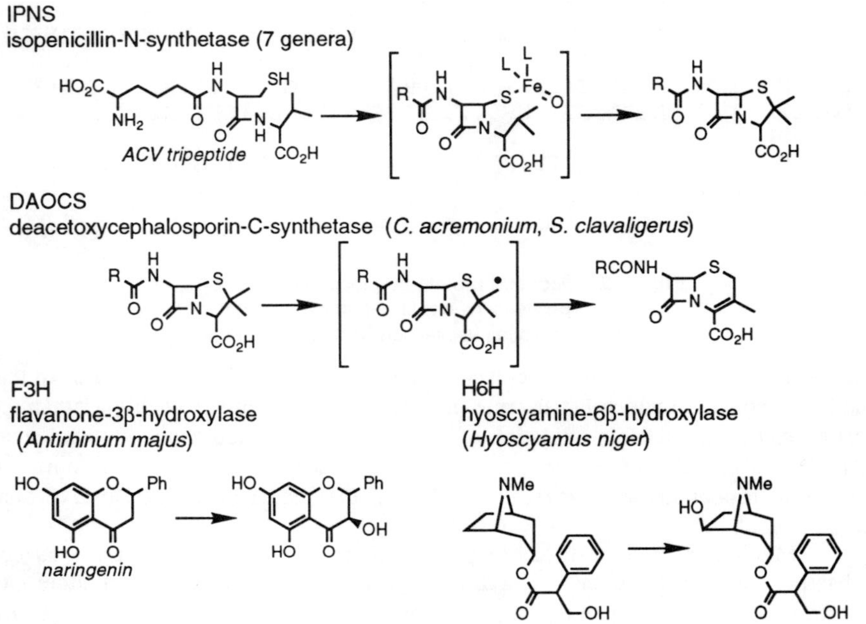

Figure 3. Biosynthetic transformations mediated by non-heme iron proteins related to the ethylene-forming enzyme.

mechanism of cytochrome P450 hydroxylation. This radical intermediate must undergo rearrangement from the penicillin to the cephalosporin ring skeleton faster than the rebound step. Otherwise, hydroxylated products would be obtained. It seems likely that rebound in the non-heme iron-oxos is slower than the extremely fast rate that has been measured in the P450 reactions.

Mechanism of Ethylene Formation

Speculation about the mechanism of the EFE has been rampant since the initial paper of Adams and Yang (*1*). In that article (eq 4), they suggested that oxygen is reduced to hydrogen peroxide, which nucleophilically attacks at C-1 of ACC to produce an intermediate carbanion. After fragmentation and protonation, an intermediate hydroperoxide decarboxylates to give formamide, the constituents of which, formate and ammonia, were at that time believed to be the products of ethylene biosynthesis. In 1979 (eq 5), Lurssen postulated the NADH-mediated reduction of a CoA-ester ACC-pyridoxal Schiff base followed by a fragmentation (*31*); this hypothesis was unsupported by evidence for the involvement of any of these cofactors in ethylene biosynthesis. In a 1981 review (*32*), Yang proposed that ACC undergoes hydroxylation by hydrogen peroxide and that N-hydroxy-ACC solvolyzes to give a nitrenium ion (eq 6) that would fragment to ethylene and cyanoformic acid. In 1982, Baldwin (*33*) proposed an unspecified covalent bonding between ACC and the EFE, oxidative decarboxylation to a cyclopropanone imine, and extrusion of ethylene (eq 7). This process would yield an isonitrile, which could hydrolyze to formate. While

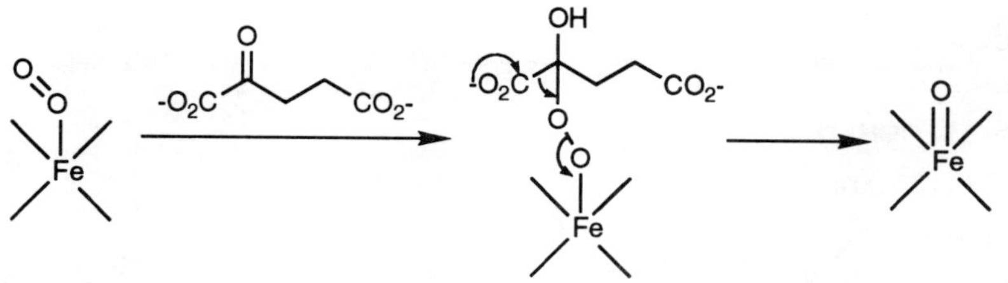

(2)

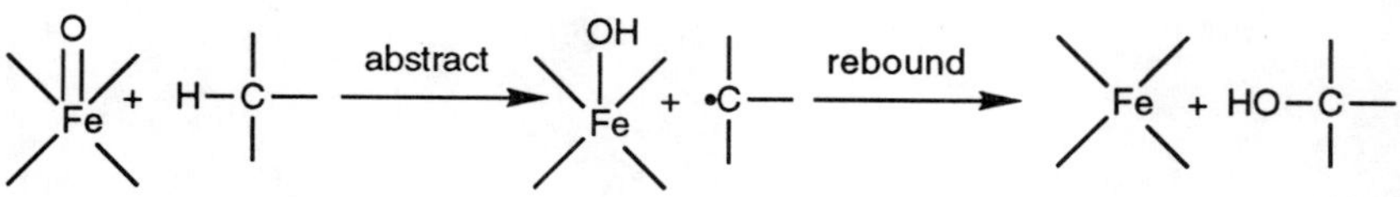

(3)

interesting, these postulates had no connection to experimental observations. A significant guide to future experimentation is that some mechanisms are expected to proceed in concerted fashion and some in stepwise fashion, issues of reaction mechanism that have fascinated organic chemists for some time.

Study of the behavior of organic intermediates postulated to inhabit enzymes is a time-tested strategy to evaluate mechanistic hypotheses since, while enzymes are capable of many amazing feats, they cannot violate laws of chemistry. In the first model system to come from our laboratory (*34*), Jerry McGeehan prepared the ACC nitrenium ion and the corresponding nitrene from azidocyclopropanecarboxylic acid and showed that both generate ethylene, carbon dioxide, and cyanide (eq 8). This product profile provided the initial experimental evidence that cyanide rather than formate should be considered as the product of C-1 of ACC in ethylene biosynthesis (*vide infra*). While direct evidence concerning the mechanism of these processes was not obtained, we argued that a 6-electron *concerted* chelotropic elimination would permit the principles of conservation of orbital symmetry to be observed. Further evidence on this point came from Adlington *et al.* who studied the stereochemistry of the oxidation of dideuterated ACC by bleach (*7*). This reaction, used for the quantitation of ACC, proceeds with retention of configuration. A likely mechanism is N-chlorination and α-elimination to generate the same nitrene or nitrenium ion prepared in our study of the azidoacid (eq 9), which undergoes *concerted* elimination of ethylene.

Shortly thereafter, the nitrenium ion mechanism fell into disfavor on the basis of experiments independently carried out in Baldwin's (*7*) and our laboratory (*35*). Each group prepared stereospecifically-dideuterated ACC derivatives. At Oxford, both *cis* and *trans* compounds were fed to apple tissue ("ripening" ethylene biosynthesis) while, at Stanford, the *cis* compound was fed to apple tissue, cantaloupe tissue, and mung bean hypocotyl segments. Concern that peroxidases known to be released when fruit is cut (the "browning" reaction) might convert ACC to ethylene non-enzymatically as well as the desire to examine "stress" ethylene in a vegetative tissue prompted our mung bean infusion studies. These were challenging because much less ethylene is produced by vegetative tissue, requiring FTIR analysis to assign the ethylene stereochemistry. In all cases, the ethylene produced from the labeled substrates is a 1:1 mixture of *cis* and *trans* isomers, making a clear case for a *non-concerted* mechanism for ACC oxidation. A model system for this non-concerted process was also developed at Stanford: the electrochemical oxidation of ACC,

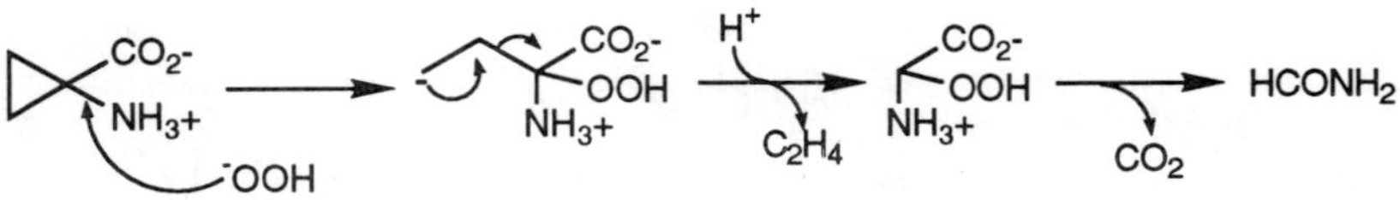
(4)

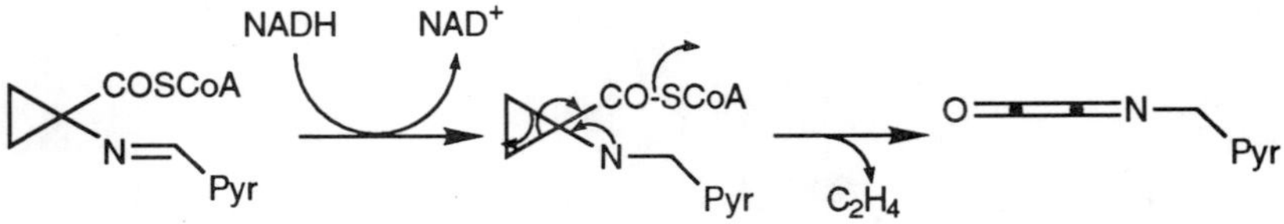
(5)

(6)

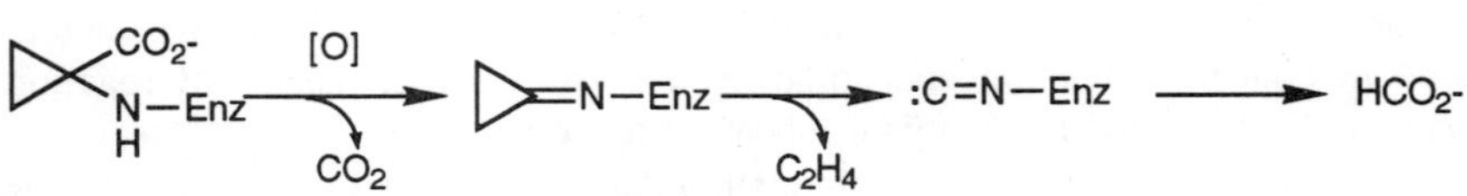
(7)

which likewise converts stereospecifically *cis* deuterated ACC to a 1:1 mixture of ethylene isomers (*35*). Furthermore, cyanide is produced in this reaction, lending further impetus to efforts to determine the product(s) of C-1 of ACC. Subsequent work in our laboratory on apple tissue (*36*) and in Yang's laboratory on mung bean segments (*37*) established that cyanide is indeed produced in ethylene biosynthesis. Because both models had predicted that cyanide is the other product, criteria other than reactants and products must be used to evaluate mechanistic proposals.

We proposed the sequential single-electron transfer mechanism for ethylene biosynthesis based on precedents of cyclopropylamine oxidations by enzymes such as cytochrome P450 (*38, 40*) and monoamine oxidase (*39, 40*). A syllogism underlying this postulate is that the electrochemical oxidation of ACC must have a close mechanistic relationship to the enzymatic oxidation because each gives the same products with the same stereochemistry. This mechanism was modified slightly by theoretical considerations (*3*) (eq 10). Key intermediates are **1**, the cyclopropylamine radical cation, and **2**, the "half-opened radical." The presence of the latter species explains, by free rotation about the $-CH_2-CH_2\bullet$ bond, the loss of stereochemistry observed with deuterated substrates. It is further consistent with cyclopropylcarbinyl

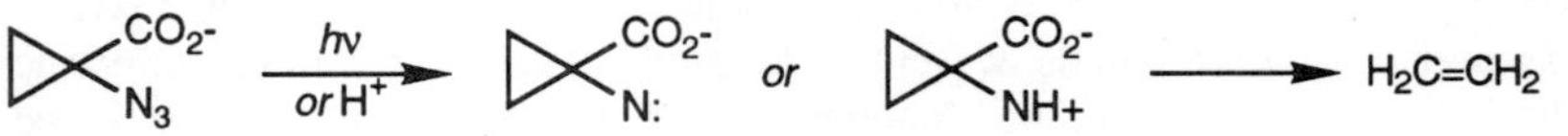
(8)

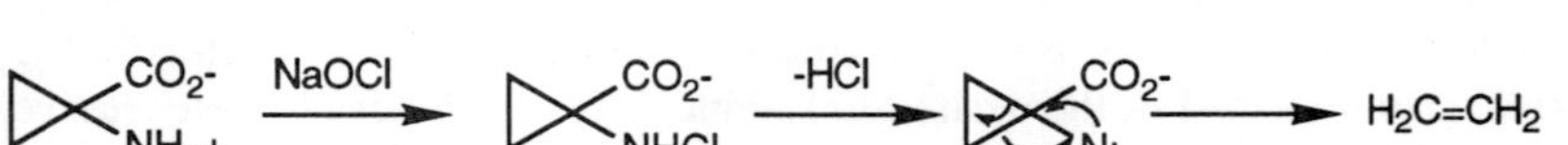
(9)

ring opening when the radical-bearing carbon is substituted by a cyclopropyl group (*6*) and the lack of an α-secondary deuterium isotope effect in biosynthetic and electrochemical ethylene production from tetradeuterated ACC (*5*). A significant isotope effect would be expected if ethylene is produced concertedly. The parent cyclopropylamine radical cation and its ring-opened isomer have now both been well-characterized by EPR spectroscopy (*41*), lending further support to the rationality of the sequential single-electron transfer mechanism. It has subsequently been used by Baldwin in interpreting biosynthetic experiments with ACC analogues substituted with both alkyl groups and deuterium (*8*) and transition metal-mediated oxidations of dideuterated ACCs (*42*). In the latter work, the first example of an ACC analogue bearing an N-O bond also was prepared. The *p*-anisaldehyde imine of dideutero-ACC methyl ester was oxidized with peracid to produce an oxaziridine (eq 11). As might be expected, this substance is quite unstable, liberating ethylene above -40 °C. However, it does so with retention of configuration, presumably in a *concerted* reaction *via* the nitrene. Oxidations of dideutero-ACC with potassium ferrate, on the other hand, give stereochemical scrambling *via* a *non-concerted* mechanism (*42*).

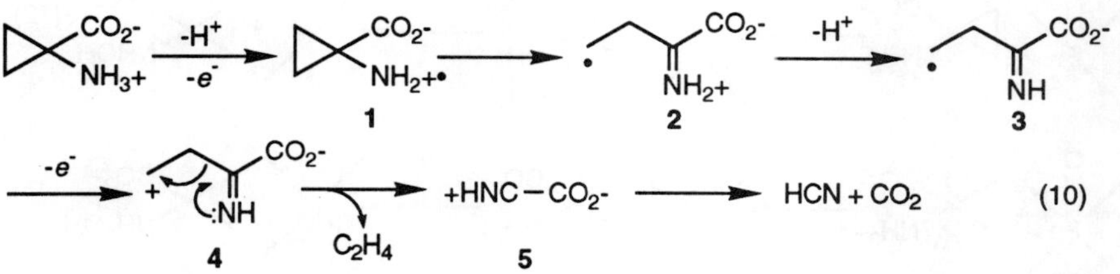

(10)

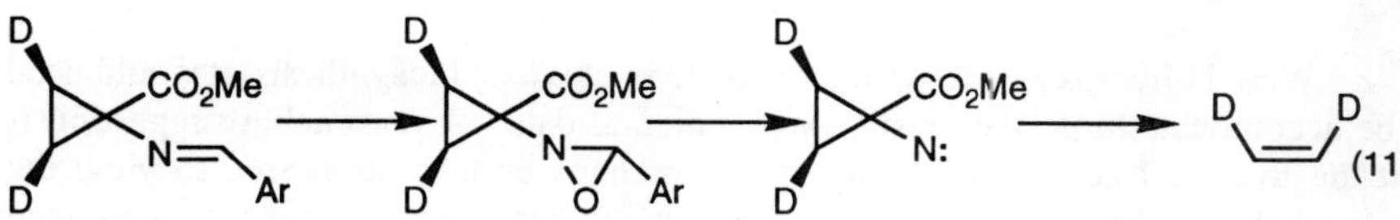

(11)

The mechanistic dichotomy for conversion of ACC to ethylene seems clear from the large body of work presented above. Formation of N-heteroatom derivatives leads to the nitrene or nitrenium ion and results in a *concerted* mechanism, while electron transfer/free radical oxidants lead to a radical cation and result in a *non-concerted* mechanism. Despite the significant evidence in favor of the radical pathway, reference to N-hydroxylation and nitrenium ion formation as a key step in ethylene biosynthesis has persisted, particularly in the plant physiology literature (*2, 43-46*). The sequence similarity of the EFE and several hydroxylase enzymes (*vide supra*) has only added fuel to this fire. However, consideration of the mechanisms for known hydroxylation processes makes the intermediacy of N-hydroxy-ACC very unlikely.

Applying the abstraction-rebound mechanism earlier discussed for iron-oxo enzymes to ACC either as its free base or zwitterion results in eqs 12 and 13. Both mechanisms produce intermediate radicals; the intermediate in eq 13 is *identical* to the cyclopropylamine radical cation generated in the electrochemical oxidation of ACC and postulated to be involved in ethylene biosynthesis. The key issue in evaluating the validity of a hydroxylation pathway is the relative rate of ring opening of such a radical compared to oxygen rebound. The best experimental data suggest that the rate of ring opening of cyclopropylamidyl radicals exceeds 5×10^8 s^{-1} (*47*). An estimate for the rate of ring-opening of cyclopropylaminyl radical is 2×10^8 s^{-1} (*48*). No data are available for ring opening of cyclopropylamine radical cations, but since aminium cation radicals are believed to be the most reactive of nitrogen radical species (aminium cation > amidyl > aminyl) (*49*), their rates are likely faster. Likewise, rates for non-heme iron-oxos rebounding to nitrogen in the hydroxylation of an amine are unknown. Because of the weak N-O bond that would be formed, rebound might be slower than the ~10^{10} s^{-1} rate observed in cytochrome P450-promoted hydroxylation of an alkane. A useful comparison can be made to the conversion of the penicillin to the cephalosporin ring system conducted by DAOCS (*vide supra*), believed to proceed by a β-thiyl radical rearrangement. This process occurs faster than rebound of the non-heme iron oxo to give the hydroxylation product. Thus, precedent exists that the rate of a molecular rearrangement can exceed the rebound rate in non-heme iron proteins. Applying this precedent in ethylene biosynthesis results in our assertion that the radical cation is formed by a "normal hydroxylation" pathway, but it ring opens *before* it can be hydroxylated.

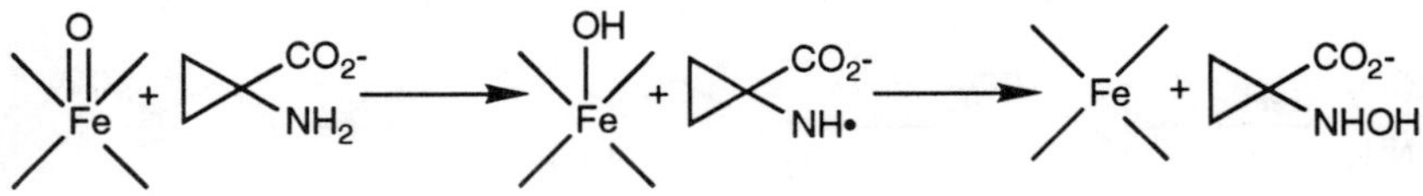

(12)

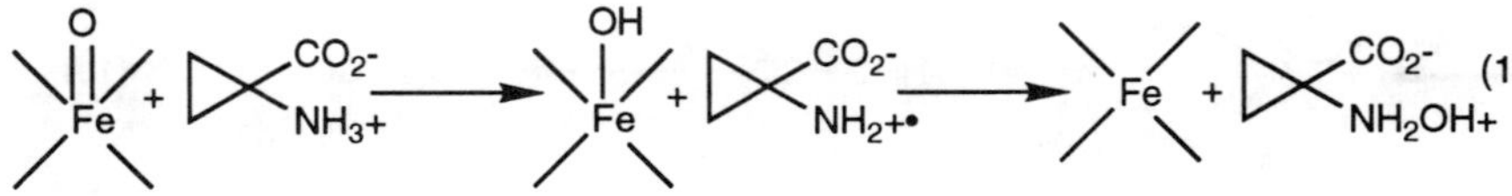

(13)

Were N-hydroxy-ACC an intermediate in ethylene biosynthesis, it would need to be accomodated into the existing experimental data. A reasonable suggestion is that the hydroxyl acts as a leaving group, perhaps upon protonation, to yield the ion (eq 14). We have already argued that this ion has a concerted reaction pathway available to it to form ethylene and cyanoformic acid, a mechanism inconsistent with experiment. In order to yield stereochemically-scrambled and cyclopropyl ring-opened products, the nitrenium ion would have to undergo ring-opening to the primary carbocation. While their great instability makes primary cations reactive intermediates rarely invoked in organic reaction mechanisms, it is conceivable that the enzyme exerts some special stabilizaton on such an ion. It would have to do so, however, without formation of a covalent linkage to the β-carbon (X-group mechanism), since that would prevent free rotation of the methylene unit.

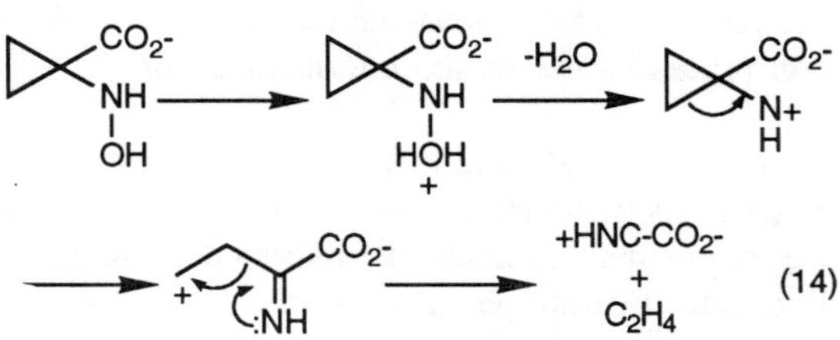

(14)

The weight of evidence is heavily in favor of radical intermediates in the production of ethylene from ACC. We advance the postulate that both the EFE and DAOCS begin their mechanisms like normal hydroxylases, but faster rearrangement of the substrate radical frustrates rebound. In such a scheme, N-hydroxy-ACC is not a discrete intermediate. Direct evidence on these points awaits the synthesis of deuterated N-hydroxy-ACCs and analysis of their chemical and enzymatic reactions. In the meantime, an attractive mechanistic scheme can be constructed by combination of the sequential single-electron transfer mechanism and a reasonable postulate for an iron enzyme (eqs 10 and 13). In this scheme (eq 15), electron transfer to an Fe-oxo produces the radical cation that ring opens. Conveniently, a basic oxide is present to deprotonate the imminium ion. Lowering of the charge on Fe^{III} through protonation facilitates oxidation of the carbon-centered radical. The resulting carbocation can fragment to ethylene and cyanoformate, and Fe^{II} is prepared to exchange ligands to cycle back to the oxo.

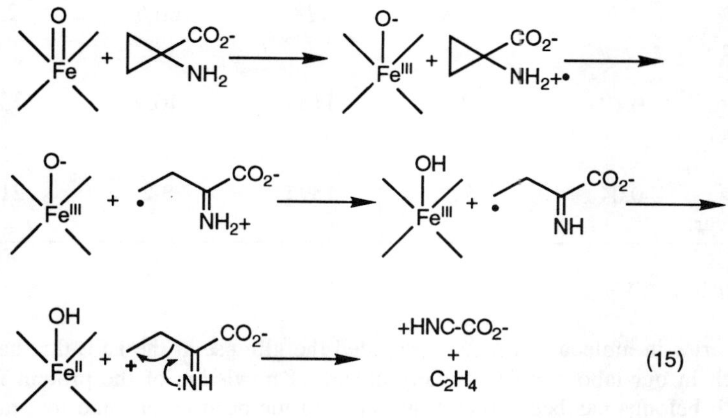

(15)

Enzyme Purification

For definitive mechanistic studies of the EFE, a homogeneous protein is required. On the basis of sequence analysis of homologous gene products, the tomato EFE is a member of a family of non-heme iron proteins (*vide supra*). Because another member of the family, the H6H protein, was purified to homogeneity and sequenced in order to obtain the cDNA clone (*24*), this work provided a model for the purification of the EFE from apple fruit. The recently-obtained apple cDNA sequence pAE12 reveals 73% DNA homology to the tomato sequence (*50*), implying that the apple enzyme is a member of the same family and should be subject to a similar purification protocol. There were already encouraging reports that cell-free extracts of EFE could be obtained from fruit (*51-53*).

Apple tissue is directly frozen in liquid nitrogen; significant activity is obtained from a storable, crude powder. A four-step extraction, precipitation, and chromatographic purification sequence provides homogeneous EFE (Table) as a 35 kDa protein. The enzyme is surprisingly abundant; in several purification procedures, only a 20- to 30-fold purification has been necessary to reach homogeneity. Some kinetic properties of the enzyme were determined: the Michaelis constant for ACC is 8.5±3 mM, and k_{cat} is $2.5 \pm 0.2 \times 10^{-3}$ s^{-1}.

Table. Purification of EFE from Apple Fruit

	TOTAL PROTEIN (mg)	ACTIVITY (nl/h)= (U)	SPECIFIC ACTIVITY (U/mg)	YIELD (%)	FOLD PURIFICATION
Crude	27	2025	75	100	-
50-80% $(NH_4)_2SO_4$	6.9	1228	178	60.6	2.4
butyl-ToyoPearl	0.19	220	1159	10.9	15.5
DEAE-ToyoPearl	0.04	64	1595	3.2	21.3

Conclusion

Discoveries in molecular genetics provided the glimpse at the EFE that has guided research in our laboratory and many others. Knowledge of the protein family to which it belongs has been crucial in obtaining the purified enzyme for study. The diversity of oxidation mechanisms utilized by this family, though, requires that each be studied individually. It would be incorrect to conclude that N-hydroxy-ACC is involved in ethylene biosynthesis only because of the similarity of the EFE sequence to known hydroxylating proteins. Pre-existing data favors a radical intermediate. A new, unified mechanism that incorporates some traits of the non-heme iron-oxo enzymes to generate the cyclopropylamine radical cation intermediate has been proposed. The availability of the purified enzyme will permit definitive studies of the mechanism by which ethylene is produced and provide opportunites to discover molecules that inhibit the process.

Acknowledgments

The most important contributors to this project were those who conducted the experimental work: Lynn M. Kaiser and Laura J. Weislo. Financial support was provided by the USDA. Thanks are due to Profs. Thomas Boller and Craig Townsend for reading the manuscript. Phaedria St. Hilaire provided invaluable advice.

Literature Cited

1. Adams, D. O.; Yang, S. F. *Proc. Natl. Acad. Sci., USA* **1979**, *76*, 170-174.
2. Yang, S. F.; Hoffman, N. E. *Annu. Rev. Plant Physiol.* **1984**, *35*, 155-189.
3. Pirrung, M. C. *J. Org. Chem.* **1987**, *52*, 4179-4184.
4. Pirrung, M. C.; Trinks, U. P *J. Chem. Soc., Chem. Commun.* **1989**, 857-859.
5. Pirrung, M. C.; McGeehan, G. M. *J. Am. Chem. Soc.* **1986**, *108*, 5647-5648.
6. Pirrung, M. C.; McGeehan, G. M. *Angew. Chem., Int. Ed. Engl.* **1985**, *24*, 1044-1045.

7. Adlington, R. M.; Baldwin, J. E; Rawlings, B. J. *J. Chem. Soc., Chem. Commun.* **1983**, 290-292.
8. Baldwin, J. E.; Adlington, R. M.; Lajoie, G. A.; Rawlings, B. J. *J. Chem. Soc., Chem. Commun.* **1985**, 1496-1498.
9. Smith, C. J. S.; Slater, A.; Grierson, D. *Planta* **1986**, *168*, 94-100.
10. Holdsworth, M. J.; Bird, C. R.; Ray, J.; Schuch, W.; Grierson, D. *Nucl. Acids Res.* **1987**, *15*, 731-739.
11. Hamilton, A. J.; Lycett, G. W.; Grierson, D. *Nature* **1990**, *346*, 284-287.
12. Spanu, P.; Reinhardt, D.; Boller, T. *EMBO J.* **1991**, *10*, 2007-2013.
13. Hoffman, N. E.; Yang, S. F.; Ichihara, A.; Sakamura, S. *Plant Physiol.* **1982**, *70*, 195-199.
14. Pirrung, M. C.; McGeehan, G. M. *J. Org. Chem.* **1986**, *51*, 2103-2106.
15. Hamilton, A. J.; Bouzayen, M.; Grierson, D. *Proc. Natl. Acad. Sci ., USA* **1991**, *88*, 7434-7437.
16. Hedden, P. *Biochem. Soc. London; Trans. 641st Ann. Mtg. Colloquium: Plant Oxygenases, Peroxidases, and Oxidases* **1992**, *20*, 373-376.
17. Abraham, E. P.; Huddleston, J. A.; Jayatilake, B. S.; O'Sulllivan, J.; White, R. L. *Spec. Publ. Chem. Soc.* **1981**, *38*, 125-134.
18. Ming, L.-J.; Que, L., Jr.; Kriauciunas, A.; Frolik, C. A.; Chen, V. J. *Inorg. Chem.* **1990**, *29*, 1111-1112.
19. Scott, R. A.; Wang, S.; Eidsness, M. K.; Kriauciunas, A.; Frolik, C. A.; Chen, V. J. *Biochemistry* **1992**, *31*, 4596-4601.
20. Chen, V. J.; Orville, A. M.; Harpel, M. R.; Frolik, C. A.; Surerus, K. K.; Munck, E.; Lipscomb, J. D. *J. Biol. Chem.* **1989**, *264*, 21677-21681.
21. Baldwin, J. E.; Adlington, R. M.; Basak, A.; Flitsch, S. L.; Forrest, A. K.; Ting, H.-H. *J. Chem. Soc., Chem. Commun.* **1984**, 1211-1213.
22. Baldwin, J. E.; Adlington, R. M.; Derome, A. E.; Ting, H.-H.; Turner, N. J. *J. Chem. Soc., Chem. Commun.* **1986**, 273-275.
23. Britsch, L. *Arch. Biochem. Biophys.* **1990**, *276*, 348-354.
24. Matsuda, J.; Okabe, S.; Hashimoto, T.; Yamada, Y. *J. Biol. Chem.* **1991**, *266*, 9460-9464.
25. Siegel, B. *Bioorg. Chem.* **1979**, *8*, 219-226.
26. Ortiz de Montellano, P. R. In *Cytochrome P-450: Structure, Mechanisms, and Biochemistry*, Ortiz de Montellano, P. R., Ed.; Plenum: New York, 1986, pp. 217-271.
27. Ortiz de Montellano, P. R. *Trends Pharmacol. Sci.* **1989**, *10*, 354-359.
28. Bowrey, V. W.; Ingold, K. U. *J. Am. Chem. Soc.* **1991**, *113*, 5699-5707.
29. Baldwin, J. E.; Adlington, R. M.; Coates, J. B.; Crabbe, M. J. C.; Crouch, N. P.; Keeping, J. W.; Knight, G. C.; Schofield, C. J.; Ting, H.-H.; Vallejo, C. A.; Abraham, E. P. *Biochem. J.* **1987**, *245*, 831-841.
30. Baldwin, J. E.; Adlington, R. M.; Kang, T. W.; Lee, E.; Schofield, C. J. *J. Chem. Soc., Chem. Commun.* **1987**, 104-106.
31. Lurssen, K.; Naumann, K.; Schroeder, R. *Naturwissenschaften* **1979**, *66*, 264-265.
32. Adams, D. O.; Yang, S. F. *Trends Biochem. Sci.* **1981**, *6*, 161-164.
33. Baldwin, J. E.; Adlington, R. M.; Aplin, R. T.; Rawlings, B. J.; Osborne, D. *J. Chem. Soc., Chem. Commun.* **1982**, 1086-1088.

34. Pirrung, M. C.; McGeehan, G. M. *J. Org. Chem.* **1983**, *48*, 5143-5144.
35. Pirrung, M. C. *J. Am. Chem. Soc.* **1983**, *105*, 7207-7209.
36. Pirrung, M. C. *Bioorg. Chem.* **1985**, *13*, 219-225.
37. Peiser, G. D.; Wang, T. T.; Hoffman, N. E.; Yang, S. F.; Liu, H.-w.; Walsh, C. T. *Proc. Natl. Acad. Sci., USA* **1984**, *81*, 3059-3063.
38. Macdonald, T.; Zirvi, K.; Burka, L.; Peyman, P.; Guengerich, F. P. *J. Am. Chem. Soc.* **1982**, *104*, 2050-2052.
39. Silverman, R. B.; Zieske, P. A. *J. Med. Chem.* **1985**, *28*, 1953-1957.
40. Tullman, R. H.; Hanzlik, R. P. *Drug Metab. Rev.* **1984**, *15*, 1163-1182.
41. Qin, X.-Z.; Williams, F. *J. Am. Chem. Soc.* **1987**, *109*, 595-597.
42. Baldwin, J. E.; Jackson, D. A.; Adlington, R. M.; Rawlings, B. J. *J. Chem. Soc., Chem. Commun.* **1985**, 206-207.
43. Yang, S. F. *Rec. Adv. Phytochem.* **1989**, *23*, 263-87.
44. Yang, S. F. *Recent Advances in the Biochemistry of Fruits and Vegetables,* Friend, J.; Rhodes, M. J. C., eds., Academic Press, 1981, p. 89-106.
45. Liu, Y.; Su, L.-y.; Yang, S. F. *Planta* **1984,** *161,* 439-443
46. Yang, S. F. *Current Topics in Plant Biochemistry and Physiology*, Vol. 4, Randall, D. D.; Blevins, D. G.; Lasson, R. L., eds., University of Missouri, Columbia, 1985, p. 126-138.
47. Sutcliffe, R.; Ingold, K. U. *J. Am. Chem. Soc.* **1982**, *104*, 6071-6075.
48. Newcomb, M.; Park, S.-U.; Kaplan, J.; Marquardt, D. J. *Tetrahedron Lett.* **1985**, *26*, 5651-5654.
49. Newcomb, M.; Deeb, T. M.; Marquardt, D. J. *Tetrahedron* **1990**, *46*, 2317-2328.
50. Dong, J. G.; Olson, D.; Silverstone, A.; Yang, S. F. *Plant Physiol.* **1992**, *98*, 1530-1531.
51. Ververidis, P., John, P. *Phytochemistry* **1991**,*30*, 725-727.
52. Fernandez-Maculet, J. C.; Yang, S. F. *Plant Physiol.* **1992**, *99*, 1751-1754.
53. McGarvey, D. J.; Christoffersen, R. E. *J. Biol. Chem.* **1992** *267*, 5964-5967.

RECEIVED April 30, 1993

Chapter 32

Inhibition of Photosynthesis by Substituted 4-Nitrophenols in Wildtype and Five Mutants of *Chlamydomonas reinhardtii* Thylakoids

Structure–Activity and Molecular Modeling Studies

W. Draber[1], U. Hilp[2], M. Schindler[1], and A. Trebst[2]

[1]Agrochemical Research, Bayer AG, 5090 Leverkusen, Germany
[2]Department of Biology, Ruhr-Universität Bochum, 4630 Bochum, Germany

Photosynthetic inhibition data were determined with thylakoids of the green alga *Chlamydomonas reinhardtii* on fourty-four 2-halo-4-nitro-6-alkyl- and 2,4-dinitro-6-alkyl-phenols. For the measurements wildtype and five different mutants of the algae were employed. The mutants had known amino acid changes in the herbicide binding protein D1. Their response to the various inhibitors was quite different from that of the wildtype. Tolerance as well as enhanced inhibitory activity was found in the mutants. This implies that the phenols are bound specifically to the D1 protein. The wealth of data encouraged structure-activity and molecular modelling studies, which were based on a model of the binding niche. It was attempted to explain the enhanced or decreased inhibitory activity of selected compounds by energy calculations.

Photosynthesis inhibitors act by displacing the secondary electron acceptor Q_B, plastoquinone, from its binding protein D1 in photosystem II as suggested first by Velthuys in 1981 (*1*). Already in 1979 we had emphasized (*2*) that there is distinction to be made between classical herbicides and phenolic inhibitors. In 1987 it was suggested (*3*) to divide the inhibitors into two chemical groups that behaved different in many biochemical respects: the classical inhibitors represented by compounds like diuron (DCMU) and atrazine, and a second group which consists mainly of phenol derivatives like the two herbicides dinoseb and bromoxynil. The first group has been called the serine family because these compounds loose their activity in plants and algae in which $serine_{264}$ in the D1 protein is replaced by alanine or glycin. The phenolic inhibitors do not show decrease of inhibitory activity in these mutants. It was assumed that they are oriented towards $histidine_{215}$ in the D1 binding niche. For that reason they were called the histidine family. In the meantime, however, experimental evidence began to show that this distinction was rather inefficient and needed to be replaced by a more detailed one. This became possible by screening a great number of inhibitors and the availibility of algae mutants, in which the amino acid sequence of the D1 protein was changed in a known manner.

0097–6156/94/0551–0449$07.00/0

Structure-activity studies have proved instrumental in describing the dependence of inhibitory activity on physicochemical and calculated descriptors. A nearly historical example of a correlation study with phenolic photosynthesis inhibitors is shown in Figure 1 (*2*). In this equation the only parameters employed were the STERIMOL parameters by Verloop et al. (*4*). In a more recent analysis of the same set of compounds (*5*) it was found that the hydrophobicity parameter π also contributed to inhibitory activity.

Computer power is now no longer a serious limitation for many applications in the area of small molecules. We have carried out studies of molecular modeling and energy calculations using a simplified model of the D1 binding niche, based on the X-ray analysis of the bacterial photosystem by Deisenhofer et al. (*6*) and using the inhibition data obtained with the phenols. The results, however, have to be regarded as preliminary.

Materials and Methods

Compounds Studied. Fourty-four 4-nitrophenols were prepared according to methods known in the literature. The structures are listed in Tables I and II. Table I contains twenty-seven 2-halo-4-nitro-6-alkylphenols, some with a methyl group in 3-position. Table II consists of seventeen 2,4-dinitro-6-alkylphenols, one of them with a methyl group in 3-position.

Chlamydomonas reinhardtii Mutants. Five different mutants with mutations in the *psb*A gene that encodes the D1 protein were obtained as described by Wildner et al. (*7*). The val219ile mutant was first described by Galloway and Mets (*8*) and the corresponding *psb*A gene sequenced by Erickson et al. (*9*). The ala251val, ser264ala and leu275phe mutants were obtained by Johanningmeier et al. (*10*) from wildtype cells after mutagenesis and metribuzin pressure. They also identified the amino acid substitutions by sequencing the *psb*A gene. The compounds were assayed for inhibitory activity in isolated thylakoid membranes by measuring photosynthetic DCPIP reduction.

Structure Activity Analyses. The pI_{50} values of wild type and mutant thylakoids were subjected to regression analyses according to the Hansch approach. Experimentally based and calculated descriptors were employed. Partition coefficients logP were obtained by using HPLC retention data that were converted by a calibration program. CLOGP and CMR were calculated with the Pomona software (*11*). The STERIMOL parameters were taken from Verloop et al. (*4,12*). Two indicator variables had to be introduced: n-R^2 which takes 0 for R^2 = H and 1 for R^2 = Me, and Ph-R^4 which takes 0 for R^4 = all alkyl and cycloalkyl groups and 1 if R^4 = phenyl or benzyl. From the many calculations that were performed, only those are presented with the best statistical criteria F, r^2, and s. From these, the standard deviation s was considered as the most important one, hence it was related to the respective range (see the last columns of Tables V and VI).

Molecular Modeling and Energy Calculations. An ESV 30/33 RISC workstation was used for molecular modeling and energy calculations with the software package SYBYL (Tripos Ass., St. Louis, USA). Figures 2 - 7 were drawn after photographs from the screen by using the software Aldus FreeHand 3.1 (Aldus Corporation).

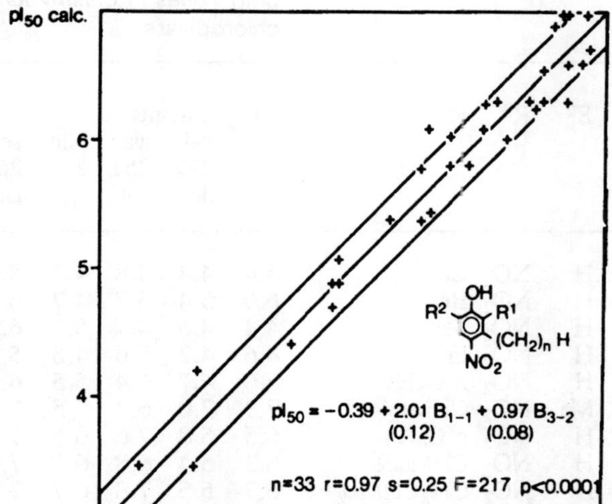

Figure 1. The First Regression Analysis of 4-Nitrophenols. (Reproduced with permission from reference 2. Copyright 1979.)

Table I. The Inhibition of Photosynthetic Electron Flow by Substituted Phenols in Wildtype and Five Mutants of *Chlamydomonas rh.*

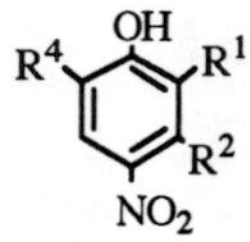

4-Nitrophenols

No.	R^1	R^2	R^3	R^4	pI_{50} values in *Chlamydomonas rh.* chloroplasts: wt	mutants: val 219 ile	ala 251 val	phe 255 tyr	ser 264 ala	leu 275 phe
1	Cl	H	NO_2	Cl	4.4	4.4	4.8	4.3	5.0	6.1
2	Br	H	NO_2	Br	5.0	5.4	5.7	4.7	6.0	6.5
3	Br	H	NO_2	Me	4.4	4.6	4.7	5.2	6.3	5.3
4	Br	H	NO_2	Et	4.6	4.2	5.6	4.6	5.7	5.6
5	Br	H	NO_2	n-C_3H_7	6.0	4.7	6.4	5.5	6.6	6.8
6	Br	Me	NO_2	$CHMe_2$	7.3	7.0	6.4	7.5	7.7	7.1
7	Br	H	NO_2	n-C_4H_9	6.5	5.8	7.6	6.6	7.3	7.5
8	Br	H	NO_2	CHMeEt	6.3	5.4	6.8	6.1	7.2	6.1
9	J	H	NO_2	CH_2CHMe_2	6.7	5.5	7.8	6.7	7.2	7.1
10	H	Me	NO_2	CMe_3	7.1	6.7	7.0	6.5	6.8	7.0
11	Cl	H	NO_2	CMe_3	6.8	6.6	7.2	6.0	6.8	5.4
12	Cl	Me	NO_2	CMe_3	8.2	8.2	7.0	7.0	8.0	7.0
13	Br	H	NO_2	CMe_3	7.2	7.1	7.1	7.2	7.5	5.8
14	Br	Me	NO_2	CMe_3	8.4	8.0	7.8	7.8	8.4	7.2
15	Br	H	NO_2	n-C_5H_{11}	7.0	5.8	7.6	7.0	7.4	7.7
16	Br	H	NO_2	n-C_6H_{13}	7.3	6.4	8.0	7.2	7.6	8.0
17	Br	H	NO_2	n-C_7H_{15}	7.5	7.0	8.1	7.8	8.0	8.5
18	Br	H	NO_2	n-C_8H_{17}	8.0	6.8	8.4	7.7	8.4	8.3
19	Br	H	NO_2	n-C_9H_{19}	8.1	7.6	8.6	7.8	8.0	8.7
20	Br	H	NO_2	n-$C_{10}H_{21}$	8.1	7.4	8.7	7.8	8.3	8.6
21	Br	H	NO_2	n-$C_{12}H_{25}$	7.6	7.1	8.0	7.2	7.8	8.0
22	Br	H	NO_2	c-C_5H_9	7.6	6.5	7.7	7.5	7.8	7.1
23	Br	H	NO_2	c-C_6H_{11}	8.0	6.4	8.1	7.2	8.1	7.0
24	Cl	H	NO_2	Ph	7.1	5.8	7.7	7.2	7.3	7.0
25	Br	H	NO_2	Ph	7.7	6.4	8.4	7.3	8.0	7.5
26	J	H	NO_2	Ph	7.5	6.0	8.2	7.0	7.5	7.5
27	Br	H	NO_2	CH_2Ph	7.2	5.5	7.4	7.2	7.6	7.1

Table II. The Inhibition of Photosynthetic Electron Flow by Substituted Phenols in Wildtype and Five Mutants of *Chlamydomonas rh.*

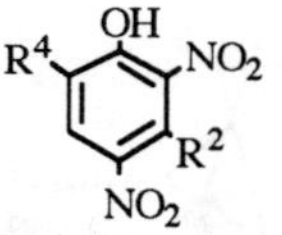

2,4-Dinitrophenols

No.	R^1	R^2	R^3	R^4	pI50 values in *Chlamydomonas rh.* chloroplasts wt	mutants val 219 ile	ala 251 val	phe 255 tyr	ser 264 ala	leu 275 phe
28	NO_2	H	NO_2	Me	4.2	4.1	4.8	5.3	4.8	5.1
29	NO_2	H	NO_2	Et	4.7	4.5	5.8	4.8	5.3	5.8
30	NO_2	H	NO_2	n-C_3H_7	5.1	4.7	6.0	5.1	5.7	5.5
31	NO_2	H	NO_2	n-C_4H_9	5.6	5.4	7.0	5.8	6.4	6.5
32	NO_2	H	NO_2	CHMeEt	5.8	5.2	6.3	5.4	6.3	5.7
33	NO_2	H	NO_2	CH_2CHMe_2	5.3	5.0	6.7	5.0	5.8	6.2
34	NO_2	H	NO_2	CMe_3	6.0	5.5	7.0	5.7	6.0	5.7
35	NO_2	Me	NO_2	CMe_3	7.1	7.3	7.4	6.4	7.1	6.7
36	NO_2	H	NO_2	n-C_5H_{11}	5.7	5.2	6.8	6.0	6.4	6.1
37	NO_2	H	NO_2	n-C_6H_{13}	6.7	6.0	7.3	6.3	6.7	7.0
38	NO_2	H	NO_2	n-C_7H_{15}	7.0	6.5	7.8	7.0	7.4	7.7
39	NO_2	H	NO_2	n-C_8H_{17}	7.4	6.7	8.0	7.0	7.7	7.8
40	NO_2	H	NO_2	n-C_9H_{19}	7.7	7.4	8.5	7.2	8.0	8.3
41	NO_2	H	NO_2	n-$C_{10}H_{21}$	7.7	7.3	8.1	7.3	8.0	8.3
42	NO_2	H	NO_2	n-$C_{12}H_{25}$	7.4	7.5	8.2	7.3	8.0	8.1
43	NO_2	H	NO_2	c-C_6H_{11}	6.5	5.7	7.3	6.3	7.3	6.2
44	NO_2	H	NO_2	Ph	6.1	4.7	7.0	6.5	6.5	6.2

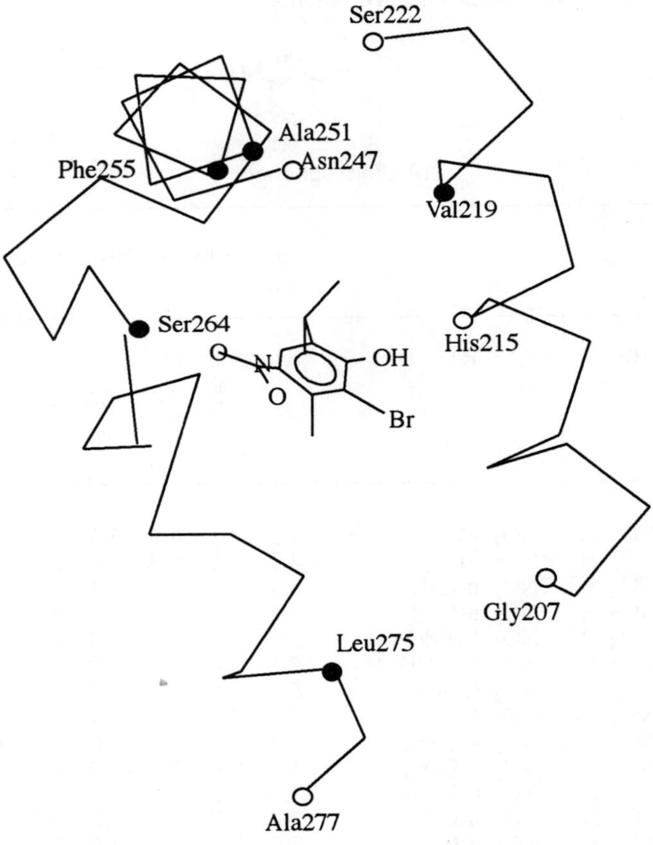

Figure 2. Compound 6 in the Wildtype Binding Niche.

In Figures 2 - 7 the Amino Acids where Mutations can Occur are Figured as Filled Circles. The Amino Acids at the Ends of the Binding Niche and His215 are Represented as Open Circles.

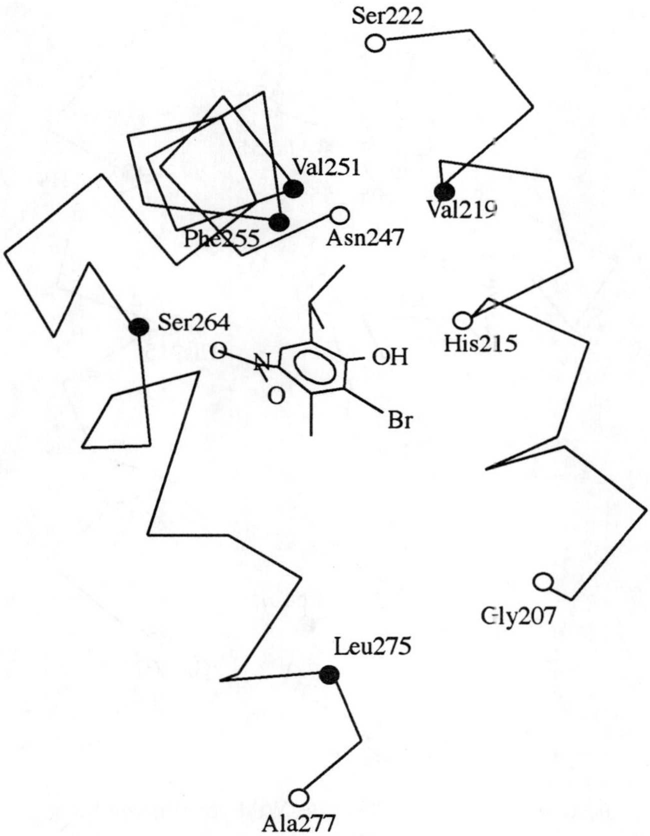

Figure 3. Compound 6 in the Ala251Val Mutant Binding Niche.

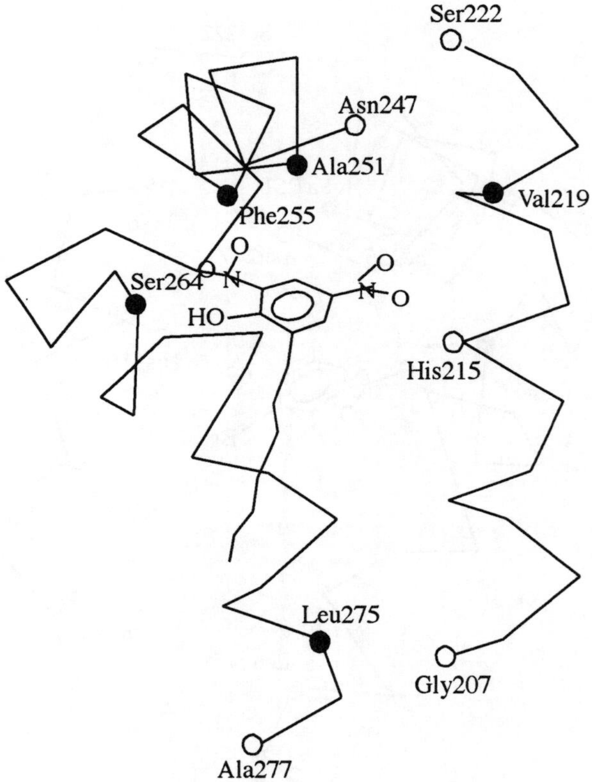

Figure 4. Compound 38 in the Wildtype Binding Niche.

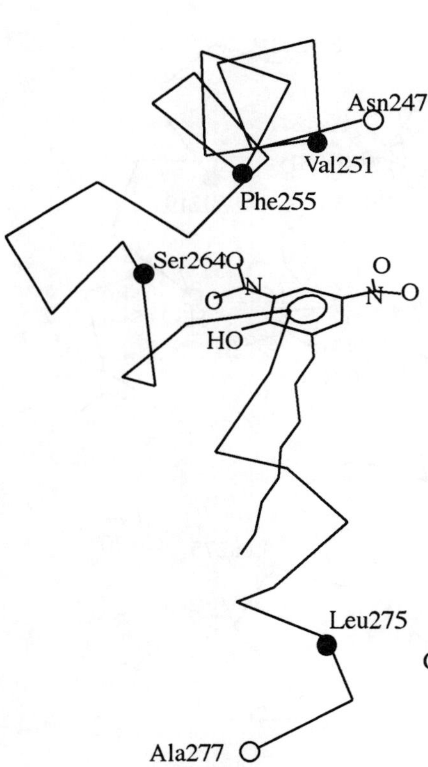

Figure 5. Compound 38 in the Ala251Val Mutant Binding Niche.

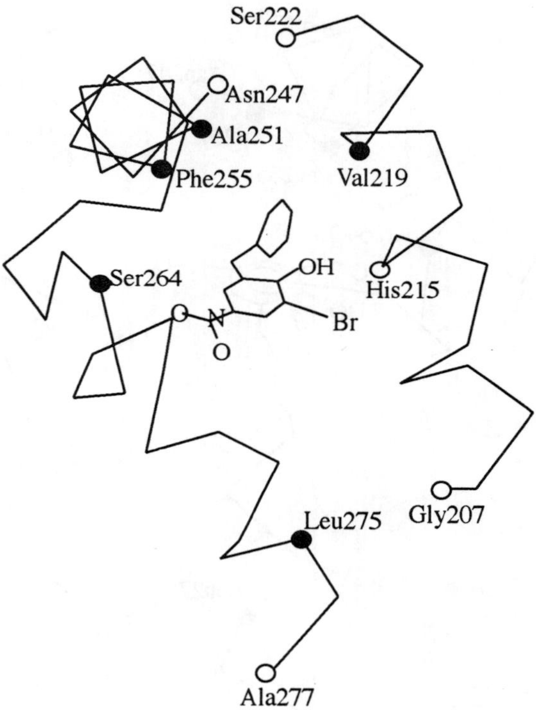

Figure 6. Compound 27 in the Wildtype Binding Niche.

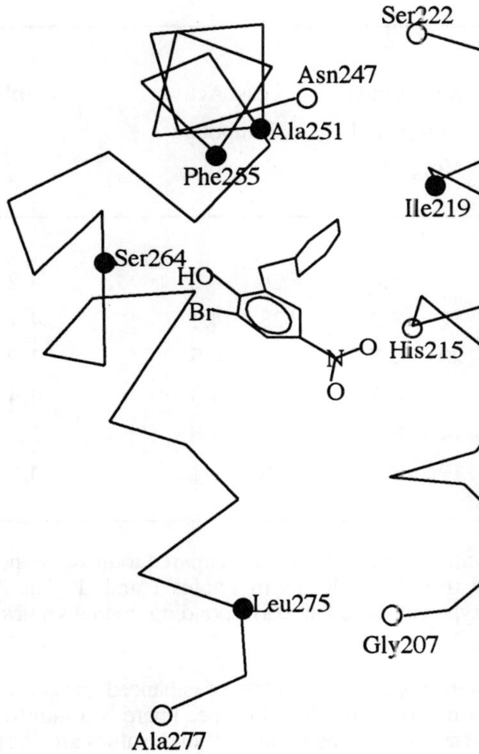

Figure 7. Compound 27 in the Val219Ile Mutant Binding Niche.

Results.

Biochemistry. From the inhibition assays with the wildtype (wt) and mutant algae it becomes evident that there are large differences in sensitivity not only to the diverse nitrophenols but also among the various mutants. The range of the pI_{50} values from Tables I and II is listed in Table III.

Table III. Spread of pI_{50} Values in Wildtype and Mutant Chloroplasts

Mutant	Most Active Compound No.	pI_{50}	Least Active Compound No.	pI_{50}	ΔpI_{50}
wt	14	8.4	28	4.2	4.2
val219ile	12	8.2	28	4.1	4.1
ala251val	20	8.7	1	4.8	3.9
phe255tyr	17	8.7	1	4.3	4.4
ser264ala	14	8.4	28	4.8	3.6
leu275phe	19	8.7	28	5.1	4.6

In Table IV the various mutant data are compared to those of the wildtype. The compounds are listed in same order as in Tables I and II. The ΔpI_{50}s are the differences of the wildtype values to the corresponding mutant values (= $pI_{50wildtype} - pI_{50mutant}$).

Table IV presents the complicated pattern of enhanced and decreased sensitivity to the 4-nitrophenols compared with the wildtype. There is no uniform response to the compounds in the sense that in one mutant all pI_{50} values are shifted in the same direction. However, a remarkable feature from Table IV is that almost all phenols show tolerance in the val219ile mutant whereas most compounds have increased sensitivity in the ser264ala mutant. There are only few exceptions to this rule (val219ile: No.s 2, 3, 42; ser264ala: 12, 19, no change: 11, 14, 26, 34, 35, 37). In this context one has to note that classical inhibitors show the opposite behaviour in the ser264ala mutant. They get tolerant to a great extent. Atrazine for example has a resistance factor (i. e. the ratio of the I_{50} values in resistant and susceptible chloroplasts) of 160 and diuron (DCMU) one of 200 (*7*) which corresponds to a $\Delta pI_{50} = 2.2$ or 2.3 respectively (*10*). In the other mutants the 4-nitrophenols display a mixed behaviour. In the ala251val mutant enhanced sensitivity prevails, whereas

Table IV. ΔpI_{50}-Values of 44 4-Nitrophenols

No.	wt	val 219 ile	ΔpI_{50}	ala 251 val	ΔpI_{50}	phe 255 tyr	ΔpI_{50}	ser 264 ala	ΔpI_{50}	leu 275 phe	ΔpI_{50}
1	4.4	4.4	**0.0**	4.8	**−0.4**	4.3	**+0.1**	5.0	**−0.6**	6.1	**−1.7**
2	5.0	5.4	**−0.4**	5.7	**−0.7**	4.7	**+0.3**	6.0	**−1.0**	6.5	**−1.5**
3	4.4	4.6	**−0.2**	4.7	**−0.3**	5.2	**−0.8**	6.3	**−1.9**	5.3	**−0.9**
3	4.6	4.2	**+0.4**	5.6	**−1.0**	4.6	**0.0**	5.7	**−1.1**	5.6	**−1.0**
5	6.0	4.7	**+1.3**	6.4	**−0.4**	5.5	**+0.5**	6.6	**−0.6**	6.8	**−0.8**
6	7.3	7.0	**+0.3**	6.4	**+0.9**	7.5	**−0.2**	7.7	**−0.4**	7.1	**+0.2**
7	6.5	5.8	**+0.7**	7.6	**−1.1**	6.6	**−0.1**	7.3	**−0.8**	7.5	**−1.0**
8	6.3	5.4	**+0.9**	6.8	**−0.5**	6.1	**+0.2**	7.2	**−0.9**	6.1	**+0.2**
9	6.7	5.5	**+1.2**	7.8	**−1.1**	6.7	**0.0**	7.2	**−0.5**	7.1	**−0.4**
10	7.1	6.7	**+0.4**	7.0	**+0.1**	6.5	**+0.6**	6.8	**+0.3**	7.0	**+0.1**
11	6.8	6.6	**+0.2**	7.2	**−0.4**	6.0	**+0.8**	6.8	**0.0**	5.4	**+1.4**
12	8.2	8.2	**0.0**	7.0	**+1.2**	7.0	**+1.2**	8.0	**+0.2**	7.0	**+1.2**
13	7.2	7.1	**+0.1**	7.1	**+0.1**	7.2	**0.0**	7.5	**−0.3**	5.8	**+1.4**
14	8.4	8.0	**+0.4**	7.8	**+0.6**	7.8	**+0.6**	8.4	**0.0**	7.2	**+1.2**
15	7.0	5.8	**+1.2**	7.6	**−0.6**	7.0	**0.0**	7.4	**−0.4**	7.7	**−0.7**
16	7.5	7.0	**+0.5**	8.0	**−0.5**	7.2	**+0.3**	7.6	**−0.1**	8.0	**−0.5**
17	7.3	6.4	**+0.9**	8.1	**−0.8**	7.8	**−0.5**	8.0	**−0.7**	8.5	**−1.2**

Continued on next page

Table IV. Contd.

No.	wt	val 219 ile	ΔpI_{50}	ala 251 val	ΔpI_{50}	phe 255 tyr	ΔpI_{50}	ser 264 ala	ΔpI_{50}	leu 275 phe	ΔpI_{50}
18	8.0	6.8	+1.2	8.4	–0.4	7.7	+0.3	8.4	–0.4	8.3	–0.3
19	8.1	7.6	+0.5	8.6	–0.5	7.8	+0.3	8.0	+0.1	8.7	–0.6
20	8.1	7.4	+0.7	8.7	–0.6	7.8	+0.3	8.3	–0.2	8.6	–0.5
21	7.6	7.1	+0.5	8.0	–0.4	7.2	+0.4	7.8	–0.2	8.0	–0.4
22	7.6	6.5	+1.1	7.7	–0.1	7.5	+0.1	7.8	–0.2	7.1	+0.5
23	8.0	6.4	+1.6	8.1	–0.1	7.2	+0.8	8.1	–0.1	7.0	+1.0
24	7.1	5.8	+1.3	7.7	–0.6	7.2	–0.1	7.3	–0.2	7.0	+0.1
25	7.7	6.4	+1.3	8.4	–0.7	7.3	+0.4	8.0	–0.3	7.5	+0.2
26	7.5	6.0	+1.5	8.2	–0.7	7.0	+0.5	7.5	0.0	7.5	0.0
27	7.2	5.5	+1.7	7.4	–0.2	7.2	0.0	7.6	–0.4	7.1	+0.1
28	4.2	4.1	+0.1	4.8	–0.6	5.3	–1.1	4.8	–0.6	5.1	–0.9
29	4.7	4.5	+0.2	5.8	–1.1	4.8	–0.1	5.3	–0.6	5.8	–1.1

30	5.1	4.7	+0.4	6.0	–0.9	5.1	0.0	5.7	–0.6	5.5	–0.4
31	5.6	5.4	+0.2	7.0	–1.4	5.8	–0.2	6.4	–0.8	6.5	–0.9
32	5.8	5.2	+0.6	6.3	–0.5	5.4	+0.4	6.3	–0.5	5.7	+0.1
33	5.3	5.0	+0.3	6.7	–1.4	5.0	+0.3	5.8	–0.5	6.2	–0.9
34	6.0	5.5	+0.5	7.0	–1.0	5.7	+0.3	6.0	0.0	5.7	+0.3
35	7.1	7.3	+0.2	7.4	–0.3	6.4	+0.7	7.1	0.0	6.7	+0.4
36	5.7	5.2	+0.5	6.8	–1.1	6.0	–0.3	6.4	–0.7	6.1	–0.4
37	6.7	6.0	+0.7	7.3	–0.6	6.3	+0.4	6.7	0.0	7.0	–0.3
38	7.0	6.5	+0.5	7.8	–0.8	7.0	0.0	7.4	–0.4	7.7	–0.7
39	7.4	6.7	+0.7	8.0	–0.6	7.0	+0.4	7.7	–0.3	7.8	–0.4
40	7.7	7.4	+0.3	8.5	–0.8	7.2	+0.5	8.0	–0.3	8.3	–0.6
41	7.7	7.3	+0.4	8.1	–0.4	7.3	+0.4	8.0	–0.3	8.1	–0.4
42	7.4	7.5	–0.1	8.2	–0.8	7.3	+0.1	8.0	–0.6	8.3	–0.9
43	6.5	5.7	+0.8	7.3	–0.8	6.3	+0.2	7.3	–0.8	6.2	+0.3
44	6.1	4.7	+1.4	7.0	–0.9	6.5	–0.4	6.5	–0.4	6.2	–0.1

decreased sensitivity is found predominantly in the phe255tyr mutant. An almost evenly distributed decreased and enhanced sensitivity is observed in the leu275phe mutant. All these results are a strong indication for a specific interaction of the phenolic inhibitors with the amino acids of the binding niche.

Quantitative Structure Activity Relationships. The goal of our structure activity studies was not the prediction of more active compounds in order to finally get new hints for the development of a herbicide. Instead, we wanted to corroborate our concept that phenols with the appropriate substitution enter the binding niche in the D1 protein just as the classical herbicidal inhibitors like atrazine and diuron do.

From earlier experience we had found it desirable to subject not too large data sets to QSAR. A wealth of data is a provision for doing so. As this was the case with the phenols, we carried out structure activity analyses on two sets of pI_{50} values. They are listed in Tables I and II.

QSAR analyses with the data set on 2-halo-4-nitrophenols in Table I (n = 27) resulted in the Equations 1 - 6 in Table V. In these equations F, r^2 and s have the usual meaning (F-test of the regression; squared correlation coefficient; standard deviation). Since the spread of the pI_{50} values for each mutant is not unimportant we have related the standard deviation to it. The numbers in parentheses below the coefficients are the corresponding t-test values.

Table VI contains the regression Equations 6 - 12 pertaining to the data set on 2,4-dinitrophenols (n = 17).

Correlation matrices with all parameters that were employed in the regressions and the wildtype pI_{50} values are compiled an Tables VII and VIII. From the matrices it is evident that some variables are not far from colinearity: logP, CLOGP, CMR and the two STERIMOL parameters L_4 and $B5_4$ in both Tables. As to be expected, L-R^4 and (L-R^4)^2 are highly colinear but they occur in all regression equations with alternate signs which indicates an optimum length for the R^4 substituent. We refrained, however, from calculating optimum values. The reason for this is that the slopes of (L-R^4)^2 in all equations are rather low (- 0.5 to - 0.008). In some Equations, B5-R^4 contributes to improve the statistics. Remarkbly, the B5-R^4 term has a negative sign in all mutants apart from the leu275phe mutant.

Another aspect is the behaviour of 2-bromo-4-nitro- and 2,4-dinitrophenols with linear alkyl chains from C_1 to C_{12}. Parameters homogeneously increasing with the C-number like the STERIMOL parameter L, logP, and CLOGP, show a good parabolic relationships to wildtype and mutant pI_{50} values in most cases. In some mutants a better correlation is obtained with the third power of L, logP, or CLOGP.

Molecular Modelling. For many compounds that exhibited extreme values of either enhanced or decreased sensitivity in the mutants, molecular modeling studies were carried out. Some examples were selected and their structures with their pI_{50} values are shown in Table IX. The models of the inhibitors sitting in the 47 amino acids of the binding niche are shown in Figures 2 to 7, again in wildtype and the corresponding mutants. Some notions become obvious already from the few models:

Table V. Regression Equations of 27 2-Halo-4-nitrophenols (from Table I)

Equ.	Mutant	pI_{50} = n-R^2	Ph-R^4	logP	L-R^4	(L-R^4)^2	B5-R^4	Intercept	F	r^2	s	ΔpI_{50}	s/ΔpI_{50}
1.	wt		+ 1.57	+ 2.59	+ 0.22	- 0.05	- 0.30	- 0.98	48	0.92	0.37	4.0	< 0.1
			(6.63)	(11.8)	(1.44)	(- 5.77)	(- 2.72)						
2.	val219ile	+ 0.95	+ 0.71	+ 1.94	+ 0.02	- 0.02	- 0.39	+ 1.01	22	0.87	0.44	4.0	> 0.1
		(2.99)	(2.51)	(6.11)	(0.09)	(- 1.83)	(- 2.96)						
3.	ala251val	- 0.19	+ 1.10	+ 1.59	+ 0.69	- 0.06	- 0.14	+ 0.09	36	0.92	0.36	3.8	< 0.1
		(- 0.74)	(4.74)	(6.06)	(4.62)	(- 6.99)	(- 1.23)						
4.	phe255tyr	+ 0.52	+ 1.17	+ 1.52	+ 0.51	- 0.05		+ 0.003	29	0.87	0.42	3.5	> 0.1
		(1.77)	(4.40)	(5.98)	(2.70)	(- 5.53)							
5.	ser264ala	+ 0.20	+ 0.96	+ 1.56	+ 0.28	- 0.04	- 0.14	+ 1.82	27	0.89	0.32	3.5	< 0.1
		(0.88)	(4.67)	(6.70)	(1.91)	(- 5.05)	(- 1.43)						
6.	leu275phe	+ 1.16	+ 0.02	- 0.42	+ 1.06	- 0.04	+ 0.22	+ 3.15	30	0.90	0.34	3.3	> 0.1
		(4.66)	(0.95)	(- 1.70)	(6.87)	(- 5.68)	(2.12)						

Table VI. Regressions Equations of 17 2,4-Dinitrophenols (from Table II)

Equ.	Mutant	pI_{50} = n-R^2	logP	L-R^4	(L-R^4)^2	B5-R^4	Intercept	F	r^2	s	ΔpI_{50}	s/ΔpI_{50}
7.	WT	+ 1.78 (5.44)	+ 1.02 (4.83)	+ 0.55 (3.80)	- 0.03 (- 3.47)	- 0.32 (- 1.96)	+ 1.13	38	0.95	0.30	3.5	< 0.1
8.	val219ile	+ 2.34 (9.06)	+0.94 (5.25)	+ 0.06 (0.66)	- 0.008 (- 1.64)		+ 1.72	85	0.97	0.24	3.4	< 0.1
9.	ala251val	+ 1.13 (3.00)	+ 0.98 (3.12)	+ 0.45 (2.68)	- 0.03 (- 3.25)	- 0.14 (- 0.72)	+ 2.07	22	0.92	0.35	3.7	< 0.1
10.	phe255tyr	+ 0.95 (2.06)	+ 0.44 (1.61)	+ 0.23 (1.47)	+ 0.002 (0.21)	- 0.31 (- 1.38)	+ 3.97	10	0.82	0.42	2.5	> 0.1
11.	ser264ala	+ 1.27 (6.02)	+ 0.94 (5.47)	+ 0.53 (5.69)	- 0.02 (- 4.46)	- 0.38 (- 3.62)	+ 2.08	79	0.97	0.19	3.2	< 0.1
12.	leu275phe	+ 1.00 (2.41)	+ 0.07 (0.21)	+ 0.15 (1.08)	- 0.01 (- 1.14)	+ 0.41 (2.04)	+ 3.71	22	0.91	0.38	3.2	> 0.1

Table VII. Correlation Matrix of 27 2-Halo-4-dinitrophenols (Table I)

WT	1.00												
n-R_2	0.28	1.00											
Ph-R_4	0.15	-0.17	1.00										
logP	0.64	-0.16	-0.22	1.00									
CLOGP	0.53	-0.16	-0.17	0.93	1.00								
CMR	0.60	-0.16	0.11	0.85	0.95	1.00							
π_1	0.07	-0.49	0.14	0.13	0.18	0.24	1.00						
L-R^1	****	-0.17	-0.03	-0.17	0.15	0.21	0.39	1.00					
B1-R^1	0.06	-0.50	0.11	0.16	0.21	0.25	0.99	0.39	1.00				
B5-R^1	0.05	-0.51	0.19	0.13	0.18	0.24	0.98	0.39	0.99	1.00			
L-R^4	0.48	-0.30	-0.05	0.93	0.92	0.87	0.17	-0.08	-0.20	0.18	1.00		
B1-R^4	0.19	0.63	-0.17	-0.23	-0.31	-0.29	-0.49	-0.12	-0.50	-0.50	-0.47	1.00	
B5-R^4	0.50	-0.24	-0.11	0.94	0.91	0.85	0.16	-0.15	-0.19	0.15	0.93	-0.43	1.00

Table VIII. Correlation Matrix of 17 2,4-Dinitrophenols (Table II)

	WT	n-R^2	Ph-R^4	logP	CLOGP	CMR	L-R^4	B1-R^4	B5-R^4
WT	1.00								
n-R^2	0.21	1.00							
Ph-R^4	****	-0.06	1.00						
logP	0.84	-0.18	-0.21	1.00					
CLOGP	0.86	-0.14	-0.17	1.00	1.00				
CMR	0.89	-0.04	-0.03	0.96	0.97	1.00			
L-R^4	0.80	-0.24	-0.07	0.96	0.97	0.94	1.00		
B1-R^4	0.10	0.64	0.00	-0.27	-0.27	-0.17	-0.38	1.00	
B5-R^4	0.77	-0.20	-0.21	0.97	0.97	0.92	0.96	-0.41	1.00

1. In two of the three reported cases, the 4-nitro group forms hydrogen bonds to histidine$_{215}$. This is consistent with most other examples not shown here. 2. The OH of serine$_{264}$ is apparently irrelevant in the binding of phenols whereas it forms a hydrogen bond to plastoquinone and the "classical" inhibitors. 3. It is not possible to tell from the comparison of wildtype and mutant models which compounds show tolerance and which supersensitivity. Therefore we attempted to carry out energy calculations.

Table IX. Molecular Modeling of Phenols

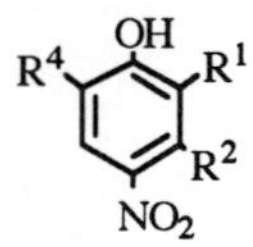

Mutant	Inhibitor No.	R^1	R^2	R^3	R^4	pI_{50wt}	$pI_{50mutant}$	ΔpI_{50}
Ala251Val	6	Br	Me	NO_2	$CHMe_2$	7.3	6.4	+ 0.9
Ala251Val	38	NO_2	H	NO_2	$(CH_2)_7H$	7.0	7.8	- 0.8
Val219Ile	27	Br	H	NO_2	CH_2Ph	7.2	5.5	+ 1.7

Energy Calculations. Figure 8 shows a rough scheme of the theoretical basis for the prediction of either enhanced or decreased sensitivity. The terms used in Fig. 8 and Table X have the following meaning:

E	=	steric energy
I	=	steric energy of the phenolic inhibitor
P_{wt}	=	steric energy of the wildtype D1 protein
P_{mut}	=	steric energy of the mutant D1 protein
ΔE_{wt}	=	gain of steric energy on formation of the complex $P_{wt} \cdot I$
ΔE_{mut}	=	gain of steric energy on formation of the complex $P_{mut} \cdot I$

The dimension of the energy data is $kcal \cdot mol^{-1}$.

These energy calculations suffer from several simplifications. First of all, they do not account for molecular dynamics, that means they are only approximations of the steric energy of the system without taking entropy into consideration. The minimizations of I, P_{wt}, P_{mut} and the corresponding complexes are not fully converged. They were broken off when the average force was $\leq$ 0.05 kcal/Mol•Å which usually took 3000 - 5000 iterations. The results for some selected compounds are shown in Table X.

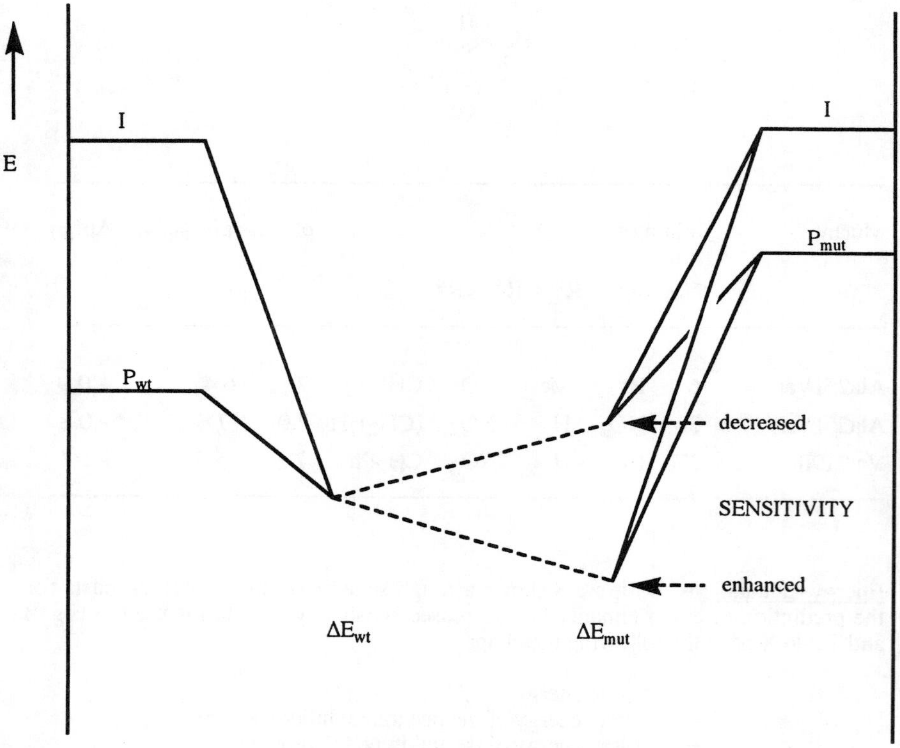

Figure 8. Energy Scheme to Symbolize the Origin of Decreased and Enhanced Sensitivity

Table X. Energy Calculations of 7 Selected Nitrophenols

Mutant	Inhibitor No.	R^1	R^2	R^3	R^4	$pI_{50}wt$	$pI_{50mutant}$	ΔpI_{50}	ΔE_{wt}	ΔE_{mut}	$\Delta E_{wt} - \Delta E_{mut}$	Accordance with Fig. 8?
Ala251Val	6	Br	Me	NO_2	$CHMe_2$	7.3	6.4	+ 0.9	- 30.25	- 27.67	- 2.58	yes
Ala251Val	38	NO_2	H	NO_2	$(CH_2)_7H$	7.0	7.8	- 0.8	- 23.62	- 24.00	+ 0.38	yes
Leu275Phe	13	Br	H	NO_2	CMe_3	7.2	5.8	+ 1.4	- 27.87	- 28.38	+ 0.51	no
Leu275Phe	17	Br	H	NO_2	$(CH_2)_7H$	7.5	8.5	- 1.0	- 28.56	- 28.21	- 0.35	no
Val219Ile	27	Br	H	NO_2	CH_2Ph	7.2	5.5	+ 1.7	- 38.03	38.67	+ 0.64	no
Ser264Ala	43	NO_2	H	NO_2	$c\text{-}C_6H_{11}$	6.5	7.3	- 0.8	- 25.76	- 26.75	+ 0.99	yes
Ser264Ala	3	Br	H	NO_2	Me	4.4	6.3	- 1.9	- 15.03	- 18.98	+ 3.95	yes

Table X are does not look too encouraging. A score of only four qualitatively correct results out of seven calculated ones shows that the simplifications are rather gross. They are probably a consequence of the negligence of entropy effects. In addition, they break-off limits are likely too high, i. e., the energy minimizations are still far from convergence. Since the procedure necessecary for calculations of ΔG is time-consuming, they will be published later.

Conclusion

By comparing the pI_{50} values measured in wildtype and five mutants in thylakoid membranes of *Chlamydomas rh.*, we have shown that also phenols enter the binding niche of the inhibitor binding D1 protein. They behave different from the classical inhibitors, especially towards the val219ile and the ser264ala mutant. In contrast to inhibitors like diuron and atrazine, many 4-nitrophenols display tolerance in the val219ile and supersensitivity in the ser264ala mutant. Structure-activity studies indicate that steric parameters are important for inhibitory activity of the phenols, especially the chain length expressed as the STERIMOL parameter L. Moreover, the partition coefficient is found to be indispensible in the regression analyses of wildtype and all mutant pI_{50} values. Molecular modelling studies suggest how the phenols fit into the binding niche of the D1 protein, though it is not possible to tell from these models which compounds are either tolerant or supersensitive in the diverse mutants. Energy calculations were carried out to answer this question, but the first ones were of limited assistance only because they involved too many simplifications.

Literature Cited

1. Velthuys, B. R. *FEBS Lett.* 1981, *126*, 277.
2. Trebst, A.; Draber, W., in *Advances in Pesticide Science*; Geissbühler, H., Ed.; Pergamon Press, Oxford, 1979, Part 3; 223-234.
3. Trebst, A. *Z. Naturforsch.* 1987, *42c*, 742.
4. Verloop, A.; Hoogenstraten, W.; Tipker, J., in *Drug Design* ; Ariens, E. J., Ed.; Academic Press, New York, 1976, Vol. VII; 165-207.
5. Tipker, J.; Verloop, A., in *Pesticide Synthesis through Rational Approaches*, Magee, P. S.; Kohn, G. K.; Menn, J.J., Eds.; ACS Symp. Ser. 255, Washington D.C., 1984; 279.
6. Deisenhofer, J., Epp, O., Miki, A., Huber, R., Michel, H. *J. Mol. Biol.* 1984, *180*, 385-398.
7. Wildner, G. F., Heisterkamp, U., Trebst, A. *Z.Naturforsch.* 1990,*45c*, 1142-1150.
8. Galloway, R. E., Mets, L. *J. Plant Physiol.* 1984, *74*, 469.
9. Erickson, J. M., Rahire, M., Rochaix, J.-D., Mets, L. *Science* 1985, *228*, 204.
10. Johanningmeier, U., Bodner, U., Wildner, G.F. *FEBS Lett.* 1987, *211*, 221.
11. Medchem Software, Daylight Chemical Information Systems, Irvine, CA 92417, USA.
12. Verloop, A. *The STERIMOL Approach in Drug Design*; Marcel Dekker, New York, 1987.
13. Trebst, A., Kluth, J.F. ,Tietjen, K., Draber, W., in *Pesticide Chemistry, Advances in International Research, Development and Legislation*, Frehse, H., Ed.; VCH Weinheim 1991, p. 111-120.
14. Tietjen, K., Kluth, J. F., Andree, R., Haug, M., Lindig, M., Müller, K.H., Wroblowsky, H. J., Trebst, A. *Pestic. Sci.* 1991, *31*, 65.

RECEIVED April 30, 1993

REGISTRATION OF BIOPESTICIDES

Chapter 33

Registration of Biopesticides

Jack R. Plimmer[1] and Richard M. Parry Jr.[2]

[1]ABC Laboratories, P.O. Box 1097, Columbia, MO 65205
[2]Agricultural Research Service, U.S. Department of Agriculture, Building 005, BARC-W, Beltsville, MD 20705

The Environmental Protection Agency (EPA) has encouraged the development of pest management techniques that offer alternatives to conventional chemical control. Many natural products and biologically derived pest management agents differ significantly from conventional chemical pesticides in chemistry, complexity, and mode of action. However, such differences are difficult to categorize in terms of relative risk and few regulatory precedents exist. This has created difficulties for the regulatory community which makes the decision-making process sometimes appear unduly slow. Progress is being made as more biopesticides are registered and experience of their use has accumulated.

Increased emphasis on pest management techniques based on substances of natural origin has been strongly advocated. However, the application of science to practical ends is subject to economic and regulatory constraints.

This discussion will be concerned with the interaction between regulation and development of newer pest control technologies based on "biochemicals". The present status of regulations and the experiences of the private sector in exploiting new technologies with particular reference to regulatory and public policy aspects are important factors in the acceptance of new techniques.

Novel pest control technologies based on natural products or products derived from biological processes appear to provide promising alternatives to conventional chemical pesticides. However, new techniques continue to receive a detailed scrutiny from regulatory authorities because of the lack of experience associated with their use. The impossibility of absolute guarantees that there would be no adverse, unanticipated effects on non-target organisms and the environment places innovators and policy makers in a difficult

0097–6156/94/0551–0474$06.00/0

situation. Although traditional techniques of risk assessment are well suited to conventional pesticide chemicals, there is no comparable well-established framework in which to assess the potential consequences of adopting novel approaches to pest management.

Legislative decisions are delayed by continued controversies over the subtle, long-term effects of using conventional pesticides because the identification and verification of such effects demand scarce resources and require a long-term commitment of expertise. The debate over safety issues has continued to intensify.

Pesticide Regulation

New scientific concepts challenge the regulator who must foresee all the implications of new technology without a substantial body of knowledge or historical experience on which predictions can be based. The challenges of biotechnology present policy makers with difficult issues that must eventually be translated into regulation. Such regulation must be framed to satisfy public concerns without inhibiting unduly the development of potentially beneficial and commercially viable technologies. As an example, the potential consequences of the escape of genetically modified organisms into the environment caused severe apprehension among many scientists and the public. The practical application of this technology must be accompanied by many safeguards to reduce the possibility of escape until there is sufficient experience to ensure that the technology is totally benevolent.

Using natural products in pest control faces several obstacles. Careful consideration of the safety issue is important because the development of new technologies and their ultimate commercialization may be impeded by the impact of regulation. There has been substantial investment in research on alternative methods of pest control. Investigators in the public sector who have often lacked resources to perform the safety tests required for registration have conducted a substantial portion of the research and developmental work. To transfer this technology successfully from the public to the private sector will depend on its role in a competitive process and ultimately on its profitability. Appropriately, US Department of Agriculture (USDA) and EPA participants in the session addressed the questions of the regulations, their current status, and the difficulties in progressing from the developmental stage to commercial exploitation.

The use of natural products as pesticides to control pests has a long history. For example, nicotine-based preparations and pyrethrum extracts have proved valuable for certain applications and they differ little in their mode of use from conventional insecticides. From a regulatory standpoint, they merit a similar approach to safety assessment. However, in recent decades approaches to pest control strategies have undergone radical changes.

New techniques for pest management have employed parasites, predators, sterile-insect release, behavioral chemicals, viruses, microorganisms, and other approaches based on new biological knowledge. The use of a variety of techniques in combination has been particularly useful in combating the growing resistance of insects to some conventional pesticides; the concept of integrated pest management (IPM) has now become a federal commitment. IPM may entail combinations of appropriate chemical, biological, cultural, and other techniques and in a particular crop or cultivation system. Its implementation is based on limiting pest damage to an economic threshold level. Thus, IPM now plays an important part in regulatory considerations and EPA considers the potential value of new pest control agents in this context.

Regulation intended to promote the use of pest control agents that act by other mechanisms than innate toxicity (biorationals: microbial and biochemical agents) was introduced in 1980.(1) Discussion during the session was, in part, a reflection of the experience of registrants and those involved in the development of candidate materials in working under regulations that were intended to facilitate the new technologies.

The major classes of biochemical agents are: semiochemicals, hormones, and natural plant regulators, allelopathic materials produced in the plant, and enzymes. Biological control agents including bacteria, fungi, viruses, and yeasts are regulated as pesticides. All these agents are generally referred to as biorational pesticides by the EPA.

Protection of the Environment

The guidelines for registering biorational pesticides evolved partly as a response to the increased understanding of biological processes. The isolation and identification of naturally occurring compounds that affected insect behavior and development at extremely low dose rates were followed by proposals that special consideration be given to such chemicals in pest management because of their negligible potential as environmental pollutants, the low amounts needed for biological efficacy, and their low mammalian toxicity.

The evolution of chemical pest control agents is paralleled by the history of regulatory measures. Initially, pesticides were regulated by the Federal Insecticide, Fungicide, and Rodenticide Act (FIFRA) (2) that set standards for pesticides moving in interstate commerce. In 1938, the passage of the Federal Food, Drug, and Cosmetic Act (3) added legislation intended to protect the public from chemical residues in food. Later, FIFRA was amended to protect not only man, but also the environment itself from any adverse effects associated with pesticide use.

During the early 1960s, as a consequence of the combination of scientific progress and increasing popular awareness of the issues, public attention became focused on the impact of pesticides on the environment. Scientific progress offered new instrumental methods of analysis, such as gas chromatography, which greatly facilitated the routine measurement of trace contaminants. At the same time, a number of popular books and articles drew wide attention to the ubiquitous presence of persistent chemicals in the environment and their potential implications.(4) Widespread use of the persistent organochlorine insecticides had distributed their residues throughout the globe and it was clear that residues were accumulating in some biota. There was considerable potential for their entry into the human diet from a variety of sources.

Until that time, policies relating to pesticide sale and use had been largely dominated by the concept of "economic poisons", and regulations were influenced by consideration of the types of compound that had been developed in the years following the Second World War. Toxicity to man and animals were the major factors weighed in assessing safety.

Data from measurements of environmental matrices demonstrated that the environmental burden of some pesticides was increasing and that action was necessary to alleviate the potential threat of increased human and animal exposure from this source. From the regulatory point of view, an initial step was taken when the PR 70-15 notice was issued by the USDA requesting that pesticide registrants provide basic information on the environmental fate of pesticides as part of the data submitted for registration.

From 1964, the Secretaries of the Interior and of the Department of Health Education and Welfare provided advice on the questions of undue hazards to fish and wildlife and to human health and, until 1972, the responsibility for pesticide registration resided in the Secretary of Agriculture. The provisions of this Act have changed as it has evolved from the 1947 version. The responsibility for implementing the regulations was transferred from the Secretary of Agriculture to the Administrator of the Environmental Protection Agency under the Federal Environment Pest Control Act of 1972 (FEPCA).(5) This Act was basically an amendment of FIFRA.

There were a number of responses to the environmental challenges. Regulators began to frame laws to protect the environment by requiring that new chemicals be subjected to more stringent testing to ensure that their potential for harmful environmental effects would be recognized. The conclusions of this assessment could then be used as a reason for denying approval or restricting their use. The law is framed in such a way that new chemicals are presumed to present substantial risk to the environment. It is the responsibility of the registrant to rebut the presumption of risk with appropriate data.

Industrial and academic scientists sought for new types of molecular structures that would be "environmentally friendly" by satisfying a number of criteria, particularly by their ability to be readily degraded by microorganisms present in soil. In addition, alternative control techniques employing microorganisms, viruses, parasitic organisms, etc., received greater attention.

Although potential combinations of techniques, whether chemical or cultural, to minimize the impact of pests on agricultural systems and IPM concepts offered promise for future pest management systems, the complexities involved created difficulties for regulators whose jurisdiction or expertise may be limited to certain components of the technology. Unfortunately, the regulatory oversight of the manufacture, distribution, and use of synthetic organic chemical pesticides became the paradigm that has dominated the formative stages of regulatory philosophies.

Technologies that did not conform to this model faced regulatory obstacles that have jeopardized their acceptance and practical application. Although many of the newer pest control agents were developed in response to the need for specifically targeted and rapidly biodegradable compounds, several considerations limited their potential for development in the rivate sector. Economic constraints generally dictated that new pesticide development should be targeted to major markets. The cost of development, registration, and marketing in a worldwide competitive market limited the number of corporations who could afford the initial investment. Costs of registration, particularly safety tests, grew to such an extent that there was little incentive for private industry to develop other than broad-spectrum pesticides for use on major crops.

Within the public sector and in industry, there had been substantial progress toward meeting some of the goals set by environmentally conscious groups; however, the prospects of economic success were hampered by the substantial costs of performing the safety tests required for registration.

"Third Generation Pesticides". Insect juvenile hormones (JHs) and semiochemicals may serve as two classes of compounds that exemplify the regulatory dilemmas. JHs are naturally occurring compounds that regulate the developmental stages of the insect. Their presence, in the correct quantities, regulate normal development of the insect as it passes from the egg to the mature adult stage. This physiological process became a target of investigation because interference with the mechanism would lead to abnormal development and interfere with the growth and reproduction of insect pests. Control methods based on such interference would act very specifically on insects as targets. The structure of the naturally occurring hormones was elucidated in a brilliant series of investigations and analogues of the natural hormones were synthesized. Some of the JH analogues had potent effects on insect development and became candidates for use as pest control agents. Such strategies were described as "biorational" approaches.

To accommodate "third generation" pesticides and chemicals derived from studies of biochemical pathways, the EPA issued guidelines (40 CFR Part

158 establishes data requirements for registration and "Subdivision M: Guidelines for Testing Biorational Pesticides" describes protocols that may be used for testing biochemical and microbial pest control agents).(6) The EPA drew several distinctions between conventional pesticides and biorationals. The latter were characterized by "unique non-toxic mode of action, low use volume, target species specificity and natural occurrence". They comprised two major categories: the "biochemical pest control agents (e.g., pheromones, natural insect and plant growth regulators and enzymes)" and the microbial pest control agents (e.g., microorganisms)". The caveat that the chemical must be naturally occurring or structurally identical with the naturally occurring material if it were a product of synthesis eliminated JH analogues and analogues of semiochemicals from consideration. Minor differences in stereochemical isomer ratios would, however, not normally eliminate a classification as a biorational unless the isomers showed significantly different toxicological properties.

Although analogues would be excluded by the above criteria, the Agency stated that these would be considered on a case-by-case basis. Chemicals similar to biorationals would be classified as biorationals or conventional pesticides based their chemical structure and mode of action in the target species. The Agency will base its decision on the chemical and toxicological significance of differences in chemical structure, mode of action of synthetic analogues compared with those of the naturally occurring compounds, and differences in toxicity (in at least Tier I screening tests for biorational pesticides). Direct toxicity to the non-target organism based on Tier I testing may or may not exclude classification as a biorational pesticide.

The biochemical agents are classified by the Agency as semiochemicals, hormones, natural plant regulators,and enzymes. Microbial pest control agents include bacteria, fungi, viruses, and protozoans. The guidelines apply to strain-improved as well as naturally occurring organisms and the Agency proposed to determine data required for genetically engineered microbials on a case-by-case basis except where Part 158 contained specific requirements.

Organisms used for pest control such as insect predators, nematodes, and macroscopic parasites do not fall within the scope of these guidelines as they are not considered biorationals and are exempt from FIFRA requirements. These organisms are subject to regulation by the Animal and Plant Health Inspection Service of the USDA (APHIS) under the Plant Pest Act.

Risk Assessment and Data Needs

Considerable knowledge of the consequences of widespread use of chemical pesticides has accumulated and has frequently been the subject of public debate. The position of the regulatory agencies must accommodate both technical needs and public opinion. The agencies have promulgated methods of assessment of risks and benefits and their guidelines have generally been recognized as providing valuable data for acceptance or denial of registration of the product. Among the pesticide industry's major criticisms of the guidelines are the costs of such test programs and the delays in bringing a new product to market.

These costs and delays are passed on to the food producer or processor as well as the consumer. New products become more costly and the reduced pace at which new pesticides are approved means that some producers will be unable to obtain pesticides to meet specialized needs. Such specialized needs are met by a USDA program that assists in registering pesticides for "minor uses". University and government laboratories are obtaining residue data that will allow the manufacturer to add additional uses to the product label.

Although the regulatory agencies have attempted to improve the reliability of methods for predicting the adverse effects from pesticide use, such assessments can never be totally adequate because of the incomplete knowledge of the biological systems in which they will be used. Decisions made are based on the state-of-the-art and future developments may reveal in hindsight that the decisions were flawed. As an example, the extent to which ground water might become contaminated by pesticides was unsuspected until extremely sensitive analytical methods were developed and systematic programs of sampling and analysis were initiated. Many questions, particularly those relating to the long-term ecological implications of pesticide use, must remain unanswered. The complexity of the factors influencing large-scale, long-term ecological changes and the quantitative and qualitative nature of such changes generally restricts such studies to the few institutions that possess considerable resources including the variety of intellectual skills, experimental techniques, and financial resources required for the investigation and the interpretation of the considerable accumulation of data that results.

With the costs and delays of registering conventional pesticides in mind, the EPA set up a scheme for registering pesticides of biological or biochemical origin. The tiered approach to testing was proposed and the EPA adopted the philosophy of maximum hazard testing. That is, if no adverse effects were revealed in the first tier of tests, the substance could be deemed safe and further tests would not be called for. This system was applied to some natural products, including plant growth regulators, semiochemicals, JHs and to all microorganisms. The original guidelines were issued as a single volume as biorational guidelines, but they have been updated and the portion on microorganisms appeared in 1989.(1) The biochemical section is still undergoing review.

Subdivision M — testing requirements.

EPA uses a maximum hazard testing concept in designing its toxicology and nontarget risk assessment strategy. The first tier of test requirements embodies the maximum hazard challenge in terms of dose, concentration, and route of administration. If the results of the first tier tests indicate no adverse effects, data from the second and third tiers will not be required.

Other data requirements include:

1. Product analysis
2. Residue chemistry
3. Toxicology
4. Nontarget organism hazard and environmental fate and expression

EPA assumes that the potential for exposure of human or nontarget organisms (hazard) to biochemicals is often very limited. EPA's criteria for reduced data requirements are low exposure pesticide formulation (traps, controlled release formulations, etc.), low rates of application ($<$20 g a.i. per acre), nonaquatic use sites (applied directly to land), and high volatility (reduced likelihood of residues on food or feed crops).

The product analysis data needed are substantially similar to those described in Subdivision D of the Pesticide Assessment Guidelines. Residue data may not be required for pheromones where rates of application are below 20 g a. i. per acre and Tier II and Tier III toxicology is not required. Pheromone products have generally been granted exemptions from tolerance requirements on a case by case basis.

The toxicity tests may be summarized:

Tier I Acute toxicity: Oral, dermal, and inhalation; Irritation — primary ocular, primary dermal; Hypersensitivity: immediate, nonimmediate; Genotoxicity; Cellular Immune Response.

Tier II Oncogenicity: Toxicity: subchronic oral, dermal, and inhalation; Cellular Immune Response; Teratogenicity

Tier III: Oncogenicity; Long term toxicity tests in two or more species.

Each tier employs different test species. Data on hazards to nontarget organisms proposed in Subdivision M cover the four areas: terrestrial wildlife, aquatic animals, plants, and beneficial insects. The testing follows a tier scheme analogous to that for toxicological tests. Tier I represents the first level hazard test and Tier II consists of environmental fate studies that will be required only if indicated by the outcome of Tier I tests.

Field tests carried out to determine efficacy trigger the requirement for an EPA Experimental Use Permit (EUP) if they are conducted on $\geq$10 cumulative acres of land or $\geq$1 surface-acre of water. Any crops grown on these sites must be destroyed or used for feeding experimental animals unless EPA grants an exemption from tolerance. Data that must be submitted for an EUP include product chemistry and toxicological testing. This acreage restriction has created difficulties for testing the efficacy of pheromones because the determination of efficacy has depended on assessments of population reduction or damage to crops or plants and it has proved difficult to obtain statistically satisfactory data within the small acreages prescribed.

Data Requirements

Categorization of the mode of action is a major factor in determining whether the registration of a naturally occurring substance should be considered under the provisions of Subdivision M or whether it must be treated as a conventional pesticide. The potential for effects on nontarget organism is much less if the compound shows extremely specific activity or its action is limited to specific sites peculiar to a limited class of organisms. Modes of action other than direct toxicity include behavioral or developmental effects. The first is exemplified by the insect semiochemicals and the latter by insect growth regulators. The inherent toxicity of the active ingredient must be considered separately and the requirements to furnish data on the chemistry of the product are the same as for conventional chemicals.

The EPA may grant data waivers at the request of the registrant where it is clearly demonstrated that the requirement is inapplicable or the substance would not be expected to present a hazard. Data waivers are granted on a product by product basis. Although this collection of information concerning waivers would be useful for guiding the development of new pest control technologies, it is not available to the public. As noted previously, pheromones are used at extremely low rates of application and, in some situations, the amounts used would contribute little more to the environment than the amount that is present naturally. This is one of the rationalizations for exemption of pheromones from registration when used in traps.

Commercial Development

Commercial Prospects. From the point of view of commercial exploitation, biopesticide sales in 1990 were estimated at $120 million, representing less than 0.5% of the world agrochemical market. Over 90% of sales are accounted for by Bacillus thuringiensis products. However, sales are estimated to be increasing at 10-25% per annum while the world agrochemical market remains static (Rodgers: unpublished data, presented at Society of Chemical Industry Symposium, London, UK, December 1992).

The Role of the Small Business. Although major agrochemical companies have expressed interest in and support for the development of biorational pesticides, their efforts have been limited. Their goals are generally control of pests of major crops. Weed control has been a primary concern and this has stimulated developments in conventional herbicides and also in biotechnology. Such strategies are within the capability of large corporations.

The market for special control techniques that are appropriate for integrated pest management systems is closer to the scope of smaller companies where large-scale production facilities are not essential. Smaller companies have found unique niches in the pest management field, based often on particular expertise or experience. Controlled-release technology is an essential component for formulation and application of pheromones, therefore small companies have entered the market on the strength of this combination of expertise. Patent protection may be obtained to cover the devices and materials used to formulate pheromones, whereas the information relating to the composition of many pheromones is available in the open literature.

Development of safer alternative technologies is a very risky business venture, small companies and venture capital investors appear to dominate the development of these products that are unproven in the market place. Although the climate appears more favorable for such ventures as chemicals are removed from the market place, the goal of business is commercial success.

Raw Materials and Production. Some factors that favor the involvement of small companies are the same factors that tend to limit the interest of major chemical companies. A constraint on the use of natural products as pest control agents (or pharmaceuticals) is the identification of sources of raw materials. It may be difficult to obtain consistent sources of botanical materials as these are often grown in less accessible regions or nontemperate climatic conditions.(7,8) Plant extracts may give varying yields of active principles, depending on climate and location in which the plants were grown. Botanical sources must also be replenished. The alternative approaches of synthesis or cell culture may be more practical, if production costs can ultimately be recovered. If the target pest is quite specific or limited to a few species, production of the active ingredient may be so costly as to preclude commercial development. This consideration also applies to the formulation of new pest control agents. If the agent is extremely labile, the formulation may represent a considerable additional development cost.

For example, the delivery of viable microorganisms or viruses to the target organisms may require special formulations. It is particularly important to protect against the action of sunlight, which rapidly reduces activity and causes variability in field performance. The problems associated with manufacture and formulation of pheromones, discussed earlier, represent additional limitations on the market potential of these chemicals.

From the regulatory standpoint, as stated above, the product chemistry requirements are described in Subdivision D of the Pesticide Assessment Guidelines.(9)

Selected Cases

Insect Growth Regulators — Neem. The registration of extracts of neem (Azadirachta indica A. Juss) is an example of the special consideration given to antifeedant and hormonal activity of the active ingredients. The effect on nontarget species, particularly beneficial insects and pollinators, is also given weight in granting registration. The neem tree is a source of biologically active materials that are finding applications in pest control. There is a long history associated with the use of this plant that grows in many tropical regions of the world and it has been investigated by many groups attempting to link its chemical composition with its biological activity. Azadirachtin is one of the principal ingredients. Products based on azadirachtin are commercially available for horticulturalist and consumer (10). The activity of these extracts is based on their insect repellent and insect growth regulatory activities. Because their mode of action is not direct toxicity, they are considered as subject to regulation as biochemicals rather than as conventional pesticides.

Behavioral Chemicals — Semiochemicals. Semiochemicals are chemicals emitted by plants or animals that modify the behavior of receptor organisms of the same or different species. This definition includes pheromones, allomones, and kairomones. The sex-attractant pheromones are emitted by the female to attract the male for mating purposes and they have been used in pest control in the following ways (11):
(1) In insect traps for monitoring or survey purposes.
(2) In a mass trapping program to reduce insect populations.
(3) To attract insects to an area treated with insecticides or to a device containing insecticide.
(4) To permeate the air and suppress insect population by disrupting mating or aggregation.

These strategies have been applied experimentally and/or practically. Those of major lepidopteran pests have been identified (12, 13) and most of these may be classified as aldehydes, acetates, or alcohols derived from long-chain unsaturated hydrocarbons. The amounts used, usually several grams/hectare, are ordinarily so small as to have little potential to cause adverse environmental effects and the molecules are generally of low mammalian toxicity since they mimic the natural compounds and are rapidly degraded in the environment.

The uses are regulated by EPA. However, pheromones and substantially similar compounds are excluded from regulation if they are to be used in insect traps solely for detection or survey purposes even if a toxicant is included in the trap. If the pheromone plus toxicant is used in a device solely to control the pest, then registration is required.(14) This leads to a curious regulatory situation exemplified by the contrasting requirements for boll weevil survey traps and boll weevil bait sticks. Both contain the boll weevil pheromone (grandlure) plus an insecticide as toxicant; however, the former are exempt from regulation, whereas the latter must be registered as a pesticide.

Field testing of semiochemical products has been hampered by acreage limitations (< 10 acres unless under an EUP) and the requirement that a tolerance must be granted for the product on each crop to which it will be applied. To adequately test efficacy, several hundred acres of crop are required to avoid "border effects" from surrounding pest populations. However, if no tolerance has been granted the crop must be purchased and destroyed, even though extremely minute quantities of product are used. This requirement has limited, and in some cases prevented, development of semiochemicals by publicly funded scientists who do not have adequate resources to access the EPA regulatory system.

During the 1970s there was rapid progress in elucidating the structure and composition of pheromones. Improved syntheses of pure pheromone components reduced production costs although they have continued to be high cost, low volume products. Stability in the field, suitable formulations and effective delivery at the target site were the initial major constraints on pheromone use. Later, the questions of demonstrating efficacy and compliance with regulatory requirements became important issues - particularly when regulatory requirements placed limitations on field testing

To demonstrate the effectiveness of pheromones required data on reduction in crop damage or reduction in mating frequency among a wild population because they do not kill insects and are intended to suppress population by interference with mating communication. Experimental recommendations to demonstrate pheromone efficacy have appeared in a report of a Task Group sponsored by the American Institute of Biological Sciences.(15)

Pheromones are volatile and readily degraded in the environment, so that they must be protected during the prolonged period over which they remain in the field. The formulation must continue to permeate the air with an adequate quantity of pheromone vapor throughout the period when the male is actively seeking to mate. Controlled-release formulations protect the pheromones by incorporating them in polymeric material in the form of capsules, hollow fibers, granules, flakes, etc. A variety of controlled-release formulations have been evaluated and types currently registered are fabricated from inert materials that are generally innocuous to the environment, including some based on polymers approved for food contact uses.

The commercial potential of pheromones was actively investigated during the late 1970s and early 1980s. They began to be marketed for trapping and for mating suppression during the 1970s and several types of proprietary controlled-release formulations became available. By 1979, several pheromones were registered (muscalure, gossyplure, disparlure, japanese beetle pheromone, and grandlure). By 1983, the list also included artichoke plume moth pheromone, western pineshoot borer pheromone, peachtree borer pheromone, elm bark beetle pheromone, and tomato budworm (Heliothis virescens) pheromone. In view of their specificity and high biological activity, the amounts used for pest control were low. This means that commercial production will never achieve the scale of conventional insecticides, costs will remain high, and production will remain the province of a few specialist manufacturers. Efficacy became a major issue since the claim that pheromones suppressed insect population was very difficult to establish convincingly without conducting large-scale, expensive field tests. Suppression of population by disrupting mating communication appeared to be a promising strategy for control (16) but the technique applied optimally to low population because mating frequency could most effectively be reduced as natural sources of pheromone (calling females) were widely dispersed.

Although the mating suppression concept was well accepted, actual conditions required for success with a particular species were difficult to define. In every case, an important prerequisite was an understanding of the behavior and reproductive cycles of the target insect. It was also necessary to understand the behavior of the formulation and its persistence under the climatic conditions in the field.

The use of pheromones for population suppression was a research-intensive activity and efforts to transfer the technology to practical use have met with mixed success. In a discussion of the registration of pheromones in practice,(17) some of the outstanding issues are summarized. The first issue is that of excessive data requirements that do not appear justifiable in light of the low rates of use of pheromones, their low toxicity, and the low potential for human exposure. It is estimated that Tier I testing costs would reach hundreds of thousands of dollars and these tests are necessary before an EUP can be obtained. It is not feasible to demonstrate efficacy unless field tests can be conducted using the target insect populations, which makes it impossible to determine whether a product will ever be marketable. Another issue is the cost of materials used for toxicology testing. Quantities of 50-100 g are required for acute toxicological tests and up to one kg may be needed for subchronic inhalation studies. The cost of materials and the amounts necessary for conducting these tests are considerable and potentially greater than those required in field experiments.

Future Outlook

Alternative Pest Control Technology. Many chemical pesticides no longer in use are likely to have become unacceptable for one of the following reasons:

1. Long-term, low-level toxicity to mammals.
2. Adverse effects on the environment or wild life.
3. Development of resistance by the target species.

These factors are now well recognized as new pesticides are discovered and are important factors in designing new active ingredients. Additionally, pesticides are expensive and the costs of developing acceptable new pesticides and maintaining the registration of current products make it unlikely that costs will be considerably lower in future.

Concerns over the presence of pesticides in the human diet are currently influencing policies and the statement that the strategy of the USDA, the EPA, and the FDA is now to "create incentives for the development of safe pesticides and to remove those pesticides that pose the greatest risk from the market place" indicates changes in the policy of the Administration in relation to the use of chemicals in food production.(18)

The release of the long awaited National Academy of Sciences/National research council report entitled "Pesticides in the Diets of Infants and Children" on June 23, 1993 has generated uncertainty among the public concerning the safety of pesticides. Although no specific hazard was identified, the perception that concern exists stimulated government officials to use all possible measures to speed development of pest control methods that leave no chemical residues in food.

Easing the Way to Registration. Industry has recommended a number of changes in the approaches to pheromone registration.(16) These include the comment that the exemption for pheromones used in traps should be extended to trap cropping in which pheromone lures are used to attract an insect to an area where it may be killed by conventional pesticides. Another recommendation was that the EPA allow the registrant to submit data relating to similar pheromones. The similarities among many common lepidopteran pheromones, as for example Z-9-tetradecenal and Z-11-hexadecenal, suggest that some representative studies may be appropriate, rather than the case-by-case basis currently adopted. A further suggestion is that testing be conducted on the end-use product rather than on the active ingredient. The high cost of these materials is cited as a major factor in recommending this approach. Tests using active ingredients represent unrealistically high-exposure levels far greater than the recommended use rates.

Delays in the approval process are also cited as major obstacles. Delays interfere with the scheduling of field tests and may critically affect the ability to have tests in place at the appropriate time of the year.

The EUP restriction of 10 cumulative acres has already been cited as a potential major hurdle in the data-gathering process. A positive move by EPA would be action to increase the EUP trigger for pheromones to 500 acres for nonfood applications and 200 acres for food use. Because pheromones act to reduce insect population by nonlethal action, large acreages are essential for adequate, statistically based demonstrations of efficacy.

Prospects for Economic Benefit. As of May 1992, EPA registrations of biorationals included 14 pheromones, 6 plant growth regulators, 13 floral lures, 5 natural insect growth regulators, and 4 others (including diallyl sulfoxide, carboxymethyl cellulose, soybean oil as a miticide).

Literature Cited

1. Pesticide Assessment Guidelines Subdivision M: (Part A Microbial) (EPA No. 540/09-89-056), March 1989; USEPA; Washington DC; 1989.
2. Federal Insecticide, Fungicide, and Rodenticide Act, Public Law 152; U.S, Statutes at Large 1; pp 331-335; April 26 1910.
3. Federal Food, Drug, and Cosmetic Act, Public Law 95-532, U.S.C.A. Title 21, 321 A,B,C.
4. Carson, R., Silent Spring, Houghton Mifflin, Boston 1962, 368 pp.
5. Public Law 92-516, 86 Stat. 973 (Oct 21, 1972): U.S. Code Title 7 Sections 136-136y, 1974 ed.
6. Pesticide Assessment Guidelines Subdivision M: Microbial and Biochemical Pest Control Agents, Guidelines for Testing Biorational Pesticides; October 1982, (EPA No. 540/09-82-028); USEPA; Washington, DC; 1982.
7. Morgan, E.D., Johnson, S., Preliminary Investigation of Azadirachtin Biosynthesis in Neem Seeds, Abstract # 51, Division of Agrochemicals 205th. ACS National Meeting, Denver CO, 1993.
8. Benke, M.D., Germplasm Improvement and Diversification, and Propagation and Cultivation Techniques for Economic Development of the Neem Tree, Abstract # 50, Division of Agrochemicals 205th. ACS National Meeting , Denver CO, 1993.
9. Walter, J.F., Barnette, D.H., Hull, C.J., Dutton, W.R., Commercialization of Margosan-O and Bioneem, Azadirachtin Based Insecticides, Abstract # 56, Division of Agrochemicals 205th. ACS National Meeting , Denver CO, 1993.
10. Code of Federal Regulations, Title 40 Section 158.190
11. R. M. Silverstein, "Pheromones: Background and Potential for Use in Insect Pest Control", Science 213, 1981, 1326-1332.
12. Arn, H., Toth, M., Priesner, E., List of Sex Pheromones of Lepidoptera and Related Attractants; IOBC-WPRS; Swiss Federal Research Station, CH-8820, Wädenswil, Switzerland; 1986; 123 pp.
13. Inscoe, M. N. In Behavior-Modifying Chemicals for Insect Management; Ridgway R. L., Inscoe, M. N., Eds.; M. Dekker Inc.; New York, NY, 1990; pp 631-715.
14. Code of Federal Regulations, Title 40 Section 162.3 (ff)(3).
15. Roelofs W. L., Ed.; Establishing Efficacy of Sex Attractants and Disruptants for Insect Control; Entomological Society of America; College Park, MD, 1979; 97pp.
16. Knipling, E. The Basic Principles of Insect Suppression and Management; Agriculture Handbook 512; USDA; Washington D. C., 1979; 421-490.

17. O'Connor, III, C.A. In Behavior-Modifying Chemicals for Insect Management; Ridgway R. L., Inscoe, M. N., Eds.; M. Dekker Inc.; New York, NY, 1990; pp 605-618.
18. EPA Press Release, June 25, 1993.

RECEIVED October 15, 1993

Chapter 34

Managing the Regulatory Process

J. H. Davis and P. S. Carlson

Crop Genetics International, 10150 Old Columbia Road, Columbia, MD 21046

Development of new biopesticides requires a thorough understanding and appreciation of the regulatory process. Products developed using the tools of recombinant DNA technology are subject to special regulations that affect not only the final product approval but all phases of research and development. Many products may be subject to both federal and state scrutiny. Close consultation with regulatory officials early in the product development process is key to insuring that research activities are focused not just on product development issues but also on developing the data base necessary for initial field testing and final product approval. Managing the regulatory process also involves creating the framework for responding to public concerns raised by individual citizens, public interest groups and the press.

Regulation of the products of genetic engineering has been the focus of considerable governmental concern and public controversy. Both industrial and academic researchers have learned that the involvement of federal and state regulatory agencies is a key factor in the development of any new genetically-engineered biological pest control agent. To insure smooth product development, particularly in the early field testing phase, it is essential to manage the regulatory process.

Regulatory management is a three-step process. First, it is essential to understand the statutory requirements and the purpose of any applicable federal and state regulations. Second, one must develop the data base and scientific rationale necessary to demonstrate the safety of a product within the regulatory framework. Finally, one must have a proactive outreach program for dealing with public concerns.

0097–6156/94/0551–0490$06.00/0

The Regulatory Framework

In the development of genetically-engineered biological pest control agents, two federal agencies play key roles: the United States Environmental Protection Agency ("EPA") and the United States Department of Agriculture's Animal and Plant Health Inspection Service ("APHIS"). In addition to these federal agencies, state agencies also have played important roles in regulating the testing and development of genetically-engineered organisms.

EPA. The Environmental Protection Agency regulates genetically-engineered microorganisms under two federal statutes: the Federal Insecticide, Fungicide and Rodenticide Act ("FIFRA") and the Toxic Substances Control Act ("TSCA"). For most products of interest to agriculture (the main exception being genetically-engineered *Rhizobia*), FIFRA is the key statute.

FIFRA is a licensing statute that provides EPA with authority to regulate the testing, distribution, sale, and use of pesticides. Under FIFRA, a pesticide must be registered before it may be distributed or sold. Obtaining registration under FIFRA involves submitting sufficient data to EPA to demonstrate that use of the pesticide will not cause unreasonable adverse effects in the environment. To obtain the data necessary to support a pesticide registration, an applicant must field test the pesticide to determine its potential environmental effects as well as to collect data on its efficacy.

Before performing any extensive field testing of a pesticide, an applicant must obtain an Experimental Use Permit ("EUP") from EPA. The EUP must be supported by sufficient data to assure EPA that the proposed test will not cause any unreasonable adverse environmental or health effects. EPA regulations generally exempt small-scale studies (defined as tests involving less than 10 acres of land or less than 1 acre of surface water) from the EUP requirement, so that applicants can collect initial field data needed for the EUP and determine whether the pesticide warrants further testing. This exemption, however, does not apply to genetically-engineered biological pest control agents. These organisms are subject to review by EPA prior to any environmental release. Specifically, EPA requires either 30 or 90 day notice (depending on the nature of the organism to be tested) prior to any field experiment. Thus it is essential for an applicant to consult with EPA in the early planning stages of a proposed field release of a genetically-engineered biopesticide to determine what laboratory and greenhouse data the agency will need to permit the proposed test.

APHIS. The Animal and Plant Health Inspection Service regulates genetically-engineered agricultural products under the authority of the Federal Plant Pest Act and the Plant Quarantine Act. APHIS has promulgated special regulations for products altered or produced through genetic engineering. Unlike EPA, APHIS does not "license" or "register" products prior to use or sale. Rather, APHIS regulates the interstate movement of regulated articles (including any environmental or field releases of genetically-engineered organisms, even if interstate movement does not take place). Specifically, APHIS requires a permit for any interstate movement or release into the environment of any organism that has been altered or produced by genetic engineering if the new organism, or any part of it, is or was associated with a plant pest (e.g. an inserted gene from a donor organism that is a plant pest or a vector derived from a plant pest). Consulting with APHIS prior to any field testing of a genetically-engineered organism is necessary to determine if APHIS regulations apply. As in the case of EPA if a permit is required, the applicant must submit sufficient laboratory or greenhouse data to APHIS to assure the agency that the proposed test does not pose a risk to agriculture.

State Agencies. In addition to the two federal agencies discussed above, consultation with state regulatory officials prior to any field release of a genetically-engineered organism is essential. Several states have established special regulatory review procedures for field release of genetically-engineered organisms, which impose additional regulatory oversight prior to testing. Even in states without special regulations for genetically-engineered organisms, consultation with state officials is necessary to insure compliance with all state agricultural and environmental regulations.

Collecting the Data

In preparation for any field testing of a genetically-engineered organism, a researcher must collect data to address three fundamental questions. First, what is the likelihood that the engineered organism will spread from the test site? Second, what is the likelihood that the organism will persist in the environment? Third, what is the environmental or public health consequence if the organism does spread or persist?

To address these questions, one must conduct a series of tests designed to understand the basic biology and ecology of the unaltered organism, the impact genetic modification will have on the organism's behavior, and the steps that could be taken to reduce the potential for adverse environmental consequences. Studies on the unaltered organism likely will need to be conducted both in laboratory and greenhouse environments as well as in the field. (It is important to remember that APHIS permits may be required for field testing even of the unaltered organism if it could be considered a plant pest.) To supplement the studies on the unaltered organism, laboratory and greenhouse studies on the engineered organism must be conducted to demonstrate the impact genetic-engineering has had on the basic biology and ecological behavior of the host organism.

Prior to any extensive data collection to support a permit application to EPA, APHIS or a state regulatory agency, it is highly advisable to consult with the relevant agencies. The purpose of these consultations is fourfold: 1) to familiarize the government agencies with the purpose and design of the proposed field tests; 2) to determine the principal concerns, if any, of the regulators with the proposed test; 3) to identify the type of data that would address those concerns; and 4) to determine what, if any, containment or monitoring requirements will be needed for the proposed test. The answers to these questions will allow an applicant to file a complete permit application and eliminate the need for submitting additional data during the review process that can delay a permit's approval.

Not only must an applicant address the concerns raised by each regulatory agency, but the applicant must coordinate the often overlapping responsibilities of these agencies. Most importantly, the applicant must insure that information submitted to one agency is shared with all agencies. If additional data are submitted to one agency in response to questions raised by that agency, the same information should be submitted immediately to the other agencies (both federal and state) involved in the review process. Similarly, if changes in the design of a proposed test are requested or mandated by one agency, it is essential that these changes be reviewed and approved by all authorities including state agencies.

Dealing with the Public

As some of the early field tests of genetically-engineered organisms demonstrated, good public relations is as much a component of field testing as good science. To insure that the public understands the nature of the testing to be conducted and the benefits to society that can result from the development of new biological control agents, one must have an active outreach program. This program must address not just the concerns of national and local environmental groups but also the concerns of the local communities where the testing is to occur.

The most useful aspect of any outreach program is face-to-face meetings. For local communities, this involves meeting with each of the neighbors surrounding a proposed field test, holding public meetings at a local civic center or Chamber of Commerce, presenting seminars at local educational institutions and generally being available to address the concerns raised by local citizens. It is also essential to meet with relevant politicians on federal, state and local levels, so that they are in a position to answer inquiries from their constituents.

It is also important to address the concerns and questions of local and national environmental groups. To accomplish this, one must be willing to share with these groups all health and safety data that have been collected on the engineered organism. Excessive claims of confidentiality, no matter how well founded, will raise concerns from these groups that the public is not being fully informed.

The time and effort involved in developing a public outreach program from the very beginning of the testing process will prove invaluable as field testing expands into full-scale product development.

Preparing for the Future

One key component of managing the regulatory process should be to minimize the impact of regulatory restrictions on product development. Indeed, in conducting the initial field trial of any genetically-engineered organism, a researcher often finds that the majority of time spent on the project is focused on monitoring,

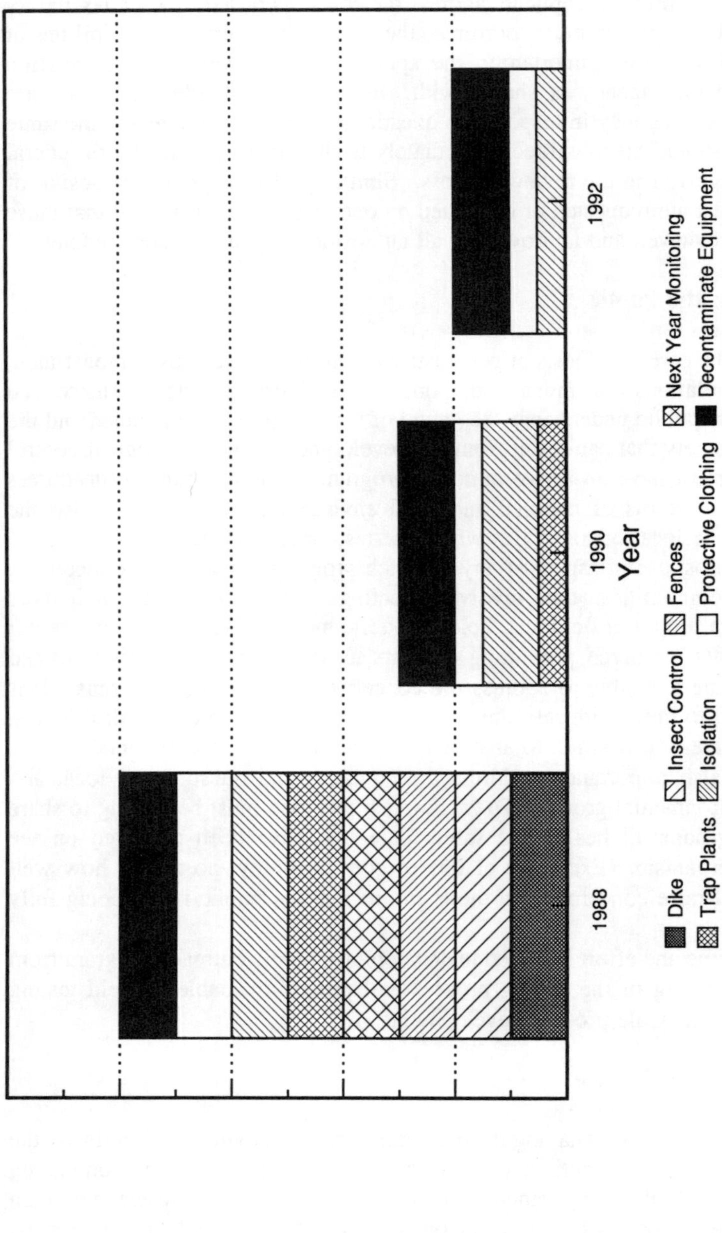

Figure 1. Environmental containment requirements for CGI field trials of Cxc/Bt - 1988-1992. Half-height bars indicate substantially reduced requirements from previous year.

decontamination, and other environmental containment activities and not on actual product evaluation and development. However, one goal of that initial field trial should be to collect sufficient data that future trials can focus more on product development and less on environmental concerns.

For example, since 1988 Crop Genetics International ("CGI") has been testing a microorganism, *Clavibacter xyli cynodontis ("Cxc")*, that has been genetically modified to produce the delta-endotoxin from *Bacillus thuringiensis ("Bt")*. This microorganism has been found to colonize the vascular system of corn plants and to protect those plants from damage caused by the European corn borer (*Ostrinia nubilalis*). During the first year of testing in 1988, CGI was required to perform a wide range of environmental containment procedures which included: 1) surrounding the plots with trap plants, fences and dikes; 2) isolating the plot from other corn growing in the area; 3) decontaminating all equipment leaving the site; 4) requiring researchers to wear protective clothing; and 5) monitoring the site the subsequent year. In addition nearly all the data collected that year were focused on environmental issues. Studies included soil persistence, presence and survival of the organism in runoff water, intraplot movement of the organism, the ability of the organism to overwinter, monitoring of trap plants, monitoring the persistence of the organism in crop residues, and the genetic stability of the transformed organism. However, based on the extensive data collected that first year, CGI was able to limit substantially the environmental containment requirements necessary for future tests and to reduce the time involved in environmental studies. Indeed as shown in Figures 1 and 2, by 1992, CGI's field studies were focused almost exclusively on product development with few regulatory requirements. The few remaining requirements are being reduced even further in 1993.

	1988	1989	1990	1991	1992
Soil Persistence	X				
Runoff Water	X				
Intraplot Spread	X	X			
Overwintering	X	X			
Trap Plants Monitoring	X	X	X		
Crop Residues	X	X	X		
Reversion	X	X	X	X	X
Product Efficacy			X	X	X

Figure 2. Field studies conducted during CGI field trials of Cxc/Bt - 1988-1992.

CGI's experience demonstrates that by working closely with the regulatory agencies in those initial field trials, one can substantially reduce the regulatory burden in future years.

Summary

Five years ago the field testing of genetically-engineered organisms was a novelty. Today field testing is becoming routine and federal agencies are moving to simplify the process. Nonetheless any field testing of a genetically-engineered organism requires careful planning and coordination. Early in the process one must identify the key agencies that will be involved in the review process and consult with them concerning the data needed to support any required permit applications. In collecting data the focus should be on the key environmental questions of spread, persistence, and potential risk if spread or persistence occurs. Once the data have been submitted to the federal and state agencies, continued coordination with all interested parties is essential. A proactive public outreach program will help to insure that the citizens and advocacy groups understand the risk and benefits to society of genetically-engineered biological pest control agents.

RECEIVED June 28, 1993

Chapter 35

Encouraging Research, Development, and Commercialization in Agricultural Biotechnology

Alvin L. Young, Daniel D. Jones, and James C. Staton

Office of Agricultural Biotechnology, U.S. Department of Agriculture, Room 1001, Rosslyn Plaza East, Washington, DC 20250–2200

The applications of biotechnology to agriculture are growing at a significant rate throughout the world. With the application of any new technology, there often arise concerns about possible economic and societal effects. In the case of agricultural biotechnology, these concerns include: the scientific underpinning of regulations; technology transfer; intellectual property protection; training/career development; consumer information/education; and research funding. The United States Department of Agriculture (USDA) is taking steps to address these issues and to encourage research, development, and commercialization of safe and effective products of agricultural biotechnology.

Agriculture is one of the oldest and most important fields of human endeavor. As the world's population continues to grow, so will the demand for food and fiber, and it is agricultural science and production which will be called upon to develop ways of growing more food and more nutritious food using fewer resources. It is the powerful and precise new tools of agricultural biotechnology which offer perhaps the best hope of satisfying this ever growing need. Indeed, the application of the tools of biotechnology to agriculture is growing at a significant rate throughout the world. With the application of any new technology, there often arise concerns about the economic and societal effects. In the case of agricultural biotechnology, these concerns include: the scientific underpinning of regulations, technology transfer, intellectual property protection, training/career development, consumer information/education, and research funding.

Science Basis of Regulations

In the United States, there are laws and regulations that are intended to assure that products of biotechnology are safe, effective, and compatible with the environment. These laws and regulations provide an effective yet workable system of oversight for product research, development, field testing, and commercialization. The Federal government has established several mechanisms which help ensure that the regulatory programs are based on sound science. For example, a United States Department of Agriculture (USDA) advisory committee developed a set of guidelines for research involving the planned introduction into the environment of genetically modified organisms. The guidelines were developed by the Agricultural Biotechnology Research Advisory Committee (ABRAC), a group made up of scientists from academia, industry, and government chosen to insure scientifically sound, non-partisan, discussion and recommendations. The guidelines established principles for assessing the safety of research with specific organisms, outlined confinement conditions to promote safety, and provided examples of the guidelines' application. They have aided researchers and institutions in the design of safe experiments with specific types of organisms conducted outside contained facilities. Indeed, the principles have provided the regulatory community with a criterion for biosafety.

USDA also supports biotechnology risk assessment research directed toward strengthening the scientific basis of oversight and regulation. Under this program, USDA awards competitive grants to support biotechnology risk assessment research to help address concerns about the effects of introducing certain biotechnology products into the environment.

Another useful approach to strengthening the scientific basis of regulation and oversight is to share and disseminate information on approved field tests and the results of field tests of genetically modified organisms. USDA maintains an electronic bulletin board under its National Biological Impact Assessment Program (NBIAP) which provides access to nearly 20 databases including Environmental Protection Agency (EPA) and Animal and Plant Health Inspection Service (APHIS) approved field tests and a catalog of field tests approved by other governments. This is available free-of-charge to the public. USDA has also sponsored and organized two international meetings which brought together researchers from around the world to present and discuss the results of field tests of genetically engineered plants and microorganisms (1,2).

Technology Transfer

The United States Department of Agriculture, and the United States Government as a whole, are committed to the process of international harmonization of standards including those for emerging products of agricultural biotechnology. This is approached by encouraging cooperation and sharing information on biosafety and ecological research in pursuing these goals.

The sharing of biotechnological developments, among researchers in the United States and in the international arena has become a major issue in the last decade. The USDA, in accordance with the Stephensen-Wydler Technology Transfer Act of 1980 and the Federal Technology Transfer Act of 1986,(3) has established Offices of Research and Technology Applications (ORTAs) at all of it's Federally owned laboratories. These offices strive to, wherever appropriate, facilitate the transfer of Federally owned or originated technology to both state and local governments and the private sector.

The Technology Transfer Act also authorized the creation of Cooperative Research and Development Agreements, CRADAs, between the Federal government and non-federal organizations. The USDA has been the biggest user of CRADAs in the Federal government thus far, with well over two hundred and fifty in existence.(3) These agreements allow the technology of the Federal government to be transferred to private inventors and businesses which may then patent them. This allows the technology to be disseminated to the people of the United States and the world more quickly and efficiently.

In 1987, President Reagan issued Executive Order 12591, "Facilitating Access to Federal Technology". This Executive Order directed U.S. agencies to encourage and facilitate cooperative research and technology transfer through their laboratories, with an end to promoting growth and economic competitiveness.(3)

In the case of transferring technology oversees, many of the agreements are negotiated by the United States Department of State, which is very supportive of both scientific and technological cooperation between the U.S. and foreign agencies, foreign universities, and firms. The different patenting processes in different nations and the issues of technology transfer, have combined with other factors to bring the topic of intellectual property protection of developments in biotechnology to the forefront of discussion groups and committees throughout the country and the world.

Intellectual Property Protection

The protection of intellectual property rights is one of the strongest incentives for investment in scientific endeavors such as agricultural biotechnology. In the United States, patents, the primary form of intellectual property protection available for biotechnological inventions, can be issued on plants, non-human animals and micro-organisms. The United States Patent and Trademark Office (PTO) issued it's first patent on an animal in 1988, and the backlog of animal patent applications is well over 150(3).

Intellectual property protection for Federally funded biotechnology developments in agriculture involves a collage of five separate issues. Those issues include: at what stage of development to file for a patent, what licensing policy should Federal agencies adopt, how much freedom with patented material should researchers be allowed, should the government extend agency application and licensing decisions to individuals and private parties, and finally, the international dimensions of intellectual property protection.

The precedents which guide agencies in these areas began largely with the Bayh-Dole Act of 1980 which required Federal agencies to adopt a "title in contractor" policy, whereby small business and non profit organizations and universities could retain title to their inventions rather than that title being held by the agency (3). In 1983, President Reagan issued the "Government Patent Policy" memorandum and extended the "title in contractor" policy to all contractors, not just small business and non-profit organizations (3). The memorandum was codified in 1984 as amendments to the Bayh-Dole Act. One question that arises during the research and development process is when to file a patent application. There are, in fact, a number of possibilities. A patent could be filed at the earliest possible time, even before a clear use or practicality of the DNA sequence or other biotechnological discovery is known. Many scientists worry that this will put an undue amount of emphasis on patenting and destroy the network of shared knowledge and research materials that has been responsible in large part for the overwhelming success of American agriculture up till now. The opposite extreme would be not to seek patent protection on biotechnology discoveries at all. However, this would contradict the government's commitment to commercializing Federally supported technology and protecting the resulting intellectual property rights.

If a middle ground is to be reached, it would have to be a policy of either seeking patents only when the structure and function of the discovery is known, or having each application reviewed on a case-by-case basis by the agencies to decide when each is appropriate for the patent application process. If the former option is chosen, it would greatly increase the risk of forfeiting foreign or U.S. patent rights through earlier publication, and this would discourage investors from making the contributions necessary to commercialize much of the technology. However, if the latter option were adopted, a great deal of variability would exist across the government. A centralized advisory body would probably have to be created to evaluate intermediate research results and suggest appropriate action which could be uniformly adopted by all Federal agencies. Which of these four options will be ultimately chosen is not yet clear.

If Federal agencies, such as the USDA, obtain patents to certain biotechnological developments and inventions, there are three different licensing policies that could implemented (4). The first is to issue an exclusive license to one individual or company. The second is to issue non-exclusive licenses to several people or groups. The third is to dedicate the knowledge to the public at large. The decision the agency makes at this point is as important as when to apply for a patent in that it has extensive effect on the actual dissemination of biotechnologies.

The use of patented materials for research does not usually demand any license currently, the existing Federal policies and common practice of companies provide sufficient protection for the patent holders. The USDA has proposed the institution of a pilot program whereby "experimental use exemption" clauses would be inserted into funding agreements with the Department (4). If this approach is successful other agencies could do the same. However, some fear that each agency developing its own form of research exemption clause would lead to inconsistent Federal policies. They urge instead a single, unified, clearly expressed, Federal policy of allowing free and uninhibited research activities using patented discoveries resulting from publicly supported research. Unfortunately, in many biotechnological fields such as that of genome research, it is often difficult to distinguish between commercially motivated and purely academic research.

How far should the policies of the Federal agencies extend? As previously discussed, Federal employees and contractors have a right to retain title to their inventions, though Federally funded. In the case of the Federal employee it is unlikely that person would take actions to commercialize the invention or any other action inconsistent with Federal policies. Government contractors and grant recipients are not so steadfast in following the Government's lead in this area. Under the current laws, it would be difficult to implement laws written for the Federal employee in relation to contractors and grantees. The adherence to Federal policies by private parties is virtually impossible to mandate. However, in that the Federal government is the largest supporter of genome research, researchers in the private sector are very likely to follow the government policies concerning patenting and licensing as a guiding principle.

The final issue concerning the protection of intellectual property in the field of agricultural biotechnology is the international dimension. The USDA and many biotechnologists in academia and industry see a need to establish an international standardization of patenting practices. Legislation was introduced in the last Congress to switch U.S. Patent Law from "first to invent" to "first to file" system. Although this would be a big step toward patenting harmonization, questions remain regarding the impact of this change in academic research.

Training/Career Development

A strong education and training system is of paramount importance if we are to continue to build a strong biotechnology program in the United States and abroad. We need to develop and fund programs to draw young people into the study of agricultural science and biotechnology, as well as programs to retrain traditional agricultural scientists in the use of the powerful new biotechnological methods and tools.

With the above mentioned goals in mind, the USDA began it's "USDA Food and Agricultural Sciences National Needs Graduate Fellowships Program" in 1984 to support the education of more scientists in the fields of food and agriculture for which there is deemed to be a national need. Since the beginning of the program, biotechnology has been designated as a national needs area. To date, a total of 205 fellows in biotechnology have been recruited and supported. 119 of those have been in plant biotechnology, 86 have been in animal biotechnology. The number of fellows supported each year however, has dropped dramatically from the onset of the program. In 1989 the total number of fellows was down to 11 from 89 five years earlier. Fortunately, those numbers are gradually increasing, in 1992 there was a total of 21 fellows being supported through this program (5). The program is administered by the Office of Higher Education Programs in the USDA's Cooperative State Research Service. The USDA Hatch Act also provides funding for predoctoral trainees as graduate research assistants available to the Agricultural Experiment Stations.

The Agricultural Research Service (ARS) of the USDA, also in 1984, established a very successful competitive postdoctoral fellowship program to support 21 people for 1 or 2 years. In the following 2 years the number of fellows underwent a significant increase and has remained a valuable tool for attracting young scientists to agricultural research (6). However, in comparison to the biomedical and basic research trainee programs of the National Institutes of Health and the National Science Foundation postgraduate work in biotechnology for agriculture has been funded at very modest levels (6).

Consumer Information/Education

Another important issue in agricultural biotechnology, as it is with any new technology, is education and public information. The commercial future of biotechnology and its promise for a better tomorrow depends upon public acceptance and trust. Public science education is essential to prevent a dangerous gap between the rapid progress the biotechnologists are making and the public's understanding of the science. It is important that the public's decisions relating to biotechnology are made on a scientific and logical basis and not merely reactions to alarmist rhetoric.

The USDA has taken many steps to aid in the education of the general public concerning biotechnology. The NBIAP and other bulletin boards\databases dealing partially and exclusively with biotechnology are accessible to the general public, and a plethora of printed information is available upon request. For example the USDA's Office of Agricultural Biotechnology (OAB) publishes a monthly newsletter entitled, "Biotechnology Notes" which enjoys a large distribution. OAB, the Agricultural Research Service (ARS), The Cooperative State Research Service (CSRS), and the Animal and Plant Health Inspection Service (APHIS), have all sponsored conferences intended to inform the public and various other special groups about the USDA's biotechnology activity. A high school curriculum has even been developed for educating our young people (7).

Research Funding

Currently, there are several Federal agencies beside USDA which fund research in agricultural biotechnology. The Agency for International Development (AID) funds research aimed at increasing the pest and disease resistance of crops and livestock, thus increasing their productivity. The Department of Commerce (DOC) supports research on the production of food and chemicals through the use of aquaculture. The Department of Energy (DOE) focuses its support on the study of plants as converters of solar energy and a renewable energy source. The Food and Drug Association (FDA) supports research relating to food science, processing, and safety. The National Aeronautics and Space Administration (NASA) promotes agricultural research relating to the closed ecosystems which would enable humans to embark on long-duration space flight without the need for major resupply. The National Science Foundation (NSF) supports a great deal of research in which the main goal is to expand our knowledge base of organisms relevant to agriculture (Committee on Life Sciences and Health; Biotechnology for the 21st Century: Realizing the Promise; in press).

USDA is the primary source of Federal funding for research in agricultural biotechnology. As shown in Figure 1, the USDA accounts for approximately 60% of all Federal spending in agricultural biotechnology. In 1992, 114.5 million dollars was spent by USDA on research in the areas of food, fiber, feed, and forestry. Funding for various research areas of biotechnology is carried out through four different agencies within the USDA.

The Agricultural Research Service is responsible for 52 percent of all USDA research expenditures. The rest comes from the Cooperative State Research Service, the Forest Service, and the Economic Research Service as shown in Figure 2.

Approximately one half of all the Federal funds committed to agricultural biotechnology goes to the study of plant systems including work with disease and pest resistant plants. The second largest funding area is animal systems, with approximately 35 percent. The remaining areas, food safety, food sciences, aquaculture, etc. together account for 15 percent of the total budget. The data in Figure 3 exemplify this breakdown pattern within the 1992 Federal agricultural biotechnology budget. Overall, about 5 percent of Federal funding for biotechnology research, shown in Figure 4, is spent on agricultural biotechnology.

Conclusion

The application of the new tools of biotechnology to agricultural research and production is growing at a significant rate. This will present many choices and challenges to the agricultural community, public policy makers, and consumers. USDA is working to ensure that decisions relating to agricultural production,

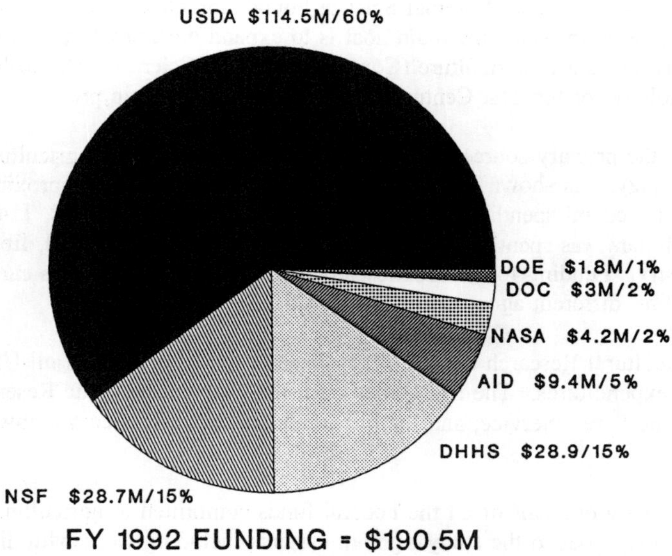

Figure 1. FY 1992 Federal Research Funding by Agencies for Agricultural Biotechnology

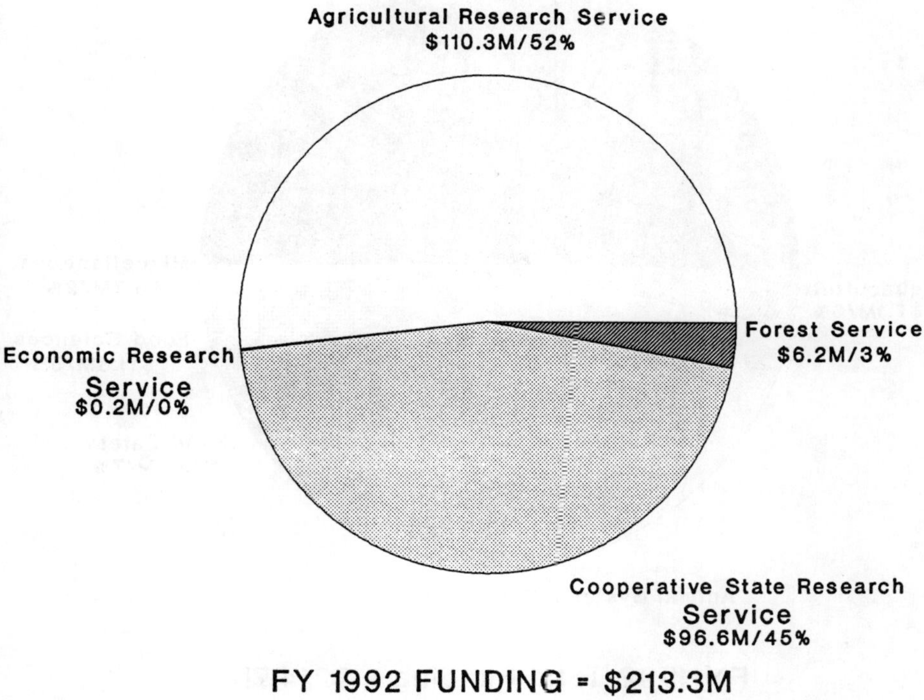

Figure 2. FY 1992 Federal Funding for Biotechnology Research USDA Science Agencies

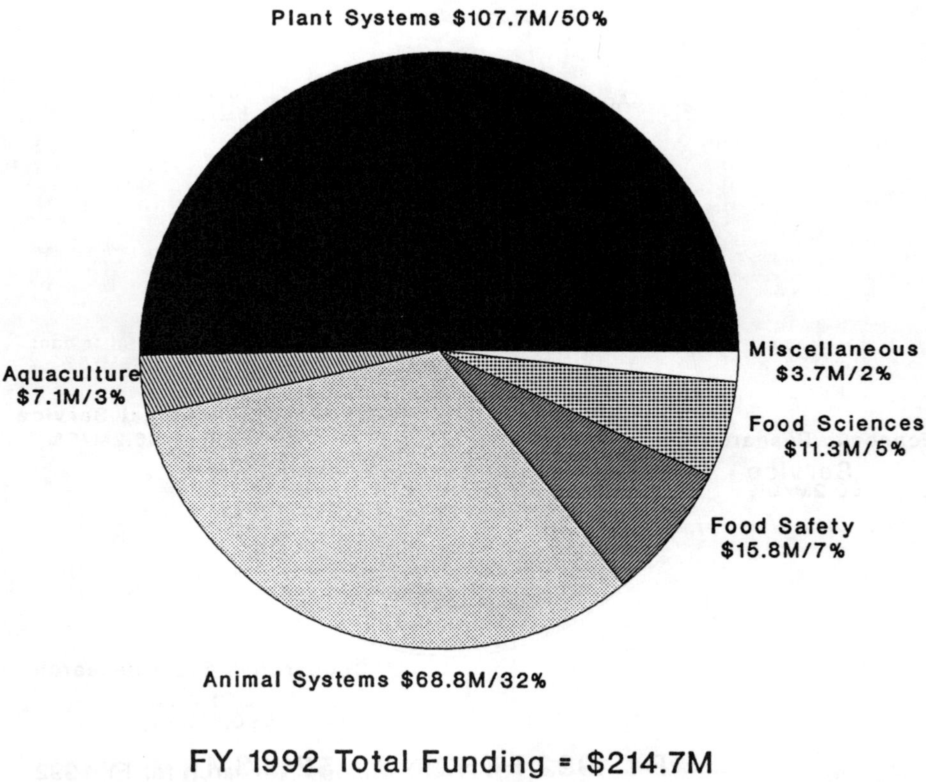

Figure 3. **USDA Funding for FY 1992 for Agricultural Biotechnology by Research Area**

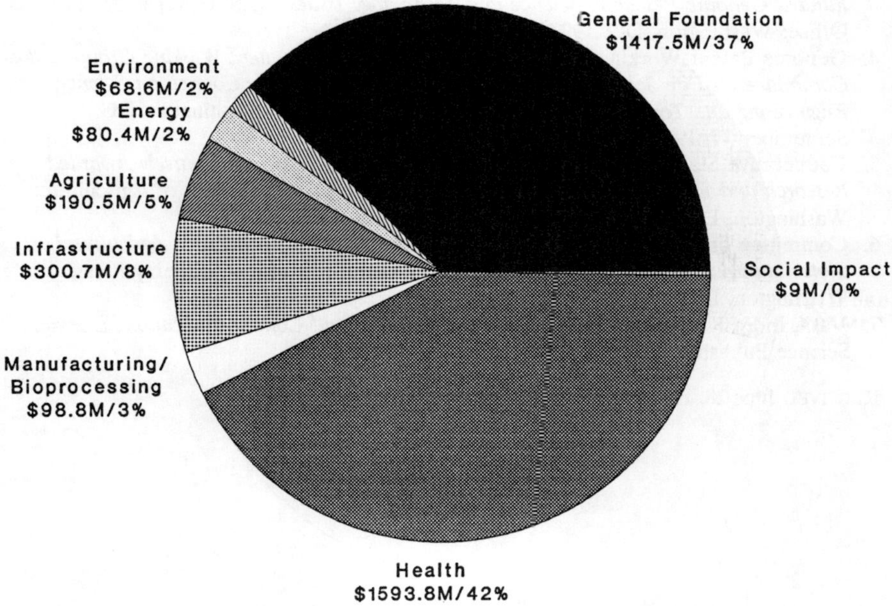

FY 1992 Total Funding = $3759.3M

Figure 4. Federal Investment in Biotechnology Research for FY 1992

product development, public policy, and consumer purchases are based on the most objective and factual information available.

Literature Cited

1. MacKenzie, D.R.; Henry, S.C., Eds.; *Biological Monitoring of Genetically Engineered Plants and Microbes;* Agricultural Research Institute: Bethesda, MD; 1991.
2. Casper, R.; Landsmann, J., Eds.; *Proceedings of the 2nd International Symposium on the Biosafety Results of Field Tests of Genetically Modified Plants and Microorganisms;* Biologische Bundesanstalt fur Land- und Forstwirtschaft: Braunschweig, Germany; 1992.
3. Genome Patent Working Group, Committee on Life Sciences and Health; *Federally Funded Genome Research, Science and Technology Issues*; U.S. Government Printing Office: Washington, DC; 1992.
4. Genome Patent Working Group; *Report of the Genome Patent Working Group to the Committee on Life Sciences and Health, Federal Coordinating Council on Science, Engineering and Technology*; National Science Foundation: Washington, DC; September 17, 1992.
5. Cooperative State Research Service; *Food and Agriculture Competitively Awarded Research and Education Grants, Fiscal Year 1992*; U.S. Department of Agriculture: Washington, DC; December, 1992.
6. Committee on a National Strategy for Biotechnology in Agriculture; *Agricultural Biotechnology: Strategies for National Competitiveness*; National Academy Press: Washington, DC; 1987; Sec. 4.
7. Vasil, Indra K.; *Biotechnology: Science, Education, and Commercialization*; Elsevier Science Publishing Co.: New York; 1990; Sec. 4

RECEIVED June 28, 1993

Chapter 36

Effect of Regulation of Pheromones as Chemical Pesticides on Their Viability in Insect Control

Terry D. Spittler

Cornell Analytical Laboratories, New York State Agricultural Experiment Station, Cornell University, Geneva, NY 14456–0462

Mating disruption of insect pests in agricultural production by application of insect sex pheromones from point source dispensers has the potential for supplanting many chemical pesticide applications, as do several other pheromone strategies. A large number of insect pheromones are being evaluated or proposed for registration as commercial pest control agents. But, each component, under current regulations, must be subjected to much the same testing and toxicity standards as potential chemical pesticide active ingredients; this threatens to price pheromone registration beyond economic reason. High species specificity, formulation complexity and low per hectare rates all limit the market viability of a given pheromone: if registration costs are also high the product is dead. Recent proposals to ease the pesticide registration requirements for "low-risk" biorational schemes are flawed in that the unequal enforcement provisions are unfair and probably illegal, plus they reinforce the pheromones/pesticide connection. Unless and until pheromones are removed from regulation as chemical pesticides, there will be no significant replacement of high-risk pest control chemicals by pheromones.

It is always a little disheartening when one of the last speakers on a program is left with the task of assessing the survival odds of the programs objectives. It would be wonderful to be able to present in-summary remarks that the business endeavors and research interests of the participants had brought us to a new era of safe, effective agricultural pest control, one that would prosper in an age of informed consumers and enlightened regulators. Sorry. Things are not so bleak that insect control using pheromones is in danger of disappearing entirely — there will always be a few token studies (funded studies) and yet another business venture; but survival is a very subjective condition, varying from minimal existence to prosperity. We are selfish, we are primarily interested in prosperity, or at least in having a chance at prosperity by building an effective and vigorous pest control option at competitive, affordable

0097–6156/94/0551–0509$06.00/0

costs per acre and with minimal adverse health and environmental effects. A return on investments would also be nice.

Perhaps I should qualify the balance of my remarks by noting that I, personally, am probable not destined for prosperity in this industry; in fact, I have only an indirect interest in its growth through the professional exercise of developing residue analytical methods for pheromone components and as a participate in the IR-4 Minor-Use Pesticide Registration Program, a program active in both conventional chemical and biologically based pest control strategies. I point this out to define my position not as a pheromone entrepreneur or biocontrol activist, but only as an interested scientist and indirect participant.

The Problem In Perspective

I don't want to parrot any one of the numerous excellent historical reviews of pheromone identification, development and deployment; even more, I don't want to abstract several and burden the world with yet another. However, I do wish to comment that as a chemist, with little agricultural or pest control training, I found it difficult to establish and maintain perspective among the many reports of specific pheromone elucidation and uses over the years. Quite frequently, I was confused as to the scale, duration and registration status of reported uses. What this did illustrate for me was the magnitude of the regulation problem, the endless combinations of simple chemical components that might be incorporated into successful pheromone insect control strategies. Not only are there hundreds of known pheromone components, many of which are common to more than one insect species or group, but they may be formulated into an infinite number of mixtures in attempts to duplicate known pheromones, or at least to elicit the desired response from the target organisms with approximations of what the natural pheromones are believed to be. Clearly, one key to establishing a successful pheromone crop protection industry will have to be the flexibility to tailor commercial preparations for rapid response to changing selectivity demands and pest behavior (*1,2*).

In the chemical pesticide arena, this situation is analogous to allowing carte blanche introduction of numerous chemical pesticide formulations composed of combinations of all known active ingredients or of compounds having similar functional groups and characteristics. Clearly, this latter would never be allowed, and for good reasons, however, for pheromones, there are few good reasons not to allow it — only policy. Overregulation is unnecessarily stifling biorational pest control well before registration for sale and use. Research and testing are delayed or made prohibitively expensive by EUP (Experimental Use Permit) requirements enacted for high-risk pesticide testing. Now they serve as de facto barriers to development of all materials except potentially high volume, high profit (high-risk?) chemical pesticides. Some token concessions have been made, but regulatory demands are still formidable, even in light of EPA's 1982 guideline establishing the tier testing scheme to limit data requirements to those on the first three tiers in the absence of adverse effects from tier one testing, and even with the provisions for waiver of data generation for information not deemed relevant for hazard evaluation of the subject product — assuming, of course, that sufficient acceptable data demonstrating the need for no data are presented to support the initial waiver request. A little more than this is needed.

Reduced Risk Approach

This talk had been in preparation only a short time when the EPA's July 14, 1992 proposal "Incentives for Development and Registration of Reduced Risk Pesticides" was released for public comment and, at first blush, took some of the urgency out of what I had intended to discuss (*3*). There is always a danger when preparing to discuss government policy that one's thoughts or points will be rendered obsolete by presentation time. Not that obsolete thoughts are forbidden (or even unusual) at these exercises, it's just that one hates to admit them at the outset. Here possibly was the answer to the registration obstacles in the path of pheromone insect control. Had a prescient answer (in the affirmative) been received to my title rhetorical question, "Will insect control using pheromones survive regulation?"

Unfortunately, reflection on this document revealed provisions that would certainly elicit unfavorable responses from several quarters. These would make its adoption unlikely. Further reflection on this document revealed consequences to the pheromone industry (and other reduced risk endeavors), as well as to agricultural in general, that would be disastrous if it were implemented. The former are more obvious:

1) It would be blatantly unfair to rank pesticides according to perceived risk or environmental correctness and then selectively and preferentially apply the FIFRA mandates to their registration, reregistration or continued existence. All pest control chemicals, once they have qualified for the definition of "pesticide" are subject to and entitled to all the provisions of FIFRA as they pertain to their regulation. While it would certainly be reasonable and allowable to waive irrelevant and/or redundant data requirements, EPA is required to review all pesticides equitably.
2) Market manipulation by private firms, either individually or by collusion, to give products an unfair competitive advantage is rarely desirable and frequently cause for intervention by one or another regulatory agency. Market manipulation by a regulatory agency, by means quite outside its legislative mandate, to achieve ends that are based on highly subjective evaluations of relative safety, would be, regardless of the purity of intent, even more objectionable.
3) Identification of any pesticide as high-risk for the purpose of identifying areas or uses for which lower risk alternatives would receive prompt and preferential consideration would elicit activist demands for its immediate removal from commerce, regardless of its other merits, regardless of whether or not a suitable replacement had been introduced. In fact, zealous attempts to replace a compound might result in the forced substitution of ineffective materials causing devastating losses of income and markets.
4) It is improbable that we can ever completely replace chemical pesticides with biorational control strategies. In fact, knowledge that several old "high-risk" control options remained available for intermittent, emergency use could give growers the security to try more biological control strategies for a major portion of their needs. Of course, there would have to remain enough of a spot market to justify continuance of a label for these emergency uses. This might require some concessions or relief from high yearly maintenance fees, if enacted. Since effective IPM programs also require occasional surgical strikes, the loss of many "high-risk" materials would make inoperative many schemes relying upon them for crisis management or threshold response (*4*). In addition, control of resistance development depends heavily on the available of alternatives.

These critical uses for our proven chemical pesticides must be preserved so that safer more biologically compatible materials can assume major (but never exclusive) roles in crop protection. Reducing risk assessment to simple terms, risk =

xicity x frequency, many of the targeted "high risk" chemicals could be scored ery benevolently if only used when absolutely necessary, and then, of course, with the exercising of all due caution against worker exposure and environmental damage. In essence, our best interests lie not in eliminating the "high-risk" compounds, but in minimizing unnecessary application, in confining their use to those situations where a "low-risk" biorational scheme falls short.

The second facet of EPA's "Incentives..." proposal, the one that I previously stated as being potentially disastrous for pheromone pest control efforts, is that by accepting a few minor concessions in the pesticide registration process, the agricultural community legitimizes the inclusion of pheromones as chemical pesticides.

Under a fundamentalist interpretation of pheromones as chemical formulations applied in acreage known to be inhabited by food producing plants, they are entitled to the full benefits of attention by the EPA Office of Pesticide Programs, Registration Division. Unfortunately, this gives pheromones equal standing with aldicarb, benomyl, carbofuran and on and on.

The inclusion of pheromones in this company is arbitrary, arbitrary because it automatically ascribes to pheromones all the toxicological and environmental detriments historically associated with classical chemical pesticides. In so doing it assigns an unreasonable registration burden to a group of potential pest control agents, having in addition to low toxicity and environmentally benign chemical characteristics, a spectrum of specificity and use strategies that sets them as far removed from chemical pesticides in eradication campaigns as phosdrin and fly swatters.

What Pheromones Are And Are Not

What is not being acknowledged is that the behavior modification mode of pest protection, a classification into which pheromone schemes fit much more closely then they do that of chemical pesticides, does not impart the directed toxicological risk or non-target environmental effects that drive the chemical pesticide registration process. Consider for a moment some similarities and differences:

1) pheromones are not applied directly to any part of the horticultural product system as are chemical insecticides, fungicides and occasionally herbicides, but are introduced as minute quantities of volatiles via non-contacting dispensers strategically placed throughout the planting: more like rows of marigolds placed along garden borders and between rows to release numerous volatile compounds that seem to minimize insect infestation.
2) in fact, insect pheromone components are probably considerably less potentially detrimental than marigold components which actually repulse insects. Pheromones are simply the essence of a communications system unique to the pest(s) being controlled. There may be a few pheromone components that have objectionable properties or would be considered dangerous or undesirable if residues were significant, but a major point about lepodopterian insect sex pheromone compounds is that almost invariably they are long-chain fatty acids, alcohols, acetates, aldehydes, saturated and unsaturated, usually straight, but occasionally with branched or annular structure: in short, they are similar or identical to naturally occurring fats, waxes, flavors and odor causing compounds found and consumed throughout the plant and animal kingdoms (i.e. our food chain) at orders of magnitude higher concentrations than man-made pheromones would ever be employed in a mating disruption program.

In our poster presentation earlier in this conference we reported finding dodecenyl alcohol and tetradecenyl alcohol in four varieties of untreated apples at concentrations ranging from 70-130 ppb. The natural occurrence of these compounds in apples had been reported in 1976, but not quantitated (*5*). The irony is that they are two of the five components in an experimental lepidopteran insect sex pheromone for which residue determinations were made of all components on treated apples. This full season application included 52 g/ha and 10 g/ha of dodecenyl alcohol and tetradecenyl alcohol, respectively. These were minor constituents, the major component was 106 g/ha of E,E-8,10-dodecadien-1-ol. No increased levels of the saturated alcohols were detected in fruit from treated orchards, and none of the latter, major component, was detectable above 10 ppb, our level of sensitivity (Spittler, T. D., Leichtweis, H. C. and Kirsch, P. Manuscript in preparation).

The specifically of the selective-ion monitoring employed in our gas chromatographic/mass spectral analyses (GC/MSD/SIM) is such that primarily the programmed pheromone compounds are quantitated and confirmed as being present or absent (*6,7*). Other, similar chemicals that might also be present are not always detected and identified, so we do not know all other compounds similar to insect pheromone components the apples are producing. However, these data do encourage speculation that some plants might be producing disruptive levels of insect sex pheromones as a defense mechanism — presumably without EPA's authorization.

Pheromones As Biological Buffers

The strategy to minimize the use of high-risk chemical pesticides by encouraging their replacement with more benign compounds through the vehicle of easing the pesticide registration requirements of biological control systems, including behavior modifying insect pheromones, is only a hollow gesture that will do little to foster the successful use of pheromones for insect control. In fact, making concessions to pheromone registration as pesticides is perhaps committing a serious error in locking them into that classification: by entrenching the concept, it becomes more difficult to remove, as a class, for example, all non-lethal biological formulations composed of natural or natural-like compounds of demonstrably insignificant toxicity, from the definition of pesticide, and into an entirely separate classification called, for sake of illustration, "species specific buffers".

Buffer would be, in fact, a much better term for a biological control strategy because the objective is rarely the active eradication of a pest species or spectrum by application of a toxic compound, but rather the intent is to cushion the impact of a biological proliferation, i.e. the target pest population, and tip it towards economically acceptable, or threshold levels. Neither pathogens, behavioral disruptions nor induced predators will bring about the eradication of a pest species. They only serve to keep populations at levels compatible with local horticultural practices. Does that make them pesticides? Yes, if we want to call them that. No, if we find it more advantageous not to.

There is no absolute that establishes what is a pesticide — there is no absolute that establishes what is not. We currently live with definitions derived as a compromise between scientific evidence, political expediency and the interests of the consumer. All of these three factors are constantly being reweighted. So why should the pesticide definition remain insulated from evidence and influence?

We Are Standing In Our Way

We would like to think that the agencies responsibility for regulating many aspects of our lives, including our food supply and environment, have the foresight and latitude to respond to changing knowledge and changing needs. Certainly we cannot recommend a disregard for legislated standards, but we live in an era of fundamentalist interpretation of statutes. Note specifically the recent interpretation that the Delaney Clause, by virtue of being the letter of the law — anachronistic as that letter now is considering its far remove from its inception — will continue to proscribe the trace residues of benomyl, captan and several other important chemical pesticides in processed foods, this despite the interpretations of negligible risk repeatedly espoused by responsible agencies and designated advisor groups.

It is precisely situations such as this that make me pessimistic that we will ever achieve real progress in biorational control as long as it is strictly regulated in concert with chemical controls. Of course, now and then a new pheromone or other biorational registration appears, but it is usually either one of the few high profile, heavily researched uses (e.g. grape berry moth pheromone) or it has been extensively subsidized by venture capital and/or public funds in yet another demonstration splash that hopes to propagate the wave of the future. I am sorry, but I see no future for insect control using pheromones ever supplanting conventional chemical controls as long as pheromone formulations are defined and regulated as pesticides.

There are hundreds, perhaps thousands, of known pheromone components that occur naturally in an infinite combination of product and proportion. Not only does each insect species require a unique mixture of known and available materials, properly employed, to have its mating, aggregation or other cycles disrupted or diminished, but the flexibility to fine tune composition for maximum efficacy is also essential. What we have brought upon ourselves is a situation where the hypothetical protection of a commodity from an insect pest using pheromones requires that toxicological data packages for several materials be completed (average three compounds per pheromone) and that residue studies for all components in each formulation be conducted in significant growing areas of the commodity.

By contrast, one broad spectrum N-methyl carbamate or versatile organophosphate material can obliterate dozens of insect species on our same hypothetically commodity for the price of one toxicology package for a parent and perhaps a metabolite and the same designated-site residue studies. Is there perhaps another price being paid here?

Literature Cited

1. Nabholz, K. In *Behaviour-Modifying Chemicals for Insect Management*; Ridgway, R. L., Silverstein, R. M. and Inscoe, M. N., Eds.; Marcel Dekker Inc.: New York and Basel, 1990, pp 553-555.
2. Tinsworth, E. F. In *Behaviour-Modifying Chemicals for Insect Management*:; Ridgway, R. L., Silverstein, R. M. and Inscoe, M. N., Eds.; Marcel Dekker Inc.: New York and Basel, 1990, pp 569-601.
3. U. S. Environmental Protection Agency, 57 Fed. Reg. 32140, (July 20, 1992).
4. Shani, A. *Role of Pheromones in Integrated Pest Management*; 1992 (unpublished survey).

5. Van Straten, S. *Volatile Compounds in Food*, 4th ed.; Central Institute for Nutrition and Food Research: Zeist, Netherlands, 1977.

6. Spittler, T. D., Leichtweis, H. C. and Dennehy, T. J. In *Biotechnology for Crop Protection*; Hedin, P., Menn, J. J. and Hollingworth, R., Eds.; ACS Symposium Series No. 379; 1988, pp 430-436.

7. Spittler, T. D., Leichtweis, H. C. and Kirsch, P. In *Insect Pheromones and Other Behaviour-Modifying Chemicals: Applications and Regulation*; Ridgway, R. L., Inscoe, M. and Arn, H. Eds.; British Crop Protection Council; 1992, 51; pp 93-108.

RECEIVED August 19, 1993

Chapter 37

Good Laboratory Practice Regulations

Applications to Field Studies

Judy Hochman[1] and Willa Garner[2]

[1]Compliance Reviews, Walkersville, MD 21793
[2]Garndal Associates, 17485 Sierra Way, Monument, CO 80132

Effective October 16, 1989, field studies became subject to the same U.S. Environmental Protection Agency (EPA) Good Laboratory Practice Standards (GLPS) regulations (*1*) as those laws regulating laboratory toxicology studies. Needless to say, applying "laboratory GLPS" to complex and unpredictable field situations gave rise to some interesting problems and many questions on how they should be handled.

A typical field study scenario that leads to the most unique GLP considerations begins when the sponsor/registrant contracts the field phase to a Field Coordinator or Study Manager or maybe to a University, but retains the study director responsibility in house. To add to the challenge, the sponsor/registrant may also elect to contract out the analytical work either directly or through the Field Coordinator/Study Manager. Figure 1 depicts possible scenarios for locations of study conduct. The analytical laboratory can be located at the sponsor's or at a contractor's facility.

The following discussion describes how the GLPS can be adapted to the typical field study components, from protocol development to field and laboratory experimental phases, and finally to report preparation and archiving of study data.

Organization and Personnel

Testing Facility Management. According to the definition set forth in §160.3, Testing Facility means "a person who actually conducts a study; i.e., uses the test substance in a test system". It appears that in our scenario, this would be the Field Coordinator/Study Manager who will be conducting the field phases of the study. However, according to §160.31, it is the testing facility management who

0097–6156/94/0551–0516$06.00/0

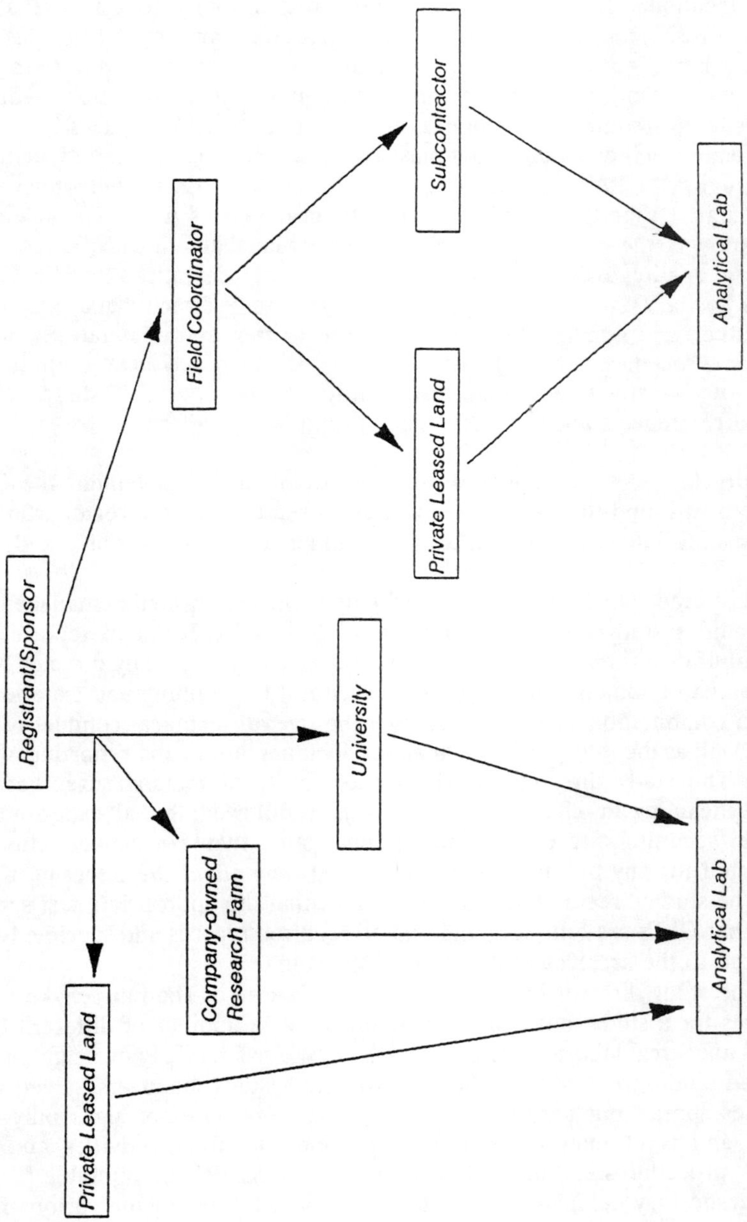

Figure 1. Possible Field Study Scenarios.

designates (and replaces, if necessary) the study director. By this definition, the sponsor/registrant would be the testing facility when the study director is at the sponsor location. It is also management's responsibility to assure that all personnel, resources, facilities, equipment, materials and methodologies are available when required to accommodate the study. Laboratory and field facilities must be of sufficient size and construction. Adequate storage areas at the test site are required for supplies, equipment, and test chemicals.

Testing facility management must also establish, implement and maintain a GLP program. EPA Office of Compliance Monitoring (OCM) personnel have verbally stated that they would like to see evidence of active management participation in the company's GLP program. Management must assure that there is a quality assurance unit (QAU); that any deviations to the GLPs reported by the QAU are communicated to the study director and corrective actions taken and documented; that there are written and adequate Standard Operating Procedures (SOPs) available for routine procedures used in the conduct of the study; and that test, control or reference substances (i.e., analytical reference standards) have been properly characterized.

Study Director. As the regulated industry attempted to implement the GLP regulations, it found that the EPA expected to see one study director who had overall responsibility for both the biological and analytical phases of a study (*2*).

As already noted, it is the responsibility of the testing facility management to designate a study director before the study is started and to replace this individual, if necessary, during the conduct of the study.The study director must be a scientist or someone with appropriate education, training and experience, or with a combination thereof, to oversee the overall technical conduct of the study, as well as the interpretation, analysis, documentation, and reporting of the results. The study director has the responsibility of making sure that the protocol, including any change, is approved and is followed; that all experimental data are accurately recorded and verified; and that corrective action is documented for any unforseen occurrence that may affect the integrity of the study. The study director also must make sure that the appropriate test system is used; that GLPs are followed; and that all required records and specimens are transferred to the archives at the close of the study.

The study director has a difficult job in that he or she must make these assurances for a study that is usually conducted at a number of different field sites and analytical laboratories over which he/she has no daily oversight and is conducted by different contract field coordinators/study managers. Even with study sites spread out across the Country, the study director signs only one protocol and is ultimately responsible for both the field activities and the analytical procedures. Although the study director is solely accountable for the study, he/she may delegate the day-to-day responsibilities to the various field coordinators or analytical managers. The titles used in this chapter are merely examples. Many different titles are used to identify the responsible field and analytical personnel. Even with portions of study conduct delegated, the study director must still approve all protocol amendments and deviations, authorize any SOP deviations, review contractor QAU inspection reports, and approve the corrective actions for any findings.

Personnel. There must be a sufficient number of personnel to conduct the study in a timely and responsible manner in accordance with the protocol. Both the study director and management are responsible for assuring that all personnel working on the study are qualified by education and/or experience to perform their designated duties. Additionally, the GLPs require documentation of these qualifications by means of a current resume or curriculum vitae, training record, and job description. These requirements apply to all personnel, whether employed by a contractor as full-time staff or as temporary help, such as a summer student or a migrant worker. If one uses a cooperator to apply the pesticide at the test site, obtain a copy of his/her applicator's license in order to meet the GLP requirements. The rule of thumb is that anyone performing a function that is required by the study must have a resume, a training record, and a job description. For example, if irrigation is required by the study design, the person performing this function must meet GLP qualification requirements. However, if a person is the subject of the study, such as in worker exposure studies, the training documentation is not required.

The GLPS also require that necessary precautions are taken with personal sanitation and that appropriate clothing is worn to avoid contamination of the test, control and reference substance and test systems. Logically, personnel conducting applications, should not be collecting Day 0 samples without first washing him/herself and changing clothing.

Another GLP requirement that rarely applies to field studies is that any individual who is found to have an illness that may adversely affect the integrity of the study is to avoid contact with the test system and report the illness to the supervisor.

Quality Assurance Unit (QAU). The testing facility must have a quality assurance unit (QAU) which is responsible for assuring management that the facilities, equipment, personnel, methods, practices, records and controls are in conformance with the GLPs. The QAU must be entirely separate from and independent of the personnel who are engaged in the conduct or direction of the study (3).

Responsibilities of the QAU include the maintenance of a copy of the master schedule, which has been indexed by the test substance and contains information on the test system, the nature of the study, the initiation date, the identity of the sponsor and the study director, and the current status for each study. Individual field and analytical contractors generally include the name of their project manager for each study listed.

Protocols must be maintained for all studies for which the QAU is responsible. In many instances field cooperators are left off of the protocol distribution list or receive only pertinent pages of the protocol. While it may not be necessary for the field personnel and their QAU to have the analytical portion of the protocol, they should at least have the title page, signature page, field study design portion, and any amendments to these sections in order to comply with the intent of §160.35(2).

The QAU must inspect each field study at least once; however, not all field sites must be inspected for each study. Additional inspections can be made at the discretion of the auditor. Many testing facility managements/study directors request inspections of all critical phases at least once, followed by periodic data reviews. During all inspections the QAU must determine that there have been no deviations from the protocol or SOPs without proper documentation and authorization by the study director.

EPA's OCM has stated that it is acceptable to use inspections conducted during other studies to provide the necessary assurances (*4*). It is also acceptable to use inspections conducted when no study is in progress to assure that methods, personnel, etc., at a particular site are in conformance with GLPs as long as the area inspected is representative of that used in the study. EPA has further clarified that protocol reviews or data reviews do not satisfy the requirement for at least one inspection (*5,6*). Additionally, for lengthy studies, one in-progress (critical phase) inspection will not be enough. Suggested inspections include chemical characterization, field aspects (application, sampling, sample storage, and shipment), and analytical procedures (sample processing, sample storage, and chromatography).

The QAU must maintain written and properly signed records of each inspection showing the date of inspection, the study inspected, the phase or segment of the inspection, the person performing the inspection, findings and problems, action recommended and taken and any scheduled date for reinspection. Any problems found during the course of an inspection which are likely to affect the integrity of the study must be brought to the attention of the study director and management immediately. Because of the organizational structure of many field studies, the QAU of a contract facility test site is often required to report findings to the sponsor study director and sponsor testing facility management as well as its own project coordinator and facility management. It is also possible that the QAU may have to report to a competitor who is acting in the capacity of study director (*7*).

The QAU must periodically submit to management and the study director written status reports on each study, noting any problems found and the corrective actions taken. Issuing each inspection report with the corrective actions taken to both the study director and study director's management satisfies the requirement for periodic reports.

Periodic facility inspections must be conducted at each test site. Most often the contractor QAU conducts these inspections at their site; however, it is recommended that off-site study director's QAU conduct a preplacement inspection at each test site to assess GLP compliance prior to study initiation.

Finally, the QAU must review the final study report to assure that it accurately reflects the raw data and SOPs used in the study. The QAU then signs a Quality Assurance Final Report Statement which specifies the dates of inspection, the phases inspected, and the dates findings were reported to the study director and study director's management. With large field studies there may be a number of separate field reports and one or more analytical reports. Generally, the QAU at each of these contractors treats their report as a final report and audits it accordingly.

Protocol

Since the EPA feels that it is possible to conduct entire field studies, including the field and analytical phases, as single studies, there can be only one protocol per study. The protocol must, therefore, include both the field and analytical portions of the study (*8*).

The protocol outlines the study design and describes what has to be done. The amount of procedural detail described in the protocol varies according to study type; however, it is best to keep the protocol more general and let the how be described in the standard operating procedures. Very specific and detail-oriented protocols can lead to numerous deviations. If possible, the QAU should review the draft protocol to assure that all GLP requirements have been addressed.

Protocol items that are required to be addressed by GLPs, as well as field specific considerations, typically include, but are not limited to:

a. A descriptive title (includes study type and name of test chemical);
b. A statement of the purpose of the study;
c. The names and addresses of the sponsor, testing facility, investigators/ cooperators test sites. The dated signatures of the study director and the sponsor;
d. The proposed experimental start and termination dates;
e. A description of the test, reference and control substances by name, CAS number, or code number. Additionally, it is suggested that, if known, the following be included:

    1. The batch/lot number, as well as the percent active ingredient of the formulation or neat chemical used;
    2. A physical description of the chemical (wettable powder, granule, emulsifiable concentrate, etc.);
    3. How and where the test chemical(s) are to be stored prior to and after use, as well as the storage location of the chemical containers; and
    4. Identity of the responsible party for the collection of the archive sample, purity analysis, and storage stability at the test site;

f. Site information, including a description of the plot size, control areas, buffer zones, etc. The test site selection process which is based on factors such as pesticide use history, crop history, soil type, precipitation and irrigation history, generally precedes protocol approval;
g. The justification for the test system selected;
h. Where applicable, the number, body weight range, sex, source of supply, species, strain, substrain and age of the test system;
i. The procedure for identification of the test system. Field site and subplot perimeter corners can be staked with the study number. Animal and crop samples are assigned unique numbers upon collection;

j. Application procedures (including method of calibration), frequency and amount, with appropriate units, and the reason for the method of application;
k. Specimen sampling procedures, including randomization techniques (for control of bias), methods and equipment to be used;
l. Storage and shipping procedures, including chain-of-custody documentation;
m. Weather data collection requirements. Irrigation requirements, if any; and
n. Records to be maintained. This statement should include where the raw data, protocol, reports, and any supporting records are to be archived. Additionally, specific documentation requirements are listed under this heading, for example:

    Maps and schematics of the test site(s)
    Application equipment calibration data
    SOPs followed
    Site maintenance
    Irrigation, weather data
    Crop and pesticide use history

Proposed statistical methods are also required by GLPs but rarely apply to the field portion of the study.

Because of the unpredictable situations that occur in field studies, frequent protocol modifications may be required. GLPs mention only the requirements for protocol "changes" or "revisions". The scientific community has, however, broken protocol changes into two categories and coined the phrases: "protocol amendments" and "protocol deviations". A protocol amendment is defined as a planned change to the protocol, whether it is to the study design or in the study personnel. A protocol deviation is an unforseen or unplanned change which the study director was not able to authorize in advance. It should be noted that protocol deviations are also GLP violations, as specified in §160.130(a) which states "The study shall be conducted in accordance with the protocol." Therefore it is of the utmost importance to inform the study director of any necessary changes and obtain his/her concurrence before proceeding if at all possible (*7*). Facsimiles and telephone records may be used as documentation. Properly documented protocol changes must include the reason for the change and be signed and dated by the study director.

Standard Operating Procedures (SOPs)

A testing facility must have standard operating procedures (SOPs) in writing setting forth study methods that management is satisfied are adequate to insure the quality and integrity of the data generated during the conduct of a study. Generally, all the operational units of a study (field, analytical) have their own

SOPs, approved by their own management. It is the study director's and his/her management's responsibility to determine if these are adequate for the study. Which facility's SOPs are to be used should be identified in the protocol.

SOPs help assure that study procedures are performed in a consistent manner and, by simply referencing a particular SOP, they eliminate the necessity of having to document repetitive procedural details in the field notebook. SOPs should be written for all common, repetitive tasks. Each SOP should describe only one procedure or operation. Published literature may be referenced and used as a supplement to SOPs. This is frequently done with equipment SOPs that have voluminous User's Manuals.

GLPs require that SOPs relative to the procedures being performed be readily available in study areas, whether the study area is a laboratory or a field site.

At a minimum, SOPs should be established for the following areas:

a. Administrative/Organizational Structure
b. Health and Safety
c. SOP Preparation
d. Study Initiation
e. Protocol Requirements
f. Chemical Management
g. Data Collection
h. Report Preparation and Storage
i. Quality Assurance Unit
j. Equipment Maintenance and Calibration
k. Computer Verification and Validation
l. Field Procedures
m. Archives

Suggested topics for individual SOPs can be found in Table I.

Each SOP should contain:

a. Title and Purpose;
b. Procedure, in outline format;
c. Definitions, Materials, Calculations, as appropriate;
d. Documentation Requirements;
e. References to Published Literature;
f. Effective Date, Date of Revision, Revision Number; and
g. Management's Dated Signature, signifying approval.

In addition to the above, equipment SOPs must address the schedules to be used in the routine maintenance, inspection, cleaning, testing, calibration/standardization and the remedial action to be taken in the event of failure or malfunction of the equipment. Additionally, the person responsible for the performance of each procedure is to be identified in the SOP.

Table I. Suggested Topics for Individual Standard Operating Procedures

UNIT I:	**ADMINISTRATIVE, FACILITY AND OPERATIONAL PROCEDURES**
CHAPTER 1	**ADMINISTRATION**
SECTION A	Organizational Structure
SECTION B	Personnel Administration
SECTION C	Security
SECTION D	Field Trial Site Inspection/Outside Agency
CHAPTER 2	**HEALTH AND SAFETY**
SECTION A	Safety Orientation
SECTION B	Safety Rules
SECTION C	Protective Clothing/Equipment
SECTION D	First Aid
SECTION E	Disposal of Hazardous Waste
CHAPTER 3	**STANDARD OPERATING PROCEDURES**
SECTION A	Writing Standard Operating Procedures
SECTION B	SOP Preparation and Maintenance
CHAPTER 4	**STUDY INITIATION**
SECTION A	Study Authorization
SECTION B	Study Number Assignment
SECTION C	Study Protocols
SECTION D	Study Personnel Assignment
SECTION E	Ordering Study Supplies
CHAPTER 5	**CHEMICAL MANAGEMENT**
SECTION A	Test Chemical Inspection and Acceptance
SECTION B	Logging-In Test Chemicals
SECTION C	Labeling of Test Chemicals
SECTION D	Test Chemical Storage
SECTION E	Final Disposition of Test Chemicals
SECTION F	Reagents and Solutions
SECTION G	Dilutions and Mixtures

Table I.—*Continued*

CHAPTER 6	**COLLECTION AND TRANSMISSION OF DATA**
SECTION A	Study Notebooks/Data Files
SECTION B	Data Forms
SECTION C	Raw Data Collection
SECTION D	Significant Figures and Arithmetic Operations
SECTION E	Data Corrections/Error Codes
SECTION F	SOP and Protocol Deviations
SECTION G	Acceptable Abbreviations
SECTION H	Transmission and Responsibility for Raw Data
SECTION I	Storage of Raw Data/Test Chemicals
CHAPTER 7	**REPORT PREPARATION AND STORAGE**
SECTION A	Interpretation of Raw Data
SECTION B	Draft/Final Report
SECTION C	Data Presentation
SECTION D	Storage of Final Report
CHAPTER 8	**QUALITY ASSURANCE**
SECTION A	Responsibility of the Quality Assurance Unit
SECTION B	QAU Records
SECTION C	Standard Operating Procedures and Protocol Review
SECTION D	QAU Master Schedule Sheet
SECTION E	Study Inspections
SECTION F	QA Auditing Procedure
SECTION G	Draft and Final Report Audits
SECTION H	Archives
SECTION I	Facility Inspection
SECTION J	QA Audit Reporting Procedures
UNIT II:	**GENERAL RESEARCH PROCEDURES**
CHAPTER 1	**PROTOCOL AND TRIAL DESIGN**
SECTION A	Protocol Development
SECTION B	Protocol Amendment/Deviation Form
SECTION C	Selection of Trial Site
SECTION D	Plot Size and Plot Plan

Continued on next page

Table I.—*Continued*

CHAPTER 2	**SITE AND ENVIRONMENT**
SECTION A	Qualification Samples
SECTION B	Meteorological Data Collection
SECTION C	Field History
SECTION D	Trial Site Location and Map
SECTION E	Soil Characterization
CHAPTER 3	**CROP MANAGEMENT**
SECTION A	Crop Management
SECTION B	Planting and Transplanting
SECTION C	Maintenance Chemicals
SECTION D	Irrigation Water Quality
SECTION E	Harvest of Test System
CHAPTER 4	**SPRAY MIXING**
SECTION A	Calibration and Maintenance of Liquid Measuring Equipment
SECTION B	Operation and Calibration of Balances
SECTION C	Mixing Test Chemicals
SECTION D	Mixing and Application Verification
SECTION E	Field Spikes
CHAPTER 5	**APPLICATION**
SECTION A	General Application Practices
SECTION B	Maintenance of Equipment
SECTION C	Cleaning of Equipment
SECTION D	Storage of Equipment
SECTION E	Calibration of Granular Application Equipment
SECTION F	Calibration of Liquid Application Equipment
SECTION G	Application Method
SECTION H	Conditions at Application
SECTION I	Tank Mix Samples
CHAPTER 6	**SAMPLING OF CROP**
SECTION A	Crop Sample Collection Procedures
SECTION B	Sampling Equipment Maintenance
SECTION C	Crop Sample Size
SECTION D	Compositing and Packaging Crop Samples
SECTION E	Identification and Labels for Crop Samples

Table I.—*Continued*

CHAPTER 7	**SAMPLING OF SOIL**
SECTION A	Soil Sampling Procedures
SECTION B	Compositing and Packaging Soil Samples
SECTION C	Identification and Labeling for Soil Samples
SECTION D	Soil Temperature Data
SECTION E	Cleaning of Equipment
CHAPTER 8	**SAMPLE STORAGE AND SHIPPING**
SECTION A	Sample Records
SECTION B	Field Handling and Cooling of Samples
SECTION C	Freezer Maintenance and Freezer Logs
SECTION D	Sample Storage
SECTION E	Sample Shipping
SECTION F	Calibration of Temperature Recording Instruments
CHAPTER 9	**OFFICE PROCEDURES**
SECTION A	Field Trial Management
SECTION B	Telephone and Correspondence Logs
SECTION C	Data Storage for Trials in Progress
SECTION D	Data Retention Period
CHAPTER 10	**COMPUTERS**
SECTION A	Computer Equipment Maintenance Logs
SECTION B	Computer Software Validation and Verification
SECTION C	Computerized Tracking Logs

SOPs should be reviewed on a periodic basis and revised as necessary. Most companies review theirs yearly. All permanent changes to a SOP must be authorized by management. One time changes or deviations to a SOP must be authorized by the study director and documented in the raw data. Documentation should include a description of the deviation, the reason for it, and the impact, if any on the study, and of course, the dated signature of the person recording the deviation. A SOP deviation does not need to be included in the final report as do protocol amendments and deviations.

A historical file of all original signature SOPs and any revisions must be retained. A responsible person should be designated by management to coordinate the preparation, revisions, distribution and maintenance of the SOPs.

Test Substance/Chemical Handling

GLPs require that there are separate areas for test, reference, and control substance receipt and storage, mixing with carrier, and storage of the mixtures. Test, reference and control substances are typically supplied by the sponsor and shipped directly to the field sites to be stored under label conditions. A computerized temperature and relative humidity monitoring device can be packaged with the test chemical prior to shipment, then down-loaded upon arrival to assure that the chemical was not subjected to conditions outside its stability range.

In order to document stability under storage conditions at the test site a max/min temperature and relative humidity recording device is needed in the storage area. Chemical storage units should be padlocked, climate controlled, and allow for the separation of chemicals in order to prevent contamination or mix-ups.

Upon receipt, the chemical label should be checked. Each storage container is to be labeled with the name, CAS number or code number, batch/lot number, expiration date, if any, and, where appropriate, the storage conditions necessary to maintain the strength, identity, purity and composition of the chemical. It is obvious that this information is always "appropriate". It is important to have an inventory log on which to document the receipt and use of the chemical (Figure 2, Ref. *9*). Documentation requirements include:

Source
Amount Received (weight, volume, number of containers, etc.)
Physical Description
Label Information
Shipping Information
Receipt Date and Condition Upon Receipt
Recipient (responsible personnel)
Storage Location and Conditions
Amount Used, Purpose, Date Dispensed, By Whom, and Amount Remaining
Final Disposition

***ptrl**-East*

PTRL RECEIPT RECORD AND INVENTORY LOG OF INCOMING TEST, REFERENCE AND/OR CONTROL ARTICLES/SYSTEMS

Receipt Date: ____________ PTRL Log Number: ____________

Name on Container Label: ____________ Formulation: ____________

Active Ingredient: ____________ Purity: ____________

Batch/Lot No.: ____________ Expiration/Re-analysis Date: ____________

Other Label Info: ____________

Shipper (Name, Address): ____________

No. of Containers: ____________ Gross Weight Received: ____________

Received By: ____________ Date: ____________

Condition on Arrival: ____________

Physical Description: ______/ Solid ______/ Liquid ______/ Solution ______/ Other

Storage Location: ____________ Storage Temp: ____________

Comments: ____________

Shipper: ____________Way Bill #: ____________

INVENTORY LOG

Project #	Purpose	Gross Wt. ( )	Amount Dispensed ( )	Adjusted Gross Wt. Remaining ( )	Initials	Date	Balance I.D. No.

Disposed: _____ Returned To: ____________ Date: ____________

Initials: ________: Date: ____________ Return Approved By: ____________

DEV 4/90

Figure 2. Chemical Receipt and Use Log. (Reproduced from Ref. 9. Copyright 1992 American Chemical Society.)

For studies of more than four weeks in experimental duration, a reserve sample of each batch of the test, reference (analytical reference standard) and control substance must be retained. As this archive sample may be retained by either the sponsor or the contract facility, the responsibility should be addressed in the protocol. Additionally, a sample of the chemical should be collected prior to each application and immediately frozen for possible analysis for "storage stability of the test, reference and control substance at the test site". If the sponsor's storage stability data are inclusive of the storage conditions recorded at the test site, the GLP requirement will have been met and the above storage stability samples will not have to be analyzed (*10*). It should be noted that all test substance containers must be retained for the duration of the study; i.e., until the study director signs the final report. A conditional exception for disposing of the containers may be obtained by writing to the OCM (*11*). OCM will set forth certain documentation requirements that will be required to account for the test substance containers.

For each test, control, or reference substance that is mixed with a carrier, such as water, appropriate analytical methods are required to determine solubility, homogeneity, and stability of the mixture. Generally, the sponsor has conducted extensive testing to meet these requirements before the chemical is used in a field study. However, the requirement to determine the uniformity of the mixture and the concentration of the test, reference or control substance in the mixture (*12*) can be addressed best by collecting tank mix samples while the tank contents are being agitated. If laboratory tests have not been conducted, tank mix samples should be taken before and after the application to demonstrate uniformity in the concentration of the mixture over the duration of the application. Documentation of the mixing time and application start and end times are necessary as the stability of some chemicals may decrease once mixed with a carrier.

Many sponsors request that field spike samples be taken. Field spikes of each sample matrix should bracket the expected range of analytical results generated from the assay of the samples. Spikes just above the screening level are necessary to validate recoveries and insure an acceptable analytical method. The number of field spikes should be completely planned and described in the protocol. These samples can be used for quality control and as samples for storage stability during shipment and laboratory storage.

All reagents and solutions, whether used for calibration or as a solvent for cleaning, must be labeled with the following information: Identity, Concentration, Expiration Date, and Storage Conditions.

Equipment Maintenance and Calibration

The GLPs state that equipment used in the generation, measurement, or assessment of data and equipment used for environmental control must be designed to perform the protocol required functions and be suitably located for operation, inspection, cleaning and maintenance. Furthermore, equipment used in the generation, measurement or assessment of data must be routinely tested, calibrated/standardized, cleaned, inspected and maintained. As noted above, the methods for maintenance and repair, as well as the schedules, must be addressed in the SOPs. Each equipment SOP should define the maintenance procedures considered to be routine.

Written records must be maintained to show the dates of the maintenance, repair, cleaning, inspection, calibration and testing, as well as a notation as to whether the SOP was followed. Documentation for repairs as a result of equipment malfunction must describe the nature of the defect, how and when the defect was discovered, and any remedial action taken to repair the defect. Documentation must be signed and dated by the responsible personnel designated in the SOP.

Typical field equipment subject to these requirements include weather stations or other weather collection equipment, temperature probes, freezers, balances, and application equipment (i.e., sprayer booms). Application equipment must be calibrated prior to its use in a GLP study, at the rate(s) to be used in the study. All calculations used in the calibration of the application equipment must be retained as raw data.

Study Conduct

Samples are to be collected in accordance with the approved protocol and labeled with the test system, study, nature, and date of collection. The GLPs state that this information may be on the specimen container or in accompanying documents. Generally, both are done unless a bar code system is used, in which case only the bar code appears on the sample container.

Additional information that needs to be recorded in the raw data includes the time of sample collection, the sampling location, the randomization scheme, if used, and the identity of sampling personnel. The methodology employed for sample collection should also be described in the raw data, i.e., the equipment that was used and the measures taken to avoid cross-contamination. A description of how the samples were handled once they were collected is required by the study guidelines. The measures employed to preserve the samples should be documented. If there was temporary storage, the storage location, including the temperature during storage, should be noted. Describe when and how samples were shipped to the analytical laboratory. A chain-of-custody form, which identifies individual samples by both description and number, must accompany the samples from the time of collection through receipt by the analytical laboratory. The chain-of-custody form is signed by both the personnel that relinquish the samples as well as those that receive them. A notation about the condition of the samples upon receipt is necessary.

All observations, procedures and measurements must be recorded promptly in ink and signed and dated by the personnel recording the data. It is recommended that a set of forms, maintained in a three-ring binder, be used, thus allowing the data to be collected in a consistent manner and prompting the field personnel to record all the required information. Original observations are considered to be the raw data. The GLPS do not allow data to be jotted down and then neatly transcribed later.

Final Report

A final report must be written for all studies, even if a study is terminated or discontinued for some reason. Reports for terminated studies need be nothing more than a memorandum to the file describing what happened with all raw data generated attached to it. The protocol should be amended to terminate it. These data packages must be archived for a period of at least two years.

For completed studies, the final report would include minimally the following:

1. The name and address of each facility that was involved in the study conduct, and the dates on which the study was initiated and completed.
2. The objectives and procedures as stated in the approved protocol, including any changes in the original protocol.
3. The statistical methods employed for analyzing the data.
4. The identity (by name, CAS number, or code number) and lot number/batch of the test, control, and reference substances. The strength, purity, composition, and solubility, if appropriate, should be provided. The stability under conditions of administration must be documented.
5. A description of the methods used.
6. A description of the test system used, and when applicable, the final report includes the number of animals used, their sex, their body weight range, the source of supply, species, strain and substrain, age, and the procedure used for identification. The source of supply, species, and procedure used for identification would be applicable to field studies regarding plants used.
7. A description of the dosage, dosage regimen, route of administration, and duration.
8. A description of all circumstances that may have affected the quality or integrity of the data. It is recommended that the protocol with its amendments and deviations be included as part of the final report. Including all lessens the perceived negative impact of listing all the changes to the protocol without any explanation.
9. A listing of the names of the study director and the other scientists or professionals, as well as all supervisory personnel, who were involved in the study.
10. A description of the transformations, calculations, or operations performed on the data, a summary and analysis of the data, and a statement of the conclusions drawn from the analysis.
11. The signed and dated reports of each of the individual scientists or other professionals involved in the study, including each person who, at the

request or direction of the testing facility or sponsor, conducted an analysis or evaluation of data or specimens from the study after data generation was complete.

12. The location where all specimens, raw data, and the final report are to be stored.
13. A statement prepared and signed and dated by the QAU which provides a listing of the audits/inspections performed, the date of the audit/inspection, and the date reported to the study director and study director's management.

The report format required by the EPA can be found in Pesticide Registration Notice 86-5 (*13*).

Literature Cited

(1) "Good Laboratory Practice Standards Under the Federal Insecticide, Fungicide, and Rodenticide Act, Final Rule", *Code of Federal Regulations, Title 40, Part 160*; Federal Register 54:58 (August 17, 1989) pp 34067-34074.

(2) "Interpretation of the Good Laboratory Practice (GLP) Regulations", GLP Regulation Advisory No. 6, U.S. Environmental Protection Agency, Washington, D.C., March 26, 1990.

(3) "Interpretation of the Good Laboratory Practice (GLP) Regulations", GLP Regulation Advisory No. 52, U.S. Environmental Protection Agency, Washington, D.C., October 16, 1992.

(4) "Interpretation of the Good Laboratory Practice (GLP) Regulations", GLP Regulation Advisory No. 13, U.S. Environmental Protection Agency, Washington, D.C., May 1, 1990.

(5) "Interpretation of the Good Laboratory Practice (GLP) Regulations", GLP Regulation Advisory No. 3, U.S. Environmental Protection Agency, Washington, D.C., December 20, 1989.

(6) "Interpretation of the Good Laboratory Practice (GLP) Regulations", GLP Regulation Advisory No. 34, U.S. Environmental Protection Agency, Washington, D.C., June 28, 1991.

(7) "Interpretation of the Good Laboratory Practice (GLP) Regulations", GLP Regulation Advisory No. 15, U.S. Environmental Protection Agency, Washington, D.C., July 16, 1990.

(8) "Interpretation of the Good Laboratory Practice (GLP) Regulations", GLP Regulation Advisory No. 8, U.S. Environmental Protection Agency, Washington, D.C., April 9, 1990.

(9) Fuller, S. R., and Garner, W. Y., Test Substance and Analytical Reference Standard Characterization and Accountability In *Good Laboratory Practice Standards: Applications for Field and Laboratory Studies* (1992) Willa Y. Garner, Maureen S. Barge, James P. Ussary, Eds.; ACS Professional Reference Book, American Chemical Society, Washington, DC.

(10) "Interpretation of the Good Laboratory Practice (GLP) Regulations", GLP Regulation Advisory No. 23, U.S. Environmental Protection Agency, Washington, D.C., November 7, 1990.
(11) "Interpretation of the Good Laboratory Practice (GLP) Regulations", GLP Regulation Advisory No. 24, U.S. Environmental Protection Agency, Washington, D.C., December 4, 1990.
(12) "Interpretation of the Good Laboratory Practice (GLP) Regulations", GLP Regulation Advisory No. 30, U.S. Environmental Protection Agency, Washington, D.C., June 28, 1991.
(13) "PR Notice 86-5", U.S. Environmental Protection Agency, Washington, DC, 1986.

RECEIVED August 28, 1993

INDEXES

Author Index

Affiliation Index

Subject Index

N

Production: C. Buzzell-Martin
Indexing: Deborah H. Steiner
Acquisition: Anne Wilson

Printed by United Book Press, Inc., Baltimore, MD